ORDINARY DIFFERENTIAL EQUATIONS

ORDINARY DIFFERENTIAL EQUATIONS

An Elementary Textbook for
Students of Mathematics, Engineering,
and the Sciences

Morris Tenenbaum
Cornell University

Harry Pollard
Purdue University

DOVER PUBLICATIONS, INC., NEW YORK

Published in Canada by General Publishing Company, Ltd., 30 Lesmill Road, Don Mills, Toronto, Ontario.

This Dover edition, first published in 1985, is an unabridged and corrected republication of the work first published by Harper & Row, Publishers, Inc., New York, in 1963.

Manufactured in the United States of America
Dover Publications, Inc., 31 East 2nd Street, Mineola, N.Y. 11501

Library of Congress Cataloging in Publication Data

Tenenbaum, Morris.
Ordinary differential equations.

Reprint. Originally published: New York : Harper & Row, 1963.
Bibliography: p.
Includes index.
1. Differential equations.   I. Pollard, Harry, 1919–   II. Title.
QA372.T4   1985   515.3'5   85-12983
ISBN 0-486-64940-7

Contents

Preface for the Teacher

IN WRITING THIS BOOK, it has been our aim to make it readable for the student, to include topics of increasing importance (such as transforms, numerical analysis, the perturbation concept) and to avoid the errors traditionally transmitted in an elementary text. In this last connection, we have abandoned the use of the terminology "general solution" of a differential equation unless the solution is in fact general, i.e., unless the solution actually contains every solution of the differential equation. We have also avoided the term "singular solution." We have exercised great care in defining function, differentials and solutions; in particular we have tried to make it clear that functions have domains.

On the other hand, this accuracy has been secondary to our main purpose: to teach the student how to use differential equations. We hope and believe that we have not overlooked any of the major applications which can be made comprehensible at this elementary level. You will find in this text an extensive list of worked examples and homework problems with answers.

We acknowledge our indebtedness to the publishers for their cooperation and willingness to let us use new pedagogical devices and to Prof. C. A. Hutchinson for his thorough editing.

M.T.
H.P.

Ithaca, New York
West Lafayette, Indiana

Preface for the Student

THIS BOOK HAS BEEN WRITTEN primarily for you, the student. We have tried to make it easy to read and easy to follow.

We do not wish to imply, however, that you will be able to read this text as if it were a novel. If you wish to derive any benefit from it, you must study each page slowly and carefully. You must have pencil and plenty of paper beside you so that you yourself can reproduce each step and equation in an argument. When we say "verify a statement," "make a substitution," "add two equations," "multiply two factors," etc., you yourself must actually perform these operations. If you carry out the explicit and detailed instructions we have given you, we can almost guarantee that you will, with relative ease, reach the conclusion.

One final suggestion—as you come across formulas, record them and their equation numbers on a separate sheet of paper for easy reference. You may also find it advantageous to do the same for Definitions and Theorems.

M.T.
H.P.

Ithaca, New York
West Lafayette, Indiana

ORDINARY DIFFERENTIAL
EQUATIONS

Chapter 1

Basic Concepts

LESSON 1. How Differential Equations Originate.

We live in a world of interrelated changing entities. The position of the earth changes with time, the velocity of a falling body changes with distance, the bending of a beam changes with the weight of the load placed on it, the area of a circle changes with the size of the radius, the path of a projectile changes with the velocity and angle at which it is fired.

In the language of mathematics, changing entities are called variables and the rate of change of one variable with respect to another a derivative. Equations which express a relationship among these variables and their derivatives are called differential equations. In both the natural and social sciences many of the problems with which they are concerned give rise to such differential equations. But what we are interested in knowing is not how the variables and their derivatives are related but only how the variables themselves are related. For example, from certain facts about the variable position of a particle and its rate of change with respect to time, we wish to determine how the position of the particle is related to the time so that we can know where the particle was, is, or will be at any time t. Differential equations thus originate whenever a universal law is expressed by means of variables and their derivatives. A course in differential equations is then concerned with the problem of determining a relationship among the variables from the information given to us about themselves and their derivatives.

We shall use an actual historical event to illustrate how a differential equation arose, how a relationship was then established between the two variables involved, and finally how from the relationship, the answer to a very interesting problem was determined. In the year 1940, a group of boys was hiking in the vicinity of a town in France named Lascaux. They suddenly became aware that their dog had disappeared. In the ensuing search he was found in a deep hole from which he was unable to climb out. When one of the boys lowered himself into the hole to help extricate the dog, he made a startling discovery. The hole was once a part of the roof of an ancient cave that had become covered with brush. On the walls of the cave there were marvellous paintings of stags, wild horses, cattle, and of a fierce-looking black beast which resembled our bull.* This accidental discovery, as you may guess, created a sensation. In addition to the wall paintings and other articles of archaeological interest, there were also found the charcoal remains of a fire. The problem we wish to solve is the following: determine from the charcoal remains how long ago the cave dwellers lived.

It is well known that charcoal is burnt wood and that with time certain changes take place in all dead organic matter. It is also known that all living organisms contain two isotopes of carbon, namely C^{12} and C^{14}. The first element is stable; the second is radioactive. Furthermore the ratio of the amounts of each present in any macroscopic piece of *living* organism remains constant. However from the moment the organism dies, the C^{14} that is lost because of radiation, is no longer replaced. Hence the amount of the unstable C^{14} present in a dead organism, as well as its ratio to the stable C^{12}, changes with time. The changing entities in this problem are therefore the element C^{14} and time. If the law which tells us how one of these changing entities is related to the other cannot be expressed without involving their derivative, then a differential equation will result.

Let t represent the elapsed time since the tree from which the charcoal came, died, and let x represent the amount of C^{14} present in the dead tree at any time t. Then the instantaneous rate at which the element C^{14} decomposes is expressed in mathematical symbols as

(1.1)
$$\frac{dx}{dt}.$$

We now make the assumption that this rate of decomposition of C^{14} varies as the first power of x (remember x is the amount of C^{14} present at any time t). Then the equation which expresses this assumption is

(1.11)
$$\frac{dx}{dt} = -kx,$$

where $k > 0$ is a proportionality constant, and the negative sign is used to indicate that x, the quantity of C^{14} present, is decreasing. Equation (1.11) is a differential equation. It states that the instantaneous rate of decomposition of C^{14} is k times the amount of C^{14} present at a moment of time. For example, if $k = 0.01$ and t is measured in years, then when $x = 200$ units at a moment in time, (1.11) tells us that the rate of decomposition of C^{14} at that moment is $1/100$ of 200 or at the rate of 2 units per

---

*You can see some of these pictures in *Primitive Art* by Erwin O. Christensen, Viking Press, 1955, and in *The Picture History of Painting* by H. W. and D. J. Jansen, Harry N. Abrams, 1957.

year. If, at another moment of time, $x = 50$ units, then (1.11) tells us that the rate of decomposition of C^{14} at that moment is $1/100$ of 50 or at the rate of $\frac{1}{2}$ unit per year.

Our next task is to try to determine from (1.11) a law that will express the relationship between the variable x (which, remember, is the amount of C^{14} present at any time t) and the time t. To do this, we multiply (1.11) by dt/x and obtain

$$(1.12) \qquad\qquad \frac{dx}{x} = -k\,dt.$$

Integration of (1.12) gives

$$(1.13) \qquad\qquad \log x = -kt + c,$$

where c is an arbitrary constant. By the definition of the logarithm, we can write (1.13) as

$$(1.14) \qquad\qquad x = e^{-kt+c} = e^c e^{-kt} = A e^{-kt},$$

where we have replaced the constant e^c by a new constant A.

Although (1.14) is an equation which expresses the relationship between the variable x and the variable t, it will not give us the answer we seek until we know the values of A and k. For this purpose, we fall back on other available information which as yet we have not used. Since time is being measured from the moment the tree died, i.e., $t = 0$ at death, we learn from (1.14) by substituting $t = 0$ in it, that $x = A$. Hence we now know, since x is the amount of C^{14} present at any time t, that A units of C^{14} were present when the tree, from which the charcoal came, died. From the chemist we learn that approximately 99.876 percent* of C^{14} present at death will remain in dead wood after 10 years and that the assumption made after (1.1) is correct. Mathematically this means that when $t = 10$, $x = 0.99876A$. Substituting these values of x and t in (1.14), we obtain

$$(1.15) \qquad 0.99876A = Ae^{-10k}, \qquad 0.99876 = e^{-10k}.$$

We can now find the value of k in either of two ways. There are tables which tell us for what value of $-10k$, $e^{-10k} = 0.99876$. Division of this value by -10 will then give us the value of k. Or if we take the natural logarithm of both sides of (1.15) there results

$$(1.2) \qquad\qquad \log 0.99876 = -10k.$$

---

*There is some difference among chemists in regard to this figure. The one used above is based on a half-life of C^{14} of 5600 years, i.e., half of C^{14} present at death will decompose in 5600 years. It is an approximate average of 5100 years, the lowest half-life figure, and 6200 years, the largest half-life figure.

From a table of natural logarithms, we find

(1.21) $-0.00124 = -10k, \quad k = 0.000124$

approximately. Equation (1.14) now becomes

(1.22) $x = Ae^{-0.000124t}$,

where A is the amount of C^{14} present at the moment the tree died.

Equation (1.22) expresses the relationship between the variable quantity x and the variable time t. We are therefore at last in a position to answer the original question: How long ago did the cave dwellers live? By a chemical analysis of the charcoal, the chemist was able to determine the ratio of the amounts of C^{14} to C^{12} present at the time of the discovery of the cave. A comparison of this ratio with the fixed ratio of these two carbons in living trees disclosed that 85.5 percent of the amount of C^{14} present at death had decomposed. Hence $0.145A$ units of C^{14} remained. Substituting this value for x in (1.22), we obtain

(1.23) $0.145A = Ae^{-0.000124t}$

$$0.145 = e^{-0.000124t}$$

$$\log 0.145 = -0.000124t$$

$$-1.9310 = -0.000124t$$

$$t = 15573.$$

Hence the cave dwellers lived approximately 15,500 years ago.

Comment 1.3. Differential equation (1.11) originated from the assumption that the rate of decomposition of C^{14} varied as the first power of the amount of C^{14} present at any time t. The resulting relationship between the variables was then verified by independent experiment. Assumptions of this kind are continually being made by scientists. From the assumption a differential equation originates. From the differential equation a relationship between variables is determined, usually in the form of an equation. From the equation certain predictions can be made. Experiments must then be devised to test these predictions. If the predictions are validated, we accept the equation as expressing a true law. It has happened in the history of science, because experiments performed were not sensitive enough, that laws which were considered as valid for many years were found to be invalid when new and more refined experiments were devised. A classical example is the laws of Newton. These were accepted as valid for a few hundred years. As long as the experiments concerned bodies which were macroscopic and speeds which were reasonable, the laws were valid. If the bodies were of the size of atoms or the speeds near that of light, then new assumptions had to be made, new equations born, new predictions foretold, and new experiments devised to test the validity of these predictions.

Comment 1.4. The method we have described for determining the age of an organic archaeological remain is known as the carbon-14 test.*

EXERCISE 1

1. The radium in a piece of lead decomposes at a rate which is proportional to the amount present. If 10 percent of the radium decomposes in 200 years, what percent of the original amount of radium will be present in a piece of lead after 1000 years?
2. Assume that the half life of the radium in a piece of lead is 1600 years. How much radium will be lost in 100 years?
3. The following item appeared in a newspaper. "The expedition used the carbon-14 test to measure the amount of radioactivity still present in the organic material found in the ruins, thereby determining that a town existed there as long ago as 7000 B.C." Using the half-life figure of C^{14} as given in the text, determine the approximate percentage of C^{14} still present in the organic material at the time of the discovery.

ANSWERS 1

1. 59.05 percent.   **2.** 4.2 percent.   **3.** Between 32 percent and 33 percent.

LESSON 2.   The Meaning of the Terms *Set* and *Function.* Implicit Functions.   Elementary Functions.

Before we can hope to solve problems in differential equations, we must first learn certain rules, methods and laws which must be observed. In the lessons that follow, we shall therefore concentrate on explaining the meaning of certain terms which we shall use and on devising methods by which certain types of differential equations can be solved. We shall then apply these methods to solving a wide variety of problems of which the one in Lesson 1 was an example.

We begin our study of differential equations by clarifying for you two of the basic notions underlying the calculus and ones which we shall use repeatedly. These are the notions of *set* and *function.*

LESSON 2A.   The Meaning of the Term *Set.*   Each of you is familiar with the word collection. Some of you in fact may have or may have had collections—such as collections of stamps, of sea shells, of coins, of butter-flies. In mathematics we call a collection of objects a **set,** and the indi-

---

*Dr. Willard F. Libby was awarded the 1960 Nobel Physics Prize for developing this method of ascertaining the age of ancient objects. His C^{14} half-life figure is 5600 years, the same as the one we used. According to Dr. Libby, the measurable age span by this test is from 1000 to 30,000 years.

vidual members of the set **elements**. A set therefore may be described by specifying what property an object must have in order to belong to it or by giving a list of the elements of the set.

Examples of Sets. 1. The collection of positive integers less than 10 is a set. Its elements are 1, 2, 3, 4, 5, 6, 7, 8, 9.

2. The collection of individuals whose surnames are Smith is a set.

3. The collection of all negative integers is a set. Its elements are $\cdots, -4, -3, -2, -1$.

Since to each point on a line, there corresponds one and only one real number, called the **coordinate** of the point, and to each real number there corresponds one and only one point on the line, we frequently refer to a point on a line by its corresponding number and vice versa.

Definition 2.1. The *set of all numbers* between any two points on a line is called an **interval** and is usually denoted by the letter I.

If the two points on a line are designated by a and b, then the notation

$$(2.11) \qquad\qquad I: \quad a < x < b$$

will mean the set of all real numbers x (or of all real values of x) which lie between the points a and b, but not including a and b. For convenience, we shall frequently omit the I and write only

$$(2.111) \qquad\qquad a < x < b$$

to represent this set of numbers. Similarly,

(2.12) $\quad I: -\infty < x < \infty$ will mean the set of all real values of x.

$I: a \leqq x \leqq b$ will mean the set of all real values of x between a and b, including the two end points.

$I: a \leqq x < b$ will mean the set of all real values of x between a and b, including a but not b.

$I: -1 < x < 3, x = 10$, will mean the set of all numbers between -1 and 3 plus the number 10.

$I: x \geqq 0$ will mean the set of all positive real values of x plus zero.

$I: x = a$ will mean the set consisting of the single number a.

LESSON 2B. The Meaning of the Term *Function of One Independent Variable.* If two variables are connected in some way so that the value of one is *uniquely* determined when a value is given to the other, we say that one is a **function** of the other. (This concept will be given a more precise meaning in Definitions 2.3 and 2.31 below.)

We shall show by examples below that the manner in which the relationship between the variables is expressed is unimportant. It may be by an equation, of the kind with which you are familiar, or by other means. It is only important for the definition of a function that there be this unambiguous relationship between the variables so that, when a value is given to one, a corresponding value to the other is thereby uniquely determined.

Example 2.2. Let l be the length of the side of a square and A its area. It is then customary to say that the area A depends on the length l, so that l is given an independent status and A a dependent one. However, there is no valid reason why l could not be considered as being dependent on A. The decision as to which variable in a problem is to be considered as dependent and which independent lies entirely within the discretion of the individual. The choice will usually be determined by convenience. It is customary to write, whenever it is possible to do so, first the dependent variable, then an equals sign, then the independent variable in a manner which expresses mathematically the relationship between the two variables.

If in this example, therefore, we express the relationship between our two variables A and l by writing

(a)
$$A = l^2;$$

we thereby give to A a dependent status and to l an independent one. Equation (a) now defines A as a function of l since for each l, it determines A uniquely. The relationship between the two variables, expressed mathematically by equation (a), is, however, not rigidly correct. It says that for each value of the length l, A, the area, is the square of l. But what if we let $l = -3$? The square of -3 is 9; yet no area exists if the side of a square has length less than zero. Hence we must place a restriction on l and say that (a) defines the area A as a function of the length l only for a *set* of positive values of l and for $l = 0$. We must therefore write

(b)
$$A = l^2, \quad l \geqq 0.$$

Example 2.21. The relationship between two variables x and y is the following. If x is between 0 and 1, y is to equal 2. If x is between 2 and 3, y is to equal $\sqrt{x}$. The equations which express the relationship between the two variables are, with the end points of the interval included,

(a)
$$y = 2, \quad 0 \leqq x \leqq 1,$$
$$y = \sqrt{x}, \quad 2 \leqq x \leqq 3.$$

These *two* equations now define y as a function of x. For each value of x in the specified intervals, a value of y is determined uniquely. The graph of this function is shown in Fig. 2.211. Note that these equations do not define y as a function of x for values of x outside the two stated intervals.

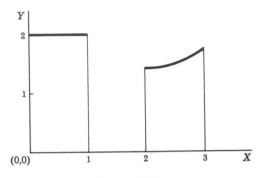

Figure 2.211

The reason is obvious. We have not been told what this relationship is. For values of x therefore, equal to say $\frac{3}{2}$ or -1 or 4, etc., we say that y is **undefined** or that the function is undefined.

Example 2.22. In Fig. 2.221, we have shown the temperature T of a body, recorded by an automatic device, in a period of 24 consecutive hours. The horizontal axis represents the time in hours; the vertical axis the temperature T at any time $t \geqq 0$. Even though we cannot express the relationship between the variables t and T by an equation, there can be no doubt that a precise, unique relationship between the two variables

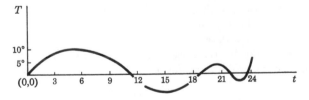

Figure 2.221

exists. For each value of the time t, the graph will give a unique value of the temperature T. Hence the graph in this case defines T as a function of t.

The main feature we wished to emphasize in the above differing examples was that for the definition of a function it was *not* essential to be able to set up the relationship between the two variables by a single equation. (Most of the functions that you have encountered thus far were of this type.) As we mentioned at the outset, what is essential for the definition of a function is that the relationship between the two variables be specific and unambiguous so that for each value taken on by an independent variable on a *specified set*, there should correspond *one and only one* value of a dependent variable. As you can verify, all the examples we gave above had this one common important property. We incorporate all the essential features of a function in the following definition.

Definition 2.3. If to each value of an independent variable x on a set E (*the set must be specified*) there corresponds one and only one real value of the dependent variable y, we say that the dependent variable y is a **function*** of the independent variable x on the set E.

It is customary to call the specified set E of values of the independent variable, the **domain** of definition of the independent variable and to call the set of resulting values of the dependent variable, the **range** of the dependent variable or the range of the function. Using this terminology, we may define a function alternately, as follows.

Definition 2.31. A function is a correspondence between a domain set D and a set R that assigns to each element of D a unique element of R.

Comment 2.32. A function is thus equivalent to a rule which tells us how to determine the unique element y of the range which is to be assigned to an element x of the domain. When we say therefore that the formula

(a) $$y = \sqrt{x - 1}, \quad x \geqq 1,$$

defines y as a function of x, we mean that the formula has given us a rule by which, for each value of the independent variable x in its domain D: $x \geqq 1$, we can determine the unique assigned value of y in the range R. (Here $R \geqq 0$.) The rule is as follows: for each x in D, subtract one and take the positive square root of the result. For example, to the element $x = 5$ of D, the function or rule defined by (a) assigns the unique value 2 of R. Conversely, we can first give the rule, and then express the rule, if possible, by a formula.

Comment 2.33. We may at times for convenience refer to an equation or formula as if it were a function. For example, we may frequently refer to the equation

(b) $$y = x^2, \quad -\infty < x < \infty,$$

as a function. What we actually mean is that the equation $y = x^2$ *defines* a function; that is $y = x^2$ gives us a rule by which to each x in D we can

---

*In advanced mathematics, the function is called a *real* function. Since we shall for the most part be considering only real functions, we shall omit the word real whenever a real function is meant.

assign a y in R. Here the rule is: for each x in D, the number y in R assigned to it is obtained by squaring x.

In view of Definition 2.3, such frequently encountered formulas as

$$y = x^2,$$

$$y = \sqrt{1 - x^2},$$

$$y = \frac{x^2 + 5x}{x - 3}$$

are meaningless because they do not specify the domain, i.e., the set of values of x, for which the formulas apply. *In practice, however, we interpret such formulas to define functions for all values of x for which they make sense.* The first equation, therefore, defines a function for all values of x, the second for values of x in the interval $-1 \leqq x \leqq 1$, and the last for all values of x except $x = 3$.

Because of the above comment, the function defined by

$$y = \sqrt{1 - x^2}$$

is not the same as the function defined by

$$y = \sqrt{1 - x^2}, \quad 0 \leqq x \leqq 1.$$

(The first is defined for all x in $-1 \leqq x \leqq 1$; the second only for x in $0 \leqq x \leqq 1$.)

It is customary and convenient to record the fact that the dependent variable y is a function of the independent variable x (these are the letters most commonly used for the dependent and independent variable respectively), by means of the symbolic expression

(2.34) $$y = f(x).$$

It is read as "y equals f of x" or "y is a function of x."

By Definition 2.3 and (2.34), we could therefore write for the temperature Example 2.22,

(2.35) $$T = f(t).$$

We then say T is a function of t and refer to the graph itself as the definition or rule which tells us which T to assign to each value of t.

Comment 2.36. We shall at times write an expression in x, say

(a) $$f(x) = x^2 + e^x, \quad -\infty < x < \infty$$

and refer to $f(x)$ as a function. What we mean is that $f(x)$, here ($x^2 + e^x$), gives a rule by which for each x, we can assign a unique value to $f(x)$.

Definition 2.4. If $f(x)$ is a function of x defined on a set E, then the symbol $f(a)$, for any a in E, means the unique value assigned to $f(x)$ obtained by substituting a for x.

Example 2.5. If

(a) $$f(x) = x^2 + 2x + 1, \quad 0 \leqq x \leqq 1,$$

find $f(0), f(1), f(\tfrac{1}{2}), f(2), f(-1)$.

Solution. By Definition 2.4 we have

(b) $$f(0) = 0^2 + 2 \cdot 0 + 1 = 1,$$
$$f(1) = 1^2 + 2 \cdot 1 + 1 = 4,$$
$$f(\tfrac{1}{2}) = (\tfrac{1}{2})^2 + 2 \cdot \tfrac{1}{2} + 1 = 2\tfrac{1}{4},$$

$f(2)$ is undefined since 2 is not in our set $E: 0 \leqq x \leqq 1$,

$f(-1)$ is undefined since -1 is not in our set E.

If several functions appear in a single context so that the use of the same letter for each would be confusing, it is permissible to replace f by other letters. Those most frequently used are g, h, G, H, F, etc. Similarly, we may use other letters in place of x and y. Those usually used are the ones at the end of the alphabet, namely u, v, w, z, s, t.

LESSON 2C. Function of Two Independent Variables. In Lesson 2B, we defined a function of one independent variable. In an analogous manner, we define a function of two independent variables x and y as follows.

Definition 2.6. If to each element (x,y) of a set E in the plane (the set must be specified) there corresponds one and only one real value of z, then z is said to be a function of x and y for the set E. In this event, x,y are called independent variables and z a dependent variable.

As in the case of one independent variable, the set E is sometimes called the **domain** of definition of the function and the set of resulting values of z, the **range** of the function.

In view of Definition 2.6, a formula such as

$$z = y\sqrt{1 - x^2}$$

is meaningless since it does not specify the domain of definition. Here again as in the case of one variable, we interpret such formulas to define functions for all values of x and y for which they make sense. In this example therefore the elements of the domain D are the points (x,y) in the plane where $-1 \leqq x \leqq 1$, $-\infty < y < \infty$. The domain, therefore, for which the formula defines z as a function of x and y consists of all points in the plane between and including the lines $x = 1$ and $x = -1$. It is the shaded area in Fig. 2.61.

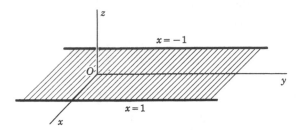

Figure 2.61

Example 2.62. Determine the domain D for which each of the following formulas define z as a function of x and y.

1. $z = \dfrac{\sqrt{1 - y^2}}{\sqrt{1 - x^2}}$.

2. $z = x + y$.

3. $z = \sqrt{x^2 + y^2 - 25}$.

4. $z = \sqrt{-(x^2 + y^2)}$.

5. $z = \sqrt{-(x^2 + y^2 + 1)}$.

Solutions. 1. The elements of the domain D for which the formula defines z as a function of x and y, consists of those points (x,y) where $-1 < x < 1, -1 \leqq y \leqq 1$. The domain D is the shaded square shown

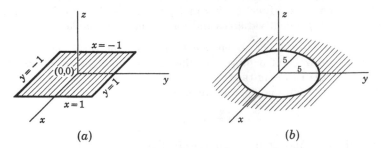

(a)   (b)

Figure 2.63

in Fig. 2.63(a). It is bounded by the lines $x = \pm 1$ and $y = \pm 1$. It includes the lines $y = \pm 1$ but not the lines $x = \pm 1$.

2. The domain is the entire plane.

3. The domain consists of those points (x,y) for which $x^2 + y^2 \geqq 25$. It is the shaded area in Fig. 2.63(b), i.e., it is the area *outside* the circle $x^2 + y^2 = 25$ plus the points on its circumference.

4. The domain consists of the single point $(0,0)$.

5. This formula does *not* define z as a function of x and y. There does not exist a domain which will determine a value of z.

It is evident from the above examples, that a two-dimensional domain may cover the whole plane or part of the plane; it may cover the whole plane with the exception of a hole in its interior; its boundaries may be circular or straight lines, or it may consist of only a finite number of points. In short, in contrast to a one-dimensional domain, a two-dimensional domain may assume a great variety of shapes and figures.

It is customary and convenient to record the fact that z is a function of x and y by means of the symbolic expression

$$(2.64) \qquad\qquad z = f(x,y).$$

It is read as "z equals f of x,y" or "z is a function of x,y."

Definition 2.65. If $f(x,y)$ is a function of two independent variables x,y, defined over a domain D, then the symbol $f(x,a)$ for any element (x,a) in D, means the function of x obtained by replacing y by a.

Example 2.66. If

(a) $f(x,y) = x^2 + xy^2 + 5y + 3, \quad -\infty < x < \infty, \quad -\infty < y < \infty,$ find $f(x,2), f(x,a), f[x,g(x)]$.

Solution. By Definition 2.65

(b) $f(x,2) = x^2 + 4x + 10 + 3 = x^2 + 4x + 13, \quad -\infty < x < \infty,$
$f(x,a) = x^2 + a^2 x + 5a + 3, \quad -\infty < x < \infty.$

And, if $g(x)$ is defined for all x,

$$f[x,g(x)] = x^2 + x[g(x)]^2 + 5g(x) + 3, \quad -\infty < x < \infty.$$

Example 2.67. If

(a) $\qquad f(x,y) = x + y, \quad -1 \leqq x \leqq 1, \quad 0 \leqq y \leqq 2,$

find $f(x,\tfrac{1}{2}), f(x,3)$.

Solution. By Definition 2.65

(b) $\qquad\qquad f(x,\tfrac{1}{2}) = x + \tfrac{1}{2}, \quad -1 \leqq x \leqq 1.$

But $f(x,3)$ is undefined since the domain of y is the interval $0 \leqq y \leqq 2$.

A special type of set, called a region, is defined as follows.

Definition 2.68. A set in the plane is called a **region** if it satisfies the following two conditions:

1. Each point of the set is the center of a circle whose entire interior consists of points of the set.
2. Every two points of the set can be joined by a curve which consists entirely of points of the set.

In Example 2.62, the domain defined in 2 is a region. If the boundary points are excluded from each set defined in 1 and 3, then each resulting domain is also a region. Each point of each set satisfies requirement 1, and every two points of each set satisfies requirement 2. On the other hand, the set consisting of the points on a line is not a region. The set satisfies requirement 2 but not 1. Also the set consisting of isolated points is not a region—the points in the set do not satisfy either of the two requirements.

Definition 2.69. A region is said to be **bounded** if there is a circle which will enclose it.

Comment 2.691. In a manner analogous to Definition 2.6, we can define a function of three or more independent variables.

LESSON 2D. Implicit Function. Consider a relationship between two variables x, y given by the formula

(2.7) $x^2 + y^2 - 25 = 0.$

Does it define a function? If $x > 5$ or $x < -5$, then the formula will not determine a value of y. For example, if $x = 7$, there is no value of y which will make the left side of (2.7) equal to zero. (Why?) However, if x lies between -5 and 5 inclusive, then there is a value of y which will make the left side of (2.7) equal to zero. To find it, we solve (2.7) for y and obtain

(2.71) $y = \pm\sqrt{25 - x^2}, \quad -5 \leqq x \leqq 5.$

When the relation between x and y is written in this form, however, we see that the formula does not define y uniquely for a value of x. Hence, by our Definition 2.3, it does not define a function. We can correct this defect by specifying which value of y is to be chosen. For example, we can choose any *one* of the following three formulas to determine y.

(2.72) $y = \sqrt{25 - x^2}, \quad -5 \leqq x \leqq 5.$

(2.73) $y = -\sqrt{25 - x^2}, \quad -5 \leqq x \leqq 5.$

(2.74) $y = \sqrt{25 - x^2}, \quad -5 \leqq x \leqq 0;$

$\qquad\qquad = -\sqrt{25 - x^2}, \quad 0 < x < 5.$

By Definition 2.3, each of these formulas now defines a function. It gives a rule which assigns a unique y to each x on the specified interval.

Now consider the formula

$$(2.75) \qquad\qquad x^2 + y^2 + 1 = 0,$$

which also connects two variables x and y, and ask of it the same question. Does it define a function? If it does, then there must be values of x for which it will determine uniquely values of y. It should be evident to you that there are no values of y for any x. (Write the equation as $x^2 + y^2 = -1$.) Hence this equation does not define a function by our Definition 2.3.

As a final example, we consider the formula

$$(2.76) \qquad\qquad x^3 + y^3 - 3xy = 0,$$

and again ask the question. Does it define a function? And if it does, for what values of x will it determine uniquely a value of y? The answer to both questions, unlike the answer to the previous formula (2.7), is not easy to give. For unlike it, (2.76) cannot be solved easily for y in terms of x. Hence we must resort to other means. The graph of equation (2.76) is shown in Fig. 2.77.

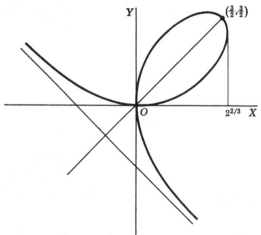

Figure 2.77

From the graph, we see that for $x \leqq 0$ and $x > 2^{2/3}$, y is uniquely determined. Hence formula (2.76) does define y as a function of x in these two intervals, but not in the interval $0 < x \leqq 2^{2/3}$. It is possible, however, to make formula (2.76) define y as a function of x for *all* x if we choose *one* of the three possible values of y for each x in the interval $0 < x < 2^{2/3}$ and choose *one* of the two possible values of y for $x = 2^{2/3}$. With these restrictions, we then would be able to assert that (2.76) defines y as a function of x for *all* x.

Whenever a relationship which exists between two variables x and y is expressed in the form (2.7) or (2.76), we write it symbolically as

$$(2.8) \qquad\qquad f(x,y) = 0.$$

It is read as "f of x,y equals zero," or as "a function of x,y equals zero."

If the relation which defines y as a function of x is expressed in the form $f(x,y) = 0$, it is customary to call y an **implicit function** of x. When we say therefore that y is an implicit function of x, we mean, as the name suggests, that the functional relationship between the two variables is not explicitly visible as when we write $y = f(x)$, but that it nevertheless *implicitly* exists. That is, there is a function, let us call it $g(x)$, which is implicitly defined by the relation $f(x,y) = 0$ and which determines uniquely a value of y for each x on a set E. Hence:

Definition 2.81. The relation

$$(2.82) \qquad\qquad f(x,y) = 0$$

defines y as an **implicit function** of x on an interval $I: a < x < b$, if there exists a function $g(x)$ defined on I such that

$$(2.83) \qquad\qquad f[x,g(x)] = 0$$

for *every* x in I.

Example 2.84. Show that

(a) $$f(x,y) = x^2 + y^2 - 25 = 0$$

defines y as an implicit function of x on the interval $I: -5 \leq x \leq 5$.

Solution. Choose for $g(x)$ any *one* of the functions (2.72), (2.73), or (2.74). If, for example, we choose (2.72), then $g(x) = \sqrt{25 - x^2}$. It is defined on I, and by (a) and Definition 2.65,

(b) $$f[x,g(x)] = x^2 + [\sqrt{25 - x^2}]^2 - 25 = 0.$$

Hence Definition 2.81 is satisfied.

Example 2.85. Show that

(a) $$f(x,y) = x^3 + y^3 - 3xy = 0$$

defines y as an implicit function of x for all x.

Solution. Here, as pointed out earlier, it is not easy to solve for y in terms of x, so that it is not easy to find the required function $g(x)$. However, if we select from Fig. 2.77 any *one* of the graphs shown in Fig. 2.86

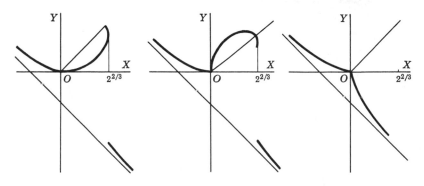

Figure 2.86

to represent the function $g(x)$, then $f[x,g(x)] = 0$ for every x. Hence Definition 2.81 is satisfied. (Note that $g(x)$ may be selected in infinitely many more ways.)

LESSON 2E. The Elementary Functions. In addition to the terms function and implicit function, we shall refer at times to a special class of functions called the **elementary functions.** These are the constants and the following functions of a variable x:

1. Powers of x: x, x^2, x^3, etc.
2. Roots of x: $\sqrt{x}$, $\sqrt[3]{x}$, etc.
3. Exponentials: e^x.
4. Logarithms: $\log x$.
5. Trigonometric functions: $\sin x$, $\cos x$, $\tan x$, etc.
6. Inverse trigonometric functions: Arc $\cos x$, etc.
7. All functions obtained by replacing x any number of times by any of the other functions 1 to 6. Examples are: $\log \sin x$, $\sin (\sin x)$, $e^{\sin x}$, e^{ax^2}, etc.
8. All functions obtained by adding, subtracting, multiplying, and dividing any of the above seven types a finite number of times. Examples are: $2x - \log x + \dfrac{e^{\sin x}}{x}$, $(\text{Arc } \cos x)^2 + \dfrac{e^{x^2+2x+1}}{\sqrt{3x}} - \log (\log 4x)$.

In the calculus course, you learned how to differentiate elementary functions and how to integrate the resulting derivatives. If you have forgotten how, it would be an excellent idea at this point to open your calculus book and review this material.

EXERCISE 2

1. Describe, in words, each of the following sets:

   (a) $x < 0$.   (b) $x \le 0$.   (c) $a < x \le b$.   (d) $-\infty < x < 5, x = 7$.

   (e) $-3 < x < -2, x > 0$.   (f) $\sqrt{2} < x < \pi$.   (g) $2\pi \le x < 3\pi$.

2. Define the area A of a circle as a function of its radius r. Which is the dependent variable and which is the independent variable? Draw a rough graph which will show how A depends on r, when r is given values between 0 and 5.

3. Under certain circumstances the pressure p of a gas and its volume V are related by the formula $pV^{3/2} = 1$. Express each variable as a function of the other.

4. Explain the difference between the function

$$y = \sqrt{x}, \quad 2 \le x \le 3,$$

   and the function

$$y = \sqrt{x}, \quad x > 0.$$

5. Let

$$\begin{aligned} F(x) &= 2 && \text{if } x < 0, \\ &= 7 && \text{if } 0 \le x \le 1, \\ &= x^3 - 2x && \text{if } x > 4. \end{aligned}$$

   Find (a) $F(-1)$,   (b) $F(0)$,   (c) $F(0.7)$,   (d) $F(4)$,   (e) $F(8)$,   (f) $F(2)$.

6. If

$$g(x) = \frac{x^2 - 2x + 1}{x - 1}, \quad x \ne 1,$$

   find (a) $g(2)$,   (b) $g(-5)$,   (c) $g(1)$,   (d) $g(u)$,   (e) $g(x^2)$,   (f) $g(x - 1)$.

7. Why is the function defined in 6 not the same as the function defined by $g(x) = x - 1$?

8. Determine the domain D for which each of the following formulas defines z as a function of x and y.

   (a) $z = \dfrac{\sqrt{1 - x^2}}{\sqrt{1 - y^2}}$.   (b) $z = x - y + 2$.

   (c) $z = \sqrt{x^2 + (y - 1)^2}$.   (d) $z = \sqrt{x^2 + y^2 - 9}$.

   (e) $z = \sqrt{-(x + y)^2}$.

   (f) $z = \sqrt{-(x^2 + y^2 + 3)}$.   (g) $z = \dfrac{1 - x}{1 - y}$.

9. Which of the domains in problem 8 are regions?

10. If $f(x,y)$ is a function of two independent variables x,y defined over a domain D, then the symbol $f(b,y)$ for any element (b,y) in D, means the function of y obtained by replacing x by b; see Definition 2.65. Let

$$f(x,y) = x^2 + 2xy + \log(xy), \quad xy > 0.$$

   Find:

   (a) $f(x,1), f(x,b), f(x,0), f(x,x^2)$.   (b) $f(1,y), f(a,y), f(0,y), f(x^2,y)$.

   (c) $f(1,1), f(2,-1), f(a,b), f(u,v)$.

11. Draw the graphs of three different functions defined by

$$x^3 + y^3 - 3xy = 0.$$

See Fig. 2.77. Is there one which is continuous for all x?

12. The following is a standard type of exercise in the calculus.

If $x^3 + y^3 - 3xy = 0$, then $3x^2 + 3y^2 \dfrac{dy}{dx} - 3x \dfrac{dy}{dx} - 3y = 0$. Therefore

$$\frac{dy}{dx} = \frac{y - x^2}{y^2 - x}, \quad y^2 \neq x$$

Explain by the use of Fig. 2.77 what this means geometrically.

13. Explain why the procedure followed in problem 12, applied to the relation $x^2 + y^2 + 1 = 0$ and yielding the result

$$\frac{dy}{dx} = -\frac{x}{y}$$

is meaningless.

14. Find the function $g(x)$ that is implicitly defined by the relation

$$\sqrt{x^2 - y^2} + \text{Arc cos } \frac{x}{y} = 0, \quad y \neq 0,$$

15. Explain why

$$\sqrt{x^2 - y^2} + \text{Arc sin } \frac{x}{y} = 0$$

does not define y as an implicit function of x.

16. Can you apply the method of implicit differentiation as taught in the calculus to the function of problem 14, of problem 15?

17. Define a function of three independent variables x_1, x_2, x_3; of n independent variables $x_1 \cdots, x_n$. *Hint.* See Definition 2.6.

ANSWERS 2

1. (a) The set of all negative values of x. (b) The set (a) plus zero. (c) The set of values of x between a and b, including b but not a. (d) The set of all values of x less than five, plus the number 7. (e) The set of all values of x between -3 and -2, plus all positive values of x.
2. $A = \pi r^2$, $r \geqq 0$.
3. $p = V^{-3/2}$, $V > 0$; $V = p^{-2/3}$, $p > 0$.
4. The first function is defined only for all x between and including two and three; the second function is defined for all x greater than zero.
5. (a) 2. (b) 7. (c) 7. (d) Undefined. (e) 496. (f) Undefined.
6. (a) 1. (b) -6. (c) Meaningless. (d) $(u^2 - 2u + 1)/(u - 1)$. $u \neq 1$. (e) $(x^4 - 2x^2 + 1)/(x^2 - 1)$, $x^2 \neq 1$. (f) $(x^2 - 4x + 4)/(x - 2)$, $x \neq 2$.
7. Function defined in 6 is meaningless when $x = 1$.
8. (a) $-1 \leqq x \leqq 1$, $-1 < y < 1$. (b) Entire plane. (c) Entire plane. (d) Area outside circle $x^2 + y^2 = 9$ plus points on the circumference of the circle. (e) Line $x + y = 0$. (f) Nonexistent. (g) $y \neq 1$.
9. (b), (c), (g). Also (a) and (d) if their boundary points are excluded.
10. (a) $x^2 + 2x + \log x$; $x^2 + 2bx + \log (bx)$; undefined; $x^2 + 2x^3 + \log x^3$. (b) $1 + 2y + \log y$; $a^2 + 2ay + \log ay$; undefined; $x^4 + 2x^2y + \log (x^2y)$. (c) 3; undefined; $a^2 + 2ab + \log (ab)$; $u^2 + 2uv + \log (uv)$.

13. $x^2 + y^2 + 1 = 0$ does not define a function.
14. $y = g(x) = x$.
15. The first term requires that $|x| \geq |y|$; the second term that $|x| \leq |y|$.
16. No, both examples.

LESSON 3. The Differential Equation.

LESSON 3A. Definition of an Ordinary Differential Equation. Order of a Differential Equation.   In the calculus, you studied various methods by which you could differentiate the elementary functions. For example, the successive derivatives of $y = \log x$ are

(a) $$y' = \frac{1}{x}, \qquad y'' = \frac{-1}{x^2}, \qquad y''' = \frac{2}{x^3}, \quad \text{etc.}$$

And if $z = x^3 - 3xy + 2y^2$, its partial derivatives with respect to x and with respect to y are respectively

(b) $$\frac{\partial z}{\partial x} = 3x^2 - 3y, \qquad \frac{\partial z}{\partial y} = -3x + 4y, \qquad \frac{\partial^2 z}{\partial x^2} = 6x, \qquad \frac{\partial^2 z}{\partial y^2} = 4, \text{ etc.}$$

Equations such as (a) and (b) which involve variables and their derivatives are called differential equations. The first involves only one independent variable x; the second two independent variables x and y. Equations of the type (a) are called ordinary differential equations; of the type (b) partial differential equations. Hence,

Definition 3.1.   Let $f(x)$ define a function of x on an interval I: $a < x < b$. By an **ordinary differential equation** we mean an equation involving x, the function $f(x)$ and one or more of its derivatives.

Note.   It is the usual custom in writing differential equations to replace $f(x)$ by y. Hence the differential equation $\dfrac{d f(x)}{dx} + x[f(x)]^2 = 0$ is usually written as $\dfrac{dy}{dx} + xy^2 = 0$; the differential equation $D_x^2[f(x)] + xD_xf(x) = e^x$ as $D_x^2 y + xD_x y = e^x$ or as $y'' + xy' = e^x$.

Examples of ordinary differential equations are:

(3.11) $$\frac{dy}{dx} + y = 0.$$

(3.12) $$y' = e^x.$$

(3.13) $$\frac{d^2 y}{dx^2} = \frac{1}{1 - x^2}.$$

(3.14) $$f'(x) = f''(x).$$

(3.15) $$xy' = 2y.$$

(3.16) $$y'' + (3y')^3 + 2x = 7.$$

(3.17) $$(y''')^2 + (y'')^4 + y' = x.$$

(3.18) $$xy^{(4)} + 2y'' + (xy')^5 = x^3.$$

Note. Since only ordinary differential equations will be considered in this text, we shall hereafter omit the word ordinary.

Definition 3.2. **The order of a differential equation** is the order of the highest derivative involved in the equation.

For the differential equations listed above, verify that (3.11), (3.12), and (3.15) are of the first order; (3.13), (3.14), and (3.16) are of the second order; (3.17) is of the third order; (3.18) is of the fourth order.

A WORD OF CAUTION. You might be tempted to assert, if you were not careful, that

$$y'' - y'' + y' - y = 0$$

is a second order differential equation because of the presence of y''. However, y'' is not really involved in the equation since it is removable. Hence the equation is of order 1.

LESSON 3B. Solution of a Differential Equation. Explicit Solution. Consider the algebraic equation

(3.3) $$x^2 - 2x - 3 = 0.$$

When we say $x = 3$ is a solution of (3.3), we mean that $x = 3$ satisfies it, i.e., if x is replaced by 3 in (3.3), the equality will hold. Similarly, when we say the function $f(x)$ defined by

(3.31) $$y = f(x) = \log x + x, \quad x > 0,$$

is a solution of

(3.32) $$x^2 y'' + 2xy' + y = \log x + 3x + 1, \quad x > 0,$$

we mean that (3.31) satisfies (3.32), i.e., if in (3.32) we substitute the function $f(x) = \log x + x$ for y, and the first and second derivatives of the function for y' and y'', respectively, the equality will hold. [Be sure to verify the assertion that (3.31) does in fact satisfy (3.32).]

We want you to note two things. First in accordance with Definition 2.3, we specified in (3.31) the values of x for which the function is defined. But even if we had not, the interval $x > 0$ would have been tacitly assumed since $\log x$ is undefined for $x \leqq 0$. Second, we also specified in (3.32) the interval for which the differential equation makes sense. Since it too contains the term $\log x$, it too is meaningless when $x \leqq 0$.

Definition 3.4. Let $y = f(x)$ define y as a function of x on an interval $I: a < x < b$. We say that the function $f(x)$ is an **explicit solution** or simply a **solution** of an ordinary differential equation involving x, $f(x)$, and its derivatives, if it satisfies the equation for *every* x in I, i.e., if we replace y by $f(x)$, y' by $f'(x)$, y'' by $f''(x)$, $\cdots$, $y^{(n)}$ by $f^{(n)}(x)$, the differential equation reduces to an identity in x. In mathematical symbols the definition says: the function $f(x)$ is a solution of the differential equation

$$(3.41) \qquad\qquad F(x,y,y',\cdots,y^{(n)}) = 0,$$

if

$$(3.42) \qquad\qquad F[x,f(x),f'(x),\cdots,f^{(n)}(x)] = 0$$

for *every* x in I.

Comment 3.43. We shall frequently use the expression, "solve a differential equation," or "find a solution of a differential equation." Both are to be interpreted to mean, find a *function* which is a solution of the differential equation in accordance with Definition 3.4. Analogously when we refer to a certain *equation as the solution* of a differential equation, we mean that the *function defined by the equation* is the solution. If the equation does not define a function, then it is not a solution of any differential equation, even though by following a formal procedure, you can show that the equation satisfies the differential equation. For example, the equation $y = \sqrt{-(1 + x^2)}$ does *not* define a function. To say, therefore, that it is a solution of the differential equation $x + yy' = 0$ is meaningless even though the formal substitution in it of $y = \sqrt{-(1 + x^2)}$ and $y' = -x/\sqrt{-(1 + x^2)}$ yields an identity. (Verify it.)

Example 3.5. Verify that the function defined by

$$(a) \qquad\qquad y = x^2, \quad -\infty < x < \infty,$$

is a solution of the differential equation

$$(b) \qquad\qquad (y'')^3 + (y')^2 - y - 3x^2 - 8 = 0.$$

Solution. By (a), the function $f(x) = x^2$. Therefore $f'(x) = 2x$, $f''(x) = 2$. Substituting these values in (b) for y, y', y'', we obtain

$$(c) \qquad\qquad 8 + 4x^2 - x^2 - 3x^2 - 8 = 0.$$

Since the left side of (c) is zero, (a), by Definition 3.4, is an explicit solution or simply a solution of (b). Note that (b) is also defined for all x.

Remark. It is the usual practice, when testing whether the function defined by the relation $y = f(x)$ on an interval I is a solution of a given differential equation, to substitute in the given equation the values of y and its derivatives. In the previous Example 3.5, therefore, if we had followed this practice, we would have substituted in (b): $y = x^2$, $y' = 2x$, $y'' = 2$. If an identity resulted, we would then say that (a) is a solution of (b). We shall hereafter, for convenience also follow this practice, but you should always remember that it is the function $f(x)$ and its derivatives which must be substituted in the given differential equation for y and its corresponding derivatives. And if $y = f(x)$ does not define a function then y or $f(x)$ cannot be the solution of any differential equation.

Example 3.51. Verify that the function defined by

(a) $$y = \log x + c, \quad x > 0$$

is a solution of

(b) $$y' = \frac{1}{x}.$$

Solution. Note first that (b) is also defined for all $x > 0$. By (a), $y' = 1/x$. Substituting this value of y' in (b) gives an identity. Hence (a) is a solution of (b) for all $x > 0$.

Example 3.52. Verify that the function defined by

(a) $y = \tan x - x, \quad x \neq (2n + 1)\dfrac{\pi}{2}, \quad n = 0, \pm 1, \pm 2, \cdots,$

is a solution of

(b) $$y' = (x + y)^2.$$

Solution. Here $y = \tan x - x$, $y' = \sec^2 x - 1 = \tan^2 x$. Substitution of these values in (b) for y and y' gives the identity

(c) $$\tan^2 x = (x + \tan x - x)^2 = \tan^2 x.$$

Hence (a) is a solution of (b) in each of the intervals specified in (a).

Comment 3.521. Note by (b) that the differential equation is defined for all x. Its solution, however, as given in (a), is not defined for all x. Hence, the interval, for which the function defined in (a) may be a solution of (b), is the smaller set of intervals given in (a).

Comment 3.53. It is also possible for a function to be defined over an interval and be the solution of a differential equation in only *part* of this interval. For example $y = |x|$ is defined for *all* x. Its graph is shown in Fig. 3.54. It has no derivative when $x = 0$. It satisfies the differential equation $y' = 1$ in the interval $x > 0$, and the differ-

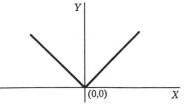

Figure 3.54

ential equation $y' = -1$ in the interval $x < 0$. But it does not satisfy any differential equation in an interval which includes the point $x = 0$.

LESSON 3C. Implicit Solution of a Differential Equation. To test whether an implicit function defined by the relation $f(x,y) = 0$ is a solution of a given differential equation, involves a much more complicated procedure than the testing of one explicitly expressed by $y = f(x)$. The trouble arises because it is usually not easy or possible to solve the equation $f(x,y) = 0$ for y in terms of x in order to obtain the needed function $g(x)$ demanded by Definition 2.81. However, whenever it can be shown that an implicit function does satisfy a given differential equation on an interval I: $a < x < b$, then the relation $f(x,y) = 0$ is called (by an unfortunate usage)* an **implicit solution** of the differential equation.

Definition 3.6. A relation $f(x,y) = 0$ will be called an **implicit solution** of the differential equation

$$(3.61) \qquad\qquad F(x,y,y',\cdots,y^{(n)}) = 0$$

on an interval I: $a < x < b$, if

1. it defines y as an implicit function of x on I, i.e., if there exists a function $g(x)$ defined on I such that $f[x,g(x)] = 0$ for *every* x in I, and if
2. $g(x)$ satisfies (3.61), i.e., if

$$(3.62) \qquad\qquad F[x,g(x),g'(x),\cdots,g^{(n)}(x)] = 0$$

for *every* x in I.

Example 3.63. Test whether

(a) $$f(x,y) = x^2 + y^2 - 25 = 0$$

is an implicit solution of the differential equation

(b) $$F(x,y,y') = yy' + x = 0$$

on the interval I: $-5 < x < 5$.

Solution. We have already shown that (a) defines y as an implicit function of x on I, if we choose for $g(x)$ any *one* of the functions (2.72), (2.73), (2.74). If we choose (2.72), then

(c) $$g(x) = \sqrt{25 - x^2}, \qquad g'(x) = -\frac{x}{\sqrt{25 - x^2}}\,; \quad -5 < x < 5.$$

Substituting in (b), $g(x)$ for y, $g'(x)$ for y', there results

(d) $$F[x,g(x), g'(x)] = \sqrt{25 - x^2}\left(-\frac{x}{\sqrt{25 - x^2}}\right) + x = 0.$$

---

*Actually $f(x,y) = 0$ is an equation and an equation is *never* a solution of a differential equation. Only a *function* can be a solution. What we really mean when we say $f(x,y) = 0$ is a solution of a differential equation is that the *function* $g(x)$ defined by the relation $f(x,y) = 0$ is the solution. See Definition 3.6, also Comment 3.43.

Since the left side of (d) is zero, the equation is an identity in x. Therefore, both requirements of Definition 3.6 are satisfied, and (a) is therefore an implicit solution of (b) on I.

Example 3.64. Test whether

(a) $\qquad f(x,y) = x^3 + y^3 - 3xy = 0, \quad -\infty < x < \infty,$

is an implicit solution of

(b) $\quad F(x,y,y') = (y^2 - x)y' - y + x^2 = 0, \quad -\infty < x < \infty.$

Solution. Unlike the previous example, it is not easy to solve for y to find the required function $g(x)$. However, we have already shown that (a) defines y as an implicit function of x, if we choose for $g(x)$ any *one* of the curves in Fig. 2.86. (Be sure to refer to the graphs shown in this figure.) If we choose the first, then $g'(x)$ does not exist when $x = 2^{2/3}$. Hence, (a) cannot be a solution of (b) for *all* x, but we shall show that (a) is a solution of (b) in any interval which excludes this point $x = 2^{2/3}$. Since we do not have an explicit expression for $g(x)$, we cannot substitute in (b), $g(x)$ for y, $g'(x)$ for y' to determine whether $F[x,g(x), g'(x)] = 0$. What we do is to differentiate (a) implicitly to obtain

(c) $\quad 3x^2 + 3y^2 y' - 3xy' - 3y = 0, \qquad (y^2 - x)y' - y + x^2 = 0.$

Since (c) now agrees with (b), we know that the slope of the function $g(x)$ implicitly defined by (a) and explicitly defined by the graph, satisfies (b) at every point x in any interval excluding $x = 2^{2/3}$. Hence both requirements of Definition 3.6 are satisfied in any interval which does not include the point $x = 2^{2/3}$. Therefore the function $g(x)$ defined by the first graph in Fig. 2.86 is an implicit solution of (a) in any interval not containing the point $x = 2^{2/3}$.

If we choose for $g(x)$ the second curve in Fig. 2.86, then $g'(x)$ does not exist when $x = 0$ and $x = 2^{2/3}$. For this $g(x)$, (a) will be a solution of (b) in any interval which excludes these two points.

If we choose for $g(x)$ the third curve in Fig. 2.86, then $g'(x)$ does not exist when $x = 0$. For this $g(x)$, (a) will be a solution of (b) in any interval which excludes this point.

Comment 3.65. The example above demonstrates the possibility of an implicit function being defined over an interval and being the solution of a differential equation in only *part* of the interval.

Comment 3.651. The standard procedure in calculus texts to prove that (a) is a solution of (b) is the following. Differentiate (a) implicitly. If it yields (b), then (a) is said to be an implicit solution of (b). If you operate blindly in this manner, then you are likely to assert that $x^2 +$

$y^2 = 0$ is an implicit solution of $x + yy' = 0$, since differentiation of the first gives the second. But $x^2 + y^2 = 0$ does not define y implicitly as a function of x on an interval. Only the point $(0,0)$ satisfies this formula. To assert, therefore, that $x^2 + y^2 = 0$ is an implicit solution of $x + yy' = 0$ because it satisfies the differential equation is meaningless.

Example 3.66. Test whether

(a) $$xy^2 - e^{-y} - 1 = 0$$

is an implicit solution of the differential equation

(b) $$(xy^2 + 2xy - 1)y' + y^2 = 0.$$

Solution. If we worked blindly and used the method of implicit differentiation as taught in the calculus, then from (a), we would obtain by differentiation

(c) $$2xyy' + y^2 + e^{-y}y' = 0, \qquad (2xy + e^{-y})y' + y^2 = 0.$$

Although (c) is not identical with (b), it can be made so if we replace e^{-y} in the second equation of (c) by its value $xy^2 - 1$ as determined from (a). Working blindly then, we would assert that (a) is an implicit solution of (b). But is it? Well, let us see. By Definition 3.6, we must first show that (a) defines y as an implicit function of x on an *interval*. If we write (a) as

(d) $$y = \pm \sqrt{\frac{1 + e^{-y}}{x}},$$

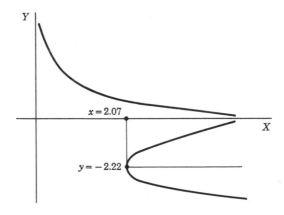

Figure 3.67

we see, since e^{-y} is always positive, that y is defined only for $x > 0$. Hence the interval for which (a) *may* be a solution of (b) must exclude values of $x \leqq 0$. Here again as in the previous example, we cannot easily solve for y explicitly in terms of x, so that we must resort to a graph to determine $g(x)$. It is given in Fig. 3.67. From the graph, we see that there are three choices of $g(x)$. If $g(x)$ is the upper branch, then (a) is an implicit solution of (b) for all $x > 0$. And if $g(x)$ is *either* of the two lower branches, one above the line $y = -2.22$, the other below this line, then (a) is an implicit solution of (b) for $I: x > 2.07$ approximately.

EXERCISE 3

1. Determine the order of the following differential equations.

   (a) $dy + (xy - \cos x)\, dx = 0.$ (b) $y'' + xy'' + 2y(y')^3 + xy = 0.$

   (c) $\left(\dfrac{d^2 y}{dx^2}\right)^3 - (y''')^4 + x = 0.$ (d) $e^{y'''} + xy'' + y = 0.$

2. Prove that the functions in the right-hand column below are solutions of the differential equations in the left-hand columns. (Be sure to state the common interval for which solution and differential equation make sense.)

   (a) $y' + y = 0$ $y = e^{-x}.$

   (b) $y' = e^x$ $y = e^x.$

   (c) $\dfrac{d^2 y}{dx^2} = \dfrac{1}{\sqrt{1 - x^2}}$ $y = x \operatorname{Arc\,sin} x + \sqrt{1 - x^2}.$

   (d) $f'(x) = f''(x)$ $y = e^x + 2.$

   (e) $xy' = 2y$ $y = x^2.$

   (f) $(1 + x^2)y' = xy$ $y = \sqrt{1 + x^2}.$

   (g) $\cos \theta \dfrac{dr}{d\theta} - 2r \sin \theta = 0$ $r = a \sec^2 \theta.$

   (h) $y'' - y = 0$ $y = ae^x + be^{-x}.$

   (i) $f'(x) = \frac{1}{3}f(x)$ $f(x) = 2e^{x/3}.$

   (j) $xy' + y = y^2$ $y = \dfrac{2}{x + 2}.$

   (k) $x + yy' = 0$ $y = \sqrt{16 - x^2}.$

3. Show that the differential equation
$$\left|\frac{dy}{dx}\right| + |y| + 1 = 0$$
   has no solutions.

4. Determine whether the equations on the right define implicit functions of x. For those which do, determine whether they are implicit solutions of the differential equations on the left.

   (a) $y^2 - 1 - (2y + xy)y' = 0$ $y^2 - 1 = (x + 2)^2.$

   (b) $e^{x-y} + e^{y-x}\dfrac{dy}{dx} = 0$ $e^{2y} + e^{2x} = 1.$

   (c) $\dfrac{dy}{dx} = -\dfrac{y}{x}$ $x^2 + y^2 + 1 = 0.$

ANSWERS 3

1. (a) 1.   (b) 2.   (c) 3.   (d) 3.

2. (a) $-\infty < x < \infty$.   (b) $-\infty < x < \infty$.   (c) $-1 < x < 1$.
  (d) $-\infty < x < \infty$.   (e) $x \neq 0$.   (f) $-\infty < x < \infty$.

  (g) $\theta \neq \pm \dfrac{\pi}{2}, \pm \dfrac{3\pi}{2}, \pm \cdots$.   (h) $-\infty < x < \infty$.

  (i) $-\infty < x < \infty$.   (j) $x \neq 0, -2$.   (k) $-4 < x < 4$.

4. (a) Yes, if one makes y single valued. Implicit solution.
  (b) Yes. Implicit solution, $x \neq 0$.
  (c) Function undefined.

LESSON 4.   The General Solution of a Differential Equation.

LESSON 4A.   Multiplicity of Solutions of a Differential Equation.
We assume at the outset that you have understood clearly the material of the previous lesson so that when we say "solve a differential equation" or "find a solution of a differential equation," or "the solution of a differential equation is," you will know what is meant (see Comment 3.43). Or if we omit intervals for which a function or a differential equation is defined, we expect that you will be able to fill in this omission yourself.

When you studied the theory of integration in the calculus, you solved some simple differential equations of the form $y' = f(x)$. For example, you learned that, if

$$(4.1) \qquad\qquad y' = e^x,$$

then its solution, obtained by a simple integration, is

$$(4.11) \qquad\qquad y = e^x + c,$$

where c can take on any numerical value. And if

$$(4.12) \qquad\qquad y'' = e^x,$$

then its solution, obtained by integrating (4.12) twice, is

$$(4.13) \qquad\qquad y = e^x + c_1 x + c_2,$$

where now c_1 and c_2 can take on arbitrary values. Finally, if

$$(4.14) \qquad\qquad y''' = e^x,$$

then its solution, obtained by integrating (4.14) three times, is

$$(4.15) \qquad\qquad y = e^x + c_1 x^2 + c_2 x + c_3,$$

where c_1, c_2, c_3 can take on any numerical values.

Two conclusions seem to stem from these examples. First, if a differential equation has a solution, it has infinitely many solutions (remember the c's can have infinitely many values). Second, if the differential equation is of the first order, its solution contains one arbitrary constant; if of the second order, its solution contains two arbitrary constants; if of the nth order, its solution contains n arbitrary constants. That both conjectures are in fact false can be seen from the following examples.

Example 4.2.   The first order differential equation

(a)
$$(y')^2 + y^2 = 0,$$

also the second order differential equation

(b)
$$(y'')^2 + y^2 = 0,$$

each has only the *one* solution $y = 0$.

Example 4.21.   The first order differential equation

(a)
$$|y'| + 1 = 0,$$

also the second order differential equation

(b)
$$|y''| + 1 = 0,$$

has *no* solution.

Example 4.22.   The first order differential equation

(a)
$$xy' = 1$$

has *no* solution if the interval I is $-1 < x < 1$. Formally one can solve (a) to obtain

(b)
$$y = \log |x| + c,$$

but this function is discontinuous at $x = 0$. By Definition 3.4, a solution must satisfy the differential equation for *every* x in I.

Remark.   If $x < 0$, then by (b) above

(c)
$$y = \log (-x) + c_1, \quad x < 0,$$

is a valid solution of (a). And if $x > 0$, then by (b)

(d)
$$y = \log x + c_2, \quad x > 0,$$

is a valid solution of (a). The line $x = 0$, therefore, divides the plane into two regions; in one (c) is valid, in the other (d) is valid. There is no solution, however, if the region includes the line $x = 0$.

Example 4.23.   The first order differential equation

(a) $$(y' - y)(y' - 2y) = 0$$

has the solution

(b) $$(y - c_1 e^x)(y - c_2 e^{2x}) = 0,$$

which has *two* arbitrary constants instead of the usual one.

These examples should warn you not to jump immediately to the conclusion that every differential equation has a solution, or if it does have a solution that this solution will contain arbitrary constants equal in number to the order of the differential equation. It should comfort you to know, however, that there are large classes of differential equations for which the above conjectures are true, and that these classes include most of the equations which you are likely to encounter. *For these classes only, then, we can assert: the solution of a differential equation of order n contains n arbitrary constants $c_1, c_2, \cdots, c_n$.*

It is customary to call a solution which contains n constants $c_1, c_2, \cdots, c_n$ an **n-parameter family of solutions,** and to refer to the constants c_1 to c_n as **parameters.** In this new notation, we would say (4.11) is a 1-parameter family of solutions of (4.1); (4.13) is a 2-parameter family of solutions of (4.12), etc.

Definition 4.3.   The *functions* defined by

(4.31) $$y = f(x, c_1, c_2, \cdots, c_n)$$

of the $n + 1$ variables, $x, c_1, c_2, \cdots, c_n$ will be called an **n-parameter family of solutions** of the nth order differential equation

(4.32) $$F(x,y,y', \cdots, y^{(n)}) = 0,$$

if for each choice of a set of values $c_1, c_2, \cdots, c_n$, the resulting *function* $f(x)$ defined by (4.31) (it will now define a function of x alone) satisfies (4.32), i.e., if

(4.33) $$F(x,f,f', \cdots, f^{(n)}) = 0.$$

For the classes of differential equations we shall consider, we can now assert: *a differential equation of the nth order has an n-parameter family of solutions.*

Example 4.34.   Show that the functions defined by

(a) $$y = f(x, c_1, c_2) = 2x + 3 + c_1 e^x + c_2 e^{2x}$$

of the three variables x, c_1, c_2, are a 2-parameter family of solutions of the second order differential equation

(b) $\qquad F(x,y,y',y'') = y'' - 3y' + 2y - 4x = 0.$

Solution. Let a,b be any two values of c_1, c_2 respectively. Then, by (a),

(c) $\qquad y = f(x) = 2x + 3 + ae^x + be^{2x}.$

[Note that (c) now defines a function only of x.] The first and second derivatives of (c) are

(d) $\qquad y' = f'(x) = 2 + ae^x + 2be^{2x}, \qquad y'' = f''(x) = ae^x + 4be^{2x}.$

Substituting in (b) the values of f, f' and f'', as found in (c) and (d), for y, y', y'', we obtain

(e) $\quad F(x,f,f',f'') = ae^x + 4be^{2x} - 6 - 3ae^x - 6be^{2x}$
$$+ 4x + 6 + 2ae^x + 2be^{2x} - 4x = 0.$$

You can verify that the left side of (e) reduces to zero. Hence by Definition 4.3, (a) is a 2-parameter family of solutions of (b).

LESSON 4B. Method of Finding a Differential Equation if Its n-Parameter Family of Solutions Is Known. We shall now show you how to find the differential equation when its n-parameter family of solutions is known. You must bear in mind that although the family will contain the requisite number of n arbitrary constants, the nth order differential equation whose solution it is, contains no such constants. In solving problems of this type, therefore, these constants must be eliminated. Unfortunately a standard method of eliminating these constants is not always the easiest to use. There are frequently simpler methods which cannot be standardized and which will depend on your own ingenuity.

Example 4.4. Find a differential equation whose 1-parameter family of solutions is

(a) $\qquad\qquad y = c \cos x + x.$

Solution. In view of what we have already said, we assume that since (a) contains one constant, it is the solution of a first order differential equation. Differentiating (a), we obtain

(b) $\qquad\qquad y' = -c \sin x + 1.$

This differential equation cannot be the one we seek since it contains the parameter c. To eliminate it, we multiply (a) by $\sin x$, (b) by $\cos x$ and add the equations. There results

(c) $\qquad y \sin x + y' \cos x = x \sin x + \cos x,$

$\qquad (y' - 1) \cos x + (y - x) \sin x = 0,$

$$y' = (x - y) \tan x + 1, \quad x \neq \pm \frac{\pi}{2}, \pm \frac{3\pi}{2}, \cdots,$$

which is the required differential equation. [We could also have solved (b) for c and substituted its value in (a).] Note that the interval for which (a) is a solution of (c) must exclude certain points even though the function (a) is defined for these points.

Example 4.5. Find a differential equation whose 2-parameter family of solutions is

(a) $\qquad\qquad\qquad y = c_1 e^x + c_2 e^{-x}.$

Solution. Since (a) contains two parameters, we assume it is the solution of a second order differential equation. We therefore differentiate (a) twice, and obtain

(b) $\qquad\qquad\qquad y' = c_1 e^x - c_2 e^{-x},$

(c) $\qquad\qquad\qquad y'' = c_1 e^x + c_2 e^{-x}.$

Because of the presence of the constants c_1 and c_2 in (c), it cannot be the differential equation we seek. A number of choices are available for eliminating c_1 and c_2. We could, for example, solve (a) and (b) simultaneously for c_1 and c_2 and then substitute these values in (c). This method is a standard one which is always available to you, provided you know how to solve the pair of equations. An easier method is to observe that the right side of (c) is the same as the right side of (a). Hence, by equating their left sides, we have

(d) $\qquad\qquad\qquad y'' - y = 0,$

which is the required differential equation.

Example 4.51. Find a differential equation whose 2-parameter family of solutions is

(a) $\qquad\qquad y = c_1 \sin x + c_2 \cos x + x^2.$

Solution. Since (a) contains two constants, we assume it is the solution of a second order differential equation. Hence we differentiate (a) twice, and obtain

(b) $\qquad\qquad y' = c_1 \cos x - c_2 \sin x + 2x,$

(c) $\qquad\qquad y'' = -c_1 \sin x - c_2 \cos x + 2$

Here again you could use the standard method of finding c_1 and c_2 by solving (a) and (b) simultaneously, and then substituting these values in (c). [Or you could solve (b) and (c) simultaneously for c_1 and c_2 and substitute these values in (a)]. An easier method is to observe from (c) that

(d) $$c_1 \sin x + c_2 \cos x = 2 - y''$$

Substitution of (d) in (a) gives

(e) $$y = 2 - y'' + x^2 \quad \text{or} \quad y'' = x^2 - y + 2,$$

which is the required differential equation.

Example 4.52.   Find a differential equation whose 1-parameter family of solutions represents a family of circles with centers at the origin.

Solution.   Here the family of solutions is not given to us in the form of a mathematical equation. However, the family of circles with center at the origin is

(a) $$x^2 + y^2 = r^2, \quad r > 0.$$

Since (a) has only 1-parameter r, we assume it is the solution of a first order differential equation. Hence we differentiate (a) once, and obtain

(b) $$x + yy' = 0,$$

which is the required differential equation. Note that in this example the parameter r was eliminated in differentiating (a), and we were thus able to obtain the required differential equation immediately.

LESSON 4C.   General Solution. Particular Solution. Initial Conditions.   An n-parameter family of solutions of an nth order differential equation has been called traditionally a "general" solution of the differential equation. And the function which results when we give a definite set of values to the constants $c_1, c_2, \cdots, c_n$ in the family has been called a "particular solution" of the differential equation.

Traditionally then, for example, $y = ce^x$ which is a 1-parameter family of solutions of $y' - y = 0$, would be called its general solution. And if we let $c = -2$, then $y = -2e^x$ would be called a particular solution of the equation. It is evident that an infinite number of particular solutions can be obtained from a general solution: one for each value of c.

A general solution, if it is to be worthy of its name, should contain *all* solutions of the differential equation, i.e., it should be possible to obtain every particular solution by giving proper values to the constants $c_1, c_2, \cdots, c_n$. Unfortunately, there are differential equations which have solutions not obtainable from the n-parameter family no matter what values are given to the constants. For example, the first order differential equation

(4.6) $$y = xy' + (y')^2$$

has for a solution the 1-parameter family

(4.61) $$y = cx + c^2.$$

Traditionally, this solution, since it contains the required one parameter, would be called the general solution of (4.6). However, it is not the general solution in the real meaning of this term since it does not include every particular solution. The function

(4.62) $$y = -\frac{x^2}{4}$$

is also a solution of (4.6). (Verify it.) And you cannot obtain this function from (4.61) no matter what value you assign to c. [(4.61) is a first degree equation; (4.62) is a second degree equation.]

Unusual solutions of the type (4.62), i.e., those which cannot be obtained from an n-parameter family or the so-called general solution, have traditionally been called **"singular solutions."** We shall show below by examples that the use of these terms—general solution and singular solution—in their traditional meanings is undesirable. Rather than being helpful in the study of differential equations, their use leads only to confusion.

Consider for example the first order differential equation

(4.63) $$y' = -2y^{3/2}.$$

Its solution is

(4.64) $$y = \frac{1}{(x + c)^2}.$$

(Verify it.) But (4.63) has another solution

(4.65) $$y = 0,$$

which cannot be obtained from (4.64) by assigning any value to c. By the traditional definition, therefore, $y = 0$ would be called a singular solution of (4.63). However, we can also write the solution of (4.63) as

(4.651) $$y = \frac{C^2.}{(Cx + 1)^2}.$$

[Now verify that (4.651) is a solution of (4.63).] In this form, $y = 0$ is not a singular solution at all. It can be obtained from (4.651) by setting $C = 0$. Hence use of the traditional definitions for general solution and singular solution in this example leads us to the uncomfortable contradiction that a solution can be both singular and nonsingular, depending on the choice of representation of the 1-parameter family.

Here is another example. The first order differential equation

$$(4.652) \qquad (y' - y)(y' - 2y) = 0$$

has the following *two distinct* 1-parameter family of solutions

$$(4.653) \qquad y = c_1 e^x,$$

$$(4.654) \qquad y = c_2 e^{2x}.$$

[Verify that each of these families satisfies (4.652).]

If we call (4.653) the general solution of (4.652), as it should be called traditionally since it contains the required one parameter, then the entire family of functions (4.654) is, in the traditional sense, singular solutions. They cannot be obtained from (4.653) by giving any values whatever to c_1. If we call (4.654) the general solution of (4.652), as well we may in the traditional sense, since it too contains the requisite one parameter, then all the functions (4.653) are singular solutions. Hence, use of the traditional definitions for general solution and singular solution again leads us, in this example, to the uncomfortable contradiction that a family of solutions can be both general and singular.

In this text, therefore, we shall not call an n-parameter family of solutions a general solution, unless we can prove that it actually contains *every* particular solution without exception. If we cannot, we shall use the term n-parameter family of solutions. In such cases, we shall make no attempt to assert that we have obtained *all* possible solutions, but shall claim only to having found an n-parameter family. Every solution of the given differential equation, in which no arbitrary constants are present, whether obtained from the family by giving values to the arbitrary constants in it or by any other means, will be called, in this text, a particular solution of a differential equation. In our meaning of the term, therefore, (4.62) is a particular solution of (4.6), not a singular solution.

Definition 4.66.   A solution of a differential equation will be called a **particular solution** if it satisfies the equation and does not contain arbitrary constants.

Definition 4.7.   An n-parameter family of solutions of a differential equation will be called a **general solution** if it contains *every* particular solution of the equation.

Since there is an infinite number of ways of choosing the n arbitrary constants $c_1, c_2, \cdots, c_n$ in an n-parameter family, one may well wonder how they are determined. What we usually want is the one solution of the infinitely many that will satisfy certain conditions. For instance, we may observe in an experiment, that at time $t = 0$ (i.e., at the start of an experiment) a body is 10 feet from an origin and is moving with a velocity of 20 ft/sec. The constants then must be so chosen that when $t = 0$, the solution will give the value 10 feet for its position and 20 ft/sec for its velocity. For example, assume the motion of the body is given by the 2-parameter family

(a) $$x = 16t^2 + c_1 t + c_2,$$

where x is the distance of the particle from an origin at time t. Its velocity, obtained by differentiating (a), is

(b) $$v = 32t + c_1.$$

Hence we must choose the constants c_1 and c_2 so that when $t = 0$, $x = 10$, and $v = 20$. Substituting these values of t, x, and v in (a) and (b), we find $c_2 = 10$, $c_1 = 20$. The particular solution, therefore, which satisfies the given conditions of this problem is

(c) $$x = 16t^2 + 20t + 10.$$

Definition 4.71. The n conditions which enable us to determine the values of the arbitrary constants $c_1, c_2, \cdots, c_n$ in an n-parameter family, if given in terms of one value of the independent variable, are called **initial conditions.**

In the example above, the given conditions were initial ones. Both the value of the function and of its derivative were given in terms of the one value $t = 0$.

Comment 4.72. Normally the number of initial conditions must equal the order of the differential equation. There are, as usual, exceptional cases where this requirement can be modified. For our classes of differential equations, however, this statement will be a true one.

Example 4.8. Find a 1-parameter family of solutions of the differential equation

(a) $$y y' = (y + 1)^2,$$

and the particular solution for which $y(2) = 0$. [This notation, $y(2) = 0$, is a shorthand way of stating the initial conditions. Here these are $x = 2$, $y = 0$. It means that the point $(2,0)$ must lie on or satisfy the particular solution.]

Solution. If $y \neq -1$, we may divide (a) by $(y + 1)^2$ and obtain

(b) $$\int \frac{y}{(y + 1)^2} \, dy = \int dx, \quad y \neq -1.$$

Performing the indicated integrations gives

(c) $$\frac{1}{y+1} + \log |y + 1| = x + c, \quad y \neq -1,$$

which is the required 1-parameter family. To find the particular solution for which $x = 2$, $y = 0$, we substitute these values in (c) and obtain

(d) $$1 = 2 + c \quad \text{or} \quad c = -1.$$

Substituting (d) in (c), there results the required particular solution,

(e) $$\frac{1}{y+1} + \log |y + 1| = x - 1, \quad y \neq -1.$$

NOTE. The function defined by $y = -1$ which we had to discard to obtain (c) is also a solution of (a). (Verify it.) Hence,

(f) $$y + 1 = 0$$

is also a particular solution of (a). It is a particular solution which cannot be obtained from the family (c) by assigning any value to the constant c.

EXERCISE 4

In problems 1–3, show that each of the functions on the left is a 2-parameter family of solutions of the differential equation on its right.

1. $y = c_1 + c_2 e^{-x} + \dfrac{x^3}{3}$, $y'' + y' - x^2 - 2x = 0$.

2. $y = c_1 e^{-2x} + c_2 e^{-x} + 2e^x$, $y'' + 3y' + 2y - 12e^x = 0$.

3. $y = c_1 x + c_2 x^{-1} + \frac{1}{2} x \log x$, $x^2 y'' + x y' - y - x = 0$.

In problems 4 and 5, show that each of the functions on the left is a 3-parameter family of solutions of the differential equation on its right.

4. $y = e^x \left(c_1 + c_2 x + c_3 x^2 + \dfrac{x^3}{6} \right)$, $y''' - 3y'' + 3y' - y - e^x = 0$.

5. $y = c_1 + c_2 e^x + c_3 e^{-x} + \left(\dfrac{1}{12} + \dfrac{9 \cos 2x - 7 \sin 2x}{520} \right) e^{2x}$,

$$y''' - y' - e^{2x} \sin^2 x = 0.$$

In each of problems 6–17, find a differential equation whose solution is the given n-parameter family.

6. $y = cx + c^3$.

7. $x^2 - cy + c^2 = 0$.

8. $y = c_1 \cos 3x + c_2 \sin 3x$.

9. $r = \theta \tan (\theta + c)$.

10. $y = cx + 3c^2 - 4c$.

11. $y = \sqrt{c_1 x^2 + c_2}$.

12. $y = c_1 e^{c_2 x}$.

13. $y = x^3 + \dfrac{c}{x}$.

14. $y = c_1 e^{2x} + c_2 e^{-2x}$.

15. $(y - c)^2 = cx$.

16. $r = a(1 - \cos \theta)$.

17. $\log y = c_1 x^2 + c_2$.

Find a differential equation whose solution is

18. A family of circles of fixed radii and centers on the x axis.

19. A family of circles of variable radii, centers on the x axis and passing through the origin.

20. A family of circles with centers at (h,k) and of fixed radius.

21. A family of circles with centers in the xy-plane and of variable radii. *Hint.* Write the equation of the family as $x^2 + y^2 - 2c_1x - 2c_2y + 2c_3 = 0$.

22. A family of parabolas with vertices at the origin and foci on the x axis.

23. A family of parabolas with foci at the origin and vertices on the x axis.

24. A family of parabolas with foci and vertices on the x axis.

25. A family of parabolas with axes parallel to the x axis and with a fixed distance $a/2$ between the vertex and focus of each parabola.

26. A family of equilateral hyperbolas whose asymptotes are the coordinate axes.

27. A family of straight lines whose y intercept is a function of its slope.

28. A family of straight lines that are tangents to the parabola $y^2 = 2x$.

29. A family of straight lines that are tangents to the circle $x^2 + y^2 = c^2$, where c is a constant.

30. Find a 1-parameter family of solutions of the differential equation $dy = y\,dx$ and the particular solution for which $y(3) = 1$.

ANSWERS 4

6. $y = xy' + (y')^3$.

7. $x^2(y')^2 - 2xyy' + 4x^2 = 0$.

8. $y'' + 9y = 0$.

9. $\theta r' = \theta^2 + r^2 + r$.

10. $y = (x - 4)y' + 3(y')^2$.

11. $xyy'' + x(y')^2 - yy' = 0$.

12. $yy'' = (y')^2$.

13. $y'x = 4x^3 - y$.

14. $y'' = 4y$.

15. $4x(y')^2 + 2xy' - y = 0$.

16. $(1 - \cos\theta)\dfrac{dr}{d\theta} = r\sin\theta$.

17. $xyy'' - yy' - x(y')^2 = 0$.

18. $(yy')^2 + y^2 = a^2$.

19. $2xyy' + x^2 - y^2 = 0$.

20. $[1 + (y')^2]^3 = a^2(y'')^2$.

21. $y'''[1 + (y')^2] = 3y'(y'')^2$.

22. $2xy' = y$.

23. $y(y')^2 + 2xy' - y = 0$.

24. $yy'' + (y')^2 = 0$.

25. $ay'' + (y')^3 = 0$.

26. $xy' + y = 0$.

27. $y = xy' + f(y')$.

28. $2x(y')^2 - 2yy' + 1 = 0$.

29. $y = xy' \pm c\sqrt{(y')^2 + 1}$.

30. $y = ce^x,\ y = e^{x-3}$.

LESSON 5. Direction Field.

LESSON 5A. Construction of a Direction Field. The Isoclines of a Direction Field.

Before beginning a formal presentation of techniques which are available for solving certain types of differential equations, we wish to emphasize the geometric significance of a solution of a first order differential equation. In many practical problems, a rough geometrical approximation to a solution, such as those we shall describe below and in later lessons, may be all that is needed. Let

$$(5.1) \qquad\qquad y = f(x) \quad \text{or} \quad f(x,y) = 0$$

define a function of x, whose derivative y' exists on an interval $I: a < x < b$.

Then y' will give the slope of the graph of this function at each point whose x coordinate is in I, i.e., y' will give the direction of the tangent to the curve at each of these points. When, therefore, we are asked to find a 1-parameter family of solutions of

$$(5.11) \qquad y' = F(x,y), \quad a < x < b,$$

we are in effect being asked the following. Find a family of curves, every member of which has at each of its points a slope given by (5.11).

Definition 5.12. If $y = f(x)$ or $f(x,y) = 0$ defines y as a function of x which satisfies (5.11) on an interval I, then the graph of this function is called an **integral curve**, i.e., it is the graph of a function which is a solution of (5.11).

Therefore even if we cannot find an elementary function which is a solution of (5.11), we can by (5.11) draw a small line element at any point (x,y), for which x is in I, to represent the slope of an integral curve. And if this line is short enough, the curve itself over that length will resemble the line. For example, let us assume that y' by (5.11) has the value 2 at the point (4,3). This means that at (4,3), the slope of an integral curve is 2. Hence we can draw a short line at this point with slope 2. In a similar manner we can draw, theoretically, such short lines over all that part of the plane for which (5.11) is valid.

These lines are called **line elements** or sometimes **lineal elements.** The totality of such lines has been given various descriptive names. We shall use the term **direction field.*** Any curve which has at each of its points one of these line elements as a tangent will satisfy (5.11), and will therefore be the graph of a particular solution.

Example 5.2. Construct a direction field for the differential equation

$$(a) \qquad y' = x + y.$$

Solution. Table 5.21 gives the values of y' for the integer coordinates from -5 to 5. In Fig. 5.22, we have drawn the line elements for these values of y' and also one integral curve. It is the graph of the particular solution

$$(b) \qquad y = e^x - x - 1$$

of (a)

The construction of line elements is unquestionably a tedious job. Further, if a sufficient number of them is not constructed in close proximity, it may be difficult or impossible to choose the correct line element

---

*Other names are **slope field, lineal element diagram.**

Table 5.21

x / y	−5	−4	−3	−2	−1	0	1	2	3	4	5
−5	−10	−9	−8	−7	−6	−5	−4	−3	−2	−1	0
−4	−9	−8	−7	−6	−5	−4	−3	−2	−1	0	1
−3	−8	−7	−6	−5	−4	−3	−2	−1	0	1	2
−2	−7	−6	−5	−4	−3	−2	−1	0	1	2	3
−1	−6	−5	−4	−3	−2	−1	0	1	2	3	4
0	−5	−4	−3	−2	−1	0	1	2	3	4	5
1	−4	−3	−2	−1	0	1	2	3	4	5	6
2	−3	−2	−1	0	1	2	3	4	5	6	7
3	−2	−1	0	1	2	3	4	5	6	7	8
4	−1	0	1	2	3	4	5	6	7	8	9
5	0	1	2	3	4	5	6	7	8	9	10

for the particular integral curve we wish to find. If such doubt exists in a certain neighborhood, it then becomes necessary to construct additional line elements in this area until the doubt is resolved. Fortunately there

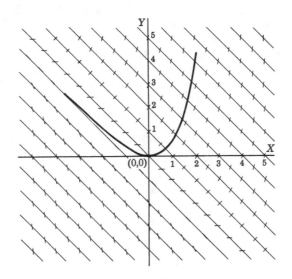

Figure 5.22

exist certain aids which can facilitate the construction of line elements. One of these is to make use of the **isoclines of a direction field.** We shall explain its meaning below.

In (a), let y' equal any value, say 3. Then (a) becomes

(c) $x + y = 3,$

which in effect says that the slope y' has the value 3 at each point where the integral curve crosses this line. [Look at the table of values given in 5.21. At all points which satisfy (c) and are therefore points on this line, as for example $(5,-2)$, $(0,3)$, $(1,2)$, etc., $y' = 3$.] Hence we can quickly draw a great many lineal elements on the line (c). All we need do is to construct at any point on it a line element with slope 3. This line therefore has been called appropriately an **isocline** of the direction field. For each different value of y', we obtain a different isocline. All the straight lines drawn in Fig. 5.22 are isoclines. In general, therefore, if

(5.23) $y' = F(x,y),$

then each curve for which

(5.24) $F(x,y) = k,$

where k is any number, will be an **isocline** of the direction field determined by (5.23). Every integral curve will cross the isocline with a slope k.

Remark. For our illustration, we chose an $F(x,y)$ which, when set equal to k, could be solved explicitly for y. We were therefore able to find the isoclines of the direction field without much trouble. We should warn you however, that in many practical cases, (5.24) may be more difficult to solve than the given differential equation itself. In such cases, we must resort to other means to find a solution.

An integral curve which has been drawn by means of a direction field may be looked upon as if it were formed by a particle moving in such a way that it is tangent to each of its line elements. Therefore the path of this particle (which remember is an integral curve) is sometimes referred to as a **streamline** of the field moving in the direction of the field. Every student of physics has witnessed the formation of a direction field when he has gently tapped a glass, covered with iron filings, which had been placed over a bar magnet. Each iron filing assumes the direction of a line element, and the imaginary curve which has the proper line elements as tangents is a streamline.

LESSON 5B. The Ordinary and Singular Points of the First Order Equation (5.11). In the example of the previous lesson, each point (x,y) in the plane determined one and only one lineal element. Now consider the following example.

Example 5.3. Construct a direction field for the differential equation

(a) $$y' = \frac{2(y-1)}{x}, \quad x \neq 0.$$

Solution. (See Fig. 5.31.) We observe from (a) that we can construct line elements at every point of the plane excepting at those points whose x coordinate is zero. If therefore we were attempting to find a particular integral curve of (a) by means of a direction field construction, we could do so as long as we did not cross the x axis. For example, if we

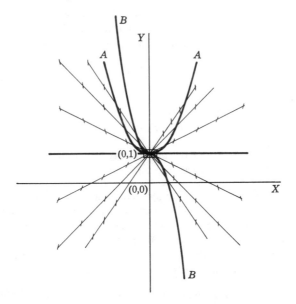

Figure 5.31

began at a point in the second quadrant of the plane and followed a streamline, we would be stopped at the point $(0,1)$ since by (a), y' is meaningless there. Even if we had concluded that an arbitrary assignment of the value zero to y' at this point would seem reasonable and give continuity to the integral curve made by the streamline, so that (a) would now read

(b) $$y' = \frac{2(y-1)}{x}, \quad x \neq 0$$
$$= 0, \qquad x = 0, y = 1,$$

we would be at a loss to know which streamline to follow after crossing $(0,1)$. If you will draw sufficient lineal elements in the neighborhood of

(0,1), it will soon become evident to you, that with the new definition of y' as given in (b), an infinite number of integral curves lies on the point (0,1) with slope zero. Hence after crossing this point, one could follow any first, third, or fourth quadrant streamline, or even another second quadrant streamline.

To overcome this difficulty, we could specify two sets of initial conditions in place of the usual one. For example, we could require our solution to lie on the point $(-3,2)$, and after crossing (0,1) to go through the point $(2,-1)$. These two initial conditions would then fix a particular integral curve. This annoying difficulty arises because of the necessity of excluding $x = 0$ from the interval of definition. Actually, there are two distinct solutions of (a), namely

(c)
$$y = c_1 x^2 + 1, \quad x < 0,$$
$$y = c_2 x^2 + 1, \quad x > 0.$$

[Be sure to verify that each function defined in (c) is a solution of (a).] Since the point (0,1) satisfies both equations in (c) and since we have agreed to define the slope y' as equal to zero at this point, we can write (c) as

(d)
$$y = c_1 x^2 + 1, \quad x \leq 0,$$
$$y = c_2 x^2 + 1, \quad x \geq 0.$$

In this form the solutions (d) include every particular solution of the given differential equation with the agreement that y' equals zero when $x = 0, y = 1$.

In the form of solution (d), we can now make the further observation that with the exception of (0,1), no integral curve lies on any point in the plane whose x coordinate is zero. For example the point (0,3) does not satisfy either equation in (d) no matter what values you assign to c_1 and c_2.

We characterize the difference between points like (0,1), (0,3), and (2,3) in the following definitions.

Definition 5.4.   An **ordinary point** of the first order differential equation (5.11) is a point in the plane which lies on *one and only one* of its integral curves.

Definition 5.41.   A **singular point** of the first order differential equation (5.11) is a point in the plane which meets the following two requirements:

1. It is not an ordinary point, i.e., it does not lie on any integral curve or it lies on more than one integral curve of (5.11).

2. If a circle of arbitrarily small radius is drawn about the point (i.e., the radius may be as small as one wishes), there is at least one ordinary point in its interior. (We describe this condition by saying the singular point is a limit of ordinary points.)

In the above Example 5.3, every circle, no matter how small, drawn about any point on the y axis, say about the point $(0,3)$, contains not only one ordinary point but also infinitely many such points.

Remark. Requirement 2 is needed to exclude extraneous points. For example, if $y' = \sqrt{1 - x^2}$, then only points whose x coordinates lie between -1 and 1 need be considered. If, therefore, we defined a singular point by requirement 1 alone, then a point like $(3,7)$ would be singular. This point, however, is extraneous to the problem.

This example has served three purposes:

1. It has shown you what an ordinary point and a singular point are.
2. It has shown you the need for specifying intervals for which a differential equation and its solution have meaning. You cannot work automatically and blindly and write as a solution of (a)

(e) $$y = cx^2 + 1, \quad -\infty < x < \infty.$$

If you did that, and even if you defined $y' = 0$ at $(0,1)$, you could get from (e) only the *parabolic curves* as solutions. One of these is shown in Fig. 5.31. It is marked A, and was obtained from (e) by setting $c = 1$ [equivalent to setting $c_1 = 1$ and $c_2 = 1$ in (d)]. If in (d), we set $c_1 = 2$ and $c_2 = -2$, we get the integral curve marked B in the graph. And if in (d) we set $c_1 = 1$ and $c_2 = -2$, we get the curve which is marked A in the second quadrant and B in the fourth quadrant.

3. It shows once more that not every first order differential equation has a 1-parameter family of solutions for its general solution. The differential equation in this example requires two 1-parameter families to include *all* possible solutions.

EXERCISE 5

1. Construct a direction field for the differential equation
$$y' = 2x.$$
Draw an integral curve.
2. Construct a direction field for the differential equation
$$y' = \frac{2y}{x}.$$

Where are its singular points? How many parameters are required to include all possible solutions? Draw the integral curve that goes through the points $(-1,2)$ and $(2,-1)$.

3. Construct a direction field for the differential equation

$$y' = \frac{x - y}{x + y}.$$

Draw an integral curve that goes through the point $(1,1)$.

4. Find the isoclines of the direction field and indicate the slope at each point where an integral curve crosses an isocline, (a) for problem 2, (b) for problem 3.

5. Describe the isoclines of the direction field, if $y' = x^2 + 2y^2$.

ANSWERS 5

2. On line $x = 0$; two parameters.

4. (a) Isoclines are the family of straight lines through the origin: $2y = cx$. Slope of an integral curve at each point where it crosses an isocline is equal to c.
(b) Isoclines: $x(1 - c) = y(1 + c)$, slope c.

5. Isoclines are the family of ellipses $x^2 + 2y^2 = c$. At each point where an integral curve crosses one of these ellipses, the slope of the integral curve is c.

Chapter 2

Special Types of Differential Equations of the First Order

Introductory Remarks. In this chapter we begin the study of formal methods of solving special types of first order differential equations. A few preliminary observations however should be instructive and helpful.

1. It is unfortunately true that only very special types of first order differential equations possess solutions (remember a solution is a function) which can be expressed in terms of the elementary functions mentioned in Lesson 2E. Most first order differential equations, in fact, one could say almost all, cannot be thus expressed.

2. There is no connection between the appearance of a differential equation and the ease or difficulty of finding its solution in terms of elementary functions. The differential equation

$$\frac{dy}{dx} = x^2 + y$$

does not look less complicated than

$$\frac{dy}{dx} = x^2 + y^2 \quad \text{or} \quad \frac{dy}{dx} = e^{x^2}.$$

Yet the first has an elementary function for its solution; the other two do not.

3. If the solution you have found can be expressed only in the implicit form $f(x,y) = 0$, it will usually be of little practical value. An implicit solution is frequently such a complicated expression that it is almost impossible to find the needed function $g(x)$ which it implicitly defines, (see Definition 2.81). And without a knowledge of the function $g(x)$ or at least a knowledge of what a rough graph of $g(x)$ looks like, the solution will not be of much use to you. While we shall show you, therefore, in the lessons which follow, formal techniques for finding solutions of a first order differential equation, keep in mind, if the solution is an implicit one,

that other forms of solutions we shall describe later, such as geometric solutions, series solutions, and numerical solutions, will be of far greater practical importance to you.

4. If you start with an algebraic equation and follow a certain procedure to find a solution for a variable x, it is possible that the value thus obtained is extraneous. For example, the usual procedure followed to solve the equation $\sqrt{x^2 + 4x - 3} = 1 - 2x$ is to square both sides and then factor the resulting equation. If you do this you will obtain the solutions $x = 2$ and $x = \frac{2}{3}$. However, both values are extraneous. Neither solution satisfies the given equation. (Verify it.) Similarly, in showing you a procedure that will lead you to a solution of a differential equation, it is possible that the function thus obtained will be extraneous. Hence, to be certain a function is a solution of a given differential equation, you should always verify that it does in fact satisfy the given equation.

5. Finally, and we cannot emphasize this point too strongly, examples generally found in textbooks are "textbook" examples. They are inserted as illustrations in order to clarify the subject matter under discussion. Hence they are carefully selected to yield "nice, relatively easy" solutions. Actual practical problems are avoided since they may require, for instance, the determination of the imaginary roots of a fourth degree equation, or the solving of a system of four or more equations in a corresponding number of unknowns—burdensome and time-consuming problems to say the least.

LESSON 6.   Meaning of the Differential of a Function. Separable Differential Equations.

LESSON 6A.   Differential of a Function of One Independent Variable.   We assume in this Lesson 6A that all functions are differentiable on an interval. Let $y = f(x)$ define y as a function of x. Then its derivative $f'(x)$ will give the slope of the curve at any point $P(x,y)$ on it, i.e., it is the slope of the tangent line drawn to the curve at P.

It is evident from Fig. 6.12, that

(6.1) $$f'(x) = \tan \alpha = \frac{dy}{\Delta x}.$$

Hence,

(6.11) $$dy = f'(x)\, \Delta x.$$

We call dy the **differential** of y, i.e., it is the differential of the function defined by $y = f(x)$. From (6.11) we note that the differential of y, namely dy, is dependent on the abscissa x (remember as the point P changes, $f'(x)$ changes), and on the size of Δx. We see, therefore, that whereas $y = f(x)$ defines y as a function of one independent variable x,

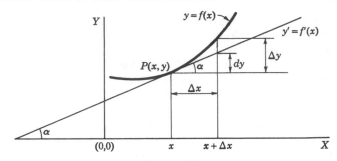

Figure 6.12

the differential dy is a function of two independent variables x and Δx. We indicate this dependence of dy on x and Δx by writing it as

$$(dy)(x,\Delta x).$$

Hence,

Definition 6.13. Let $y = f(x)$ define y as a function of x on an interval I. The **differential of y,** written as dy (or df) is defined by

$$(6.14) \qquad (dy)(x,\Delta x) = f'(x)\ \Delta x.$$

Note. We shall want to apply Definition 6.13 to the function defined by $y = x$. Therefore, in order to distinguish between the *function* defined by $y = x$ and the *variable x,* we place the symbol $\wedge$ over the x so that

$$(6.15) \qquad y = \hat{x}$$

will define the function that assigns to each value of the independent variable x the same unique value to the dependent variable y.

Theorem 6.2. *If*

$$(6.21) \qquad y = \hat{x},$$

then

$$(6.22) \qquad (dy)(x,\Delta x) \equiv (d\hat{x})(x,\Delta x) = \Delta x.$$

Proof. Since

$$(a) \qquad y = \hat{x}$$

defines the function that assigns to each value of the independent variable x the same unique value to the dependent variable y, its graph is the straight line whose slope is given by

$$(b) \qquad y' \equiv f'(x) = 1.$$

Substituting $f'(x) = 1$ in (6.14), we obtain (6.22).

Comment 6.23. If in (6.14) we replace Δx by its value as given in (6.22), it becomes

(6.24) $$(dy)(x,\Delta x) = f'(x)(d\hat{x})(x,\Delta x).$$

In words (6.24) says that if $y = f(x)$ defines y as a function of x, then the differential of y is the product of the derivative of the function f and the differential of the function defined by $y = \hat{x}$. The relation (6.24) is the correct one, but in the course of time, it became customary to write (6.24) in the more familiar form

(6.25) $$dy = f'(x)\, dx, \qquad \frac{dy}{dx} = f'(x).$$

Example 6.26. If

(a) $$y = x^2,$$

defines y as a function of x, find dy.

Solution. Here $f(x) = x^2$. Therefore $f'(x) = 2x$. Hence by (6.24)

(b) $$(dy)(x,\Delta x) = 2x\, d\hat{x}(x,\Delta x),$$

which is customarily written as

(c) $$dy = 2x\, dx.$$

The importance of the definition of the differential as given in 6.13 lies in the following theorem.

Theorem 6.3. *If $y = f(x)$ defines y as a function of x and $x = g(t)$, $y = f[g(t)] = F(t)$, define x and y as functions of t, then*

(6.31) $$(dy)(t,\Delta t) = f'(x)(dx)(t,\Delta t).$$

Proof. Since $x = g(t)$ defines x as a function of t, we have by (6.24)

(a) $$(dx)(t,\Delta t) = g'(t)\, d\hat{t}(t,\Delta t),$$

where $x = \hat{t}$ defines the function which assigns to each value of the independent variable t, the same value to the dependent variable x. By hypothesis $y = f[g(t)]$ defines y as a function of t. Therefore by (6.24) and the chain rule of differentiation,

(b) $$(dy)(t,\Delta t) = f'[g(t)][g'(t)\, d\hat{t}(t,\Delta t)].$$

In (b) replace the last expression in brackets on the right, by its equal left side of (a) and replace $g(t)$ by its equal x. The result is (6.31).

To summarize:

1. If $y = f(x)$, then $(dy)(x,\Delta x) = f'(x) \, d\hat{x}(x,\Delta x)$.

2. If $y = f(x)$ and $x = g(t)$ so that $y = f[g(t)]$, then $(dy)(t,\Delta t) = f'(x)(dx)(t,\Delta t)$, where dy and dx are differentials of y and x respectively. The first is the differential of $f[g(t)]$; the second is the differential of $g(t)$.

In both cases 1 and 2, we shall follow the usual custom and write

(6.32) $$dy = f'(x) \, dx \quad \text{or} \quad \frac{dy}{dx} = f'(x).$$

Comment 6.33. If $y = f(x)$ and $x = g(t)$, then $y = f[g(t)]$ defines y as a function of t. The independent variable is therefore t; the dependent variables x and y. In general if y is a dependent variable, the increment $\Delta y \neq dy$, see Fig. 6.12. It follows therefore that $\Delta x \neq dx$ since here x is also a dependent variable. Thus there is no justification in replacing an increment Δx by dx in "$dy = f'(x) \, \Delta x$." However, if both dy and dx are differentials as defined in 6.13, then as we proved in Theorem 6.3, "$dy = f'(x) \, dx$" even when x is itself dependent on a third variable t.

LESSON 6B.   Differential of a Function of Two Independent Variables. Let $z = f(x,y)$ define z as a function of the two independent variables x and y. Then, following the analogy of the one independent variable treatment, we define the differential of z as follows.

Definition 6.4. Let $z = f(x,y)$ define z as a function of x and y. The **differential of z**, written as dz or df, is defined by

(6.41) $$(dz)(x,y,\Delta x,\Delta y) = \frac{\partial f(x,y)}{\partial x} \Delta x + \frac{\partial f(x,y)}{\partial y} \Delta y.$$

Note that whereas z is a function of two independent variables, the differential of z is a function of four independent variables.

Theorem 6.42. *If*

(6.43) $$z = f(x,y) = \hat{x},$$

where $\hat{x}$ has the usual meaning, then

(6.44) $$(dz)(x,y,\Delta x,\Delta y) \equiv d\hat{x}(x,y,\Delta x,\Delta y) = \Delta x.$$

Proof. Here $z = f(x,y) = \hat{x}$. Hence

$$\frac{\partial f(x,y)}{\partial x} \equiv \frac{\partial \hat{x}}{\partial x} = 1, \qquad \frac{\partial \hat{x}}{\partial y} = 0.$$

Substituting these values in (6.41), we obtain (6.44).

Similarly, it can be shown, if $z = f(x,y) = \hat{y}$, that

(6.45) $$(d\hat{y})(x,y,\Delta x,\Delta y) = \Delta y,$$

where $\hat{y}$ has the usual meaning. Substituting (6.44) and (6.45) in (6.41), we obtain

(6.46)
$$(dz)(x,y,\Delta x,\Delta y) = \frac{\partial f(x,y)}{\partial x} (d\hat{x})(x,y,\Delta x,\Delta y) + \frac{\partial f(x,y)}{\partial y} (d\hat{y})(x,y,\Delta x,\Delta y).$$

This relation (6.46) is the correct one. However, in the course of time, it became the custom to write (6.46) as

(6.47)
$$dz = \frac{\partial f(x,y)}{\partial x} dx + \frac{\partial f(x,y)}{\partial y} dy.$$

Keep in mind that dx and dy mean $d\hat{x}$ and $d\hat{y}$ and are therefore differentials, not increments. With this understanding of the meaning of dx and dy we shall now state, but not prove, an important theorem analogous to Theorem 6.3 in the case of one independent variable. The theorem asserts that (6.47) is valid even when x and y are both dependent on other variables.

Theorem 6.5. *If $z = f(x,y)$ defines z as a function of x and y, and $x = x(r,s,\cdots)$, $y = y(r,s,\cdots)$, $z = f[x(r,s,\cdots), y(r,s,\cdots)] = F(r,s,\cdots)$ define x, y, and z as functions of r, s, and a finite number of other variables (indicated by the dots after s), then*

(6.51) $(dz)(r,s,\cdots,\Delta r,\Delta s,\cdots) = \dfrac{\partial f(x,y)}{\partial x} (dx)(r,s,\cdots,\Delta r,\Delta s\cdots)$

$$+ \frac{\partial f(x,y)}{\partial y} (dy)(r,s,\cdots,\Delta r,\Delta s,\cdots).$$

Here also we shall follow the usual custom and write (6.51) as

(6.52)
$$dz = \frac{\partial f(x,y)}{\partial x} dx + \frac{\partial f(x,y)}{\partial y} dy.$$

Example 6.53. Find dz if

(a)
$$z = f(x,y) = x^3 + 3x^2 y + y^3 + 5.$$

Solution. Here

(b)
$$\frac{\partial f(x,y)}{\partial x} = 3x^2 + 6xy, \quad \frac{\partial f(x,y)}{\partial y} = 3x^2 + 3y^2.$$

Hence by (6.52)

(c)
$$dz = (3x^2 + 6xy)\, dx + (3x^2 + 3y^2)\, dy.$$

LESSON 6C. Differential Equations with Separable Variables. The first order differential equations we shall study in this chapter will be those which can be written in the form

(6.6) $$Q(x,y) \frac{dy}{dx} + P(x,y) = 0.$$

Written in this form it is assumed that y is the dependent variable and x is the independent variable. If we multiply (6.6) by dx, it becomes

(6.61) $$P(x,y) \, dx + Q(x,y) \, dy = 0.$$

Written in this form, either x or y may be considered as being the dependent variable. In both cases, however, dy and dx are *differentials* and not increments.

Although (6.6) and (6.61) are not the most general equations of the first order, they are sufficiently inclusive to cover most of the applications which you will meet. Examples of such equations are

(a) $$\frac{dy}{dx} = 2xy + e^x,$$

(b) $$y' = \log x + y,$$

(c) $$(x - 2y) \, dx + (x + 2y + 1) \, dy = 0,$$

(d) $$e^x \cos y \, dx + x \sin y \, dy = 0.$$

If it is possible to rewrite (6.6) or (6.61) in the form

(6.62) $$f(x) \, dx + g(y) \, dy = 0,$$

so that the coefficient of dx is a function of x alone and the coefficient of dy is a function of y alone, then the variables are called **separable**. And after they have been put in the form (6.62), they are said to be **separated**. A 1-parameter family of solutions of (6.62) is then

(6.63) $$\int f(x) \, dx + \int g(y) \, dy = C,$$

where C is an arbitrary constant.

Example 6.64. Find a 1-parameter family of solutions of

(a) $$2x \, dx - 9y^2 \, dy = 0.$$

Solution. A comparison of (a) with (6.62) shows that the variables are separated. Hence, by (6.63), its solution is

(b) $$x^2 - 3y^3 = C.$$

Example 6.65. Find a 1-parameter family of solutions of

(a) $$\sqrt{1 - x^2} \, dx + \sqrt{5 + y} \, dy = 0, \quad -1 \leqq x \leqq 1, y > -5.$$

Solution. A comparison of (a) with (6.62) shows that the variables are separated. Hence its solution by (6.63) is

(b) $\frac{1}{2}x\sqrt{1-x^2}+\frac{1}{2}$ Arc sin x | $\frac{2}{3}(5+y)^{3/2}-C$, $-1 \leq x \leq 1$, $y > -5$.

Comment 6.651. Because of the presence of the inverse sine, (b) implicitly defines a multiple-valued function. By our definition of a function it must be single-valued, i.e., each value of x should determine one and only one value of y. For this reason we have written the inverse sine with a capital A to indicate that we mean only its principal values, namely those values which lie between $-\pi/2$ and $\pi/2$.

Example 6.66. Find a 1-parameter family of solutions of

(a) $$x\sqrt{1-y}\,dx - \sqrt{1-x^2}\,dy = 0;$$

also a particular solution not obtainable from the family.

Solution. We note first that (a) makes sense only if $y \leq 1$ and $-1 \leq x \leq 1$. Further if $y \neq 1$, $x \neq \pm 1$, we can divide (a) by $\sqrt{1-y}\,\sqrt{1-x^2}$ and obtain

(b) $$\frac{x\,dx}{\sqrt{1-x^2}} - \frac{dy}{\sqrt{1-y}} = 0, \quad -1 < x < 1, \ y < 1.$$

This equation is now of the form (6.62). A 1-parameter family of solutions by (6.63) is

(c) $$\sqrt{1-x^2} - 2\sqrt{1-y} = C, \quad -1 < x < 1, \ y < 1.$$

The function $y = 1$, which we had to exclude to obtain (c) also satisfies (a) for values of x between -1 and 1. (Be sure to verify it.) It is a particular solution of (a) that cannot be obtained from the family (c).

Remark. In Fig. 6.67 we have indicated the set in the plane for which the solution (c) is valid. It is bounded on the top by the line $y = 1$, and on the sides by the lines $x = 1$ and $x = -1$. By (c), when $x = \pm 1$ and $y < 1$, a unique value of the constant C and therefore a unique particular solution of (a) is determined. However, if $y < 1$,

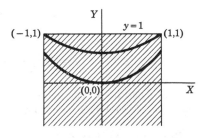

Figure 6.67

$dy/dx \to \pm\infty$ as $x \to \pm 1$. Hence if $y < 1$, the lines $x = \pm 1$ are tangents to the family of integral curves. The corner points $(1,1)$ and $(-1,1)$ can be made part of the set. For these two points, we obtain from (c) the solution

(d) $$\sqrt{1-x^2} - 2\sqrt{1-y} = 0.$$

Squaring (d) and taking the derivative of the resulting function, we obtain

(e) $$y' = \frac{x}{2}.$$

By (a), y' is meaningless at the two corner points $(1,1)$ and $(-1,1)$. However, because of (e) we are led to define the slope of the integral curve at these two corner points as $\frac{1}{2}$ and $-\frac{1}{2}$. We see now that every particular solution of (a) obtained from (c) lies in the region below the line $y = 1$. The line $y = 1$ however, is, as we observed previously, also a particular integral curve of (a) and is part of the set for which solutions of (a) are valid. In Fig. 6.67 we have drawn the integral curve (d) and an integral curve of (c) through $(0,0)$.

Example 6.68. Find a 1-parameter family of solutions of

(a) $$x \cos y \, dx + \sqrt{x + 1} \sin y \, dy = 0;$$

also a particular solution not obtainable from the family.

Solution. We note first that (a) makes sense only if $x > -1$. Further, if $x \neq -1, y \neq \pm \dfrac{\pi}{2}, \pm \dfrac{3\pi}{2}, \cdots$, we can divide (a) by $\sqrt{x + 1} \cos y$ and obtain

(b) $$\frac{x}{\sqrt{x + 1}} \, dx + \frac{\sin y}{\cos y} \, dy = 0, \quad x > -1, \quad y \neq \pm \frac{\pi}{2}, \quad \pm \frac{3\pi}{2} \pm, \cdots.$$

The equation is now of the form (6.62). A 1-parameter family, by (6.63), is

(c) $$\frac{2(x - 2)}{3} \sqrt{x + 1} - \log |\cos y| = C,$$
$$x > -1, \quad y \neq \pm \frac{\pi}{2}, \quad \pm \frac{3\pi}{2} \pm, \cdots.$$

The functions

(d) $$y = \pm \frac{\pi}{2}, \quad \pm \frac{3\pi}{2}, \quad \pm, \cdots,$$

which we had to exclude to obtain (c), also satisfy (a). They are particular solutions of (a) which cannot be obtained from the family (c).

Remark. In Fig. 6.69 we have indicated the set in the plane for which the solutions (c) are valid. It is bounded on the left by the line $x = -1$, and excludes the lines $y = \pm \dfrac{\pi}{2}.$ These last two lines, however, also are particular solutions of (a) not obtainable from (c), and are therefore also part of the set for which solutions of (a) are valid. Each line $y = 3\pi/2, -3\pi/2$, etc., is also a solution of (a).

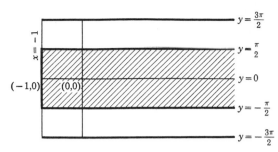

Figure 6.69

Example 6.7. Find a particular solution of

(a) $$xy^2\,dx + (1 - x)\,dy = 0,$$

for which $y(2) = 1$.

Solution. If $y \neq 0$ and $x \neq 1$, we can obtain from (a)

(b) $$\frac{x}{1 - x}\,dx + y^{-2}\,dy = 0, \quad x \neq 1, \quad y \neq 0,$$

which we can write as

(c) $$\left(\frac{1}{1 - x} - 1\right) dx + y^{-2}\,dy = 0, \quad x \neq 1, y \neq 0.$$

By (6.63), a family of solutions of (a) is

(d) $$\log|1 - x| + x + \frac{1}{y} = C, \quad x \neq 1, y \neq 0.$$

To find the particular solution for which $x = 2$, $y = 1$, we substitute these values in (d) and obtain $0 + 2 + 1 = C$, or $C = 3$. Hence (d) becomes

(e) $$\log|1 - x| + x + \frac{1}{y} = 3, \quad x \neq 1, y \neq 0.$$

EXERCISE 6

Find a 1-parameter family of solutions of each of the differential equations 1–16 listed below. Be careful to justify all steps used in obtaining a solution and to indicate intervals for which the differential equation and the solution are valid. Also try to discover particular solutions which are not members of the family of solutions.

1. $y' = y$.   2. $x\,dy - y\,dx = 0$.   3. $\dfrac{dr}{d\theta} = -\sin\theta$.

4. $(y^2 + 1)\,dx - (x^2 + 1)\,dy = 0$.   5. $\dfrac{dr}{d\theta}\cot\theta - r = 2$.

6. $yx^2\, dy - y^3\, dx = 2x^2\, dy.$ **7.** $(y^2 - 1)\, dx - (2y + xy)\, dy = 0.$

8. $x \log x\, dy + \sqrt{1 + y^2}\, dx = 0.$ **9.** $e^{x+1} \tan y\, dx + \cos y\, dy = 0.$

10. $x \cos y\, dx + x^2 \sin y\, dy = a^2 \sin y\, dy.$

11. $\dfrac{dr}{d\theta} = r \tan \theta.$

12. $(x - 1) \cos y\, dy = 2x \sin y\, dx.$

13. $y' = y \log y \cot x.$

14. $x\, dy + (1 + y^2)\, \text{Arc tan } y\, dx = 0.$

15. $dy + x(y + 1)\, dx = 0.$

16. $e^{y^2}(x^2 + 2x + 1)\, dx + (xy + y)\, dy = 0.$

Find a particular solution satisfying the initial condition, of each of the following differential equations 17–21. The initial condition is indicated alongside each equation.

17. $\dfrac{dy}{dx} + y = 0, \quad y(1) = 1.$

18. $\sin x \cos 2y\, dx + \cos x \sin 2y\, dy = 0, \quad y(0) = \pi/2.$

19. $(1 - x)\, dy = x(y + 1)\, dx, \quad y(0) = 0.$

20. $y\, dy + x\, dx = 3xy^2\, dx, \quad y(2) = 1.$

21. $dy = e^{x+y}\, dx, \quad y(0) = 0.$

22. Define the differential of a function of three independent variables; of n independent variables.

ANSWERS 6

1. $y = ce^x.$ **2.** $y = cx.$ **3.** $r = \cos \theta + c.$

4. Arc tan $x =$ Arc tan $y + c$ or $x - y = C(1 + xy)$, where $C = \tan c.$

5. $r = c \sec \theta - 2, \quad r \neq -2, \quad \theta \neq \dfrac{\pi}{2} + n\pi; \quad r = -2.$

6. $(cx + 1)y^2 = (y - 1)x, \quad x \neq 0, \quad y \neq 0; \quad y = 0.$

7. $x + 2 = c\sqrt{y^2 - 1}, \quad x \neq -2, \quad y \neq \pm 1; \quad y = \pm 1.$

8. $\log |x|(y + \sqrt{y^2 + 1}) = c, \quad x \neq 0, \quad x \neq 1.$

9. $e^{x+1} + \log (\csc y - \cot y) + \cos y = c, \quad y \neq n\pi; \quad y = n\pi.$

10. $a^2 - x^2 = c \cos^2 y, \quad x^2 \neq a^2, \quad y \neq \dfrac{\pi}{2} + n\pi; \quad y = \dfrac{\pi}{2} + n\pi.$

11. $r \cos \theta = c, \quad r \neq 0, \quad \theta \neq \dfrac{\pi}{2} + n\pi; \quad r = 0.$

12. $\sin y = (x - 1)^2 e^{2x+c}, \quad x \neq 1, \quad y \neq n\pi; \quad y = n\pi.$

13. $y = e^{c \sin x}, \quad x \neq n\pi, \quad y \neq 0.$

14. $y = \tan (c/x), \quad x \neq 0.$

15. $y = ce^{-x^2/2} - 1, \quad y \neq -1; \quad y = -1.$

16. $x^2 + 2x = e^{-y^2} + c, \quad x \neq -1.$

17. $y = e^{1-x}.$

18. $\cos^2 x \cos 2y = -1.$

19. $(y + 1)(1 - x) = e^{-x}.$

20. $3y^2 = 1 + 2e^{3x^2-12}.$

21. $e^x + e^{-y} = 2.$

LESSON 7.   First Order Differential Equation with Homogeneous Coefficients.

LESSON 7A.   Definition of a Homogeneous Function.

Definition 7.1.   Let $z = f(x,y)$ define z as a function of x and y in a region R. The function $f(x,y)$ is said to be **homogeneous of order n** if it can be written as

$$(7.11) \qquad f(x,y) = x^n g(u),$$

where $u = y/x$ and $g(u)$ is a function of u; or alternately if it can be written as

$$(7.12) \qquad f(x,y) = y^n h(u),$$

where $u = x/y$ and $h(u)$ is a function of u.

Example 7.13.   Determine whether the function

(a) $$f(x,y) = x^2 + y^2 \log \frac{y}{x}, \quad R: x > 0, y > 0,$$

is homogeneous. If it is, give its order.

Solution.   We can write the right side of (a) as

(b) $$x^2 \left(1 + \frac{y^2}{x^2} \log \frac{y}{x}\right).$$

If we now let $u = y/x$, it becomes

(c) $$x^2(1 + u^2 \log u) = x^2 g(u).$$

Hence, by Definition 7.1, (a) is a homogeneous function. Comparing (c) with (7.11), we see that it is of order 2.

Or if we wished, we could have written (a) as

(d) $$y^2 \left(\frac{x^2}{y^2} + \log \frac{y}{x}\right) = y^2(u^2 - \log u) = y^2 h(u),$$

where $u = x/y$. Hence by Definition 7.1, (a) is homogeneous of order 2.

Example 7.14.   Determine whether the function

(a) $$f(x,y) = \sqrt{y} \sin \left(\frac{x}{y}\right)$$

is homogeneous. If it is, give its order.

Solution.   With $u = x/y$, we can write the right side of (a) as

(b) $$y^{1/2} \sin u = y^{1/2} h(u).$$

Hence, by Definition 7.1, (a) is homogeneous of order $\frac{1}{2}$.

Follow the procedure used in Examples 7.13 to 7.14 to check the accuracy of the answers given in the examples below.

Answers

1. $e^{y/x} + \tan(y/x)$. Homogeneous of order zero.
2. $x^2 + \sin x \cos y$. Nonhomogeneous.
3. $\sqrt{x + y}$. Homogeneous of order $\frac{1}{2}$.
4. $\sqrt{x^2 + 3xy + 2y^2}$. Homogeneous of order 1.
5. $x^4 - 3x^3 y + 5y^2 x^2 - 2y^4$. Homogeneous of order 4.

Comment 7.15. An alternate definition of a homogeneous function is the following. A function $f(x,y)$ is said to be **homogeneous of order n** if

$$(7.16) \qquad f(tx,ty) = t^n f(x,y),$$

where $t > 0$ and n is a constant. By using this definition, (a) of Example 7.13 becomes

(a)
$$\begin{aligned} f(tx,ty) &= t^2 x^2 + t^2 y^2 \log \frac{ty}{tx} \\ &= t^2 \left(x^2 + y^2 \log \frac{y}{x} \right) \\ &= t^2 f(x,y). \end{aligned}$$

Hence the given function $x^2 + y^2 \log(y/x)$ is homogeneous of order two. By using this definition, (a) of Example 7.14 becomes

(b)
$$\begin{aligned} f(tx,ty) &= (ty)^{1/2} \sin \left(\frac{tx}{ty} \right) = t^{1/2} \left(y^{1/2} \sin \frac{x}{y} \right) \\ &= t^{1/2} f(x, y). \end{aligned}$$

Hence the given function is of order $\frac{1}{2}$. As an exercise, use (7.16) to test the accuracy of the answers given for the functions 1 to 5 after Example 7.14.

LESSON 7B. Solution of a Differential Equation in Which the Coefficients of dx and dy Are Each Homogeneous Functions of the Same Order.

Definition 7.2. The differential equation

$$(7.3) \qquad P(x,y)\, dx + Q(x,y)\, dy = 0,$$

where $P(x,y)$ and $Q(x,y)$ are each homogeneous functions of order n is called a **first order differential equation with homogeneous coefficients.**

We shall now prove that the substitution in (7.3) of

$$(7.31) \qquad y = ux, \qquad dy = u\,dx + x\,du$$

will always lead to a differential equation in x and u in which the variables are separable and hence solvable for u by Lesson 6. The solution y will then be obtainable by (7.31). The proof is incorporated in the following theorem.

Theorem 7.32. *If the coefficients in (7.3) are each homogeneous functions of order n, then the substitution in it of (7.31) will lead to an equation in which the variables are separable.*

Proof. By hypothesis $P(x,y)$ and $Q(x,y)$ are each homogeneous functions of order n. Hence by Definition 7.1 with $u = y/x$, each can be written as

$$(a) \qquad P(x,y) = x^n g_1(u), \qquad Q(x,y) = x^n g_2(u).$$

Substituting in (7.3) the value of dy as given in (7.31) and the values of $P(x,y)$, $Q(x,y)$ as given in (a), we obtain

$$(b) \qquad x^n g_1(u)\,dx + x^n g_2(u)(u\,dx + x\,du) = 0,$$

which simplifies to

$$(c) \qquad [g_1(u) + u g_2(u)]\,dx + x g_2(u)\,du = 0,$$

$$\frac{dx}{x} + \frac{g_2(u)}{g_1(u) + u g_2(u)}\,du = 0, \quad x \neq 0,\ g_1(u) + u g_2(u) \neq 0,$$

an equation in which the variables x and u have been separated.

Prove as an exercise that the substitution in (7.3) of

$$(7.33) \qquad x = uy, \qquad dx = u\,dy + y\,du$$

will also lead to a separable equation in u and y.

Remark. If the differential equation (7.3) is written in the form

$$(7.4) \qquad \frac{dy}{dx} = \frac{P(x,y)}{Q(x,y)} = F(x,y),$$

then the statement that $P(x,y)$ and $Q(x,y)$ are each homogeneous of order n is equivalent to saying $F(x,y)$ is homogeneous of order 0. For by (7.4) and Definition 7.1

$$(7.41) \qquad F(x,y) = \frac{P(x,y)}{Q(x,y)} = \frac{x^n g_1(u)}{x^n g_2(u)} = x^0 G(u).$$

Example 7.5. Find a 1-parameter family of solutions of

(a) $$(\sqrt{x^2 - y^2} + y)\,dx - x\,dy = 0;$$

also any particular solution not obtainable from the family.

Solution. We observe first that (a) makes sense only if $|y| \leqq |x|$, or $|y/x| \leqq 1$, $x \neq 0$. Second we note by Definition 7.1 that (a) is a differential equation with homogeneous coefficients of order one. We have a choice therefore of either of the substitutions (7.31) or (7.33). By experimenting with both, you quickly will discover that the first is preferable. By using this substitution in (a), we obtain

(b) $$(\sqrt{x^2 - u^2 x^2} + ux)\,dx - x(u\,dx + x\,du) = 0,$$

$$|u| \doteq \left|\frac{y}{x}\right| \leqq 1, \quad x \neq 0.$$

Since $x \neq 0$, we can divide (b) by it to obtain after simplification

(c) $$\pm\sqrt{1 - u^2}\,dx - x\,du = 0, \quad x \neq 0, \quad |u| = \left|\frac{y}{x}\right| \leqq 1,$$

where the $+$ sign is to be used if $x > 0$; the $-$ sign if $x < 0$.* Further if $u \neq \pm 1$, we may divide (c) by $\sqrt{1 - u^2}$. Therefore (c) becomes

(d) $$\frac{dx}{x} = \pm\frac{du}{\sqrt{1 - u^2}}, \quad x \neq 0, \quad |u| = \left|\frac{y}{x}\right| < 1.$$

The variables are now separated. Hence, by (6.63), a 1-parameter family of solutions of (d) is

(e) $$\log x = \text{Arc sin } u + c, \quad |u| < 1, x > 0,$$

$$-\log(-x) = \text{Arc sin } u + c, \quad |u| < 1, x < 0.$$

Replacing u in (e) by its value as given in (7.31), we have

(f) $$\log x = \text{Arc sin } \frac{y}{x} + c, \quad \left|\frac{y}{x}\right| < 1, x > 0,$$

$$-\log(-x) = \text{Arc sin } \frac{y}{x} + c, \quad \left|\frac{y}{x}\right| < 1, x < 0.$$

In obtaining the solution (f), we had to exclude the values $|u| = |y/x| = 1$. This means we had to exclude the functions $y = \pm x$. You can and should verify that these two functions also satisfy (a). They are particular solutions of (a) not obtainable from the family (f).

---

*For real x, $\sqrt{x^2} = x$ if $x \geqq 0$ and $\sqrt{x^2} = -x$ if $x \leqq 0$. For example, if $x = 2$, $\sqrt{2^2} = 2$ and if $x = -2$, $\sqrt{(-2)^2} = -(-2) = 2$.

EXERCISE 7

1. Prove that the substitution in (7.3) of

$$x = uy, \qquad dx = u\, dy + y\, du, \quad y \neq 0,$$

leads to a separable equation.

Find a 1-parameter family of solutions of each of the following equations. Assume in each case that the coefficient of $dy \neq 0$.

2. $2xy\, dx + (x^2 + y^2)\, dy = 0$.
3. $(x + \sqrt{y^2 - xy})\, dy - y\, dx = 0$.
4. $(x + y)\, dx - (x - y)\, dy = 0$.
5. $xy' - y - x \sin(y/x) = 0$.
6. $(2x^2y + y^3)\, dx + (xy^2 - 2x^3)\, dy = 0$.
7. $y^2\, dx + (x\sqrt{y^2 - x^2} - xy)\, dy = 0$.
8. $\dfrac{y}{x} \cos \dfrac{y}{x}\, dx - \left(\dfrac{x}{y} \sin \dfrac{y}{x} + \cos \dfrac{y}{x} \right) dy = 0$.
9. $y\, dx + x \log \dfrac{y}{x}\, dy - 2x\, dy = 0$.
10. $2ye^{x/y}\, dx + (y - 2xe^{x/y})\, dy = 0$.
11. $\left(xe^{y/x} - y \sin \dfrac{y}{x} \right) dx + x \sin \dfrac{y}{x}\, dy = 0$.

Find a particular solution, satisfying the initial condition, of each of the following differential equations.

12. $(x^2 + y^2)\, dx = 2xy\, dy, \quad y(-1) = 0$.
13. $(xe^{y/x} + y)\, dx = x\, dy, \quad y(1) = 0$.
14. $y' - \dfrac{y}{x} + \csc \dfrac{y}{x} = 0, \quad y(1) = 0$.
15. $(xy - y^2)\, dx - x^2\, dy = 0, \quad y(1) = 1$.

ANSWERS 7

2. $3x^2y + y^3 = c$.
3. $y = ce^{-2\sqrt{1 - x/y}}, \quad y > 0, x < y; \quad y = ce^{2\sqrt{1 - x/y}}, \quad y < 0, \quad x > y$.
4. Arc tan $(y/x) - \frac{1}{2} \log (x^2 + y^2) = c$.
5. $y = 2x$ Arc tan cx.
6. $\dfrac{x^2}{y^2} + \log xy = c, \quad x \neq 0, \quad y \neq 0$.
7. $y^2 - cx = y\sqrt{y^2 - x^2}$, or equivalently, $c(y + \sqrt{y^2 - x^2}) = xy, y^2 > x^2$.
8. $y \sin \dfrac{y}{x} = c$.
9. $y = c(1 + \log x/y)$.
10. $2e^{x/y} + \log y = c$.
11. $\log x^2 - e^{-y/x} \left(\sin \dfrac{y}{x} + \cos \dfrac{y}{x} \right) = c$.
12. $y^2 = x^2 + x$.   14. $\log x - \cos \dfrac{y}{x} + 1 = 0$.
13. $\log x + e^{-y/x} = 1$.   15. $x = e^{(x/y)-1}$.

LESSON 8.   Differential Equations with Linear Coefficients.

LESSON 8A.   A Review of Some Plane Analytic Geometry. The first degree equation $ax + by + c = 0$ represents a straight line. For this reason it is called a **linear equation.** (NOTE. The presence of the constant c in the equation prevents the function defined by it from being homogeneous.) If the coefficients of x and y in one linear equation are proportional to the x and y coefficients in another, the two lines they represent are parallel. For example, the two lines

$$3x - 2y + 7 = 0,$$

$$6x - 4y + 3 = 0,$$

are parallel since $3 : -2 = 6 : -4$. If the three constants in one linear equation are proportional to the three constants respectively in a second linear equation, the two lines coincide, i.e., they are the same line. For example, the two lines

$$2x + 3y + 1 = 0,$$

$$4x + 6y + 2 = 0,$$

are coincident. (Do you see why?)

Another concept of analytic geometry that we shall need for this lesson is that of "translation of axes." Let (x,y) be the coordinates of a point P with respect to an origin $(0,0)$ (Fig. 8.1), and let us translate the origin to

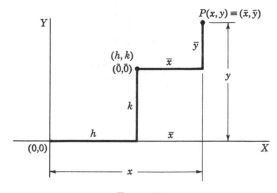

Figure 8.1

a new position whose x and y distances from $(0,0)$ are h and k respectively. To distinguish the new origin from the old one, we call its coordinates $(\bar{0},\bar{0})$. The point P will then have two sets of coordinates, one with respect to $(\bar{0},\bar{0})$, which we designate by $(\bar{x},\bar{y})$, and the other with respect to

(0,0), which we have already designated as (x,y). This means that if a point is measured from (0,0), its coordinates have no bars over them; if it is measured from $(\overline{0},\overline{0})$ its coordinates have bars over them. The question we now ask and whose answer we seek is this. What is the relationship between the two sets of coordinates (x,y) and $(\overline{x},\overline{y})$?

If you will examine Fig. 8.1 carefully, you will see that

(8.11) $$x = \overline{x} + h, \qquad y = \overline{y} + k.$$

Hence by (8.11),

(8.12) $$\overline{x} = x - h, \qquad \overline{y} = y - k.$$

These are the equations of translation. Their purpose, you may recall, is to change more complicated second degree equations into simpler ones by eliminating the first degree terms. We shall now demonstrate how a translation of axes can help solve a differential equation with linear coefficients.

LESSON 8B. Solution of a Differential Equation in Which the Coefficients of dx and dy Are Linear, Nonhomogeneous, and When Equated to Zero Represent Nonparallel Lines. Consider the differential equation

(8.2) $$(a_1x + b_1y + c_1)\,dx + (a_2x + b_2y + c_2)\,dy = 0,$$

in which the coefficients of dx and dy are linear and when equated to zero represent nonparallel lines. We assume also that both c_1 and c_2 are not zero. (If both $c_1 = 0$ and $c_2 = 0$, then (8.2) is a differential equation with homogeneous coefficients which can be solved by the method of Lesson 7.) Since the coefficients in (8.2) are assumed to define nonparallel lines, the pair of equations

(8.21) $$a_1x + b_1y + c_1 = 0,$$
$$a_2x + b_2y + c_2 = 0,$$

formed with them, have a unique point of intersection and therefore a unique solution for x and y. Let us call this point (h,k). If we now translate the origin to (h,k), then by (8.11), (8.2) becomes, with respect to this new origin $(\overline{0},\overline{0})$,

(8.22) $$[a_1(\overline{x} + h) + b_1(\overline{y} + k) + c_1]\,d\overline{x}$$
$$+ [a_2(\overline{x} + h) + b_2(\overline{y} + k) + c_2]\,d\overline{y} = 0,$$

which simplifies to

(8.23) $$[a_1\overline{x} + b_1\overline{y} + (a_1h + b_1k + c_1)]\,d\overline{x}$$
$$+ [a_2\overline{x} + b_2\overline{y} + (a_2h + b_2k + c_2)]\,d\overline{y} = 0.$$

But (h,k) is the point of intersection of the two lines in (8.21) and therefore lies on both of them. Hence the term in the parentheses in each bracket of (8.23) is zero. This equation therefore reduces to

(8.24) $$(a_1\bar{x} + b_1\bar{y})\, d\bar{x} + (a_2\bar{x} + b_2\bar{y})\, d\bar{y} = 0,$$

which is now a homogeneous type solvable for $\bar{x}$ and $\bar{y}$ by the method of Lesson 7. By (8.11) we can then find solutions in terms of x and y.

Note.

1. The left-hand members of the system (8.21) by which h and k are determined are the coefficients in the given differential equation (8.2).

2. Equation (8.24) which is equivalent to (8.2) with respect to a new origin translated to the point (h,k), can be easily obtained from (8.2). Omit the constants c_1 and c_2 and place bars over x and y.

Example 8.25. Find a 1-parameter family of solutions of

(a) $$(2x - y + 1)\, dx + (x + y)\, dy = 0.$$

Solution. The coefficient of dx is linear but nonhomogeneous, and the two lines defined by the coefficients of dx and dy are nonparallel. Hence the procedure outlined above applies. Solving simultaneously the two equations determined by the coefficients of dx and dy, namely,

(b) $$2x - y + 1 = 0,$$
$$x + y = 0,$$

we find that their point of intersection is $(-\frac{1}{3}, \frac{1}{3})$. Hence, in (8.11) $h = -\frac{1}{3}$, $k = \frac{1}{3}$. Translating the origin to the point $(-\frac{1}{3}, \frac{1}{3})$, the equations of translation are by (8.11) and (8.12)

(c) $\quad x = \bar{x} - \frac{1}{3}, \qquad y = \bar{y} + \frac{1}{3}, \qquad \bar{x} = x + \frac{1}{3}, \qquad \bar{y} = y - \frac{1}{3}.$

By (8.24), (a) becomes with respect to this new origin (see Note 2 above)

(d) $$(2\bar{x} - \bar{y})\, d\bar{x} + (\bar{x} + \bar{y})\, d\bar{y} = 0.$$

To solve it, we apply the method of Lesson 7. Let

(e) $$\bar{y} = u\bar{x}, \qquad d\bar{y} = u\, d\bar{x} + \bar{x}\, du,$$

By substituting (e) in (d) and following the procedure outlined in Lesson 7, we obtain

(f) $\quad \log |\bar{x}| = c_1 - \dfrac{1}{\sqrt{2}} \text{ Arc tan } \dfrac{u}{\sqrt{2}} - \dfrac{1}{2} \log |2 + u^2|, \quad \bar{x} \neq 0,$

which is a 1-parameter family of solutions of (d). In (f), we replace u by its value in (e) and multiply by 2. There results

(g) $$\log \left| \overline{x}^2 \left(2 + \frac{\overline{y}^2}{\overline{x}^2} \right) \right| = c - \sqrt{2} \text{ Arc tan } \frac{\overline{y}}{\sqrt{2}\,\overline{x}}, \quad \overline{x} \neq 0.$$

Substituting the last two equations of (c) in (g) gives finally

(h) $$\log \left| 2 \left(\frac{3x+1}{3} \right)^2 + \left(\frac{3y-1}{3} \right)^2 \right| = c - \sqrt{2} \text{ Arc tan } \frac{3y-1}{\sqrt{2}\,(3x+1)},$$

$$x \neq -\frac{1}{3}.$$

Comment 8.26. The solution (h) above is an excellent example of the point made at the beginning of this chapter in introductory remark No. 3. Here is a solution written in implicit form, which has little value for practical purposes. To try to find the function $g(x)$ implicitly defined by this relation would be an extremely laborious if not a hopeless task. In general a 1-parameter family of solutions of a differential equation with linear coefficients or with homogeneous coefficients will usually be a complicated expression of this kind. In these cases, more important for practical purposes than an implicit solution is a knowledge of the approximate behavior of the integral curves. There are fortunately means available by which it is possible to determine the character of these integral curves from the *differential equation (8.2) itself*, without the need to solve it. Since differential equations with homogeneous or linear coefficients arise in practical problems when trying to find an approximation to the behavior of the motion of a particle whose velocities in the x and y directions are given by the two differential equations

$$\frac{dx}{dt} = f(x,y),$$

$$\frac{dy}{dt} = g(x,y),$$

we have deferred to Lesson 32 a further discussion of this important topic. We shall show there, how it is possible to find an approximation of the particle's motion by changing the *two* equations into *one* equation with linear coefficients, and then showing how more useful information can be obtained from the resulting *differential equation* itself than from its usual complicated *implicit solution*. We shall thus be able to learn what the solution (h) in the above example approximately looks like, not from this solution, but from the given differential equation (a); see Example 32.44.

LESSON 8C.   A Second Method of Solving the Differential Equation (8.2) with Nonhomogeneous Coefficients.   In (8.2), let

$$(8.3) \qquad u = a_1 x + b_1 y + c_1,$$
$$v = a_2 x + b_2 y + c_2.$$

Therefore

$$(8.31) \qquad du = a_1\, dx + b_1\, dy,$$
$$dv = a_2\, dx + b_2\, dy.$$

Now solve (8.31) for dx and dy. The substitution in (8.2) of (8.3) and these values of dx and dy will also lead to a differential equation with homogeneous coefficients solvable by the method of Lesson 7.

Example 8.32.   Find a 1-parameter family of solutions of

$$(a) \qquad (2x - y + 1)\, dx + (x + y)\, dy = 0.$$

Solution.   As indicated in (8.3), we let

$$(b) \qquad u = 2x - y + 1, \qquad v = x + y.$$

Therefore

$$(c) \qquad du = 2dx - dy,$$
$$dv = dx + dy,$$

The solution of (c) for dx and dy is

$$(d) \qquad dx = \frac{du + dv}{3}, \qquad dy = -\frac{du - 2\, dv}{3}.$$

Substituting (b) and (d) in (a), we obtain

$$(e) \qquad u\left(\frac{du + dv}{3}\right) - v\left(\frac{du - 2\, dv}{3}\right) = 0,$$

which simplifies to

$$(f) \qquad (u - v)\, du + (u + 2v)\, dv = 0.$$

This equation is now of thē type with homogeneous coefficients. Following the method of Lesson 7, we let

$$(g) \qquad u = tv, \qquad du = t\, dv + v\, dt.$$

Substituting these values in (f), we obtain

(h) $$(tv - v)(t\,dv + v\,dt) + (tv + 2v)\,dv = 0,$$

which reduces to

(i) $$\frac{dv}{v} + \frac{t - 1}{t^2 + 2}\,dt = 0, \quad v \neq 0.$$

Its solution is

(j) $$\log |v| + \tfrac{1}{2} \log (t^2 + 2) - \frac{1}{\sqrt{2}} \operatorname{Arc\,tan} \frac{t}{\sqrt{2}} = c, \quad v \neq 0,$$

$$\log [v^2(t^2 + 2)] = C + \sqrt{2} \operatorname{Arc\,tan} \frac{t}{\sqrt{2}}, \quad v \neq 0.$$

By (g) and (b),

(k) $$t = \frac{u}{v} = \frac{2x - y + 1}{x + y}, \quad x + y \neq 0.$$

Substituting (k) in (j), we obtain

(l) $$\log [(2x - y + 1)^2 + 2(x + y)^2] = C + \sqrt{2} \operatorname{Arc\,tan} \frac{2x - y + 1}{\sqrt{2}\,(x + y)},$$

$$x + y \neq 0.$$

LESSON 8D. Solution of a Differential Equation in Which the Coefficients of dx and dy Define Parallel or Coincident Lines. If the lines defined by the coefficients of dx and dy in (8.2) are parallel, the method of Lesson 8B will not work. Parallel lines do not have a point of intersection and therefore (8.21) has no solution for x and y. In this case we must resort to a different substitution. It is illustrated in the following example.

Example 8.4. Find a 1-parameter family of solutions of

(a) $$(2x + 3y - 1)\,dx + (4x + 6y + 2)\,dy = 0,$$

also any particular solution not obtainable from the family.

Solution. We observe that the lines defined by the coefficients of dx and dy are parallel but not coincident lines. In *all* such cases, the substitution of a new variable for the coefficient of dx *or* of dy will transform the equation into one which is separable. We therefore let

(b) $u = 2x + 3y - 1, \qquad du = 2dx + 3dy, \qquad dx = \dfrac{du - 3dy}{2}.$

Then by (b)

(c) $$2u + 4 = 4x + 6y + 2.$$

Substituting (b) and (c) in (a), we obtain

(d) $$u\left(\frac{du - 3dy}{2}\right) + (2u + 4)\, dy = 0,$$

which simplifies to

(e) $$u\, du + (u + 8)\, dy = 0,$$

an equation whose variables are separable. If $u \neq -8$, (e) can be written as

(f) $$\frac{u}{u + 8}\, du + dy = 0, \quad u \neq -8.$$

Integration of (f) gives

(g) $$u - 8 \log |u + 8| + y = c, \quad u \neq -8.$$

Finally, replace in (g) the value of u as given in (b), noting at the same time that the exclusion of $u = -8$ implies the exclusion of the line $2x + 3y + 7 = 0$. Hence (g) becomes

(h) $$2x + 3y - 1 - 8 \log |2x + 3y + 7| + y = c, \quad 2x + 3y + 7 \neq 0,$$

which is a 1-parameter family of solutions of (a).

The function defined by

(i) $$2x + 3y + 7 = 0,$$

which had to be excluded in obtaining (h) also satisfies (a). (Be sure to verify it.) It is a particular solution not obtainable from the family (h).

Example 8.41. Find a 1-parameter family of solutions of

(a) $$(2x + 3y + 2)\, dx + (4x + 6y + 4)\, dy = 0;$$

also any particular solution not obtainable from the family.

Solution. We observe that the coefficients in (a) define the same line. If we exclude values of x and y for which

(b) $$2x + 3y + 2 = 0,$$

we may divide (a) by it and obtain

(c) $$dx + 2dy = 0.$$

Its solution is

(d) $$x + 2y = c,$$

which is valid for those values of x and y which do not lie on the line $2x + 3y + 2 = 0$. It is the required 1-parameter family. However, the function defined by

(e) $$2x + 3y + 2 = 0$$

also satisfies (a). It is a particular solution not obtainable from the family.

EXERCISE 8

Find a 1-parameter family of solutions of each of the following equations.

1. $(x + 2y - 4)\, dx - (2x - 4y)\, dy = 0$.
2. $(3x + 2y + 1)\, dx - (3x + 2y - 1)\, dy = 0$.
3. $(x + y + 1)\, dx + (2x + 2y + 2)\, dy = 0$.
4. $(x + y - 1)\, dx + (2x + 2y - 3)\, dy = 0$.
5. $(x + y - 1)\, dx - (x - y - 1)\, dy = 0$.
6. $(x + y)\, dx + (2x + 2y - 1)\, dy = 0$.
7. $(7y - 3)\, dx + (2x + 1)\, dy = 0$.
8. $(x + 2y)\, dx + (3x + 6y + 3)\, dy = 0$.
9. $(x + 2y)\, dx + (y - 1)\, dy = 0$.
10. $(3x - 2y + 4)\, dx - (2x + 7y - 1)\, dy = 0$.

Find a particular solution, satisfying the initial condition, of each of the following differential equations.

11. $(x + y)\, dx + (3x + 3y - 4)\, dy = 0, \quad y(1) = 0$.
12. $(3x + 2y + 3)\, dx - (x + 2y - 1)\, dy = 0, \quad y(-2) = 1$.
13. $(y + 7)\, dx + (2x + y + 3)\, dy = 0, \quad y(0) = 1$.
14. $(x + y + 2)\, dx - (x - y - 4)\, dy = 0, \quad y(1) = 0$.

ANSWERS 8

1. $\log [4(y - 1)^2 + (x - 2)^2] - 2 \operatorname{Arc\,tan} \dfrac{2y - 2}{x - 2} = c$.

2. $\log |15x + 10y - 1| + \frac{5}{2}(x - y) = c$.

3. $x + 2y = c$.

4. $x + 2y + \log |x + y - 2| = c$.

5. $(x - 1)^2 + y^2 = ce^{2\,\operatorname{Arc\,tan}\,[y/(x-1)]}$.

6. $x + 2y + \log |x + y - 1| = c$.

7. $7 \log |2x + 1| + 2 \log |7y - 3| = c$.

8. $x + 3y - 3 \log |x + 2y + 3| = c$.

9. $[(x + 2)/(x + y + 1)] + \log |x + y + 1| = c$.

10. $3x^2 - 4xy + 8x - 7y^2 + 2y = c$.

11. $x + 3y + 2 \log (2 - x - y) = 1$.

12. $(2x + 2y + 1)(3x - 2y + 9)^4 = -1$.

13. $(y + 7)^2 (3x + y + 1) = 128$.

14. $\log [(x - 1)^2 + (y + 3)^2] + 2 \operatorname{Arc\,tan} \dfrac{x - 1}{y + 3} = 2 \log 3$.

LESSON 9. Exact Differential Equations.

Before beginning a study of this type of equation, we shall review those concepts from the theory of integration which we shall need.

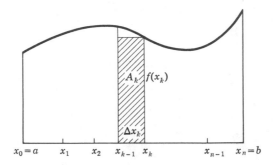

Figure 9.1

1. Let $f(x)$ be a function of x defined on an interval $I: a \leqq x \leqq b$. Let I be divided into n subintervals and call Δx_k the width of the kth subinterval, (Fig. 9.1). Then if

$$(9.11) \qquad\qquad \lim_{n \to \infty} \sum_{k=1}^{n} f(x_k) \, \Delta x_k$$

exists as the number of subintervals increases in such a manner that the largest subinterval approaches zero, we say that

$$(9.12) \qquad\qquad \lim_{n \to \infty} \sum_{k=1}^{n} f(x_k) \, \Delta x_k = \int_a^b f(x) \, dx.$$

This limit is called the **Riemann integral** of $f(x)$ over I. If this limit does not exist, we say $f(x)$ is not Riemann-integrable over I.

2. If $f(x)$ is a continuous function of x on an interval $I: a \leqq x \leqq b$, and

$$(9.13) \qquad\qquad F(x) = \int_{x_0}^{x} f(u) \, du,$$

then by the *fundamental theorem of the calculus*

$$(9.14) \qquad\qquad F'(x) = f(x), \quad a < x < b,$$

or equivalently

$$(9.15) \qquad\qquad \frac{d}{dx} \int_{x_0}^{x} f(u) \, du = f(x).$$

3. Let $P(x,y)$ and $Q(x,y)$ be functions of two independent variables, x,y, both functions being defined on a common domain D. In Lesson 2C (we suggest your rereading this lesson), we showed that a two-dimensional domain may assume various shapes. Hence if we wish to perform, for example, the following integration, the first with respect to x (y constant), and the second with respect to y (x constant),

$$(9.16) \qquad \int_{x_0}^{x} P(x,y)\,dx + \int_{y_0}^{y} Q(x,y)\,dy,$$

we must be sure that (x_0,y_0) is a point of D and that the rectangle determined by the line segments joining the points (x_0,y_0), (x,y_0) and (x_0,y_0), (x_0,y) lies entirely in D. If the domain D is either of the regions shown

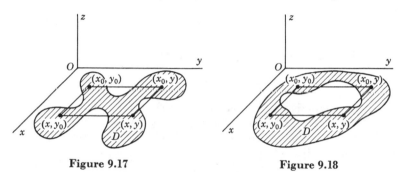

Figure 9.17   **Figure 9.18**

in the shaded areas in Fig. 9.17 or 9.18, then you cannot integrate (9.16) along the straight line from x_0 to x or from y_0 to y because part of these lines are not in D.

There is also another factor to be considered. If the domain D is the region shown in Fig. 9.17, and the rectangle determined by the line segments joining the points (x_0,y_0), (x,y_0) and (x_0,y_0), (x_0,y) lies entirely in D, then (9.16) can be integrated. If, however, the domain D is a region with a hole in it such as in Fig. 9.18, then even if the rectangle were entirely in D, (9.16) cannot be integrated because of this hole in its interior. Therefore, we must in some way distinguish between these two types of regions. The region in Fig. 9.17, i.e., the one which has no hole in it, is called a simply connected region. Its formal definition is the following.

Definition 9.19.   A region is called a **simply connected region** if *every* simple *closed* curve lying entirely in the region *encloses only points of the region.*

To summarize: We can perform the integrations called for in (9.16) only if:

(a) The common domain of definition of the functions is a simply connected region R.

(b) The point (x_0, y_0) is in R.

(c) The rectangle determined by the lines joining the points (x_0, y_0), (x, y_0) and (x_0, y_0), (x_0, y) lies entirely in R.

LESSON 9A. Definition of an Exact Differential and of an Exact Differential Equation. We showed in Lesson 6B that if, for example,

$$(9.2) \qquad z = f(x,y) = 3x^2 y + 5xy + y^3 + 5,$$

then the differential of z [see (6.47)] is

$$(9.21)$$
$$dz = \frac{\partial f(x,y)}{\partial x}\, dx + \frac{\partial f(x,y)}{\partial y}\, dy = (6xy + 5y)\, dx + (3x^2 + 5x + 3y^2)\, dy.$$

If therefore, we had started with the differential expression

$$(9.22) \qquad (6xy + 5y)\, dx + (3x^2 + 5x + 3y^2)\, dy,$$

we would know that it was the **total differential** of the function $f(x,y)$ defined in (9.2). Differential expressions of the type (9.22); i.e., those which are the total differentials of a function $f(x,y)$, are called **exact differentials.** Hence:

Definition 9.23. A differential expression

$$(9.24) \qquad P(x,y)\, dx + Q(x,y)\, dy$$

is called an **exact differential** if it is the total differential of a function $f(x,y)$, i.e., if

$$(9.241) \qquad P(x,y) = \frac{\partial}{\partial x} f(x,y) \quad \text{and} \quad Q(x,y) = \frac{\partial}{\partial y} f(x,y).$$

Setting the differential expression (9.22) equal to zero, we obtain the differential equation

$$(9.25) \qquad (6xy + 5y)\, dx + (3x^2 + 5x + 3y^2)\, dy = 0,$$

whose solution, by (9.2), is

$$(9.26) \qquad f(x,y) = 3x^2 y + 5xy + y^3 = c,$$

valid for all values of x for which (9.26) defines y as an implicit function of x and for which dy/dx exists. [If you have any doubt that (9.26) is an implicit solution of (9.25) or that $\partial f(x,y)/\partial x$ is the dx coefficient in (9.25) or that $\partial f(x,y)/\partial y$ is the dy coefficient in (9.25), verify these statements.]

A differential equation of the type (9.25), i.e., one whose dx coefficient is the partial derivative with respect to x of a function $f(x,y)$ and whose dy

coefficient is the partial derivative with respect to y of the *same* function, is called an exact differential equation. Hence:

Definition 9.27. The differential equation

$$(9.28) \qquad P(x,y)\, dx + Q(x,y)\, dy = 0$$

is called **exact** if there exists a function $f(x,y)$ such that its partial derivative with respect to x is $P(x,y)$ and its partial derivative with respect to y is $Q(x,y)$. In symbolic notation, the definition says that (9.28) is an **exact differential equation** if there exists a function $f(x,y)$ such that

$$(9.29) \qquad \frac{\partial f(x,y)}{\partial x} = P(x,y), \qquad \frac{\partial f(x,y)}{\partial y} = Q(x,y).$$

A 1-parameter family of solutions of the exact differential equation (9.28) is then

$$(9.291) \qquad f(x,y) = c.$$

In the example used above, we knew in advance that (9.25) was exact and that the function defined in (9.26) was its solution, because we started with this function $f(x,y)$ and then set its total differential equal to zero to obtain the differential equation (9.25). In general, however, if a first order differential equation were selected at random, two questions would present themselves. First, how would we know it was exact, and second, if it were exact, how could we find the solution $f(x,y) = c$? The answer to both questions is incorporated in the theorem and proof which follow.

LESSON 9B. Necessary and Sufficient Condition for Exactness and Method of Solving an Exact Differential Equation.

Theorem 9.3. *A necessary and sufficient condition that the differential equation*

$$(9.31) \qquad P(x,y)\, dx + Q(x,y)\, dy = 0$$

be exact is that

$$(9.32) \qquad \frac{\partial}{\partial y} P(x,y) = \frac{\partial}{\partial x} Q(x,y),$$

where the functions defined by $P(x,y)$ and $Q(x,y)$, the partial derivatives in (9.32) and $\partial P(x,y)/\partial x$, $\partial Q(x,y)/\partial y$ exist and are continuous in a simply connected region R.

Note. Although the theorem is valid as stated, the proof will be given only for a rectangular domain contained entirely within the simply connected region R.

Proof of necessary condition, i.e., given (9.31) is exact, to prove (9.32).
Since (9.31) is exact, it follows from Definition 9.27 that there is a function $f(x,y)$ such that

$$(9.33) \qquad \frac{\partial}{\partial x} f(x,y) = {}^{\backprime}P(x,y), \qquad \frac{\partial}{\partial y} f(x,y) = Q(x,y).$$

Because of the assumptions about the functions P and Q stated after
(9.32) and by a theorem in analysis, we are permitted to assert that

1. $\dfrac{\partial}{\partial y} \left(\dfrac{\partial}{\partial x} f(x,y) \right)$ and $\dfrac{\partial}{\partial x} \left(\dfrac{\partial}{\partial y} f(x,y) \right)$ exist.

2. $\dfrac{\partial}{\partial y} \dfrac{\partial}{\partial x} f(x,y) = \dfrac{\partial}{\partial x} \dfrac{\partial}{\partial y} f(x,y),$

i.e., the order in which we take the first and second partial derivatives of
$f(x,y)$ is immaterial. Substituting (9.33) in 2, above, we obtain

$$(9.34) \qquad \frac{\partial}{\partial y} P(x,y) = \frac{\partial}{\partial x} Q(x,y),$$

which is (9.32).

Proof of sufficient condition, i.e., given (9.32), to prove (9.31) is exact.
By Definition 9.27, the proof that (9.31) is exact is equivalent to proving
the existence of a function $f(x,y)$ such that $\partial f/\partial x = P(x,y)$ and $\partial f/\partial y = Q(x,y)$. [In the proof of this sufficient condition, we shall at the same
time discover the method of finding $f(x,y)$.] Hence the function $f(x,y)$, if
it exists, must have the property that

$$(9.35) \qquad \frac{\partial f(x,y)}{\partial x} = P(x,y).$$

Therefore, *with y constant, $f(x,y)$*, by (9.14) and (9.13), must be a function
such that

$$(9.36) \qquad f(x,y) = \int_{x_0}^{x} P(x,y) \, dx + R(y),$$

where x_0 is a constant and $R(y)$ stands for the arbitrary constant of integration. [Remember that in going from (9.36) to (9.35), $R(y)$ and y are
constants.]

But this function $f(x,y)$ must also have the property, by Definition 9.27,
that

$$(9.37) \qquad \frac{\partial}{\partial y} f(x,y) = Q(x,y).$$

Therefore, differentiating (9.36) with respect to y and setting the result equal to $Q(x,y)$, we obtain in symbolic notation

$$(9.38) \qquad \frac{\partial}{\partial y} \int_{x_0}^{x} P(x,y)\, dx + R'(y) = Q(x,y).$$

By hypotheses $P(x,y)$ is continuous. Hence, by a theorem in analysis, we can, in (9.38), put the symbol $\partial/\partial y$ inside the integral sign. It will then read

$$(9.381) \qquad \int_{x_0}^{x} \frac{\partial}{\partial y}\, P(x,y)\, dx + R'(y) = Q(x,y).$$

By (9.32), we can write (9.381) as

$$(9.39) \qquad \int_{x_0}^{x} \frac{\partial}{\partial x}\, Q(x,y)\, dx + R'(y) = Q(x,y).$$

Study the integral in (9.39) carefully. In words, it says: differentiate $Q(x,y)$ with respect to x, with y fixed, and then integrate this result with y still fixed. The net result is to get back the function $Q(x,y)$. [Try it for the function $Q(x,y) = x^2 y$ with y constant.] Hence (9.39) becomes

$$(9.4) \qquad Q(x,y)|_{x_0}^{x} + R'(y) = Q(x,y),$$

which simplifies to [remember $Q(x,y)|_{x_0}^{x} = Q(x,y) - Q(x_0,y)$]

$$(9.41) \qquad R'(y) = Q(x_0,y).$$

Integration of (9.41) gives, by (9.14) and (9.13),

$$(9.42) \qquad R(y) = \int_{y_0}^{y} Q(x_0, y)\, dy,$$

where y_0 is a constant. Substituting (9.42) in (9.36), we obtain finally

$$(9.43) \qquad f(x,y) = \int_{x_0}^{x} P(x,y)\, dx + \int_{y_0}^{y} Q(x_0,y)\, dy,$$

where (x_0,y_0) is a point in R, and the line segments joining the points (x_0,y_0), (x,y_0) and (x_0,y_0), (x_0,y) lie entirely in R.

This function $f(x,y)$ we shall now show is the one we seek. Since the second integral in (9.43) is a function of y, we have, by (9.13) and (9.14),

$$(9.44) \qquad \frac{\partial f(x,y)}{\partial x} = \frac{\partial}{\partial x} \int_{x_0}^{x} P(x,y)\, dx + 0 = P(x,y).$$

And by following the steps from (9.38) to (9.4), we obtain

(9.441) $$\frac{\partial}{\partial y} \int_{x_0}^{x} P(x,y) \, dx = Q(x,y) - Q(x_0,y).$$

By (9.13) and (9.14)

(9.442) $$\frac{\partial}{\partial y} \int_{y_0}^{y} Q(x_0,y) \, dy = Q(x_0,y).$$

Hence, by (9.43), (9.441) and (9.442), it follows that

(9.443) $$\frac{\partial}{\partial y} f(x,y) = Q(x,y).$$

We have thus not only proved the theorem, but have shown at the same time how to find a 1-parameter family of solutions of (9.31). It is, by (9.291) and (9.43),

(9.45) $$f(x,y) = \int_{x_0}^{x} P(x,y) \, dx + \int_{y_0}^{y} Q(x_0,y) \, dy = c,$$

where (x_0,y_0) is a point in R and the rectangle determined by the line segments joining the points (x_0,y_0), (x,y_0) and (x_0,y_0), (x_0,y) lies entirely in R.

Prove as an exercise that if in place of (9.35) we had started with

(9.46) $$\frac{\partial}{\partial y} f(x,y) = Q(x,y),$$

we would have obtained for the solution of (9.31)

(9.47) $$f(x,y) = \int_{y_0}^{y} Q(x,y) \, dy + \int_{x_0}^{x} P(x,y_0) \, dx = c.$$

Remark. Both (9.45) and (9.47) will give a 1-parameter family of solutions of (9.31). You may use whichever you find easiest in a particular problem.

Example 9.5. Show that the following differential equation is exact and find a 1-parameter family of solutions.

(a) $$\cos y \, dx - (x \sin y - y^2) \, dy = 0.$$

Solution. Comparing (a) with (9.31) we see that $P(x,y) = \cos y$ and $Q(x,y) = -x \sin y + y^2$. Therefore $\partial P(x,y)/\partial y = -\sin y$ and $\partial Q(x,y)/\partial x = -\sin y$. Since $\partial P/\partial y = \partial Q/\partial x$, the equation, by Theorem 9.3, is exact and since $P(x,y)$ and $Q(x,y)$ are defined for all x,y, the region R is the whole plane. Hence we may take $x_0 = 0$ and $y_0 = 0$. With $x_0 = 0$,

$Q(x_0,y) = Q(0,y) = y^2$. Thus (9.45) becomes

(b) $$\int_0^x \cos y \, dx + \int_0^y y^2 \, dy = c.$$

Integration of (b) gives (remember in the first integration y is a constant)

(c) $$x \cos y + \frac{y^3}{3} = c,$$

which is the required 1-parameter family.

Example 9.51.   Show that the following differential equation is exact and find a particular solution $y(x)$ for which $y(1) = 0$.

(a) $$(x - 2xy + e^y) \, dx + (y - x^2 + xe^y) \, dy = 0.$$

Solution.   Comparing (a) with (9.31) we see that $P(x,y) = x - 2xy + e^y$ and $Q(x,y) = y - x^2 + xe^y$. Therefore $\partial P/\partial y = -2x + e^y$ and $\partial Q/\partial x = -2x + e^y$. Since $\partial P/\partial y = \partial Q/\partial x$, the equation, by Theorem 9.3, is exact. And since the region R is the whole plane, we may take $x_0 = 0$ and $y_0 = 0$. With $x_0 = 0$, $Q(x_0,y) = y$. Hence (9.45) becomes

(b) $$\int_0^x (x - 2xy + e^y) \, dx + \int_0^y y \, dy = c.$$

Integration of (b) gives

(c) $$\frac{x^2}{2} - x^2y + xe^y + \frac{y^2}{2} = c.$$

To obtain a particular solution for which $x = 1$, $y = 0$, we substitute these values in (c) and find $c = \frac{3}{2}$. Hence the required particular solution is

(d) $$x^2 - 2x^2y + 2xe^y + y^2 = 3.$$

Example 9.52.   Show that the following equation is exact, and find a 1-parameter family of solutions.

(a) $$(x^3 + xy^2 \sin 2x + y^2 \sin^2 x) \, dx + (2xy \sin^2 x) \, dy = 0.$$

Solution.   Here $\partial P/\partial y = 2xy \sin 2x + 2y \sin^2 x$ and

$$\frac{\partial Q}{\partial x} = 2y(2x \sin x \cos x + \sin^2 x) = 2xy \sin 2x + 2y \sin^2 x.$$

Since $\partial P/\partial y = \partial Q/\partial x$, the equation by Theorem 9.3 is exact. In this case, it will be found easier to use (9.47). Since the region R is the whole plane, we may take $x_0 = 0$ and $y_0 = 0$ so that $P(x,0) = x^3$. Hence (9.47) becomes

(b)
$$\int_0^y 2xy \sin^2 x \, dy + \int_0^x x^3 \, dx = c.$$

Integration of (b) gives (remember that this time x is a constant in the first integration)

(c)
$$xy^2 \sin^2 x + \frac{x^4}{4} = c.$$

which is the required 1-parameter family.

Remark. Formulas have the advantage of enabling one to obtain a result relatively easily, but have the disadvantage of being easily forgotten. We therefore outline a method by which you can solve any of the above differential equations directly from Definition 9.27. In our opinion, this method is the preferable one.

Example 9.6. Solve Example 9.5, using Definition 9.27.

Solution. The differential equation is

(a)
$$\cos y \, dx - (x \sin y - y^2) \, dy = 0.$$

We have already proved (a) is exact. By Definition 9.27, therefore, there exists a function $f(x,y)$ such that

(b)
$$\frac{\partial f(x,y)}{\partial x} = P(x,y) = \cos y.$$

Hence integrating (b) *with respect to* x, we obtain

(c)
$$f(x,y) = \int \cos y \, dx + R(y) = x \cos y + R(y).$$

Again by Definition 9.27, this function $f(x,y)$ must also have the property that

(d)
$$\frac{\partial f(x,y)}{\partial y} = Q(x,y) = -x \sin y + y^2.$$

Hence differentiating the last expression in (c) partially with respect to y and substituting this value in (d) we obtain

(e)
$$-x \sin y + R'(y) = -x \sin y + y^2,$$

which simplifies to

(f)
$$R(y) = \int y^2 \, dy = \frac{y^3}{3}.$$

Substituting (f) in (c) gives

(g) $$f(x,y) = x \cos y + \frac{y^3}{3}.$$

By (9.291), a 1-parameter family of solutions of (a) is therefore

(h) $$x \cos y + \frac{y^3}{3} = c,$$

just as we found previously.

As an exercise, solve the other two Examples 9.51 and 9.52 directly from the definition as we did above.

EXERCISE 9

1. Prove formula (9.47).
2. Solve Example 9.51 by the method of Example 9.6.
3. Solve Example 9.52 by the method of Example 9.6.

Show that each of the following differential equations 4–13 is exact and find a 1-parameter family of solutions using formula (9.45) or (9.47), and also the method outlined in Example 9.6.

4. $(3x^2y + 8xy^2)\, dx + (x^3 + 8x^2y + 12y^2)\, dy = 0$.

5. $\left(\dfrac{2xy + 1}{y}\right) dx + \left(\dfrac{y - x}{y^2}\right) dy = 0$.

6. $2xy\, dx + (x^2 + y^2)\, dy = 0$.

7. $(e^x \sin y + e^{-v})\, dx - (xe^{-v} - e^x \cos y\, dy) = 0$.

8. $\cos y\, dx - (x \sin y - y^2)\, dy = 0$.

9. $(x - 2xy + e^v)\, dx + (y - x^2 + xe^v)\, dy = 0$.

10. $(x^2 - x + y^2)\, dx - (e^v - 2xy)\, dy = 0$.

11. $(2x + y \cos x)\, dx + (2y + \sin x - \sin y)\, dy = 0$.

12. $x\sqrt{x^2 + y^2}\, dx - \dfrac{x^2 y}{y - \sqrt{x^2 + y^2}}\, dy = 0$.

13. $(4x^3 - \sin x + y^3)\, dx - (y^2 + 1 - 3xy^2)\, dy = 0$.

14. If $f(x,y)$ is the function defined in (9.43), prove (9.44).

Find a particular solution, satisfying the initial condition, of each of the following differential equations.

15. $e^x(y^3 + xy^3 + 1)\, dx + 3y^2(xe^x - 6)\, dy = 0$, $y(0) = 1$.

16. $\sin x \cos y\, dx + \cos x \sin y\, dy = 0$, $y(\pi/4) = \pi/4$.

17. $(y^2 e^{xy^2} + 4x^3)\, dx + (2xye^{xy^2} - 3y^2)\, dy = 0$, $y(1) = 0$.

ANSWERS 9

4. $x^3y + 4x^2y^2 + 4y^3 = c$.

5. $x^2 + \dfrac{x}{y} + \log |y| = c$.

6. $3x^2y + y^3 = c$.

7. $e^x \sin y + xe^{-v} = c$.

8. $3x \cos y + y^3 = c$.

9. $x^2 - 2x^2y + y^2 + 2xe^v = c$.

10. $2x^3 - 3x^2 + 6xy^2 - 6e^y = c.$ **11.** $x^2 + y \sin x + y^2 + \cos y = c.$
12. $(x^2 + y^2)^{3/2} + y^3 = c.$
13. $3x^4 + 3 \cos x + 3y^3 x - y^3 - 3y = c.$
15. $xe^x y^3 + e^x - 6y^3 = -5.$ **16.** $2 \cos x \cos y = 1.$
17. $e^{xy^2} + x^4 - y^3 = 2.$

LESSON 10.   Recognizable Exact Differential Equations. Integrating Factors.

LESSON 10A.   Recognizable Exact Differential Equations. It is sometimes possible to recognize the solution $f(x,y) = c$ of an exact differential equation without the necessity of resorting to the methods of Lesson 9b. For example, if the exact differential equation is

(10.1) $$2xy^2 \, dx + 2x^2 y \, dy = 0,$$

you might be able to recognize that its solution is

(10.11) $$x^2 y^2 = c.$$

If you cannot, you must of course use the method of solution outlined in the previous lesson.

We list below a number of exact differential equations and their solutions. It will be profitable for you to verify some of them by taking the total differential of the function on the right and seeing if it yields the differential equation on the left.

	Exact Differential Equation	*Solution*
(10.2)	$y \, dx + x \, dy = 0$	$xy = c$
(10.21)	$2xy \, dx + x^2 \, dy = 0$	$x^2 y = c$
(10.22)	$y^2 \, dx + 2xy \, dy = 0$	$xy^2 = c$
(10.23)	$2xy^2 \, dx + 2x^2 y \, dy = 0$	$x^2 y^2 = c$
(10.24)	$3x^2 y^3 \, dx + 3x^3 y^2 \, dy = 0$	$x^3 y^3 = c$
(10.25)	$3x^2 y \, dx + x^3 \, dy = 0$	$x^3 y = c$
(10.26)	$y \cos x \, dx + \sin x \, dy = 0$	$y \sin x = c$
(10.27)	$\sin y \, dx + x \cos y \, dy = 0$	$x \sin y = c$
(10.28)	$ye^{xy} \, dx + xe^{xy} \, dy = 0$	$e^{xy} = c$
(10.29)	$\dfrac{dx}{x} + \dfrac{dy}{y} = 0$	$\log (xy) = c$
(10.3)	$\dfrac{y \, dx - x \, dy}{y^2} = 0$	$\dfrac{x}{y} = c$
(10.31)	$-\dfrac{y \, dx - x \, dy}{x^2} = 0$	$\dfrac{y}{x} = c$
(10.32)	$\dfrac{2xy \, dy - y^2 \, dx}{x^2} = 0$	$\dfrac{y^2}{x} = c$

	Exact Differential Equation	*Solution*
(10.33)	$\dfrac{2xy\,dx - x^2\,dy}{y^2} - 0$	$\dfrac{x^2}{y} - c$
(10.34)	$\dfrac{x\,dy - y\,dx}{x^2 + y^2} = 0$	$\text{Arc tan } \dfrac{y}{x} = c$
(10.35)	$2\,\dfrac{x\,dy - y\,dx}{x^2 - y^2} = 0$	$\log \dfrac{x+y}{x-y} = c$
(10.36)	$-\,\dfrac{3x^2y\,dx - x^3\,dy}{x^6} = 0$	$\dfrac{y}{x^3} = c$
(10.37)	$\dfrac{y^2\,dx - 2xy\,dy}{y^4} = 0$	$\dfrac{x}{y^2} = c$
(10.38)	$-\,\dfrac{y\,dx + x\,dy}{x^2y^2} = 0$	$\dfrac{1}{xy} = c$
(10.39)	$-\,\dfrac{y^3\,dx - xy^2\,dy}{x^4} = 0$	$\dfrac{y^3}{3x^3} = c$
(10.391)	$\dfrac{x\,dx + y\,dy}{\sqrt{x^2 + y^2}} = 0$	$\sqrt{x^2 + y^2} = c$
(10.392)	$e^{3x}\,dy + 3e^{3x}y\,dx = 0$	$e^{3x}y = c$

The exact differential equations on the left are sometimes referred to as **integrable combinations.** We encourage you to add to this list whenever you discover the solution of a new integrable combination.

If a differential equation is exact, but no integrable combination is readily apparent to you, you can always fall back on the method of solution outlined in Lesson 9B. However, it is sometimes possible, if an equation is exact, to solve it more readily and easily by a judicious rearrangement of terms so as to take advantage of any integrable combinations it may possess.

Note. Hereafter when we use the word *solve,* in connection with first order equations, we shall mean "find a 1-parameter family of solutions of the given differential equation."

Example 10.4. Solve

(a) $$\frac{xy + 1}{y}\,dx + \frac{2y - x}{y^2}\,dy = 0, \quad y \neq 0.$$

Solution. By (9.31), $P(x,y)$, $Q(x,y)$ are the respective coefficients of dx, dy. Therefore,

(b) $$\frac{\partial P}{\partial y} = -\frac{1}{y^2} \quad \text{and} \quad \frac{\partial Q}{\partial x} = -\frac{1}{y^2}.$$

Hence the equation is exact. A solution therefore can be obtained by Theorem 9.3. However, if we rewrite the equation as

(c) $$x\,dx + \frac{1}{y}\,dx + \frac{2}{y}\,dy - \frac{x}{y^2}\,dy = 0, \quad y \neq 0,$$

which can be put in the form

(d) $$x \, dx + \frac{2}{y} \, dy + \frac{y \, dx - x \, dy}{y^2} = 0, \quad y \neq 0,$$

we observe that the last term of the left side is, by (10.30), $d(x/y)$. Each of the other terms can be integrated individually. Hence integration of (d) yields

(e) $$\frac{x^2}{2} + 2 \log |y| + \frac{x}{y} = c, \quad y \neq 0,$$

which is the required solution.

Example 10.41. Solve

(a) $$(3e^{3x}y - 2x) \, dx + e^{3x} \, dy = 0.$$

Solution. You can easily verify that (a) is exact. It therefore can be solved by the method of Lesson 9B. However, if we rewrite the equation as

(b) $$3e^{3x}y \, dx + e^{3x} \, dy - 2x \, dx = 0,$$

we observe that the sum of the first two terms is by (10.392), $d(e^{3x}y)$, and that the third term can be integrated individually. Hence integration of (b) yields the solution

(c) $$e^{3x}y - x^2 = c.$$

Example 10.42. Solve

(a) $$(x - 2xy + e^y) \, dx + (y - x^2 + xe^y) \, dy = 0.$$

Solution. We have already shown (see Example 9.51) that this equation is exact, and solved it by use of Lesson 9B. However, if we rewrite it as

(b) $$x \, dx - (2xy \, dx + x^2 \, dy) + (e^y \, dx + xe^y \, dy) + y \, dy = 0,$$

the second expression, by (10.21), is $d(x^2y)$ and the third expression is recognizable as $d(xe^y)$. The solution of (b) is therefore

(c) $$\frac{x^2}{2} - x^2y + xe^y + \frac{y^2}{2} = c.$$

just as we found previously.

LESSON 10B. Integrating Factors.

Definition 10.5. A multiplying factor which will convert an inexact differential equation into an exact one is called an **integrating factor.**

For example, the equation $(y^2 + y)\,dx - x\,dy = 0$ is not exact. If, however, we multiply it by y^{-2}, the resulting equation

$$\left(1 + \frac{1}{y}\right) dx - \frac{x}{y^2}\,dy = 0, \quad y \neq 0,$$

is exact. (Verify it.) Hence by Definition 10.5, y^{-2} is an integrating factor.

Remark. Theoretically an integrating factor exists for every differential equation of the form $P(x,y)\,dx + Q(x,y)\,dy = 0$, but no general rule is known to discover it. Methods have been devised for finding integrating factors for certain special types of differential equations, but the types are so special that the methods are of little practical value. It is evident that if a standard method of finding integrating factors were available then every first order equation of this form would be solvable by this means. Unfortunately this is not the case. However, in the next lesson we shall discuss a special, important, first order differential equation for which an integrating factor is known and for which a standard method for finding it exists.

In the meantime, we shall show by a few examples how an integrating factor, if you are shrewd enough to discover one, can help you solve a differential equation.

Example 10.51. Solve

(a) $$(y^2 + y)\,dx - x\,dy = 0.$$

Solution. We proved to you above that y^{-2} is an integrating factor of (a). Hence multiplication of (a) by y^{-2} will convert it into the exact differential equation

(b) $$\left(1 + \frac{1}{y}\right) dx - \frac{x}{y^2}\,dy = 0, \quad y \neq 0.$$

We can now solve (b) by means of Lesson 9B or 10A. If we rearrange the terms to read

(c) $$dx + \frac{y\,dx - x\,dy}{y^2} = 0,$$

the second term by (10.3) is $d(x/y)$. Hence by integration of (c) we obtain the solution

(d) $$x + \frac{x}{y} = c; \quad y = \frac{x}{c - x}, \quad y \neq 0.$$

NOTE. The curve $y = 0$ also satisfies (a). It is a particular solution not obtainable from the family (d).

Example 10.52. Solve

(a) $\qquad y \sec x \, dx + \sin x \, dy = 0, \quad x \neq \dfrac{\pi}{2}, \dfrac{3\pi}{2}, \cdots.$

Solution. First verify that (a) is not exact. However, $\sec x$ is an integrating factor. Multiplication of (a) by it will therefore yield the exact equation

(b) $\qquad y \sec^2 x \, dx + \tan x \, dy = 0, \quad x \neq \dfrac{\pi}{2}, \dfrac{3\pi}{2}, \cdots.$

[Now verify that (b) is exact.] The solution of (b) by means of Lesson 9B or by recognizing that its left side is an integrable combination is

(c) $\qquad y \tan x = c \quad \text{or} \quad y = c \cot x.$

LESSON 10C. Finding an Integrating Factor. As noted in the remark following Definition 10.5, a standard method of finding an integrating factor is known only for certain very special types of differential equations. We shall discuss some of these special types below.

We assume that

(10.6) $\qquad P(x,y) \, dx + Q(x,y) \, dy = 0$

is not an exact differential equation and that h is an integrating factor of (10.6), where h is an unknown function which we wish to determine. Hence, by Definition 10.5,

(10.61) $\qquad hP(x,y) \, dx + hQ(x,y) \, dy = 0$

is exact. It therefore follows, by Theorem 9.3, that

(10.62) $\qquad \dfrac{\partial}{\partial y} [hP(x,y)] = \dfrac{\partial}{\partial x} [hQ(x,y)].$

We consider five possibilities.

1. h is a function only of x, i.e., $h = h(x)$. In this case we obtain from (10.62),

(10.63) $\qquad h(x) \dfrac{\partial}{\partial y} P(x,y) = h(x) \dfrac{\partial}{\partial x} Q(x,y) + Q(x,y) \dfrac{dh(x)}{dx},$

which we can write as

(10.64) $\qquad \dfrac{dh(x)}{h(x)} = \dfrac{\left[\dfrac{\partial}{\partial y} P(x,y) - \dfrac{\partial}{\partial x} Q(x,y) \right]}{Q(x,y)} \, dx.$

In the special case that the coefficient of dx in (10.64) also simplifies to a function only of x, let us call it $F(x)$, so that

(10.65)
$$F(x) = \frac{\frac{\partial}{\partial y} P(x,y) - \frac{\partial}{\partial x} Q(x,y)}{Q(x,y)},$$

then, by (10.64) and (10.65), $\log [h(x)] = \int F(x) \, dx$. Hence,

(10.66)
$$h(x) = e^{\int F(x) \, dx},$$

where we have omitted the constant of integration, is an integrating factor of (10.6).

Example 10.661.   First show that the differential equation

(a)
$$(e^x - \sin y) \, dx + \cos y \, dy = 0$$

is not exact and then find an integrating factor.

Solution.   Comparing (a) with (10.6), we see that

(b)
$$P(x,y) = e^x - \sin y, \qquad Q(x,y) = \cos y.$$

Therefore

(c)
$$\frac{\partial P(x,y)}{\partial y} = -\cos y, \qquad \frac{\partial Q(x,y)}{\partial x} = 0.$$

Hence, by Theorem 9.3, (a) is not exact. By (b), (c), and (10.65),

(d)
$$F(x) = \frac{-\cos y}{\cos y} = -1.$$

Therefore, by (d) and (10.66),

(e)
$$h(x) = e^{\int -dx} = e^{-x}$$

is an integrating factor of (a). (Verify it.)

2. h is a function only of y, i.e., $h = h(y)$.   In this case we obtain from (10.62),

(10.67)
$$h(y) \frac{\partial}{\partial y} P(x,y) + P(x,y) \frac{dh(y)}{dy} = h(y) \frac{\partial}{\partial x} Q(x,y),$$

which we can write as

(10.68)
$$\frac{dh(y)}{h(y)} = \frac{\frac{\partial}{\partial x} Q(x,y) - \frac{\partial}{\partial y} P(x,y)}{P(x,y)} \, dy.$$

If the coefficient of dy in (10.68) also simplifies to a function only of y—

let us call it $G(y)$—so that

(10.69)
$$G(y) = \frac{\dfrac{\partial}{\partial x} Q(x,y) - \dfrac{\partial}{\partial y} P(x,y)}{P(x,y)},$$

then, by (10.68) and (10.69), $\log [h(y)] = \int G(y)\, dy$. Hence,

(10.7)
$$h(y) = e^{\int G(y)\, dy},$$

is an integrating factor of (10.6).

Example 10.701. First show that the differential equation

(a)
$$xy\, dx + (1 + x^2)\, dy = 0$$

is not exact and then find an integrating factor.

Solution. Comparing (a) with (10.6), we see that

(b)
$$P(x,y) = xy, \qquad Q(x,y) = 1 + x^2.$$

Therefore

(c)
$$\frac{\partial P(x,y)}{\partial y} = x, \qquad \frac{\partial Q(x,y)}{\partial x} = 2x.$$

Hence, by Theorem 9.3, (a) is not exact. By (b), (c), and (10.69),

(d)
$$G(y) = \frac{2x - x}{xy} = \frac{1}{y}.$$

Therefore, by (d) and (10.7),

(e)
$$h(y) = e^{\int \frac{1}{y}\, dy} = e^{\log y} = y$$

is an integrating factor of (a). (Verify it.)

3. h is a function of xy, i.e., $h = h(u)$, where $u = xy$. In this case we obtain from (10.62),

(10.71)
$$h(u) \frac{\partial}{\partial y} P(x,y) + P(x,y) \left[\frac{\partial}{\partial y} h(u) \right] = h(u) \frac{\partial}{\partial x} Q(x,y) + Q(x,y) \left[\frac{\partial}{\partial x} h(u) \right].$$

Since $u = xy$, $\dfrac{\partial u}{\partial y} = x$. Therefore, $\dfrac{\partial}{\partial y} h(u) = h'(u) \dfrac{\partial u}{\partial y} = x \dfrac{d}{du} h(u)$.

Similarly, $\dfrac{\partial u}{\partial x} = y$ and $\dfrac{\partial}{\partial x} h(u) = h'(u) \dfrac{\partial u}{\partial x} = y \dfrac{d}{du} h(u)$. Substituting in

(10.71) these values of $\frac{\partial}{\partial y} h(u)$ and $\frac{\partial}{\partial x} h(u)$ and simplifying the result, we obtain

(10.72) $$\frac{d[h(u)]}{h(u)} = \frac{\frac{\partial}{\partial y} P(x,y) - \frac{\partial}{\partial x} Q(x,y)}{yQ(x,y) - xP(x,y)} du.$$

If the coefficient of du in (10.72) also simplifies to a function of $u = xy$—let us call it $F(u) \equiv F(xy)$—so that

(10.73) $$F(u) = \frac{\frac{\partial}{\partial y} P(x,y) - \frac{\partial}{\partial x} Q(x,y)}{yQ(x,y) - xP(x,y)},$$

then, by (10.72) and (10.73), $\log [h(u)] = \int F(u) \, du$. Hence,

(10.74) $$h(u) = e^{\int F(u) \, du},$$

is an integrating factor of (10.6), where $u = xy$.

Example 10.741.   First show that the differential equation

(a) $(y^3 + xy^2 + y) \, dx + (x^3 + x^2 y + x) \, dy = 0$

is not exact and then find an integrating factor.

Solution.   Comparing (a) with (10.6), we see that

(b) $P(x,y) = y^3 + xy^2 + y, \quad Q(x,y) = x^3 + x^2 y + x.$

Therefore

(c) $\frac{\partial P(x,y)}{\partial y} = 3y^2 + 2xy + 1, \quad \frac{\partial Q(x,y)}{\partial x} = 3x^2 + 2xy + 1.$

Hence, by Theorem 9.3, (a) is not exact. By (b), (c), and (10.73),

(d) $F(u) = \dfrac{3y^2 + 2xy + 1 - 3x^2 - 2xy - 1}{x^3 y + x^2 y^2 + xy - xy^3 - x^2 y^2 - xy} = \dfrac{-3(x^2 - y^2)}{xy(x^2 - y^2)}$

$= -\dfrac{3}{u}.$

Therefore, by (d) and (10.74),

(e) $h(u) = e^{\int -\frac{3}{u} \, du} = e^{-3 \log u} = u^{-3} = (xy)^{-3}$

is an integrating factor of (a). (Verify it.)

4. h is a function of x/y, i.e., $h = h(u)$, where $u = x/y$. Here $u = x/y$ so that $\partial u/\partial y = -x/y^2$ and $\partial u/\partial x = 1/y$. Therefore

$$(10.75) \qquad \frac{\partial}{\partial y} h(u) = h'(u) \frac{\partial u}{\partial y} = -\frac{x}{y^2} \frac{d}{du} h(u),$$

$$\frac{\partial}{\partial x} h(u) = h'(u) \frac{\partial u}{\partial x} = \frac{1}{y} \frac{d}{du} h(u).$$

Substituting (10.75) in (10.71) and simplifying the result, we obtain

$$(10.76) \qquad \frac{d[h(u)]}{h(u)} = \frac{y^2 \left[\dfrac{\partial P(x,y)}{\partial y} - \dfrac{\partial Q(x,y)}{\partial x} \right]}{x P(x,y) + y Q(x,y)} du.$$

If the coefficient of du in (10.76) also simplifies to a function of $u = x/y$—let us call it $G(u) \equiv G(x/y)$—so that

$$(10.77) \qquad G(u) = \frac{y^2 \left[\dfrac{\partial P(x,y)}{\partial y} - \dfrac{\partial Q(x,y)}{\partial x} \right]}{x P(x,y) + y Q(x,y)},$$

then, by (10.76) and (10.77), $\log [h(u)] = \int G(u)\, du$. Hence,

$$(10.78) \qquad h(u) = e^{\int G(u)\, du},$$

is an integrating factor of (10.6), where $u = x/y$.

Example 10.781. First show that the differential equation

(a) $$3y\, dx - x\, dy = 0$$

is not exact and then find an integrating factor.

Solution. Comparing (a) with (10.6), we see that

(b) $$P(x,y) = 3y, \qquad Q(x,y) = -x.$$

Therefore

(c) $$\frac{\partial P(x,y)}{\partial y} = 3, \qquad \frac{\partial Q(x,y)}{\partial x} = -1,$$

Hence, by Theorem 9.3, (a) is not exact. By (b), (c), and (10.77),

(d) $$G(u) = \frac{y^2(3+1)}{3xy - xy} = 2\frac{y}{x} = \frac{2}{u} \cdot$$

Therefore, by (d) and (10.78),

(e) $$h(u) = e^{\int \frac{2}{u}\, du} = e^{\log u^2} = u^2 = \frac{x^2}{y^2}$$

is an integrating factor of (a). (Verify it.)

5. h is a function of y/x, i.e., $h = h(u)$, where $u = y/x$. In this case, we leave it to you as an exercise—follow the method used in **4**—to show that an integrating factor of (10.6) is

(10.79) $$h(u) = e^{\int K(u)\, du},$$

where $u = y/x$ and

(10.8) $$K(u) = \frac{x^2\left[\dfrac{\partial Q(x,y)}{\partial x} - \dfrac{\partial P(x,y)}{\partial y}\right]}{xP(x,y) + yQ(x,y)}.$$

Example 10.81. First show that the differential equation

(a) $$y\, dx - 3x\, dy = 0$$

is not exact and then find an integrating factor.

Solution. Comparing (a) with (10.6), we see that

(b) $$P(x,y) = y, \qquad Q(x,y) = -3x.$$

Therefore

(c) $$\frac{\partial P(x,y)}{\partial y} = 1, \qquad \frac{\partial Q(x,y)}{\partial x} = -3.$$

Hence, by Theorem 9.3, (a) is not exact. By (b), (c), and (10.8),

(d) $$K(u) = \frac{x^2[-3 - 1]}{xy - 3xy} = \frac{2x}{y} = \frac{2}{u}.$$

Therefore, by (d) and (10.79),

(e) $$h(u) = e^{\int \frac{2}{u}\, du} = e^{\log u^2} = u^2 = \frac{y^2}{x^2}$$

is an integrating factor of (a). (Verify it.)

If a differential equation can be put in the special form

(10.82) $$y(Ax^p y^q + Bx^r y^s)\, dx + x(Cx^p y^q + Dx^r y^s)\, dy = 0,$$

where A, B, C, D are constants, then it can be shown that an integrating factor of (10.82) has the form $x^a y^b$ where a and b are suitably chosen constants. We illustrate, by an example, the method of finding an integrating factor of (10.82).

Example 10.83. First show that the differential equation

(a) $$y(2x^2 y^3 + 3)\, dx + x(x^2 y^3 - 1)\, dy = 0$$

is not exact and then find an integrating factor.

Solution. Comparing (a) with (10.6), we see that

(b) $\qquad P(x,y) = 2x^2y^4 + 3y, \qquad Q(x,y) = x^3y^3 - x.$

Therefore

(c) $\qquad \dfrac{\partial P(x,y)}{\partial y} = 8x^2y^3 + 3, \qquad \dfrac{\partial Q(x,y)}{\partial x} = 3x^2y^3 - 1.$

Hence, by Theorem 9.3, (a) is not exact. Since (a) has the form of (10.82), an integrating factor will have the form $x^a y^b$. Multiplying (a) by $x^a y^b$, we have

(d) $\qquad (2x^{a+2}y^{b+4} + 3x^a y^{b+1})\, dx + (x^{a+3}y^{b+3} - x^{a+1}y^b)\, dy = 0.$

By Theorem 9.3, (d) will be exact if

(e) $\quad 2(b+4)x^{a+2}y^{b+3} + (b+1)3x^a y^b = (a+3)x^{a+2}y^{b+3} - (a+1)x^a y^b.$

Multiplying (e) by $1/(x^a y^b)$, we obtain

(f) $\qquad (2b+8)x^2y^3 + 3b + 3 = (a+3)x^2y^3 - (a+1).$

Equation (f) will be an equality if we choose a and b so that

(g) $\qquad 2b + 8 = a + 3, \qquad 3b + 3 = -a - 1.$

Solving (g) for a and b, we find

(h) $\qquad a = \tfrac{7}{5}, \qquad b = -\tfrac{9}{5}.$

Hence, $x^{7/5}y^{-9/5}$ is an integrating factor of (a). (Verify it.)

EXERCISE 10

Test each of the following equations 1–19 for exactness. If it is not exact, try to find an integrating factor. (Integrating factors for nonexact equations are given in the answers.) After the equation is made exact, solve by looking for integrable combinations. If you cannot find any, use method of Lesson 9.

1. $(2xy + x^2)\, dx + (x^2 + y^2)\, dy = 0.$
2. $(x^2 + y\cos x)\, dx + (y^3 + \sin x)\, dy = 0.$
3. $(x^2 + y^2 + x)\, dx + xy\, dy = 0.$
4. $(x - 2xy + e^y)\, dx + (y - x^2 + xe^y)\, dy = 0.$
5. $(e^x \sin y + e^{-y})\, dx - (xe^{-y} - e^x \cos y)\, dy = 0.$
6. $(x^2 - y^2 - y)\, dx - (x^2 - y^2 - x)\, dy = 0.$
7. $(x^4 y^2 - y)\, dx + (x^2 y^4 - x)\, dy = 0.$
8. $y(2x + y^3)\, dx - x(2x - y^3)\, dy = 0.$
9. $\left(\text{Arc tan } xy + \dfrac{xy - 2xy^2}{1 + x^2 y^2}\right) dx + \dfrac{x^2 - 2x^2 y}{1 + x^2 y^2}\, dy = 0.$

10. $e^x(x+1) \, dx + (ye^y - xe^x) \, dy = 0.$

11. $\dfrac{xy+1}{y} \, dx + \dfrac{2y-x}{y^2} \, dy = 0.$

12. $(y^2 - 3xy - 2x^2) \, dx + (xy - x^2) \, dy = 0.$

13. $y(y + 2x + 1) \, dx - x(2y + x - 1) \, dy = 0.$

14. $y(2x - y - 1) \, dx + x(2y - x - 1) \, dy = 0.$

15. $(y^2 + 12x^2y) \, dx + (2xy + 4x^3) \, dy = 0.$

16. $3(y + x)^2 \, dx + x(3y + 2x) \, dy = 0.$

17. $y \, dx - (y^2 + x^2 + x) \, dy = 0.$

18. $2xy \, dx + (x^2 + y^2 + a) \, dy = 0.$

19. $(2xy + x^2 + b) \, dx + (y^2 + x^2 + a) \, dy = 0.$

ANSWERS 10

1. $3x^2y + x^3 + y^3 = c.$

2. $4x^3 + 3y^4 + 12y \sin x = c.$

3. $3x^4 + 4x^3 + 6x^2y^2 = c;$ integrating factor $x.$

4. $x^2 - 2x^2y + y^2 + 2xe^y = c.$

5. $e^x \sin y + xe^{-y} = c.$

6. $x - y + \log \sqrt{x+y} - \log \sqrt{x-y} = c;$ integrating factor $1/(x^2 - y^2).$

7. $x^4y + xy^4 - cxy = -3;$ integrating factor $1/x^2y^2.$

8. $x^2 + xy^3 = cy^2;$ integrating factor $1/y^3.$

9. $x \operatorname{Arc\,tan} xy - \log(1 + x^2y^2) = c.$

10. $2xe^{x-y} + y^2 = c;$ integrating factor $e^{-y}.$

11. $x^2 + \dfrac{2x}{y} + 4 \log |y| = c.$

12. $x^2y^2 - 2x^3y - x^4 = c;$ integrating factor $2x.$

13. $(y - x + 1)^3 = cxy;$ integrating factor $(xy)^{-4/3}.$

14. $(x + y + 1)^3 = cxy;$ integrating factor $x^{-1}y^{-1}(x + y + 1)^{-1}.$

15. $4x^3y + xy^2 = c.$

16. $6x^2y^2 + 8x^3y + 3x^4 = c;$ integrating factor $x.$

17. $y + \operatorname{Arc\,tan} \dfrac{y}{x} = c;$ integrating factor $1/(x^2 + y^2).$

18. $y^3 + 3x^2y + 3ay = c.$

19. $y^3 + x^3 + 3(x^2y + ay + bx) = c.$

LESSON 11. The Linear Differential Equation of the First Order. Bernoulli Equation.

LESSON 11A. Definition of a Linear Differential Equation of the First Order. The important differential equation which we are about to discuss has many theoretical and practical applications. It is a special type of first order differential equation in which both the *dependent variable and its derivative are of the first degree.* An equation of this type is called a linear differential equation of the first order. Hence:

Definition 11.1. A **linear differential equation of the first order** is one which can be written as

$$(11.11) \qquad \frac{dy}{dx} + P(x)y = Q(x),$$

where $P(x)$ and $Q(x)$ are continuous functions of x over the intervals for which solutions are sought. (Note that y and its derivative both have exponent one.)

For this differential equation we shall prove in Lesson 11B that the 1-parameter family of solutions we shall obtain is actually a true general solution as we defined the term in 4.7. *Every particular solution* of (11.11) will be obtainable from this 1-parameter family of solutions.

LESSON 11B.   Method of Solution of a Linear Differential Equation of the First Order.   As mentioned in Lesson 10B, an integrating factor is known for this type of equation (11.11). It is

$$(11.12) \qquad e^{\int P(x)dx},$$

where the constant of integration is taken to be zero.

The motivation and means by which this rather terrifying looking integrating factor was obtained have been deferred to Lesson 11C. In the meantime let us verify that (11.12) is indeed an integrating factor for (11.11). Multiplying (11.11) by (11.12) and changing the order in which the terms appear, we obtain

$$(11.13) \qquad [P(x)e^{\int P(x)dx}y - Q(x)e^{\int P(x)dx}]\,dx + e^{\int P(x)dx}\,dy = 0.$$

The terms $P(x)e^{\int P(x)dx}$ and $Q(x)e^{\int(x)dx}$ are functions of x. Hence the partial derivative with respect to y of the coefficient of dx in (11.13) is

$$(11.14) \qquad P(x)e^{\int P(x)dx}.$$

By (9.15) the partial derivative with respect to x of the coefficient of dy in (11.13) is $\left[\text{remember } \dfrac{d}{dx}e^u = e^u\dfrac{du}{dx} \text{ ; here } u = \int P(x)\,dx\right]$

$$(11.15) \qquad P(x)e^{\int P(x)dx}.$$

Since the functions in (11.14) and (11.15) are the same, by Theorem 9.3, (11.13) is exact. Hence by Definition 10.5, $e^{\int P(x)dx}$ is an integrating factor.

Let us now rewrite (11.13) in the form

$$(11.16) \qquad e^{\int P(x)dx}\,dy + P(x)e^{\int P(x)dx}y\,dx = Q(x)e^{\int P(x)dx}\,dx.$$

Since we know (11.16) is exact, we can solve it either by the method of Lesson 9B, or by trying to discover an integrable combination. A shrewd observer now discerns that the left side of (11.16) is indeed

$$(11.17) \qquad d(e^{\int P(x)dx}y).$$

[Be sure to verify this statement. Remember $d(uy) = u\,dy + y\,du$. Here $u = e^{\int P(x)dx}$.] Hence (11.16) becomes

$$(11.18) \qquad d(e^{\int P(x)dx}y) = e^{\int P(x)dx}Q(x)\,dx$$

whose solution is

$$(11.19) \qquad e^{\int P(x)dx}y = \int e^{\int P(x)dx}Q(x)\,dx + c.$$

Proof that (11.19) is a true general solution of (11.11). The argument proceeds as follows. Since $e^{\int P(x)dx} \neq 0$, [for proof see Lesson 18, (18.86)], (11.11) holds if and only if (11.13) holds; (11.13) holds if and only if (11.16) holds; (11.16) holds if and only if (11.18) holds. Finally (11.18) holds if and only if (11.19) holds. We have thus demonstrated that (11.11) is true if and only if (11.19) is true. And since $e^{\int P(x)dx} \neq 0$, we can in (11.19) divide by this factor to obtain a 1-parameter family of solutions of (11.11) in the explicit form

$$(11.191) \qquad y = e^{-\int P(x)dx}\int e^{\int P(x)dx}Q(x)\,dx + ce^{-\int P(x)dx}.$$

Hence (11.11) holds if and only if (11.191) holds. The "if and only if" clause is equivalent to saying that if (11.11) holds, then (11.191) holds, and if (11.191) holds, then (11.11) holds. This means that if (11.11) is true, then its solutions are (11.191), and if (11.191) is true (i.e., true for each c), then (11.11) is satisfied.

Example 11.2. Find the general solution of

$$(a) \qquad y' - 2xy = e^{x^2}.$$

Solution. By comparing (a) with (11.11), we see that the equation is linear, $P(x) = -2x$ and $Q(x) = e^{x^2}$. By (11.12) an integrating factor of (a) is therefore

$$(b) \qquad e^{\int -2x\,dx} = e^{-x^2}.$$

Multiplication of (a) by e^{-x^2} gives

$$(c) \qquad e^{-x^2}\,dy - 2e^{-x^2}xy\,dx = dx,$$

an equation which now corresponds to (11.16). Hence by (11.17), the left side of (c) should be (and is)

$$(d) \qquad d(e^{-x^2}y).$$

Replacing the left side of (c) by (d), and integrating the resulting expression, we obtain for the general solution of (a),

(e) $$e^{-x^2}y = \int dx = x + c,$$

which we can write as

(f) $$y = e^{x^2}(x + c).$$

Comment 11.21. After the integrating factor e^{-x^2} has been determined, we could by (11.19) go immediately from (a) to (e). For (11.19) says, "y times the integrating factor $= \int$ (integrating factor) $Q(x)\,dx + c$."

Example 11.3. Find the particular solution of

(a) $$x\frac{dy}{dx} + 3y = \frac{\sin x}{x^2}, \quad x \neq 0,$$

for which $y(\pi/2) = 1$.

Solution. Since $x \neq 0$, we may divide (a) by it and obtain

(b) $$\frac{dy}{dx} + \frac{3}{x}y = \frac{\sin x}{x^3}, \quad x \neq 0.$$

Comparing (b) with (11.11) we see that (b) is linear, $P(x) = 3/x$ and $Q(x) = \sin x/x^3$. Hence by (11.12) an integrating factor is

(c) $$e^{\int \frac{3}{x}\,dx} = e^{3\log x} = e^{\log x^3} = x^3.$$

(By taking the logarithm of both sides, you can show that $e^{\log u} = u$.) Multiplication of (b) by the integrating factor x^3 will give, by (11.19) [see Comment (11.21)],

(d) $$x^3 y = \int \sin x\,dx + c = -\cos x + c.$$

To find the particular solution for which $x = \pi/2$, $y = 1$, we substitute these values in (d) and obtain

(e) $$c = \frac{\pi^3}{8}.$$

Replacing (e) in (d), the required solution is

(f) $$yx^3 + \cos x = \frac{\pi^3}{8}.$$

LESSON 11C. Determination of the Integrating Factor $e^{\int P(x)\,dx}$. We outline below a method by which the integrating factor for (11.11) can be determined. First we rewrite (11.11) as

(11.4) $$[P(x)y - Q(x)]\,dx + dy = 0,$$

and then multiply it by $u(x)$. There results

(11.41) $[u(x)P(x)y - u(x)Q(x)] \, dx + u(x) \, dy = 0.$

We now ask ourselves this question. What must $u(x)$ look like if it is to be an integrating factor for (11.11)? We know by Definition 10.5 that it will be an integrating factor if it makes (11.41) exact. And by Theorem 9.3, we know that (11.41) will be exact if

(11.42) $\dfrac{\partial}{\partial y} [u(x)P(x)y - u(x)Q(x)] = \dfrac{\partial}{\partial x} u(x).$

Taking these derivatives [observe that $u(x)$, $P(x)$, and $Q(x)$ are functions only of x], we obtain

(11.43) $u(x)P(x) = \dfrac{d}{dx} u(x),$

which we can write as

(11.44) $P(x) \, dx = \dfrac{du(x)}{u(x)} \cdot$

Integration of (11.44), with the constant of integration taken to be zero, gives $\left(\text{remember } \displaystyle\int \dfrac{du}{u} = \log u \right)$

(11.45) $\log u(x) = \displaystyle\int P(x) \, dx,$

which is equivalent to

(11.46) $u(x) = e^{\int P(x)dx}.$

Hence if $u(x)$ has the value $e^{\int P(x)dx}$, it will be an integrating factor for (11.11).

LESSON 11D. Bernoulli Equation. A special type of first order differential equation, named for the Swiss mathematician James Bernoulli (1654–1705), and solvable by the methods of this lesson is the following.

(11.5) $\dfrac{dy}{dx} + P(x)y = Q(x)y^n.$

If $n = 1$, (11.5) can be written as $dy/dx = [Q(x) - P(x)]y$, an equation in which the variables are separable and therefore solvable by the method of Lesson 6C. Hence we assume $n \neq 1$. Note also that the presence of y^n prevents the equation from being linear. If we multiply (11.5) by

(11.51) $(1 - n)y^{-n},$

we obtain

(11.52) $(1 - n)y^{-n} \dfrac{dy}{dx} + (1 - n)P(x)(y^{1-n}) = (1 - n)Q(x).$

The first term in (11.52) is $\dfrac{d}{dx}(y^{1-n})$. Hence (11.52) can be written as

(11.53) $\dfrac{d}{dx}(y^{1-n}) + (1-n)P(x)(y^{1-n}) = (1-n)Q(x).$

If we now think of y^{1-n} as the dependent variable instead of the usual y [or if you prefer you can replace y^{1-n} by a new variable u so that (11.53) becomes $\dfrac{du}{dx} + (1-n)P(x)u = (1-n)Q(x)$], then, by Definition 11.1, (11.53) is linear in y^{1-n} (or u). It can therefore be solved by the method of Lesson 11B.

Example 11.54. Solve

(a) $y' + xy = \dfrac{x}{y^3}, \quad y \neq 0.$

Solution. Comparing (a) with (11.5), we see that (a) is a Bernoulli equation with $n = -3$. Hence, by (11.51), we must multiply (a) by $4y^3$. There results

(b) $4y^3y' + 4xy^4 = 4x.$

Because of the sentence after (11.52), we know that the first term of (b) should be (and is)

(c) $\dfrac{d}{dx}y^4.$

Hence we can write (b) as

(d) $\dfrac{d}{dx}[y^4] + 4x[y^4] = 4x,$

an equation which is now linear in the variable y^4. An integrating factor for (d) is therefore, by (11.12),

(e) $e^{\int 4x\,dx} = e^{2x^2}.$

After multiplying (d) by e^{2x^2}, we can take advantage of (11.19) to write immediately

(f) $e^{2x^2}y^4 = 4\int xe^{2x^2}\,dx = e^{2x^2} + c.$

Therefore,

(g) $y^4 = 1 + ce^{-2x^2}$

is the required solution.

EXERCISE 11

Find the general solution of each of the following.

1. $xy' + y = x^3$.
2. $y' + ay = b$.
3. $xy' + y = y^2 \log x$.

4. $\dfrac{dx}{dy} + 2yx = e^{-y^2}$. *Hint.* Consider x as the dependent variable.

5. $\dfrac{dr}{d\theta} = (r + e^{-\theta}) \tan \theta$.

6. $\dfrac{dy}{dx} - \dfrac{2xy}{x^2 + 1} = 1$.

7. $y' + y = xy^3$.

8. $(1 - x^3) \dfrac{dy}{dx} - 2(1 + x)y = y^{5/2}$.

9. $\tan \theta \, \dfrac{dr}{d\theta} - r = \tan^2 \theta$.

10. $L \dfrac{di}{dt} + Ri = E \sin kt$. (This is the equation of a simple electric circuit containing an inductor, a resistor, and an applied electromotive force. For the meaning of these terms and for a more complete discussion of electric circuits, see Lessons 30 and 33C.)

11. $y' + 2y = 3e^{-2x}$.   15. $y' + y \cos x = \frac{1}{2} \sin 2x$.
12. $y' + 2y = \frac{3}{4}e^{-2x}$.   16. $xy' + y = x \sin x$.
13. $y' + 2y = \sin x$.   17. $xy' - y = x^2 \sin x$.
14. $y' + y \cos x = e^{2x}$.   18. $xy' + xy^2 - y = 0$.
19. $xy' - y(2y \log x - 1) = 0$.
20. $x^2(x - 1)y' - y^2 - x(x - 2)y = 0$.

Find a particular solution of each of the differential equations 21–24.

21. $y' - y = e^x$, $y(0) = 1$.

22. $y' + \dfrac{1}{x} y = \dfrac{y^2}{x}$, $y(-1) = 1$.

23. $2 \cos x \, dy = (y \sin x - y^3) \, dx$, $y(0) = 1$.
24. $(x - \sin y) \, dy + \tan y \, dx = 0$, $y(1) = \pi/6$.
25. The differential equation

(11.6) $$y' = f_0(x) + f_1(x)y + f_2(x)y^2, \quad f_2(x) \neq 0,$$

is called a **Riccati equation**. If $y_1(x)$ is a particular solution of this equation, show that the substitution

(11.61) $$y = y_1 + \dfrac{1}{u}, \qquad y' = y_1' - \dfrac{1}{u^2} u',$$

will transform the equation into the first order linear equation

(11.62) $$u' + [f_1(x) + 2f_2(x)y_1]u = -f_2(x).$$

    Hint. Since y_1 is a particular solution of the given equation, $y_1' = f_0(x) + f_1(x)y_1 + f_2(x)y_1^2$.

With the aid of problem 25 above, find the general solution of each of the following Riccati equations.

26. $y' = x^3 + \dfrac{2}{x} y - \dfrac{1}{x} y^2, \quad y_1(x) = -x^2.$

27. $y' = 2 \tan x \sec x - y^2 \sin x, \quad y_1(x) = \sec x.$

28. $y' = \dfrac{1}{x^2} - \dfrac{y}{x} - y^2, \quad y_1(x) = \dfrac{1}{x}.$

29. $y' = 1 + \dfrac{y}{x} - \dfrac{y^2}{x^2}, \quad y_1(x) = x.$

ANSWERS 11

1. $4xy = x^4 + c.$

2. $y = \dfrac{b}{a} + ce^{-ax}.$

3. $y \log x + y + cxy = 1.$

4. $x = e^{-y^2}(y + c).$

5. $2r = c \sec \theta - e^{-\theta}(\tan \theta + 1).$

6. $y = (\text{Arc} \tan x + c)(x^2 + 1).$

7. $\dfrac{1}{y^2} = ce^{2x} + x + \dfrac{1}{2}.$

8. $y^{-3/2} = -\dfrac{3}{4(1 + x + x^2)} + \dfrac{c(1 - x)^2}{1 + x + x^2}.$

9. $r = \sin \theta[\log (\sec \theta + \tan \theta)] + c \sin \theta.$

10. $i = ce^{-Rt/L} + \dfrac{E(R \sin kt - kL \cos kt)}{R^2 + k^2 L^2}.$

11. $y = 3xe^{-2x} + ce^{-2x}.$

12. $y = \frac{3}{4}xe^{-2x} + ce^{-2x}.$

13. $y = \frac{1}{5}(2 \sin x - \cos x) + ce^{-2x}.$

14. $y = e^{-\sin x}\left(c + \displaystyle\int e^{2x + \sin x}\, dx\right).$

15. $y = \sin x - 1 + ce^{-\sin x}.$

16. $y = \dfrac{\sin x}{x} - \cos x + \dfrac{c}{x}.$

17. $y = x(c - \cos x).$

18. $y = \dfrac{2x}{x^2 + c}, \quad y = 0.$

19. $1 - 2y(1 + \log x) = cxy.$

20. $y = \dfrac{x^2}{(x - 1)c + 1}.$

21. $y = e^x(x + 1).$

22. $y = 1.$

23. $\sec x = y^2(\tan x + 1).$

24. $8x \sin y = 4 \sin^2 y + 3.$

26. $u' + \left(\dfrac{2}{x} + 2x\right) u = \dfrac{1}{x}, \quad y = -x^2 + \dfrac{2x^2 e^{x^2}}{e^{x^2} + c}.$

27. $u' - 2u \tan x = \sin x, \quad y = \dfrac{1}{\cos x} + \dfrac{3 \cos^2 x}{c - \cos^3 x}.$

28. $u' - \dfrac{3}{x} u = 1, \quad y = \dfrac{1}{x} + \dfrac{2}{cx^3 - x}.$

29. $u' - \dfrac{1}{x} u = \dfrac{1}{x^2}, \quad y = x + \dfrac{2x}{cx^2 - 1}.$

LESSON 12.   Miscellaneous Methods of Solving a First Order Differential Equation.

LESSON 12A.   Equations Permitting a Choice of Method. If a differential equation is selected at random, it may be solvable by more than one method. The one selected will depend ultimately on your ingenuity in determining which will most readily lead to a solution. The following examples will illustrate this point.

Example 12.1.   Solve

(a) $x\,dy - y\,dx = y^2\,dx.$

Solution. As was shown in Example 10.51, multiplication by y^{-2}, $y \neq 0$, will make (a) exact. It will then be solvable by the method of Lesson 9B, or by means of (10.30). Finally if we divide by x, ($x \neq 0$), the equation becomes

(b) $y' - \dfrac{y}{x} = \dfrac{y^2}{x}, \quad x \neq 0,$

which one recognizes as a Bernoulli equation. Hence it is solvable by the method of Lesson 11D. Of the three choices available to solve (a), you will find that use of (10.30) is the easiest.

Example 12.11.   Solve

(a) $(x^2 + y^2)\,dy + 2xy\,dx = 0.$

Solution. This equation is of the type with homogeneous coefficients and is therefore solvable by the method of Lesson 7B. It is also exact and therefore solvable by the method of Lesson 9B. However, the easiest method is to rewrite (a) as

(b) $x^2\,dy + 2xy\,dx + y^2\,dy = 0,$

and then make use of (10.21).

Example 12.12.   Solve

(a) $(3e^{3x}y - 2x)\,dx + e^{3x}\,dy = 0.$

Solution. The equation is exact and is therefore solvable by the method of Lesson 9B. If we divide (a) by $e^{3x}\,dx$, we obtain

(b) $\dfrac{dy}{dx} + 3y = 2xe^{-3x},$

an equation which is linear and hence solvable by the method of Lesson

11B. If, however, (a) is rewritten as

(c) $$3e^{3x}y\,dx + e^{3x}\,dy - 2x\,dx = 0,$$

it can be solved most easily by making use of (10.392).

Example 12.13. Solve

(a) $$x^2\,dy - (xy + y\sqrt{x^2 + y^2})\,dx = 0.$$

Solution. The equation is of the type with homogeneous coefficients and therefore is solvable by the method of Lesson 7B. However, the presence of the combination $x(x\,dy - y\,dx)$ leads one to try to make use of (10.3) or (10.31). We therefore divide (a) by x, $(x \neq 0)$ and rewrite it to read

(b) $$x\,dy - y\,dx = \frac{y}{x}\sqrt{x^2 + y^2}\,dx, \quad x \neq 0.$$

We then divide (b) by y^2 to obtain

(c) $$\frac{x\,dy - y\,dx}{y^2} = \frac{1}{xy}\sqrt{x^2 + y^2}\,dx = \frac{1}{x}\sqrt{\frac{x^2}{y^2} + 1}\,dx, \quad x \neq 0,\, y \neq 0.$$

By (10.3) the left side of (c) is $-d(x/y)$. Hence (c) can be written as

(d) $$-\frac{d(x/y)}{\sqrt{1 + (x/y)^2}} = \frac{dx}{x}, \quad x \neq 0,\, y \neq 0.$$

Integration of (d) now gives

(e) $$-\log\left|\frac{x}{y} + \sqrt{1 + (x/y)^2}\right| = \log|x| + \log|c|, \quad x \neq 0,\, y \neq 0,$$

which can be written as

(f) $$\frac{y}{x + \sqrt{x^2 + y^2}} = cx, \quad x \neq 0,\, y \neq 0.$$

The solution of (a) is therefore

(g) $$y = cx(x + \sqrt{x^2 + y^2}), \quad x \neq 0,\, y \neq 0.$$

Comment 12.14. The function $y = 0$ which we had to exclude to obtain (g) also satisfies (a). Moreover every member of the family of solutions (g) goes through the point (0,0). Observe from (a), however, that dy/dx is undefined when $x = 0$. The point (0,0) is one of those singular points we discussed in Lesson 5B. Every member of the family (g) lies on it, but no member goes through any other point on the line $x = 0$. This solution, therefore, actually should have been written with two parameters instead of one, namely

$$y = c_1 x(x + \sqrt{x^2 + y^2}), \quad x \leqq 0$$
$$y = c_2 x(x + \sqrt{x^2 + y^2}), \quad x \geqq 0.$$

LESSON 12B. Solution by Substitution and Other Means. A first order differential equation need not come under any of the headings mentioned heretofore. This fact should not be too surprising. You yourself could easily write a first order differential equation which would not fit any of the types thus far discussed. It is possible in some cases to solve a differential equation by means of a *shrewd substitution*, or by discovering an integrating factor, or by some other ingenious method. We give below a number of examples which do not, as they stand, lend themselves to standardization. You should keep in mind that these examples have been specially designed to yield a solution in terms of elementary functions. It is easily conceivable, if a differential equation were selected at random, that one could spend hours and days using every known method and device at one's disposal and still fail to find an explicit or implicit solution in terms of the elementary functions. More than likely no such solution exists.

Example 12.2. Solve

(a) $$(y - y^2 - x^2)\, dx - x\, dy = 0.$$

The equation as it stands cannot be solved by any of the methods outlined thus far. However, the presence of the combination $y\, dx - x\, dy$ leads one to try to make use of (10.3) or (10.31). Rearranging terms and dividing by x^2 gives

(b) $$\frac{y\, dx - x\, dy}{x^2} = \left[1 + \left(\frac{y}{x}\right)^2\right] dx, \quad x \neq 0.$$

By (10.31), the left side of (b) is $-d(y/x)$. Hence (b) can be written as

(c) $$-\frac{d\left(\frac{y}{x}\right)}{1 + \left(\frac{y}{x}\right)^2} = dx.$$

Integration of (c) gives

(d) $$\text{Arc tan } \frac{y}{x} = -(x + c) \quad \text{or} \quad \frac{y}{x} = -\tan(x + c), \quad x \neq 0.$$

Hence the solution of (a) is

(e) $$y = -x \tan(x + c), \quad x \neq 0.$$

NOTE. In regard to the line $x = 0$ and the point $(0,0)$, we refer you to our Comment 12.14.

Example 12.21. Solve

(a) $\qquad (2 \cos y)\, y' + \sin y = x^2 \csc y, \quad y \neq 0.$

Solution. The equation as it stands cannot be solved by any of the methods outlined thus far. If we multiply it by $\sin y$, we obtain

(b) $\qquad (2 \sin y \cos y)\, y' + \sin^2 y = x^2, \quad y \neq 0.$

The first term is equal to $\dfrac{d}{dx}\,(\sin^2 y)$. We therefore can write (b) as

(c) $\qquad \dfrac{d}{dx}\,(\sin^2 y) + (\sin^2 y) = x^2, \quad y \neq 0,$

an equation which is now linear in the variable $\sin^2 y$. The integrating factor by (11.12) is found to be e^x. Hence by (11.19)

(d) $\qquad e^x \sin^2 y = \displaystyle\int x^2 e^x \, dx = e^x(x^2 - 2x + 2) + c, \quad y \neq 0.$

The solution of (a) is therefore

(e) $\qquad \sin^2 y = (x^2 - 2x + 2) + ce^{-x}, \quad y \neq 0.$

Example 12.22. Solve

(a) $\qquad y' + 2x = 2(x^2 + y - 1)^{2/3}.$

Solution. The equation as it stands cannot be solved by any of the methods outlined thus far. A powerful and useful method frequently used by mathematicians to integrate functions is that of substitution. An attempt is made to simplify the integrand by using a new variable to represent a function of the given variable. In some cases this method of substitution can also be used profitably to find solutions of differential equations. For this example, we try the substitution

(b) $\qquad u = x^2 + y - 1,$

and hope it will yield a differential equation in u which we can solve in terms of elementary functions. Differentiating (b) we obtain

(c) $\qquad \dfrac{du}{dx} = 2x + \dfrac{dy}{dx}\,; \quad \dfrac{dy}{dx} = \dfrac{du}{dx} - 2x.$

We now substitute (b) and (c) in (a). There results

(d) $\qquad \dfrac{du}{dx} = 2u^{2/3}, \qquad u^{-2/3}\, du = 2\, dx, \quad u \neq 0.$

By integration of (d) we obtain

(e) $$3u^{1/3} = 2x + c, \quad u \neq 0.$$

Replacing u by its value in (b) and cubing, we have

(f) $$x^2 + y - 1 = \frac{1}{27}(2x + c)^3, \quad x^2 + y - 1 \neq 0.$$

The solution of (a) is therefore

(g) $$y = 1 - x^2 + \frac{(2x + c)^3}{27}, \quad x^2 + y - 1 \neq 0.$$

NOTE. The function $y = 1 - x^2$ which had to be excluded in obtaining (g) also satisfies (a). It is a particular solution of (a) not obtainable from the family (g).

EXERCISE 12

Solve each of the following differential equations.

1. $2xy \dfrac{dy}{dx} + (1 + x)y^2 = e^x.$

2. $\cos y \dfrac{dy}{dx} + \sin y = x^2.$ Hint. $\dfrac{d}{dx}(\sin y) = \cos y \dfrac{dy}{dx}.$

3. $(x + 1) \, dy - (y + 1) \, dx = (x + 1)\sqrt{y + 1} \, dx.$ Let $u = y + 1.$

4. $e^y(y' + 1) = e^x.$ Hint. $\dfrac{d}{dx} e^y = e^y y'.$

5. $y' \sin y + \sin x \cos y = \sin x.$ Hint. $\dfrac{d}{dx} \cos y = -(\sin y)y'.$

6. $(x - y)^2 \dfrac{dy}{dx} = 4.$ Let $u = x - y.$

7. $x \dfrac{dy}{dx} - y = \sqrt{x^2 + y^2}.$

8. $(3x + 2y + 1) \, dy + (4x + 3y + 2) \, dx = 0.$
9. $(x^2 - y^2) \, dy = 2xy \, dx.$
10. $y \, dx + (1 + y^2 e^{2x}) \, dy = 0.$ Let $y = e^{-x}u.$
11. $(x^2 y + y^2) \, dx + x^3 \, dy = 0.$
12. $(y^2 e^{xy^2} + 4x^3) \, dx + (2xy e^{xy^2} - 3y^2) \, dy = 0.$
13. $y' = (x^2 + 2y - 1)^{2/3} - x.$ Let $u = x^2 + 2y - 1.$

14. $x \dfrac{dy}{dx} + y = x^2(1 + e^x)y^2.$

MISCELLANEOUS PROBLEMS

In the following set of problems, classify each differential equation by type before attempting to find a 1-parameter family of solutions. Some may fall into more than one type; some may not fall into any. In the answers, you will find hints as to methods of solving. Always first try to solve the equation before looking at these hints. Also keep in mind that there may be other methods in addition to the ones we have given.

15. $(2y - xy \log x)\, dx - 2x \log x\, dy = 0.$
16. $y' + ay = ke^{bx}.$
17. $y' = (x + y)^2.$
18. $y' + 8x^3y^3 + 2xy = 0.$
19. $(xy\sqrt{x^2 - y^2} + x)y' = y - x^2\sqrt{x^2 - y^2}.$
20. $y' + ay = b \sin kx.$
21. $xy' - y^2 + 1 = 0.$
22. $(y^2 + a \sin x)\, \dfrac{dy}{dx} = \cos x.$
23. $xy' = xe^{y/x} + x + y.$
24. $y' + y \cos x = e^{-\sin x}.$
25. $xy' - y(\log xy - 1) = 0.$
26. $x^3y' - y^2 - x^2y = 0.$
27. $xy' + ay + bx^n = 0, \quad x > 0.$
28. $xy' - x \sin \dfrac{y}{x} - y = 0.$
29. $(xy - x^2)y' + y^2 - 3xy - 2x^2 = 0.$
30. $(6xy + x^2 + 3)y' + 3y^2 + 2xy + 2x = 0.$
31. $x^2y' + y^2 + xy + x^2 = 0.$
32. $(x^2 - 1)y' + 2xy - \cos x = 0.$
33. $(x^2y - 1)y' + xy^2 - 1 = 0.$
34. $(x^2 - 1)y' + xy - 3xy^2 = 0.$
35. $(x^2 - 1)y' - 2xy \log y = 0.$
36. $(x^2 + y^2 + 1)y' + 2xy + x^2 + 3 = 0.$
37. $y' \cos x + y + (1 + \sin x) \cos x = 0.$
38. $(2xy + 4x^3)y' + y^2 + 12x^2y = 0.$
39. $(x^2 - y)y' + x = 0.$
40. $(x^2 - y)y' - 4xy = 0.$
41. $xyy' + x^2 + y^2 = 0.$
42. $2xyy' + 3x^2 - y^2 = 0.$
43. $(2xy^3 - x^4)y' + 2x^3y - y^4 = 0.$
44. $(xy - 1)^2xy' + (x^2y^2 + 1)y = 0.$
45. $(x^2 + y^2)y' + 2x(2x + y) = 0.$
46. $3xy^2y' + y^3 - 2x = 0.$
47. $2y^3y' + xy^2 - x^3 = 0.$
48. $(2xy^3 + xy + x^2)y' - xy + y^2 = 0.$
49. $(2y^3 + y)y' - 2x^3 - x = 0.$
50. $y' - e^{x-y} + e^x = 0.$

ANSWERS 12

1. $2y^2x = e^x + ce^{-x}.$ 4. $y = \log\left(\tfrac{1}{2}e^x + ce^{-x}\right).$
2. $\sin y = x^2 - 2x + 2 + ce^{-x}.$ 5. $\cos y = 1 + ce^{-\cos x}.$
3. $\sqrt{y + 1} = x + 1 + c\sqrt{x + 1}.$
6. $y = \log\left|\dfrac{x - y - 2}{x - y + 2}\right| + c.$
7. $y + \sqrt{x^2 + y^2} = cx^2,$ or equivalently, $y = C + \sqrt{x^2 + y^2}.$
8. $3xy + y^2 + y + 2x^2 + 2x = c.$

9. $x^2 + y^2 + cy = 0$.
10. $e^{-2x} = 2y^2(\log |y| - c)$.
11. $3x^2 + y = cx^3y$.

12. $e^{xy^2} + x^4 - y^3 = c$.
13. $3(x^2 + 2y - 1)^{1/3} = 2x + c$.
14. $xy(x + e^x + c) + 1 = 0$.

ANSWERS FOR MISCELLANEOUS PROBLEMS

15. Separable. $x + 2 \log |y| - 2 \log (\log |x|) = c$.

16. Linear.
$$y = \frac{k}{a+b} e^{bx} + ce^{-ax}, \quad a + b \neq 0,$$
$$y = kxe^{bx} + ce^{-ax}, \quad a + b = 0.$$

17. Let $u = x + y$. Resulting equation is separable. $x + y = \tan (x + c)$.

18. Bernoulli. $y^{-2} = ce^{2x^2} - 4x^2 - 2$.

19. Let $y = ux$. Resulting equation can be made exact.
$$y = x \sin [c - \tfrac{1}{2}(x^2 + y^2)], \quad x > 0$$
$$y = x \sin [c + \tfrac{1}{2}(x^2 + y^2)], \quad x < 0.$$

20. Linear. $y = \dfrac{b}{a^2 + k^2} (a \sin kx - k \cos kx) + ce^{-ax}$.

21. Separable. $y = (1 - cx^2)/(1 + cx^2)$.

22. Let y be the independent variable. Bernoulli in $\sin x$; see problem 2 above.
$$\sin x = ce^{ay} - \left(\frac{y^2}{a} + \frac{2y}{a^2} + \frac{2}{a^3}\right).$$

23. Homogeneous. $(1 - cx)e^{y/x} = cx$.

24. Linear. $y = (x + c)e^{-\sin x}$.

25. Let $u = xy$. Resulting equation is separable. $xy = e^{cx}$.

26. Bernoulli. Also $1/x^2y^2$ is an integrating factor. $x^2 - y = cxy$.

27. Linear
$$y = cx^{-a} - \frac{b}{a+n} x^n, \quad a \neq -n.$$
$$y = cx^{-a} - bx^{-a} \log x, \quad a = -n.$$

28. Homogeneous. $\csc \dfrac{y}{x} - \cot \dfrac{y}{x} = cx$.

29. Integrating factor x. $x^2y^2 - 2x^3y - x^4 = c$.

30. Exact. $3xy^2 + x^2y + 3y + x^2 = c$.

31. Homogeneous. $x = (y + x)(\log |x| + c)$.

32. Linear. Also exact. $(x^2 - 1)y = \sin x + c$.

33. Exact. $x^2y^2 - 2(x + y) = c$.

34. Bernoulli. $y^{-1} = 3 + c\sqrt{|x^2 - 1|}$.

35. Separable. $y = e^{c(x^2-1)}$.

36. Exact. $x^3 + y^3 + 3(x^2y + y + 3x) = c$.

37. Linear. $(1 + \sin x)y = \cos x \left(\sin x - 2 \log \dfrac{1 + \sin x}{\cos^2 x} + c\right)$.

38. Integrable combinations. $4x^3y + xy^2 = c$.

39. Bernoulli, y independent variable. $2x^2 = 2y - 1 + ce^{-2y}$.

40. Let $y = ux^2$. Resulting equation is separable. $(x^2 + y)^2 = cy$.

41. Homogeneous. $x^2(x^2 + 2y^2) = c$.

42. Bernoulli. $y^2 = cx - 3x^2$.
43. Integrating factor $1/x^2y^2$. $x^3 + y^3 = cxy$.
44. Let $u = xy$. Resulting equation is separable. $y^2 = ce^{[xy-(1/xy)]}$.
45. Homogeneous. Also exact. $y^3 + 4x^3 + 3x^2y = c$.
46. Bernoulli. $xy^3 = x^2 + c$.
47. Homogeneous. Also $x^2 + y^2$ is an integrating factor.

$$(x^2 + y^2)^2(2y^2 - x^2) = c.$$

48. Integrating factor $1/xy^2$. $y^3 + y \log |xy| - x = cy$.
49. Exact. Also separable. $y^4 + y^2 = x^4 + x^2 + c$.
50. Let $u = e^y$. Resulting equation is linear. $e^y = 1 + ce^{-e^x}$.

Chapter 3

Problems Leading to Differential Equations of the First Order

LESSON 13. Geometric Problems.

We are at last ready to study a wide variety of problems which lead to differential equations of the first order. We consider first certain geometric problems in which we seek the equation of a curve whose derivative y' has certain preassigned properties.

Example 13.1. Find the family of curves which has the property that the segment of a tangent line drawn between a point of tangency and the y axis is bisected by the x axis.

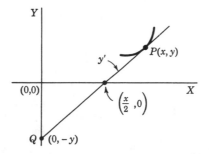

Figure 13.11

Solution (Fig. 13.11). Let $P(x,y)$ be a point on a curve of the required family and let PQ be a segment of the tangent line drawn at this point. By hypothesis, PQ is bisected by the x axis. Hence the coordinates of the point Q are $(0,-y)$. $\Big[$ The mid-point formula for two points (x_1,y_1) and (x_2,y_2) is $\left(\dfrac{x_1 + x_2}{2},\ \dfrac{y_1 + y_2}{2}\right).\Big]$ The equation of the line PQ is, therefore, given by

(a) $$\frac{2y}{x} = y' = \frac{dy}{dx}.$$

By Lesson 6C, the solution of (a) is

(b) $$y = cx^2, \quad x \neq 0, y \neq 0.$$

We have thus shown that if there exists a family of curves which meets the requirements of our problem, it must satisfy (a) and hence must satisfy (b).

Conversely, we must show that if a family of curves satisfies (b), it will meet the requirements of our problem. Let $P(x_0, y_0)$ be a point on a curve of (b). Then by (b)

(c) $$y_0 = cx_0^2, \quad y'_{(x_0, y_0)} = 2cx_0.$$

The equation of the line at $P(x_0, y_0)$ with slope $2cx_0$ is

(d) $$y - y_0 = 2cx_0(x - x_0).$$

When $x = 0$, we obtain from (d)

(e) $$y = y_0 - 2cx_0^2,$$

which is the y coordinate of the intersection of the line (d) with the y axis, (Q in Fig. 13.11). The coordinates of the mid-point of PQ are therefore

(f) $$\left(\frac{x_0}{2}, y_0 - cx_0^2\right).$$

Replacing y_0 in (f) by its value in (c), we obtain for the coordinates of the mid-point of PQ

(g) $$\left(\frac{x_0}{2}, 0\right).$$

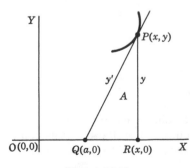

Figure 13.21

We have thus proved that the tangent at any point $P(x_0, y_0)$ of a member of the family (b), drawn to the y axis is bisected by the x axis.

Example 13.2 (Fig. 13.21). Find the family of curves with the property that the area of the region bounded by the x axis, the tangent line drawn at a point $P(x,y)$ of a curve of the family and the projection of the tangent line on the x axis has a constant value A.

Solution. At $P(x,y)$ draw a tangent to a curve of the required family. Call $Q(a,0)$ the point of inter-

section of this line with the x axis, $R(x,0)$ the intersection of the projection of the tangent line with the x axis. The equation of the tangent line is

(a) $$\frac{y}{x - a} = y', \quad x \neq a.$$

Solving this equation for a, we obtain

(b) $$a = x - \frac{y}{y'},$$

which represents the distance $0Q$. Hence the distance QR is

(c) $$x - \left(x - \frac{y}{y'} \right) = \frac{y}{y'}.$$

The region whose area is A, is therefore given by the formula

(d) $$A = \frac{1}{2} y \left(\frac{y}{y'} \right) = \frac{y^2}{2y'}.$$

Hence,

(e) $$y' = \frac{y^2}{2A},$$

whose solution, by Lesson 6C, is

(f) $$\frac{1}{y} = -\frac{x}{2A} + c \quad \text{or} \quad y = \frac{2A}{2Ac - x}, \quad x \neq 2Ac.$$

We have thus shown that if there exists a family of curves which meets
the requirements of our problems it must satisfy (a) and hence must
satisfy (f).

Conversely, we must show that if a family of curves satisfies (f), it will
meet the requirements of our problem. Let $P(x_0,y_0)$ be a point on a
curve of (f). Then by (f)

(g) $$y_0 = \frac{2A}{2Ac - x_0}, \quad y'_{(x_0,y_0)} = \frac{2A}{(2Ac - x_0)^2}.$$

The equation of the line at $P(x_0,y_0)$ with slope given in (g) is

(h) $$y - y_0 = (x - x_0)\frac{2A}{(2Ac - x_0)^2}.$$

When $y = 0$, we obtain from (h),

(i) $$x = \frac{2Ax_0 - y_0(2Ac - x_0)^2}{2A},$$

which is the x coordinate of the intersection of the line (h) with the x axis

(Q in Fig. 13.21). The distance QR is therefore

(j) $$x_0 - \frac{2Ax_0 - y_0(2Ac - x_0)^2}{2A} = \frac{y_0(2Ac - x_0)^2}{2A}.$$

Hence the area of the region A is

(k) $$\tfrac{1}{2}y_0 \left[\frac{y_0(2Ac - x_0)^2}{2A} \right].$$

Substituting in (k) the value of $(2Ac - x_0)^2$ as determined by the first equation in (g), we obtain

(l) $$\tfrac{1}{2}y_0 \left(\frac{y_0 4A^2}{2Ay_0^2} \right),$$

which simplifies to the constant A. Hence our family (f) meets the requirements of the problem.

Example 13.3. Find the family of curves such that the angle from a tangent to a normal at any point of a curve of the family is bisected by the radius vector at that point. (In problems involving radius vectors, it is usually preferable to use polar coordinates.)

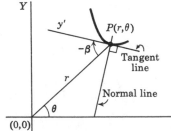

Figure 13.31

Solution (Fig. 13.31). Let $P(r,\theta)$ be the polar coordinates of a point on a curve of the required family. Call β the angle measured from the radius vector r counterclockwise to the tangent line at P. Then by a theorem in the calculus

(a) $$\tan \beta = r \frac{d\theta}{dr}.$$

By hypothesis r bisects the angle between the normal and tangent lines. Hence $\beta = 45°$ or $-45°$. In Fig. 13.31 it is $-45°$. Therefore (a) becomes (using $\beta = 45°$)

(b) $$1 = r \frac{d\theta}{dr}, \qquad \frac{dr}{r} = d\theta,$$

whose solution is

(c) $$\log r = \theta + c', \qquad r = e^{\theta+c'} = e^\theta e^{c'} = ce^\theta.$$

We leave it to you as an exercise to solve (a) when $\beta = -45°$ and to prove the converse, i.e., that if a family of curves satisfies (c), then it meets the requirements of our problem.

Example 13.4.   Find the family of curves with the property that the area of the region bounded by a curve of the family, the x axis, the lines $x = a$, $x = x$ is proportional to the length of the arc included between these two vertical lines.

Solution (Fig. 13.41).   The formula for the *arc length of a curve* between the points A,B whose abscissas are $x = a$ and $x = x$ is

(a) $s = \displaystyle\int_a^x \sqrt{1 + \left(\dfrac{dy}{dx}\right)^2}\, dx.$

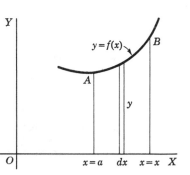

The area of the region bounded by the lines, $x = a$, $x = x$, the x axis, and the arc of the curve $y = f(x)$ between these lines, is

(b) $A = \displaystyle\int_a^x f(x)\, dx.$

By hypothesis A is proportional to s; therefore, by (a) and (b),

Figure 13.41

(c) $\displaystyle\int_a^x f(x)\, dx = k\int_a^x \sqrt{1 + \left(\dfrac{dy}{dx}\right)^2}\, dx,$

where $k > 0$ is a proportionality constant.   Differentiation of (c) with respect to x, and replacing $f(x)$ by its equivalent y, give [see (9.15)]

(d) $y = k \sqrt{1 + \left(\dfrac{dy}{dx}\right)^2},\qquad y^2 = k^2\left[1 + \left(\dfrac{dy}{dx}\right)^2\right],\quad y > 0,\ k > 0,$

which simplifies to

(e) $\pm\sqrt{\dfrac{y^2 - k^2}{k^2}} = \dfrac{dy}{dx},\qquad \dfrac{dx}{k} = \dfrac{\pm dy}{\sqrt{y^2 - k^2}},\quad y \neq k.$

By integration of (e), we obtain

(f) $\dfrac{x}{k} = \log\left(y \pm \sqrt{y^2 - k^2}\right) - \log c,$

which can be written as

(g) $ce^{x/k} = y \pm \sqrt{y^2 - k^2},\qquad y = \dfrac{1}{2c}\left(c^2 e^{x/k} + k^2 e^{-x/k}\right).$

We leave it to you as an exercise to prove the converse, i.e., if a family of curves satisfies (g), then it meets the requirements of our problem.

EXERCISE 13

1. Let $P(x,y)$ be a point on the curve $y = f(x)$. At P draw a tangent and a normal to the curve. The slope of the curve at P is therefore y'; the slope of the normal is $-1/y'$. Prove (see Fig. 13.5) each of the following.

(a) $x - y/y'$ is the x intercept of the tangent line.
(b) $y - xy'$ is the y intercept of the tangent line.
(c) $x + yy'$ is the x intercept of the normal line.
(d) $y + x/y'$ is the y intercept of the normal line.
(e) $|y/y'|$ is the length AC of the projection on the x axis, of the segment of the tangent AP. The length AC is called the **subtangent.**
(f) $|yy'|$ is the length CB of the projection on the x axis of the segment of the normal BP. The length CB is called the **subnormal.**

(g) The length of the tangent segment $AP = \left| y \sqrt{\dfrac{1}{(y')^2} + 1} \right|$.

(h) The length of the tangent segment $DP = |x\sqrt{1 + (y')^2}|$.

(i) The length of the normal segment $PB = |y\sqrt{1 + (y')^2}|$.

(j) The length of the normal segment $PE = \left| x \sqrt{1 + \dfrac{1}{(y')^2}} \right|$.

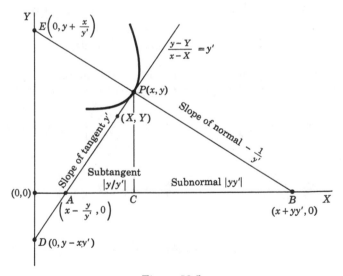

Figure 13.5

In each of the following problems 2–20, use rectangular coordinates to find the equation of the family of curves that satisfies the property described.

2. The slope of the tangent at each point of a curve is equal to the sum of the coordinates of the point. Find the particular curve through the origin.
3. The subtangent is a positive constant k for each point of the curve. *Hint.* See 1(e).

4. The subnormal is a positive constant k for each point of the curve. *Hint.* See 1(f).
5. The subnormal is proportional to the square of the abscissa.
6. The segment of a normal line between the curve and the y axis is bisected by the x axis. Find the particular curve through the point $(4,2)$, with this property.
7. The x intercept of a tangent line is equal to the ordinate. *Hint.* See 1(a).
8. The length of a tangent segment from point of contact to the x intercept is a constant. *Hint.* See 1(g).
9. Change the word tangent in 8 above to normal. *Hint.* See 1(i).
10. The area of the right triangle formed by a tangent line, the x axis and the ordinate of the point of contact of tangent and curve, has constant area 8. *Hint.* See 1(e).
11. The area of the region bounded by a curve $y = f(x)$, the x axis, and the lines $x = 2$, $x = x$, is one-half the length of the arc included between these vertical lines.
12. The slope of the curve is equal to the square of the abscissa of the point of contact of tangent line and curve. Find the particular curve through the point $(-1,1)$.
13. The subtangent is equal to the sum of the coordinates of the point of contact of tangent and curve. *Hint.* See 1(e).
14. The length of a tangent segment from point of contact to the x intercept is equal to the x intercept of the tangent line. *Hint.* See (1g) and 1(a).
15. The normal and the line drawn to the origin from point of contact of normal and curve form an isosceles triangle with the x axis.
16. Change the word normal in 15 above to tangent.
17. The point of contact of tangent and curve bisects the segment of the tangent line between the coordinate axis.
18. Change the word tangent in 17 above to normal.
19. The length of arc between $x = a$ and $x = x$ is equal to $x^2/2$.
20. The area of the region bounded by the curve $y = f(x)$, the lines $x = a$, $x = x$, and the x axis, is proportional to the difference of the ordinates.
21. Let $y = f(x)$ define a curve that passes through the origin. Find a family of curves with the property that the volume of the region bounded by $y = f(x)$, $x = 0$, $x = x$ rotated about the x axis is equal to the volume of the region bounded by $y = f(x)$, $y = 0$, $y = y$ rotated about the y axis.
22. Let $P(r,\theta)$ be a point on the curve $r = r(\theta)$. At P draw a tangent to the curve. Prove (see Fig. 13.6) each of the following.

(a) $\dfrac{dA}{d\theta} = \tfrac{1}{2}r^2$, where A is the area of the region bounded by an arc of the curve and two radii vectors.

(b) $\dfrac{ds}{d\theta} = \sqrt{\left(\dfrac{dr}{d\theta}\right)^2 + r^2}$, where s is the length of arc of the curve between two radii vectors.

(c) $\tan \beta = r \dfrac{d\theta}{dr}$. See (a) of Example 13.3.

In each of the following problems 23–25, use polar coordinates to find the equation of the family of curves that satisfies the property described.

23. The radius vector r and the tangent at $P(r,\theta)$ intersect in a constant angle β.

24. The radius vector r and the tangent at $P(r,\theta)$ intersect in an angle which is k times the polar angle θ.

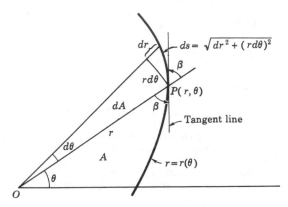

Figure 13.6

25. The area of the region bounded by an arc of a curve and the radii vectors to the end points of the arc is equal to one-half the length of the arc. *Hint.* See 22(a) and (b).

ANSWERS 13

2. $y = e^x - x - 1$.

3. $y^k = ce^x$, $(y/y') > 0$; $y^k = ce^{-x}$, $(y/y') < 0$.

4. $y^2 = 2kx + c$, $yy' > 0$; $y^2 = -2kx + c$, $yy' < 0$.

5. $3y^2 = 2kx^3 + c$, $yy' > 0$; $3y^2 = -2kx^3 + c$, $yy' < 0$.

6. $x^2 + 2y^2 = 24$.   **7.** $x + y \log y + cy = 0$.

8. $\pm x + c = \sqrt{k^2 - y^2} - k \log \left(\dfrac{k + \sqrt{k^2 - y^2}}{y} \right)$.

9. $x = \pm\sqrt{k^2 - y^2} + c$.

10. $xy = cy - 16$, $y' > 0$; $xy = cy + 16$, $y' < 0$.

11. $8y = 4ce^{\pm 2x} + c^{-1} e^{\mp 2x}$.   **12.** $3y = x^3 + 4$.

13. $x = y \log |cy|$, $(y/y') > 0$; $2xy + y^2 = c$, $(y/y') < 0$.

14. $x^2 + y^2 = cy$.   **17.** $xy = c$.

15. $x^2 - y^2 = c$.   **18.** $x^2 - y^2 = c$.

16. $xy = c$.

19. $y = \pm \left[\dfrac{x}{2} \sqrt{x^2 - 1} - \tfrac{1}{2} \log (x + \sqrt{x^2 - 1}) \right] + c$.

20. $y^k = ce^x$.   **24.** $r^k = c \sin k\theta$.

21. $x - y = cxy$.   **25.** $r = 1$; $r = \sec (\theta + c)$, $r \neq 1$.

23. $r = ce^{\theta \cot \beta}$.

LESSON 14. Trajectories.

LESSON 14A. Isogonal Trajectories. When two curves intersect in a plane, the angle between them is defined to be the angle made by their respective tangents drawn at their point of intersection. Since these lines determine two angles, it is customary to specify the particular one desired

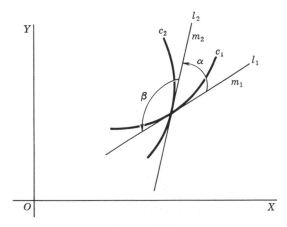

Figure 14.1

by stating from which tangent line we are to proceed in a counterclockwise direction to reach the other. In Fig. 14.1, α is the positive angle from the curve c_1 with tangent line l_1 to the curve c_2 with tangent line l_2; β is the positive angle from the curve c_2 to the curve c_1. If we call m_1 the slope of l_1 and m_2 the slope of l_2, then by a formula in analytic geometry

$$(14.11) \qquad \tan \alpha = \frac{m_2 - m_1}{1 + m_1 m_2} \; ; \quad \tan \beta = \frac{m_1 - m_2}{1 + m_1 m_2} \cdot$$

Definition 14.12. A curve which cuts every member of a given 1-parameter family of curves in the *same angle* is called an **isogonal trajectory of the family.**

If two 1-parameter families have the property that every member of one family cuts every member of the other family in the same angle, then each family may be said to be a 1-parameter family of isogonal trajectories of the other, i.e., the curves of either family are isogonal trajectories of the other.

An interesting problem is to find a family of isogonal trajectories that makes a predetermined angle with a given 1-parameter family of curves. If we call y_1' the slope of a curve of a given 1-parameter family, y' the slope of an isogonal trajectory of the family, and α their angle of intersection measured from the tangent line with slope y' to the tangent line with slope y_1', then by (14.11)

(14.13) $$\tan \alpha = \frac{y_1' - y'}{1 + y'y_1'}.$$

Example 14.14. Given the 1-parameter family of parabolas

(a) $$y = ax^2,$$

find a 1-parameter family of isogonal trajectories of the family if the angle of intersection α, measured from the required trajectory to the given family, is $\pi/4$.

Solution. Differentiation of (a) gives

(b) $$y' = 2ax.$$

From (a) again, we obtain

(c) $$a = y/x^2, \quad x \neq 0.$$

Substitution of this value in (b) gives

(d) $$y' = \frac{2y}{x}, \quad x \neq 0,$$

which is the slope of the given family at any point (x,y), $x \neq 0$. It therefore corresponds to the y_1' in (14.13). Since by hypothesis $\alpha = \pi/4$, $\tan \alpha = 1$. Hence (14.13) becomes

(e) $$1 = \frac{\dfrac{2y}{x} - y'}{1 + \dfrac{2y}{x} y'} = \frac{2y - xy'}{x + 2yy'}.$$

Simplification of (e) gives

(f) $$(x - 2y)\, dx + (x + 2y)\, dy = 0,$$

whose solution, by Lesson 7, is

(g) $$\log \sqrt{2y^2 - xy + x^2} + \frac{3}{\sqrt{7}} \operatorname{Arc\,tan} \frac{4y - x}{\sqrt{7}\, x} = c, \quad x \neq 0.$$

Comment 14.15. Note that before we used (14.13), we eliminated the parameter in (b) to obtain (d). **This elimination is essential.**

Comment 14.16. Because of the presence of the inverse tangent, (g) implicitly defines a multiple-valued function. By our definition of a function it must be single-valued, i.e., each value of x should determine one and only one y. For this reason we have written the inverse tangent with a capital A to indicate that we mean only its principal values, namely those values which lie between $-\pi/2$ and $\pi/2$. By placing this restriction on the inverse tangent, we have thus excluded infinitely many solutions, for example solutions for which the arc tan lies between $\pi/2$ and $3\pi/2$, between $3\pi/2$ and $5\pi/2$, etc.

LESSON 14B.   Orthogonal Trajectories.

Definition 14.2.   A curve which cuts every member of a given 1-parameter family of curves in a 90° angle is called an **orthogonal trajectory** of the family.

If two 1-parameter families have the property that every member of one family cuts every member of the other family in a right angle, then each family may be said to be an orthogonal trajectory of the other. Orthogonal trajectory problems are of special interest since they occur in many physical fields.

Let y_1' be the slope of a given family and let y' be the slope of an orthogonal trajectory family. Then by a theorem in analytic geometry

(14.21) $$y_1'y' = -1, \quad y' = -\frac{1}{y_1'}.$$

Example 14.22.   Find the orthogonal trajectories of the 1-parameter family of curves

(a) $$y = cx^5.$$

Solution (Fig. 14.23).   Differentiation of (a) gives

(b) $$y' = 5cx^4.$$

From (a) again we obtain

(c) $$c = y/x^5, \quad x \neq 0$$

Substituting this value in (b), we have

(d) $$y' = 5\frac{y}{x}, \quad x \neq 0,$$

which is the slope of the given family at any point (x,y), $x \neq 0$. It corresponds therefore to the y_1' in (14.21). Hence, by (14.21), the slope y' of an orthogonal family is

(e) $$y' = -\frac{x}{5y}, \quad x \neq 0, y \neq 0.$$

Its solution, by Lesson 6C, is

(f) $$x^2 + 5y^2 = k, \quad x \neq 0, y \neq 0.$$

which is a 1-parameter family of ellipses.

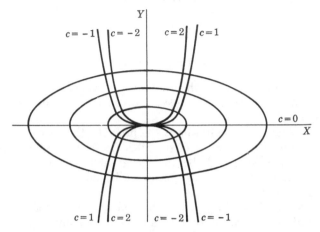

Figure 14.23

Comment 14.24.   Note that before we used (14.21), we eliminated the parameter in (b) to obtain (d). **This elimination is essential.**

LESSON 14C.   Orthogonal Trajectory Formula in Polar Coordinates.   Call $P(r,\theta)$ (Fig. 14.3) the point of intersection in polar coordi-

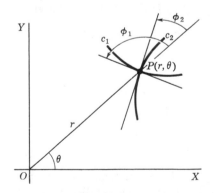

Figure 14.3

nates of two curves c_1, c_2 which are orthogonal trajectories of each other. Call ϕ_1 and ϕ_2 the respective angle the tangent to each curve c_1 and c_2

makes with the radius vector r (measured from the radius vector counterclockwise to the tangent). Since the two tangents are orthogonal, it is evident from the figure that

$$\phi_1 = \phi_2 + \frac{\pi}{2}.$$

Therefore

(14.31) $$\tan \phi_1 = \tan \left(\phi_2 + \frac{\pi}{2} \right) = - \frac{1}{\tan \phi_2}.$$

As remarked previously in Example 13.3, in polar coordinates

(14.32) $$\tan \phi_2 = r \frac{d\theta}{dr}.$$

Therefore (14.31) becomes

(14.33) $$\tan \phi_1 = - \frac{dr}{r \, d\theta}.$$

Comparing (14.32) with (14.33), we see that if two curves are orthogonal, then $r \dfrac{d\theta}{dr}$ of one is the negative reciprocal of $r \dfrac{d\theta}{dr}$ of the other. Conversely, if one of two curves satisfies (14.32) and the other satisfies (14.33), then the curves are orthogonal. Hence to find an orthogonal family of a given family, we proceed as follows. Calculate $r \dfrac{d\theta}{dr}$ of the given family. Replace $r \dfrac{d\theta}{dr}$ by its negative reciprocal $- \dfrac{dr}{r \, d\theta}$. The family of solutions of this new resulting differential equation is orthogonal to the given family.

Example 14.34. Find, in polar form, the orthogonal trajectories of the family of curves given by

(a) $$r = k \sec \theta.$$

Solution. Differentiation of (a) gives

(b) $$\frac{dr}{d\theta} = k \sec \theta \tan \theta.$$

From (a), again, we obtain

(c) $$k = \frac{r}{\sec \theta}.$$

Substituting this value in (b), we have

(d) $$\frac{dr}{d\theta} = r \tan \theta, \qquad r \frac{d\theta}{dr} = \frac{1}{\tan \theta}.$$

By (14.33), therefore, the differential equation of the orthogonal family is

(e) $$-\frac{dr}{r\,d\theta} = \frac{1}{\tan\theta} \quad \text{or} \quad -\frac{dr}{r} = \cot\theta\,d\theta,$$

whose solution is

(f) $$r\sin\theta = c.$$

Comment 14.35. We remark once more that before we could use (14.33), we had to eliminate the parameter in (b) to obtain (d).

EXERCISE 14

For each of the following family of curves, 1–4, find a 1-parameter family of isogonal trajectories of the family, where the angle of intersection, measured from the required trajectory to the given family, is the angle shown alongside each problem.

1. $y^2 = 4kx$, $\alpha = 45°$.
2. $x^2 + y^2 = k^2$, $\alpha = 45°$.
3. $y = kx$, $\tan\alpha = \frac{1}{2}$.
4. $y^2 + 2xy - x^2 = k$, $\alpha = 45°$.
5. Given that the differential equation of a family of curves is $M(x,y)\,dx + N(x,y)\,dy = 0$. Find the differential equation of a family of isogonal trajectories which makes an angle $\alpha \neq \pi/2$ with the given family, where α is measured from trajectory family to given family.
6. Given that the differential equation of a family of curves is of the homogeneous type. Prove that the differential equation of a family of isogonal trajectories which makes an angle $\alpha \neq \pi/2$ with the given family is also homogeneous. *Hint.* You will find a proof in Case 3–2 of Lesson 32.

Find in rectangular form the orthogonal trajectories of each of the following family of curves.

7. $x^2 + 2xy - y^2 = k$.
8. $x^2 + (y - k)^2 = k^2$.
9. $(x - k)^2 + y^2 = k^2$.
10. $x - y = ke^x$.
11. $y^2 = 4px$.
12. $xy = kx - 1$.
13. $x^2y = k$.
14. $x^2 - y^2 = k^2$.
15. $y^2 = kx^3$.
16. $e^x \cos y = k$.
17. $\sin y = ke^{x^2}$.

Find in rectangular form the orthogonal trajectories of each of the following family of curves.

18. A family of straight lines through the origin.
19. A family of circles with variable radii, centers on the x axis, and passing through the origin.
20. A family of ellipses with centers at the origin and vertices at $(\pm1,0)$.
21. A family of equilateral hyperbolas whose asymptotes are the coordinate axes.

22. A family of parabolas with vertices at the origin and foci on the x axis.

If the differential equation of a given family of curves remains unchanged when y' is replaced by $-1/y'$, the family is called **self orthogonal**, i.e., a self-orthogonal family has the property that a curve orthogonal to a member of the family also belongs to the family. Show that each of the following families is self orthogonal.

23. The family of parabolas that have a common focus and axis. *Hint.* See answer.

24. The family of central conics that have common foci and axis. *Hint.* See answer.

Find in polar form the orthogonal trajectories of each of the following family of curves.

25. $r = k \cos \theta$.

26. $r\theta = k$.

27. $r = k(1 + \sin \theta)$.

28. $r = \sin \theta + k$.

29. $r = k \sin 2\theta$.

30. $r^2 = k(r \sin \theta - 1)$.

31. $r^2 = k \cos 2\theta$.

32. $r^{-1} = \sin^2 \theta + k$.

33. $r^n \sin n\theta = k$.

34. $r = e^{k\theta}$.

ANSWERS 14

1. $\log |2x^2 + xy + y^2| + \dfrac{6}{\sqrt{7}} \operatorname{Arc\,tan} \dfrac{x + 2y}{x\sqrt{7}} = c$.

2. $\log |c(x^2 + y^2)| - 2 \operatorname{Arc\,tan}(y/x) = 0$.

3. $\log |c(x^2 + y^2)| + 4 \operatorname{Arc\,tan}(y/x) = 0$.

4. $xy = c$.

5. $y' = (N \tan \alpha + M)/(M \tan \alpha - N)$.

7. $x^2 - 2xy - y^2 = c$.

8. $x^2 + y^2 = cx$.

9. $x^2 + y^2 = cy$.

10. $ce^y = y - x + 2$.

11. $2x^2 + y^2 = c$.

12. $x^3 + 3y = c$.

13. $2y^2 - x^2 = c$.

14. $xy = c$.

15. $3y^2 + 2x^2 = c$.

16. $e^x \sin y = c$.

17. $x = c \cos^2 y$.

18. $x^2 + y^2 = c$.

19. See problem 9.

20. $x^2 = ce^{x^2 + y^2}$.

21. $x^2 - y^2 = c$.

22. See problem 11.

23. Equation of family is $y^2 = 4p(x + p)$.

24. Equation of family is $\dfrac{x^2}{a^2} + \dfrac{y^2}{a^2 - c^2} = 1$, where $(\pm c, 0)$ are the coordinates of the foci.

25. $r = c \sin \theta$.

26. $cr = e^{\theta^2/2}$.

27. $r = c(1 - \sin \theta)$.

28. $e^{1/r} = c(\sec \theta + \tan \theta)$.

29. $r^4 = c \cos 2\theta$.

30. $r = 2 \sin \theta + c \cos \theta$.

31. $r^2 = c \sin 2\theta$.

32. $\tan \theta = ce^{2r}$.

33. $r^n \cos n\theta = c$.

34. $r = e^{\pm\sqrt{c - \theta^2}}$.

LESSON 15.   Dilution and Accretion Problems.   Interest
Problems.   Temperature Problems.
Decomposition and Growth Problems.
Second Order Processes.

LESSON 15A.   Dilution and Accretion Problems.

In this type of problem, we seek a formula which will express the amount of a substance in solution as a function of the time t, where this amount is changing instantaneously with time.

Example 15.1. A tank contains 100 gallons of water. In error 300 pounds of salt are poured into the tank instead of 200 pounds. To correct this condition, a stopper is removed from the bottom of the tank allowing 3 gallons of the brine to flow out each minute. At the same time 3 gallons of fresh water per minute are pumped into the tank. If the mixture is kept uniform by constant stirring, how long will it take for the brine to contain the desired amount of salt?

Solution. The usual procedure in solving problems of this type is to let a variable, say x, represent the number of pounds of salt in solution at *any time t*. Then an equation is set up which will reflect the approximate change in x in an arbitrarily small time interval Δt. In this problem, 3 gallons of brine flow out each minute. Hence $3\Delta t$ gallons of brine will flow out in Δt minutes. Since the 100 gallons of solution are thoroughly mixed, and x represents the number of pounds of salt present in the solution at any instant of time t, we may assume that of the $3\Delta t$ gallons of brine flowing out, $\dfrac{3\Delta t}{100}x$ will be the approximate loss of salt from the solution. (For example, if 250 pounds of salt are in the 100-gallon solution of brine at time t, and if Δt is sufficiently small, then $\dfrac{3\Delta t}{100}250$ will be the approximate loss of salt in time Δt.)

If, therefore, Δx represents the approximate loss of salt in time Δt, then

$$\Delta x \approx -\frac{3\Delta t}{100}x, \qquad \frac{\Delta x}{\Delta t} \approx -\frac{3}{100}x$$

(the symbol $\approx$ means approximately equal to). The negative sign is necessary to indicate that x is decreasing. Since no salt enters the solution, we are led to the differential equation

(a) $$\frac{dx}{dt} = -\frac{3x}{100}, \qquad \frac{dx}{x} = -0.03\,dt.$$

The solution of (a) is

(b) $$\log x = -0.03t + c', \qquad x = ce^{-0.03t}.$$

Inserting in (b) the initial conditions $x = 300$, $t = 0$, we obtain $c = 300$. Hence (b) becomes

(c) $$x = 300e^{-0.03t},$$

an equation which gives the amount of salt in solution as a function of the time t. And when $x = 200$, we obtain from (c)

(d) $$\tfrac{2}{3} = e^{-0.03t}, \qquad \log \tfrac{3}{2} = 0.03t,$$

from which we find

(e) $$t = 13.5 \text{ min},$$

i.e., it will take 13.5 minutes for the amount of salt in the solution to be reduced to 200 pounds.

Comment 15.11. A much shorter and more desirable method of solving the above problem is to insert in the second equation of (a) the initial and final conditions as limits of integration. We would thus obtain

(f) $$\int_{x=300}^{200} \frac{dx}{x} = -0.03 \int_{t=0}^{t} dt.$$

Integration of (f) gives immediately

(g) $$\log \tfrac{2}{3} = -0.03t, \qquad \log \tfrac{3}{2} = 0.03t,$$

which is the same as (d) above.

Example 15.12. A tank contains 100 gallons of brine whose salt concentration is 3 pounds per gallon. Three gallons of brine whose salt concentration is 2 pounds per gallon flow into the tank each minute, and at the same time 3 gallons of the mixture flow out each minute. If the mixture is kept uniform by constant stirring, find the salt content of the brine as a function of the time t.

Solution. Let x represent the number of pounds of salt in solution at any time t. By hypothesis 6 pounds of salt enter and 3 gallons of brine leave the tank each minute. In time Δt, therefore, $6\Delta t$ pounds of salt flow in and approximately $\dfrac{x}{100}(3\Delta t)$ pounds flow out. Hence in time Δt, the approximate change of the salt content in the solution is

(a) $$\Delta x \approx 6\Delta t - \frac{x}{100} 3\Delta t, \qquad \frac{\Delta x}{\Delta t} \approx 6 - 0.03x.$$

We are thus led to the differential equation $dx/dt = 6 - 0.03x$, which we write as

(b) $$\frac{dx}{6 - 0.03x} = dt, \qquad \frac{dx}{0.03x - 6} = -dt.$$

As suggested in Comment 15.11, we integrate (b) and insert the initial condition as a limit of integration. We thus obtain

(c)
$$\int_{x=300}^{x} \frac{dx}{0.03x - 6} = \int_{t=0}^{t} - dt,$$

whose solution is

(d)
$$x = 100(2 + e^{-0.03t}).$$

Note. We also could have solved (b) by the method of Lesson 11B.

Example 15.13. Same problem as in Example 15.12, excepting that 3 gallons of fresh water flow into the tank instead of brine and 5 gallons of the mixture flow out in place of 3.

Solution. Since liquid is flowing into the tank at the rate of 3 gallons per minute and the mixture is flowing out at the rate of 5 gallons per minute, there will be $(100 - 2t)$ gallons of the brine in the tank at the end of t minutes. If x represents the number of pounds of salt in solution at time t, then the approximate change in the salt content of the brine in a sufficiently small time interval Δt is

$$\Delta x \approx - \frac{x}{100 - 2t} (5\Delta t), \qquad \frac{\Delta x}{\Delta t} \approx - \frac{5x}{100 - 2t}.$$

We are thus led to the differential equation

(a)
$$\frac{dx}{dt} = - \frac{5x}{100 - 2t}, \qquad \frac{dx}{5x} = - \frac{dt}{100 - 2t}, \qquad 0 \leq t < 50.$$

Following the suggestion in comment 15.11, we integrate the second equation in (a) and insert the initial condition as a limit of integration. We thus obtain

(b)
$$\frac{1}{5} \int_{x=300}^{x} \frac{dx}{x} = \int_{t=0}^{t} - \frac{dt}{100 - 2t}.$$

Its solution is

(c)
$$x = \tfrac{3}{1000}(100 - 2t)^{5/2}.$$

EXERCISE 15A

It is assumed in the problems below that all mixtures are kept uniform by constant stirring.

1. A tank initially holds 100 gal of brine containing 30 lb of dissolved salt. Fresh water flows into the tank at the rate of 3 gal/min and brine flows out at the same rate. (a) Find the salt content of the brine at the end of 10 min. (b) When will the salt content be 15 lb?

2. Solve problem 1 if 2 gal/min of fresh water enter the tank instead of 3 gal/min.
3. Solve problem 1 if 4 gal/min of fresh water enter the tank instead of 3 gal/min.
4. A tank initially contains 200 gal of brine whose salt concentration is 3 lb/gal. Brine whose salt concentration is 2 lb/gal flows into the tank at the rate of 4 gal/min. The mixture flows out at the same rate. (a) Find the salt content of the brine at the end of 20 min. (b) When will the salt concentration be reduced to 2.5 lb/gal?
5. A tank initially contains 100 gal of brine whose salt concentration is $\frac{1}{2}$ lb/gal. Brine whose salt concentration is 2 lb/gal flows into the tank at the rate of 3 gal/min. The mixture flows out at the rate of 2 gal/min. Find the salt content of the brine and its concentration at the end of 30 min. *Hint.* After 30 min, the tank contains 130 gal of brine.
6. A tank initially contains 100 gal of brine whose salt concentration is 0.6 lb/gal. Brine whose salt concentration is 1 lb/gal flows into the tank at the rate of 2 gal/min. The mixture flows out at the rate of 3 gal/min. Find the salt content of the brine and its concentration at the end of 60 min. *Hint.* After 60 min, the tank contains 40 gal of brine.
7. A tank initially contains 200 gal of fresh water. Brine whose salt concentration is 2 lb/gal flows into the tank at the rate of 2 gal/min. The mixture flows out at the same rate.

   (a) Find the salt content of the brine at the end of 100 min.
   (b) At what time will the salt concentration reach 1 lb/gal?
   (c) Could the salt content of the brine ever reach 400 lb?

8. A tank initially contains 100 gal of fresh water. Brine whose salt concentration is 1 lb/gal flows into the tank at the rate of 2 gal/min. The mixture flows out at the rate of 1 gal/min. (a) Find the salt content of the brine and its concentration at the end of 60 min.
9. A tank initially contains 200 gal of fresh water. It receives brine of an unknown salt concentration at the rate of 2 gal/min. The mixture flows out at the same rate. At the end of 120 min, 280 lb of salt are in the tank. Find the salt concentration of the entering brine.
10. Two tanks, A and B, each contain 5000 gal of water. To each tank 150 gal of a chemical should be added, but in error the entire 300 gal are poured into the A tank. Pumps are set to work to circulate the liquid through the two

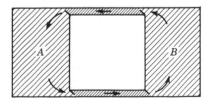

Figure 15.14·

tanks at the rate of 100 gal/min (Fig. 15.14). (a) How long will it take for tank A to contain 200 gal of the chemical and tank B to contain 100 gallons? (b) Is it theoretically possible for each tank to contain 150 gal?
11. The CO_2 content of the air in a 5000-cu-ft room is 0.3 percent. Fresh air containing 0.1 percent CO_2 is pumped into the room at the rate of 1000 ft^3/min. (a) Find the percentage of CO_2 in the room after 30 min. When will the CO_2 content be 0.2 percent?

12. The CO_2 content of the air in a 7200-cu-ft room is 0.2 percent. What volume of fresh air containing 0.05 percent CO_2 must be pumped into the room each minute in order to reduce the CO_2 content to 0.1 percent in 15 min?

ANSWERS 15A

1. (a) $x = 30e^{-0.3} = 22.2$ lb.   (b) $e^{-0.03t} = \frac{1}{2}$, $t = 23.1$ min.
2. (a) $x = 30(1 - 0.01t)^3 = 30(0.9)^3 = 21.87$ lb.   (b) $t = 20.6$ min.
3. (a) $x = 30e^{-0.286} = 22.5$ lb.   (b) $t = 26.0$ min.
4. (a) $x = 200(2 + e^{-0.4}) = 534.1$ lb.   (b) $t = 50 \log 2 = 34.7$ min.
5. 171 lb, 1.32 lb/gal.
6. 37.4 lb, 0.94 lb/gal.
7. (a) 252.8 lb.   (b) 69.3 min.   (c) Salt content approaches 400 lb as t increases without limit.
8. 98 lb, 0.61 lb/gal.
9. 2 lb/gal.
10. (a) 27.5 min.   (b) No. Tank B contains 150 gal of the chemical only if t is infinite.
11. (a) 0.10 percent.   (b) 3.47 min.
12. 527 ft^3/min.

LESSON 15B. Interest Problems. Let $\$A$ be invested at 6 percent per annum. Then the principal P at the end of one year will be

(a) $P = A(1 + 0.06)$ if interest is compounded annually,

$$P = A\left(1 + \frac{0.06}{2}\right)^2 \text{ if interest is compounded semiannually,}$$

$$P = A\left(1 + \frac{0.06}{4}\right)^4 \text{ if interest is compounded quarterly,}$$

$$P = A\left(1 + \frac{0.06}{12}\right)^{12} \text{ if interest is compounded monthly.}$$

And, in general, the principal P at the end of one year will be

(b) $$P = A\left(1 + \frac{r}{m}\right)^m,$$

if the interest rate is r percent per annum compounded m times per year. At the end of n years, it will be

(c) $$P = A\left[\left(1 + \frac{r}{m}\right)^m\right]^n.$$

If the number m of compoundings in one year, increases without limit, then

(d) $$P = \lim_{m \to \infty} A\left[\left(1 + \frac{r}{m}\right)^m\right]^n = \lim_{m \to \infty} A\left[\left(1 + \frac{r}{m}\right)^{m/r}\right]^{nr}.$$

But $\lim\limits_{m \to \infty} \left(1 + \dfrac{r}{m}\right)^{m/r} = e$. Hence (d) becomes

(e) $$P = Ae^{nr}.$$

Finally, replacing n by t, we obtain

(15.2) $$P = Ae^{rt},$$

which gives the principal at the end of time t if $\$A$ are compounded *instantaneously* or *continuously* at r percent per annum. The differential equation of which (15.2) is the 1-parameter family, is

(15.21) $$\frac{dP}{dt} = rP.$$

(Verify it.)

Example 15.22. How long will it take for $\$1.00$ to double itself if it is compounded continuously at 4 percent annum.

Solution. By (15.21) with $r = 0.04$, we obtain

(a) $$\frac{dP}{P} = 0.04dt.$$

Integrating (a) and inserting the initial and final conditions as limits of integration, we have

(b) $$\int_{P=1}^{2} \frac{dP}{P} = 0.04 \int_{t=0}^{t} dt, \qquad \log 2 = 0.04t, \qquad t = 17\tfrac{1}{3} \text{ years,}$$

approximately.

Remark 1. We could have solved this problem by using (15.2) directly with $P = 2$, $A = 1$, $r = 0.04$.

Remark 2. At 4 percent per annum, compounded *semiannually*, $\$1.00$ will double itself in $17\tfrac{1}{2}$ years. Compounded continuously, as we saw above, it doubles in $17\tfrac{1}{3}$ years. The continuous compounding of interest property therefore is not as powerful as one might have believed.

EXERCISE 15B

It is assumed in the problems below that interest is compounded continuously unless otherwise stated.

1. In how many years will $\$1.00$ double itself at 5 percent per annum?
2. At what interest rate will $\$1.00$ double itself in 12 years?
3. How much will $\$1000.00$ be worth at $4\tfrac{1}{2}$ percent interest after 10 years?

4. In a will, a man left a few million dollars, to be divided among several trusts. The will provided that the money was to be deposited in savings institutions and held for 500 years before being distributed to the designated legatees. The will was contested by the government on the grounds that the monetary wealth of the nation would be concentrated in these trusts. If the money earned an average of 4 percent interest, approximately how much would only $1,000,000 amount to at the end of the 500 years?

5. How much money would you need to deposit in a bank at 5 percent interest in order to be able to withdraw $3600.00 per year for 20 years if you wish the entire principal to be consumed at the end of this time: (a) if the money is also being withdrawn continuously from the date of deposit as, for example, withdrawing $3600/365 each day; (b) if the money is being withdrawn at the rate of $300.00 per month beginning with the first month after the deposit. *Hint.* After 1 month, money left in the bank equals $Ae^{0.05/12} - 300$, where A is the amount at the beginning of the month.

(c) Try to solve this problem assuming the more realistic situation of interest being credited quarterly and $900.00 withdrawn quarterly, beginning with the first quarter after the deposit. NOTE. This problem no longer involves a differential equation. *Hint.* At the end of the first quarter, money left in the bank equals $A\left(1 + \dfrac{0.05}{4}\right) - 900$.

6. You plan to retire in 30 years. At the end of that time, you wish to have $45,500.00, the approximate amount needed—see problem 5(b)—in order to withdraw $300.00 monthly for 20 years after retirement. (a) What amount must you deposit monthly at 5 percent? *Hint.* At the end of one month $P = A$; at the end of two months $P = Ae^{0.05/12} + A$, where A is the monthly deposit.

(b) What amount must you deposit semiannually if interest is credited semiannually at 5 percent instead of continuously. NOTE. This problem no longer involves a differential equation.

ANSWERS 15B

1. 13.86 years. (Compounded semiannually at 5 percent interest, $1.00 doubles itself in 14 years.)

2. 5.78 percent approximately.

3. $1568.31.

4. $485,165,195,400,000.

5. (a) $dP = (rP)\,dt - (3600)\,dt$, $A = \$45,513$.

(b) $A = 300\left(\dfrac{1 - e}{e - e^{12.05/12}}\right) = \$45,418$.

(c) $A = 900\left[\dfrac{1 - (1.0125)^{-80}}{0.0125}\right] = \$45,348$.

6. (a) $\$45,500 = A\left[\dfrac{1 - e^{1.5}}{1 - e^{0.05/12}}\right]$, $A = \$54.57$ monthly.

(b) $\$45,500 = A\left[\dfrac{(1.025)^{60} - 1}{0.025}\right]$, $A = \$334.58$ semiannually.

LESSON 15C. Temperature Problems. It has been proved experimentally that, under certain conditions, the rate of change of the temperature of a body, immersed in a medium whose temperature (kept constant) differs from it, is proportional to the difference in temperature between it and the medium. In mathematical symbols, this statement is written as

(15.3) $$\frac{dT_B}{dt} = -k(T_B - T_M),$$

where $k > 0$ is a proportionality constant, T_B is the temperature of the body at any time t, and T_M is the constant temperature of the medium.

Comment 15.31. In solving problems in which a proportionality constant k is present, it is necessary to know another condition in addition to the initial condition. In the temperature problem, for example, we shall need to know, in addition to the initial condition, the temperature of the body at some future time t. With these two sets of conditions, it will then be possible to determine the values of the proportionality constant k and the arbitrary constant of integration c. And if we wish to take advantage of Comment 15.11 we must use (15.3) *twice*, once to find k, the second time to find the desired answer.

Example 15.32. A body whose temperature is 180° is immersed in a liquid which is kept at a constant temperature of 60°. In one minute, the temperature of the immersed body decreases to 120°. How long will it take for the body's temperature to decrease to 90°?

Solution. Let T represent the temperature of the body at any time t. Then by (15.3) with $T_M = 60$,

(a) $$\frac{dT}{dt} = -k(T - 60), \qquad \frac{dT}{T - 60} = -k\,dt,$$

where the negative sign is used to indicate a decreasing T. Writing (a) twice as suggested in Comment 15.31, integrating both equations, and inserting all given conditions, we obtain

(b) $$\int_{T=180}^{120} \frac{dT}{T - 60} = -k\int_{t=0}^{1} dt; \qquad \int_{T=180}^{90} \frac{dT}{T - 60} = -k\int_{t=0}^{t} dt.$$

From the first integral equation, we obtain

(c) $$\log 0.5 = -k, \quad k = \log 2.$$

With the proportionality constant k known, we find from the second integral equation

(d) $\log 0.25 = -(\log 2)t$, $t = \dfrac{\log 4}{\log 2} = \dfrac{2 \log 2}{\log 2} = 2.$

Hence it will take 2 minutes for the body's temperature to decrease to 90°

EXERCISE 15C

In the problems below, assume that the rate of change of the temperature of a body obeys the law given in (15.3).

1. A body whose temperature is 100° is placed in a medium which is kept at a constant temperature of 20°. In 10 min the temperature of the body falls to 60°. (a) Find the temperature T of the body as a function of the time t. (b) Find the temperature of the body after 40 min. (c) When will the body's temperature be 50°?

2. The temperature of a body differs from that of a medium, whose temperature is kept constant, by 40°. In 5 min, this difference is 20°. (a) What is the value of k in (15.3)? (b) In how many minutes will the difference in temperature be 10°?

3. A body whose temperature is 20° is placed in a medium which is kept at a constant temperature of 60°. In 5 min the body's temperature has risen to 30°. (a) Find the body's temperature after 20 min. (b) When will the body's temperature be 40°?

4. The temperature in a room is 70°F. A thermometer which has been kept in it is placed outside. In 5 min the thermometer reading is 60°F. Five minutes later, it is 55°F. Find the outdoor temperature.

The **specific heat** of a substance is defined as the ratio of the quantity of heat required to raise a unit weight of the substance 1° to the quantity of heat required to raise the same unit weight of water 1°. For example, it takes 1 calorie to change the temperature of 1 gram of water 1°C (or 1 British thermal unit to change the temperature of 1 lb of water 1°F). If, therefore, it takes only $\frac{1}{10}$ of a calorie to change the temperature of 1 gram of a substance 1°C (or $\frac{1}{10}$ of a British thermal unit to change 1 lb of the substance 1°F), then the specific heat of the substance is $\frac{1}{10}$.

In problems 5–7 below, assume that the only exchanges of heat occur between the body and water.

5. A 50-lb iron ball is heated to 200°F and is then immediately plunged into a vessel containing 100 lb of water whose temperature is 40°F. The specific heat of iron is 0.11. (a) Find the temperature of the body as a function of time. *Hint.* The quantity of heat lost by the iron body in time t is 50(0.11) $(200 - T_B)$, where T_B is its temperature at the end of time t. The quantity of heat gained by the water—remember the specific heat of water is one— is 100(1)$(T_W - 40)$, where T_W is the temperature of the water at the end of time t. Since the heat gained by the water is equal to the heat lost by the ball, 50(0.11)$(200 - T_B) = 100(T_W - 40)$. Solve for T_W and substitute this value for T_M in (15.3). Solve for T_B. (b) Find the common temperature approached by body and water as $t \to \infty$.

6. The specific heat of tin is 0.05. A 10-lb body of tin, whose temperature is 100°F, is plunged into a vessel containing 50 lb of water at 10°F. (a) Find the temperature T of the tin as a function of time. (b) Find the common temperature approached by tin and water as $t \to \infty$.

7. The temperature of a 100-lb body whose specific heat is $\frac{1}{10}$ is 200°F. It is plunged into a 40-lb liquid whose specific heat is $\frac{1}{2}$ and whose temperature is 50°. (a) Find the temperature T of the body as a function of time. (b) To what temperature will the body eventually cool?

ANSWERS 15C

1. (a) $T = 20(1 + 4e^{-0.06931 t})$.   (b) 25°.   (c) 14.2 min.
2. (a) $k = -\frac{1}{5} \log (0.5) = 0.1386$.   (b) $t = 10$ min.
3. (a) 47.3°.   (b) 12 min.
4. 50°F.

5. (a) $T_B = \dfrac{1}{1.055} (51 + 160e^{-1.055kt})$.   (b) 48.3°F.

6. (a) $T = \dfrac{1}{1.01} (11 + 90e^{-1.01kt})$.   (b) 10.9°F.

7. (a) $T = 100(1 + e^{-3kt/2})$.   (b) 100°F.

LESSON 15D.   Decomposition and Growth Problems.   These problems will also involve a proportionality constant and will therefore require an additional reading after an interval of time t. The method of solution is essentially the same as that used to solve the problem discussed in Lesson 1

Example 15.4.   The number of bacteria in a yeast culture grows at a rate which is proportional to the number present. If the population of a colony of yeast bacteria doubles in one hour, find the number of bacteria which will be present at the end of $3\frac{1}{2}$ hours.

Solution.   Let x equal the number of bacteria present at any time t. Then in mathematical symbols, the first sentence of the problem states

(a) $$\frac{dx}{dt} = kx, \qquad \frac{dx}{x} = k \, dt,$$

where k is a proportionality constant. Writing (a) twice, integrating both equations, and inserting all the given conditions, we obtain (omitting percent signs)

(b) $$\int_{x=100}^{200} \frac{dx}{x} = k \int_{t=0}^{1} dt; \qquad \int_{x=100}^{x} \frac{dx}{x} = k \int_{t=0}^{7/2} dt.$$

From the first integral equation, we obtain

(c) $$k = \log 2,$$

and from the second integral equation

(d) $\log(x/100) = \frac{7}{2}\log 2, \qquad \dfrac{x}{100} = 2^{7/2}, \quad x = 1131.$

Hence 1131 percent or 11.31 times the initial number of bacteria will be present at the end of $3\frac{1}{2}$ hours.

Example 15.41. The death rate of an ant colony is proportional to the number present. If no births were to take place, the population at the end of one week would be reduced by one-half. However because of births, the rate of which is also proportional to the population present, the ant population doubles in 2 weeks. Determine the birth rate of the colony per week.

Solution. In this problem, we must determine two proportionality constants, one for births which we call k_1, the other for deaths which we call k_2. Using first the fact that the death rate is proportional to the number present and that deaths without births would reduce the colony in one week by one-half, we have

(a) $\dfrac{dx}{dt} = -k_2 x, \qquad \displaystyle\int_{x=1}^{0.5} \dfrac{dx}{x} = -k_2 \int_{t=0}^{1} dt,$

where x represents the population of the colony at any time t, and $x = 1$ stands for 100 percent. The solution of (a) is

(b) $k_2 = \log 2.$

Hence, the differential equation which takes into consideration both births and deaths of the colony is

(c) $\dfrac{dx}{dt} = k_1 x - (\log 2)x, \qquad \dfrac{dx}{x} = (k_1 - \log 2)\, dt.$

Integrating (c) and inserting the given conditions which reflect the net change in the population, we obtain

(d) $\displaystyle\int_{x=1}^{2} \dfrac{dx}{x} = (k_1 - \log 2)\int_{t=0}^{2} dt.$

The solution of (d) is

(e) $\log 2 = 2k_1 - \log 4, \qquad k_1 = \frac{1}{2}\log 8 = 1.0397.$

Hence the birth rate is 103.97 percent per week.

EXERCISE 15D

In problems 1 10 below, assume that the decomposition of a substance is proportional to the amount of the substance *remaining* and that the growth of population is proportional to the number *present*. (A suggestion: review Lesson 1.)

1. The population of a colony doubles in 50 days. In how many days will the population triple?
2. Assume that the half life of the radium in a piece of lead is 1500 years. How much radium will remain in the lead after 2500 years?
3. If 1.7 percent of a substance decomposes in 50 years, what percentage of the substance will remain after 100 years? How many years will be required for 10 percent to decompose?
4. The bacteria count in a culture is 100,000. In $2\frac{1}{2}$ hours, the number has increased by 10 percent. (a) In how many hours will the count reach 200,000? (b) What will the bacteria count be in 10 hours?
5. The population of a country doubles in 50 years. Its present population is 20,000,000. (a) When will its population reach 30,000,000? (b) What will its population be in 10 years?
6. Ten percent of a substance disintegrates in 100 years. What is its half life?
7. The bacteria count in a culture doubles in 3 hours. At the end of 15 hours, the count is 1,000,000. How many bacteria were in the count initially?
8. By natural increase, a city, whose population is 40,000, will double in 50 years. There is a net addition of 400 persons per year because of people leaving and moving into the city. Estimate its population in 10 years. *Hint*. First find the natural growth proportionality factor.
9. Solve problem 8, if there is a net decrease in the population of 400 persons per year.
10. A culture of bacteria whose population is N_0 will, by natural increase, double in $4 \log 2$ days. If bacteria are extracted from the colony at the uniform rate of R per day, find the number of bacteria present as a function of time. Show that the population will increase if $R < N_0/4$, will remain stationary if $R = N_0/4$, will decrease if $R > N_0/4$.
11. The rate of loss of the volume of a spherical substance, for example a moth ball, due to evaporation, is proportional to its surface area. Express the radius of the ball as a function of time.
12. The volume of a spherical raindrop increases as it falls because of the adhesion to its surface of mist particles. Assume it retains its spherical shape during its fall and that the rate of change of its volume with respect to the distance y it has fallen, is proportional to the surface area at that distance. Express the radius of the raindrop as a function of y.

ANSWERS 15D

1. 79 days.
2. 31 percent.
3. 96.6 percent; 307 years.
4. (a) 18.2 hours.   (b) 146,400.
9. 41,660.
10. $x = 4\left[R + \left(\dfrac{N_0}{4} - R\right) e^{t/4}\right].$

5. (a) 29 years.   (b) 22,970,000 approx.
6. 658 years.
7. 31,250.
8. 50,240.

11. $r = r_0 - kt$, where r_0 is the initial radius and k is a positive proportionality factor.

12. $r = r_0 + ky$ where r_0 is the initial radius and k is a positive proportionality factor.

LESSON 15E. Second Order Processes. A new substance C is sometimes formed from two given substances A and B by taking something away from each; the growth of the new substance being jointly proportional to the amount *remaining* of each of the original substances. Let s_1 and s_2 be the respective amounts of A and B present initially and let x represent the number of units of the new substance C formed in time t. If, for example, one unit of C is formed by combining 2 units of substance A with three units of substance B, then when x units of the new substance are present at time t

$$(s_1 - 2x) \text{ is the amount of } A \text{ remaining at time } t,$$

$$(s_2 - 3x) \text{ is the amount of } B \text{ remaining at time } t.$$

By the first sentence above, therefore, the differential equation which represents the rate of change of C at any time t is given by

$$\frac{dx}{dt} = k(s_1 - 2x)(s_2 - 3x),$$

where k is a proportionality constant. In general, if one unit of C is formed by combining m units of A and n units of B, then the differential equation becomes

$$(15.5) \qquad \frac{dx}{dt} = k(s_1 - mx)(s_2 - nx),$$

where s_1 and s_2 are the respective number of units of A and B present initially and x is the number of units of C present in time t.

A substance may also be dissolved in a solution, its rate of dissolution being jointly proportional to:

1. The amount of the substance which is still undissolved.
2. The difference between the concentration of the substance in a saturated solution and the actual concentration of the substance in the solution. For example, if 10 gallons of water can hold a maximum of 30 pounds of salt, it is said to be saturated when it holds this amount of salt. The concentration of salt in a saturated solution is then 3 pounds per gallon. When therefore the solution contains only 15 pounds of salt, the actual concentration of the salt in solution is 1.5 pounds per gallon or 50 percent of saturation.

Let x represent the amount of the substance *undissolved* at any time t, x_0 the initial amount of the substance, and v the volume of the solution.

Then at any time t,

$(x_0 - x)$ is the amount of the substance *dissolved* in the solution,

$\dfrac{x_0 - x}{v}$ is the concentration of the substance in the solution.

If c represents the concentration of the substance in a saturated solution, then the differential equation which expresses mathematically conditions 1 and 2 above is

(15.51) $$\frac{dx}{dt} = kx\left(c - \frac{x_0 - x}{v}\right).$$

Problems which involve joint proportionality factors are known as **second order processes**.

Example 15.52. A new substance C is to be formed by removing two units from each of two substances whose initial quantities are 10 and 8 units respectively. Assume that the rate at which the new substance is formed is jointly proportional to the amount remaining of each of the original substances. If x is the number of units of C formed at any time t and $x = 1$ unit when $t = 5$ minutes, find x when $t = 10$ minutes.

Solution. In (15.5), $s_1 = 10$, $s_2 = 8$, $m = n = 2$. Hence (15.5) becomes

(a) $$\frac{dx}{dt} = k(10 - 2x)(8 - 2x) = 4k(5 - x)(4 - x).$$

Therefore

(b) $$4k \, dt = \frac{dx}{(5 - x)(4 - x)} = \text{(see Lesson 26)} \left(\frac{1}{4 - x} - \frac{1}{5 - x}\right) dx.$$

Writing (b) twice, integrating both equations and inserting all the given conditions, we obtain

(c) $$4k \int_{t=0}^{5} dt = \int_{x=0}^{1} \left(\frac{1}{4 - x} - \frac{1}{5 - x}\right) dx;$$

$$4k \int_{t=0}^{10} dt = \int_{x=0}^{x} \left(\frac{1}{4 - x} - \frac{1}{5 - x}\right) dx.$$

From the first integral equation, we find

(d) $$k = \frac{1}{20}\left(\log \frac{5 - x}{4 - x}\right)_0^1 = \frac{1}{20} \log \frac{16}{15},$$

and from the second integral equation,

(e) $$4\left(\frac{1}{20} \log \frac{16}{15}\right)(10) = \log \frac{5 - x}{4 - x} - \log \frac{5}{4} = \log \frac{4}{5}\left(\frac{5 - x}{4 - x}\right).$$

Simplification of (e) gives

(f) $\left(\dfrac{16}{15}\right)^2 = \dfrac{4}{5}\left(\dfrac{5-x}{4-x}\right), \quad \dfrac{5-x}{4-x} = \dfrac{64}{45}, \quad 19x = 31, \quad x = 1.63.$

Hence 1.63 units of the substance x are formed in 10 minutes.

Example 15.53. Six grams of sulfur are placed in a solution of 100 cc of benzol which when saturated will hold 10 grams of sulfur. If 3 grams of sulfur are in the solution in 50 minutes, how many grams will be in the solution in 250 minutes?

Solution. Let x represent the number of grams of sulfur not yet dissolved at any time t. Then, at time t, $(6 - x)$ is the amount of sulfur dissolved and $(6 - x)/100$ is the concentration of sulfur in benzol. Here c, the concentration of sulfur in a saturated solution of benzol, is given as $10/100 = 0.1$, the initial amount x_0 of the substance is given as 6 and $v = 100$. Hence (15.51) becomes

(a) $\qquad \dfrac{dx}{dt} = kx\left(0.1 - \dfrac{6-x}{100}\right) = \dfrac{kx(4+x)}{100}.$

Therefore

(b) $\qquad \dfrac{k}{100}\, dt = \dfrac{dx}{x(x+4)} = \dfrac{1}{4}\left(\dfrac{dx}{x} - \dfrac{dx}{x+4}\right).$

Writing (b) twice, integrating both equations, and inserting all the given conditions, we obtain

(c) $\qquad \dfrac{k}{25}\displaystyle\int_{t=0}^{50} dt = \int_{x=6}^{3}\left(\dfrac{1}{x} - \dfrac{1}{x+4}\right)dx,$

$\qquad \dfrac{k}{25}\displaystyle\int_{t=0}^{250} dt = \int_{x=6}^{x}\left(\dfrac{1}{x} - \dfrac{1}{x+4}\right)dx.$

From the first integral equation, we find

(d) $\qquad k = \dfrac{1}{2}\log\dfrac{x}{x+4}\Big|_6^3 = \dfrac{1}{2}\left(\log\dfrac{3}{7} - \log\dfrac{6}{10}\right) = \dfrac{1}{2}\log\dfrac{5}{7},$

and from the second integral equation,

(e) $\qquad \dfrac{1}{25}\left(\dfrac{1}{2}\log\dfrac{5}{7}\right)250 = \log\dfrac{x}{x+4}\Big|_6^x$

$\qquad\qquad = \log\dfrac{x}{x+4} - \log\dfrac{3}{5} = \log\dfrac{5}{3}\left(\dfrac{x}{x+4}\right).$

Simplification of (e) gives

(f) $\qquad \left(\dfrac{5}{7}\right)^5 = \dfrac{5}{3}\left(\dfrac{x}{x+4}\right), \quad x = 0.5$ gram approximately,

which is the amount of sulfur not yet dissolved at the end of 250 minutes. Therefore since 6 grams of sulfur were undissolved in the solution originally, 5.5 grams are in the solution at the end of 250 minutes.

EXERCISE 15E

In problems 1–8, assume all reactions are governed by formulas (15.5) or (15.51), with the exception of problem 4 which is a modified version of (15.51).

1. In (15.5) take $s_1 = 10$, $s_2 = 10$, $m = 1$, $n = 1$. If 5 units of C are formed in 10 min, determine the number of C units formed in 50 min.
2. In (15.5) take $s_1 = 10$, $s_2 = 8$, $m = 1$, $n = 1$. If 1 unit of C is formed in 5 min, determine the number of C units formed in 10 min.
3. In (15.5) take $m = 1$, $n = 1$. (a) Solve for x as a function of time when $s_1 \neq s_2$ and when $s_1 = s_2$. (b) Show that as $t \to \infty$, $x \to s_1$ if $s_2 \geqq s_1$ and $x \to s_2$ if $s_2 \leqq s_1$.
4. In a certain chemical reaction, substance A, initially weighing 12 lb, is converted into substance B. The rate at which B is formed is proportional to the amount of A *remaining*. At the end of 2.5 min, 4 lb of B have been formed.

   (a) How much of the B substance will be present after 6 min?
   (b) How much time will be required to convert 60 percent of A?

   Work this problem in two ways:

       1. Letting x represent amount of A remaining at time t.
       2. Letting x represent amount of B formed at time t.

   Chemical reactions of this type are called **first order processes.**
5. A new substance C is to be formed from two given substances A and B by combining one unit of A with two units of B. Initially A weighs 20 lb and B weighs 40 lb. (a) If 12 lb of C are formed in $\frac{1}{3}$ hr, express x as a function of time in hours, where x is the number of units of C formed in time t. (b) What is the maximum possible value of x?
6. A saturated solution of salt water will hold approximately 3 lb of salt per gallon. A block of salt weighing 60 lb is placed into a vessel containing 100 gal of water. In 5 min, 20 lb of salt are dissolved.

   (a) How much salt will be dissolved in 1 hr?
   (b) When will 45 lb of salt be dissolved?
7. Five grams of a chemical A are placed in a solution of 100 cc of a liquid B which, when saturated, will hold 10 g of A. If 2 g of A are in the solution in 1 hr, how many grams of A will be in the solution in 2 hr?
8. Fifteen grams of a chemical A are placed into 50 cc of water, which when saturated will hold 25 g of A. If 5 g of A are dissolved in 2 hr, how many grams of A will be dissolved in 5 hr?
9. A substance containing 10 lb of moisture is placed in a sealed room, whose volume is 2000 cu ft and which when saturated can hold 0.015 lb of moisture per cubic foot. Initially the relative humidity of the air is 30 percent. If the

substance loses 4 lb of moisture in 1 hr, how much time is required for the substance to lose 80 percent of its moisture content? Assume the substance loses moisture at a rate that is proportional to its moisture content and to the difference between the moisture content of saturated air and the moisture content of the air.

ANSWERS 15E

1. $8\frac{1}{3}$ units.   2. 1.80 units.

3. $x = \dfrac{s_1 s_2 [e^{k(s_1 - s_2)t} - 1]}{s_1 e^{k(s_1 - s_2)t} - s_2}$, $x = \dfrac{s_1{}^2 kt}{1 + s_1 kt}$.

4. (a) 7.47 lb.   (b) 5.6 min.

5. (a) $x = 180t/(2 + 9t)$.   (b) 20 lb.

6. (a) 59.2 lb.   (b) 18.2 min.

7. 3.04.   8. 8.9 g.   9. $dx/dt = kx[30 - (19 - x)]$; 3.8 hr.

LESSON 16.   Motion of a Particle Along a Straight Line— Vertical, Horizontal, Inclined.

In this lesson we discuss a wide variety of problems involving the motion of a particle along a straight line. In Lesson 34, we shall discuss the motion of a particle moving in a plane.

By Newton's first law of motion, a body at rest will remain at rest, and a body in motion will maintain its velocity, (i.e., its *speed* and *direction*), unless acted upon by an outside force. By his second law, the rate of change of the momentum of a body (momentum = mass × velocity) is proportional to the resultant external force F acting upon it. In mathematical symbols, the second law says

(a) $$F = km\frac{dv}{dt}$$

where m is the mass of the body, v its velocity, and $k > 0$ is a proportionality constant whose value depends on the units used. If these are foot for distance, pound for force, slug for mass (= 1/32 pound), second for time, then $k = 1$ and (a) becomes

(16.1) $$F = m\frac{dv}{dt} = ma = m\frac{d^2 s}{dt^2},$$

where a is the rate of change in velocity, commonly called the **acceleration** of the particle, and s is the distance the particle has moved from a fixed point. A force of 1 lb therefore will give a mass of 1 slug an acceleration of 1 ft/sec^2. Remember that F, a, and v are vector quantities, i.e., they not only have magnitude but also direction. (For a discussion of a vector quantity, see Lesson 16C.) *Hence it is always essential in a problem to indicate the positive direction.*

If we write

(b)
$$\frac{dv}{dt} = \frac{dv}{ds}\frac{ds}{dt}$$

and recognize that $v = ds/dt$, then (b) becomes

(16.11)
$$\frac{dv}{dt} = v\frac{dv}{ds}.$$

Hence we can also write (16.1) as

(16.111)
$$F = mv\frac{dv}{ds}.$$

Newton also gave us the law of attraction between bodies. If m_1 and m_2 are the masses of two bodies whose centers of gravity are r distance apart, the force of attraction between them is given by

(16.12)
$$F = k\frac{m_1 m_2}{r^2},$$

where $k > 0$ is a proportionality constant.

LESSON 16A.   Vertical Motion.   Let, see Fig. 16.13,

M = mass of the earth, assumed to be a sphere,
m = mass of a body in the earth's gravitational field,
R = the radius of the earth,
y = the distance of the body above the earth's surface.

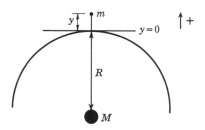

Figure 16.13

By (16.12) the force of attraction between earth and body is (we assume their masses are concentrated at their respective centers)

(16.14)
$$F = -G\frac{Mm}{(R + y)^2}.$$

The proportionality constant G which we have used in place of k is called the **gravitational constant.** The negative sign is necessary because the resulting force acts downward toward the earth's center, and our positive direction is upward. If the distance y of the body above the earth's surface is small compared to the radius R of the earth, then the error in writing (16.14) as

(16.15) $$F = - \frac{GMm}{R^2}$$

is also small. [$R = 4000$ miles approximately so that even if y is as high as 1 mile above the earth, the difference between using $(4000 \times 5280)^2$ feet and $(4001 \times 5280)^2$ feet in the denominator is relatively negligible.] By (16.1) with y replacing s, we can write (16.15) as

(16.16) $$m \frac{d^2y}{dt^2} = - \frac{GMm}{R^2}.$$

Since G, M, and R are constants, we may replace GM/R^2 by a new constant which we call g. We thus finally obtain for the differential equation of motion of a falling body in the gravitational field of the earth,

(16.17) $$m \frac{d^2y}{dt^2} = -gm, \qquad m \frac{dv}{dt} = -gm,$$

where $v = dy/dt$. The minus sign is necessary because we have taken the upward direction as positive (see Fig. 16.13) and the force of the earth's attraction is downward. From (16.17), we have

(16.18) $$\frac{d^2y}{dt^2} = -g.$$

The constant g is thus the acceleration of a body due to the earth's attractive force, commonly known as the **force of gravity**. Its value varies slightly for different locations on the earth and for different heights. For convenience we shall use the value 32 ft/sec^2.

Integration of (16.18) gives the velocity equation

(16.19) $$v \left(= \frac{dy}{dt} \right) = -gt + c_1.$$

And by integration of (16.19), we obtain the distance equation

(16.2) $$y = - \frac{gt^2}{2} + c_1 t + c_2.$$

Example 16.21. A ball is thrown upward from a building which is 64 feet above the ground, with a velocity of 48 ft/sec. Find:

1. How high the ball will rise.
2. How long it will take the ball to reach the ground.
3. The velocity of the ball when it touches the ground.

Solution (Fig. 16.22).   By (16.2), with $g = 32$,

(a) $$y = -16t^2 + c_1 t + c_2.$$

Differentiation of (a) gives

(b) $$v = -32t + c_1.$$

If the origin is taken at ground level, the initial conditions are $t = 0$, $y = 64$, $v - 48$. Inserting these values in (a) and (b), we obtain

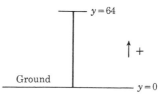

Figure 16.22

(c) $$c_2 = 64, \qquad c_1 = 48.$$

Hence (a) and (b) become respectively

(d) $$y = -16t^2 + 48t + 64, \qquad v = -32t + 48.$$

The   ball will continue to rise until its velocity is zero. By (d), when $v = 0$, $t = 1.5$ seconds, and when $t = 1.5$ seconds, $y = 100$ feet. Hence the ball will rise 100 feet above the ground.

When the ball is at ground level, $y = 0$, and by (d) when $y = 0$, $t = 4$ seconds. Hence the ball will reach the ground in 4 seconds. Its velocity at that moment will then be, by the second equation in (d),

$$v = (-32)(4) + 48 = -80 \text{ ft/sec.}$$

The negative sign indicates that the ball is moving in a downward direction.

Comment 16.23.   In the above example, we ignored the very important factor of air resistance. In a real situation, this factor cannot be thus ignored. Air resistance varies, among other things, with air density and with the speed of the object. Furthermore, air density itself changes with height and with time. It is different for different heights and may be different from day to day. The factor of air resistance in a real problem is thus a complicated one.

When, therefore, we assume in the examples which follow, a constant atmosphere and an air resistance which is dependent only on the speed of the object, we have simplified the practical problem enormously. And when in addition we suppose that this simplified air resistance is proportional to an integral power of the speed, we have simplified the problem considerably further. There is no valid reason why air resistance may not be proportional to the logarithm of the speed or to the square root of the speed, etc.

In all cases, however, air resistance always acts in a direction to oppose the motion.

Example 16.24. A body of mass m slugs is dropped from a height of 5000 feet. Find the velocity and the distance it will fall in time t. Assume that the force of the air resistance is proportional to the first power of the velocity, the proportionality constant being $m/40$.

Solution (Fig. 16.241). The force of the air resistance is given as $(m/40)v$. The downward force due to the weight of the mass m is mg pounds. Hence the differential equation of motion (16.17) must be modified to read, with the *positive direction downward* (remember mass $\times$ acceleration of a body = the net forces acting upon it),

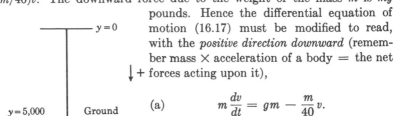

(a) $$m\frac{dv}{dt} = gm - \frac{m}{40}v.$$

Figure 16.241

Note that the force of gravity gm is now positive since it acts in the chosen positive direction. This equation can be solved by the method of Lesson 6C or 11B. Using the latter method, we write (a) as

(b) $$\frac{dv}{dt} + \frac{1}{40}v = g.$$

The integrating factor by (11.12) is $e^{t/40}$. The solution of (b) is therefore

(c) $$v = 40g + c_1 e^{-t/40}.$$

Integration of (c) gives

(d) $$y = 40gt - 40c_1 e^{-t/40} + c_2.$$

If the origin is taken at the point where the body is dropped, then the initial conditions are $t = 0$, $v = 0$, $y = 0$. Substituting these values in (c) and (d), we find

(e) $$c_1 = -40g, \qquad c_2 = -1600g.$$

Hence the two required equations are

(f) $$v = 40g(1 - e^{-t/40}), \qquad y = 40g(t + 40e^{-t/40} - 40).$$

Comment 16.25. We see from (f), that as $t \to \infty$, the velocity $v \to 40g$. This means that when a resisting force is present, the velocity does not increase indefinitely with time but approaches a limiting value beyond which it will not increase. This limiting velocity is called the **terminal velocity** of the falling body. In this example, it is $40g$ ft/sec.

Example 16.26.   The problem and initial conditions are the same as in Example 16.24 excepting that the force of the air resistance is assumed to be proportional to the second power of the velocity. Find the velocity of the body as a function of time and also the terminal velocity of the body.

Solution.   Here the force of the air resistance is $(m/40)v^2$. Hence (a) of Example 16.24 must be modified to read

(a) $m\dfrac{dv}{dt} = mg - \dfrac{m}{40}v^2, \qquad \dfrac{dv}{dt} = \dfrac{40g - v^2}{40}, \qquad \dfrac{dv}{40g - v^2} = \dfrac{1}{40}\,dt.$

Integrating the last equation in (a) and inserting the initial conditions as limits of integration, we obtain

(b) $$\int_{v=0}^{v} \frac{dv}{(2\sqrt{10g})^2 - v^2} = \frac{1}{40}\int_{t=0}^{t} dt.$$

Its solution is

(c) $\dfrac{1}{4\sqrt{10g}}\log\left(\dfrac{2\sqrt{10g} + v}{2\sqrt{10g} - v}\right) = \dfrac{1}{40}\,t, \qquad \dfrac{2\sqrt{10g} + v}{2\sqrt{10g} - v} = e^{(\sqrt{10g}/10)t}.$

Solving the last equation for v, we obtain

(d) $$v = 2\sqrt{10g}\left(\frac{e^{(\sqrt{10g}/10)t} - 1}{e^{(\sqrt{10g}/10)t} + 1}\right),$$

which gives the velocity of the body as a function of t.

As $t \to \infty$, we see from (d) that $v \to 2\sqrt{10g}$. This is the terminal velocity of the body.

Example 16.27.   A raindrop falls from a motionless cloud. Find its velocity as a function of the distance it falls. Assume it is subject to a resisting force which is proportional to the second power of the velocity. Also find its terminal velocity.

Solution.   Taking the *downward direction as positive*, the differential equation of motion (16.17) must be modified to read

(a) $$m\frac{dv}{dt} = mg - kv^2.$$

where $k > 0$ is a proportionality constant. Since we wish to find v as a function of the distance y, we replace dv/dt by its equal as given in (16.11). Hence (a) becomes

(b) $$mv\frac{dv}{dy} = mg - kv^2, \qquad \frac{v\,dv}{mg - kv^2} = \frac{dy}{m}.$$

If the cloud is taken as the origin, then the initial conditions are $y = 0$,

$v = 0$. Integration of (b) and insertion of the initial conditions give

(c)
$$\int_{v=0}^{v} \frac{v\,dv}{mg - kv^2} = \frac{1}{m} \int_{y=0}^{y} dy,$$

whose solution is

(d)
$$-\frac{1}{2k} \log\left(\frac{mg - kv^2}{mg}\right) = \frac{y}{m},$$

$$mg - kv^2 = mge^{-2ky/m},$$

$$v^2 = \frac{mg}{k}\left(1 - e^{-2ky/m}\right).$$

As $t \to \infty$, the distance y the raindrop falls approaches infinity, and as $y \to \infty$, we see from (d) that $v^2 \to mg/k$. Hence the terminal velocity is

(e)   T. V. $= \sqrt{mg/k}$.

Note. Since the body is falling and the downward direction is positive, the *positive* square root must be taken for the velocity in the last equation of (d).

Comment 16.28. The terminal or limiting velocity has no y in it, and is therefore independent of the height from which the raindrop falls. It is also independent of the initial velocity. From actual experience we know that a raindrop reaches its limiting velocity in a finite and not in an infinite time. This is because other factors also operate to slow the raindrop's velocity.

Comment 16.29. *A body falling in water* encounters a resistance just as does the body falling in air. If the magnitude of the velocity is small, the resistance of the water is approximately proportional to the first power of the velocity. The differential equation of motion (16.17) therefore becomes, with the *downward direction positive,*

(a) $$m\frac{dv}{dt} = mg - kv,$$

which is similar to (a) of Example 16.24.

Example 16.3. A man with a parachute jumps at a great height from an airplane moving horizontally. After 10 seconds, he opens his parachute. Find his velocity at the end of 15 seconds and his terminal velocity (i.e., the approximate velocity with which he will float to the ground). Assume that the combined weight of man and parachute is 160 pounds, and the force of the air resistance is proportional to the first power of the velocity, equaling $\frac{1}{2}v$ when the parachute is closed and $10v$ when it is opened.

Solution. For the first 10 seconds of fall, the differential equation of motion (16.17) of the man is, with *positive direction downward*,

(a) $$m\frac{dv}{dt} = mg - \tfrac{1}{2}v.$$

Here the downward force mg is equal to 160 pounds and the mass $m = 160/32$. Hence (a) becomes

(b) $$\frac{160}{32}\frac{dv}{dt} = 160 - \tfrac{1}{2}v, \qquad \frac{dv}{dt} + \frac{1}{10}v = 32.$$

Its solution, by the method of Lesson 11B, is

(c) $$v = 320 + ce^{-0.1t}.$$

If we take the origin at the point of jump, then $t = 0$, $v = 0$. Hence by (c), we find $c = -320$ so that

(d) $$v = 320(1 - e^{-0.1t}).$$

When $t = 10$

(e) $$v = 320(1 - e^{-1}) = 320(0.6321) = 202.3 \text{ ft/sec.}$$

Starting with the tenth second, the differential equation (16.17) becomes (remember the resistance is now $10v$)

(f) $$\frac{160}{32}\frac{dv}{dt} = 160 - 10v, \qquad \frac{dv}{dt} + 2v = 32,$$

whose solution is

(g) $$v = 16 + ce^{-2t}.$$

Inserting in (g) the initial condition which, by (e), is $t = 0$, $v = 202.3$, we find $c = 186.3$. Hence (g) becomes

(h) $$v = 16 + 186.3e^{-2t}.$$

When $t = 5$, i.e., 5 seconds after the parachute opens and 15 seconds after his jump,

(i) $$v = 16 + 186.3e^{-10} = 16 + 186.3(0.000045)$$
$$= 16 + 0.008 = 16.008 \text{ ft/sec.}$$

The terminal velocity is [in (h) let $t \to \infty$] 16 ft/sec. We see from (i), therefore, that only 5 seconds after the parachute is opened, the man is already floating to earth with a practically steady velocity of 16 ft/sec.

Comment 16.31. In deriving formula (16.17) for a vertically falling body, we ignored the distance y of the object above the earth's surface, since we assumed it to be relatively small in comparison with the radius

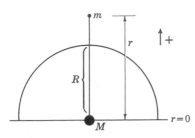

R of the earth. If, however, the distance of the object is very far above the earth's surface, then this distance cannot be thus ignored. In this case (16.14) becomes

$$(16.32) \qquad F = -G\frac{Mm}{r^2},$$

Figure 16.33

where M is the mass of the earth considered as being concentrated at its center and r is the distance of the body of mass m from this *center* (Fig.

16.33). Replacing in (16.32) the value of F as given in (16.1), we obtain

$$(16.34) \qquad m\frac{dv}{dt} = -G\frac{Mm}{r^2}, \qquad \frac{dv}{dt} = -\frac{GM}{r^2}.$$

Since G and M are constants, we can replace GM by a new constant k. There results

$$(16.35) \qquad \frac{dv}{dt} = -\frac{k}{r^2}, \qquad \frac{d^2r}{dt^2} = -\frac{k}{r^2},$$

where $v = dr/dt$. From (16.35) we deduce that the acceleration of a body in the gravitational field of the earth varies inversely as the square of the distance of the body from the center of the earth.

Example 16.36. A body is shot straight up from the surface of the earth with an initial velocity v_0. Assuming no air resistance, find:

1. The velocity v of the body as a function of the distance r from the center of the earth.
2. Its velocity when it is 4000 miles above the earth's surface.
3. How high the body will rise.
4. The magnitude of the initial velocity v_0 in order that the body may escape the earth, i.e., in order that it may never return to the earth.
5. The time t as a function of the distance r of the body from the earth's center.

Solution (Fig. 16.361). We take the origin at the center of the earth, and call R the radius of the earth. Then, by (16.35),

$$(a) \qquad a = \frac{dv}{dt} = -\frac{k}{r^2}.$$

Substituting in (a), the initial conditions $r = R$, $a = -g$, we find $k = gR^2$. Hence (a) becomes

(b)
$$\frac{dv}{dt} = -\frac{gR^2}{r^2}.$$

Since we wish to find v as a function of the distance r, we replace dv/dt by its equivalent value as given in (16.11). Hence (b) becomes

(c)
$$v\frac{dv}{dr} = -\frac{gR^2}{r^2}, \qquad v\,dv = -\frac{gR^2}{r^2}\,dr.$$

Integration of (c) and insertion of the initial conditions $v = v_0$, $r = R$, give

(d)
$$\int_{v=v_0}^{v} v\,dv = -gR^2 \int_{r=R}^{r} \frac{dr}{r^2},$$

whose solution is

(e)
$$v^2 = v_0{}^2 + \frac{2gR^2}{r} - 2gR = v_0{}^2 + 2gR\left(\frac{R}{r} - 1\right).$$

Hence the answer to question 1 is

(f)
$$v = \pm\sqrt{v_0{}^2 + 2gR\left(\frac{R}{r} - 1\right)};$$

the positive sign is to be used when the body is rising, the negative sign when it is falling. When the body is 4000 miles above the earth's surface,

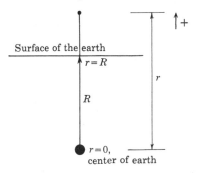

Figure 16.361

$r = 8000$ ($R = 4000$ miles approximately). Inserting this value in (f), we obtain

(g)
$$v = \pm\sqrt{v_0{}^2 + 2gR\left(\frac{R}{8000} - 1\right)} = \pm\sqrt{v_0{}^2 - 4000g},$$

which is the answer to question 2.

The body will continue to rise until $v = 0$. Hence by (f)

(h) $$0 = v_0{}^2 + 2gR\left(\frac{R}{r} - 1\right), \quad r = \frac{2gR^2}{2gR - v_0{}^2},$$

which is the distance the body will rise above the *center* of the earth if fired with an initial velocity v_0. Subtracting R from this value will give the distance the body will rise above the earth's *surface*. This is the answer to question 3.

The body will escape the earth, i.e., it will never return to the earth, if r increases with time. This means we want r to become infinite as the velocity v of the body approaches zero. By (e) we see that if $v_0{}^2 = 2gR$, then $r \to \infty$ as $v \to 0$. Hence the answer to question 4 is

(i) $$v_0 = \sqrt{2gR} = \sqrt{(2)(32)(4000)(5280)}$$
$$= 36{,}765 \text{ ft/sec} = 7 \text{ mi/sec,}^* \text{ approx.,}$$
$$= 25{,}100 \text{ mi/hr, approx.,}$$

which is the escape velocity of a body if air resistance is ignored.

The answer to question 5 is somewhat more difficult to obtain. In (e) replace v by dr/dt and let

(j) $$a = 2gR^2, \quad b = v_0{}^2 - 2gR.$$

Hence (e) becomes

(k) $$v = \frac{dr}{dt} = \pm \sqrt{\frac{a}{r} + b} = \pm \sqrt{\frac{a + br}{r}} = \pm \frac{1}{r} \sqrt{ar + br^2}.$$

When the body is rising, the velocity is positive and we can, therefore, write (k) as

(l) $$dt = \frac{r\,dr}{\sqrt{ar + br^2}} = \frac{1}{2b} \frac{(a + 2br)\,dr}{\sqrt{ar + br^2}} - \frac{a}{2b} \frac{dr}{\sqrt{ar + br^2}}.$$

If we assume $b < 0$, i.e., if we assume [see (j)] $v_0{}^2 < 2gR$, so that the body cannot escape the earth, then integration of (l) gives

(m) $$t = c + \frac{1}{b} \sqrt{ar + br^2} - \frac{a}{2b\sqrt{-b}} \text{ Arc sin}\left(\frac{-2br - a}{a}\right), \quad b < 0,$$

where a,b have the values given in (j).

Substituting in (m) the initial conditions $t = 0$, $r = R$, we obtain

(n) $$c = -\frac{1}{b} \sqrt{aR + bR^2} + \frac{a}{2b\sqrt{-b}} \text{ Arc sin}\left(\frac{-2bR - a}{a}\right), \quad b < 0.$$

With this value of c, (m) defines t as a function of r for a **rising** body.

---

*With $g = 32$ ft/sec^2, $v_0 = 6.96$ mi/sec instead of 7 mi/sec. However, g is closer to 32.17 ft/sec^2. With this value of g, $v_0 = 6.98$ mi/sec.

Remark. When the body is **falling,** v is negative. Hence for a falling body, we must, in (k), take

(o) $$v = -\sqrt{\frac{a}{r} + b},$$

in order to arrive at an equation comparable to (l) above. Or, if you wish, you may use formula (d) of Example 16.38 following. It gives the time of a falling body as a function of r with initial conditions $t = 0$, $v = 0$, $r = r_0$.

If we assume that $b = 0$, i.e., if we assume $v_0{}^2 = 2gR$ [see (j)] so that the body will escape the earth, then (e) becomes

(p) $\quad v = \dfrac{dr}{dt} = \dfrac{\sqrt{2g}\,R}{r^{1/2}}\,,\qquad r^{1/2}\,dr = \sqrt{2g}\,R\,dt,\qquad \tfrac{2}{3}r^{3/2} = \sqrt{2g}\,Rt + C.$

When $t = 0$, $r = R$. Hence $c = \tfrac{2}{3}R^{3/2}$. Therefore the last equation in (p) becomes

(q) $$t = \frac{2}{3R\sqrt{2g}}\,(r^{3/2} - R^{3/2}).$$

Comment 16.37. 1. Note from (16.35) that, as $r \to \infty$, the acceleration d^2r/dt^2 due to the gravitational force of the earth approaches zero. This means that the influence of the earth's gravitational field, although never zero, becomes insignificant.

2. From (e) we observe that when $v_0{}^2 = 2gR$, the escape velocity of the body, the velocity equation reduces to $v^2 = 2gR^2/r$. Hence as r gets larger, the velocity of the body will continue to get smaller until such time as it enters the gravitational field of another heavenly body. And if $v_0{}^2 > 2gR$ so that $v_0{}^2 - 2gR$ equals a positive constant k^2, then the velocity equation (e) reduces to $v^2 = k^2 + \dfrac{2gR^2}{r}$. Hence as $r \to \infty$, $v \to k$.

3. From equation (q) above, which expresses time as a function of r with $v_0{}^2 = 2gR$, we see that t also approaches infinity as $r \to \infty$.

Example 16.38. A body falls from interstellar space at a distance r_0 from the center of the earth. Find:

1. Its velocity v as a function of the distance r, where r is measured from the center of the earth.
2. Its velocity when it reaches the surface of the earth.
3. The time t as a function of the distance r.

Take the earth's center as the origin and the outward direction as positive.

Solution. The differential equation of motion of the body is the same as that of (c) in the previous example. Integration of this equation and insertion of the initial conditions gives

(a) $$\int_{v=0}^{v} v \, dv = -gR^2 \int_{r=r_0}^{r} \frac{dr}{r^2}.$$

Its solution is

(b) $$v^2 = 2gR^2 \left(\frac{1}{r} - \frac{1}{r_0} \right),$$

which is the answer to question 1.

When $r = R$, i.e., when the body is at the surface of the earth, its velocity is, by (b),

(c) $$v = - \sqrt{2gR - \frac{2gR^2}{r_0}}.$$

The negative sign is needed because the body is moving toward the earth and the outward direction is positive. This is the answer to question 2. From (c) we see that if r_0 is very large, i.e., if the body is extremely far away from the earth, then $2gR^2/r_0$ is very close to zero, and $|v|$ is extremely close to, but less than $\sqrt{2gR}$. This means that a body falling from outer space can never exceed a velocity equal to $\sqrt{2gR}$. As we saw in (i) of Example 16.36, $\sqrt{2gR} = 25{,}100$ miles/hour. Hence, if *air resistance is ignored*, a body falling from interstellar space will have a velocity at the earth's surface which differs extremely little from 25,100 miles/hour.

We leave it to you as an exercise to show that the answer to question 3 is

(d) $$t = \frac{\sqrt{r_0}}{R\sqrt{2g}} \left[\sqrt{r_0 r - r^2} + \frac{r_0}{2} \left(\frac{\pi}{2} - \text{Arc} \sin \frac{2r - r_0}{r_0} \right) \right]$$

$$= \frac{\sqrt{r_0}}{R\sqrt{2g}} \left(\sqrt{r_0 r - r^2} + \frac{r_0}{2} \text{Arc} \cos \frac{2r - r_0}{r_0} \right).$$

Hint. Start with (b) and follow the method we used in Example 16.36 to find the answer to question 5. You do not need to make the substitutions (j) of Example 16.36.

Comment 16.39. In solving the two previous problems, we assumed no air resistance. Since there is air resistance, the escape velocity v_0 would have to be sufficiently greater than $\sqrt{2gR} = 25{,}100$ miles/hour to overcome this resistance. If, however, the body emerged from the earth's atmosphere, which is rare 100 miles above its surface,* with a velocity equal to or perhaps very slightly more than 25,100 miles/hour, it would escape the earth. Conversely, the formula for the velocity of a body falling from outer space will give fairly accurate results until the object reaches the earth's atmosphere or about 100 miles from its surface. Con-

---

*A calculation made from an analysis of one of our satellite's orbits shows that the density of atmosphere at 932 miles above the earth is one thousand million millionths the density of air at sea level.

sidering the great distances involved, 100 miles is relatively insignificant, but its importance is tremendous. It complicates the whole problem of exit and reentry of satellites.

EXERCISE 16A

1. Verify the accuracy of the answer as given in the text to question 3 of Example 16.38.

In problems 2–5, assume **no air resistance** and that the object is **near** the earth's surface.

2. A ball is thrown vertically upward from the ground with an initial velocity of 80 ft/sec.
   - (a) Find its velocity and distance equations as functions of time. Take the origin at the point where the ball is thrown and the upward direction as positive.
   - (b) What are its velocity and height at the end of 1 sec?
   - (c) How long and how high will it rise?
   - (d) When will it reach the ground and with what velocity?
   - (e) Would the results differ if the ball were a projectile weighing 5 tons?

3. A man leans over the side of a bridge and drops a stone. His stop watch shows that the stone touched the water in 2.1 seconds. How high is the bridge above the water?

4. A ball is given a downward velocity of 8 ft/sec from a height of 120 ft above the ground.
   - (a) When will it reach the ground and with what velocity will it strike the ground?
   - (b) How much lower must one stand in order to drop a ball and have it reach the ground at the same time as the first ball?
   - (c) With what velocity will the second ball strike the ground?
   - (d) What is the significance of the negative sign of -3 seconds obtained in (a)? *Hint.* Substitute $t = -3$ in your velocity equation. Then solve this problem: If a ball is thrown upward from the ground with a velocity of 88 ft/sec, how high will it go; at what height will its velocity be 8 ft/sec *downward;* when will it reach this height and velocity?

5. A person, 81 ft above the ground, drops an object. With what velocity must a second person 180 ft above the ground throw an object straight down in order that both objects reach the ground at the same time?

In the problems below, weight in pounds is equal to mass in slugs times the acceleration of gravity in feet per second per second, i.e.,

(16.391) $$W = mg, \qquad m = W/g.$$

6. A man weighs 160 lb on earth.
   - (a) What is his mass?
   - (b) The acceleration of gravity on the surface of the moon is approximately one-sixth that of the earth. What will he weigh on the surface of the moon?
   - (c) Find formulas comparable to the velocity and distance equations (16.19) and (16.2) for the surface of the moon.

7. A ball thrown vertically upward from the surface of the earth with a velocity of 64 ft/sec will reach a maximum height of 64 ft in 2 sec (verify it). If thrown with the same velocity on the surface of the moon, find, by means of the formulas developed in 6(c), comparable figures for the maximum height reached and the time needed to attain this height.

8. If a man can high-jump 5 ft on earth, how high will he jump on the moon and how much longer will he be in the air as compared with the time in the air on the earth? Assume his center of gravity is 2 ft from the top of the bar. *Hint*. See problem 7.

9. A man whose weight is 160 lb is in an elevator which is descending with an acceleration of 2 ft/sec^2. What is his weight while riding in the elevator? *Hint*. Use (16.391); remember his mass is constant, and when the elevator accelerates down, he accelerates up.

In problems 10–17 and 20, 21, assume that **the force R of the air resistance is proportional to the first power of the velocity,** i.e., $R = kv$, and that the falling or rising body is **near** the earth's surface.

10. A body of mass m is dropped from a great height.

   (a) Find its velocity and the distance it falls as functions of time. Take the positive direction as downward and the origin at the point where the body is dropped. *Hint*. In (a) of Example 16.24 replace $m/40$ by k. Use your results to check the accuracy of the answers given in (f).

   (b) What is its terminal velocity?

11. Solve problem 10 if the body initially is given a downward velocity of v_0 ft/sec. What is its terminal velocity? Compare with 10(b) above. Note that the initial velocity does not affect the terminal velocity.

12. A body weighing 192 lb is dropped from a great height. The proportionality constant k of the air resistance is 12.

   (a) Find its velocity and the distance it falls as a function of time. Solve independently. Use results obtained in 10 only as a check.

   (b) What is its terminal velocity?

   (c) How far does it fall in 10 sec? What is its velocity at that moment?

13. Solve problem 12 if the body is given an initial downward velocity of 170 ft/sec instead of being dropped. Solve independently. Use the results obtained in 11 only as a check.

14. When a paratrooper falls freely from a great height before opening his chute, his terminal velocity is approximately 175 ft/sec. Assume a paratrooper and his chute together weigh 200 lb.

   (a) Find the proportionality factor k of the air resistance. *Hint*. Use the formula for T.V. found in 10(b).

   (b) Find his velocity and the distance he falls as a function of time.

   (c) What is his velocity at the end of $8\frac{3}{4}$ sec, $17\frac{1}{2}$ sec, $26\frac{1}{4}$ sec, $32\frac{5}{6}$ sec?

   (d) How far has he fallen in $32\frac{5}{6}$ sec?

15. Assume the paratrooper of problem 14 opens his chute when he has reached his terminal velocity of 175 ft/sec, and that his chute is designed to give him a safe landing speed of 16 ft/sec.

   (a) What is the new value of k? *Hint*. Use the formula for T.V. found in problem 11.

   (b) Find his velocity and the distance he falls after he opens his chute as functions of time.

(c) What is his velocity at the end of 1 sec, 2 sec, 3 sec, 4 sec, 5 sec?

(d) How far has he fallen in 5 sec?

(e) Do your answers in (c) and (d) suggest a safe height at which he can open his chute?

(f) If he opens his chute at a height of 1040 ft, in how many seconds does he reach the ground?

16. A paratrooper jumps from a plane flying horizontally at a great height. When he feels that he has reached a steady velocity (i.e., when he has reached his terminal velocity), he opens his chute. Assume this steady velocity is 180 ft/sec.

(a) Find his velocity and the distance he falls as a function of time before the chute opens. *Hint.* Use formula for T.V. found in 10(b) and solve for m/k.

(b) Calculate his velocity at the end of $11\frac{1}{4}$ sec, $22\frac{1}{2}$ sec, $33\frac{3}{4}$ sec, 45 sec.

(c) How far has he fallen in 45 sec?

17. If the force of the air resistance is 50 lb when a body is falling at a velocity of 25 ft/sec, what is the value of the proportionality constant k of the air resistance? For this k, find the terminal velocity of a falling body weighing 100 lb. If the body has an initial velocity of 20 ft/sec, find its velocity and distance equations as functions of time.

18. A body weighing 96 lb begins to sink as soon as it is placed in water. Two forces act on it to oppose its motion, an upward force due to the buoyancy of the object and a force due to the resistance of the water. Assume the buoyant force is 12 lb and the resistance of the water is $6v$. Take the origin on the surface of the water and downward direction as positive.

(a) Find the velocity and position of the body as functions of time.

(b) Find its terminal velocity.

19. The **specific gravity** of a body is defined as the ratio of its weight to the weight of an equal volume of water. Assume a body is released from the surface of a medium whose specific gravity is one-fourth that of the body and that the resistance offered by the medium is $mv/3$.

(a) Find the velocity of the body as a function of time. *Hint.* The specific gravity of the medium equal to $\frac{1}{4}$ that of the body, implies that the medium's upward buoyant force is one-fourth the weight of the body.

(b) Find its terminal velocity.

20. A body of mass m is shot straight up from the ground with an initial velocity of v_0 ft/sec.

(a) Find the velocity and position of the body as functions of time. Take the positive direction upwards and the origin on the ground.

(b) How high will the body rise and when will it reach this maximum height?

21. An object weighing 64 lb is shot straight up from the ground with an initial velocity of 96 ft/sec. Assume the force of the air resistance is $4v$.

(a) Find the velocity and position of the object as functions of time.

(b) How high will the body rise and when will it reach this maximum height? Check the accuracy of your answers in (a) and (b) with the formulas obtained in problem 20.

(c) When and with what velocity will it strike the ground? Note that the down trip takes longer than the up trip, whereas when there is no air resistance both times are the same. Note also that the return velocity is smaller than the initial velocity.

(d) Do you believe you would get the same answer for the velocity of the falling body when it strikes the ground and for the time of the down trip if you used the formulas for a falling body as found in problem 10, with y having the value determined in (b)? Try it.

In problems 22–30, assume that the force R of air resistance is proportional to the second power of the velocity, i.e., $R = kv^2$, and that the falling or rising body is **near** the earth's surface.

22. In Example 16.27, we found the velocity v of a falling body as a function of the distance fallen. Starting with equation (a) of this example, find the velocity and distance of a falling body as functions of time. Find its terminal velocity. Compare with (e) of Example 16.27.

23. A body of mass m falls from a great height. Its initial velocity is v_0 ft/sec. Find:

(a) Its velocity as a function of the distance fallen. Take the positive direction downwards and the origin at the point of fall.

(b) Its velocity as a function of time.

(c) Its terminal velocity. Compare with (e) of Example 16.27 and with problem 22. Note that the initial velocity does not affect the terminal velocity.

24. A body of mass m is fired vertically upward from the ground at an initial velocity of v_0 ft/sec.

(a) Find its velocity as a function of its height. Take positive direction upwards and origin on ground.

(b) Find its velocity as a function of time.

(c) Find its height as a function of time.

(d) When will it reach maximum height?

(e) How high will it rise?

Note. *These formulas are valid only while the body is rising.* When it begins to fall, formulas developed in Example 16.27 and problem 22 must be used. Compare with problem 21 where we used one formula to calculate the time for a round trip.

25. With what velocity will the body of problem 24 return to the earth and how long will it take for its descent? *Hint.* Read note in problem 24. To find the velocity, use (d) of Example 16.27 with y equal to the value of the maximum height as found in problem 24(e). Note that the returning velocity is less than the initial velocity v_0. To find the time of descent, use the formula for y as found in problem 22. Here it is not easy to see that the time of descent is longer than the time of ascent.

26. The velocity of a parachutist at the moment his chute opens is 160 ft/sec. The force of the air resistance is $mv^2/8$. Find:

(a) His subsequent velocity and distance as functions of time. Take the origin at the point where the chute opens and the positive direction downwards.

(b) His velocity 1 sec after his chute opens; 2 sec after.

(c) His terminal velocity.

(d) How far he falls in the first second; in the second second.

(e) Approximately when he reaches the ground if he is 1064 ft above the earth when his chute opens.

27. A body of mass m is dropped from a plane flying horizontally 1 mile above the earth. The force of the air resistance is $2mk^2v^2$. The terminal velocity of the body is 100 ft/sec. Find:

  (a) The value of the constant k. (*Hint.* T.V. $= \sqrt{mg/k}$; replace k by $2mk^2$.)
  (b) The velocity of the body as a function of time.
  (c) The velocity of the body at the end of 3 sec.
  (d) When the body reaches a velocity of 60 ft/sec.

28. A body falls from a great height. Its terminal velocity is 10 ft/sec. (a) Find its velocity and distance equations as functions of time. *Hint.* T.V. $= \sqrt{mg/k}$. Solve for k/m. (b) Find its velocity equation as a function of distance.

29. A paratrooper and his chute, which together weigh 192 lb, drop from an airplane moving horizontally. He opens his chute at the end of 10 sec. Assuming the proportionality constant of air resistance is 1/120 when the chute is closed and 4/3 when it is open, find:

  (a) His velocity as a function of time before the chute is opened.
  (b) His terminal velocity before the chute is opened.
  (c) His velocity at the end of the first 10 sec.
  (d) His velocity as a function of time after the chute is opened.
  (e) His terminal velocity after the chute is opened.
  (f) His velocity at the end of 15 sec, i.e., his velocity 5 sec after the chute is opened.

  Solve independently and then check your results with the formulas found in problems 22 and 23.

30. A man and his parachute weigh 192 lb. Assume that a safe landing velocity is 16 ft/sec and that air resistance is proportional to the square of the velocity, equaling $\frac{1}{2}$ lb for each square foot of cross-sectional area of the parachute when it is moving at 20 ft/sec at right angles to the direction of motion. What must the cross-sectional area of a parachute be in order that the paratrooper land safely? *Hint.* First find the force of the air resistance. Then find k such that T.V. $= 16$. Then find the number of square feet of parachute that will make the force of the air resistance equal to kv^2.

In problems 31–39, assume **no air resistance** and that the object is **far** enough from the earth so that equation (16.35) applies. Use the following data: $R = 4000$ miles, $g = 32$ ft/sec^2, $\sqrt{2gR} = 6.96$ mi/sec.

31. With what velocity must a rocket be fired in order to reach a height of 400 mi above the earth; 4000 mi above the earth? Solve independently. Check your results with (h) of Example 16.36.

32. The "air" 200 mi above the earth is so thin that it will hardly slow a space vehicle. It is called the **F-2 region** of the atmosphere. What velocity should a rocket have at 200 miles above the earth, if all its fuel is exhausted at that point, in order to go another 3800 mi?

33. A body is shot to a height of 400 mi and then starts to fall. What is its velocity (a) when it has fallen 200 mi and (b) when it is at the surface of the earth? *Hint.* Use (a) of Example 16.38 with $r_0 = R + 400 = 4400$. Note that the velocity, when it reaches the ground, is the same as the velocity required to propel it 400 miles upward. See problem 31.

34. Assume a body falls from rest at a distance of $61R$, i.e., at a distance of 244,000 mi from the center of the earth (equivalent to the moon's distance from the center of the earth). With what velocity and in how many hours will it reach the earth?

35. A body is fired straight up with an initial velocity equal to the escape velocity $\sqrt{2gR}$. Find:

(a) The velocity of the projectile as a function of the distance r from the center of the earth.

(b) When it will have reached 244,000 mi, the distance of the moon from the center of the earth.

Hint. Use (e) and (q) of Example 16.36.

36. A body is fired straight up with an initial velocity v_0 whose magnitude is less than escape velocity. When will it reach its maximum height? *Hint.* The maximum height the body will reach is given in (h) of Example 16.36. Express this value of r in terms of a and b as defined in (j). Then use (m) and (n). Note. This time equation is valid only for a *rising* particle. If you wish to compute its return time, you must use the time equation given in (d) of Example 16.38, with r_0 equal to the height from which it begins its fall.

37. Show that if v_0 is very much less than the escape velocity $\sqrt{2gR}$, then the time for the object to reach its maximum height, as given in problem 36, is approximately v_0/g. *Hint.* Replace the Arc sin function by its series expansion, Arc sin $x = x + x^3/6 + \cdots$. Then eliminate $v_0{}^2$ and higher powers of v_0. To see some justification for this elimination, let $v_0 = \frac{1}{30}$ mi/sec in the time equation of problem 36.

38. A body is shot straight up from the surface of the moon with an initial velocity v_0. (a) Find the velocity v of the body as a function of its distance r_m from the center of the moon. (b) Find the escape velocity of the body. The radius of the moon is approximately 1080 mi; the acceleration of gravity on the surface of the moon is approximately one-sixth that of the earth. Take the outward direction from the moon as positive.

39.   (a) Prove that if a particle were placed approximately nine-tenths of the distance D from the center of the earth to the center of the moon on a line connecting moon to earth, Fig. 16.392, the particle would be at rest, i.e., the gravitational pulls of moon and earth on a particle placed nine-tenths of the distance from the center of earth to the center of the moon are equal. Assume the mass of the moon is 1/81 the mass of the earth. *Hint.* Apply (16.32) to both earth and moon. Then equate the two forces.

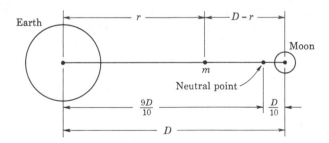

Figure 16.392

(b) Set up the differential equation of motion of a particle shot from the surface of the earth toward the moon, considered fixed, taking into account both the earth's and moon's gravitational attractions. Solve the equation with $t = 0$, $v = v_0$. Take the positive direction as outward from the earth, and the orgin at the earth's center.

(c) At what velocity must the particle be fired in order to reach the neutral point? *Hint.* In the velocity equation found in (b), you want $v = 0$, when $r = \frac{9}{10}D$. Assume $R_m{}^2 = 6R^2/81$, $D = 61R + R/4$, $g_m = g/6$, where R_m is the radius of the moon and g_m is the acceleration due to the force of gravity of the moon. Note the small effect of the moon's gravitational attraction.

(d) At what velocity must a projectile be fired in order to reach the moon?

40. If a body were dropped in a hole bored through the center of the earth, it would be attracted toward the center with a force directly proportional to the distance of the body from the center.

(a) With what velocity will it pass the center?
(b) When will it reach the other end of the hole?

NOTE. The motion of the body is known as simple harmonic motion. A fuller discussion of this motion can be found in Lesson 28.

ANSWERS 16A

2. (a) $v = -32t + 80$, $y = -16t^2 + 80t$.   (b) 48 ft/sec, 64 ft.
(c) $2\frac{1}{2}$ sec, 100 ft.   (d) 5 sec, -80 ft/sec. (e) No.

3. 70.56 ft.

4. (a) $2\frac{1}{2}$ sec, 88 ft/sec.   (b) 20 ft.   (c) 80 ft/sec.
(d) If 3 sec ago, the ball were thrown upward from the ground with a velocity of 88 ft/sec, it would reach a height of 121 ft and have a velocity of 8 ft/sec as it passed the 120 ft point on its way down.

5. 44 ft/sec.

6. (a) 5 slugs.   (b) $26\frac{2}{3}$ lb.   (c) $v = -\dfrac{16t}{3} + c_1$, $y = -\frac{8}{3}t^2 + c_1 t + c_2$.

7. 384 ft, 12 sec.

8. 15 ft if one assumes he scales the bar horizontally so that he raises his center of gravity 2 ft on earth; 6 times as long.

9. 150 lb.

10. (a) $v = \dfrac{mg}{k}(1 - e^{-kt/m})$, $y = \dfrac{mg}{k}t + \dfrac{m^2 g}{k^2}(e^{-kt/m} - 1)$.
(b) T.V. $= mg/k$.

11. $v = \dfrac{mg}{k}\left(1 - e^{-kt/m}\right) + v_0 e^{-kt/m}$,

$y = \dfrac{mg}{k}t - \left(\dfrac{m^2 g}{k^2} - \dfrac{mv_0}{k}\right)(1 - e^{-kt/m})$.
T.V. $= mg/k$, same as in 10(b).

12. (a) $v = 16(1 - e^{-2t})$, $y = 16t + 8(e^{-2t} - 1)$.   (b) 16 ft/sec.
(c) $y = 8(19 + e^{-20}) = 152$ ft, $v = 16(1 - e^{-20}) = 16$ ft/sec.

13. (a) $v = 16(1 - e^{-2t}) + 170e^{-2t} = 16 + 154e^{-2t}$, $y = 16t + 77(1 - e^{-2t})$.
(b) 16 ft/sec.   (c) 237 ft, 16 ft/sec.

14. (a) $k = \frac{8}{7}$.

(b) $v = 175\left(1 - e^{-32t/175}\right)$, $y = 175t + \dfrac{175^2}{32}(e^{-32t/175} - 1)$.

(c) 139.7 ft/sec, 167.9, 173.6, 174.6.   (d) 4791 ft.

15. (a) 12.5.   (b) $v = 16(1 - e^{-2t}) + 175e^{-2t}$, $y = 16t + \dfrac{159}{2}(1 - e^{-2t})$.

    (c) 37.5 ft/sec, 18.9, 16.4, 16.05, 16.01.   (d) 159.5 ft.

    (e) Over 160 ft.   (f) Approx. 60 sec.

16. (a) $v = 180(1 - e^{-8t/45})$, $y = 180t + \dfrac{180^2}{32}(e^{-8t/45} - 1)$.

    (b) $v(11\frac{1}{4}) = 180(1 - e^{-2}) = 155.6$ ft/sec, 176.7, 179.6, 179.9.

    (c) Approx. 7100 ft.

17. $k = 2$; T.V. $= 50$ ft/sec; $v = 50 - 30e^{-gt/50} = 50 - 30e^{-0.64t}$;

$$y = 50t - \frac{1500}{g}(1 - e^{-gt/50}) = 50t - \frac{375}{8}(1 - e^{-0.64t}).$$

18. (a) $v = 14(1 - e^{-2t})$, $y = 14[t - \frac{1}{2}(1 - e^{-2t})]$.

    (b) T.V. $= 14$ ft/sec.

19. (a) $v = 72(1 - e^{-t/3})$.   (b) T.V. $= 72$ ft/sec.

20. (a) $v = \dfrac{gm}{k}(e^{-kt/m} - 1) + v_0 e^{-kt/m}$,

$$y = \left(\frac{gm^2}{k^2} + \frac{mv_0}{k}\right)(1 - e^{-kt/m}) - \frac{gm}{k}t.$$

    (b) $y = \dfrac{m}{k}\left(v_0 - \dfrac{gm}{k}\log\dfrac{kv_0 + gm}{gm}\right)$, $t = \dfrac{m}{k}\log\dfrac{kv_0 + gm}{gm}$.

21. (a) $v = 16(7e^{-2t} - 1)$, $y = 56(1 - e^{-2t}) - 16t$.

    (b) $y = 8(6 - \log 7) = 32.4$ ft, $t = \frac{1}{2}\log 7 = 0.97$ sec.

    (c) 3.5 sec approx., -16 ft/sec.   (d) Same answer.

22. $v = \sqrt{mg/k}\,\tanh\sqrt{gk/m}\,t = \sqrt{mg/k}\,\dfrac{e^{2\sqrt{kg/m}\,t} - 1}{e^{2\sqrt{kg/m}\,t} + 1}$,

$y = \dfrac{m}{k}\log\cosh\sqrt{gk/m}\,t = \dfrac{m}{k}\log\dfrac{e^{\sqrt{gk/m}\,t} + e^{-\sqrt{gk/m}\,t}}{2}$,

T.V. $= \sqrt{mg/k}$ ft/sec.

For definitions of cosh and tanh, see (18.91) and (18.92).

23. (a) $v^2 = \dfrac{mg}{k}(1 - e^{-2ky/m}) + v_0^2 e^{-2ky/m}$. Since the positive direction is

downward, the positive square root must be taken for v when the body is falling.

    (b) $\dfrac{\sqrt{k}\,v + \sqrt{mg}}{\sqrt{k}\,v - \sqrt{mg}} = \dfrac{\sqrt{k}\,v_0 + \sqrt{mg}}{\sqrt{k}\,v_0 - \sqrt{mg}}\,e^{2\sqrt{gk/m}\,t}$.

    (c) T.V. $= \sqrt{mg/k}$ ft/sec.

24. (a) $v^2 = \dfrac{mg}{k}(e^{-2ky/m} - 1) + v_0^2 e^{-2ky/m}$. Since the positive direction is

upward, the positive square root must be taken for v when the body is rising.

    (b) $t = \sqrt{m/kg}\,(\text{Arc tan}\sqrt{k/mg}\,v_0 - \text{Arc tan}\sqrt{k/mg}\,v)$,

    or $v = \sqrt{gm/k}\,\tan(c - \sqrt{kg/m}\,t)$, where $c = \text{Arc tan}\sqrt{k/gm}\,v_0$.

    The formula for the time t is valid only for a rising body.

(c) $y = \dfrac{m}{k} \log \left[\dfrac{\cos (c - \sqrt{kg/m}\, t)}{\cos c} \right]$.

(d) $t = c\sqrt{m/kg}$.

(e) $y = \dfrac{m}{k} \log \sec c$.

25. $v = \sqrt{mg/(kv_0^2 + mg)}\; v_0$ ft/sec,

$t = \sqrt{m/gk} \cosh^{-1} \sec c$ or $\cosh \sqrt{gk/m}\, t = \sec c$,

where $c = \text{Arc tan} \sqrt{k/gm}\; v_0$. For definition of cosh, see (18.91).

26. (a) $v = 16 \dfrac{11 + 9e^{-4t}}{11 - 9e^{-4t}}$, $y = 16t + 8 \log \left(5.5 - 4.5e^{-4t}\right)$.

(b) 16.49 ft/sec, 16.009 ft/sec.   (c) 16 ft/sec.

(d) 29.5 ft., 16.1 ft.   (e) 66 sec.

27. (a) $k = 1/25$.

(b) $v = 100 \dfrac{e^{0.64t} - 1}{e^{0.64t} + 1}$.

(c) 74.4 ft/sec.   (d) 2.2 sec.

28. (a) $v = 10 \tanh (3.2t) = 10 \dfrac{e^{6.4t} - 1}{e^{6.4t} + 1}$,

$y = \dfrac{100}{g} \log \cosh 3.2t = \dfrac{100}{g} \log \dfrac{e^{3.2t} + e^{-3.2t}}{2}$.

(b) $v = 10\sqrt{1 - e^{-0.64y}}$.

29. (a) $v = 151.8 \dfrac{e^{2t\sqrt{10}/15} - 1}{e^{2t\sqrt{10}/15} + 1}$.

(b) T.V. $= 151.8$ ft/sec.   (c) 147.4 ft/sec.

(d) $v = 12 \dfrac{1.18e^{16t/3} + 1}{1.18e^{16t/3} - 1}$.

(e) 12 ft/sec.   (f) 12 ft/sec.

30. 600 sq ft.

31. 2.10 mi/sec, 4.92 mi/sec.

32. 4.7 mi/sec.

33. (a) 1.45 mi/sec.   (b) 2.10 mi/sec.

34. 6.9 mi/sec, 119 hours.

35. (a) $v = R\sqrt{2g/r}$,   (b) 50.6 hours.

36. $t = -\dfrac{1}{b} \sqrt{aR + bR^2} - \dfrac{a}{2b\sqrt{-b}} \left[\dfrac{\pi}{2} - \text{Arc sin} \dfrac{-2bR - a}{a} \right]$.

$t = \dfrac{gR^2}{(2gR - v_0^2)^{3/2}} \left[\dfrac{v_0\sqrt{2gR - v_0^2}}{gR} + \text{Arc cos} \left(\dfrac{gR - v_0^2}{gR} \right) \right]$

$= \dfrac{gR^2}{(2gR - v_0^2)^{3/2}} \left[\dfrac{v_0\sqrt{2gR - v_0^2}}{gR} + 2 \, \text{Arc sin} \dfrac{v_0}{\sqrt{2gR}} \right]$.

38. (a) $v^2 = v_0{}^2 + \dfrac{gR_m}{3}\left(\dfrac{R_m}{r_m} - 1\right)$, where R_m is the radius of the moon.

(b) $v_0 = \sqrt{gR_m/3} = 1.48$ mi/sec.

39. (b) $v\,\dfrac{dv}{dr} = \dfrac{-gR^2}{r^2} + \dfrac{g_m R_m{}^2}{(D - r)^2}$,

$$v^2 = 2g\,\frac{R^2}{r} + \frac{2g_m R_m{}^2}{D - r} + v_0{}^2 - 2gR - \frac{2g_m R_m{}^2}{D - R}.$$

(c) $v_0 = \sqrt{2gR}\,(0.99) = 99$ percent of the escape velocity of the earth when the gravitational pull of the moon is ignored.

(d) The velocity must have a positive value when it reaches the neutral point. See (c).

40. (a) 4.9 mi/sec.   (b) 42.5 min.

LESSON 16B.   Horizontal Motion.   If a body moves in a horizontal direction, as on a table or platform, a frictional force develops, which operates to stop the progress of the motion. This frictional force is due to

1. The gravitational force of the earth pressing the body to the platform.
2. The smoothness or roughness of the surface of the platform. This quality of the surface, i.e., its roughness or smoothness, is characterized by means of a letter μ, called the **coefficient of friction** of the surface.

Definition 16.4.   The **frictional force of a body** *moving* on a horizontal surface is, by definition, equal to the product of the coefficient of friction μ of the body and the gravitational force mg, i.e., frictional force $= \mu(mg)$.

Comment 16.41.   From experience we know that it requires a greater force to begin the movement of an object than it does to keep it moving. There are thus two coefficients of friction, one called **static friction,** which operates at the start of the motion, the other called **sliding friction,** which operates after the motion has begun.

In addition to the frictional force, a body also may be subject to a resisting force due to the air or other medium in which it moves.

Example 16.411.   An object on a sled is pulled by a force of 10 pounds across a frozen pond. Object and sled weigh 64 pounds. The coefficients of static and sliding friction are negligible. However, the force of the air resistance is twice the velocity of the sled. If the sled starts from rest, find its velocity at the end of 5 seconds and the distance it has traveled in that time. What is its terminal velocity?

Solution.   Here $m = 64/32 = 2$ and the force of the air resistance is given as $2v$ pounds. Hence the differential equation of motion of the sled is (remember mass × acceleration of a body = net force acting upon it)

(a) $$2\frac{dv}{dt} = 10 - 2v, \qquad \frac{dv}{dt} + v = 5.$$

Its solution is

(b) $$v = 5 + c_1 e^{-t}.$$

The initial condition is $t = 0$, $v = 0$. Hence $c_1 = -5$, and (b) becomes

(c) $$v = 5(1 - e^{-t}).$$

When $t = 5$,

(d) $\quad v = 5(1 - e^{-5}) = 5(1 - 0.0067) = 5(0.9933) = 4.97$ ft/sec.

By (c) the terminal velocity is found to be 5 ft/sec.

To find the distance x traveled in time t, we integrate (c) to obtain

(e) $$x = 5(t + e^{-t}) + c_2.$$

When $t = 0$ and $x = 0$ (taking the origin at the starting position), we find, from (e), $c_2 = -5$. Hence (e) becomes

(f) $$x = 5(t + e^{-t} - 1).$$

And when $t = 5$,

(g) $\quad x = 5(5 + e^{-5} - 1) = 5(4 + 0.0067) = 20$ ft approx.

Example 16.42. A boy weighing 75 pounds runs for a slide and reaches it at a velocity of 10 ft/sec. If the coefficient of sliding friction μ between his shoes and the ice is $1/25$, how far will he slide? Ignore wind resistance.

Solution. By Definition 16.4, the frictional force is $1/25 \cdot 75 = 3$ pounds. Since this is the only force which is opposing the motion and the mass of the boy is $75/32$, the differential equation of motion becomes

(a) $$\frac{75}{32}\frac{dv}{dt} = -3, \qquad \frac{dv}{dt} = -\frac{32}{25}.$$

By (16.11), we can write (a) as

(b) $$v\frac{dv}{dx} = -\frac{32}{25}, \qquad v\,dv = -\frac{32}{25}\,dx,$$

where x is the distance measured from the beginning of the slide. Integration of (b) and insertion of the initial and final conditions results in

(c) $$\int_{v=10}^{0} v\,dv = -\frac{32}{25}\int_{x=0}^{x} dx.$$

Its solution is

(d) $$50 = \tfrac{32}{25}x, \quad x = 39 \text{ ft approx.}$$

Example 16.43. Solve the previous problem if a wind is blowing against the boy with a force equal to his velocity.

Solution. The differential equation (a) above must now be modified to read [with dv/dt replaced by its equal $v(dv/dx)$, — see (16.11)],

(a) $$\frac{75}{32} v \frac{dv}{dx} = -3 - v, \quad \frac{v\,dv}{v+3} = -\frac{32}{75}\,dx,$$

$$\int_{v=10}^{0} \left(1 - \frac{3}{v+3}\right) dv = -\frac{32}{75} \int_{x=0}^{x} dx.$$

The solution of (a) is

(b) $$-3 \log 3 - 10 + 3 \log 13 = -\tfrac{32}{75}x, \quad x = 13.1 \text{ ft approx.}$$

Example 16.44. A boat is being towed at the rate of 18 ft sec. At the instant when the towing line is cast off, a man takes up the oars and begins to row with a force of 20 pounds in the direction of the moving boat. If man and boat together weigh 480 pounds, and the resistance is equal to $\tfrac{7}{4}v$ pounds, find the speed of the boat at the end of 30 seconds.

Solution. The net force acting on the boat at the instant $t = 0$ when the towing line is cast off is the man's force of 20 pounds, less the resisting force of $\tfrac{7}{4}v$ pounds. The mass of man and boat is $480/32 = 15$. Hence the differential equation of motion is

(a) $$15 \frac{dv}{dt} = 20 - \frac{7}{4}v, \quad \frac{dv}{dt} + \frac{7}{60}v = \frac{4}{3}.$$

Its solution by Lesson 11B is

(b) $$v = ce^{-7t/60} + \tfrac{80}{7}.$$

Substituting in (b) the initial conditions $t = 0$, $v = 18$, we find $c = 46/7$. Hence (b) becomes

(c) $$v = \tfrac{46}{7}e^{-7t/60} + \tfrac{80}{7}.$$

When $t = 30$,

(d) $$v = \tfrac{46}{7}e^{-7/2} + \tfrac{80}{7} = 11.6 \text{ ft/sec.}$$

EXERCISE 16B

1. A boy pulls a sled, on which a parcel has been placed, with a constant force of F lb. The sled and parcel together weigh mg lb. The frictional force of the ice on the runners is negligible. However, the force of the air resistance is k times the velocity of the sled. If the sled starts from rest, find:

   (a) Its velocity as a function of time.
   (b) Its distance as a function of time.
   (c) Its distance as a function of velocity.
   (d) Its terminal velocity.

2. In problem 1, assume sled and parcel weigh 96 lb, that air resistance is $\frac{1}{9}$ times the velocity and that the boy is moving the sled at a constant rate of 4.5 ft/sec (i.e., the terminal velocity of the sled is 4.5 ft/sec).

   (a) Find the constant force which the boy is applying to the sled.
   (b) Find the velocity and distance equations as functions of time.
   (c) How far has the sled moved in 5 min?

   Solve independently and then check your answers with formulas found in problem 1.

3. A boy weighing mg lb runs for a slide and reaches it with a velocity of v_0 ft/sec. The coefficient of sliding friction between his shoes and the ice is r. (a) Find his velocity as a function of distance. Ignore air resistance. (b) How far will he slide?

4. A boy weighing 80 lb runs for a slide and reaches it with a velocity of 12 ft/sec. The coefficient of sliding friction between his shoes and the ice is 1/20 and the wind blows against him with a force equal to twice his velocity. (a) Find his distance as a function of his velocity. (b) How far will he slide?

5. A 32,000-ton ship starting from rest begins to move because of the actions of its propellers that exert a forward force of 120,000 lb. The force of the water resistance is $5000v$. Find the velocity of the ship as a function of time and its terminal velocity.

6. The brakes are applied to a car, traveling on a slippery road, when it has slowed down to a speed of 6 mi/hr = 8.8 ft/sec. It slides 80 ft before coming to a stop. Compute the coefficient of sliding friction between tires and street. Neglect the force of air resistance.

7. A body weighing 50 lb rests on a table whose coefficient of sliding friction is 1/25. The body is attached by a string to a weight of 14 lb that hangs vertically over the table. At the moment the system is released, the 50-lb body is 15 ft from the edge of the table. Assume no other forces are operating. When and with what velocity does the body leave the table? *Hint.* The total mass of the system is 64/32.

8. The forward thrust of an airplane due to its propellers is F lb. The air resistance is kv^2. Find the terminal velocity of the plane.

9. An 8-lb body starting from rest is being pulled along a surface, whose coefficient of sliding friction is $\frac{1}{4}$, by a force that is equal to twice the distance of the body from its starting point $x = 0$. Air resistance is $v^2/8$. Find the velocity of the body as a function of its distance. *Hint.* The resulting differential equation is linear in v^2.

10. A man and his boat weigh 320 lb. The man exerts a force of 16 lb on the oars. The resistance of the water is twice the speed. Find:
(a) The velocity of the boat as a function of time.
(b) Its speed after 5 sec.
(c) Its terminal velocity.

11. A man and his boat weigh 400 lb. At the moment the man picks up his oars to row, the boat is moving at the rate of 22 ft/sec. If the resistance of the water is $2v$ lb and the oars exert a constant force of 15 lb, find the velocity of the boat as a function of time; also find the terminal velocity of the boat.

ANSWERS 16B

1. (a) $v = \dfrac{F}{k}(1 - e^{-kt/m})$.

  (b) $x = \dfrac{F}{k}\left(t + \dfrac{me^{-kt/m}}{k} - \dfrac{m}{k}\right)$.

  (c) $x = \dfrac{m}{k}\left(-v - \dfrac{F}{k}\log\dfrac{F - kv}{F}\right)$.

  (d) T.V. $= F/k$ ft/sec.

2. (a) $\frac{1}{2}$ lb.

  (b) $v = \frac{9}{2}(1 - e^{-t/27})$, $x = \frac{9}{2}(t + 27e^{-t/27} - 27)$.

  (c) 1228.5 ft.

3. (a) $v^2 = v_0{}^2 - 2rgx$.   (b) $x = v_0{}^2/2rg$.

4. (a) $v - 2\log\dfrac{2 + v}{14} = 12 - \frac{4}{5}x$.   (b) 10.1.

5. $v = 24(1 - e^{-t/400})$, T.V. $= 24$ ft/sec.

6. 0.015.   **8.** T.V. $= \sqrt{F/k}$ ft/sec.

7. $t = 2.2$ sec, $v = 13.4$ ft/sec.   **9.** $v^2 = 16(x - 2 + 2e^{-x})$.

10. (a) $v = 8(1 - e^{-0.2t})$.   (b) $v = 5.1$ ft/sec.   (c) T.V. $= 8$ ft/sec.

11. $v = \frac{1}{2}(15 + 29e^{-4t/25})$; T.V. $= 15/2$ ft/sec.

LESSON 16C. Inclined Motion. Quantities that have both magnitude and direction, such as force, velocity, acceleration, are called **vector**

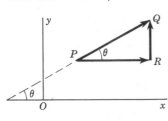

quantities. It is the usual custom to represent a vector quantity by a line with an arrowhead at one end. The magnitude of the vector is given by the length of the line and its direction by the inclination of the line. In Fig. 16.5, PQ represents a vector quantity. With PQ as hypotenuse, we construct a right triangle whose sides are parallel to the x and y axes. Then the vectors **PR** and **RQ** are called respec-

Figure 16.5

tively the **x component** and the **y component** of the vector **PQ**. Their respective magnitudes are given by

(16.51) $\qquad |\mathbf{PR}| = |PQ \cos \theta|, \qquad |\mathbf{RQ}| = |PQ \sin \theta|,$

and their respective directions by the directions of the arrowheads.

Comment 16.511. A vector formerly was written with an arrow over it to distinguish it from a line segment. The current practice, and the one we shall adopt, is to use bold face type.

A vector may also be broken up into two or more components in *any* directions. In Fig. 16.52, we have broken up the vector **PQ** into four component vectors **PR, RS, ST, TQ**. Note that each vector begins where the other leaves off and that the final vector's arrowhead touches the original vector's arrowhead.

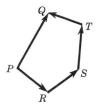

If a body moves along a line which is inclined to the horizontal, the effective force causing it to move downhill is that component of the gravitational force acting in a direction *parallel* to the motion. The forces opposing the motion are the frictional force and the wind resistance. Here the frictional force is equal to

Figure 16.52

the product of the coefficient of friction μ and the component of the gravitational force acting in a direction *perpendicular* to the motion.

Example 16.53. A toboggan with two people on it weighs 520 pounds. It moves down a slope whose gradient is 5/12. If the coefficient of sliding friction is 1/50, and the force of the wind resistance is 5 times the velocity, find the time it will take the toboggan to reach the bottom of a 650-ft long incline. What would the terminal velocity be if the slope were of infinite length?

Solution. Let α be the angle of incline of the slide. Since its gradient $= 5/12$, $\tan \alpha = 5/12$. Hence (Fig. 16.54a) $\sin \alpha = 5/13$ and $\cos \alpha = 12/13$. The gravitational force, which is the combined weight of sled

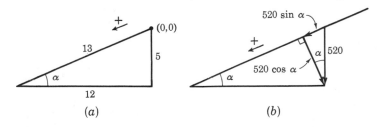

Figure 16.54

and two people, equals 520 pounds. Therefore the magnitude of the component of the gravitational force in the direction of the incline (Fig. 16.54b) is $520 \sin \alpha = 520(5/13) = 200$ pounds. The magnitude of the component of the gravitational force perpendicular to the incline is $520 \cos \alpha = 520(12/13) = 480$ pounds. Since the coefficient of sliding friction is $1/50$, the sliding frictional force is $480(1/50) = 9.6$ pounds. The mass of the body is $520/32$. Hence the differential equation of motion is (remember mass $\times$ acceleration $=$ net force acting on the body)

(a) $\quad \dfrac{520}{32} \dfrac{dv}{dt} = 200 - 9.6 - 5v = 190.4 - 5v, \qquad \dfrac{dv}{dt} + \dfrac{4}{13} v = 11.7.$

Its solution by the method of Lesson 11B is

(b) $\qquad\qquad\qquad v = 38 + c_1 e^{-4t/13}.$

The initial conditions are $t = 0$, $v = 0$. Hence $c_1 = -38$, and (b) becomes

(c) $\qquad\qquad\qquad v = 38(1 - e^{-4t/13}).$

Integration of (c) gives

(d) $\qquad\qquad\qquad s = 38(t + \tfrac{13}{4} e^{-4t/13}) + c_2.$

If we take the origin at the starting point of the toboggan, then $t = 0$, $s = 0$. Therefore by (d), $c_2 = -123.5$. Hence (d) becomes

(e) $\qquad\qquad\qquad s = 38(t + \tfrac{13}{4} e^{-4t/13}) - 123.5.$

When $s = 650$, we obtain from (e)

(f) $\qquad\qquad\qquad 20.4 = t + \tfrac{13}{4} e^{-4t/13},$

whose solution is $t = 20.4$ seconds approximately. This is the time it will take the toboggan to reach the bottom of the incline.

From (b), the terminal velocity is 38 ft/sec (let $t \to \infty$).

EXERCISE 16C

In the problems below, take the origin at the start of the motion and the positive direction in the initial direction of motion.

1. A toboggan with four boys on it weighs 400 lb. It slides down a slope with a 30° incline. Find the position and velocity of the toboggan as functions of time if it starts from rest and the coefficient of sliding friction is $\tfrac{1}{50}$. If the slope is 123.2 ft long, what is the velocity of the toboggan when it reaches the bottom? Neglect air resistance.

2. A body weighing 400 lb is shot up a 30° incline with an initial velocity of 50 ft/sec. The coefficient of sliding friction is $\tfrac{1}{10}$. (a) Find the position and velocity of the body as functions of time. (b) How long and how far will the body move before coming to rest? Neglect air resistance.

3. Two particles start from rest and from the same point on the circumference of a circle in a vertical plane. One moves along a vertical diameter; the other along a chord of the circle (any chord will do). If the only force acting on the particles is that of gravity, prove that both particles reach the circumference of the circle at the same time.

4. A body weighing W lb slides down a slope with an $\alpha°$ incline. Its initial velocity is v_0 ft/sec. The coefficient of sliding friction is r. Ignore air resistance.

(a) Express the velocity and the distance of the body as functions of time.
(b) Express its velocity as a function of distance.
(c) Express the time as a function of the distance.

Use these formulas to verify the accuracy of your answers to 1.

5. A body weighing W lb is shot up a slope with an $\alpha°$ incline. Its initial velocity is v_0 ft/sec. The coefficient of sliding friction is r. Ignore air resistance.

(a) Find the position and velocity of the body as functions of time.
(b) How long and how far will the body move before coming to rest?

Use these formulas to verify the accuracy of your answers to 2.

6. A body weighing W lb, starting from rest, slides down a slope with an $\alpha°$ incline. The coefficient of sliding friction is r and the force of the air resistance is kv lb.

(a) Express the velocity and the distance of the body as functions of time.
(b) Express its distance as a function of the velocity.
(c) What is its terminal velocity?

7. A boy and his sled weigh 96 lb. Starting from rest, he slides down an incline whose gradient is $\frac{3}{4}$. The coefficient of sliding friction is $\frac{1}{24}$, and the force of the air resistance is twice the velocity.

(a) Find the velocity and the distance of the sled as functions of time.
(b) What is his terminal velocity?
(c) When will he reach the bottom of the slope if it is 272 ft long?
(d) What is his velocity at the bottom?

Solve independently. Use the formulas found in problem 6 to verify the accuracy of your answers.

8. A body weighing W lb is shot up an $\alpha°$ incline with a velocity of v_0 ft/sec. The coefficient of sliding friction is r and the force of the air resistance is kv lb.

(a) Express the velocity and the distance of the body as functions of time.
(b) How long and how far will it go?

9. A body weighing 64 lb is shot up an incline, whose gradient is $\frac{3}{4}$, with a velocity of 96 ft/sec. The coefficient of sliding friction is $\frac{1}{4}$ and the force of the air resistance is $\frac{1}{10}v$ lb.

(a) Find the velocity and the distance of the body as functions of time.
(b) How long and how far will it go?

Solve independently. Use the formulas found in problem 8 to verify the accuracy of your answers.

10. A toboggan with two people on it weighs 300 lb. It starts from rest down a slope, $\frac{1}{4}$ mile long, from a height 200 ft above a horizontal level. The coefficient of sliding friction is $\frac{3}{100}$ and the force of the wind resistance is proportional to the square of the velocity. When the velocity is 30 ft/sec, this force is 6 lb.

(a) Find the velocity of the toboggan as a function of the distance and of the time.
(b) With what velocity will the toboggan reach the bottom of the slide?
(c) When will it reach the bottom?
(d) What would its terminal velocity be if the slide were infinite in length?

ANSWERS 16C

1. $v = 15.4t$, $s = 7.7t^2$, $v = 61.7$ ft/sec.

2. (a) $v = 50 - 18.77t$, $s = 50t - 9.39t^2$.
   (b) $t = 2.7$ sec, $s = 66.6$ ft.

4. (a) $v = v_0 + g(\sin \alpha - r \cos \alpha)t$; $s = v_0t + \dfrac{g}{2}(\sin \alpha - r \cos \alpha)t^2$.

   (b) $v^2 = v_0{}^2 + 2g(\sin \alpha - r \cos \alpha)s$.

   (c) $t = \dfrac{\sqrt{v_0{}^2 + 2gs(\sin \alpha - r \cos \alpha)} - v_0}{g(\sin \alpha - r \cos \alpha)}$.

5. (a) $v = v_0 - g(\sin \alpha + r \cos \alpha)t$, $s = v_0t - \dfrac{g}{2}(\sin \alpha + r \cos \alpha)t^2$.

   (b) $t = v_0/g(\sin \alpha + r \cos \alpha)$ sec, $s = v_0{}^2/2g(\sin \alpha + r \cos \alpha)$.

6. (a) $v = \dfrac{W}{k}(\sin \alpha - r \cos \alpha)(1 - e^{-ktl/m})$,

   $s = \dfrac{W}{k}(\sin \alpha - r \cos \alpha)\left(t + \dfrac{m}{k}e^{-ktl/m} - \dfrac{m}{k}\right)$.

   (b) $s = \dfrac{m}{k}\left(-v - \dfrac{A}{k}\log\dfrac{A - kv}{A}\right)$, where $A = mg(\sin \alpha - r \cos \alpha)$.

   (c) T.V. $= \dfrac{W}{k}(\sin \alpha - r \cos \alpha)$.

7. (a) $v = 27.2(1 - e^{-2t/3})$, $s = 27.2(t + \frac{3}{2}e^{-2t/3} - \frac{3}{2})$.
   (b) T.V. $= 27.2$ ft/sec.   (c) 11.5 sec.   (d) 27.2 ft/sec.

8. (a) $v = \dfrac{W(\sin \alpha + r \cos \alpha)}{k}(e^{-ktl/m} - 1) + v_0e^{-ktl/m}$,

   $s = \dfrac{W(\sin \alpha + r \cos \alpha)}{k}\left[\dfrac{m}{k}(1 - e^{-ktl/m}) - t\right] + \dfrac{mv_0}{k}(1 - e^{-ktl/m})$.

   (b) $t = \dfrac{m}{k}\log\left(1 + \dfrac{kv_0}{W(\sin \alpha + r \cos \alpha)}\right)$ sec,

   $s = \dfrac{mv_0}{k} - \dfrac{m^2g(\sin \alpha + r \cos \alpha)}{k^2}\log\left(1 + \dfrac{kv_0}{W(\sin \alpha + r \cos \alpha)}\right)$ ft.

9. In the formulas: $W = 64$, $r = \frac{1}{4}$, $k = \frac{1}{10}$, $\sin \alpha = \frac{3}{5}$, $\cos \alpha = \frac{4}{5}$,
   $v_0 = 96$, $m = 64/32 = 2$.

10. (a) $v = 74.1\,\dfrac{e^{0.105t} - 1}{e^{0.105t} + 1}$, $v^2 = 5484(1 - e^{-0.0014s})$.

    (b) 68 ft/sec.   (c) 30 sec, approx.   (d) 74.1 ft/sec.

LESSON 17. Pursuit Curves. Relative Pursuit Curves.

LESSON 17A. Pursuit Curves. The path traced by a body which always moves in the direction of a fixed point or of another moving object is called a **pursuit curve.**

Example 17.1. A pilot always keeps the nose of his plane pointed toward a city T due west of his starting point. If his speed is v miles per hour and a wind is blowing from the south at the rate of w miles per hour, find the equation of the plane's path. Assume that it starts from a flying field which is at a distance a miles from T.

Solution (Fig. 17.11).   Let $P(x,y)$ be the position of the plane at any time t. The vector representing its speed has magnitude v and is pointed toward T. Call θ the angle this vector makes with the horizontal line connecting the flying field and T. The wind vector points due north and has a magnitude w. The diagonal of the parallelogram formed by the vectors $\mathbf{v}$

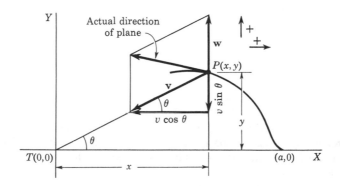

Figure 17.11

and $\mathbf{w}$ then represents the actual direction and magnitude of the plane's velocity at time t. Since this direction is changing instantaneously, it must be tangent to the path of the plane at P. It must therefore be the slope dy/dx of the desired curve. Our problem then is to find an equation which expresses dy/dx as a function of x and y.

The respective components of the airplane's velocity in the x and y directions are

(a) $$\frac{dx}{dt} = -v \cos \theta, \qquad \frac{dy}{dt} = -v \sin \theta.$$

Hence the effective velocity of the plane in the y direction, taking the wind's velocity into account, is

(b) $$\frac{dy}{dt} = -v \sin \theta + w.$$

From Fig. 17.11, we see that

(c) $$\sin \theta = \frac{y}{\sqrt{x^2 + y^2}}, \qquad \cos \theta = \frac{x}{\sqrt{x^2 + y^2}}.$$

Substituting these values in the first equation of (a) and in (b) we obtain

(d) $\qquad \dfrac{dx}{dt} = \dfrac{-vx}{\sqrt{x^2 + y^2}}, \qquad \dfrac{dy}{dt} = -\dfrac{vy}{\sqrt{x^2 + y^2}} + w.$

Division of the second equation in (d) by the first gives

(e) $\qquad \dfrac{dy}{dx} = \dfrac{-vy + w\sqrt{x^2 + y^2}}{-vx} = \dfrac{y - \dfrac{w}{v}\sqrt{x^2 + y^2}}{x}.$

For convenience, we let

(f) $\qquad\qquad\qquad\qquad k = \dfrac{w}{v},$

so that k represents the ratio of the speeds of wind and plane. Hence (e) now becomes

(g) $\qquad\qquad\qquad x\, dy = (y - k\sqrt{x^2 + y^2})\, dx.$

This equation is of the homogeneous type discussed in Lesson 7. It can therefore be solved by the method outlined there. Easier perhaps is to make use of the integrable combination (10.31). The necessary steps are outlined below without further comment. (Initial conditions are $t = 0$, $x = a$, $y = 0$, $y' = y/x = 0$.)

(h) $\quad \dfrac{x\, dy - y\, dx}{x^2} = -\dfrac{k}{x}\sqrt{1 + (y/x)^2}\, dx, \qquad \dfrac{d(y/x)}{\sqrt{1 + (y/x)^2}} = -\dfrac{k}{x}\, dx,$

(i) $\qquad\qquad \displaystyle\int_{y/x=0}^{y/x} \dfrac{d(y/x)}{\sqrt{1 + (y/x)^2}} = -\int_{x=a}^{x} \dfrac{k}{x}\, dx,$

(j) $\qquad\qquad \log\left(\dfrac{y}{x} + \sqrt{1 + (y/x)^2}\right) = -k \log \dfrac{x}{a}.$

Let

(k) $\qquad\qquad\qquad\qquad u = \dfrac{y}{x}.$

Then (j) becomes

(l) $\qquad\qquad \log\left(u + \sqrt{1 + u^2}\right) = -k \log \dfrac{x}{a},$

(m) $\qquad\qquad u + \sqrt{1 + u^2} = \left(\dfrac{x}{a}\right)^{-k},$

(n) $\qquad\qquad 1 + u^2 = \left(\dfrac{x}{a}\right)^{-2k} - 2\left(\dfrac{x}{a}\right)^{-k} u + u^2,$

(o) $\qquad\qquad 2\left(\dfrac{x}{a}\right)^{-k} u = \left(\dfrac{x}{a}\right)^{-2k} - 1,$

(p)
$$u = \frac{1}{2}\left[\left(\frac{x}{a}\right)^{-k} - \left(\frac{x}{a}\right)^{k}\right].$$

Replacing u by its value in (k), we obtain

(q)
$$y = \frac{x}{2}\left[\left(\frac{x}{a}\right)^{-k} - \left(\frac{x}{a}\right)^{k}\right] = \frac{a}{2}\left[\left(\frac{x}{a}\right)^{1-k} - \left(\frac{x}{a}\right)^{1+k}\right].$$

Replacing k by its value in (f), we have finally

(r)
$$y = \frac{a}{2}\left[\left(\frac{x}{a}\right)^{1-(w/v)} - \left(\frac{x}{a}\right)^{1+(w/v)}\right],$$

as the equation of the path.

Comment 17.111. Equation (r) enables us to obtain interesting conclusions in regard to the path of the plane.

Case 1. *Wind speed w = plane speed v.* In this case (r) becomes

(s)
$$y = \frac{a}{2}\left(1 - \frac{x^2}{a^2}\right), \qquad x^2 = -2a\left(y - \frac{a}{2}\right),$$

which is the equation of a parabola, Fig. 17.12. Note that the plane will never reach its destination T.

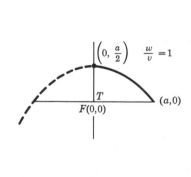

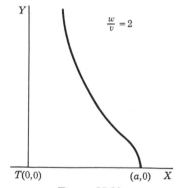

Figure 17.12   **Figure 17.13**

Case 2. *Wind speed w > plane speed v.* In this case $w/v > 1$ so that $1 - (w/v) < 0$. Hence as $x \to 0$, $\left(\dfrac{x}{a}\right)^{1-(w/v)}\left[= \left(\dfrac{a}{x}\right)^{(w/v)-1}\right] \to \infty$. We see from (r), therefore, that as $x \to 0$, $y \to \infty$. Again the plane will never reach T. A rough graph of its path is shown in Fig. 17.13 with $w = 2v$.

Case 3. *Wind speed w < plane speed v.* In this case $w/v < 1$, so that $1 - (w/v) > 0$. We see from (r), therefore that when $x = 0$, $y = 0$. Hence the plane will reach the town T. A rough graph of its path is shown in Fig. 17.14, with $v = 2w$.

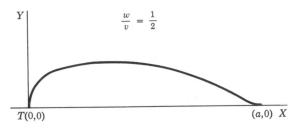

$$\frac{w}{v} = \frac{1}{2}$$

Y

$T(0,0)$ $(a,0)$ X

Figure 17.14

Example 17.2. Solve the problem of Example 17.1 by using polar coordinates.

Solution (Fig. 17.21). The components of the wind velocity **w** in the radial and transverse directions are respectively [for the derivation of these formulas, see Lesson 34, (34.2)].

(a) $$\frac{dr}{dt} = w \sin \theta, \qquad r\frac{d\theta}{dt} = w \cos \theta.$$

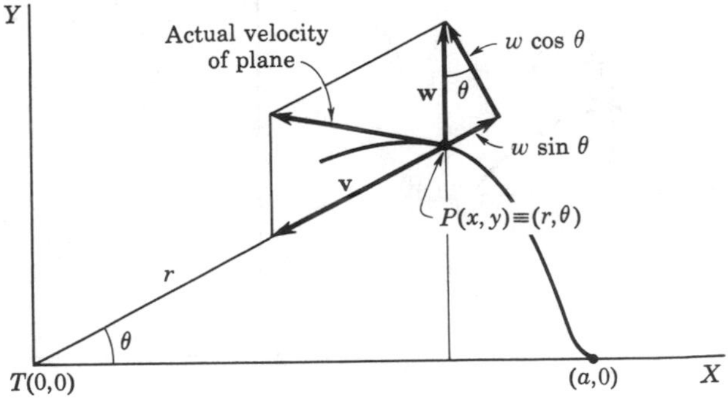

Y

Actual velocity of plane

$w \cos \theta$

w θ

$w \sin \theta$

v

$P(x,y) \equiv (r,\theta)$

r

θ

$T(0,0)$ $(a,0)$ X

Figure 17.21

Hence the effective velocity of the plane in the radial direction, taking into account the wind's speed w and the plane's speed v, is

(b) $$\frac{dr}{dt} = -v + w \sin \theta.$$

Dividing (b) by the second equation in (a), we obtain

(c) $$\frac{dr}{r\,d\theta} = \frac{-v + w\sin\theta}{w\cos\theta} = -\frac{v}{w}\sec\theta + \tan\theta.$$

Let

(d) $$k = -\frac{v}{w}.$$

Then (c) becomes

(e) $$\frac{dr}{r} = (k\sec\theta + \tan\theta)\,d\theta.$$

Its solution, by integration, is

(f) $$\log r + \log c = k\log(\sec\theta + \tan\theta) - \log\cos\theta,$$

which can be written as

(g) $$cr = (\sec\theta + \tan\theta)^k \sec\theta.$$

At $t = 0, r = a, \theta = 0$. Substituting these values in (g), we have

(h) $$ca = 1, \qquad c = 1/a.$$

The equation of the path is, therefore, after replacing k by its value in (d),

(i) $$r = a(\sec\theta + \tan\theta)^{-v/w}\sec\theta,$$

or

(j) $$r\cos\theta = a(\sec\theta + \tan\theta)^{-v/w}.$$

Example 17.3. Solve the problem of Example 17.1 if the wind is blowing with a velocity **w** in a direction which makes an angle α with the vertical, Fig. 17.31.

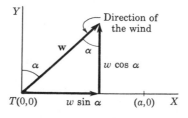

Figure 17.31

Solution. We shall give two methods by which this problem may be solved.

Method 1. Choose the axes so that the direction of the wind becomes the y axis (Fig. 17.32). The initial conditions at $t = 0$, therefore, become

(a) $$x = a \cos \alpha, \qquad y = a \sin \alpha.$$

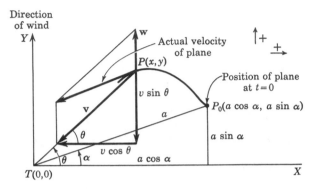

Figure 17.32

The differential equations of motion, however, are exactly the same as those in Example 17.1, namely

(b) $$\frac{dx}{dt} = -v \cos \theta, \qquad \frac{dy}{dt} = -v \sin \theta + w.$$

Proceeding just as we did in Example 17.1, we obtain,

(c) $$\log (u + \sqrt{1 + u^2}) = -k \log x + \log c, \qquad k = w/v.$$

But now at $t = 0$, $x = a \cos \alpha$, $u = y/x = (a \sin \alpha)/(a \cos \alpha) = \tan \alpha$. Substituting these values in (c), and solving for c, we obtain

(d) $$c = (\tan \alpha + \sec \alpha)(a \cos \alpha)^k.$$

Hence (c) becomes, after replacing u by its value y/x,

(e) $$\log \left(\frac{y}{x} + \sqrt{\frac{x^2 + y^2}{x^2}} \right) + \log x^k = \log [(\tan \alpha + \sec \alpha)(a \cos \alpha)^k].$$

Simplification of (e) gives

(f) $$x^{k-1}(y + \sqrt{x^2 + y^2}) = (\tan \alpha + \sec \alpha)(a \cos \alpha)^k, \qquad k \doteq w/v.$$

Replacing k by its value w/v, we obtain finally as the equation of the plane's path

(g) $$x^{(w/v)-1}(y + \sqrt{x^2 + y^2}) = (\tan \alpha + \sec \alpha)(a \cos \alpha)^{w/v}.$$

We leave it to you as an exercise to draw rough graphs of its path as we did in Comment 17.11. Remember $\tan \alpha$, $\sec \alpha$, $\cos \alpha$ and a are constants.

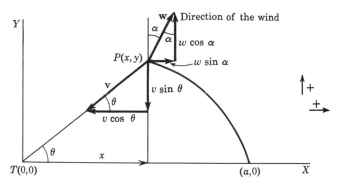

Figure 17.33

Method 2 (Fig. 17.33). From Fig. 17.33, we see that the resultant of the effective velocities of wind and plane in the x and y directions are respectively

(h) $\qquad \dfrac{dx}{dt} = -v \cos \theta + w \sin \alpha, \qquad \dfrac{dy}{dt} = -v \sin \theta + w \cos \alpha.$

Dividing the second equation in (h) by the first we obtain

(i) $\qquad\qquad\qquad \dfrac{dy}{dx} = \dfrac{-v \sin \theta + w \cos \alpha}{-v \cos \theta + w \sin \alpha}\,.$

In (i) replace $\sin \theta$ by its value $y/\sqrt{x^2 + y^2}$ and $\cos \theta$ by $x/\sqrt{x^2 + y^2}$. It then simplifies to

(j) $\qquad \left(\dfrac{-vy}{\sqrt{x^2 + y^2}} + w \cos \alpha\right) dx = \left(-\dfrac{vx}{\sqrt{x^2 + y^2}} + w \sin \alpha\right) dy,$

which is a homogeneous equation. Remember v, $w \cos \alpha$, and $w \sin \alpha$ are constants. Hence it can be solved by the method of Lesson 7. The initial conditions are $x = a$, $y = 0$. We leave it to you as an exercise to complete.

EXERCISE 17A

1. A man swims across a river 100 ft wide, always heading for a tree directly across from his starting point. He can swim at the rate of 3 ft/sec.

   (a) Find the equation of his path if the current is carrying him downstream at the rate of 1 ft/sec; 3 ft/sec; 4 ft/sec.
   (b) Draw the graph of each equation.

Solve independently and check your answers with (r) of Example 17.1.

2. In problem 1 show that the man will reach the tree on the opposite bank if the current is 1 ft/sec, that he will reach a point on the opposite shore 50 ft from the tree if the current is 3 ft/sec, that he will never reach the opposite shore, if the current is 4 ft/sec.

3. Solve problem 1 by using polar coordinates. Solve independently and check your answers with (j) of Example 17.2.

4. An insect steps on the edge of a turntable of radius a that is rotating at a constant angular velocity α. It moves straight toward the center of the table at a constant velocity v_0. Find the equation of its path in polar coordinates, relative to axes fixed in space. (*Hint.* If θ is the angle through which the turntable has rotated in time t and r is the distance of the insect from the center at that moment, then $d\theta/dt = \alpha$, $r(d\theta/dt) = r\alpha$.) Draw a graph of the equation if $a = 10$ ft, $\alpha = \pi/4$ radians/sec, $v_0 = 1$ ft/sec. How many revolutions will the table have made by the time the insect reaches the center?

5. Assume in problem 4 that the insect always moves in a direction parallel to the diameter drawn through the point where he steps on the table.

(a) Find the equation of its path relative to axes fixed in space.

(b) What kind of curve is it? *Hint.* Change the equation to rectangular coordinates if the polar form is unfamiliar to you.

6. A boy stands at A, Fig. 17.34. By means of a string l ft long, he holds a boat, which is in the water at B. He starts walking in a direction perpendicular to AB, always keeping the string taut. Find the equation of the boat's path. Ignore the height of the boy above the horizontal and assume that the string is always tangent to the path. The resulting curve is known as a **tractrix**. *Hint.* The slope of the tractrix is: $\tan \theta = dy/dx$.

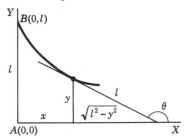

Figure 17.34

7. A pilot always keeps the nose of his plane pointed toward a city which is 400 mi due west of him. A wind is blowing from the south at the rate of 20 mi/hr. His speed is 300 mi/hr. Find the equation of his path. Solve independently, by using both rectangular and polar coordinates. Check the accuracy of your answers with (r) of Example 17.1 and (j) of Example 17.2.

8. A pilot always keeps the nose of his plane pointed toward a city which is a mi due north of him. His speed is v mi/hr and a wind is blowing from the west at w mi/hr. Find the equation of his path.

9. Solve problem 8 if the city is a mi due south of the pilot.

10. A pilot always keeps the nose of his plane pointed toward a city which is 300 mi due west of him. A 25-mi/hr wind is blowing in a direction whose

slope is $\frac{4}{3}$. His speed is 200 mi/hr. Find the equation of his path. Solve independently. Use both methods outlined in this lesson. Check the accuracy of your differential equations with (b) and (j) of Example 17.3.

11. Solve problem 10 if the wind is blowing in a direction whose slope is $-\frac{4}{3}$. Use method 2 of Example 17.3. Solve independently and then check the accuracy of your differential equation with (j) of this example.

12. Assume in problem 4 that the insect moves straight toward a light which is fixed in space directly above the end of the diameter drawn through the point where it steps on the table. Find the differential equation of its path in polar coordinates.

ANSWERS 17A

1. $y = 50\left[\left(\dfrac{x}{100}\right)^{2/3} - \left(\dfrac{x}{100}\right)^{4/3}\right]$; $x^2 = -200(y - 50)$;

$y = 50\left[\left(\dfrac{x}{100}\right)^{-1/3} - \left(\dfrac{x}{100}\right)^{7/3}\right]$.

3. $r = 100(1 - \sin\theta)/(1 + \sin\theta)^2$; $r = 100/(1 + \sin\theta)$;
$r\cos\theta = 100(\sec\theta + \tan\theta)^{-3/4}$.

4. $r = a - \dfrac{v_0}{\alpha}\theta$ (origin at center). $1\frac{1}{4}$ revolutions.

5. (a) $2v_0(r\sin\theta) = \alpha(a^2 - r^2)$.   (b) Circle: center $(0, -v_0/\alpha)$,
radius $\sqrt{a^2 + v_0^2/\alpha^2}$.

6. $x = l\log\dfrac{l + \sqrt{l^2 - y^2}}{y} - \sqrt{l^2 - y^2}$.

7. $y = 200\left[\left(\dfrac{x}{400}\right)^{14/15} - \left(\dfrac{x}{400}\right)^{16/15}\right]$; $r\cos\theta = 400(\sec\theta + \tan\theta)^{-15}$.

8. $x = \dfrac{a}{2}\left[\left(\dfrac{y}{a}\right)^{1-(w/v)} - \left(\dfrac{y}{a}\right)^{1+(w/v)}\right]$ with positive directions to the east and to the south.

9. Same as 8 with positive directions to the east and to the north.

10. First method: solution is $y + \sqrt{x^2 + y^2} = 2(240)^{1/8}x^{7/8}$.
Second method: Differential equation is
$$(200y - 20\sqrt{x^2 + y^2})\,dx - (200x - 15\sqrt{x^2 + y^2})\,dy = 0.$$

11. Differential equation is
$$(200y - 20\sqrt{x^2 + y^2})\,dx - (200x + 15\sqrt{x^2 + y^2})\,dy = 0.$$

12. $\dfrac{dr}{d\theta} = -\dfrac{v_0 r\cos(\theta - \psi)}{r\alpha + v_0\sin(\theta - \psi)}$, where ψ is the angle made by the original

diameter and a line drawn from its end (i.e., where the light is) to the insect's position.

LESSON 17B.   Relative Pursuit Curve.   If the origin of a coordinate system is not fixed on the ground but is attached to a pursued object, then the path described by a pursuer, moving always in the direction of the object he is pursuing, is called a **relative pursuit curve**. It is, in effect, the path traced by the pursuing body as seen by an observer in the pursued object.

Example 17.4. A fighter plane whose speed is V_F is chasing a bomber plane whose speed is V_B. The nose of the fighter plane is always pointed toward the bomber which is flying in a direction making an angle β with the horizontal. Find the path traced by the fighter as plotted by an observer in the bomber.

Solution (Fig. 17.41). Call (0,0) the origin of a coordinate system fixed on the ground, and $(\overline{0},\overline{0})$ the origin of a coordinate system which is moving with the bomber plane. The position of the fighter plane at any

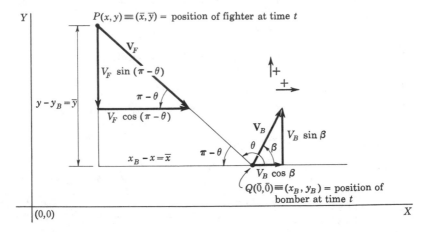

Figure 17.41

time t is therefore given by two sets of coordinates, one (x,y) with respect to the fixed ground origin (0,0) and the other, $(\overline{x},\overline{y})$ with respect to the moving origin $(\overline{0},\overline{0})$.

As measured from an observer on the ground, the effective velocities of the fighter plane in the x and y directions are (see Fig. 17.41)

(a)
$$\frac{dx}{dt} = V_F \cos(\pi - \theta) = -V_F \cos\theta,$$

$$\frac{dy}{dt} = -V_F \sin(\pi - \theta) = -V_F \sin\theta.$$

If two planes are approaching each other along a straight line, then to an observer in one plane, it will seem as if he were standing still and the other plane were coming toward him with a speed equal to the sum of the speeds of the two planes. If the two planes are moving away from each other along a straight line, then it will seem to an observer in one plane as if he were standing still and the other plane were going away from him with a speed equal to the sum of the speeds of the two planes. If, however, the planes are going in the same direction along a straight line, then to an observer in a pursued plane it will seem, if his speed is slower than the other, as if he were standing still and the other plane were coming toward him with a speed equal to the difference of the speeds of the two planes. If his speed is greater than the other, then it will seem to him as if he were standing still and the other plane were going away from him with a speed equal to the difference of the speeds of the two planes.

Here the horizontal component of fighter and bomber velocities are in the same direction. Hence (see Fig. 17.41), to an observer in the bomber (we assume the fighter's component is greater than the bomber's), it is as if he were standing still, and the fighter plane were coming toward him in the positive x direction with a velocity equal to the difference of their two x components of velocity, i.e., the rate of change of $\bar{x}$ to an observer in the bomber is

(b) $$\frac{d\bar{x}}{dt} = -V_F \cos \theta - V_B \cos \beta.$$

The vertical components of fighter and bomber velocities are in opposite directions and pointed toward each other. Hence, the velocity of the fighter relative to an observer in the bomber, is in the negative y direction and is the sum of their two vertical velocities, i.e., the rate of change of $\bar{y}$ is

(c) $$\frac{d\bar{y}}{dt} = -(V_F \sin \theta + V_B \sin \beta).$$

Dividing (c) by (b), we obtain

(d) $$\frac{d\bar{y}}{d\bar{x}} = \frac{V_F \sin \theta + V_B \sin \beta}{V_F \cos \theta + V_B \cos \beta},$$

which has the same form as (i) of Example 17.3. Remember V_F, $V_B \sin \beta$, and $V_B \cos \beta$ are constants. Hence it can be solved by the method suggested there. We leave it to you as an exercise to complete. The initial conditions will depend on the distance and direction of the fighter plane from the bomber plane at $t = 0$, i.e., at the moment when the pursuit began.

Example 17.5. Solve the problem of Example 17.4 by using polar coordinates.

Solution (Fig. 17.51). To an observer in the bomber, when he moves to the right it appears to him as if he were standing still and the fighter plane were moving to the left. Hence, to an observer in the bomber, when he moves with a velocity $\mathbf{V}_B$, represented by the lower vector in Fig. 17.51, it appears to him as if he were standing still and the fighter is

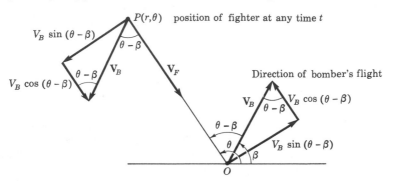

$P(r,\theta)$ position of fighter at any time t

Direction of bomber's flight

Figure 17.51

moving with a velocity $\mathbf{V}_B$ represented by the upper vector in this figure. The radial and transverse components of the upper $\mathbf{V}_B$ are

(a) $\dfrac{dr}{dt} = -V_B \cos(\theta - \beta), \qquad r\dfrac{d\theta}{dt} = V_B \sin(\theta - \beta).$

As seen from an observer in the bomber, therefore, the effective velocity of the fighter in the radial direction is (remember it is the sum of the two velocities in the same negative r direction)

(b) $\dfrac{dr}{dt} = -V_F - V_B \cos(\theta - \beta),$

and in the transverse direction is

(c) $r\dfrac{d\theta}{dt} = V_B \sin(\theta - \beta).$

Dividing (b) by (c), we obtain

(d) $\dfrac{dr}{r\,d\theta} = -\dfrac{\cos(\theta - \beta)}{\sin(\theta - \beta)} - \dfrac{V_F}{V_B}\csc(\theta - \beta).$

Its solution is

(e) $\log r + \log c$
$$= -\log \sin(\theta - \beta) - \dfrac{V_F}{V_B}\log[\csc(\theta - \beta) - \cot(\theta - \beta)],$$

which simplifies to

(f)
$$cr = \frac{[\csc(\theta - \beta) - \cot(\theta - \beta)]^{-V_F/V_B}}{\sin(\theta - \beta)}$$

$$= \frac{1}{\sin(\theta - \beta)} \left[\frac{1 - \cos(\theta - \beta)}{\sin(\theta - \beta)} \right]^{V_F/V_B}$$

$$= \frac{1}{[\sin(\theta - \beta)]^{1-(V_F/V_B)}[1 - \cos(\theta - \beta)]^{V_F/V_B}} .$$

NOTE. The initial conditions are $t = 0$, $r = r_0$, $\theta = \theta_0$.

Comment 17.52. As viewed by an observer in the bomber, it is as if he were standing still, and the path of the fighter as given in (f) (which is the resultant of both planes' motions) were due entirely to the fighter plane's movements.

Comment 17.521. By means of (f), we can draw a rough graph of the fighter plane's path, as seen from the observer in the bomber.

Case 1. If $V_B = V_F$, i.e., if the ground speed of both planes are the same, then (f) reduces to

(g)
$$cr = \frac{1}{1 - \cos(\theta - \beta)} ,$$

which is the equation of a parabola [see (34.68) and comment following].

A rough graph of the path of the fighter plane as viewed from the bomber is given for this case in Fig. 17.53. It is evident that the fighter will never reach the bomber.

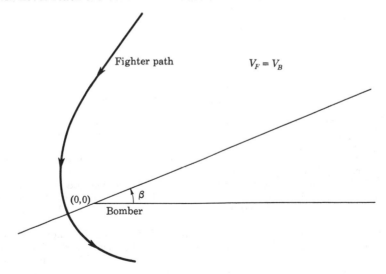

Figure 17.53

Case 2. If $V_F > V_B$, then $1 - (V_F/V_B) < 0$, and (f) can be written as

(h)
$$cr = \frac{[\sin(\theta - \beta)]^{(V_F/V_B)-1}}{[1 - \cos(\theta - \beta)]^{V_F/V_B}}.$$

A rough graph of the fighter plane as viewed from the bomber is given for this case in Fig. 17.54 (with $V_F = 2V_B$).

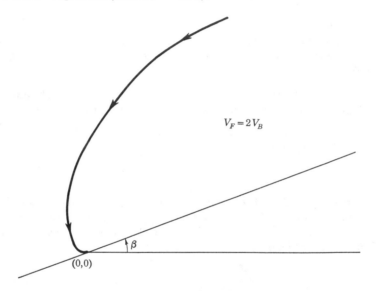

$$V_F = 2V_B$$

Figure 17.54

EXERCISE 17B

1. An airplane A is flying with a speed of 200 mi/hr in a direction whose slope is $\frac{3}{4}$. A plane B, 50 miles due north of him, starts in pursuit. Plane B, whose speed is 300 mi/hr, always keeps the nose of his plane pointed toward A. Find the equation of B's path as observed from A. Use rectangular and polar coordinates. *Hint.* In polar coordinates at $t = 0$, $\theta = \pi/2$, $\tan \beta = \frac{3}{4}$, $r = 50$.
2. Solve problem 1, in polar coordinates only, if A is flying in a southeasterly direction and plane B, at the instant he starts in pursuit, is 50 miles northeast of A. *Hint.* At $t = 0$, $\theta = \pi/4$, $\beta = -\pi/4$, $r = 50$. Make a figure corresponding to 17.51. Be very careful of signs.
3. Solve problem 1, in polar coordinates only, if A is flying in a northeasterly direction and plane B, at the instant he starts in pursuit, is 50 miles southeast of A. *Hint.* At $t = 0$, $\theta = -\pi/4$, $\beta = \pi/4$, $r = 50$. Make a figure corresponding to Fig. 17.51. Be very careful of signs.

ANSWERS 17B

1. Differential equation in rectangular coordinates is

$$\frac{d\bar{y}}{d\bar{x}} = \frac{300\bar{y} + 120\sqrt{\bar{x}^2 + \bar{y}^2}}{300\bar{x} + 160\sqrt{\bar{x}^2 + \bar{y}^2}} \quad \text{with } t = 0, \ \bar{x} = 0, \ \bar{y} = 50.$$

Solution in polar coordinate is:

$$r = \frac{50\sqrt{2}\,(4\sin\theta - 3\cos\theta)^{1/2}}{(5 - 4\cos\theta - 3\sin\theta)^{3/2}}.$$

2. $r = \dfrac{50\sqrt{2}\,(\sin\theta + \cos\theta)^{1/2}}{(\sqrt{2} - \cos\theta + \sin\theta)^{3/2}}.$

3. $r = \dfrac{50\sqrt{2}\,(\cos\theta - \sin\theta)^{1/2}}{(\sqrt{2} - \cos\theta - \sin\theta)^{3/2}}.$

LESSON 17M.

MISCELLANEOUS TYPES OF PROBLEMS LEADING TO EQUATIONS OF THE FIRST ORDER

A. Flow of Water Through an Orifice. When water flows from a tank through a small hole in its bottom, it has been proved that the rate of flow of water is proportional to the area of the hole and the square root of the height of the water in the tank. Hence

(17.6) $$\frac{dV}{dt} = -ka\sqrt{h}, \qquad dV = -ka\sqrt{h}\,dt,$$

where

       V is the number of cubic feet of water in the tank at time t sec,
       k is a positive proportionality constant,
       a is the area of the hole in square feet,
       h is the height in feet of the water above the hole at time t.

The minus sign is necessary because the volume of water is decreasing. If A is the cross-sectional area of the water surface at time t, then $dV = A\,dh$. Substituting this value of dV in the second equation of (17.6) and taking $k = 4.8$, a figure determined experimentally as valid under certain conditions, we obtain

(17.61) $A\,dh = -4.8a\sqrt{h}\,dt, \qquad Ah^{-1/2}\,dh = -4.8\pi r^2\,dt,$

where

       A is the cross-sectional area in square feet of the water surface at time t sec,
       h is the height in feet of the water level above the hole or orifice at time t,
       r is the radius in feet of the orifice.

With the aid of (17.61), solve the following problems.

1. A tank, whose cross section measures 3 ft $\times$ 4 ft, is filled with water to a height of 9 ft. It has a hole at the bottom of radius 1 in. (a) When will the tank be empty? (b) When will the water be 5 ft high? *Hint.* At $t = 0$, $h = 9$ and in (17.61) $r = \frac{1}{12}$, $A = 12$.
2. A water container, whose circular cross section is 6 ft in diameter and whose height is 8 ft, is filled with water. It has a hole at the bottom of radius 1 in. When will the tank be empty?
3. Solve problem 2 if the tank rests on supports above ground so that its 8 ft height is now in a horizontal direction, Fig. 17.62, and the hole of radius 1 in. is in its bottom.

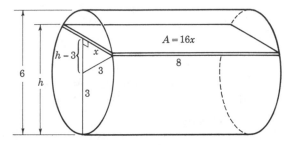

Figure 17.62

4. A water tank, in the shape of a conical funnel, with its apex at the bottom and vertical axis, is 12 ft across the top and 18 ft high. It has a hole at its apex of radius 1 in. When will the tank be empty if initially it is filled with water?
5. A water tank, in the shape of a paraboloid of revolution, measures 6 ft in diameter at the top and is 3 ft deep. A hole at the bottom is 1 in. in diameter. When will the tank be empty if initially it is filled with water?
6. A cylindrical tank is 12 ft in diameter and 9 ft high. Water flows into the tank at the rate of $\pi/10$ ft^3/sec. It has a hole of radius $\frac{1}{2}$ in. at the bottom. When will the tank be full if initially it is empty? *Hint.* Adjust the first equation in (17.6) to reflect the increase in volume due to the water intake. Remember $dV = A\,dh$. After simplification, you should obtain

$$\int_{h=0}^{9} \frac{dh}{12 - \sqrt{h}} = \frac{1}{4320} \int_{t=0}^{t} dt.$$

If you have trouble integrating, see hint given in answer.

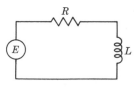

Figure 17.621

B. First Order Linear Electric Circuit. The differential equation which results when an applied electromotive force E, an inductor L, and a resistor R are connected in series is, Fig. 17.621,

$$(17.63) \qquad L\frac{di}{dt} + Ri = E.$$

For an understanding of this equation and for the meaning of these terms, read Lesson 30. The units we shall adopt are

ohms for the coefficient of resistance of the resistor,
henries for the coefficient of inductance of the inductor,
volts for the electromotive force (also written as emf),
amperes for the current.

7. Solve the differential equation (17.63), if $E = E_0$ and at $t = 0$, the current $i = i_0$. What is the limiting current as $t \to \infty$?
8. Solve (17.63) if $E = E_0 \sin \omega t$ and at $t = 0$, $i = i_0$.
9. An inductance of 3 henries and a resistance of 30 ohms are connected in series with an emf of 150 volts. At $t = 0$, $i = 0$. Find the current when $t = 0.01$ sec.
10. An inductance of 2 henries and a resistance of 20 ohms are connected in series with an emf of $100 \sin 150t$ volts. At $t = 0$, $i = 0$. Find the current when $t = 0.01$ sec.
11. When an emf is disconnected from a circuit in which a current is flowing, i.e., at $t = 0$, $E = 0$, the current is called an **induced current**. The term $L\, di/dt$ is then referred to as an **induced electromotive force**. At the moment an emf is disconnected from a circuit in which there is a resistance of 30 ohms and an inductance of 6 henries, the current $i = 15$ amp. (a) Find the current equation as a function of time. (b) When will the current be 7.5 amp?

C. Steady State Flow of Heat. When the inner and outer walls of a body, as for example the inner and outer walls of a house or of a pipe, are maintained at *different constant* temperatures, heat will flow from the warmer wall to the colder one. When each surface parallel to a wall has attained a constant temperature, we say that the flow of heat has reached a **steady state.** In a steady state flow of heat, therefore, each surface parallel to a wall, because its temperature is now constant, is called an **isothermal surface.** Isothermal surfaces at different distances from an interior wall will, of course, have different temperatures. In many cases the temperature of an isothermal surface is a function only of its distance x from an interior wall, and the rate of flow of heat in a unit time across such a surface is proportional both to the area of the surface and to dT/dx, where T is the temperature of the isothermal surface. Hence

$$(17.64) \qquad Q = -kA\,\frac{dT}{dx},$$

where

Q is the rate of flow of calories* of heat in 1 sec across an isothermal surface,

k, the proportionality constant, is called the **thermal conductivity** of the material that is between the walls,

---

*A calorie is equal to the amount of heat required to change the temperature of 1 gram of water 1 degree centigrade.

A is the area in square centimeters of an isothermal surface,

T is the temperature in centigrade degrees of the isothermal surface,

x is the distance in centimeters of the isothermal surface from an interior wall.

The negative sign is used to indicate that heat flows from the interior wall of higher temperature to the exterior wall of lower temperature.

With the help of (17.64), solve the following problems.

12. The inner and outer radii of a hollow spherical shell are 4 cm and 9 cm respectively. The thermal conductivity of the material between the walls is 0.75 cal/deg-cm-sec. The inner surface is kept at a constant temperature of 100°C and the outer surface at 0°C. Find:

(a) The rate of heat loss per second flowing outward through the exterior of the shell.

(b) The temperature T of a surface 5 cm from the center.

Hint. In (17.64) $A = 4\pi r^2$, where r, which replaces x in the formula, is the radius of an isothermal surface. Initial conditions are: $r = 4$, $T = 100$; $r = 9$, $T = 0$.

Note, from answer, that Q has a constant value. In this example, it is 2160π cal/sec. Hence, through each isothermal surface, this much heat escapes each second. If a surface is nearer the center of the sphere, so that its area is smaller than the area of a surface farther away, more heat per *unit area* will escape each second from the nearer surface than from the farther one. The total loss of heat through each surface is, however, the same.

13. A steam pipe of negligible thickness has an inside radius of 6 cm. It is insulated with 3 cm coating of magnesia, whose thermal conductivity is 0.000175. The interior of the pipe is maintained at 100°C and the outer surface at 0°C. Find:

(a) The rate of heat loss per second flowing outward from each *meter* length of pipe.

(b) The temperature of an isothermal surface whose radius is 8 cm.

Hint. In (17.64), $A = 2\pi r(100)$, where r, which replaces x in the formula, is the radius of an isothermal surface.

D. Pressure—Atmospheric and Oceanic. We consider a column of air of cross-sectional area A, of height dh and at a distance h units from the surface of the earth, Fig. 17.65. For simplicity, we assume the earth is a plane, the air is at rest, and is unaffected by changes of latitude, longitude, and temperature. The positive direction is upwards. Let p be the pressure* on this column of air, due to the weight of all the air above it. Hence the total force P on a column of air of cross-sectional area A is $P = Ap$. By differentiation we obtain

(a) $$dP = A \, dp,$$

---

*Pressure is the force acting perpendicularly on a *unit area* of surface.

which gives the change of the total force P for a change in the pressure p. Let ρ be the weight of a unit volume of air. Therefore the weight of a column of air of height dh and cross-sectional area A is

(b) $(A\,dh)\rho$.

The decrease in total force as one goes from the height h to the height $h + dh$ must be equal to the weight of the column of air of thickness dh. Hence equating the negative of (a) with (b), we obtain

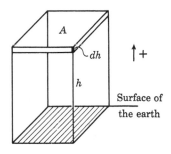

Figure 17.65

(17.66) $-A\,dp = \rho A\,dh,\qquad \dfrac{dp}{dh} = -\rho,$

which states that the rate of change of air pressure with height is equal to the negative of the weight of a unit volume of air at that height. The units we shall use are: pounds per cubic foot for ρ, pounds per square foot for p, and feet for height.

If p is oceanic pressure and h is the depth below sea level, then (17.66) becomes

(17.67) $\dfrac{dp}{dh} = \rho,$

where ρ is the weight per cubic foot of ocean water.

With the aid of (17.66) and (17.67), solve the following problems.

14. Assume that the air is an isothermal gas. Therefore by Boyle's law, its pressure and density are related by the formula

(17.671) $\rho = kp.$

   (a) Determine the pressure of the air as a function of the height h above the earth if at sea level the air pressure is 14.7 lb/sq in. *Hint.* In (17.66), replace ρ by kp.

   (b) What is the value of the constant k in (17.671) if $\rho = 0.081$ lb/cu ft at sea level?

   (c) What is the air pressure at 10,000 ft, at 15,000 ft, at 70,000 ft, at 50 mi?

   (d) Show that the pressure is zero only when h is infinite.

   (e) Express the density of air as a function of height. *Hint.* In (17.66) replace dp by $d\rho/k$ as obtained from (17.671).

   (f) What is the weight of the air at an altitude of 1000 ft, of 5000 ft, of 10,000 ft, of 50,000 ft?

15. If air expands adiabatically, i.e., without gaining or losing heat, then $p = k\rho^{1.4}$.

   (a) Find the pressure of the air as a function of height. Assume that at sea level $p = 14.7$ lb/sq in. and $\rho = 0.081$ lb/cu ft. *Hint.* First find the value of k. Then in (17.66) replace ρ by $(p/k)^{5/7}$.

(b) How high is the atmosphere, i.e., at what height is $\rho = 0$? *Hint.* In (17.66) replace dp by $1.4k\rho^{0.4} \, dp$. Solve for ρ. Then find h when $\rho = 0$.

16. Assume that the weight ρ of a cubic foot of sea water, under a pressure of p lb/ft^2 is given by the formula

$$(17.672) \qquad \rho = k(1 + 2 \cdot 10^{-8} p) \text{ lb/ft}^3,$$

and is 64 lb/ft^3 at sea level.

(a) Find the value of k. *Hint.* At sea level $p = 0$.
(b) Find the pressure of sea water as a function of its depth below sea level. *Hint.* In (17.67), replace ρ by its value as given in (17.672).
(c) Find the weight per cubic foot of sea water as a function of depth. *Hint.* In (17.67), replace dp by $10^8 \, d\rho/2k$ as determined from (17.672).
(d) What is the pressure and density of sea water at 20,000 ft below sea level?

17. If the earth is assumed spherical instead of planar, show that (17.66) becomes

$$(17.68) \qquad \frac{dp}{dh} = -\rho - \frac{2p}{R+h},$$

where R is the radius of the earth. *Hint.* See Fig. 17.69. The volume of a thin spherical shell of air at a distance r units from the center of the earth and thickness dr is $4\pi r^2 \, dr$. Its weight, therefore, is $(4\pi r^2 \, dr)\rho$. The total

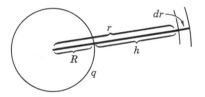

Figure 17.69

force P on this spherical surface due to the weight of the air outside it is $P = 4\pi r^2 p$. Hence $dP = 4\pi(r^2 \, dp + 2rp \, dr)$. Set $-dP$, which is the decrease in the total force P as one goes from the distance r to the distance $r + dr$, equal to the weight of the spherical shell.

E. Rope or Chain Around a Cylinder. When a rope or chain, assumed uniform, flexible, and inextensible, is wound around a rough cylindrical post, a small force applied at one end of the rope or chain can control a much larger force applied at the other end. For example, a man can hold in check a large weight by winding a rope attached to it, a sufficient number of times about a pole.

For a post, whose axis is horizontal, it has been proved that the differential equation

$$(17.7) \qquad \frac{dT}{d\theta} = \delta r(\cos \theta + \mu \sin \theta) + \mu T$$

expresses the tension T in the rope or chain at a point P on it, when the rope or chain is just on the verge of slipping, see Fig. 17.71. In equation (17.7),

T is the pull or tension on the rope at any point P on it, in pounds,
r is the radius of the cylinder, in feet,
μ is the coefficient of friction between rope and post,
δ is the weight of the rope or chain in pounds per foot,
θ is the radial angle of P.

With the help of (17.7), solve the following problems.

18. Show that the solution of (17.7) is

$$(17.72) \qquad T = \frac{\delta r}{1 + \mu^2}\,[(1 - \mu^2)\sin\theta - 2\mu\cos\theta] + ce^{\mu\theta}.$$

19. A chain weighs δ lb/ft. It hangs over a circular cylinder with horizontal axis and radius r ft. One end of the chain is at A, Fig. 17.71. How far must the other end extend below D, so that the chain is on the verge of slipping. *Hint.* The initial conditions are: $\theta = 0$, $T = 0$, and $\theta = \pi$, $T = l\delta$, where l is the length of the portion of the chain overhanging at D.

20. A chain weighs δ lb/ft. It hangs over a circular cylinder with horizontal axis and radius r ft. One end of the chain is at B, Fig. 17.71, the other end just reaches to D. What is the least value of μ so that the chain will not slip? *Hint.* The initial conditions are: $\theta = \pi/2$, $T = 0$, and if the chain is not to slip, then at $\theta = \pi$, $T = 0$.

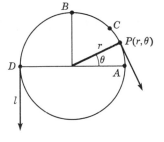

Figure 17.71

21. A chain weighs 1 lb/ft. It hangs over a circular cylinder with horizontal axis and radius $\frac{1}{2}$ ft. One end of the chain reaches three quarters around the top to C, Fig. 17.71; the other end extends below D. The coefficient of friction between chain and cylinder is $\frac{1}{2}$. If the chain is on the verge of slipping, find the length l of the overhang.

22. If the axis of the cylinder is vertical, then the rope's or chain's weight which now acts vertically so that it does not press down upon the cylinder, has little effective force. Hence, in formula (17.7), δ which is the weight per unit length of rope, may be taken to be zero. The formula then simplifies to

$$(17.73) \qquad \frac{dT}{d\theta} = \mu T.$$

With the help of (17.73), solve the following problem. A longshoreman is holding a ship by means of a hawser wound around a vertical post. The ship is pulling on the rope with a force of 5 tons. If the coefficient of friction between rope and post is $\frac{1}{3}$, and the man exerts a force of 50 lb to hold the ship, approximately how many turns of rope is he using? *Hint.* Initial condition is $\theta = 0$, $T = 50$. We want θ when $T = 10,000$.

F. Motion of a Complex System. Solve problems 23–27 below, by use of Newton's law of motion $F = ma = mv(dv/dy)$, where F is the algebraic sum of all the forces acting on a body of mass m, and a is the acceleration of the center of gravity of the body.

23. A 24-ft chain weighing δ lb/ft hangs over a frictionless support, which is more than 24 ft above the floor. Initially the chain is held at rest with 10 ft overhanging on one side of the support and 14 ft on the other side, Fig. 17.74.

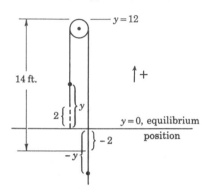

Figure 17.74

How long after its release and with what velocity will the chain leave the support? *Hint.* When 12 ft of chain overhang on each side of the support, the chain is in equilibrium. Call this position $y = 0$. Then if the distance of one end of the chain from equilibrium is y, the distance of the other end is $-y$. The effective force moving the chain is therefore $2y\delta$. The mass of the chain is $24\delta/32$. Initial conditions are $t = 0$, $v = 0$, $y = 2$.

24. In problem 23, assume that the 14-ft overhang just touches the floor. Compute the velocity with which the other end will leave the support.

25. A chain is 12 ft long. Six feet of the chain are held extended on a frictionless flat table which is more than 12 ft above the ground, the other 6 ft hangs over the table. When and with what velocity will the end of the chain leave the table after its release? *Hint.* $m = 12\delta/32$, $F = y\delta$, where y is the distance of the overhanging part of the chain from the edge of the table and δ is its weight per foot. Initial conditions are $t = 0$, $y = 6$, $v = 0$.

26. Assume, in problem 25, that the table is only 4 ft above the ground. With what velocity will the chain of problem 25 leave the table?

27. It has been proved that when a mass particle slides without friction down a fixed curved path whose equation is $y = f(x)$, its differential equation of motion, with upward direction positive, is

$$(17.75) \qquad\qquad v\,dv = -g\,dy,$$

where g is the acceleration due to gravity, v is the velocity of the particle *along the curve*, and y is the *vertical* position of the particle at time t. Therefore, $v = ds/dt$, where s is the distance the particle has moved along the curved path. Show that the solution of (17.75), with initial conditions $t = 0$, $v = v_0$, $y = y_0$, is

$$(17.76) \qquad\qquad v^2 = \left(\frac{ds}{dt}\right)^2 = v_0{}^2 + 2g(y_0 - y).$$

By means of the substitution in (17.76) of $ds/dt = \sqrt{1 + (dy/dx)^2}\, dx/dt$, the equation of the path $y = f(x)$ and the initial condition $y_0 = f(x_0)$, show that (17.76) becomes

$$(17.77) \qquad \left(\frac{dx}{dt}\right)^2 = \frac{v_0^2 + 2g[f(x_0) - f(x)]}{1 + [f'(x)]^2}.$$

The solution of (17.77) will give x as a function of t. With x known, we can determine y by means of the given equation of the path $y = f(x)$.

Use (17.77) to solve the following problem. A particle moves along a smooth wire, shaped in the form of the parabola $x^2 = -y$. Initially it is at the origin and has a velocity of 4 ft/sec. Find the position of the particle at the end of 5 sec. *Hint.* $f(x) = -x^2, f(x_0) = 0, f'(x) = -2x$.

G. Variable Mass. Rocket Motion. In the straight line motion problems thus far considered, the mass of a particle or of a body remained constant throughout the motion. If, however, the mass itself is also changing with time, decreasing or increasing, then Newton's second law of motion no longer holds and must be modified. It has been proved, in the case of a body of variable mass moving in a straight line, that the differential equation governing its motion is given by

$$(17.78) \qquad m\frac{dv}{dt} = F + u\frac{dm}{dt}$$

where

    m is the mass of the body at time t,

    v is the velocity of the body at time t,

    F is the algebraic sum of all the forces acting on the body at time t,

    dm is the mass joining or leaving the body in the time interval dt,

    u is the velocity of dm at the moment it joins or leaves the body, relative to an observer stationed on the body.

Note that (17.78) differs from Newton's second law of motion by the term $u\, dm/dt$.

If a mass dm **leaves** the system in the time interval dt, dm/dt will be a negative quantity; if it **joins** the system, dm/dt will be a positive quantity.

With the aid of (17.78), solve the following problems.

28. A rocket, which weighs $32M$ lb and contains fuel weighing $32m_0$ lb, is propelled straight up from the surface of the earth by burning $32k$ lb of fuel per second and expelling it backwards at a constant velocity of A ft/sec relative to an observer on the rocket. Assume that the only force acting on the rocket is that of gravity. Find the velocity of the rocket and the distance it travels as functions of time. Take positive direction upward. *Hint.* In (17.78), the variable mass m at time t is $m = M + m_0 - kt$; therefore $dm/dt = -k$. The relative velocity u of dm is $-A$. The force of gravity at time t is $F = -(M + m_0 - kt)g$. Answers are

$$(17.79) \quad v = -gt - A \log\left(1 - \frac{k}{M + m_0}t\right), \quad 0 \leqq t < \frac{M + m_0}{k},$$

$$(17.8) \quad y = At - \tfrac{1}{2}gt^2 + \frac{A}{k}(M + m_0 - kt)\log\left(1 - \frac{k}{M + m_0}t\right),$$

$$0 \leqq t < \frac{M + m_0}{k}.$$

Note. If the rocket *moves in free space* so that it is not subject to the gravitational force of the earth, then $F = 0$ in (17.78) and $g = 0$ in (17.79) and (17.8).

29. (a) Show that, when the rocket's fuel of problem 28 is exhausted, it has reached a theoretical height of

$$(17.81) \qquad y = \frac{Am_0}{k} - \frac{g}{2}\left(\frac{m_0}{k}\right)^2 + \frac{AM}{k}\log\left(\frac{M}{M + m_0}\right).$$

Hint. The mass m_0 of the fuel will be exhausted in time $t = m_0/k$. Substitute this value in (17.8).

(b) Show that its velocity at that moment is

$$(17.82) \qquad v = -\frac{gm_0}{k} - A\log\left(\frac{M}{M + m_0}\right).$$

30. A rocket of mass M, containing fuel of mass m_0, falls to the earth from a great height. It burns an amount k of its mass per second and ejects it downward with a constant velocity relative to an observer on the rocket of A ft/sec. Find the distance it falls in time t. Take positive downward direction. *Hint.* See problem 28.

31. A rocket and its fuel have mass m_0. At the moment it starts to burn an amount k of its mass per second, it is moving with a velocity v_0. The fuel is ejected backwards with just enough velocity, so that the ejected fuel is motionless in space. Find the subsequent velocity and distance equations of the rocket as functions of time. *Hint.* For the ejected fuel to be motionless relative to an observer on the earth, the backward velocity of the fuel must equal the forward velocity of the rocket; remember, the fuel at the instant of ejection has the same forward velocity v as the rocket itself. Relative to an observer on the rocket, however, it will seem to him as if he were standing still and the ejected fuel moving away from him at the rate of $-v$ ft/sec, where $+v$ ft/sec is his own velocity in the positive direction. Therefore u in (17.78) is $-v$. The variable mass $m = m_0 - kt$ and $dm/dt = -k$.

32. Solve problem 31 if the rocket were moving in free space. *Hint.* $F = 0$ in (17.78).

33. A body moves in a straight line in free space with a velocity of v_0 ft/sec. Initially its mass is m_0 and as it moves it adds to its mass k slugs per second. Find its velocity and distance equations as functions of time. *Hint.* In (17.78), $F = 0$ and assuming the added mass dm is stationary in space, then its velocity relative to an observer on the body is $-v$, where v is the velocity of the body. See problem 31.

34. A spherical raindrop falls under the influence of gravity. Its mass increases by the addition of stationary moisture particles at a rate which is proportional to its surface area. Initially its radius $r = r_0$. Find its acceleration, velocity, and distance equations as functions of time. Take positive direction downward. Show that if initially $r_0 = 0$, the acceleration has the constant

value $g/4$. For hints, see answer section. First try to solve without making
use of these hints.

35. A chain unwinds from a coil held at rest. It falls straight down under the
influence of gravity, which is the only acting force. Initially l ft of the chain
are unwound. Find its velocity as a function of time. Take positive direction
downward. For hints see answer section.

36. Solve problem 35, if the chain must first slide along a frictionless plane inclined at an angle θ with the horizontal before dropping straight down.
Assume that the coil is held at rest at a distance l ft from one end of the plane
and that initially one end of the chain is at the end of the plane. For hints,
see answer section.

H. Rotation of the Liquid in a Cylinder.

37. A vessel of water is rotated about a vertical axis with a constant angular
velocity ω. Show that when the water is motionless relative to the vessel, the
surface of the water assumes the form of a paraboloid of revolution. Find
the equation of the curve made by a vertical cross section through the axis of
the cylinder. *Hint.* See Fig. 17.9. Two forces act on a particle of water at P:

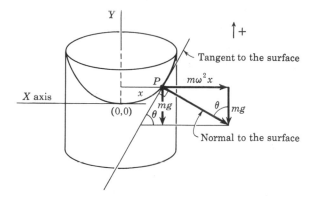

Figure 17.9

one downward due to the weight mg of the particle; the other $m\omega^2x$ due to
the centrifugal force* of revolution. Since the particle of water is motionless,
the resultant of these two forces must be perpendicular to the surface of the
water at P; if it were not, then the resultant would itself have a component
of force in a direction tangent to the curve and thus cause the particle of
water at P to move. You can therefore equate $\tan \theta = dy/dx$ with $m\omega^2x/mg$.

38. Assume that the rotating vessel of problem 37 is a cylinder containing a liquid
whose weight per unit volume is ρ. If the pressure on the axis of the vessel is
p_0, show that the pressure at the surface of the liquid at a distance r from
the axis is given by the differential equation $dp/dr = \rho\omega^2r/g$. Then show
that its solution is $p = p_0 + (\rho\omega^2r^2/2g)$. *Hint.* As in problem 37, use the
fact that an element of volume is in equilibrium under the pressures on its
surfaces and centrifugal force.

---

*For a definition of centrifugal force see Lesson 30M, A.

39. Assume the cylinder of problem 38 contains a gas whose weight per unit volume is ρ and which obeys Boyle's law $p = k\rho$, where p is pressure and k a constant. Show that $p = p_0 e^{\omega^2 r^2/2kg}$. *Hint.* Replace ρ by p/k in the formula given in problem 38.

ANSWERS 17M.

1. (a) $36/\pi$ min.   (b) $12(3 - \sqrt{5})/\pi$ min.

2. $18\sqrt{2}$ min.   **4.** 30.5 min.

3. $32\sqrt{6}/\pi$ min.   **5.** $12\sqrt{3}$ min.

6. Make the substitution $u = 12 - \sqrt{h}$; 65 min.

7. $i = \dfrac{E_0}{R}(1 - e^{-Rt/L}) + i_0 e^{-Rt/L}$; $i = E_0/R$.

8. $i = \dfrac{E_0}{\omega^2 L^2 + R^2}[R \sin \omega t - \omega L \cos \omega t + \omega L e^{-Rt/L}] + i_0 e^{-Rt/L}$.

9. 0.476 amp.   **11.** (a) $i = 15e^{-5t}$.   (b) 0.14 sec.

10. 0.299 amp.   **12.** (a) $Q = 2160\pi$ cal/sec.   (b) 64°C.

13. (a) 8.6π cal/sec.   (b) 29°C.

14. (a) $p = 14.7e^{-kh}$ lb/sq in. $= 2117e^{-kh}$ lb/sq ft.

    (b) 0.000038.

    (c) 10.0 lb/sq in., 8.3 lb/sq in., 1.0 lb/sq in., 0.00060 lb/sq in.

    (e) $\rho = 0.081e^{-0.000038h}$.

    (f) 0.078 lb/cu ft, 0.067, 0.055, 0.012.

15. (a) $p^{2/7} = p_0^{2/7} - \frac{2}{7}k^{-5/7}h$, where $p_0 = 14.7$ lb/sq in. $= 2116.8$ lb/sq ft and $k = 2116.8(0.081)^{-7/5}$.

    (b) $h = \frac{7}{2}(p_0/\rho_0) = \frac{7}{2}(2116.8/0.081) = 91{,}500$ ft $= 17\frac{1}{3}$ mi.

16. (a) $k = 64$.   (b) $p = 5 \cdot 10^7(e^{128h/10^8} - 1)$.   (c) $\rho = 64e^{128h/10^8}$.

    (d) 1,296,500 lb/ft², 65.7 lb/ft³.

19. $l = \dfrac{2r\mu}{1 + \mu^2}(1 + e^{\tau\mu})$. Note that for a fixed μ, l is a function only of r.

20. $2\mu = (1 - \mu^2)e^{\tau\mu/2}$, $\mu = 0.7324$.

21. $l = 0.63$ ft.

22. $\theta = 15.9$ radians $=$ slightly more than $2\frac{1}{2}$ turns of rope about the post.

23. $v = -\frac{4}{3}\sqrt{210}$ ft/sec; $t = \sqrt{\frac{3}{8}}\log(6 + \sqrt{35})$ sec.

24. 18.1 ft/sec.

25. 17.0 ft/sec; $\sqrt{\frac{3}{8}}\log(2 + \sqrt{3})$ sec.

26. 15.3 ft/sec.

27. $x = 20$ ft, $y = -400$ ft, or $x = -20$ ft, $y = -400$ ft.

30. $y = \frac{1}{2}gt^2 - At - \dfrac{A}{k}(M + m_0 - kt)\log\left(1 - \dfrac{kt}{M + m_0}\right)$,

    $0 \leqq t < \dfrac{M + m_0}{k}$.

31. $v = \dfrac{g}{2k}(m_0 - kt) + \dfrac{2km_0v_0 - gm_0^2}{2k(m_0 - kt)}$,

    $y = \dfrac{g}{4k^2}(2m_0kt - k^2t^2) - \dfrac{2km_0v_0 - gm_0^2}{2k^2}\log\left(1 - \dfrac{k}{m_0}t\right)$.

32. $v = m_0v_0/(m_0 - kt)$, $y = \dfrac{m_0v_0}{k}\log\dfrac{m_0}{m_0 - kt}$.

33. $v = m_0v_0/(m_0 + kt)$, $y = \dfrac{m_0v_0}{k}\log\dfrac{m_0 + kt}{m_0}$.

34. First show, see Exercise 15D, 12, that $r = r_0 + kt/\delta$, where r is the radius of the raindrop at time t, k is a proportionality constant and δ is the mass per cubic foot of water. Hence $dr = k\,dt/\delta$. The velocity u of the particle dm is $-v$ where v is the velocity of the raindrop at time t, since relative to an observer moving with the raindrop, it is as if he were stationary and the particle were moving to him, in the negative upward direction, with a velocity equal to his own velocity at the moment the moisture particle attaches itself to the raindrop. The variable mass m at time t is $m = 4\pi r^3 \delta/3$; $dm/dt = k4\pi r^2$. Answers are

$$\text{Acceleration } a = \frac{g}{4}\left(1 + \frac{3r_0^4}{r^4}\right),$$

$$\text{Velocity } v = \frac{\delta}{k}\frac{g}{4}\left(r - \frac{r_0^4}{r^3}\right),$$

$$\text{Distance } y = \frac{g}{8}\left(\frac{\delta}{k}\right)^2\left(r^2 + \frac{r_0^4}{r^2} - 2r_0^2\right).$$

Replacing r by $r_0 + kt/\delta$ in the above equations will give acceleration, velocity, and distance equations as functions of time.

35. The mass of chain at time t is $(l + y)\delta$, where δ is the mass of chain per unit length and $l + y$ is the distance fallen in time t. Therefore $dm/dt = \delta\,dy/dt = \delta v$. The velocity of a piece dm as it leaves the spool, relative to a person moving with the chain is $-v$, where v is the velocity of the chain at time t. Change dv/dt in (17.78) to $v\,dv/dy$. Answer is

$$v^2(l + y)^2 = \frac{2g}{3}[(l + y)^3 - l^3].$$

36. See problem 35. Acting force now is $[(l\sin\theta)\delta + y\delta]g$, where y is the vertical distance the chain has fallen in time t. Answer is

$$v^2(l + y^2) = \frac{g}{3}[3ly\sin\theta(2l + y) + y^2(3l + 2y)].$$

37. $y = \omega^2 x^2/2g$, if the origin is taken at the lowest point of the parabola.

Chapter 4

Linear Differential Equations of Order Greater Than One

Introductory Remarks. In Lesson 11, we introduced the important linear differential equation of the first order. The linear differential equation of higher order which we shall discuss in this chapter has even greater importance. Motions of pendulums, of elastic springs, of falling bodies, the flow of electric currents, and many more such types of problems are intimately related to the solution of a linear differential equation of order greater than one.

Definition 18.1. A **linear differential equation of order** n is an equation which can be written in the form

(18.11) $f_n(x)y^{(n)} + f_{n-1}(x)y^{(n-1)} + \cdots + f_1(x)y' + f_0(x)y = Q(x),$

where $f_0(x), f_1(x), \cdots, f_n(x)$, and $Q(x)$ are each continuous functions of x defined on a common interval I and $f_n(x) \not\equiv 0$ in I.*

Note that in a linear differential equation of order n, y and each of its derivatives have exponent one. It cannot have, for example, terms such as y^2 or $(y')^{1/2}$ or $[y^{(n)}]^3$.

Definition 18.12. If $Q(x) \not\equiv 0$ on I, (18.11) is called a **nonhomogeneous linear differential equation of order** n. If, in (18.11), $Q(x) \equiv 0$ on I, the resulting equation

(18.13) $f_n(x)y^{(n)} + f_{n-1}(x)y^{(n-1)} + \cdots + f_1(x)y' + f_0(x)y = 0$

is called a **homogeneous linear differential equation of order** n.

---

*The notation "$f_n(x) \equiv 0$ in I" (or on I or over I) means $f_n(x) = 0$ for *every* x in I (or on I or over I) and is read as "$f_n(x)$ is identically zero in I (or on I or over I)." The symbol $f_n(x) \not\equiv 0$ in I means $f_n(x)$ is not equal to zero for *every* x in I (or on I or over I), although it may equal zero for *some* x in I. It is read as "$f_n(x)$ is not identically zero in I (or on I or over I)."

Remark. Do not confuse the term homogeneous as used in Lesson 7 with the above use of the term.

For a clearer understanding of the solutions of linear differential equations of order n you should be familiar with:

1. Complex numbers and complex functions.
2. The meaning of the linear independence of a set of functions.

We shall, therefore, briefly discuss these topics in this and the next lesson.

LESSON 18.   Complex Numbers and Complex Functions.

LESSON 18A.   Complex Numbers.   We shall not attempt to enter into a lengthy discussion of the theory of the number system, but will indicate only briefly the important facts we shall need. The **real number system** consists of:

1. The rational numbers; examples are the positive integers, zero, the negative integers, fractions formed with integers such as $\frac{2}{3}$, $-\frac{5}{7}$, and $\frac{3}{4}$.
2. The irrational numbers; examples are $\sqrt{2}$, $\sqrt[3]{5}$, π, and e.

Definition 18.2.   A **pure imaginary number** is the product of a real number and a number i which is defined by the relation $i^2 = -1$.

Examples of pure imaginary numbers are $3i$, $-5i$, $\sqrt{2}\,i$, and $\sqrt[3]{\pi}\,i$.

Definition 18.21.   A **complex number** is one which can be written in the form $a + bi$, where a and b are real numbers.

It is evident that the *complex numbers include the real numbers* since every real number can be written as $a + 0i$. They also include the pure imaginary numbers, since every pure imaginary number can be written as $0 + bi$.

Definition 18.22.   The a in the complex number $z = a + bi$ is called the **real part of** z; the b the **imaginary part of** z.

Note that b, the imaginary part of the complex number, is itself real.

Remark.   The complex numbers developed historically because of the necessity of solving equations of the type

$$x^2 + 1 = 0, \qquad x^2 + 2x + 2 = 0.$$

If we require x to be real, then these equations have no solutions; if, however, x may be complex, then the solutions are respectively $x = \pm i$ and $x = -1 \pm i$. In fact, if we admit complex numbers, we can make the following very important assertion. It is so important that it has been labeled the "Fundamental Theorem of Algebra."

Theorem 18.23.   *Every equation of the form*

$$(18.24) \qquad a_n x^n + a_{n-1} x^{n-1} + \cdots + a_1 x + a_0 = 0, \quad a_n \neq 0,$$

where the a's are complex numbers has at least one root and not more than n distinct roots. Its left side can be written as

$$(18.25) \qquad a_n(x - r_1)(x - r_2) \cdots (x - r_n),$$

where the r's are complex numbers which need not be distinct.

Comment 18.26.   Although Theorem 18.23 tells us that (18.24) has n roots, it does not tell us how to find them. If the coefficients are real numbers, then you may have learned Horner's or Newton's method for finding approximate *real* roots. Or you can set y equal to the left side of (18.24) and draw a careful graph. The approximate values of the coordinates of the points where the curve crosses the x axis will also give the real roots of (18.24).

To find the *imaginary* roots of (18.24) is a more difficult matter. However, there are methods available for approximating such roots.*

Definition 18.3.   If $z = a + bi$, then the **conjugate of z,** written as $\bar{z}$, is $\bar{z} = a - bi$.

To form the conjugate of a complex number, change the sign of the coefficient of i. For example if $z = 2 + 3i$, $\bar{z} = 2 - 3i$; if $z = 2 - 3i$, $\bar{z} = 2 + 3i$.

Definition 18.31.   If $z = a + bi$, then the **absolute value of z,** written as $|z|$, is $|z| = \sqrt{a^2 + b^2}$.

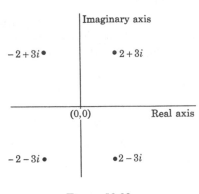

Figure 18.32

For example, if $z = 2 + 3i$, then $|z| = \sqrt{2^2 + 3^2} = \sqrt{13}$; if $z = 2 - 3i$, then $|z| = \sqrt{2^2 + (-3)^2} = \sqrt{13}$. (Note. By our definition $|z|$ is always a non-negative number.)

If, in a rectangular coordinate system, we label the x axis as the real axis and the y axis as the imaginary axis, we may represent any complex number $z = a + bi$ graphically. We plot a along the real axis and b along the imaginary axis. In Fig. 18.32, we have plotted the numbers $2 + 3i$, $2 - 3i$, $-2 + 3i$, $-2 - 3i$.

By the graphical method of representing complex numbers, it is easy to establish relationships between a complex number and the polar coordi-

---

*W. E. Milne, *Numerical Calculus*, Princeton University Press, 1949.

nates of its point. In Fig. 18.33, we have represented graphically the complex number $z = x + iy$ and the polar coordinate (r, θ) of its point. If

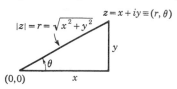

Figure 18.33

you examine this figure carefully, you will have no trouble establishing the following relationships.

(18.4) $$|z| = r = \sqrt{x^2 + y^2},$$

(18.41) $$x = r \cos \theta,$$

(18.42) $$y = r \sin \theta.$$

Definition 18.5. If in

(18.51) $$z = x + yi,$$

(called the **rectangular form of z**), we replace x by (18.41), and y by (18.42), we obtain

(18.52) $$z = r \cos \theta + ir \sin \theta = r(\cos \theta + i \sin \theta),$$

(called the **polar form of z**).

Definition 18.53. The polar angle θ in Fig. 18.33 is called the **Argument of z**, written as **Arg z**. In general, Arg z is defined to be the smallest positive angle satisfying the two equalities

(18.54) $$\cos \theta = \frac{x}{|z|}, \qquad \sin \theta = \frac{y}{|z|}.$$

Formulas (18.4) to (18.42) plus (18.52) and (18.54) enable us to change a complex number from its rectangular form to its polar form and vice versa.

Remark. Note that we have written arg z with a capital A to designate the smallest or **principal value of θ**. If written with a small a, then

$$\arg z = \text{Arg } z \pm 2n\pi, \; n = 1,2,3\ldots$$

Example 18.55. Find Arg z and the polar form of z, if the rectangular form of z is

(a) $$z = 1 - i.$$

Solution. Comparing (a) with (18.51), we see that $x = 1$ and $y = -1$. Therefore by (18.4)

(b) $$|z| = r = \sqrt{1^2 + (-1)^2} = \sqrt{2}.$$

By (18.54)

(c) $$\cos \theta = \frac{1}{\sqrt{2}}, \quad \sin \theta = \frac{-1}{\sqrt{2}}.$$

Hence θ is a fourth quadrant angle. By Definition 18.53, therefore,

(d) $$\operatorname{Arg} z = \frac{7\pi}{4}.$$

Finally by (18.52), and making use of (b) and (d), we have

(e) $$z = \sqrt{2}\left(\cos \frac{7\pi}{4} + i \sin \frac{7\pi}{4}\right),$$

which is the polar form of z.

Example 18.56. Find the rectangular form of z if its polar form is

(a) $$z = \sqrt{2}\left(\cos \frac{\pi}{3} + i \sin \frac{\pi}{3}\right).$$

Solution. Comparing (a) with (18.52), we see that

(b) $$r = \sqrt{2}, \quad \theta = \frac{\pi}{3}.$$

Hence by (18.41) and (18.42),

(c) $$x = \sqrt{2}\cos \frac{\pi}{3} = \frac{\sqrt{2}}{2} \quad \text{and} \quad y = \sqrt{2}\sin \frac{\pi}{3} = \frac{\sqrt{6}}{2}.$$

By (18.51), the rectangular form of z is therefore

(d) $$z = \tfrac{1}{2}(\sqrt{2} + \sqrt{6}\,i).$$

LESSON 18B. Algebra of Complex Numbers. Complex numbers would not be of much use if rules were not available by which we could add, subtract, multiply, and divide them. The rules which have been laid down for these operations follow the ordinary rules of algebra with the number i^2 replaced by its agreed value -1. If $z_1 = a + bi$ and $z_2 = c + di$ are two complex numbers, then by definition,

(18.6)　　$z_1 + z_2 = (a + bi) + (c + di) = (a + c) + (b + d)i,$

(18.61)　$z_1 - z_2 = (a + bi) - (c + di) = (a - c) + (b - d)i,$

(18.62) $\quad z_1 z_2 = (a + bi)(c + di) = ac + adi + bci + bdi^2$
$\qquad\qquad = (ac - bd) + (ad + bc)i,$

(18.63) $\quad \dfrac{z_1}{z_2} = \dfrac{a + bi}{c + di} \cdot \dfrac{c - di}{c - di} = \dfrac{(ac + bd) + (bc - ad)i}{c^2 + d^2}$

$\qquad\qquad = \dfrac{ac + bd}{c^2 + d^2} + \dfrac{bc - ad}{c^2 + d^2}\, i, \qquad c^2 + d^2 \neq 0.$

Example 18.64.   If $z_1 = 3 + 5i$ and $z_2 = 1 - 3i$, compute $z_1 + z_2$, $z_1 - z_2$, $z_1 z_2$, z_1/z_2.

Solutions.

$z_1 + z_2 = (3 + 5i) + (1 - 3i) = 4 + 2i.$

$z_1 - z_2 = (3 + 5i) - (1 - 3i) = 2 + 8i.$

$z_1 z_2 = (3 + 5i)(1 - 3i) = (3 + 15) + (-9 + 5)i = 18 - 4i.$

$\dfrac{z_1}{z_2} = \dfrac{3 + 5i}{1 - 3i} \cdot \dfrac{1 + 3i}{1 + 3i} = \dfrac{(3 - 15) + (9 + 5)i}{1 + 9} = -\dfrac{12}{10} + \dfrac{14}{10}\, i.$

LESSON 18C.   Exponential, Trigonometric, and Hyperbolic Functions of Complex Numbers. If x represents a real number, i.e., if x can take on only real values, then the Maclaurin series expansions (which you studied in the calculus) for e^x, $\sin x$ and $\cos x$ are

(18.7) $\qquad e^x = 1 + \dfrac{x}{1!} + \dfrac{x^2}{2!} + \dfrac{x^3}{3!} + \cdots, \quad -\infty < x < \infty,$

(18.71) $\qquad \sin x = x - \dfrac{x^3}{3!} + \dfrac{x^5}{5!} - \cdots, \quad -\infty < x < \infty,$

(18.72) $\qquad \cos x = 1 - \dfrac{x^2}{2!} + \dfrac{x^4}{4!} - \cdots, \quad -\infty < x < \infty.$

Each of these series is valid for all values of x, i.e., each series converges for all x. If z represents a complex number, i.e., if z can take on complex values, we define

(18.73) $\qquad\qquad e^z = 1 + \dfrac{z}{1!} + \dfrac{z^2}{2!} + \dfrac{z^3}{3!} + \cdots,$

(18.74) $\qquad\qquad \sin z = z - \dfrac{z^3}{3!} + \dfrac{z^5}{5!} - \cdots,$

(18.75) $\qquad\qquad \cos z = 1 - \dfrac{z^2}{2!} + \dfrac{z^4}{4!} - \cdots.$

It has been proved that each of these series also converges for all z.

Example 18.76. Find the series expansion for $\cos i$.

Solution. By (18.75) with $z = i$ and $i^2 = -1$, we have

(a) $$\cos i = 1 + \frac{1}{2!} + \frac{1}{4!} + \frac{1}{6!} + \cdots.$$

Note that $\cos i$ turns out to be a *real* number.

By means of (18.73) to (18.75), we can prove the following identities. These are the ones you will meet most frequently, and the ones we shall need.

(18.8) $$e^0 = 1,$$

(18.81) $$e^{z_1}e^{z_2} = e^{z_1 + z_2},$$

(18.82) $$e^{iz} = \cos z + i \sin z,$$

(18.83) $$e^{-iz} = \cos z - i \sin z,$$

(18.84) $$\sin z = \frac{1}{2i}(e^{iz} - e^{-iz}),$$

(18.85) $$\cos z = \tfrac{1}{2}(e^{iz} + e^{-iz}),$$

(18.86) $$e^z \neq 0 \text{ for any value of } z.$$

Proofs of (18.8) to (18.86). We give below an outline of the proofs of (18.8) to (18.86). Rigorous proofs would be beyond the scope of this text.

Proof of (18.8). In (18.73) replace z by zero.

Proof of (18.81). By (18.73)

$$e^{z_1} = 1 + \frac{z_1}{1!} + \frac{z_1^2}{2!} + \frac{z_1^3}{3!} + \cdots,$$

$$e^{z_2} = 1 + \frac{z_2}{1!} + \frac{z_2^2}{2!} + \frac{z_2^3}{3!} + \cdots$$

Since each series also converges absolutely, we may multiply them to obtain

$$
\begin{aligned}
e^{z_1}e^{z_2} &= 1 + (z_1 + z_2) + \left(\frac{z_1^2}{2!} + z_1 z_2 + \frac{z_2^2}{2!}\right) \\
&\quad + \left(\frac{z_1^3}{3!} + \frac{z_1^2 z_2}{2!} + \frac{z_1 z_2^2}{2!} + \frac{z_2^3}{3!}\right) + \cdots \\
&= 1 + \frac{(z_1 + z_2)}{1!} + \frac{(z_1 + z_2)^2}{2!} + \frac{(z_1 + z_2)^3}{3!} + \cdots \\
&= e^{z_1 + z_2}.
\end{aligned}
$$

Proof of (18.82). Replace z by iz in (18.73). There results, with $i^2 = -1$,

$$e^{iz} = 1 + \frac{iz}{1!} + \frac{i^2 z^2}{2!} + \frac{i^3 z^3}{3!} + \frac{i^4 z^4}{4!} + \cdots$$

$$= 1 + iz - \frac{z^2}{2!} - \frac{iz^3}{3!} + \frac{z^4}{4!} + \cdots$$

$$= \left(1 - \frac{z^2}{2!} + \frac{z^4}{4!} + \cdots\right) + i\left(z - \frac{z^3}{3!} + \cdots\right)$$

$$= \cos z + i \sin z \quad \text{[by (18.74) and (18.75)]}.$$

Proof of (18.83). Replace z by $-iz$ in (18.73) and proceed as above. Or replace z by $-z$ in (18.82) and then note by (18.74) and (18.75) that $\sin(-z) = -\sin z;\ \cos(-z) = \cos z$.

Proof of (18.84). Subtract (18.83) from (18.82) and solve for $\sin z$.

Proof of (18.85). Add (18.82) and (18.83) and solve for $\cos z$.

Proof of (18.86). By (18.81) and (18.8)

$$e^z e^{-z} = e^{z+(-z)} = e^0 = 1.$$

Since the product $e^z e^{-z} = 1$, e^z cannot have the value zero for any z.

The combinations

$$\frac{e^z - e^{-z}}{2} \quad \text{and} \quad \frac{e^z + e^{-z}}{2}$$

appear so frequently in problems that, for convenience, symbols have been introduced to represent them. These are

$$(18.9) \qquad \sinh z = \frac{e^z - e^{-z}}{2},$$

$$(18.91) \qquad \cosh z = \frac{e^z + e^{-z}}{2},$$

$$(18.92) \qquad \tanh z = \frac{\sinh z}{\cosh z} = \frac{e^z - e^{-z}}{e^z + e^{-z}}.$$

These three functions are called respectively the **hyperbolic sine,** the **hyperbolic cosine,** and the **hyperbolic tangent** of z.

EXERCISE 18

1. Find the conjugate of each of the following complex numbers. (a) $1 + i$. (b) 3. (c) $-3i$. (d) $5 - 6i$. (e) -2. (f) $2i$. (g) $-3 - 4i$.

2. Find the absolute value of each of the complex numbers in 1 above.

3. Find Arg z and the polar form of z of each of the complex numbers in 1 above.
4. Find the rectangular form of z, if its polar form is:

(a) $2\left(\cos\dfrac{\pi}{4}+i\sin\dfrac{\pi}{4}\right).$

(e) $\sqrt{2}\,(\cos\frac{5}{4}\pi+i\sin\frac{5}{4}\pi).$

(b) $3\left(\cos\dfrac{2\pi}{3}+i\sin\dfrac{2\pi}{3}\right).$

(f) $\sqrt{3}\,(\cos\frac{3}{2}\pi+i\sin\frac{3}{2}\pi).$

(c) $4(\cos\pi+i\sin\pi).$

(g) $\sqrt{3}\,(\cos\frac{5}{3}\pi+i\sin\frac{5}{3}\pi).$

(d) $5\left(\cos\dfrac{\pi}{2}+i\sin\dfrac{\pi}{2}\right).$

5. Given $z_1=-2+i$, $z_2=3-2i$. Find: (a) z_1+z_2. (b) z_1-z_2.
   (c) z_1z_2. (d) z_1/z_2.

6. If z_1 and z_2 are two complex numbers, prove:

(a) $\overline{z_1+z_2}=\overline{z_1}+\overline{z_2}.$
(b) $\overline{-z_1}=-\overline{z_1}.$

(c) $\overline{z_1z_2}=\overline{z_1}\,\overline{z_2}.$
(d) $\overline{z_1/z_2}=\overline{z_1}/\overline{z_2},\ z_2\neq 0.$

7. With the help of 6, prove that if z_1, z_2, z_3 are complex numbers and

$$z=z_1z_2+z_1(z_2+z_3)+z_1+z_2-z_3+z_1/z_2,\quad z_2\neq 0,$$

then

$$\bar{z}=\overline{z_1}\,\overline{z_2}+\overline{z_1}(\overline{z_2}+\overline{z_3})+\overline{z_1}+\overline{z_2}-\overline{z_3}+\overline{z_1}/\overline{z_2}.$$

NOTE. If you have proved 7, then you have proved that if z is obtained by adding, subtracting, multiplying, and dividing complex numbers z_1, z_2, $z_3, \cdots$, then $\bar{z}$ can be obtained by performing the same arithmetic operations on $\overline{z_1}$, $\overline{z_2}$, $\overline{z_3}, \cdots$.

8. With the help of 7, prove that if r is a root of

(a)
$$a_nx^n+a_{n-1}x^{n-1}+\cdots+a_1x+a_0=0,$$

then $\bar{r}$ is a root of

(b)
$$\bar{a}_nx^n+\bar{a}_{n-1}x^{n-1}+\cdots+\bar{a}_1x+\bar{a}_0=0.$$

9. With the help of 8, prove that if the coefficients in 8(a) are real and r is a root of 8(a), then $\bar{r}$ is also a root of 8(a). NOTE. If you have succeeded in proving 9, then you have proved the following very important theorem. *If the coefficients in 8(a) are real, then its imaginary roots must occur in conjugate pairs.*

10. Prove that the product of two conjugate complex numbers is a positive real number.

ANSWERS 18

1. (a) $1-i$.   (b) 3.   (c) $3i$.   (d) $5+6i$ (e) -2 (f) $-2i$.
   (g) $-3+4i$.
2. (a) $\sqrt{2}$.   (b) 3.   (c) 3.   (d) $\sqrt{61}$.   (e) 2.   (f) 2.   (g) 5.
3. (a) $\dfrac{\pi}{4}$, $\sqrt{2}\left(\cos\dfrac{\pi}{4}+i\sin\dfrac{\pi}{4}\right).$

(b) 0, $3(\cos 0° + i \sin 0°)$.

(c) $\dfrac{3\pi}{2}$, $3\left(\cos \dfrac{3\pi}{2} + i \sin \dfrac{3\pi}{2}\right)$.

(d) $309°48'$, $\sqrt{61}(\cos 309°48' + i \sin 309°48')$.

(e) π, $2(\cos \pi + i \sin \pi)$.

(f) $\dfrac{\pi}{2}$, $2\left(\cos \dfrac{\pi}{2} + i \sin \dfrac{\pi}{2}\right)$.

(g) $233°8'$, $5(\cos 233°8' + i \sin 233°8')$.

4. (a) $\sqrt{2} + i\sqrt{2}$.　(b) $-\frac{3}{2} + i\frac{3}{2}\sqrt{3}$.　(c) -4.　(d) $5i$.

 (e) $-1 - i$.　(f) $-\sqrt{3}\,i$.　(g) $\frac{1}{2}\sqrt{3} - i\frac{3}{2}$.

5. (a) $1 - i$.　(b) $-5 + 3i$.　(c) $-4 + 7i$.　(d) $-\frac{8}{13} - i\frac{1}{13}$.

LESSON 19.   Linear Independence of Functions. The Linear Differential Equation of Order n.

LESSON 19A.   Linear Independence of Functions.

Definition 19.1.   A set of functions $f_1(x)$, $f_2(x)$, $\cdots$, $f_n(x)$, each defined on a common interval I, is called **linearly dependent** on I, if there exists a set of constants c_1, c_2, $\cdots$, c_n, not *all* zero, such that

$$(19.11) \qquad c_1 f_1(x) + c_2 f_2(x) + \cdots + c_n f_n(x) = 0,$$

for *every* x in I. If no such set of constants c_1, c_2, $\cdots$, c_n exists, then the set of functions is called **linearly independent**.

Definition 19.12.   The left side of (19.11) is called a **linear combination** of the set of functions $f_1(x)$, $f_2(x)$, $\cdots$, $f_n(x)$.

Example 19.13.   For each of the following sets of functions, determine whether it is linearly dependent or independent.

1. x, $-2x$, $-3x$, $4x$,　$I: -\infty < x < \infty$.

2. x^p, x^q,　$p \neq q$,　$x > 0$, $x < 0$.

3. e^{px}, e^{qx},　$p \neq q$,　$I: -\infty < x < \infty$.

4. e^x, 0, $\sin x$, 1,　$I: -\infty < x < \infty$.

Solution.   First we observe that all the functions in each set are defined on their respective intervals. Second we form for each set the linear combination called for in (19.11).

The linear combination of the set of functions in 1, equated to zero, is

(a) $\qquad c_1 x + c_2(-2x) + c_3(-3x) + c_4(4x) = 0$.

We must now ask and answer this question. Does a set of c's exist, not all zero, which will make (a) a true equation for *every* x in the interval

$-\infty < x < \infty$? Rewriting (a) as

(b) $$(c_1 - 2c_2 - 3c_3 + 4c_4)x = 0,$$

we see immediately there are an infinite number of sets of c's which will satisfy (b), and therefore (a), for every x in I. For example,

(c) $$c_1 = 2, \quad c_2 = 1, \quad c_3 = 0, \quad c_4 = 0;$$
$$c_1 = 5, \quad c_2 = 3, \quad c_3 = 1, \quad c_4 = 1,$$

are two such sets. Hence by Definition 19.1 the set of functions in 1 is linearly dependent.

The linear combination of the set of functions in 2, equated to zero, is

(d) $$c_1 x^p + c_2 x^q = 0.$$

If we assume x^p and x^q are linearly dependent functions in I, then by Definition 19.1, there must exist constants c_1 and c_2 both not zero such that (d) is an identity in x. Since c_1 and c_2 are not both zero, we may choose one of them, say $c_1 \neq 0$. Dividing (d) by $c_1 x^q$, we obtain

(e) $$x^{p-q} = -\frac{c_2}{c_1}, \quad p \neq q.$$

The value of the left side of (e) varies with each x in the interval $x > 0$, $x < 0$. The right side, however, is a constant for fixed values of c_1 and c_2. Hence to assume that x^p, x^q are linearly dependent leads to a contradiction. They must therefore be linearly independent.

The proof that the functions in 3, $e^{px}, e^{qx}, p \neq q$ are linearly independent is practically identical with the proof that x^p and x^q, $p \neq q$ are linearly independent. It has therefore been left to you as an exercise; see Exercise 19,1.

The linear combination of the set of functions in 4, equated to zero, is

(f) $$c_1 e^x + c_2 0 + c_3 \sin x + c_4 \cdot 1 = 0.$$

Choose $c_1 = c_3 = c_4 = 0$ and $c_2 = $ any number not zero. Then (f) will read

(g) $$0 \cdot e^x + c_2 \cdot 0 + 0 \cdot \sin x + 0 \cdot 1 = 0,$$

which is a true equation for every x in I. Hence the given set is linearly dependent.

Comment 19.14. Whenever a set of functions contains zero for one of its members, the set must be linearly dependent. All you have to do is to choose for every c the value zero, except the one which is the coefficient of zero.

If a set of functions were selected at random, one could work a long time trying to find a set of constants $c_1, c_2, \cdots, c_n$, not all zero, which would make (19.11) true. However, a failure to find such a set would not of itself permit one to assert that the given set of functions was independent, since the possible values assignable to the set of c's are infinite in number. It is conceivable that such a set of constants exist, not easily discoverable, for which (19.11) holds. Fortunately there are tests available to determine whether a set of functions is linearly dependent or independent. A discussion of these tests will be found in Lesson 63B and Exercise 63,5.

You may be wondering what linear dependence and linear independence of a set of functions has to do with solving a linear differential equation of order n. It turns out that a homogeneous linear differential equation has as many linearly independent solutions as the order of its equation. (For its proof, see Theorems 19.3 and 65.4.) If the homogeneous linear differential equation, therefore, is of order n, we must find not only n solutions but must also be sure that these n solutions are linearly independent. And as we shall show, the linear combination of these n solutions will be its true general solution. For instance, if we solved a fourth order homogeneous linear differential equation and thought the four functions x, $-2x$, $-3x$, $4x$ in Example 19.13–1 were its four distinct and different solutions, we would be very much mistaken. All of them can be included in the one solution $y = cx$. Hence we would have to hunt for three more solutions. On the other hand, the functions x and x^2, which we proved were linearly independent (see Example 19.13–2 with $p = 1$, $q = 2$), could be two distinct solutions of a homogeneous linear differential equation of order two. In that event, the linear combination $c_1 x + c_2 x^2$ would be its general solution.

Hence in solving an nth order homogeneous linear differential equation, we must not only find n solutions, but must also show that they are linearly independent.

LESSON 19B. The Linear Differential Equation of Order n. We have already defined in 18.1 the linear differential equation of order n. In connection with this equation, there are two extremely important theorems. The proof of the first one, Theorem 19.2 below, is too complicated to be given at this stage and would only lead us too far afield. In order, therefore, not to delay the study of methods of solving nth order linear differential equations more than necessary, we have postponed its proof to Lesson 65. The proof of the second one, however, Theorem 19.3 below, is given in this lesson.

Theorem 19.2. If $f_0(x), f_1(x), \cdots, f_n(x)$ and $Q(x)$ are each continuous functions of x on a common interval I, and $f_n(x) \neq 0$ when x is in I, then *the linear differential equation*

(19.21) $f_n(x)y^{(n)} + f_{n-1}(x)y^{(n-1)} + \cdots + f_1(x)y' + f_0(x)y = Q(x)$

has one and only one solution,

(19.22) $y = y(x)$,

satisfying the set of initial conditions

(19.23) $y(x_0) = y_0, \quad y'(x_0) = y_1, \quad \cdots, \quad y^{(n-1)}(x_0) = y_{n-1}$,

where x_0 is in I, and $y_0, y_1, \cdots, y_{n-1}$ are constants.

This theorem is an *existence and a uniqueness theorem*. It is an existence theorem because it gives the conditions under which a solution of (19.21) satisfying (19.23) must exist. It is also a uniqueness theorem because it gives conditions under which the solution of (19.21) satisfying (19.23) is unique. As remarked previously, the proof of this *very important* theorem has been postponed to Lesson 65; see Theorem 65.2.

Right now we shall content ourselves with the proofs of three important properties of linear differential equations that we shall need for a clearer understanding of the remaining lessons of this chapter. These properties are incorporated in the following theorem.

Theorem 19.3. *If $f_0(x), f_1(x), \cdots, f_n(x)$ and $Q(x)$ are each continuous functions of x on a common interval I and $f_n(x) \neq 0$ when x is in I, then:*

1. *The homogeneous linear differential equation*

(19.31) $f_n(x)y^{(n)} + f_{n-1}(x)y^{(n-1)} + \cdots + f_1(x)y' + f_0(x)y = 0$

has n linearly independent solutions $y_1(x), y_2(x), \cdots, y_n(x)$.

2. *The linear combination of these n solutions*

(19.32) $y_c(x) = c_1y_1(x) + c_2y_2(x) + \cdots + c_ny_n(x)$,

where $c_1, c_2, \cdots, c_n$ is a set of n arbitrary constants, is also a solution of (19.31). It is an n-parameter family of solutions of (19.31). (We shall explain later the significance of the subscript c in y_c.)

3. *The function*

(19.33) $y(x) = y_c(x) + y_p(x)$,

where $y_c(x)$ is defined in (19.32) and $y_p(x)$ is a particular solution of the nonhomogeneous linear differential equation corresponding to (19.31), namely

(19.34) $f_n(x)y^{(n)} + f_{n-1}(x)y^{(n-1)} + \cdots + f_1(x)y' + f_0(x)y = Q(x)$,

is an n-parameter family of solutions of (19.34).

Proof of 1. We shall, for the present, limit the proof of 1 to the special case where the coefficients $f_0(x), \cdots, f_n(x)$ are constants. The proof will consist in actually showing, in lessons 20 to 22 which follow, how to find these n linearly independent solutions. The proof, for the case where the coefficients are not constants, has been deferred to Lesson 65, Theorem 65.4.

Proof of 2. By hypothesis each function $y_1, y_2, \cdots, y_n$ is a solution of (19.31). Hence each satisfies (19.31). This means, by Definition 3.4 and the Remark at the bottom of page 22, that

(a) $f_n(x)y_1^{(n)} + f_{n-1}(x)y_1^{(n-1)} + \cdots + f_1(x)y_1' + f_0(x)y_1 = 0,$
$f_n(x)y_2^{(n)} + f_{n-1}(x)y_2^{(n-1)} + \cdots + f_1(x)y_2' + f_0(x)y_2 = 0,$
$\cdots \cdots \cdots \cdots \cdots \cdots \cdots \cdots \cdots \cdots \cdots \cdots \cdots \cdots \cdots \cdots$
$f_n(x)y_n^{(n)} + f_{n-1}(x)y_n^{(n-1)} + \cdots + f_1(x)y_n' + f_0(x)y_n = 0.$

Multiply the first equation of (a) by c_1, the second by c_2, $\cdots$, the last by c_n, where $c_1, c_2, \cdots, c_n$ are n arbitrary constants, and add them. The result is

(b) $f_n(x)[c_1y_1^{(n)} + c_2y_2^{(n)} + \cdots + c_ny_n^{(n)}]$
$+ f_{n-1}(x)[c_1y_1^{(n-1)} + c_2y_2^{(n-1)} + \cdots + c_ny_n^{(n-1)}] + \cdots$
$+ f_0(x)[c_1y_1 + c_2y_2 + \cdots + c_ny_n] = 0.$

This last equation can be written as*

(c) $f_n(x)[c_1y_1 + c_2y_2 + \cdots + c_ny]^{(n)}$
$+ f_{n-1}(x)[c_1y_1 + \cdots + c_ny_n]^{(n-1)} + \cdots$
$+ f_0(x)[c_1y_1 + \cdots + c_ny_n] = 0.$

By (19.32) each quantity inside the brackets is y_c. Hence (c) is

(d) $f_n(x)y_c^{(n)} + f_{n-1}(x)y_c^{(n-1)} + \cdots + f_0(x)y_c = 0,$

which says that y_c satisfies (19.31). It is therefore a solution. Since it is a linear combination of n independent solutions and contains n parameters, it is an n-parameter family of solutions.

Proof of 3. By hypothesis, y_p is a particular solution of (19.34). Hence, by Definition 3.4,

(e) $f_n(x)y_p^{(n)} + f_{n-1}(x)y_p^{(n-1)} + \cdots + f_1(x)y_p' + f_0(x)y_p = Q(x).$

---

*For example,
$$u''(x) + v''(x) = \frac{d^2u(x)}{dx^2} + \frac{d^2v(x)}{dx^2} = \frac{d^2(u+v)}{dx^2} = (u+v)''.$$

Adding (d) and (e) we obtain

(f) $\qquad f_n(x)(y_c + y_p)^{(n)} + f_{n-1}(x)(y_c + y_p)^{(n-1)}$
$\qquad\qquad + \cdots + f_0(x)(y_c + y_p) = Q(x).$

This last equation says that $(y_c + y_p)$ satisfies (19.34) and is therefore a solution. Further, since $y = y_c + y_p$ contains n parameters, it is an n-parameter family of solutions of (19.34).

Definition 19.4. The solution $y_c(x)$ in (19.32) of the homogeneous equation (19.31) is called the **complementary function** of (19.34). Hence the use of the subscript c in $y_c(x)$.

Remark. The subscript p in $y_p(x)$ of (19.33) is used to distinguish it from the y_c part of the n-parameter family of solutions.

Comment 19.41. We shall prove in Lesson 65, see Theorem 65.5, that the function $y_c(x)$ of (19.32), which is an n-parameter family of solutions of (19.31), is in fact its true *general* solution in accordance with our Definition 4.7. *Every* particular solution of (19.31) can be obtained from it by properly choosing the n arbitrary constants. We therefore shall refer to this n-parameter family of solutions as a *general solution*. We infer further from this theorem that if one person has obtained the n independent solutions $y_1, y_2, \cdots, y_n$ (whose linear combination is the general solution) by using one method, then it is not possible for a second person using another method to find a general solution which is essentially different from it. The second person's solution will be obtainable from the first by a proper choice of the constants $c_1, c_2, \cdots, c_n$ in (19.32).

A similar statement can be made for the function $y(x)$ in (19.33). The proof will be found in Theorem 65.6. We shall therefore refer to it as the general solution of (19.34).

EXERCISE 19

1. Prove that if $p \neq q$, the functions e^{px} and e^{qx} are linearly independent. *Hint.* Follow the method used in proving the linear independence of the functions in Example 19.13–2.
2. Prove that the functions e^{ax} and xe^{ax} are linearly independent. *Hint.* Make use of Example 19.13–2 with $p = 0$, $q = 1$, and the fact that $e^z \neq 0$ for all z.
3. Prove that the functions $\sin x$, 0, $\cos x$ are linearly dependent.
4. Prove that the functions $3e^{2x}$ and $-2e^{2x}$ are linearly dependent.

It is extremely important that you prove the statements in Exercises 5 to 7 below. Follow the method used to prove statements 2 and 3 of Theorem 19.3.

5. If y_p is a solution of

(19.5) $\qquad f_n(x)y^{(n)} + \cdots + f_1(x)y' + f_0(x)y = Q(x),$

then Ay_p is a solution of (19.5) with $Q(x)$ replaced by $AQ(x)$.

6. **Principle of Superposition.** (Also see Comment 24.25.) If y_{p_1} is a solution of (19.5) with $Q(x)$ replaced by $Q_1(x)$ and y_{p_2} is a solution of (19.5) with $Q(x)$ replaced by $Q_2(x)$, then $y_p = y_{p_1} + y_{p_2}$ is a solution of

$$f_n(x)y^{(n)} + \cdots + f_1(x)y' + f_0(x)y = Q_1(x) + Q_2(x).$$

7. If $y_p(x) = u(x) + iv(x)$ is a solution of

(19.51) $\qquad f_n(x)y^{(n)} + \cdots + f_1(x)y' + f_0(x)y = R(x) + iS(x),$

where $f_0(x), \cdots, f_n(x)$ are real functions of x, then

(a) the real part of y_p, i.e., $u(x)$, is a solution of

$$f_n(x)y^{(n)} + \cdots + f_1(x)y' + f_0(x)y = R(x),$$

(b) the imaginary part of y_p, i.e., $v(x)$, is a solution of

$$f_n(x)y^{(n)} + \cdots + f_1(x)y' + f_0(x)y = S(x).$$

Hint. Two complex numbers are equal if and only if their real parts are equal and their imaginary parts are equal.

8. Assume that $y_p = x^2$ is a solution of $y'' + y' - 2y = 2(1 + x - x^2)$. Use 5 above to find a particular solution of $y'' + y' - 2y = 6(1 + x - x^2)$. Answer: $y_p = 3x^2$. Verify the correctness of this result.

9. Assume that $y_{p_1} = 1 + x$ is a solution of

$$y'' - y' + y = x,$$

and $y_{p_2} = e^{2x}$ is a solution of

$$y'' - y' + y = 3e^{2x}.$$

Use 6 above to find a particular solution of $y'' - y' + y = x + 3e^{2x}$. Answer: $y_p = 1 + x + e^{2x}$. Verify the correctness of this result.

10. Assume that $y_p = (\frac{1}{10}\cos x - \frac{3}{10}\sin x) + i(\frac{1}{10}\sin x + \frac{3}{10}\cos x)$ is a solution of $y'' - 3y' + 2y = e^{ix} = \cos x + i\sin x$. Use 7 above to find a particular solution of (a) $y'' - 3y' + 2y = \cos x$, (b) $y'' - 3y' + 2y = \sin x$. *Ans.* (a) $y(x) = \frac{1}{10}\cos x - \frac{3}{10}\sin x$. (b) $y(x) = \frac{1}{10}\sin x + \frac{3}{10}\cos x$. Verify the correctness of each of these results.

11. Prove that two functions are linearly dependent if one is a constant multiple of the other.

12. Assume f_1, f_2, f_3 are three linearly independent functions. Show that the addition to the set of one of these functions, say f_1, makes the new set linearly dependent.

13. Prove that a set of functions $f_1, f_2, \cdots, f_n$, is linearly dependent if two functions of the set are the same.

14. Prove that a set of functions, $f_1, f_2, \cdots, f_n$, is linearly dependent, if a subset, i.e., if a part of the set, is linearly dependent.

LESSON 20.   Solution of the Homogeneous Linear Differential Equation of Order n with Constant Coefficients.

LESSON 20A. General Form of Its Solutions. In actual practice, equations of the type (19.31), where the coefficients are functions of x with no restrictions placed on their simplicity or complexity, do not usually have solutions expressible in terms of elementary functions. And even when they do, it is in general extremely difficult to find them. If, however, each coefficient in (19.31) is a constant, then solutions in terms of elementary functions can be readily obtained. For the next few lessons, therefore, we shall concentrate on solving the differential equation

$$(20.1) \qquad a_n y^{(n)} + a_{n-1} y^{(n-1)} + \cdots + a_1 y' + a_0 y = 0,$$

where $a_0, a_1, \cdots, a_n$ are *constants* and $a_n \neq 0$.

Without going into the question of motivation, let us guess that a possible solution of (20.1) has the form

$$(20.11) \qquad\qquad\qquad y = e^{mx}.$$

We now ask ourselves this question. For what value of m will (20.11) be a solution of (20.1)? By Definition 3.4, it must be a value for which

$$(20.12) \quad a_n \frac{d^n}{dx^n} e^{mx} + a_{n-1} \frac{d^{n-1}}{dx^{n-1}} e^{mx} + \cdots + a_1 \frac{d}{dx} e^{mx} + a_0 e^{mx} = 0.$$

Since the kth derivative of $e^{mx} = m^k e^{mx}$, we may rewrite (20.12) as

$$(20.13) \quad a_n m^n e^{mx} + a_{n-1} m^{n-1} e^{mx} + \cdots + a_1 m e^{mx} + a_0 e^{mx} = 0.$$

By (18.86), $e^{mx} \neq 0$ for all m and x. We therefore can divide (20.13) by it to obtain

$$(20.14) \qquad a_n m^n + a_{n-1} m^{n-1} + \cdots + a_1 m + a_0 = 0.$$

We at last have the answer to our question. Each value of m for which (20.14) is true will make $y = e^{mx}$ a solution of (20.1).

But (20.14) is an algebraic equation in m of degree n, and therefore, by the fundamental theorem of algebra (see Theorem 18.23), it has at least one and not more than n distinct roots. Let us call these n roots m_1, $m_2, \cdots, m_n$, where the m's need not all be distinct. Then each function

$$(20.15) \qquad y_1 = e^{m_1 x}, \qquad y_2 = e^{m_2 x}, \qquad \cdots, \qquad y_n = e^{m_n x}$$

is a solution of (20.1).

Definition 20.16. Equation (20.14) is called the **characteristic equation** of (20.1).

Note. The characteristic equation (20.14) is easily obtainable from (20.1). Replace y by m and the order of the derivative by a numerically equal exponent.

In solving the characteristic equation (20.14), the following three possibilities may occur.

     1. All its roots are distinct and real.
     2. All its roots are real but some of its roots repeat.
     3. All its roots are imaginary.

We shall discuss each of the above three possibilities separately. Other possibilities may also occur as, for example, when all roots are distinct but some are real and some are imaginary. It will be made apparent why such other possible combinations do not require special consideration.

Remark. We have already commented, see Comment 18.26, in regard to the difficulty, in general, of finding the n roots of the characteristic equation (20.14). If an equation of this type, of degree greater than two, were written at random, the probability is very high that it would have irrational or complex roots which would be difficult and laborious to find. Since this is a differential equations text and not an algebra text, the examples used for illustration and exercises have been carefully chosen so that their roots can be readily found. In practical problems, however, finding the roots of the resulting characteristic equation may not be and usually will not be an easy task. We cannot emphasize this point too strongly.

LESSON 20B.   Roots of the Characteristic Equation (20.14) Real and Distinct. If the n roots $m_1, m_2, \cdots, m_n$ of the characteristic equation (20.14) are distinct, then the n solutions of (20.1), namely,

(20.2) $$y_1 = e^{m_1 x}, \qquad y_2 = e^{m_2 x}, \cdots, y_n = e^{m_n x}$$

are linearly independent functions. (For proof when $n = 2$, see Exercise 19,1. For proof when $n > 2$, see Example 64.2.) Hence by Theorem 19.3 [see in particular (19.32)] and Comment 19.41,

(20.21) $$y_c = c_1 e^{m_1 x} + c_2 e^{m_2 x} + \cdots + c_n e^{m_n x}$$

is the *general solution* of (20.1).

Example 20.22.   Find the *general solution* of

(a) $$y''' + 2y'' - y' - 2y = 0.$$

Solution.   By Definition 20.16, the characteristic equation of (a) is

(b) $$m^3 + 2m^2 - m - 2 = 0,$$

whose roots are

(c) $$m_1 = 1, \qquad m_2 = -1, \qquad m_3 = -2.$$

Hence by (20.21) the general solution of (a) is

(d) $$y_c = c_1 e^x + c_2 e^{-x} + c_3 e^{-2x}.$$

Example 20.23.   Find the particular solution $y(x)$ of

(a) $$y'' - 3y' + 2y = 0,$$

for which $y_c(0) = 1$, $y_c'(0) = 0$.

Solution. By Definition 20.16, the characteristic equation of (a) is

(b) $$m^2 - 3m + 2 = 0,$$

whose roots are

(c) $$m_1 = 1, \qquad m_2 = 2.$$

Hence by (20.21) the general solution of (a) is

(d) $$y_c = c_1 e^x + c_2 e^{2x}.$$

By differentiation of (d), we obtain

(e) $$y_c' = c_1 e^x + 2c_2 e^{2x}.$$

Substituting the given initial conditions $x = 0$, $y_c = 1$, $y_c' = 0$ in (d) and (e), there results

(f) $$1 = c_1 + c_2,$$
$$0 = c_1 + 2c_2.$$

Solving (f) simultaneously for c_1 and c_2, we find $c_1 = 2$, $c_2 = -1$. Substituting these values in (d), we obtain the required particular solution

(g) $$y = 2e^x - e^{2x}.$$

LESSON 20C. Roots of Characteristic Equation (20.14) Real but Some Multiple. If two or more roots of the characteristic equation (20.14) are alike, then the functions (20.15) formed with each of these n roots are not linearly independent. For example, the characteristic equation of

(a) $$y'' - 4y' + 4y = 0$$

is

(b) $$m^2 - 4m + 4 = 0,$$

which has the double root $m = 2$. It is easy to show that the two functions $y_1 = e^{2x}$ and $y_2 = e^{2x}$ are linearly dependent. Form the linear combination

(c) $$c_1 e^{2x} + c_2 e^{2x} = 0$$

and take $c_1 = 1$, $c_2 = -1$. Hence the general solution of (a) could not be $y = c_1 e^{2x} + c_2 e^{2x}$. (Remember, a general solution of a second order linear differential equation is a linear combination of two linearly independent solutions.) Actually we can write the solution as

(d) $$y = c_1 e^{2x} + c_2 e^{2x} = (c_1 + c_2)e^{2x} = ce^{2x},$$

from which we see that we really have only one solution and not two. We must therefore search for a second solution, independent of e^{2x}.

To generalize matters for the second order linear differential equation, we confine our attention to the equation

$$(20.3) \qquad y'' - 2ay' + a^2 y = 0,$$

whose characteristic equation

$$(20.31) \qquad m^2 - 2am + a^2 = 0$$

has the double root $m = a$. Let

$$(20.32) \qquad y_c = ue^{ax},$$

where u is a function of x. We now ask ourselves the usual question. What must u look like for (20.32) to be a solution of (20.3)? We know, by Definition 3.4, that (20.32) will be a solution of (20.3) if

$$(20.33) \qquad (ue^{ax})'' - 2a(ue^{ax})' + a^2(ue^{ax}) = 0.$$

Performing the indicated differentiations in (20.33), we obtain

$$(20.34) \qquad e^{ax}(u'' + 2au' + a^2 u - 2au' - 2a^2 u + a^2 u) = 0,$$

which simplifies to

$$(20.35) \qquad e^{ax}u'' = 0.$$

By (18.86), $e^{ax} \neq 0$. Hence, (20.35) will be true if and only if

$$(20.36) \qquad u'' = 0.$$

Integration of (20.36) twice gives

$$(20.37) \qquad u = c_1 + c_2 x.$$

We now have the answer to our question. If, in (20.32), u has the value (20.37), then

$$(20.38) \qquad y_c = (c_1 + c_2 x)e^{ax}$$

will be a solution (20.3). It will be the general solution of (20.3) provided the two functions e^{ax} and xe^{ax} are linearly independent. The proof that they are indeed linearly independent was left to you as an exercise; see Exercise 19,2. Hence by Theorem 19.3 and Comment 19.41, (20.38) is the general solution of (20.3).

In general it can be shown that if the characteristic equation (20.14) has a root $m = a$, which repeats n times, then the general solution of (20.1) is

$$(20.4) \qquad y_c = (c_1 + c_2 x + c_3 x^2 + \cdots + c_n x^{n-1})e^{ax}.$$

And if, for example, the characteristic equation (20.14) can be written in the form

$$(20.41) \qquad m^2(m-a)^3(m+b)^4(m+c) = 0,$$

which implies that its roots are

$$m = 0 \text{ twice}, \quad m = a \text{ three times}, \quad m = -b \text{ four times}, \quad m = -c \text{ once},$$

then the general solution of its related differential equation is

$$(20.42) \qquad y_c = c_1 + c_2 x + (c_3 + c_4 x + c_5 x^2)e^{ax}$$
$$+ (c_6 + c_7 x + c_8 x^2 + c_9 x^3)e^{-bx} + c_{10}e^{-cx}.$$

Observe that with each n-fold root p, e^{px} is multiplied by a linear combination of powers of x beginning with x^0 and ending with x^{n-1}.

Example 20.43. Find the general solution of

(a) $$y^{(4)} - 3y'' + 2y' = 0.$$

Solution. The characteristic equation of (a) is

(b) $$m^4 - 3m^2 + 2m = 0,$$

whose roots are

(c) $$m = 0, \quad m = 1, \quad m = 1, \quad m = -2.$$

Since the root 1 appears twice, the general solution of (a), by (20.42), is

(d) $$y_c = c_1 + (c_2 + c_3 x)e^x + c_4 e^{-2x}.$$

Example 20.44. Find the particular solution of

(a) $$y'' - 2y' + y = 0,$$

for which $y_c(0) = 1$, $y_c'(0) = 0$.

Solution. The characteristic equation of (a) is

(b) $$m^2 - 2m + 1 = 0,$$

whose roots are $m = 1$ twice. Hence the general solution of (a), by (20.38), is

(c) $$y_c = (c_1 + c_2 x)e^x.$$

Differentiation of (c) gives

(d) $$y_c' = (c_1 + c_2 + c_2 x)e^x.$$

To find the particular solution for which $x = 0$, $y_c = 1$, $y_c' = 0$, we substitute these values in (c) and (d). The result is

(e) $$1 = c_1,$$
$$0 = c_1 + c_2.$$

The simultaneous solution of (e) for c_1 and c_2 gives $c_1 = 1$, $c_2 = -1$. Substituting these values in (c), we obtain the required particular solution

(f) $$y = (1 - x)e^x.$$

LESSON 20D. Some or All Roots of the Characteristic Equation (20.14) Imaginary. If the constant coefficients in the characteristic equation (20.14) are real, then (see Exercise 18,9) any imaginary roots it may have must occur in conjugate pairs. Hence if $\alpha + i\beta$ is one root, another root must be $\alpha - i\beta$. Assume now that $\alpha + i\beta$ and $\alpha - i\beta$ are two imaginary roots of the characteristic equation of a second order linear differential equation. Then its general solution by (20.21) is

(20.5) $$y_c = c_1'e^{(\alpha+i\beta)x} + c_2'e^{(\alpha-i\beta)x}$$
$$= c_1'e^{\alpha x}e^{i\beta x} + c_2'e^{\alpha x}e^{-i\beta x}$$
$$= e^{\alpha x}(c_1'e^{i\beta x} + c_2'e^{-i\beta x}).$$

By (18.82) and (18.83)

(20.51) $$e^{i\beta x} = \cos \beta x + i \sin \beta x; \qquad e^{-i\beta x} = \cos \beta x - i \sin \beta x.$$

Substituting (20.51) in the last equation of (20.5) and simplifying the result gives

(20.52) $$y_c = e^{\alpha x}[(c_1' + c_2') \cos \beta x + i(c_1' - c_2') \sin \beta x].$$

We now replace the constant $c_1' + c_2'$ by a new constant c_1 and the constant $i(c_1' - c_2')$ by a new constant c_2. Then (20.52) becomes

(20.53) $$y_c = e^{\alpha x}(c_1 \cos \beta x + c_2 \sin \beta x),$$

which is a second form of the general solution (20.5).

A third form of writing the general solution (20.5), more useful for practical purposes, is obtained as follows. Write the equality

(20.54) $$c_1 \cos \beta x + c_2 \sin \beta x$$
$$= \sqrt{c_1^2 + c_2^2} \left(\frac{c_1}{\sqrt{c_1^2 + c_2^2}} \cos \beta x + \frac{c_2}{\sqrt{c_1^2 + c_2^2}} \sin \beta x \right).$$

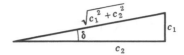

Figure 20.55

We see from Fig. 20.55 that

(20.56) $\qquad \sin \delta = \dfrac{c_1}{\sqrt{c_1{}^2 + c_2{}^2}}, \qquad \cos \delta = \dfrac{c_2}{\sqrt{c_1{}^2 + c_2{}^2}}.$

The substitution of these values in the right side of (20.54) gives

(20.57)
$$c_1 \cos \beta x + c_2 \sin \beta x = \sqrt{c_1{}^2 + c_2{}^2}\,(\sin \delta \cos \beta x + \cos \delta \sin \beta x)$$
$$= \sqrt{c_1{}^2 + c_2{}^2}\,\sin(\beta x + \delta).$$

Prove as an exercise that if we had interchanged the positions of c_1 and c_2 in Fig. 20.55, equation (20.54) would have become

(20.58)
$$c_1 \cos \beta x + c_2 \sin \beta x = \sqrt{c_1{}^2 + c_2{}^2}\,(\cos \delta \cos \beta x + \sin \delta \sin \beta x)$$
$$= \sqrt{c_1{}^2 + c_2{}^2}\,\cos(\beta x - \delta).$$

Replacing in (20.57) and in (20.58) a new constant c for the constant $\sqrt{c_1{}^2 + c_2{}^2}$, we may write the general solution (20.53) in either of the forms

(20.59) $\qquad y_c = ce^{\alpha x}\sin(\beta x + \delta) \quad$ or $\quad y_c = ce^{\alpha x}\cos(\beta x - \delta).$

NOTE. The δ in the first equation is not the same as the δ in the second equation.

Hence in the case of complex roots, the general solution of a linear differential equation of order two, whose characteristic equation has the conjugate roots $\alpha + \beta i$ and $\alpha - \beta i$, can be written in any of the following forms:

(20.6) $\qquad$ (a) $y_c = c_1 e^{(\alpha + i\beta)x} + c_2 e^{(\alpha - i\beta)x},$

$\qquad\qquad\qquad$ (b) $y_c = e^{\alpha x}(c_1 \cos \beta x + c_2 \sin \beta x),$

$\qquad\qquad\qquad$ (c) $y_c = ce^{\alpha x}\sin(\beta x + \delta),$

$\qquad\qquad\qquad$ (d) $y_c = ce^{\alpha x}\cos(\beta x - \delta).$

The significance of the two arbitrary constants c and δ which appear in (20.6) (c) and (d) will be discussed in Lesson 28.

Example 20.61.   Find the general solution of

(a) $$y'' - 2y' + 2y = 0.$$

Solution.   The characteristic equation of (a) is

(b) $$m^2 - 2m + 2 = 0,$$

whose roots are $1 \pm i$. Hence in (20.6), $\alpha = 1$, $\beta = 1$. The general solution of (a), therefore, can be written in any of the following forms:

(c)
$$y_c = c_1 e^{(1+i)x} + c_2 e^{(1-i)x},$$
$$y_c = e^x(c_1 \cos x + c_2 \sin x),$$
$$y_c = c e^x \sin (x + \delta),$$
$$y_c = c e^x \cos (x - \delta).$$

If the linear differential equation

(20.7) $$a_4 y^{(4)} + a_3 y''' + a_2 y'' + a_1 y' + a_0 y = 0, \quad a_4 \neq 0,$$

has a conjugate pair of repeated imaginary roots, i.e., if $\alpha + i\beta$ and $\alpha - i\beta$ each occurs twice as a root, then by (20.38) the general solution of (20.7) is

(20.71) $$y_c = (c_1 + c_2 x)e^{(\alpha + i\beta)x} + (c_3 + c_4 x)e^{(\alpha - i\beta)x}.$$

The equivalent forms are

(20.72)
$$y_c = e^{\alpha x}[(c_1 + c_2 x) \cos \beta x + (c_3 + c_4 x) \sin \beta x],$$
$$y_c = e^{\alpha x}[c_1 \sin (\beta x + \delta_1) + c_2 x \sin (\beta x + \delta_2)],$$
$$y_c = e^{\alpha x}[c_1 \cos (\beta x - \delta_1) + c_2 x \cos (\beta x - \delta_2)].$$

Example 20.73.   Find the general solution of

(a) $$y^{(4)} + 2y'' + 1 = 0.$$

Solution.   The characteristic equation of (a) is

(b) $$m^4 + 2m^2 + 1 = 0; \quad (m^2 + 1)^2 = 0,$$

whose roots are $m = \pm i$ twice. Hence in (20.71) and (20.72) $\alpha = 0$, $\beta = 1$. The general solution of (a) therefore can be written in any of the following forms:

(c)
$$y_c = (c_1 + c_2 x)e^{ix} + (c_3 + c_4 x)e^{-ix},$$
$$y_c = (c_1 + c_2 x) \cos x + (c_3 + c_4 x) \sin x,$$
$$y_c = c_1 \sin (x + \delta_1) + c_2 x \sin (x + \delta_2),$$
$$y_c = c_1 \cos (x - \delta_1) + c_2 x \cos (x - \delta_2).$$

Example 20.74.   Find the general solution of

(a) $$y''' - 12y'' + 22y' - 20y = 0$$

Solution.   The characteristic equation of (a) is

(b) $$m^3 - 8m^2 + 22m - 20 = 0$$

whose roots are 2, 3 $\pm$ *i*. The general solution of (a), therefore, can be written in any of the following forms:

(c)
$$y_c = c_1 e^{2x} + c_2 e^{(3+i)x} + c_3 e^{(3-i)x},$$
$$y_c = c_1 e^{2x} + e^{3x}(c_2 \cos x + c_3 \sin x),$$
$$y_c = c_1 e^{2x} + c_2 e^{3x} \cos(x - \delta),$$
$$y_c = c_1 e^{2x} + c_2 e^{3x} \sin(x + \delta).$$

EXERCISE 20

Find the general solution of each of the following equations.

1. $y'' + 2y' = 0$.
2. $y'' - 3y' + 2y = 0$.
3. $y'' - y = 0$.
4. $y''' + y'' - 6y' = 0$.

5. $6y'' - 11y' + 4y = 0$.
6. $y'' + 2y' - y = 0$.
7. $y''' + y'' - 10y' - 6y = 0$.
8. $y^{(4)} - y''' - 4y'' + 4y' = 0$.

9. $y^{(4)} + 4y''' + y'' - 4y' - 2y = 0$.

10. $y^{(4)} - a^2 y = 0$, $a > 0$.
11. $y'' - 2ky' - 2y = 0$.
12. $y'' + 4ky' - 12k^2 y = 0$.
13. $y^{(4)} = 0$.
14. $y'' + 4y' + 4y = 0$.

15. $3y''' + 5y'' + y' - y = 0$.
16. $y'''' - 6y''' + 12y' - 8y = 0$.
17. $y'' - 2ay' + a^2 y = 0$.
18. $y^{(4)} + 3y''' = 0$.
19. $y^{(4)} - 2y'' = 0$.

20. $y^{(4)} + 2y''' - 11y'' - 12y' + 36y = 0$.

21. $36y^{(4)} - 37y'' + 4y' + 5y = 0$.
22. $y^{(4)} - 8y'' + 16y = 0$.
23. $y'' - 2y' + 5y = 0$.
24. $y'' - y' + y = 0$.
25. $y^{(4)} + 5y'' + 6y = 0$.

26. $y'' - 4y' + 20y = 0$.
27. $y^{(4)} + 4y'' + 4y = 0$.
28. $y''' + 8y = 0$.
29. $y^{(4)} + 4y'' = 0$.
30. $y^{(5)} + 2y''' + y' = 0$.

For each of the following equations find a particular solution which satisfies the given initial conditions.

31. $y'' = 0$, $y(1) = 2$, $y'(1) = -1$.
32. $y'' + 4y' + 4y = 0$, $y(0) = 1$, $y'(0) = 1$.
33. $y'' - 2y' + 5y = 0$, $y(0) = 2$, $y'(0) = 4$.
34. $y'' - 4y' + 20y = 0$, $y(\pi/2) = 0$, $y'(\pi/2) = 1$.
35. $3y''' + 5y'' + y' - y = 0$, $y(0) = 0$, $y'(0) = 1$, $y''(0) = -1$.

ANSWERS 20

1. $y = c_1 + c_2 e^{-2x}$.
2. $y = c_1 e^x + c_2 e^{2x}$.
3. $y = c_1 e^x + c_2 e^{-x}$.

4. $y = c_1 + c_2 e^{2x} + c_3 e^{-3x}$.
5. $y = c_1 e^{x/2} + c_2 e^{4x/3}$.
6. $y = c_1 e^{(-1+\sqrt{2})x} + c_2 e^{(-1-\sqrt{2})x}$.

7. $y = c_1 e^{3x} + c_2 e^{(-2+\sqrt{2})x} + c_3 e^{(-2-\sqrt{2})x}$.

8. $y = c_1 + c_2 e^x + c_3 e^{2x} + c_4 e^{-2x}$.

9. $y = c_1 e^x + c_2 e^{-x} + c_3 e^{(-2+\sqrt{2})x} + c_4 e^{(-2-\sqrt{2})x}$.

10. $y = c_1 e^{\sqrt{a}x} + c_2 e^{-\sqrt{a}x} + c_3 \cos \sqrt{a}\, x + c_4 \sin \sqrt{a}\, x$.

11. $y = c_1 e^{(k+\sqrt{k^2+2})x} + c_2 e^{(k-\sqrt{k^2+2})x}$.

12. $y = c_1 e^{-6kx} + c_2 e^{2kx}$.   **16.** $y = (c_1 + c_2 x + c_3 x^2)e^{2x}$.

13. $y = c_1 + c_2 x + c_3 x^2 + c_4 x^3$.   **17.** $y = (c_1 + c_2 x)e^{ax}$.

14. $y = (c_1 + c_2 x)e^{-2x}$.   **18.** $y = c_1 + c_2 x + c_3 x^2 + c_4 e^{-3x}$.

15. $y = c_1 e^{x/3} + (c_2 + c_3 x)e^{-x}$.   **19.** $y = c_1 + c_2 x + c_3 e^{\sqrt{2}x} + c_4 e^{-\sqrt{2}x}$.

20. $y = (c_1 + c_2 x)e^{2x} + (c_3 + c_4 x)e^{-3x}$.

21. $y = c_1 e^{-x} + c_2 e^{x/2} + c_3 e^{5x/6} + c_4 e^{-x/3}$.

22. $y = (c_1 + c_2 x)e^{2x} + (c_3 + c_4 x)e^{-2x}$.

23. $y = e^x(c_1 \cos 2x + c_2 \sin 2x)$.

24. $y = c_1 e^{\frac{1+\sqrt{3}i}{2}x} + c_2 e^{\frac{1-\sqrt{3}i}{2}x} = e^{x/2}\left(c_3 \cos \dfrac{\sqrt{3}}{2}\, x + c_4 \sin \dfrac{\sqrt{3}}{2}\, x\right)$.

25. $y = c_1 \cos \sqrt{3}\, x + c_2 \sin \sqrt{3}\, x + c_3 \cos \sqrt{2}\, x + c_4 \sin \sqrt{2}\, x$.

26. $y = e^{2x}(c_1 \cos 4x + c_2 \sin 4x)$.

27. $y = (c_1 + c_2 x) \cos \sqrt{2}\, x + (c_3 + c_4 x) \sin \sqrt{2}\, x$.

28. $y = c_1 e^{-2x} + e^x[c_2 \cos \sqrt{3}\, x + c_3 \sin \sqrt{3}\, x]$.

29. $y = c_1 + c_2 x + c_3 \cos 2x + c_4 \sin 2x$.

30. $y = c_1 + c_2 \sin x + c_3 \cos x + c_4 x \sin x + c_5 x \cos x$.

31. $y = 3 - x$.

32. $y = (1 + 3x)e^{-2x}$.

33. $y = e^x(2 \cos 2x + \sin 2x)$.

34. $y = \frac{1}{4} e^{2x-\pi} \sin 4x$.

35. $y = \dfrac{9}{16} e^{x/3} + \left(\dfrac{x}{4} - \dfrac{9}{16}\right) e^{-x}$.

LESSON 21. Solution of the Nonhomogeneous Linear Differential Equation of Order n with Constant Coefficients.

LESSON 21A. Solution by the Method of Undetermined Coefficients. By Theorem 19.3 and Comment 19.41, the general solution of the differential equation

$$(21.1) \qquad a_n y^{(n)} + a_{n-1} y^{(n-1)} + \cdots + a_1 y' + a_0 y = Q(x),$$

where $a_n \neq 0$ and $Q(x) \not\equiv 0$ in an interval I, is

$$(21.11) \qquad\qquad y(x) = y_c(x) + y_p(x),$$

where $y_c(x)$, the complementary function, is the general solution of the related homogeneous equation of (21.1) and $y_p(x)$ is a particular solution of (21.1). In Lesson 20, we showed how to find y_c. There remains the problem of finding y_p.

The procedure we are about to describe for finding y_p is called the **method of undetermined coefficients.** It can be used only if $Q(x)$ consists of a sum of terms each of which has a finite number of linearly independent derivatives. This restriction implies that $Q(x)$ can only contain terms such as a, x^k, e^{ax}, $\sin ax$, $\cos ax$, and combinations of such terms, where a is a constant and k is a positive integer. See Exercise 21,2. For example, the successive derivatives of $\sin 2x$ are

$$2 \cos 2x, \ -4 \sin 2x, \ -8 \cos 2x, \ \text{etc.}$$

However, only the set consisting of $\sin 2x$ and $2 \cos 2x$ is linearly independent. The addition of any succeeding derivative makes the set linearly dependent. Verify it. The linearly independent derivatives of x^3 are

$$3x^2, \ 6x, \ 6.$$

The addition to this set of the next derivative, which is zero, makes the set linearly dependent. See Comment 19.14. However the function x^{-1}, for example, has an infinite number of linearly independent derivatives.

To find y_p by the method of undetermined coefficients, it is necessary to compare the terms of $Q(x)$ in (21.1) with those of the complementary function y_c. In making this comparison, a number of different possibilities may occur, each of which we consider separately in the cases below.

Case 1. *No term of $Q(x)$ in (21.1) is the same as a term of y_c.* In this case, a particular solution y_p of (21.1) will be a linear combination of the terms in $Q(x)$ and *all* its linearly independent derivatives.

Example 21.2. Find the general solution of

(a) $$y'' + 4y' + 4y = 4x^2 + 6e^x.$$

The complementary function of (a) is

(b) $$y_c = (c_1 + c_2 x)e^{-2x}.$$

(Verify it.) Since $Q(x)$, which is the right side of (a), has no term in common with y_c this case applies. A particular solution y_p will therefore be a linear combination of $Q(x)$ and all its linearly independent derivatives. These are, ignoring constant coefficients, x^2, x, 1, e^x. Hence the trial solution y_p must be a linear combination of these functions, namely

(c) $$y_p = Ax^2 + Bx + C + De^x,$$

where A, B, C, D are to be determined. Successive derivatives of (c) are

(d) $$y_p{}' = 2Ax + B + De^x,$$

(e) $$y_p{}'' = 2A + De^x.$$

As we have repeatedly remarked, (c) will be a solution of (a) if the substitution of (c), (d), and (e) in (a) will make it an identity in x. Hence (c) will be a solution of (a) if

(f) $2A + De^x + 4(2Ax + B + De^x)$
$$+ 4(Ax^2 + Bx + C + De^x) \equiv 4x^2 + 6e^x.$$

Simplification of (f) gives

(g) $4Ax^2 + (8A + 4B)x + (2A + 4B + 4C) + 9De^x \equiv 4x^2 + 6e^x.$

We now ask ourselves the question: What values shall we assign to A, B, C, and D to make (g) an identity in x? We proved in Example 19.13, that x, x^2 are linearly independent functions. The proof can be extended to show that x^0, x, x^2 are also linearly independent functions. Hence the answer to our question is: values which will make each coefficient of like powers of x zero. This means that the following equalities must hold.

(h)
$$4A = 4,$$
$$8A + 4B = 0,$$
$$2A + 4B + 4C = 0,$$
$$9D = 6.$$

Solving (h) simultaneously, there results

(i) $A = 1, \quad B = -2, \quad C = \tfrac{3}{2}, \quad D = \tfrac{2}{3}.$

Substituting these values in (c), we obtain

(j) $y_p = x^2 - 2x + \tfrac{3}{2} + \tfrac{2}{3}e^x,$

which is a particular solution of (a). Hence by (21.11), the general solution of (a) is (b) + (j), namely,

(k) $y = (c_1 + c_2 x)e^{-2x} + x^2 - 2x + \tfrac{3}{2} + \tfrac{2}{3}e^x.$

Example 21.21.   Find the general solution of

(a) $y'' - 3y' + 2y = 2xe^{3x} + 3\sin x.$

Solution.   The complementary function of (a) is

(b) $y_c = c_1 e^x + c_2 e^{2x}.$

(Verify it.) Since $Q(x)$, which is the right side of (a), has no term in common with y_c, a particular solution y_p will be a linear combination of $Q(x)$ and all its linearly independent derivatives. These are, ignoring constant coefficients, xe^{3x}, e^{3x}, $\sin x$, $\cos x$. Therefore the trial solution y_p must be of the form

(c) $$y_p = Axe^{3x} + Be^{3x} + C \sin x + D \cos x.$$

Successive derivatives of (c) are

(d) $$y_p' = 3Axe^{3x} + Ae^{3x} + 3Be^{3x} + C \cos x - D \sin x,$$

(e) $$y_p'' = 9Axe^{3x} + 6Ae^{3x} + 9Be^{3x} - C \sin x - D \cos x.$$

The function defined by (c) will be a solution of (a) if the substitution of (c), (d), and (e) in (a) will result in an identity in x. Making these substitutions and simplifying the resulting expression, we obtain

(f) $2Axe^{3x} + (3A + 2B)e^{3x} + (C + 3D) \sin x$
$$+ (D - 3C) \cos x \equiv 2xe^{3x} + 3 \sin x.$$

Equation (f) will be an identity in x if the coefficients of like terms on each side of the equal sign have the same value. Hence we must have

(g) $$2A = 2,$$
$$3A + 2B = 0,$$
$$C + 3D = 3,$$
$$-3C + D = 0.$$

Solving (g) simultaneously, there results

(h) $$A = 1, \quad B = -\tfrac{3}{2}, \quad C = \tfrac{3}{10}, \quad D = \tfrac{9}{10}.$$

Substituting these values in (c), we obtain

(i) $$y_p = xe^{3x} - \tfrac{3}{2}e^{3x} + \tfrac{3}{10} \sin x + \tfrac{9}{10} \cos x.$$

Hence by (21.11) the general solution of (a) is (b) + (i), namely

(j) $$y = c_1e^x + c_2e^{2x} + xe^{3x} - \tfrac{3}{2}e^{3x} + \tfrac{3}{10} \sin x + \tfrac{9}{10} \cos x.$$

Case 2. $Q(x)$ *in (21.1) contains a term which, ignoring constant coefficients, is x^k times a term $u(x)$ of y_c, where k is zero or a positive integer.* In this case a particular solution y_p of (21.1) will be a linear combination of $x^{k+1}u(x)$ and all its linearly independent derivatives (ignoring constant coefficients). If in addition $Q(x)$ contains terms which belong to Case 1, then the proper terms called for by this case must be included in y_p.

Example 21.3. Find the general solution of

(a) $$y'' - 3y' + 2y = 2x^2 + 3e^{2x}.$$

Solution. The complementary function of (a) is

(b) $y_c = c_1 e^x + c_2 e^{2x}$.

(Verify it.) Comparing $Q(x)$, which is the right side of (a), with (b), we
see that $Q(x)$ contains the term e^{2x} which, ignoring constant coefficients,
is x^0 times the same term in y_c. Hence for this term, y_p must contain a
linear combination of $x^{0+1} e^{2x}$ and all its linearly independent derivatives.
$Q(x)$ also has the term x^2 which belongs to Case 1. For this term, therefore, y_p must include a linear combination of it and all its linearly independent derivatives. In forming the linear combination of these functions
and their linearly independent derivatives, we may omit the function e^{2x}
since it already appears in y_c; see Exercise 21,1. Hence the trial solution
y_p must be of the form

(c) $y_p = Ax^2 + Bx + C + Dxe^{2x}$.

Successive derivatives of (c) are

(d) $y_p{}' = 2Ax + B + 2Dxe^{2x} + De^{2x}$,

(e) $y_p{}'' = 2A + 4Dxe^{2x} + 4De^{2x}$.

Substituting (c), (d), (e) in (a) and simplifying, we see that (c) will be a
solution of (a) if

(f) $2Ax^2 + (2B - 6A)x + (2A - 3B + 2C) + De^{2x} \equiv 2x^2 + 3e^{2x}$.

Equation (f) will be an identity in x if the coefficients of like terms on
each side of the equal sign have the same value. Hence we must have

(g) $2A = 2$, $2B - 6A = 0$, $2A - 3B + 2C = 0$, $D = 3$.

From (g) we find

(h) $A = 1$, $B = 3$, $C = \frac{7}{2}$, $D = 3$.

Substituting these values in (c), there results

(i) $y_p = x^2 + 3x + \frac{7}{2} + 3xe^{2x}$.

Combining this solution with (b), we obtain for the general solution of (a)

(j) $y = x^2 + 3x + \frac{7}{2} + 3xe^{2x} + c_1 e^x + c_2 e^{2x}$.

Example 21.31.   Find a general solution of

(a) $y'' - 3y' + 2y = xe^{2x} + \sin x$.

Solution. The complementary function of (a) is

(b) $$y_c = c_1 e^x + c_2 e^{2x}.$$

Comparing $Q(x)$ which is the right side of (a), with (b), we see that $Q(x)$ contains a term xe^{2x} which, ignoring constant coefficients, is x times a term e^{2x} in y_c. For this term, therefore, y_p must be a linear combination of $x^{1+1}e^{2x} = x^2 e^{2x}$ and all its independent derivatives. In addition we note $Q(x)$ contains a term $\sin x$ which belongs to Case 1. For this term, therefore, y_p must include a linear combination of it and its independent derivatives. In forming the linear combination of all these functions and their independent derivatives, we may omit the function e^{2x} since it already appears in y_c. Hence y_p must be of the form

(c) $$y_p = Ax^2 e^{2x} + Bxe^{2x} + C \sin x + D \cos x.$$

Successive derivatives of (c) are

(d) $\quad y_p' = 2Ax^2 e^{2x} + 2Axe^{2x} + 2Bxe^{2x} + Be^{2x} + C \cos x - D \sin x,$

(e) $\quad y_p'' = 4Ax^2 e^{2x} + 8Axe^{2x} + 2Ae^{2x} + 4Bxe^{2x} + 4Be^{2x}$
$$- C \sin x - D \cos x.$$

Substituting (c), (d), (e) in (a) and simplifying the result, we see that (c) will be a solution of (a) if

(f) $\quad 2Axe^{2x} + (2A + B)e^{2x} + (C + 3D) \sin x$
$$+ (D - 3C) \cos x \equiv xe^{2x} + \sin x.$$

Equating the coefficients of like terms on each side of the equal sign, we find that

(g) $\quad 2A = 1, \quad 2A + B = 0, \quad C + 3D = 1, \quad D - 3C = 0.$

From (g), we obtain

(h) $\qquad A = \tfrac{1}{2}, \quad B = -1, \quad C = \tfrac{1}{10}, \quad D = \tfrac{3}{10}.$

Substituting these values in (c), there results

(i) $$y_p = \tfrac{1}{2}x^2 e^{2x} - xe^{2x} + \tfrac{1}{10} \sin x + \tfrac{3}{10} \cos x.$$

The general solution of (a) is therefore the sum of (b) and (i).

Example 21.32. Find a general solution of

(a) $$y'' + y = \sin^3 x.$$

Solution. The complementary function of (a) is

(b) $$y_c = c_1 \sin x + c_2 \cos x.$$

By (18.84),

(c) $\quad \sin^3 x = \left(\dfrac{e^{ix} - e^{-ix}}{2i} \right)^3 = \dfrac{e^{3ix} - e^{-3ix}}{-8i} + \dfrac{3(e^{ix} - e^{-ix})}{8i}$

$\qquad = -\frac{1}{4} \sin 3x + \frac{3}{4} \sin x.$

Comparing $Q(x)$, which is the right side of (c), with (b), we see that $Q(x)$ contains a term, which, ignoring constant coefficients, is x^0 times a term $\sin x$ in y_c. Hence the trial solution y_p must be of the form

(d) $\qquad y_p = A \sin 3x + B \cos 3x + Cx \sin x + Dx \cos x.$

Successive derivatives of (d) are

(e) $\qquad y_p{}' = 3A \cos 3x - 3B \sin 3x + Cx \cos x + C \sin x$

$\qquad\qquad - Dx \sin x + D \cos x.$

$\qquad y_p{}'' = -9A \sin 3x - 9B \cos 3x - Cx \sin x + 2C \cos x$

$\qquad\qquad - Dx \cos x - 2D \sin x.$

Substituting (c), (d), (e) in (a), we obtain

(f) $\quad -8A \sin 3x - 8B \cos 3x + 2C \cos x - 2D \sin x$

$\qquad\qquad\qquad = -\frac{1}{4} \sin 3x + \frac{3}{4} \sin x.$

Equating coefficients of like terms on each side of the equal sign, there results

(g) $\qquad -8A = -\frac{1}{4}, \qquad B = 0, \qquad C = 0, \qquad D = -\frac{3}{8}.$

From (g), we have

(h) $\qquad\qquad\qquad A = \frac{1}{32}, \qquad D = -\frac{3}{8}.$

Substituting these values in (d), we obtain

(i) $\qquad\qquad\qquad y_p = \frac{1}{32} \sin 3x - \frac{3}{8} x \cos x.$

A general solution of (a) is therefore the sum of (b) and (i).

Case 3. *This case is applicable only if* **both** *of the following conditions are fulfilled.*

A. The characteristic equation of the given differential equation (21.1) has an r multiple root.

B. $Q(x)$ contains a term which, ignoring constant coefficients, is x^k times a term $u(x)$ in y_c, where $u(x)$ was obtained from the r multiple root.

In this case, a particular solution y_p will be a linear combination of $x^{k+r} u(x)$ and all its linearly independent derivatives. If in addition $Q(x)$

contains terms which belong to Cases 1 and 2, then the proper terms called for by these cases must also be added to y_p.

Example 21.4. Find the general solution of

(a) $$y'' + 4y' + 4y = 3xe^{-2x}.$$

Solution. The complementary function of (a) is

(b) $$y_c = c_1 e^{-2x} + c_2 x e^{-2x}.$$

We observe first that the characteristic equation of (a), namely $m^2 + 4m + 4 = 0$, has a multiple root, $m = -2$. Secondly we observe that $Q(x)$ which is the right side of (a), contains the term xe^{-2x} which is x times the term e^{-2x} in y_c (or alternately xe^{-2x} of $Q(x)$ is x^0 times the term xe^{-2x} in y_c), and that this term in y_c came from a multiple root. Hence, by the above remarks under Case 3, $r = 2$, $k = 1$, and $r + k = 3$, (or alternately $r = 2$, $k = 0$ and $r + k = 2$). Therefore, y_p must be a linear combination of $x^3 e^{-2x}$ and all its linearly independent derivatives [or alternately $x^2(xe^{-2x})$, which yields the same $x^3 e^{-2x}$, and its derivatives]. In forming this linear combination, we may omit the functions e^{-2x} and xe^{-2x} since they already appear in y_c. Hence y_p must be of the form

(c) $$y_p = Ax^3 e^{-2x} + Bx^2 e^{-2x}.$$

The successive derivatives of (c) are

(d) $$y_p' = -2Ax^3 e^{-2x} + 3Ax^2 e^{-2x} - 2Bx^2 e^{-2x} + 2Bxe^{-2x},$$

(e) $$y_p'' = 4Ax^3 e^{-2x} - 12Ax^2 e^{-2x} + 6Axe^{-2x} + 4Bx^2 e^{-2x}$$
$$- 8Bxe^{-2x} + 2Be^{-2x}.$$

Substituting (c), (d), (e) in (a) and simplifying the result, we see that (c) will be a solution of (a) if

(f) $$6Axe^{-2x} + 2Be^{-2x} \equiv 3xe^{-2x}.$$

Equating coefficients of like terms on each side of the equal sign, there results

(g) $$A = \tfrac{1}{2}, \qquad B = 0.$$

Substituting these values in (c), we obtain

(h) $$y_p = \tfrac{1}{2}x^3 e^{-2x}.$$

The general solution of (a) is, therefore, the sum of (b) and (h), namely

(i) $$y = \tfrac{1}{2}x^3 e^{-2x} + c_1 e^{-2x} + c_2 x e^{-2x}.$$

Example 21.41. Find the general solution of

(a) $$y'' + 4y' + 4y = 3e^{-2x}.$$

Solution. The complementary function of (a) is

(b) $$y_c = c_1e^{-2x} + c_2xe^{-2x}.$$

The characteristic equation of (a), namely $m^2 + 4m + 4 = 0$, has a multiple root $m = -2$. The function $Q(x)$, which is the right side of (a), contains the term e^{-2x} which is x^0 times a term in y_c (or alternately x^{-1} times the term xe^{-2x} in y_c). Since this term in y_c came from a multiple root, this Case 3 applies. Hence by the remarks under Case 3, $r = 2$, $k = 0$, and $r + k = 2$ (or alternately $r = 2$, $k = -1$ and $r + k = 1$). Therefore y_p must be a linear combination of x^2e^{-2x} and all its linearly independent derivatives [or alternately $x(xe^{-2x})$, which yields the same x^2e^{-2x}]. In forming this linear combination, we may omit the terms e^{-2x} and xe^{-2x} since they already appear in y_c. Hence y_p must be of the form

(c) $$y_p = Ax^2e^{-2x}, \quad y_p' = 2Axe^{-2x} - 2Ax^2e^{-2x},$$
$$y_p'' = 2Ae^{-2x} - 8Axe^{-2x} + 4Ax^2e^{-2x}.$$

Substituting (c) in (a) and simplifying the result, we see that y_p will be a solution of (a) if

(d) $$2Ae^{-2x} = 3e^{-2x}, \quad A = \tfrac{3}{2}.$$

Substituting this value of A in the first equation of (c), we obtain

(e) $$y_p = \tfrac{3}{2}x^2e^{-2x}.$$

The general solution of (a) is therefore the sum of (b) and (e).

Comment 21.42. The function which we have labeled y_p, since it does not contain arbitrary constants, has been correctly called, by our Definition 4.66, a particular solution of (21.1). There are, of course, infinitely many other particular solutions of the differential equation, one for each set of values of the arbitrary constants in the y_c part of the general solution $y = y_c + y_p$. These constants are determined in the usual way by inserting the initial conditions in the *general* solution y. Do not confuse, therefore, a particular solution obtained by the method of undetermined coefficients and the particular solution which will satisfy given initial conditions. See example below.

Example 21.43. Find the particular solution of

(a) $$y'' - 3y' + 2y = 6e^{-x}$$

for which $y(0) = 1$, $y'(0) = 2$.

Solution. By the methods outlined previously, verify that

(b) $$y_c = c_1 e^x + c_2 e^{2x}, \quad y_p = e^{-x}.$$

However this y_p is not the particular solution which satisfies the initial conditions. To find it, we must first write the general solution, and then substitute the initial conditions in it and in its derivative. The general solution of (a) and its derivative are, by (b),

(c) $$y = c_1 e^x + c_2 e^{2x} + e^{-x}, \quad y' = c_1 e^x + 2c_2 e^{2x} - e^{-x}.$$

Substituting in (c) the initial conditions $x = 0$, $y = 1$, $y' = 2$, we obtain

(d) $$1 = c_1 + c_2 + 1,$$
$$2 = c_1 + 2c_2 - 1,$$

whose solutions are $c_1 = -3$, $c_2 = 3$. Hence the particular solution of (a) which satisfies the given initial conditions is, by (c) and these values of c_1, c_2,

(e) $$y = 3e^{2x} - 3e^x + e^{-x}.$$

LESSON 21B. Solution by the Use of Complex Variables. There is another way of solving certain types of nonhomogeneous linear equations with constant coefficients. If in

(21.5) $$a_n y^{(n)} + a_{n-1} y^{(n-1)} + \cdots + a_1 y' + a_0 y = Q(x), \quad a_n \neq 0,$$

the a's are real, $Q(x)$ a complex-valued function (i.e., a function which can take on complex values) and $y_p(x)$ is a solution of (21.5), then (see Exercise 19,7):

1. The real part of y_p is a solution of (21.5) with $Q(x)$ replaced by its real part.
2. The imaginary part of y_p is a solution of (21.5) with $Q(x)$ replaced by its imaginary part.

Remark. Statements 1 and 2 above would still be valid if the coefficients in (21.5) were real, continuous functions of x instead of constants.

Example 21.51. Find a particular solution of

(a) $$y'' - 3y' + 2y = \sin x.$$

Solution. Instead of solving (a), let us solve the differential equation

(b) $$y'' - 3y' + 2y = e^{ix}.$$

By (18.82), $e^{ix} = \cos x + i \sin x$. Its imaginary part is, therefore, $\sin x$. Hence by 2 above, the imaginary part of a particular solution y_p of (b) will be a solution of (a). A particular solution of (b), using the method of undetermined coefficients is [take for the trial solution $y_p = (A + Bi)e^{ix}$],

(c)
$$y_p = \tfrac{1}{10}e^{ix} + \tfrac{3}{10}ie^{ix}$$

$$= \tfrac{1}{10}(\cos x + i \sin x) + \frac{3i}{10}(\cos x + i \sin x)$$

$$= \tfrac{1}{10}\cos x - \tfrac{3}{10}\sin x + i(\tfrac{1}{10}\sin x + \tfrac{3}{10}\cos x).$$

The imaginary part of the solution y_p is $\tfrac{1}{10}\sin x + \tfrac{3}{10}\cos x$. Hence a particular solution of (a) is

(d)
$$y_p = \tfrac{1}{10}\sin x + \tfrac{3}{10}\cos x.$$

Question. What would a particular solution of (a) be, if in it $\sin x$ were replaced by $\cos x$? [*Ans.* The real part of y_p, namely $y_p = \tfrac{1}{10}\cos x - \tfrac{3}{10}\sin x$.]

EXERCISE 21

1. Prove that any term which is in the complementary function y_c need not be included in the trial solution y_p. (*Hint.* Show that the coefficients of this term will always add to zero.)
2. Prove that if $F(x)$ is a function with a finite number of linearly independent derivatives, i.e., if $F^{(n)}(x)$, $F^{(n-1)}(x)$, $\cdots$, $F'(x)$, $F(x)$ are linearly independent functions, where n is a finite number, then $F(x)$ consists only of such terms as a, x^k, e^{ax}, $\sin ax$, $\cos ax$, and combinations of such terms, where a is a constant and k is a positive integer. *Hint.* Set the linear combination of these functions equal to zero, i.e., set

$$C_n F^{(n)}(x) + C_{n-1} F^{(n-1)}(x) + \cdots + C_1 F'(x) + C_0 F(x) = 0,$$

where the C's are not all zero, and then show, by Lesson 20, that the only functions $F(x)$ that can satisfy this equation are those stated.

Find the general solution of each of the following equations.

3. $y'' + 3y' + 2y = 4$.
4. $y'' + 3y' + 2y = 12e^x$.
5. $y'' + 3y' + 2y = e^{ix}$.
6. $y'' + 3y' + 2y = \sin x$.
7. $y'' + 3y' + 2y = \cos x$.
8. $y'' + 3y' + 2y = 8 + 6e^x + 2\sin x$.
9. $y'' + y' + y = x^2$.

10. $y'' - 2y' - 8y = 9xe^x + 10e^{-x}$.
11. $y'' - 3y' = 2e^{2x}\sin x$.
12. $y^{(4)} - 2y'' + y = x - \sin x$.
13. $y'' + y' = x^2 + 2x$.
14. $y'' + y' = x + \sin 2x$.
15. $y'' + y = 4x\sin x$.
16. $y'' + 4y = x\sin 2x$.

17. $y'' + 2y' + y = x^2 e^{-x}$.
18. $y''' + 3y'' + 3y' + y = 2e^{-x} - x^2 e^{-x}$.
19. $y'' + 3y' + 2y = e^{-2x} + x^2$.
20. $y'' - 3y' + 2y = xe^{-x}$.
21. $y'' + y' - 6y = x + e^{2x}$.
22. $y'' + y = \sin x + e^{-x}$.

23. $y''' - 3y'' + 3y' - y = e^x$.
24. $y'' + y = \sin^2 x$. *Hint.* $\sin^2 x = \tfrac{1}{2} - \tfrac{1}{2}\cos 2x$.
25. $y''' - y' = e^{2x}\sin^2 x$.
26. $y^{(5)} + 2y''' + y' = 2x + \sin x + \cos x$. *Hint.* Solve $y^{(5)} + 2y''' + y' = 2x + e^{ix}$; see Example 21.51.
27. $y'' + y = \sin 2x \sin x$. *Hint.* $\sin 2x \sin x = \tfrac{1}{2}\cos x - \tfrac{1}{2}\cos 3x$.

For each of the following equations, find a particular solution which satisfies the given initial conditions.

28. $y'' - 5y' - 6y = e^{3x}$, $y(0) = 2$, $y'(0) = 1$.

29. $y'' - y' - 2y = 5 \sin x$, $y(0) = 1$, $y'(0) = -1$.

30. $y''' - 2y'' + y' = 2e^x + 2x$, $y(0) = 0$, $y'(0) = 0$, $y''(0) = 0$.

31. $y'' + 9y = 8 \cos x$, $y(\pi/2) = -1$, $y'(\pi/2) = 1$.

32. $y'' - 5y' + 6y = e^x(2x - 3)$, $y(0) = 1$, $y'(0) = 3$.

33. $y'' - 3y' + 2y = e^{-x}$, $y(0) = 1$, $y'(0) = -1$.

ANSWERS 21

3. $y = c_1 e^{-2x} + c_2 e^{-x} + 2$.

4. $y = c_1 e^{-2x} + c_2 e^{-x} + 2e^x$.

5. $y = c_1 e^{-2x} + c_2 e^{-x} + \frac{1}{10}(e^{ix} - 3ie^{ix})$.

6. $y = c_1 e^{-2x} + c_2 e^{-x} + \frac{1}{10}(\sin x - 3 \cos x)$.

7. $y = c_1 e^{-2x} + c_2 e^{-x} + \frac{1}{10}(3 \sin x + \cos x)$.

8. $y = c_1 e^{-2x} + c_2 e^{-x} + 4 + e^x + \frac{1}{5}(\sin x - 3 \cos x)$.

9. $y = e^{-x/2}\left(c_1 \cos \frac{\sqrt{3}}{2} x + c_2 \sin \frac{\sqrt{3}}{2} x\right) + x^2 - 2x$.

10. $y = c_1 e^{4x} + c_2 e^{-2x} - xe^x - 2e^{-x}$.

11. $y = c_1 + c_2 e^{3x} - \frac{e^{2x}}{5}(3 \sin x + \cos x)$.

12. $y = (c_1 + c_2 x)e^x + (c_3 + c_4 x)e^{-x} + x - \frac{\sin x}{4}$.

13. $y = c_1 + c_2 e^{-x} + \frac{x^3}{3}$.

14. $y = c_1 + c_2 e^{-x} + \frac{x^2}{2} - x - \frac{1}{10}(2 \sin 2x + \cos 2x)$.

15. $y = c_1 \cos x + c_2 \sin x - x(x \cos x - \sin x)$.

16. $y = c_1 \cos 2x + c_2 \sin 2x - \frac{x}{16}(2x \cos 2x - \sin 2x)$.

17. $y = c_1 e^{-x} + c_2 x e^{-x} + \frac{x^4 e^{-x}}{12}$.

18. $y = c_1 e^{-x} + c_2 x e^{-x} + c_3 x^2 e^{-x} + \frac{x^3 e^{-x}}{60}(20 - x^2)$.

19. $y = c_1 e^{-2x} + c_2 e^{-x} + \frac{x^2}{2} - \frac{3x}{2} + \frac{7}{4} - xe^{-2x}$.

20. $y = c_1 e^{2x} + c_2 e^x + \frac{1}{36}(6xe^{-x} + 5e^{-x})$.

21. $y = c_1 e^{-3x} + c_2 e^{2x} - \frac{x}{6} - \frac{1}{36} + \frac{xe^{2x}}{5}$.

22. $y = c_1 \cos x + c_2 \sin x - \frac{x}{2} \cos x + \frac{1}{2}e^{-x}$.

23. $y = \left(c_1 + c_2 x + c_3 x^2 + \frac{x^3}{6}\right)e^x$

24. $y = c_1 \cos x + c_2 \sin x + \frac{1}{2} + \frac{\cos 2x}{6}$.

25. $y = c_1 + c_2 e^x + c_3 e^{-x} + \left(\frac{1}{12} + \frac{9 \cos 2x - 7 \sin 2x}{520}\right)e^{2x}$.

26. $y = c_1 + c_2 \sin x + c_3 \cos x + c_4 x \sin x + c_5 x \cos x + x^2$
$\quad + \dfrac{x^2}{8}(\cos x - \sin x).$

27. $y = c_1 \cos x + c_2 \sin x + \dfrac{x}{4}\sin x + \dfrac{\cos 3x}{16}.$

28. $y = \frac{10}{21}e^{6x} + \frac{45}{28}e^{-x} - \frac{1}{12}e^{3x}.$

29. $y = \frac{1}{3}e^{2x} + \frac{1}{6}e^{-x} - \frac{3}{2}\sin x + \frac{1}{2}\cos x.$

30. $y = x^2 + 4x + 4 + (x^2 - 4)e^x.$

31. $y = \cos x + \frac{2}{3}\cos 3x + \sin 3x.$

32. $y = e^{2x} + xe^x.$

33. $6y = -10e^{2x} + 15e^x + e^{-x}.$

LESSON 22. Solution of the Nonhomogeneous Linear Differential Equation by the Method of Variation of Parameters.

LESSON 22A. Introductory Remarks. In the previous lessons of this chapter, we showed how to solve the linear differential equation

$$(22.1) \quad a_n y^{(n)} + a_{n-1} y^{(n-1)} + \cdots + a_1 y' + a_0 y = Q(x), \quad a_n \neq 0,$$

where:

1. The coefficients are constants.
2. $Q(x)$ is a function which has a finite number of linearly independent derivatives.

You may be wondering whether either or both of these restrictions may be removed.

In regard to the first restriction there are very few types of linear equations with nonconstant coefficients whose solutions can be expressed in terms of elementary functions and for which standard methods of obtaining them, if they do exist, are available. In Lesson 23, we shall describe a method by which a general solution of a second order linear differential equation with nonconstant coefficients can be found provided *one* solution is known. Again therefore, the equation must be of a special type so that the needed one solution can be discovered.

As for the second restriction, it is possible to solve (22.1) even when $Q(x)$ has an infinite number of linearly independent derivatives. The method used is known by the name of "variation of parameters" and is discussed below.

LESSON 22B. The Method of Variation of Parameters. For convenience and clarity, we restrict our attention to the second order linear equation with constant coefficients,

$$(22.2) \quad a_2 y'' + a_1 y' + a_0 y = Q(x), \quad a_2 \neq 0,$$

where $Q(x)$ is a continuous function of x on an interval I and is $\neq 0$ on I.

If the two linearly independent solutions of the related homogeneous equation

$$(22.21) \qquad a_2 y'' + a_1 y' + a_0 y = 0$$

are known, then it is possible to find a particular solution of (22.2) by a method called **variation of parameters,** even when $Q(x)$ contains terms whose linearly independent derivatives are infinite in number. In describing this method, we assume therefore, that you would have no trouble in finding the two linearly independent solutions y_1 and y_2 of (22.21). With them we form the equation

$$(22.22) \qquad y_p(x) = u_1(x) y_1(x) + u_2(x) y_2(x),$$

where u_1 and u_2 are *unknown* functions of x which are to be determined.

The successive derivatives of (22.22) are

$$(22.23) \qquad \begin{aligned} y_p' &= u_1 y_1' + u_1' y_1 + u_2' y_2 + u_2 y_2' \\ &= (u_1 y_1' + u_2 y_2') + (u_1' y_1 + u_2' y_2), \end{aligned}$$

$$(22.24) \quad y_p'' = (u_1 y_1'' + u_2 y_2'') + (u_1' y_1' + u_2' y_2') + (u_1' y_1 + u_2' y_2)'.$$

Substituting the above values of y_p, y_p', and y_p'' in (22.2), we see that y_p will be a solution of (22.2) if

$$(22.25) \qquad \begin{aligned} a_2 (u_1 y_1'' + u_2 y_2'') &+ a_2 (u_1' y_1' + u_2' y_2') \\ &+ a_2 (u_1' y_1 + u_2' y_2)' + a_1 (u_1 y_1' + u_2 y_2') \\ &+ a_1 (u_1' y_1 + u_2' y_2) + a_0 (u_1 y_1 + u_2 y_2) = Q(x). \end{aligned}$$

This equation can be written as

$$(22.26) \qquad \begin{aligned} u_1 (a_2 y_1'' + a_1 y_1' + a_0 y_1) &+ u_2 (a_2 y_2'' + a_1 y_2' + a_0 y_2) \\ &+ a_2 (u_1' y_1' + u_2' y_2') + a_2 (u_1' y_1 + u_2' y_2)' \\ &+ a_1 (u_1' y_1 + u_2' y_2) = Q(x). \end{aligned}$$

Since y_1 and y_2 are assumed to be solutions of (22.21), the quantities in the first two parentheses in (22.26) equal zero. The remaining three terms will equal $Q(x)$ if we choose u_1 and u_2 such that

$$(22.27) \qquad u_1' y_1 + u_2' y_2 = 0,$$

$$u_1' y_1' + u_2' y_2' = \frac{Q(x)}{a_2}.$$

The pair of equations in (22.27) can be solved for u_1' and u_2' in terms of the other functions by the ordinary algebraic methods with which you are familiar. Or, if you are acquainted with determinants (see Lessons 31 and 63), the solutions of (22.27) are

$$(22.28) \qquad u_1' = \frac{\begin{vmatrix} 0 & y_2 \\ \dfrac{Q(x)}{a_2} & y_2' \end{vmatrix}}{\begin{vmatrix} y_1 & y_2 \\ y_1' & y_2' \end{vmatrix}}, \qquad u_2' = \frac{\begin{vmatrix} y_1 & 0 \\ y_1' & \dfrac{Q(x)}{a_2} \end{vmatrix}}{\begin{vmatrix} y_1 & y_2 \\ y_1' & y_2' \end{vmatrix}}.$$

These equations (22.28) will always give solutions for u_1' and u_2' provided the denominator determinant $\neq 0$. We shall prove in Lesson 64 that, if y_1 and y_2 are linearly independent solutions of (22.21), then this denominator is never zero.

Integration of (22.28) will enable us to determine u_1 and u_2. The substitution of these values in (22.22) will give a particular solution y_p of (22.2).

Comment 22.29. Since we seek a particular solution y_p, constants of integration may be omitted when integrating u_1' and u_2'.

Comment 22.291. If the nonhomogeneous linear differential equation is of order $n > 2$, then it can be shown that

$$(22.3) \qquad y_p = u_1 y_1 + u_2 y_2 + \cdots + u_n y_n$$

will be a particular solution of the equation, where $y_1, y_2, \cdots, y_n$ are the n independent solutions of its related homogeneous equation, and $u_1', u_2', \cdots, u_n'$ are the functions obtained by solving simultaneously the following set of equations:

$$(22.31) \qquad u_1' y_1 + u_2' y_2 + \cdots + u_n' y_n = 0,$$
$$u_1' y_1' + u_2' y_2' + \cdots + u_n' y_n' = 0,$$
$$\cdots \cdots \cdots \cdots \cdots \cdots \cdots \cdots \cdots \cdots \cdots \cdots$$
$$u_1' y_1^{(n-1)} + u_2' y_2^{(n-1)} + \cdots + u_n' y_n^{(n-1)} = \frac{Q(x)}{a_n}.$$

In (22.31), a_n is the coefficient of $y^{(n)}$ in the given differential equation. Again we remark that since we seek a particular solution y_p, arbitrary constants may be omitted when integrating $u_1', u_2', \cdots, u_n'$ to find $u_1, u_2, \cdots, u_n$.

Comment 22.32. The method of variation of parameters can also be used when $Q(x)$ has a finite number of linearly independent derivatives. In the above description of this method, the only requirement placed on $Q(x)$ is that it be a continuous function of x. You will find however, that if $Q(x)$ has a finite number of linearly independent derivatives, the method of undetermined coefficients explained in Lesson 21 will usually be easier to use. To show you, however, that the method of variation of parameters will also work in this case, we have included in the examples below one which was solved previously by the method of undetermined coefficients.

Comment 22.33. The proof we gave above to arrive at (22.27) would also have been valid if the constant coefficients in (22.2) and (22.21) were replaced by continuous functions of x. The method of variation of parameters can be used, therefore, to find a particular solution of the equation

$$(22.34) \qquad f_2(x)y'' + f_1(x)y' + f_0(x)y = Q(x),$$

provided we know two independent solutions y_1 and y_2 of the related homogeneous equation

$$(22.35) \qquad f_2(x)y'' + f_1(x)y' + f_0(x)y = 0.$$

Example 22.4. Find the general solution of

(a) $$y'' - 3y' + 2y = \sin e^{-x}.$$

(NOTE. This equation cannot be solved by the method of undetermined coefficients explained in Lesson 21. Here $Q(x) = \sin e^{-x}$, which has an infinite number of linearly independent derivatives.)

Solution. The roots of the characteristic equation of (a) are $m = 1$, $m = 2$. Hence the complementary function of (a) is

(b) $$y_c = c_1 e^x + c_2 e^{2x}.$$

The two linearly independent solutions of the related homogeneous equation of (a) are therefore

(c) $$y_1 = e^x \quad \text{and} \quad y_2 = e^{2x}.$$

Substituting these values and their derivatives in (22.27), we obtain, with $a_2 = 1$, (remember a_2 is the coefficient of y'')

(d) $$u_1'e^x + u_2'e^{2x} = 0,$$
$$u_1'e^x + u_2'(2e^{2x}) = \sin e^{-x}.$$

Solving (d) for u_1' and u_2', there results

(e) $$u_1' = -e^{-x} \sin e^{-x}, \qquad u_2' = e^{-2x} \sin e^{-x}.$$

Therefore

(f) $$u_1 = \int \sin e^{-x}(-e^{-x})\, dx, \qquad u_2 = -\int e^{-x} \sin e^{-x}(-e^{-x})\, dx.$$

Hence (in the integrands, let $u = e^{-x}$, $du = -e^{-x}\,dx$)

(g) $\qquad u_1 = -\cos e^{-x}, \qquad u_2 = \sin e^{-x} \mid e^{-x} \cos e^{-x}.$

Substituting (c) and (g) in (22.22), we obtain

(h) $\qquad y_p = -(\cos e^{-x})e^x + (e^{-x} \cos e^{-x} - \sin e^{-x})e^{2x}$
$\qquad\qquad = -e^{2x} \sin e^{-x}.$

Combining (b) and (h) gives

(i) $\qquad\qquad y = c_1 e^x + c_2 e^{2x} - e^{2x} \sin e^{-x},$

which is the general solution of (a).

Example 22.41.   Find the general solution of

(a) $\qquad\qquad y'' + 4y' + 4y = 3xe^{-2x}.$

(NOTE. We have already solved this example by the method of undetermined coefficients. See Example 21.4.)

Solution.   The complementary function of (a) is

(b) $\qquad\qquad y_c = c_1 e^{-2x} + c_2 x e^{-2x}.$

Therefore the two independent solutions of the related homogeneous equation of (a) are

(c) $\qquad\qquad y_1 = e^{-2x}, \qquad y_2 = xe^{-2x}.$

Substituting these values and their derivatives in (22.27), we obtain

(d) $\qquad\qquad u_1' e^{-2x} + u_2'(xe^{-2x}) = 0,$
$\qquad u_1'(-2e^{-2x}) + u_2'(-2xe^{-2x} + e^{-2x}) = 3xe^{-2x}.$

Solving (d) for u_1' and u_2', there results

(e) $\qquad\qquad u_1' = -3x^2, \qquad u_2' = 3x.$

Hence,

(f) $\qquad\qquad u_1 = -x^3, \qquad u_2 = \tfrac{3}{2}x^2.$

Substituting (c) and (f) in (22.22), we have

(g) $\qquad\qquad y_p = -x^3 e^{-2x} + \dfrac{3x^2}{2}(xe^{-2x}) = \tfrac{1}{2}e^{-2x}x^3.$

Combining (b) and (g) we obtain for the general solution of (a)

(h) $\qquad\qquad y = c_1 e^{-2x} + c_2 x e^{-2x} + \tfrac{1}{2}e^{-2x}x^3,$

which is the same as the one we found previously in Example 21.4.

Remark. The method of variation of parameters has one advantage over the method of undetermined coefficients. In the variation of parameter method, there is no need to concern oneself with the different cases encountered in the method of undetermined coefficients. In the above example, the characteristic equation has a repeated root and $Q(x)$ contains a term xe^{-2x}, which is x times the term e^{-2x} of y_c.

Example 22.42. Find the general solution of

(a) $$y'' + y = \tan x, \qquad -\frac{\pi}{2} < x < \frac{\pi}{2}.$$

(Note. This equation cannot be solved by the method of Lesson 21. Here $Q(x) = \tan x$ which has an infinite number of linearly independent derivatives.)

Solution. The complementary function of (a) is

(b) $$y_c = c_1 \cos x + c_2 \sin x.$$

The two independent solutions of the related homogeneous equation of (a) are therefore

(c) $$y_1 = \cos x, \qquad y_2 = \sin x.$$

Substituting these values and their derivatives in (22.27), we obtain

(d) $$u_1' \cos x + u_2' \sin x = 0,$$
$$u_1'(-\sin x) + u_2' \cos x = \tan x.$$

Solving (d) for u_1' and u_2', there results

(e) $$u_1' = -\frac{\sin^2 x}{\cos x} = -\frac{1 - \cos^2 x}{\cos x} = -\sec x + \cos x, \quad u_2' = \sin x.$$

Hence

(f) $$u_1 = -\log(\sec x + \tan x) + \sin x, \quad u_2 = -\cos x.$$

Substituting (c) and (f) in (22.22), we have

(g) $$y_p = -\cos x \log(\sec x + \tan x) + \sin x \cos x - \sin x \cos x$$
$$= -\cos x \log(\sec x + \tan x), \quad -\frac{\pi}{2} < x < \frac{\pi}{2}.$$

Combining (b) and (g), we obtain for the general solution of (a),

(h)
$$y = c_1 \cos x + c_2 \sin x - \cos x \log(\sec x + \tan x), \quad -\frac{\pi}{2} < x < \frac{\pi}{2}.$$

In Fig. 22.43, we have plotted a graph, using polar coordinates, with x as the polar angle and y as the radius vector, of a particular solution of (a) obtained by setting $c_1 = c_2 = 0$ in (h).

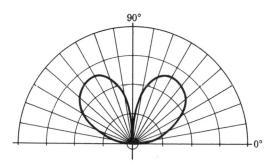

Figure 22.43

Example 22.44. If

(a) $$y_1 = x \quad \text{and} \quad y_2 = x^{-1}$$

are two solutions of the differential equation

(b) $$x^2 y'' + xy' - y = 0,$$

find the general solution of

(c) $$x^2 y'' + xy' - y = x, \quad x \neq 0.$$

Solution. (See Comment 22.33.) Substituting the given two solutions and their derivatives in (22.27), it becomes

(d) $$xu_1' + x^{-1}u_2' = 0,$$
$$u_1' - x^{-2}u_2' = \frac{x}{x^2} = \frac{1}{x}.$$

Solutions of (d) for u_1' and u_2', by any method you wish to choose, are

(e) $$u_1' = \frac{1}{2x}, \qquad u_2' = -\frac{x}{2}.$$

Hence

(f) $$u_1 = \frac{1}{2} \log x, \qquad u_2 = -\frac{x^2}{4}.$$

Substituting (a) and (f) in (22.22) gives

(g) $$y_p = \frac{x}{2} \log x - \frac{x}{4}.$$

Combining (g) with (a), we obtain for the general solution of (c),

(h) $$y = c_1'x + c_2x^{-1} + \frac{x}{2}\log x - \frac{x}{4},$$

which simplifies to

(i) $$y = c_1x + c_2x^{-1} + \frac{x}{2}\log x.$$

EXERCISE 22

Use the method of variation of parameters to find the general solution of each of the following equations.

1. $y'' + y = \sec x.$
2. $y'' + y = \cot x.$
3. $y'' + y = \sec^3 x.$
4. $y'' - y = \sin^2 x.$
5. $y'' + y = \sin^2 x.$
6. $y'' + 3y' + 2y = 12e^x.$
7. $y'' + 2y' + y = x^2e^{-x}.$
8. $y'' + y = 4x\sin x.$

9. $y'' + 2y' + y = e^{-x}\log x.$
10. $y'' + y = \csc x.$
11. $y'' + y = \tan^2 x.$
12. $y'' + 2y' + y = e^{-x}/x.$
13. $y'' + y = \sec x \csc x.$
14. $y'' - 2y' + y = e^x\log x.$
15. $y'' - 3y' + 2y = \cos e^{-x}.$

Use the method of variation of parameters to find the general solution of each of the following equations. Solutions for the related homogeneous equation are shown alongside each equation.

16. $x^2y'' - xy' + y = x;$ $y_1 = x,\ y_2 = x\log x.$

17. $y'' - \dfrac{2}{x}y' + \dfrac{2}{x^2}y = x\log x;$ $y_1 = x,\ y_2 = x^2.$

18. $x^2y'' + xy' - 4y = x^3;$ $y_1 = x^2,\ y_2 = \dfrac{1}{x^2}.$

19. $x^2y'' + xy' - y = x^2e^{-x};$ $y_1 = x,\ y_2 = x^{-1}.$

20. $2x^2y'' + 3xy' - y = x^{-1};$ $y_1 = x^{1/2},\ y_2 = x^{-1}.$

ANSWERS 22

1. $y = c_1\cos x + c_2\sin x + x\sin x + \cos x\log\cos x.$
2. $y = c_1\cos x + c_2\sin x + \sin x\log(\csc x - \cot x).$
3. $y = c_1\cos x + c_2\sin x + \frac{1}{2}\tan x\sin x.$
4. $y = c_1e^x + c_2e^{-x} - \frac{1}{5}\sin^2 x - \frac{2}{5}.$
5. $y = c_1\cos x + c_2\sin x + \frac{1}{6}\cos 2x + \frac{1}{2}.$
6. $y = c_1e^{-2x} + c_2e^{-x} + 2e^x.$
7. $y = c_1e^{-x} + c_2xe^{-x} + \dfrac{x^4e^{-x}}{12}.$
8. $y = c_1\cos x + c_2\sin x - x^2\cos x + x\sin x.$
9. $y = \frac{1}{4}x^2e^{-x}(2\log x - 3) + (c_1 + c_2x)e^{-x}.$
10. $y = c_1\cos x + c_2\sin x + \sin x\log\sin x - x\cos x.$
11. $y = c_1\cos x + c_2\sin x + \sin x\log(\sec x + \tan x) - 2.$

12. $y = e^{-x}(c_1 + c_2 x + x \log x)$.

13. $y = c_1 \cos x + c_2 \sin x + \sin x \log (\csc x - \cot x) -$
$\cos x \log (\sec x + \tan x)$.

14. $y = (c_1 + c_2 x)e^x + x^2 e^x (\frac{1}{2} \log x - \frac{3}{4})$.

15. $y = c_1 e^x + c_2 e^{2x} - e^{2x} \cos e^{-x}$.

16. $y = c_1 x + c_2 x \log x + \dfrac{x}{2}(\log x)^2$.

17. $y = c_1 x + c_2 x^2 + \frac{1}{2}x^3 \log x - \frac{3}{4}x^3$.

18. $y = c_1 x^2 + c_2 x^{-2} + \dfrac{x^3}{5}$.

19. $y = c_1 x + c_2 x^{-1} + e^{-x}(1 + x^{-1})$.

20. $y = c_1 x^{1/2} + c_2 x^{-1} - \frac{1}{3}x^{-1} \log x$.

LESSON 23.   Solution of the Linear Differential Equation with Nonconstant Coefficients. Reduction of Order Method.

LESSON 23A.   Introductory Remarks.   We are at last ready to examine the general linear differential equation

(23.1) $f_n(x)y^{(n)} + f_{n-1}(x)y^{(n-1)} + \cdots + f_1(x)y' + f_0(x)y = Q(x)$,

and its related homogeneous equation

(23.11) $f_n(x)y^{(n)} + f_{n-1}(x)y^{(n-1)} + \cdots + f_1(x)y' + f_0(x)y = 0$,

where $f_0(x), f_1(x), \cdots, f_n(x), Q(x)$ are each continuous functions of x on a common interval I and $f_n(x) \neq 0$ when x is in I.

We have already remarked that in most cases the solutions of (23.1) will not be expressible in terms of elementary functions. However, even when they are, no standard method is known of finding them, as is the case when the coefficients in (23.1) or (23.11) are constants, unless the coefficient functions $f_n(x)$ are of a very special type; see, for example, Exercise 23,18. For an unrestricted nth order equation (23.11) that has a solution expressible in terms of elementary functions, the best you can hope for by the use of a standard method is to find *one* independent solution, if the *other $n-1$ independent solutions* are known. And for (23.1), the best you can obtain from a standard method is to find one independent solution of (23.11) and a particular solution of (23.1), again provided the other $n-1$ independent solutions of (23.11) are known.

You can see, therefore, that even in showing you a standard method for finding a general solution of only the second order equation

(23.12) $f_2(x)y'' + f_1(x)y' + f_0(x)y = Q(x)$,

it is essential that the functions $f_0(x), f_1(x), f_2(x)$ be of such a character that the needed first solution of its related homogeneous equation can be discovered.

Comment 23.13. The function $y \equiv 0$ always satisfies (23.11). Since this solution is of no value, it has been appropriately called the **trivial solution.**

LESSON 23B. Solution of the Linear Differential Equation with Nonconstant Coefficients by the Reduction of Order Method. As remarked in Lesson 23A, we assume that we have been able to find a nontrivial solution y_1 of the homogeneous equation

(23.14) $\qquad f_2(x)y'' + f_1(x)y' + f_0(x)y = 0.$

The method by which we shall obtain a second independent solution of (23.14), as well as a particular solution of the related nonhomogeneous equation

(23.15) $\qquad f_2(x)y'' + f_1(x)y' + f_0(x)y = Q(x),$

is called the **reduction of order** method.

Let $y_2(x)$ be a second solution of (23.14) and assume that it will have the form

(23.2) $\qquad y_2(x) = y_1(x) \int u(x)\, dx,$

where $u(x)$ is an unknown function of x which is to be determined. The derivatives of (23.2) are

(23.21) $\qquad y_2'(x) = y_1 u + y_1' \int u(x)\, dx,$

(23.22) $\qquad y_2''(x) = y_1 u' + y_1' u + y_1' u + y_1'' \int u(x)\, dx.$

Substituting the above values for y_2, y_2', and y_2'' in (23.14), we see, by Definition 3.4, that y_2 will be a solution of (23.14) if

(23.23) $\qquad f_2(x)\left[y_1 u' + 2y_1' u + y_1'' \int u(x)\, dx \right]$

$$+ f_1(x)\left[y_1 u + y_1' \int u(x)\, dx \right] + f_0(x)\left[y_1(x) \int u(x)\, dx \right] = 0.$$

We can rewrite this expression as

(23.24) $\qquad \left(f_2(x)y_1'' + f_1(x)y_1' + f_0(x)y_1 \right) \int u(x)\, dx$

$$+ f_2(x)y_1 u' + [2f_2(x)y_1' + f_1(x)y_1]u = 0.$$

Since we have assumed that y_1 is a solution of (23.14), the quantity in the first parenthesis of (23.24) is zero. Hence (23.24) reduces to

(23.25) $\qquad f_2(x)y_1 u' + [2f_2(x)y_1' + f_1(x)y_1]u = 0.$

Multiplying (23.25) by $dx/[uf_2(x)y_1]$, we obtain

(23.26) $$\frac{du}{u} + \frac{2\,dy_1}{y_1} = \frac{-f_1(x)}{f_2(x)}\,dx.$$

Integration of (23.26) gives

(23.27) $$\log u + 2\log y_1 = -\int \frac{f_1(x)}{f_2(x)}\,dx,$$

$$\log(uy_1{}^2) = -\int \frac{f_1(x)}{f_2(x)}\,dx,$$

$$uy_1{}^2 = e^{-\int \frac{f_1(x)}{f_2(x)}\,dx},$$

$$u = e^{-\int \frac{f_1(x)}{f_2(x)}\,dx}/y_1{}^2.$$

Substituting this value of u in (23.2), we obtain for the second solution of (23.12)

(23.28) $$y_2 = y_1 \int \frac{e^{-\int \frac{f_1(x)}{f_2(x)}\,dx}}{y_1{}^2}\,dx$$

We shall prove in Lesson 64, see Example 64.22, that this second solution y_2 is linearly independent of y_1.

Comment 23.29. We do not wish to imply that the integral in (23.28) will yield an elementary function. In most cases it will not, since in most cases, we repeat, the given differential equation does not have a solution which can be expressed in terms of elementary functions. Further the solution y_1 need not itself be an elementary function, although in the examples below and in the exercises, we have carefully chosen differential equations which have at least one solution expressible in terms of the elementary functions.

Example 23.291. Find the general solution of

(a) $$x^2 y'' + xy' - y = 0, \quad x \neq 0,$$

given that $y = x$ is a solution of (a).

Solution. Comparing (a) with (23.14), we see that $f_2(x) = x^2$, $f_1(x) = x$. Hence by (23.28) with $y_1 = x$, we obtain

(b) $$y_2 = x \int \frac{e^{-\int \frac{x}{x^2}\,dx}}{x^2}\,dx = x \int \frac{e^{-\log x}}{x^2}\,dx = x \int x^{-3}\,dx$$

$$= x\frac{x^{-2}}{-2} = -\frac{x^{-1}}{2}.$$

Hence the general solution of (a) is

(c) $$y = c_1 x + c_2 x^{-1}.$$

Comment 23.292. Although formulas are useful, it is not necessary to memorize (23.28). We shall now solve the same Example 23.29 by using the substitution (23.2). With $y_1 = x$, (23.2) becomes

(d) $$y_2 = x \int u(x)\, dx.$$

Differentiating (d) twice, we obtain

(e) $$y_2' = xu + \int u(x)\, dx, \qquad y_2'' = xu' + 2u.$$

Substituting (d) and (e) in (a), we have

(f) $$x^2(xu' + 2u) + x\left[xu + \int u(x)\, dx\right] - x\int u\, dx = 0,$$

which simplifies to the separable equation

(g) $$xu' + 3u = 0, \quad x \neq 0.$$

Its solution is

(h) $$u = x^{-3}.$$

Substituting (h) in (d), there results

(i) $$y_2 = x\int x^{-3}\, dx = x\,\frac{x^{-2}}{-2} = \frac{x^{-1}}{-2}.$$

Hence the general solution of (a) is, as found previously,

(j) $$y = c_1 x + c_2 x^{-1}.$$

The same substitution (23.2), namely

(23.3) $$y(x) = y_1(x)\int u(x)\, dx,$$

in the nonhomogeneous equation (23.15) will yield not only a second independent solution of (23.14) but also a particular solution of (23.15). As before, we differentiate (23.3) twice and substitute the values of y, y', and y'' in (23.15). The left side of the resulting equation will be exactly the same as (23.25) found previously. Its right side, however, will have $Q(x)$ in it instead of zero. Hence in place of (23.25), we would obtain

(23.31) $$f_2(x)y_1 u' + [2f_2(x)y_1' + f_1(x)y_1]u = Q(x),$$

an equation which is now linear in u and therefore solvable by the method of Lesson 11B. The substitution of this value of u in (23.3), will make $y(x)$ a solution of (23.15).

Comment 23.311. Two integrations will be involved in this procedure; one when solving for u in (23.31), the other when integrating u in (23.3). There will, therefore, be two constants of integration. If we include these two constants, we shall obtain not only a second independent solution $y_2(x)$ of (23.14) and a particular solution $y_p(x)$ of (23.15), but also the first solution $y_1(x)$ with which we started. Hence by this method the substitution (23.3) will give the general solution of (23.15).

Example 23.32. Given that $y = x$ is a solution of

$$x^2 y'' + xy' - y = 0, \quad x \neq 0,$$

find the general solution of

(a) $$x^2 y'' + xy' - y = x, \quad x \neq 0.$$

Solution. By (23.3), with one solution $y_1(x) = x$, we have

(b) $$y(x) = x \int u(x)\, dx.$$

Its derivatives are

(c) $$y'(x) = xu + \int u\, dx,$$

$$y''(x) = xu' + u + u.$$

Substituting (b) and (c) in (a) gives

(d) $$x^2(xu' + 2u) + x\left(xu + \int u\, dx\right) - x\int u\, dx = x,$$

which simplifies to

(e) $$xu' + 3u = x^{-1}, \qquad u' + \frac{3}{x}u = x^{-2}.$$

This equation is linear in u. By Lesson 11B, its integrating factor is $e^{\int 3x^{-1}dx} = e^{\log x^3} = x^3$. Hence, by (11.19),

(f) $$ux^3 = \int x\, dx = \frac{x^2}{2} + c_1',$$

$$u = \frac{x^{-1}}{2} + c_1' x^{-3}.$$

Substituting this value of u in (b), we obtain

(g) $\quad y(x) = x \int \left(\dfrac{x^{-1}}{2} + c_1' x^{-3} \right) dx = x \left[\tfrac{1}{2} \log x - c_1' \dfrac{x^{-2}}{2} + c_2 \right]$

$\qquad = \dfrac{x}{2} \log x - c_1' \dfrac{x^{-1}}{2} + c_2 x,$

which is equivalent to

(h) $\qquad\qquad y(x) = c_1 x^{-1} + c_2 x + \dfrac{x}{2} \log x.$

Comment 23.33. We could, if we had wished, have obtained (e) directly from (23.31). Comparing (a) with (23.15), we see that $f_2(x) = x^2$, $f_1(x) = x$ and $Q(x) = x$. The y_1 in this formula is the given solution x. Substituting these values in (23.31) will yield (e). Verify it.

EXERCISE 23

Use the reduction of order method to find the general solution of each of the following equations. One solution of the homogeneous equation is shown alongside each equation.

1. $x^2 y'' - xy' + y = 0,\quad y_1 = x.$

2. $y'' - \dfrac{2}{x} y' + \dfrac{2}{x^2} y = 0,\quad y_1 = x.$

3. $(2x^2 + 1)y'' - 4xy' + 4y = 0,\quad y_1 = x.$

4. $y'' + (x^2 - x)y' - (x - 1)y = 0,\quad y_1 = x.$

5. $y'' + \left(\dfrac{x}{2} - \dfrac{1}{x} \right)y' - y = 0,\quad y_1 = x^2.$

6. $2x^2 y'' + 3xy' - y = 0,\quad y_1 = x^{1/2}.$

7. $y'' + [f(x) - 1]y' - f(x)y = 0,\quad y_1 = e^x.$

8. $y'' + xf(x)y' - f(x)y = 0,\quad y_1 = x.$

9. $x^2 y'' - xy' + y = x,\quad y_1 = x.$

10. $y'' - \dfrac{2}{x} y' + \dfrac{2}{x^2} y = x \log x,\quad y_1 = x.$

11. $x^2 y'' + xy' - 4y = x^3,\quad y_1 = x^2.$

12. $x^2 y'' + xy' - y = x^2 e^{-x},\quad y_1 = x.$

13. $2x^2 y'' + 3xy' - y = \dfrac{1}{x},\quad y_1 = x^{1/2}.$

14. $x^2 y'' - 2y = 2x,\quad y_1 = x^2.$

15. $y'' + (x^2 - 1)y' - x^2 y = 0,\quad y_1 = e^x.$

16. $x^2 y'' - 2y = 2x^2,\quad y_1 = x^2.$

17. $x^2 y'' + xy' - y = 1,\quad y_1 = x.$

18. The differential equation

(23.4) $\quad a_3(x - x_0)^3 y''' + a_2(x - x_0)^2 y'' + a_1(x - x_0)y' + a_0 y = Q(x),\quad x \neq x_0,$

where a_0, a_1, a_2, a_3, x_0 are constants, and $Q(x)$ is a continuous function of x, is known as an **Euler** equation. Prove that the substitution

$$x - x_0 = e^u, \quad u = \log (x - x_0),$$

will transform (23.4) into a linear equation with constant coefficients. *Hint.*

$$\frac{dy}{du} = \frac{dy}{dx}\frac{dx}{du} = (x - x_0)\frac{dy}{dx},$$

$$\frac{d^2y}{du^2} - \frac{dy}{du} = (x - x_0)^2 \frac{d^2y}{dx^2},$$

$$\frac{d^3y}{du^3} - 3\frac{d^2y}{du^2} + 2\frac{dy}{du} = (x - x_0)^3 \frac{d^3y}{dx^3}.$$

Use the substitution given in problem 18 above to solve equations 19–21.

19. $(x - 3)^2 y'' + (x - 3)y' + y = x, \quad x \neq 3$.
20. $x^2 y'' + xy' = 0, \quad x \neq 0$.
21. $x^3 y''' + 2x^2 y'' - xy' + y = x, \quad x \neq 0$.
22. Prove that the substitutions

$$y = e^{\int y_1(x)\,dx}, \; y' = y_1 e^{\int y_1\,dx}, \text{ and } y'' = y_1^2 e^{\int y_1\,dx} + y_1' e^{\int y_1\,dx}$$

will transform the linear equation

$$g_2(x)y'' + g_1(x)y' + g_0(x)y = 0$$

into the **Riccati** equation (see also Exercise 11,25)

$$y_1' = -\frac{g_0(x)}{g_2(x)} - \frac{g_1(x)}{g_2(x)}y_1 - y_1^2.$$

Hence if y_1 is a solution of the Riccati equation, $y = e^{\int y_1\,dx}$ is a solution of the linear equation.

23. Use the substitutions given in 22 to transform the linear equation $xy'' - y' - x^3 y = 0$ into a Riccati equation. Find, by trial, a solution of the Riccati equation; then find a solution of the linear equation. With one solution of the linear equation known, find its general solution.

24. Prove that the substitution

$$y = -\frac{1}{f_2(x)u}\frac{du}{dx}, \quad f_2(x) \neq 0,$$

$$y' = -\frac{1}{f_2 u}u'' + u'\left(\frac{f_2 u' + f_2' u}{f_2^2 u^2}\right),$$

will transform the Riccati equation,

$$y' = f_0(x) + f_1(x)y + f_2(x)y^2, \quad f_2(x) \neq 0,$$

into the second order linear equation

$$f_2(x)u'' - [f_2'(x) + f_1(x)f_2(x)]u' + f_0(x)[f_2(x)]^2 u = 0.$$

25. A second order *linear* differential equation

(23.5) $$f_2(x)y'' + f_1(x)y' + f_0(x)y = Q(x),$$

is said to be **exact** if it can be written as

(23.51) $$\frac{d}{dx}\left[M(x)\frac{dy}{dx} + N(x)y \right] = Q(x),$$

i.e., if its left side can be written as the derivative of a first order *linear* expression. A necessary and sufficient condition that the equation be exact is that

(23.52) $$\frac{d^2 f_2}{dx^2} - \frac{df_1}{dx} + f_0 \equiv 0.$$

If the equation is exact, then $M(x)$ and $N(x)$ of (23.51) are given by

(23.53) $$M(x) = f_2(x) \quad \text{and} \quad N(x) = f_1(x) - f_2'(x).$$

Show that the following equation is exact and solve.

$$(x^2 - 2x)y'' + 4(x - 1)y' + 2y = e^{2x}.$$

Ans. $(x^2 - 2x)y = \dfrac{e^{2x}}{4} + c_1 x + c_2.$

26. A function $h(x)$ is called an **integrating factor** of an inexact differential equation, if after multiplication of the equation by it, the resulting equation is exact. A necessary and sufficient condition that $h(x)$ be an integrating factor of the linear differential equation

(23.6) $$f_2(x)y'' + f_1(x)y'''f_0(x)y = Q(x)$$

is that it be a solution of the differential equation

(23.61) $$\frac{d^2}{dx^2}(f_2 h) - \frac{d}{dx}(f_1 h) + f_0 h = 0.$$

Find an integrating factor of the following equation, then multiply the equation by the integrating factor and use (23.52) to prove that the resulting equation is exact.

$$x^3 y'' - (2x^3 - 6x^2)y' - 3(x^3 + 2x^2 - 2x)y = 0.$$

Ans. $h(x) = e^x$; also $h(x) = e^{-3x}.$

ANSWERS 23

1. $y = c_1 x + c_2 x \log x.$

2. $y = c_1 x + c_2 x^2.$

3. $y = c_1 x + c_2(2x^2 - 1).$

4. $y = c_1 x + c_2 x \displaystyle\int \frac{e^{-\frac{x^3}{3} + \frac{x^2}{2}}}{x^2}\, dx.$

5. $y = c_1 x^2 + c_2 x^2 \displaystyle\int \frac{e^{-x^2/4}}{x^3}\, dx.$

6. $y = c_1 x^{1/2} + c_2 x^{-1}.$

7. $y = c_1 e^x + c_2 e^x \int e^{-x - \int f(x) dx} \, dx.$

8. $y = c_1 x + c_2 x \int \dfrac{e^{-\int x f(x) dx}}{x^2} \, .$

9. $y = c_1 x + c_2 x \log x + \dfrac{x}{2} (\log x)^2.$

10. $y = c_1 x + c_2 x^2 + \frac{1}{2} x^3 \log x - \frac{3}{4} x^3.$

11. $y = c_1 x^2 + c_2 x^{-2} + \dfrac{x^3}{5} \, .$

12. $y = c_1 x + c_2 x^{-1} + e^{-x}(1 + x^{-1}).$

13. $y = c_1 x^{1/2} + c_2 x^{-1} - \frac{1}{3} x^{-1} \log x.$

14. $y = c_1 x^2 + c_2 x^{-1} - x.$

15. $y = c_1 e^x + c_2 e^x \int e^{-x - (x^3/3)} \, dx.$

16. $y = c_1 x^2 + c_2 x^{-1} + \frac{2}{3} x^2 \log x.$

17. $y = c_1 x - c_2 x^{-1} - 1.$

19. $\dfrac{d^2 y}{du^2} + y = e^u + 3,$

   $y = c_1 \cos [\log (x - 3)] + c_2 \sin [\log (x - 3)] + \dfrac{x}{2} + \dfrac{3}{2} \, .$

20. $\dfrac{d^2 y}{du^2} = 0, \quad y = c_1 \log x + c_2.$

21. $\dfrac{d^3 y}{du^3} - \dfrac{d^2 y}{du^2} - \dfrac{dy}{du} + y = e^u, \quad y = c_1 x + c_2 x \log x + c_3 x^{-1} + \dfrac{x}{4} (\log x)^2.$

23. Riccati equation: $y_1' = x^2 + \dfrac{y_1}{x} - y_1^2 \, ; y_1 = x, y = e^{x^2/2}.$

   General solution: $y = c_1 e^{x^2/2} + c_2 e^{-x^2/2}.$

Chapter 5

Operators and Laplace Transforms

In the next and succeeding lessons we shall prove certain theorems by means of mathematical induction. We digress momentarily, therefore, to explain the meaning of a **"proof by mathematical induction."**

Suppose we wish to prove a statement about the positive integers as, for example, that the sum of the first n odd integers is n^2, i.e., suppose we wish to prove

(a) $$1 + 3 + 5 + 7 + \cdots + (2n - 1) = n^2.$$

By direct substitution we can verify that (a) is true when $n = 1$, $n = 2$, $n = 3$, for then (a) reduces to the respective identities $1 = 1^2$, $1 + 3 = 2^2$, $1 + 3 + 5 = 3^2$. Since it would be impossible to continue in this manner to check the accuracy of (a) for *every* n (there are just too many of them), we reason as follows. If by assuming (a) is valid when $n = k$, where k is an integer, we can then prove that (a) is valid when $n = k + 1$, it will follow that (a) will be true for *every* n. For then the validity of (a) when $n = 3$ will insure its validity when $n = 4$. The validity of (a) when $n = 4$ will insure its validity when $n = 5$, etc., ad infinitum. And since we have already shown by direct substitution that (a) is valid when $n = 3$, it follows that (a) is valid when $n = 4$, etc., ad infinitum.

We shall now use this reasoning to prove that (a) is true for every n. We have already shown that (a) is true when $n = 1$, $n = 2$, and $n = 3$. We now assume that (a) is true when $n = k$. Hence we assume that

(b) $$1 + 3 + 5 + 7 + \cdots + (2k - 1) = k^2$$

is a true equality. Let us add to both sides of (b) the next odd number in the series, namely $(2k - 1 + 2) = 2k + 1$. Equation (b) then becomes

(c) $$1 + 3 + 5 + 7 + \cdots + (2k - 1) + (2k + 1)$$
$$= k^2 + 2k + 1 = (k + 1)^2.$$

Since we have assumed (b) is true, it follows that (c) is true. But if in (a), we let $n = k + 1$, we obtain

(d) $\qquad 1 + 3 + 5 + 7 + \cdots + (2k + 1) = (k + 1)^2$

which agrees with (c). Therefore, **if** (b) is true so also is (d). Hence we
have shown that **if** (a) is true when $n = k$, it is true when $n = k + 1$.
But (a) **is** true when $n = 3$. It is therefore true when $n = 4$. Since
(a) is true when $n = 4$, it is therefore true when $n = 5$, etc., ad infinitum.

A proof which uses the above method of reasoning is called a **proof by
induction.**

Comment 24.1. Every proof by induction must consist of these two
parts. First you must show that the assertion is true when $n = 1$ (or
for whatever other letter appears in the formula). Second you must show
that if the assertion is true when $n = k$, where k is an integer, it will
then be true when $n = k + 1$.

LESSON 24.   Differential and Polynomial Operators.

**LESSON 24A.   Definition of an Operator.   Linear Property of
Polynomial Operators.** An operator is a mathematical device which
converts one function into another. For example, the operation of differentiation is an operator since it converts a differentiable function $f(x)$ into
a new function $f'(x)$. The operation of integrating $\int_{x_0}^{x} f(t)\, dt$ is also an
operator. It converts an integrable function $f(t)$ into a new function $F(x)$.

Because the derivative at one time was known as a differential coefficient, the letter D, which we now introduce to denote the operation of
differentiation, is called a **differential operator.** Hence, if y is an nth
order differentiable function, then

(24.11) $\quad D^0 y = y, \qquad Dy = y', \qquad D^2 y = y'', \cdots, D^n y = y^{(n)}.$

(Other letters may also be used in place of y.) For example, if $y(x) = x^3$,
then $D^0 y = x^3$, $Dy = dy/dx = 3x^2$, $D^2 y = d^2y/dx^2 = 6x$, $D^3 y =
d^3y/dx^3 = 6$, $D^4 y = d^4y/dx^4 = 0$; if $r(\theta) = \sin \theta + \theta^2$, then $D^0 r =
\sin \theta + \theta^2$, $Dr = dr/d\theta = \cos \theta + 2\theta$; $D^2 r = d^2r/d\theta^2 = -\sin \theta + 2$; if
$x(t) = t^2$, then $D^0 x = t^2$, $Dx = dx/dt = 2t$, $D^2 x = d^2x/dt^2 = 2$, $D^3 x = 0$.

By forming a linear combination of differential operators of orders 0 to
n, we obtain the expression

(24.12) $\quad P(D) = a_0 + a_1 D + a_2 D^2 + \cdots + a_n D^n, \quad a_n \neq 0,$

where $a_0, a_1, \cdots, a_n$ are constants.

Because of the resemblance of $P(D)$ to a polynomial, we shall refer to
it as a **polynomial operator of order n.** Its meaning is given in the
following definition.

Definition 24.13. Let $P(D)$ be the polynomial operator (24.12) of order n and let y be an nth order differentiable function. Then we define $P(D)y$ to mean

$$(24.14) \qquad P(D)y = (a_n D^n + \cdots + a_1 D + a_0)y$$
$$= a_n D^n y + \cdots + a_1 D y + a_0 y.$$

By (24.11), we can also write (24.14) as

$$(24.15) \qquad P(D)y = a_n y^{(n)} + \cdots + a_1 y' + a_0 y, \quad a_n \neq 0.$$

Hence, by (24.15), we can write the linear equation with constant coefficients,

$$(24.16) \quad a_n y^{(n)} + a_{n-1} y^{(n-1)} + \cdots + a_1 y' + a_0 y = Q(x), \quad a_n \neq 0,$$

as

$$(24.17) \qquad\qquad P(D)y = Q(x),$$

where $P(D)$ is the polynomial operator (24.12).

Theorem 24.2. *If $P(D)$ is the polynomial operator (24.12) and y_1, y_2 are two nth order differentiable functions, then*

$$(24.21) \qquad P(D)(c_1 y_1 + c_2 y_2) = c_1 P(D)y_1 + c_2 P(D)y_2,$$

where c_1 and c_2 are constants.

Proof. In the left side of (24.21) replace $P(D)$ by its value as given in (24.12). There results

$$(a) \qquad (a_n D^n + a_{n-1} D^{n-1} + \cdots + a_1 D + a_0)(c_1 y_1 + c_2 y_2).$$

By Definition 24.13, we can write (a) as

$$(b) \quad a_n D^n(c_1 y_1 + c_2 y_2) + a_{n-1} D^{n-1}(c_1 y_1 + c_2 y_2) + \cdots$$
$$+ a_1 D(c_1 y_1 + c_2 y_2) + a_0(c_1 y_1 + c_2 y_2).$$

By (24.11) and the property of derivatives,

$$(c) \qquad D^k(c_1 y_1 + c_2 y_2) = c_1 y^{(k)} + c_2 y^{(k)} = c_1 D^k y_1 + c_2 D^k y_2,$$

for each $k = 0, 1, 2, \cdots, n$. Hence (b) becomes

$$(d) \quad a_n(c_1 D^n y_1 + c_2 D^n y_2) + a_{n-1}(c_1 D^{n-1} y_1 + c_2 D^{n-1} y_2)$$
$$+ \cdots + a_1(c_1 D y_1 + c_2 D y_2) + a_0(c_1 y_1 + c_2 y_2)$$
$$= c_1(a_n D^n + a_{n-1} D^{n-1} + \cdots + a_1 D + a_0)y_1$$
$$+ c_2(a_n D^n + a_{n-1} D^{n-1} + \cdots + a_1 D + a_0)y_2$$
$$= c_1 P(D)y_1 + c_2 P(D)y_2,$$

which is the same as the right side of (24.21). We have thus shown that the left side of (24.21) is equal to its right side.

By repeated application of (24.21), it can be proved that

(24.22) $P(D)(c_1 y_1 + c_2 y_2 + \cdots + c_n y_n)$
$$= c_1 P(D)y_1 + c_2 P(D)y_2 + \cdots + c_n P(D)y_n,$$

where $c_1, c_2, \ldots, c_n$ are constants, and each of $y_1, y_2, \cdots, y_n$ is an nth order differentiable function.

Example 24.221.   Use Definition 24.13 to evaluate

(a) $$(D^2 - 3D + 5)(2x^3 + e^{2x} + \sin x).$$

Solution.   Here $P(D) = D^2 - 3D + 5$ and the function y of (24.14) is $2x^3 + e^{2x} + \sin x$. Therefore, by (24.14) and (24.11),

(b) $(D^2 - 3D + 5)(2x^3 + e^{2x} + \sin x)$
$$= D^2(2x^3 + e^{2x} + \sin x) - 3D(2x^3 + e^{2x} + \sin x)$$
$$+ 5(2x^3 + e^{2x} + \sin x)$$
$$= 12x + 4e^{2x} - \sin x - 18x^2 - 6e^{2x} - 3\cos x$$
$$+ 10x^3 + 5e^{2x} + 5\sin x$$
$$= 10x^3 - 18x^2 + 12x + 3e^{2x} + 4\sin x - 3\cos x.$$

Example 24.222.   Use (24.22) to evaluate

(a) $$(D^2 - 3D + 5)(2x^3 + e^{2x} + \sin x).$$

NOTE. This example is the same as Example 24.221 above.

Solution.   By (24.22), with $P(D) = D^2 - 3D + 5$,

(b) $(D^2 - 3D + 5)(2x^3 + e^{2x} + \sin x)$
$$= (D^2 - 3D + 5)(2x^3) + (D^2 - 3D + 5)e^{2x}$$
$$+ (D^2 - 3D + 5)\sin x$$
$$= 2(6x - 9x^2 + 5x^3) + (4e^{2x} - 6e^{2x} + 5e^{2x})$$
$$+ (-\sin x - 3\cos x + 5\sin x)$$
$$= 10x^3 - 18x^2 + 12x + 3e^{2x} + 4\sin x - 3\cos x,$$

the same result obtained previously.

Definition 24.23.   An operator which has the property (24.21) is called a **linear operator**. Hence the polynomial operator (24.12) is **linear**.

Comment 24.24. With the aid of the linear property of the polynomial operator $P(D)$ of (24.12) we can easily prove the following two assertions, proved previously in Theorem 19.3.

1. If $y_1, y_2, \cdots, y_n$ are n solutions of the homogeneous linear equation $P(D)y = 0$, then $y_c = c_1y_1 + c_2y_2 + \cdots + c_ny_n$ is also a solution.
2. If y_c is a solution of the homogeneous linear equation $P(D)y = 0$, and y_p is a particular solution of the nonhomogeneous equation $P(D)y = Q(x)$, then $y = y_c + y_p$ is a solution of $P(D)y = Q(x)$.

Proof of 1. Since $y_1, y_2, \cdots, y_n$ are each solutions of $P(D)y = 0$, we have

(a) $P(D)y_1 = 0, \qquad P(D)y_2 = 0, \cdots, P(D)y_n = 0.$

Hence also

(b) $c_1P(D)y_1 = 0, \qquad c_2P(D)y_2 = 0, \cdots, c_nP(D)y_n = 0,$

where $c_1, c_2, \cdots, c_n$ are constants. Adding all equations in (b) and making use of (24.22), we obtain

(c) $P(D)(c_1y_1 + c_2y_2 + \cdots + c_ny_n) = 0,$

which implies that $y_c = c_1y_1 + c_2y_2 + \cdots + c_ny_n$ is a solution of $P(D)y = 0$.

Proof of 2. By hypothesis y_c is a solution of $P(D)y = 0$, and y_p is a solution of $P(D)y = Q(x)$. Therefore

(d) $P(D)y_c = 0 \quad \text{and} \quad P(D)y_p = Q(x).$

Adding the two equations in (d) and making use of (24.21), we obtain

(e) $P(D)(y_c + y_p) = Q(x),$

which implies that $y = y_c + y_p$ is a solution of $P(D)y = Q(x)$.

Comment 24.25. Principle of Superposition. In place of the linear differential equation

(a) $P(D)y = Q_1 + Q_2 + \cdots + Q_n,$

where $P(D)$ is a polynomial operator (24.12), let us write the n equations

(b) $P(D)y = Q_1, \qquad P(D)y = Q_2, \cdots, P(D)y = Q_n.$

Let $y_{1p}, y_{2p}, \cdots, y_{np}$ be respective particular solutions of the n equations of (b). Therefore

(c) $P(D)y_{1p} = Q_1, \qquad P(D)y_{2p} = Q_2, \cdots, P(D)y_{np} = Q_n.$

Adding all the equations in (c) and making use of (24.22), there results

(d) $P(D)(y_{1p} + y_{2p} + \cdots + y_{np}) = Q_1 + Q_2 + \cdots + Q_n,$

which implies that

(e) $y_p = y_{1p} + y_{2p} + \cdots + y_{np}$

is a solution of (a).

We have thus shown that a particular solution y_p of (a) can be obtained by summing the particular solutions $y_{1p}, y_{2p}, \cdots, y_{np}$ of the n equations of (b). The principle used in this method of obtaining a particular solution of (a) is known as the **principle of superposition**.

Example 24.26.   Use the principle of superposition to find a particular solution of the equation

(a) $(D^2 + 1)y = x^2 + xe^{2x} + 3.$

Solution.   By Comment 24.25, a solution y_p of (a) is the sum of the particular solutions of each of the following equations.

(b) $(D^2 + 1)y = x^2,$ $(D^2 + 1)y = xe^{2x},$ $(D^2 + 1)y = 3.$

Particular solutions of each of these equations are respectively—write $(D^2 + 1)y$ as $y'' + y$ and use the method of Lesson 21—

(c) $y_{1p} = x^2 - 2,$ $y_{2p} = \frac{1}{5}(xe^{2x} - \frac{4}{5}e^{2x}),$ $y_{3p} = 3.$

Hence a particular solution of (a) is

(d) $y_p = x^2 - 2 + \frac{1}{5}(xe^{2x} - \frac{4}{5}e^{2x}) + 3 = x^2 + 1 + \frac{1}{5}(xe^{2x} - \frac{4}{5}e^{2x}).$

LESSON 24B.   Algebraic Properties of Polynomial Operators. Whenever $P_1(D)$ and $P_2(D)$ appear in this lesson, we assume that

(24.3) $P_1(D) = a_n D^n + a_{n-1}D^{n-1} + \cdots + a_1 D + a_0,$

           $P_2(D) = b_m D^m + b_{m-1}D^{m-1} + \cdots + b_1 D + b_0,$

are two polynomial operators of orders n and m respectively, $n \geq m$. Whenever y appears we assume it is an nth order differentiable function, defined on an interval I.

Definition 24.31.   The sum of two polynomial operators $P_1(D)$ and $P_2(D)$ is defined by the relation

(24.32) $(P_1 + P_2)y = P_1 y + P_2 y.$

Theorem 24.33. *Let $P_1(D)$ and $P_2(D)$ be two polynomial operators of the forms (24.3). Then P_1 and P_2 can be added just as if they were ordinary polynomials, i.e., we can add the coefficients of like orders of D.*

Proof. By (24.32), (24.3) and the rules of differentiation,

(a) $(P_1 + P_2)y = (a_n D^n + \cdots + a_2 D^2 + a_1 D + a_0)y$
$\qquad\qquad + (b_m D^m + \cdots + b_2 D^2 + b_1 D + b_0)y$
$\qquad = (a_n D^n + \cdots + b_m D^m + \cdots + a_2 D^2 + b_2 D^2$
$\qquad\qquad + a_1 D + b_1 D + a_0 + b_0)y$
$\qquad = [a_n D^n + \cdots + b_m D^m + \cdots + (a_2 + b_2)D^2$
$\qquad\qquad + (a_1 + b_1)D + (a_0 + b_0)]y.$

Example 24.331. If

(a) $P_1(D) = 3D^3 + D^2 + 3D - 1, \qquad P_2(D) = 5D^2 - 7D + 3,$

find $P_1(D) + P_2(D)$.

Solution. By Theorem 24.33,

(b) $P_1(D) + P_2(D) = 3D^3 + 6D^2 - 4D + 2.$

As an exercise, prove that the following identities follow from Definition 24.31.

(24.34) $P_1(D) + P_2(D) = P_2(D) + P_1(D)$
$\qquad\qquad\qquad\qquad$ (commutative law of addition).
$\qquad P_1(D) + [P_2(D) + P_3(D)] = [P_1(D) + P_2(D)] + P_3(D)$
$\qquad\qquad\qquad\qquad = P_1(D) + P_2(D) + P_3(D)$
$\qquad\qquad\qquad\qquad$ (associative law of addition).

Definition 24.35. The product of a function $h(x)$ by a polynomial operator $P(D)$ is defined by the relation

(24.36) $[h(x)P(D)]y = h(x)[P(D)y].$

Example 24.37. Evaluate

(a) $[2x^2(3D^2 + 1)]e^{3x}.$

Solution. Comparing (a) with (24.36), we see that $h(x) = 2x^2$, $P(D) = 3D^2 + 1, y(x) = e^{3x}$. Therefore, by (24.36), (24.14) and (24.11),

(b) $[2x^2(3D^2 + 1)]e^{3x} = 2x^2[(3D^2 + 1)e^{3x}]$
$\qquad\qquad\qquad = 2x^2[3(9e^{3x}) + e^{3x}] = 2x^2(28e^{3x}) = 56x^2 e^{3x}.$

Example 24.38.   Evaluate

(a) $\qquad (3x^3 + 2)[(D^2 + 3D + 2) + (2D^2 - 1)]e^{2x}.$

Solution.   By Theorem 24.33 and (24.36), we can write (a) as

(b) $\qquad (3x^3 + 2)[(3D^2 + 3D + 1)e^{2x}].$

Carrying out the indicated differentiations in (b), we obtain

(c) $\qquad (3x^3 + 2)(12 + 6 + 1)e^{2x} = 19e^{2x}(3x^3 + 2).$

Comment 24.381.   By (24.32), we may also write

(24.39) $\quad h(x)[P_1(D) + P_2(D)]y = h(x)[P_1(D)y + P_2(D)y]$
$$= h(x)[P_1(D)y] + h(x)[P_2(D)y].$$

Example 24.4.   Evaluate (a) of Example 24.38, by use of (24.39).

Solution.

(a) $\quad (3x^3 + 2)[(D^2 + 3D + 2) + (2D^2 - 1)]e^{2x}$
$$= (3x^3 + 2)[(D^2 + 3D + 2)e^{2x}] + (3x^3 + 2)[(2D^2 - 1)e^{2x}]$$
$$= (3x^3 + 2)(4 + 6 + 2)e^{2x} + (3x^3 + 2)(8 - 1)e^{2x}$$
$$= 19e^{2x}(3x^3 + 2).$$

Definition 24.41.   The product of two polynomial operators $P_1(D)$ and $P_2(D)$ is defined by the relation

(24.42) $\qquad [P_1(D)P_2(D)]y = P_1(D)[P_2(D)y].$

Example 24.43.   Evaluate

(a) $\qquad [(3D^2 - 5D + 3)(D^2 - 2)](x^3 + 2x).$

Solution.   By (24.42), (24.14) and (24.11),

(b) $\quad (3D^2 - 5D + 3)[(D^2 - 2)(x^3 + 2x)]$
$$= (3D^2 - 5D + 3)(6x - 2x^3 - 4x)$$
$$= 3(-12x) - 5(6 - 6x^2 - 4) + 18x - 6x^3 - 12x$$
$$= -6x^3 + 30x^2 - 30x - 10.$$

Prove as exercises that the following identity follows from Definition 24.41,

(24.44) $\qquad P_1(D)[P_2(D)P_3(D)] = [P_1(D)P_2(D)]P_3(D)$
$$= P_1(D)P_2(D)P_3(D)$$
$$\text{(associative law of multiplication),}$$

and the following identity from Definitions 24.31 and 24.41,

(24.45) $P_1(D)[P_2(D) + P_3(D)] = P_1(D)P_2(D) + P_1(D)P_3(D)$
                                        (distributive law of multiplication).

Theorem 24.46.   *If*

(24.47) $P(D) = a_n D^n + a_{n-1} D^{n-1} + \cdots + a_1 D + a_0, \quad a_n \neq 0,$

where $a_0, a_1, \cdots, a_n$ *are constants, then*

(24.48) $P(D) = a_n(D - r_1)(D - r_2) \cdots (D - r_n),$

where $r_1, r_2, \cdots, r_n$ *are the real or imaginary roots of the characteristic equation (20.14) of* $P(D)y = 0$, *i.e., a polynomial operator with constant coefficients can be factored just as if it were an ordinary polynomial.*

Proof.   We shall prove the theorem only for $n = 2$, i.e., we shall prove

(a) $D^2 - (r_1 + r_2)D + r_1 r_2 = (D - r_1)(D - r_2).$

In the equalities which follow, we have indicated the reason after each step. Be sure to refer to these numbers. We start with the right side of (a) and show that it yields the left side.

	By
(b) $[(D - r_1)(D - r_2)]y$	
$\quad = (D - r_1)[(D - r_2)y]$	(24.42),
$\quad = (D - r_1)(Dy - r_2 y)$	(24.14),
$\quad = (D - r_1)Dy - (D - r_1)(r_2 y)$	(24.21),
$\quad = D^2 y - r_1 Dy - r_2 Dy + r_1 r_2 y$	(24.14),
$\quad = D^2 y - (r_1 + r_2)Dy + r_1 r_2 y$	Theorem 24.33,
$\quad = [D^2 - (r_1 + r_2)D + r_1 r_2]y$	(24.14).

Corollary 24.481.   *If* $P(D)$ *is the polynomial operator (24.47), then*

(24.482) $P(D) = P_1(D)P_2(D),$

where $P_1(D)$ *and* $P_2(D)$ *may be composite factors of* $P(D)$, *i.e.,* P_1 *and* P_2 *may be products of factors of (24.48).*

The proof follows from Theorem 24.46.

Theorem 24.49.   *The commutative law of multiplication is valid for polynomial operators, i.e.,*

(24.5) $(D - r_1)(D - r_2) = (D - r_2)(D - r_1).$

Proof. In the proof of Theorem 24.46, interchange the subscripts 1 and 2 of r_1 and r_2. Since the final formula on the right of (b) in the proof will remain the same, the equality (24.5) follows.

Example 24.51. Evaluate $(D^2 - 2D - 3)(\sin x + x^2)$.

Solution. *Method 1.* By application of Theorem 24.2.

(a) $(D^2 - 2D - 3)(\sin x + x^2)$
$$= (D^2 - 2D - 3)\sin x + (D^2 - 2D - 3)x^2$$
$$= -\sin x - 2\cos x - 3\sin x + 2 - 4x - 3x^2$$
$$= 2 - 4x - 3x^2 - 4\sin x - 2\cos x.$$

Method 2. By application of Theorem 24.46 and (24.42).

(b) $(D^2 - 2D - 3)(\sin x + x^2)$
$$= [(D + 1)(D - 3)](\sin x + x^2)$$
$$= (D + 1)[(D - 3)(\sin x + x^2)]$$
$$= (D + 1)(\cos x + 2x - 3\sin x - 3x^2)$$
$$= -\sin x + 2 - 3\cos x - 6x + \cos x$$
$$\quad + 2x - 3\sin x - 3x^2$$
$$= 2 - 4x - 3x^2 - 4\sin x - 2\cos x.$$

Definition 24.52. If

(24.521) $$P(D) = a_n D^n + \cdots + a_1 D + a_0$$

is a polynomial operator of order n, then

(24.522) $$P(D + a) = a_n(D + a)^n + \cdots + a_1(D + a) + a_0,$$

where a is a constant, i.e., (24.522) is the polynomial operator obtained by replacing D in (24.521) by $D + a$.

Summary 24.523.* Polynomial operators can be added, multiplied, factored, and multiplied by a constant, just as if they were ordinary polynomials. Furthermore, the

(24.53) Commutative law: $P_1 P_2 = P_2 P_1$,
(24.54) Associative law: $P_1(P_2 P_3) = (P_1 P_2)P_3 = P_1 P_2 P_3$,
(24.55) Distributive law: $P_1(P_2 + P_3) = P_1 P_2 + P_1 P_3$

are all valid.

---

*Some of the properties here summarized do not apply to polynomial operators with *nonconstant* coefficients. See, for example, Exercise 24, 17 and 18.

LESSON 24C.   Exponential Shift Theorem for Polynomial Operators. If the function to be operated on has the special form ue^{ax}, where u is an nth order differentiable function of x defined on an interval I, then the following theorem, called the **exponential shift theorem** because the formula we shall obtain shifts the position of the exponential e^{ax}, will aid materially in evaluating $P(D)ue^{ax}$.

Theorem 24.56.   (Exponential Shift Theorem) *If*

$$(24.57)\qquad P(D) = a_n D^n + a_{n-1}D^{n-1} + \cdots + a_1 D + a_0,$$

$a_n \neq 0$, *is a polynomial operator with constant coefficients and $u(x)$ is an nth order differentiable function of x defined on an interval I, then*

$$(24.58)\qquad\qquad P(D)(ue^{ax}) = e^{ax}P(D + a)u,$$

where a is a constant.

Proof. We shall first prove by induction that the theorem is true for the special polynomial operator $P(D) = D^k$. By Comment 24.1 we must show that:

1. (24.58) is true when $k = 1$, i.e., when $P(D) = D$.
2. If (24.58) is valid when $k = n$, then it is true when $k = n + 1$, i.e., if (24.58) is valid when $P(D) = D^n$, then it is true when $P(D) = D^{n+1}$

Proof of 1. When $P(D) = D$, the left side of (24.58) is $D(ue^{ax})$. By (24.11) and (24.14),

$$(a)\qquad D(ue^{ax}) = aue^{ax} + e^{ax}u' = e^{ax}(u' + au) = e^{ax}(D + a)u.$$

By Definition 24.52, if $P(D) = D$, then $P(D + a) = D + a$. Substituting these values of D and $D + a$ in the first and last terms of (a), we obtain (24.58).

Proof of 2. We assume that (24.58) is true when $P(D) = D^n$. We must then prove that the theorem is true when $P(D) = D^{n+1}$. By Definition (24.52), if $P(D) = D^n$, $P(D + a) = (D + a)^n$. Substituting these values of $P(D)$ and $P(D + a)$ in (24.58), we obtain

$$(b)\qquad\qquad D^n(ue^{ax}) = e^{ax}(D + a)^n u,$$

which, by our assumption, is a true equality. Operating on (b) with D, we obtain

		By
(c) $D[D^n(ue^{ax})]$	$= D[e^{ax}(D + a)^n u]$	(b),
	$= e^{ax}D(D + a)^n u + ae^{ax}(D + a)^n u$	(24.11),
	$= e^{ax}[D(D + a)^n + a(D + a)^n]u$	(24.39),
	$= e^{ax}[(D + a)^n(D + a)]u$	Corollary 24.481,
	$= e^{ax}(D + a)^{n+1}u.$	Corollary 24.481.

The left side of (c) is $D^{n+1}(ue^{ax})$. Hence we have shown by (c) that the validity of (b) leads to the validity of

(d) $$D^{n+1}(ue^{ax}) = e^{ax}(D + a)^{n+1}u.$$

It therefore follows by Comment 24.1, that

(24.59) $$D^k(ue^{ax}) = e^{ax}(D + a)^k u,$$

for *every* k.

Let $P(D)$ be the operator (24.57). Then

(e) $P(D)(ue^{ax})$

$\quad = a_n D^n(ue^{ax}) + \cdots + a_1 D(ue^{ax}) + a_0 ue^{ax}$ (24.14),

$\quad = e^{ax}[a_n(D + a)^n + \cdots + a_1(D + a) + a_0]u$ (24.59),

$\quad = e^{ax}P(D + a)u$ Definition 24.52.

By

Corollary 24.6.

(24.61) $$(D - a)^n(ue^{ax}) = e^{ax}D^n u.$$

Proof. Let $P(D) = (D - a)^n$. Then by Definition 24.52, $P(D + a) = (D + a - a)^n = D^n$. Substituting these values of $P(D)$ and $P(D + a)$ in (24.58), we obtain (24.61).

Corollary 24.7. *If c is a constant, and $P(D)$ is the polynomial operator* (24.57), *then*

(24.71) $$P(D)(ce^{ax}) = ce^{ax}P(a).$$

Proof. By (24.58) and Definition 24.52,

(a) $P(D)ce^{ax} = e^{ax}P(D + a)c$

$\quad = e^{ax}[a_n(D + a)^n + a_{n-1}(D + a)^{n-1} + \cdots + a_0]c.$

By Theorem 24.46, (24.42) and (24.14),

(b) $(D + a)^k c = (D + a)^{k-1}(D + a)c = (D + a)^{k-1}ac$

$\quad = (D + a)^{k-2}(D + a)(ac) = (D + a)^{k-2}(a^2 c)$

$\quad \cdots\cdots\cdots\cdots\cdots\cdots\cdots\cdots\cdots\cdots$

$\quad = (D + a)(a^{k-1}c) = a^k c.$

Hence, by (b), we can write the second equation in (a) as

(c) $\quad P(D)ce^{ax} = e^{ax}[a_n a^n + a_{n-1}a^{n-1} + \cdots + a_1 a + a_0]c.$

By Definition 24.52, (c) is equivalent to

$$P(D)ce^{ax} = e^{ax}P(a)c = ce^{ax}P(a).$$

Example 24.8.   Evaluate

(a) $$(D^2 + 2D + 3)(e^{2x} \sin x).$$

Solution. Comparing (a) with (24.58) we see that $P(D) = D^2 + 2D + 3$, $a = 2$, $u = \sin x$. Hence by Definition 24.52, $P(D + 2) = (D + 2)^2 + 2(D + 2) + 3 = D^2 + 6D + 11$. Therefore by (24.58),

(b) $$\begin{aligned}(D^2 + 2D + 3)(e^{2x} \sin x) &= e^{2x}(D^2 + 6D + 11) \sin x \\ &= e^{2x}(10 \sin x + 6 \cos x).\end{aligned}$$

Example 24.81.   Evaluate

(a) $$(D^2 - D + 3)(x^3 e^{-2x}).$$

Solution. Comparing (a) with (24.58), we see that $P(D) = D^2 - D + 3$, $a = -2$, $u = x^3$. Hence by Definition 24.52, $P(D - 2) = (D - 2)^2 - (D - 2) + 3 = D^2 - 5D + 9$. Therefore, by (24.58),

(b) $$\begin{aligned}(D^2 - D + 3)(x^3 e^{-2x}) &= e^{-2x}(D^2 - 5D + 9)x^3 \\ &= e^{-2x}(6x - 15x^2 + 9x^3).\end{aligned}$$

Example 24.82.   Evaluate

(a) $$(D - 2)^3(e^{2x} \sin x).$$

Solution. Comparing (a) with (24.61), we see that $a = 2$, $u = \sin x$, $n = 3$. Hence by (24.61),

(b) $$(D - 2)^3(e^{2x} \sin x) = e^{2x}D^3 \sin x = -e^{2x} \cos x.$$

Example 24.83.   Evaluate

(a) $$(D^3 - 3D^2 + 2)(5e^{-4x}).$$

Solution. Comparing (a) with (24.71), we see that $P(D) = D^3 - 3D^2 + 2$, $c = 5$, $a = -4$. Therefore $P(-4) = -64 - 48 + 2 = -110$. Hence by (24.71)

(b) $$(D^3 - 3D^2 + 2)(5e^{-4x}) = 5e^{-4x}(-110) = -550e^{-4x}.$$

LESSON 24D. Solution of a Linear Differential Equation with Constant Coefficients by Means of Polynomial Operators. In Lessons 21 and 22 we outlined methods for finding the complementary function y_c and a particular solution y_p of the nonhomogeneous linear equation,

(24.9) $a_n y^{(n)} + a_{n-1} y^{(n-1)} + \cdots + a_1 y' + a_0 = Q(x)$, $a_n \neq 0$.

In this lesson we shall solve (24.9) by means of polynomial operators. We illustrate the method by means of examples.

Example 24.91. Find the general solution of

(a) $$y''' + 2y'' - y' - 2y = e^{2x}.$$

Solution. In operator notation, (a) can be written as

(b) $$(D^3 + 2D^2 - D - 2)y = e^{2x}.$$

By Theorem 24.46, (b) is equivalent to

(c) $$(D - 1)(D + 1)(D + 2)y = e^{2x}.$$

Let

(d) $$u = (D + 1)(D + 2)y.$$

Then (c) becomes

(e) $$(D - 1)u = e^{2x}, \qquad u' - u = e^{2x},$$

which is a first order linear differential equation in u. Its solution, by Lesson 11B, is

(f) $$u = e^{2x} + c_1 e^x.$$

Substituting this value of u in (d) gives

(g) $$(D + 1)(D + 2)y = e^{2x} + c_1 e^x.$$

Let

(h) $$v = (D + 2)y.$$

Then (e) can be written as

(i) $$(D + 1)v = e^{2x} + c_1 e^x; \qquad v' + v = e^{2x} + c_1 e^x,$$

an equation linear in v. By Lesson 11B, its solution is

(j) $$v = \tfrac{1}{3}e^{2x} + \frac{c_1}{2} e^x + c_2 e^{-x}.$$

Substituting this value of v in (h) gives

(k) $$y' + 2y = \tfrac{1}{3}e^{2x} + \frac{c_1}{2} e^x + c_2 e^{-x},$$

an equation linear in y. Its solution, by Lesson 11B, is

(l) $$y = \tfrac{1}{12}e^{2x} + \frac{c_1}{6} e^x + c_2 e^{-x} + c_3 e^{-2x},$$

which can be written as

(m) $$y = \tfrac{1}{12}e^{2x} + C_1 e^x + c_2 e^{-x} + c_3 e^{-2x}.$$

Comment 24.92. The solution (m) could have been obtained much more easily if we had found y_c by means of Lesson 20, and used the above method to find only the particular solution $e^{2x}/12$. The solution (f) would then have read

$$u = e^{2x};$$

the solution (j)

$$v = \tfrac{1}{3}e^{2x},$$

and the solution (l),

$$y_p = \tfrac{1}{12}e^{2x}.$$

The roots of the characteristic equation are, by (c), 1, −1, −2. We could, therefore, easily have written the complementary function

$$y_c = c_1 e^x + c_2 e^{-x} + c_3 e^{-2x}.$$

Example 24.93. Find the general solution of

(a) $$y'' + y = e^x.$$

Solution. In operator notation, (a) can be written as

(b) $$(D^2 + 1)y = e^x.$$

By Theorem 24.46, (b) is equivalent to

(c) $$(D + i)(D - i)y = e^x.$$

Let

(d) $$u = (D - i)y.$$

Then (c) becomes

(e) $$(D + i)u = e^x, \qquad u' + iu = e^x,$$

an equation linear in u. Its solution, by Lesson 11B, is

(f) $$u = \frac{1}{1 + i}e^x + c_1'e^{-ix}.$$

Hence (d) becomes

(g) $$y' - iy = \frac{1}{1 + i}e^x + c_1'e^{-ix},$$

whose solution is

(h) $$y = \tfrac{1}{2}e^x + \tfrac{1}{2}c_1'ie^{-ix} + c_2'e^{ix}.$$

This last equation can be written with new parameters as

(i) $$y = \tfrac{1}{2}e^x + c_1 e^{-ix} + c_2 e^{ix}.$$

Again we remark that (i) could have been obtained more easily, if we had found y_c by means of Lesson 20 and used the above method to find only the particular solution $e^x/2$.

In general, if the nonhomogeneous linear differential equation (24.9) of order n is expressed as

$$(24.94) \qquad (D - r_1)(D - r_2) \cdots (D - r_n)y = Q(x),$$

where $r_1, r_2, \cdots, r_n$ are the roots of its characteristic equation, then a general solution (or, if arbitrary constants of integration are ignored, a particular solution) can be obtained as follows. Let

$$(24.95) \qquad u = (D - r_2) \cdots (D - r_n)y.$$

Then (24.94) can be written as

$$(24.96) \qquad (D - r_1)u = Q(x),$$

an equation linear in u. If its solution $u(x)$ can be found, substituting it in (24.95) will give

$$(24.97) \qquad (D - r_2)(D - r_3) \cdots (D - r_n)y = u(x).$$

Let

$$(24.98) \qquad v = (D - r_3) \cdots (D - r_n)y.$$

Then (24.97) becomes

$$(24.99) \qquad (D - r_2)v = u(x),$$

an equation linear in v. If a solution for $v(x)$ can be found, substituting this value in (24.98) will give

$$(24.991) \qquad (D - r_3) \cdots (D - r_n)y = v(x).$$

The repetition of the above process an additional $(n - 2)$ times will eventually lead to a solution for y.

EXERCISE 24

1. Prove by induction that
$$1^2 + 2^2 + 3^2 + \cdots + n^2 = \frac{n(n + 1)(2n + 1)}{6}.$$

2. Find D^0y, Dy, D^2y, D^3y for each of the following:

   (a) $y(x) = 3x^2$,　　(b) $y(x) = 3 \sin 2x$,　　(c) $y(x) = \sqrt{x}$.

3. Find D^0r, Dr, D^2r for each of the following:

   (a) $r(\theta) = \cos \theta + \tan \theta$.　　(b) $r(\theta) = \theta^2 + \sin \theta$.

   (c) $r(\theta) = \dfrac{1}{\theta^2}$.

4. Find D^0x, Dx, D^2x for each of the following:

    (a) $x(t) = t^2 + 3t + 1$.   (b) $x(t) = a \cos \theta t + b \sin \theta t$.
    (c) $x(t) = a \cos (\theta t + b)$.

5. Use Definition 24.13 to evaluate each of the following:

    (a) $(D^2 - 2D - 3) \cos 2x$.   (b) $(D^2 - 6D + 5)2e^{3x}$
    (c) $(D^4 - 2D^2)4x^3$.   (d) $(D^2 - 4D + 4)(x^2 + x + 1)$.

6. Use (24.22) to evaluate each of the following:

    (a) $D^3(\cos ax + \sin bx)$.   (b) $(D^2 + D)(3e^x + 2x^3)$.
    (c) $(D^2 - 2D + 4)(xe^x + 5x^2 + 2)$.

7. Use the principle of superposition to find a particular solution of each of the following equations.

    (a) $y'' + 3y' + 2y = 8 + 6e^x + 2 \sin x$.
    (b) $y'' - 2y' - 8y = 9xe^x + 10e^{-x}$.
    (c) $y'' + 2y' + 10y = e^x + 2$.

8. Evaluate $P_1(D) + P_2(D)$, where

    (a) $P_1(D) = D^2 + 2D - 1$, $P_2(D) = 3D^3 + 4D^2 - D + 3$.
    (b) $P_1(D) = D^4 - 2D^2$, $P_2(D) = D^2 - 6D + 5$.

9. Evaluate $P_1(D) + P_2(D) + P_3(D)$, where

    (a) $P_1(D) = D^2 - 2D - 3$, $P_2(D) = D^2 - D$,
       $P_3(D) = D^4 + 2D^3 - 2D^2 + 3D + 4$.
    (b) $P_1(D) = 3D^3 + 2D^2 + 1$, $P_2(D) = 2D^4 + D^3 + 2D^2 - 3D + 2$,
       $P_3(D) = 3D^5 - 4D^2 + 7$.

10. Evaluate by means of (24.36)

    (a) $[x^2(D^2 + 1)](2e^x)$.   (b) $[(x - 1)(D^3 + D^2)](e^{2x} + x^2)$.
    (c) $[e^{-x}(3D + 4)](\tan x + 1)$.   (d) $[(x^2 + 3x + 4)(D^3 + 1)](x^3 + 2)$.

11. Evaluate two ways, by Theorem 24.33 and by (24.39),

    (a) $x^2[(D^2 + 1) + D](2e^{-2x})$.
    (b) $(x - 1)[(D^3 + D^2) + (2D^2 - 6D + 5)](x^3 + 2)$.
    (c) $\sin x[(2D^2 - D + 1) + (D^2 + D - 1)] \cot x$.

12. Evaluate two ways, by (24.42) and by Theorem 24.46,

    (a) $(D - 1)(D + 1)(3e^{-x})$.
    (b) $(D^2 - D + 1)(D + 1)(3x^3 + 2x^2 - 5x - 6)$.
    (c) $(D^2 + 1)(D - 3)(\cos x + 2e^{3x})$.

13. Evaluate, by means of Theorem 24.56,

    (a) $(D^2 - 2D - 3)(e^{2x} \cos 2x)$.   (b) $(D^2 - D + 3)(3x^2e^{-2x})$.
    (c) $(D - 2)^2(e^{-x} \tan x)$.   (d) $(D^2 + 2D)(3e^{2x} \csc x)$.
    (e) $(D^2 - 2D + 6)(e^{-3x} \log x)$.

14. Evaluate, by means of Corollary 24.6,

    (a) $(D - 1)e^x \sin x$.   (b) $(D + 1)e^{-x} \cos x$.
    (c) $(D - 2)^2(e^{2x} \log x)$.   (d) $(D + 2)^2(e^{-2x} \tan x)$.
    (e) $(D - 3)^2(e^{3x} \text{ Arc} \sin x)$.   (f) $(D + 3)^2(e^{-3x} \cot x)$.

15. Evaluate, by means of Corollary 24.7,
   (a) $(D^3 + 3D + 1)(5e^{2x})$. (b) $(D^3 + 3D + 1)(2e^{-2x})$.
   (c) $(D - 2)^3(3e^{4x})$. (d) $(D^2 - 3D + 7)5e^{3x}$.

16. Prove Theorem 24.46 for $n = 3$. *Hint.* $(D - r_1)(D - r_2)(D - r_3) = D^3 - (r_1 + r_2 + r_3)D^2 + (r_1r_2 + r_1r_3 + r_2r_3)D - r_1r_2r_3$.

17. Prove Theorem 24.46 is false if the coefficients of $P(D)$ in (24.47) are functions of x. *Hint.* Show, for example, that $(x^2D^2 - 1) \neq (xD - 1)(xD + 1)$ by evaluating $(x^2D^2 - 1)e^{2x}$ and $(xD - 1)[(xD + 1)e^{2x}]$.

18. Prove the commutative law (24.5) is false if the coefficients of $P(D)$ in (24.47) are functions of x. *Hint.* Show, for example, that $(D^2 + xD)(3D)y \neq 3D(D^2 + xD)y$, where y is a second order differentiable function of x.

Find the general solution of each of the following differential equations. Follow the method of Lesson 24D.

19. $y'' - y' - 2y = e^x$.
20. $y'' + 3y' + 2y = 12e^x$.
21. $y'' + 3y' + 2y = e^{ix}$.
22. $y'' + 3y' + 2y = \sin x$.
23. $y'' + 3y' + 2y = \cos x$.
24. $y'' + 3y' + 2y = 8 + 6e^x + 2\sin x$.
25. $y'' - 2y' - 8y = 10e^{-x} + 9xe^x$.
26. $y'' - 3y' = 2e^{2x}\sin x$.
27. $y^{(4)} - 2y'' + y = x - \sin x$.
28. $y'' + y' = x^2 + 2x$.

29. $y''' - 3y' + 2y = e^{-x}$.
30. $y'' + 4y = 4x\sin 2x$.
31. $y''' - 3y'' + 3y' - y = e^x$.
32. $y''' - y' = e^{2x}\sin^2 x$.
33. $4y'' - 5y' = x^2e^{-x}$.
34. $y''' - y'' + y' - y = 4e^{-3x} + 2x^4$.
35. $y''' - 5y'' + 8y' - 4y = 3e^{2x}$.
36. Prove the equalities in (24.34).
37. Prove the equality in (24.44).
38. Prove the equality in (24.45).

ANSWERS 24

2. (a) $3x^2$, $6x$, 6, 0. (b) $3\sin 2x$, $6\cos 2x$, $-12\sin 2x$, $-24\cos 2x$.
   (c) $\sqrt{x}$, $\frac{1}{2}x^{-1/2}$, $-\frac{1}{4}x^{-3/2}$, $\frac{3}{8}x^{-5/2}$.
3. (a) $\cos\theta + \tan\theta$, $-\sin\theta + \sec^2\theta$, $-\cos\theta + 2\sec^2\theta\tan\theta$.
   (b) $\theta^2 + \sin\theta$, $2\theta + \cos\theta$, $2 - \sin\theta$. (c) θ^{-2}, $-2\theta^{-3}$, $6\theta^{-4}$.
4. (a) $t^2 + 3t + 1$, $2t + 3$, 2. (b) $a\cos\theta t + b\sin\theta t$, $-a\theta\sin\theta t + b\theta\cos\theta t$, $-a\theta^2\cos\theta t - b\theta^2\sin\theta t$.
   (c) $a\cos(\theta t + b)$, $-a\theta\sin(\theta t + b)$, $-a\theta^2\cos(\theta t + b)$.
5. (a) $4\sin 2x - 7\cos 2x$. (b) $-8e^{3x}$. (c) $-48x$. (d) $4x^2 - 4x + 2$.
6. (a) $a^3\sin ax - b^3\cos bx$. (b) $6e^x + 6x^2 + 12x$.
   (c) $3xe^x + 18 - 20x + 20x^2$.
7. (a) $y_p = e^x + 4 + \frac{1}{5}(\sin x - 3\cos x)$.
   (b) $y_p = -xe^x - 2e^{-x}$. (c) $y_p = x^3/3$.
   (c) $y_p = \frac{1}{13}e^x + \frac{1}{5}$.
8. (a) $3D^3 + 5D^2 + D + 2$. (b) $D^4 - D^2 - 6D + 5$.
9. (a) $D^4 + 2D^3 + 1$. (b) $3D^5 + 2D^4 + 4D^3 - 3D + 10$.
10. (a) $4x^2e^x$. (b) $2(x - 1)(6e^{2x} + 1)$. (c) $e^{-x}(3\sec^2 x + 4\tan x + 4)$.
   (d) $(x^2 + 3x + 4)(x^3 + 8)$.
11. (a) $6x^2e^{-2x}$. (b) $(x - 1)(5x^3 - 18x^2 + 18x + 16)$.
   (c) $6\cot x\csc x$.
12. (a) 0. (b) $3x^3 + 2x^2 - 5x + 12$. (c) 0.
13. (a) $e^{2x}(-4\sin 2x - 7\cos 2x)$. (b) $3e^{-2x}(2 - 10x + 9x^2)$.
   (c) $e^{-x}(2\sec^2 x\tan x - 6\sec^2 x + 9\tan x)$.
   (d) $3e^{2x}\csc x(\csc^2 x + \cot^2 x - 6\cot x + 8)$.
   (e) $e^{-3x}(21\log x - x^{-2} - 8x^{-1})$.

14. (a) $e^x \cos x$.   (b) $-e^{-x} \sin x$.   (c) $-e^{2x}x^{-2}$.
     (d) $2e^{-2x} \sec^2 x \tan x$.   (e) $xe^{3x}(1 - x^2)^{-3/2}$.
     (f) $2e^{-3x} \csc^2 x \cot x$.

15. (a) $75e^{2x}$.   (b) $-26e^{-2x}$.   (c) $24e^{4x}$.   (d) $235e^{8x}$.

19. $y = c_1 e^{-x} + c_2 e^{2x} - \frac{1}{2}e^x$.

20. $y = 2e^x + c_1 e^{-2x} + c_2 e^{-x}$.

21. $y = \frac{1}{10}(e^{ix} - 3ie^{ix}) + c_1 e^{-2x} + c_2 e^{-x}$.

22. $y = \frac{1}{10}(\sin x - 3 \cos x) + c_1 e^{-2x} + c_2 e^{-x}$.

23. $y = \frac{1}{10}(3 \sin x + \cos x) + c_1 e^{-2x} + c_2 e^{-x}$.

24. $y = 4 + e^x + \frac{1}{5}(\sin x - 3 \cos x) + c_1 e^{-2x} + c_2 e^{-x}$.

25. $y = c_1 e^{4x} + c_2 e^{-2x} - xe^x - 2e^{-x}$.

26. $y = -\dfrac{e^{2x}}{5}(3 \sin x + \cos x) + c_1 + c_2 e^{3x}$.

27. $y = (c_1 + c_2 x)e^x + (c_3 + c_4 x)e^{-x} + x - \dfrac{\sin x}{4}$.

28. $y = c_1 + c_2 e^{-x} + \dfrac{x^3}{3}$.

29. $y = (c_1 + c_2 x)e^x + c_3 e^{-2x} + \frac{1}{4}e^{-x}$.

30. $y = c_1 \cos 2x + c_2 \sin 2x - \dfrac{x}{4}(2x \cos 2x - \sin 2x)$.

31. $y = \left(c_1 + c_2 x + c_3 x^2 + \dfrac{x^3}{6}\right)e^x$.

32. $y = c_1 + c_2 e^x + c_3 e^{-x} + \left(\dfrac{1}{12} + \dfrac{9 \cos 2x - 7 \sin 2x}{520}\right)e^{2x}$.

33. $y = c_1 + c_2 e^{5x/4} + (81x^2 + 234x + 266)\dfrac{e^{-x}}{729}$.

34. $y = c_1 e^x + c_2 \sin x + c_3 \cos x - \left(\dfrac{e^{-3x}}{10} + 2x^4 + 8x^3 + 48\right)$.

35. $y = (c_1 + c_2 x)e^{2x} + c_3 e^x + \dfrac{3x^2 e^{2x}}{2}$.

LESSON 25. Inverse Operators.

In Lesson 21, we found, by the method of undetermined coefficients, a particular solution of the nth order linear differential equation

(25.1) $$P(D)y = Q(x),$$

where $P(D)$ is the polynomial operator

(25.11) $$P(D) = a_n D^n + \cdots + a_1 D + a_0, \quad a_n \neq 0,$$

and $Q(x)$ is a function which consists only of such terms as b, x^k, e^{ax}, $\sin ax$, $\cos ax$, and a finite number of combinations of such terms. Here a and b are constants and k is a positive integer. In this lesson we shall show how inverse operators may furnish a relatively easy and quick method for obtaining this same particular solution.

Let

(25.12) $$y_c = c_1 y_1 + \cdots + c_n y_n$$

be the complementary function of (25.1), i.e., let y_c be the general solution of $P(D)y = 0$, and let y_p be a particular solution of (25.1). Therefore the general solution of (25.1), by Theorem 19.3, is

$$(25.13) \qquad\qquad y = y_c + y_p.$$

By following the method of undetermined coefficients as outlined in Lesson 21, we obtained a particular solution y_p that contained no term which was a constant multiple of a term in y_c. However, there are infinitely many other particular solutions of (25.1). By Definition 4.66, each solution which satisfies (25.1) and does not contain arbitrary constants is a particular solution of (25.1). For example, the complementary function of the differential equation

(a) $$(D^2 - 1)y = x^2$$

is

(b) $$y_c = c_1 e^x + c_2 e^{-x}.$$

A particular solution of (a), found by the method of undetermined coefficients, is

(c) $$y_p = -x^2 - 2.$$

Therefore the general solution of (a) is

(d) $$y = -x^2 - 2 + c_1 e^x + c_2 e^{-x}.$$

You can verify that the following solutions, obtained by assigning arbitrary values to the constants c_1 and c_2 of (d) are, by Definition 4.66, also particular solutions of (a).

(e) $\quad y_p = -x^2 - 2, \qquad y_p = -x^2 - 2 - 3e^x,$

$\qquad y_p = -x^2 - 2 + 6e^x - e^{-x}, \qquad y_p = -x^2 - 2 + 3e^{-x}$, etc.

Note, however, that every two functions differ from each other by terms which are constant multiples of terms in y_c. In the proof of Theorem 65.6, we show that this observation holds for all particular solutions of (25.1).

In the remainder of this lesson and the next, whenever we refer to a particular solution y_p of (25.1), we shall mean that particular solution from which all constant multiples of terms in y_c have been eliminated. For instance, in the above example, our particular solution of (a) is $-x^2 - 2$.

LESSON 25A. Meaning of an Inverse Operator.

Definition 25.2. Let $P(D)y = Q(x)$, where $P(D)$ is the polynomial operator (25.11) and $Q(x)$ is the special function consisting only of such terms as b, x^k, e^{ax}, $\sin ax$, $\cos ax$, and a finite number of combinations of such terms,* where a, b are constants and k is a positive integer. Then the **inverse operator** of $P(D)$, written as $P^{-1}(D)$ or $1/P(D)$, is defined as an operator which, when operating on $Q(x)$, will give the particular solution y_p of (25.1) that contains no constant multiples of a term in the complementary function y_c, i.e.,

(25.21) $\qquad P^{-1}(D)Q(x) = y_p \quad \text{or} \quad \dfrac{1}{P(D)} Q(x) = y_p,$

where y_p is the particular solution of $P(D)y = Q(x)$ that contains no constant multiple of a term in y_c.

Comment 25.22. If we are given $P(D)$ and $Q(x)$, we now know how to find $P^{-1}(D)Q(x)$. By Definition 25.2, it is the particular solution y_p of $P(D)y = Q(x)$ that contains no constant multiples of terms in y_c.

Example 25.23. Evaluate

(a) $\qquad\qquad\qquad (D^2 - 3D + 2)^{-1}x.$

Solution. By Definition 25.2, $(D^2 - 3D + 2)^{-1}x = y_p$, where y_p is a particular solution of

(b) $\qquad (D^2 - 3D + 2)y = x, \qquad y'' - 3y' + 2y = x,$

that contains no constant multiples of terms in y_c.

By the method of Lesson 21 or 24D, we obtain the particular solution

(c) $\qquad\qquad\qquad y_p = \dfrac{x}{2} + \dfrac{3}{4}.$

Hence

(d) $\qquad\qquad (D^2 - 3D + 2)^{-1}x = \dfrac{x}{2} + \dfrac{3}{4}.$

Comment 25.24. By Definition 25.2, we conclude that

(25.25) $\qquad D^{-n}Q(x) \equiv$ integrating $Q(x)$ n times and ignoring constants of integration.

Proof. By Definition 25.2, $D^{-n}Q(x) = y_p$, where y_p is the particular solution of

(a) $\qquad\qquad\qquad D^n y = Q(x)$

that contains no constant multiples of terms in the complementary func-

---

*This restriction on $Q(x)$ is a drastic one. Our definition, however, would not be meaningful if $Q(x)$ were not thus restricted, since it might then be difficult to exhibit y_p explicitly or implicitly in terms of elementary functions.

tion y_c of (a). The complementary function of (a) is

(b) $$y_c = c_1 + c_2 x + c_3 x^2 + \cdots + c_{n-1} x^{n-1},$$

and these terms result from retaining the constants of integration when integrating (a) n times.

Example 25.251. Evaluate

(a) $$D^{-2}(2x + 3).$$

Solution. By (25.25),

(b) $$D^{-2}(2x + 3) = D^{-1} \int (2x + 3)\, dx = D^{-1}(x^2 + 3x) = \frac{x^3}{3} + \frac{3x^2}{2}.$$

You can verify that $x^3/3 + 3x^2/2$ is a particular solution of

(c) $$D^2 y = 2x + 3, \qquad y'' = 2x + 3,$$

and that the complementary function of (c) is $y_c = c_1 + c_2 x$.

Comment 25.26. We draw another important conclusion from Definition 25.2, namely that if $P(D)y = 0$, then

(25.27) $$y_p = P^{-1}(D)(0) = 0 \quad \text{or} \quad y_p = \frac{1}{P(D)}\,(0) = 0.$$

Proof. By Definition 25.2, $P^{-1}(D)(0) = y_p$, where y_p is the particular solution of

(a) $$P(D)y = 0$$

that contains no constant multiple of a term in the complementary function y_c of (a). This particular solution is $y_p = 0$.

Theorem 25.28.

(25.29) $$P(D)[P^{-1}(D)Q] = Q \quad \text{or} \quad P(D)\left[\frac{1}{P(D)}\,Q\right] = Q.$$

Proof. Let y_p be a particular solution of

(a) $$P(D)y = Q.$$

Therefore

(b) $$P(D)y_p = Q.$$

By (a) and Definition 25.2,

(c) $$y_p = \frac{1}{P(D)}\,Q.$$

In (b) replace y_p by its value in (c). The result is (25.29).

LESSON 25B. Solution of (25.1) by Means of Inverse Operators.
As stipulated at the beginning of Lesson 25, the function $Q(x)$ of (25.1)
may contain only such terms as b, x^k, e^{ax}, sin ax, cos ax, or a finite combination of such terms, where a and b are constants and k is a positive
integer. We shall first consider each of these functions individually and
then combinations of them.

    1. If $Q(x) = bx^k$ and $P(D) = D - a_0$, then (25.1) becomes

(a) $(D - a_0)y = bx^k$, $y' - a_0 y = bx^k$, $a_0 \neq 0$.

The complementary function of (a) is $y_c = ce^{a_0 x}$. Hence a trial solution
y_p, by the method of Lesson 21A, is

(b) $y_p = A_1 x^k + A_2 x^{k-1} + A_3 x^{k-2} + A_4 x^{k-3} + \cdots + A_k x + A_{k+1}$.

Differentiation of (b) gives

(c) $y_p' = A_1 k x^{k-1} + A_2(k-1)x^{k-2} + A_3(k-2)x^{k-3} + \cdots + A_k$.

Therefore, by (b) and (c), y_p will be a solution of (a) if

(d) $y_p' - a_0 y_p = -a_0 A_1 x^k + (A_1 k - a_0 A_2)x^{k-1}$
$$+ [A_2(k-1) - a_0 A_3]x^{k-2}$$
$$+ [A_3(k-2) - a_0 A_4]x^{k-3} + \cdots + (A_k - a_0 A_{k+1})$$
$$= bx^k.$$

Equation (d) will be an identity in x, if

(e) $-a_0 A_1 = b$, $A_1 = -\dfrac{b}{a_0}$.

$$A_1 k - a_0 A_2 = 0, \quad A_2 = \frac{A_1 k}{a_0} = -\frac{bk}{a_0^2}.$$

$$A_2(k-1) - a_0 A_3 = 0, \quad A_3 = \frac{A_2(k-1)}{a_0} = -\frac{bk(k-1)}{a_0^3}.$$

$$A_3(k-2) - a_0 A_4 = 0, \quad A_4 = \frac{A_3(k-2)}{a_0} = -\frac{bk(k-1)(k-2)}{a_0^4}.$$

. . . .

$$A_k - a_0 A_{k+1} = 0, \quad A_{k+1} = \frac{A_k}{a_0} = -\frac{bk!}{a_0^{k+1}}.$$

Substituting (e) in (b), we have

(f) $y_p = -\dfrac{b}{a_0}\left[x^k + \dfrac{k}{a_0} x^{k-1} + \dfrac{k(k-1)}{a_0^2} x^{k-2} + \cdots + \dfrac{k!}{a_0^k} \right]$, $a_0 \neq 0$.

We shall now prove that the same particular solution results if we formally expand $1/(D - a_0)$ in ascending powers of D and then perform the necessary differentiations. By Definition 25.2, a particular solution of (a) is

(g) $$y_p = \frac{1}{D - a_0}(bx^k) = \frac{1}{-a_0\left(1 - \dfrac{D}{a_0}\right)}(bx^k)$$

$$= -\frac{1}{a_0}\left[1 + \frac{D}{a_0} + \frac{D^2}{a_0{}^2} + \frac{D^3}{a_0{}^3} + \cdots + \frac{D^k}{a_0{}^k}\right](bx^k),$$

where the last series was obtained by ordinary division. Note that in making this division, *it is not necessary to go beyond the $D^k/a_0{}^k$ term since $D^{k+1}x^k = 0$.* Performing the indicated differentiations, we have

(h) $$y_p = -\frac{b}{a_0}\left[x^k + \frac{k}{a_0}x^{k-1} + \frac{k(k - 1)}{a_0{}^2}x^{k-2} + \cdots + \frac{k!}{a_0{}^k}\right], \quad a_0 \neq 0,$$

which is the same as (f). In general, it has been proved that if

(i) $$P(D)y = (a_nD^n + \cdots + a_1D + a_0)y = bx^k,$$

then

(25.3) $$y_p = \frac{1}{P(D)}(bx^k)$$

$$= \frac{1}{a_0\left(1 + \dfrac{a_1}{a_0}D + \dfrac{a_2}{a_0}D^2 + \cdots + \dfrac{a_n}{a_0}D^n\right)}(bx^k)$$

$$= \frac{b}{a_0}(1 + b_1D + b_2D^2 + \cdots + b_kD^k)x^k, \quad a_0 \neq 0.$$

where $(1 + b_1D + b_2D^2 + \cdots + b_kD^k)/a_0$ is the series expansion of the inverse operator $1/P(D)$ obtained by ordinary division.

If $k = 0$, then (i) becomes $P(D)y = b$. Hence, by (25.3),

(25.31) $$y_p = \frac{1}{P(D)}b = \frac{b}{a_0}, \quad a_0 \neq 0.$$

Example 25.32.   Find a particular solution of

(a) $$y'' - 2y' - 3y = 5, \quad (D^2 - 2D - 3)y = 5.$$

Solution.   Comparing (a) with (i) above we see that $a_0 = -3$, $b = 5$, $k = 0$. Therefore, by (25.31),

(b) $$y_p = \frac{-5}{3}.$$

Example 25.33.   Find a particular solution of

(a) $4y'' - 3y' + 9y = 5x^2,$ $(4D^2 - 3D + 9)y = 5x^2.$

Solution.   Here $P(D) = 4D^2 - 3D + 9$, $a_0 = 9$, $b = 5$. Hence, by (25.3),

(b) $$y_p = \frac{1}{9\left(1 - \dfrac{D}{3} + \dfrac{4}{9}D^2\right)}(5x^2)$$

$$= \frac{5}{9}\left(1 + \frac{D}{3} - \frac{D^2}{3}\right)x^2 = \tfrac{5}{9}(x^2 + \tfrac{2}{3}x - \tfrac{2}{3}).$$

Note that we did not need to go beyond the D^2 term since $D^3(x^2) = 0$.

2. If $Q(x) = bx^k$ and $P(D) = a_nD^n + \cdots + a_1D$, so that $a_0 = 0$, then D is a factor of $P(D)$. Therefore, by Theorem 24.46, we can write $P(D) = D(a_nD^{n-1} + \cdots + a_2D + a_1)$, where $a_1 \neq 0$. If both $a_0 = 0$ and $a_1 = 0$, then D^2 is a factor of $P(D)$, so that we can write $P(D) = D^2(a_nD^{n-2} + \cdots + a_3D + a_2)$. In general, let D^r be a factor of $P(D)$. Then $P(D)y = bx^k$ can be written as

(a) $P(D)y = D^r(a_nD^{n-r} + \cdots + a_{r+1}D + a_r)y = bx^k,$ $a_r \neq 0.$

Therefore, by Definition 25.2,

(b) $$y_p = \frac{1}{D^r(a_nD^{n-r} + \cdots + a_{r+1}D + a_r)}(bx^k),$$ $a_r \neq 0.$

We shall now stipulate that the inverse operator in (b) means

(25.34) $$y_p = \frac{1}{D^r}\left[\frac{1}{a_nD^{n-r} + \cdots + a_{r+1}D + a_r}(bx^k)\right],$$ $a_r \neq 0.$

Comment 25.35.   Since polynomial operators commute, we could also have written, in place of (b) above,

(c) $$y_p = \frac{1}{(a_nD^{n-r} + \cdots + a_{r+1}D + a_r)D^r}(bx^k)$$

$$= \frac{1}{a_nD^{n-r} + \cdots + a_{r+1}D + a_r}\left[\frac{1}{D^r}(bx^k)\right],$$ $a_r \neq 0.$

In effect we would now be integrating first, see (25.25), and then differentiating. Although no harm results, following this order may introduce terms in the solution y_p that are constant multiples of terms in y_c. In that event, we merely eliminate such terms. As exercises, follow the order of procedure given in (c) to find a particular solution of each of the two examples below. In both cases you will obtain a term in y_p that is a constant multiple of a term in y_c.

Example 25.36.   Find a particular solution of

(a) $$y'' - 2y' = 5, \qquad (D^2 - 2D)y = 5.$$

Solution.   Here $P(D) = D^2 - 2D = D(D - 2)$. Therefore by Definition 25.2 and (25.34)

(b) $$y_p - \frac{1}{D}\left[\frac{1}{D - 2}\,(5)\right].$$

By (25.31), with $b = 5$, $a_0 = -2$,

(c) $$\frac{1}{D - 2}\,(5) = -\tfrac{5}{2}.$$

Substituting (c) in (b), and then applying (25.25) to the result, we obtain

(d) $$y_p = \frac{1}{D}\left(-\frac{5}{2}\right) = -\tfrac{5}{2}x.$$

Example 25.37.   Find a particular solution of

(a) $$y^{(5)} - y^{(3)} = 2x^2, \qquad (D^5 - D^3)y = 2x^2.$$

Solution.   Here $P(D) = D^5 - D^3 = D^3(D^2 - 1)$. Therefore, by Definition 25.2 and (25.34),

(b) $$y_p = \frac{1}{D^3}\left[\frac{1}{D^2 - 1}\,(2x^2)\right].$$

By (25.3), with $a_0 = -1$, $b = 2$,

(c) $$\frac{1}{D^2 - 1}\,(2x^2) = \frac{1}{-(1 - D^2)}\,(2x^2) = -2(1 + D^2)x^2$$
$$= -2(x^2 + 2).$$

Substituting (c) in (b), and then applying (25.25) to the result, we obtain

(d) $$y_p = \frac{1}{D^3}\,[-2(x^2 + 2)] = -2\left(\frac{x^5}{60} + \frac{x^3}{3}\right).$$

3. If $Q(x) = be^{ax}$, then (25.1) becomes $P(D)y = be^{ax}$. We shall now prove that a particular solution of this equation is

(25.4) $$y_p = \frac{1}{P(D)}\,be^{ax} = \frac{be^{ax}}{P(a)}, \qquad P(a) \neq 0.$$

Note that the a in $P(a)$ is the same as the exponent a in e^{ax}.

Proof. By (25.1) and (25.11), $P(D)y = be^{ax}$ is equivalent to

(a) $\qquad (a_n D^n + a_{n-1} D^{n-1} + \cdots + a_1 D + a_0)y = be^{ax}$.

Since $P(a) \neq 0$, $(D - a)$ cannot be a factor of $P(D)$. This means that a cannot be a root of the characteristic equation of (a). This in turn implies that the complementary function of (a) cannot have a term e^{ax} in it. Hence the trial solution y_p of (a), by Case 1 of Lesson 21A, is

(b) $\qquad\qquad\qquad y_p = Ae^{ax}$.

Differentiating (b) n times, we obtain

(c) $\qquad\qquad y_p{}' = aAe^{ax}, \qquad y_p{}'' = a^2 Ae^{ax},$
$$y_p{}''' = a^3 Ae^{ax}, \cdots, y_p{}^{(n)} = a^n Ae^{ax}.$$

The substitution of (b) and (c) in (a) gives

(d) $\quad a_n a^n Ae^{ax} + a_{n-1} a^{n-1} Ae^{ax} + \cdots + a_1 a Ae^{ax} + a_0 Ae^{ax} = be^{ax},$
$$Ae^{ax}(a_n a^n + a_{n-1} a^{n-1} + \cdots + a_1 a + a_0) = be^{ax}.$$

By Definition 24.52, the quantity in parenthesis is $P(a)$. The last equation in (d) therefore simplifies to

(e) $\qquad\qquad\qquad AP(a) = b, \quad A = \dfrac{b}{P(a)}.$

Substituting this value of A in (b) gives the expression on the extreme right of (25.4). Hence $be^{ax}/P(a)$ is a particular solution of $P(D)y = be^{ax}$.

Example 25.41. Find a particular solution of

(a) $\quad y''' - y'' + y' + y = 3e^{-2x}, \qquad (D^3 - D^2 + D + 1)y = 3e^{-2x}.$

Solution. Here $P(D) = D^3 - D^2 + D + 1$. Therefore, by (25.4), with $b = 3$, $a = -2$,

(b) $\quad y_p = \dfrac{1}{P(D)} 3e^{-2x} = \dfrac{3e^{-2x}}{P(-2)} = \dfrac{3e^{-2x}}{(-2)^3 - (-2)^2 - 2 + 1}$
$$= -\tfrac{3}{13} e^{-2x}.$$

4. If $Q(x) = b \sin ax$ or $b \cos ax$, no special difficulty arises. For, by (18.84) and (18.85), we can change these functions to their exponential equivalents and use (25.4). Easier, perhaps, is to apply the method of Lesson 21B. We shall use this latter method to solve the following problem.

Example 25.42. Find a particular solution of

(a) $\quad y'' - 3y' + 2y = 3 \sin 2x, \qquad (D^2 - 3D + 2)y = 3 \sin 2x.$

Solution. Since $e^{2ix} = \cos 2x + i \sin 2x$, the imaginary part of a particular solution of

(b) $$(D^2 - 3D + 2)y = 3e^{2ix}$$

will be a solution of (a). Here $P(D) = D^2 - 3D + 2$. Therefore, by (25.4), with $b = 3$, $a = 2i$,

(c) $$y_p = \frac{3e^{2ix}}{P(2i)} = \frac{3e^{2ix}}{(2i)^2 - 6i + 2} = \frac{3e^{2ix}}{-2(1 + 3i)} \cdot \frac{(1 - 3i)}{(1 - 3i)}$$

$$= \frac{3(1 - 3i)}{-20} e^{2ix} = -\tfrac{3}{20}(1 - 3i)(\cos 2x + i \sin 2x)$$

$$= -\tfrac{3}{20}[(\cos 2x + 3 \sin 2x) + i(\sin 2x - 3 \cos 2x)].$$

The imaginary part of (c) is

(d) $$y_p = \tfrac{3}{20}(3 \cos 2x - \sin 2x),$$

which is a particular solution of (a).

5. Exponential Shift Theorem for Inverse Operators.

Theorem 25.5. *If* $P(D)y = ue^{ax}$, *where* $P(D)$ *is a polynomial operator of order* n *and* u *is a polynomial in* x, *then*

(25.51) $$y_p = \frac{1}{P(D)} ue^{ax} = e^{ax} \frac{1}{P(D + a)} u.$$

Proof.

(a) $$ue^{ax} = e^{ax}u.$$

By applying (25.29) to each side of (a), we can write (a) as

(b) $$P(D)\left[\frac{1}{P(D)}(ue^{ax})\right] = e^{ax}P(D + a)\left[\frac{1}{P(D + a)}u\right].$$

By (24.58), $e^{ax}P(D + a)u = P(D)(e^{ax}u)$. Applying this equality to the right side of (b), with $\dfrac{1}{P(D + a)} u$ playing the role of u, we obtain, by (b),

(c) $$P(D)\left[\frac{1}{P(D)}(ue^{ax})\right] = P(D)\left[e^{ax} \frac{1}{P(D + a)} u\right].$$

By (24.21), we can write (c) as

(d) $$P(D)\left[\frac{1}{P(D)} ue^{ax} - e^{ax} \frac{1}{P(D + a)} u\right] = 0.$$

Let y represent the quantity in brackets. Then, by (d) and Comment 25.26,

(e) $\quad y_p = \left[\dfrac{1}{P(D)}(ue^{ax}) - e^{ax}\dfrac{1}{P(D+a)}u\right]_p = \dfrac{1}{P(D)}(0) = 0,$

from which we deduce,

(f) $\qquad\qquad y_p = \dfrac{1}{P(D)}(ue^{ax}) = e^{ax}\dfrac{1}{P(D+a)}u.$

Example 25.52. Find a particular solution of

(a) $\quad y'' - 2y' - 3y = x^2e^{2x}, \qquad (D^2 - 2D - 3)y = x^2e^{2x}.$

Solution. Here $P(D) = D^2 - 2D - 3$. Therefore

$$\begin{array}{ll}
\text{(b)} \quad y_p = \dfrac{1}{P(D)}(x^2e^{2x}) = \dfrac{1}{D^2 - 2D - 3}(x^2e^{2x}) & \text{By Def. 25.2,}\\[2ex]
\qquad = e^{2x}\dfrac{1}{P(D+2)}x^2 & \text{(25.51),}\\[2ex]
\qquad = e^{2x}\dfrac{1}{(D+2)^2 - 2(D+2) - 3}x^2 & \text{Def. 24.52,}\\[2ex]
\qquad = e^{2x}\dfrac{1}{D^2 + 2D - 3}x^2 & \text{Theorems 24.46 and 24.33,}\\[2ex]
\qquad = e^{2x}\dfrac{1}{-3\left(1 - \dfrac{2D}{3} - \dfrac{D^2}{3}\right)}x^2 & \\[2ex]
\qquad = -\dfrac{e^{2x}}{3}(1 + \tfrac{2}{3}D + \tfrac{7}{9}D^2)x^2 & \text{(25.3),}\\[2ex]
\qquad = -\dfrac{e^{2x}}{3}(x^2 + \tfrac{4}{3}x + \tfrac{14}{9}) & \text{Def. 24.13 and (24.11)}
\end{array}$$

6. **Formula (25.4) can be used only if $P(a) \neq 0$. What if $P(a) = 0$?** If $P(a) = 0$, then $(D - a)$ is a factor of $P(D)$. Assume $(D - a)^r$ is a factor of $P(D)$. Therefore we can write

$$P(D) = (D - a)^r F(D), \quad F(a) \neq 0.$$

We shall now prove that if $P(D) = (D - a)^r F(D)$, where $F(a) \neq 0$, and $P(D)y = be^{ax}$, then

(25.6) $\quad y_p = \dfrac{1}{P(D)}(be^{ax}) \equiv \dfrac{1}{(D - a)^r F(D)}(be^{ax}) = \dfrac{bx^r e^{ax}}{r!F(a)}, \quad F(a) \neq 0.$

Proof. We stipulate that the inverse operator in (25.6) shall mean

(a) $\qquad\qquad\qquad \dfrac{1}{(D - a)^r}\left[\dfrac{1}{F(D)}(be^{ax})\right].$

(Remark. As commented in 25.35, no harm results, other than more labor, if the order of performing the inverse operations is interchanged. As remarked there, using this inverted order may introduce terms in y_p that are constant multiples of terms in y_c. In that event we merely eliminate such terms.) By (25.4),

(b) $$\frac{1}{F(D)} be^{ax} = \frac{be^{ax}}{F(a)}, \quad F(a) \neq 0.$$

Substituting (b) in (a), we obtain

(c) $$\frac{1}{(D-a)^r} \left[\frac{b}{F(a)} e^{ax} \right].$$

By Definition 24.52, if $P(D) = (D-a)^r$, then $P(D+a) = (D+a-a)^r = D^r$. In (25.51), let $P(D) = (D-a)^r$ and $u = b/F(a)$. Then (c) is equivalent to

(d) $$e^{ax} \frac{1}{D^r} \left[\frac{b}{F(a)} \right].$$

By (25.25), $D^{-r}[b/F(a)]$ means integrate $[b/F(a)]$ r times, ignoring constants of integration. The first integration gives $bx/F(a)$, the second $bx^2/2!F(a)$, the third $bx^3/3!F(a)$, and finally the rth integration gives $bx^r/r!F(a)$. Hence (d) becomes

(e) $$\frac{e^{ax}bx^r}{r!F(a)}, \quad F(a) \neq 0,$$

which is the same as the last expression in (25.6).

Example 25.61.   Find a particular solution of

(a) $y''' - 5y'' + 8y' - 4y = 3e^{2x}, \qquad (D^3 - 5D + 8D - 4)y = 3e^{2x}.$

Solution.   Here $P(D) = D^3 - 5D^2 + 8D - 4$. Therefore, by Definition 25.2,

(b) $$y_p = \frac{1}{D^3 - 5D^2 + 8D - 4} (3e^{2x}).$$

If we now attempted to apply (25.4) to the right side of (b), we would find that $P(a)$, which here equals $P(2)$, is zero. Hence we must resort to (25.6). Since $(D^3 - 5D^2 + 8D - 4) = (D-2)^2(D-1)$, we have by (b),

(c) $$y_p = \frac{1}{(D-2)^2(D-1)} 3e^{2x}.$$

Comparing (c) with (25.6), we see that $b = 3$, $a = 2$, $r = 2$, $F(D) =$

$D - 1$ so that $F(a) = F(2) = 2 - 1 = 1$. Hence by (25.6), (c) becomes

(d) $$y_p = \frac{3x^2 e^{2x}}{2!}.$$

Comment 25.62. The above solution (d) could also have been found by the method of Lesson 21, Case 3. A good way to discover how much easier the above method is, is to try to obtain (d) by means of that lesson.

Remark. As an exercise, show that if in place of (c), we had inverted the order of the inverse operators and written

(e) $$y_p = \frac{1}{D - 1}\left[\frac{1}{(D - 2)^2}(3e^{2x})\right],$$

and solved it by first applying (25.51) and then (25.4), the solution y_p would have contained additional terms that were constant multiples of terms in y_c.

Example 25.63. Find a particular solution of

(a) $$y'' + 4y' + 4y = 5e^{-2x}.$$

Solution. Here $P(D) = D^2 + 4D + 4 = (D + 2)^2$. Therefore, by Definition 25.2,

(b) $$y_p = \frac{1}{(D + 2)^2 (1)}(5e^{-2x}).$$

If we now attempted to apply (25.4) to the right side of (b), we would find $P(a) = P(-2) = 0$. Hence we must resort to (25.6). Comparing (b) with (25.6), we see that $a = -2$, $r = 2$, $F(D) = 1$, $b = 5$. Hence, by (25.6) and (b),

(c) $$y_p = \frac{5x^2 e^{-2x}}{2}.$$

7. Let $Q(x) = Q_1(x) + Q_2(x) + \cdots + Q_n(x)$. Therefore $P(D)y = Q(x) = Q_1(x) + Q_2(x) + \cdots + Q_n(x)$. We showed in comment 24.25, that a solution y_p of $P(D)y = Q(x)$ is the sum of the respective particular solutions of $P(D)y = Q_1$, $P(D)y = Q_2$, $\cdots$, $P(D)y = Q_n$. It follows, therefore, that if $P(D)y = Q$, then

(25.7) $$y_p = \frac{1}{P(D)}Q \equiv \frac{1}{P(D)}Q_1 + \frac{1}{P(D)}Q_2 + \cdots + \frac{1}{P(D)}Q_n.$$

Because of (25.7) and the rules developed in this lesson, we are now in a position to solve a linear differential equation

(25.71) $$a_n y^{(n)} + \cdots + a_1 y' + a_0 y = Q(x),$$

where $Q(x)$ may contain such terms as b, x^k, e^{ax}, $\sin ax$, $\cos ax$, and combinations of such terms. Here a and b are constants and k is a positive integer.

Example 25.72.   Find the general solution of

(a) $\qquad y'' - y = x^3 + 3x - 4$, $\qquad (D^2 - 1)y = x^3 + 3x - 4$.

Solution.   Here $P(D) = D^2 - 1$. Therefore by Definition 25.2 and (25.3)

(b) $\qquad y_p = \dfrac{1}{D^2 - 1}(x^3 + 3x - 4) = -(1 + D^2)(x^3 + 3x - 4)$.

By (25.7), we may apply $1/(D^2 - 1) = -(1 + D^2)$ to each of the terms in $Q(x)$, namely to x^3, $3x$, -5. Hence, by (b) and (25.7),

(c) $\qquad y_p = -(x^3 + 3x - 4 + 6x) = -(x^3 + 9x - 4)$.

The complementary function of (a) is

(d) $\qquad\qquad\qquad\qquad y_c = c_1 e^x + c_2 e^{-x}$.

The general solution of (a) is therefore the sum of (c) and (d).

Example 25.73.   Find the general solution of

(a) $\qquad\qquad\qquad y''' - y'' + y' - y = 4e^{-3x} + 2x^4$,
$\qquad\qquad (D^3 - D^2 + D - 1)y = 4e^{-3x} + 2x^4$.

Solution.   Here $P(D) = D^3 - D^2 + D - 1$. Therefore by (25.7)

(b) $\quad y_p = \dfrac{1}{D^3 - D^2 + D - 1}(4e^{-3x}) + \dfrac{1}{D^3 - D^2 + D - 1}(2x^4)$.

By (25.4)

(c) $\quad \dfrac{1}{D^3 - D^2 + D - 1} 4e^{-3x} = \dfrac{4e^{-3x}}{(-3)^3 - (-3)^2 + (-3) - 1}$

$\qquad\qquad\qquad\qquad\qquad = -\dfrac{e^{-3x}}{10}\,.$

By (25.3)

(d) $\quad \dfrac{1}{-(1 - D + D^2 - D^3)}(2x^4) = -(1 + D + D^4)(2x^4)$

$\qquad\qquad\qquad\qquad\qquad = -(2x^4 + 8x^3 + 48)$.

Therefore, by (b), (c), and (d),

(e) $\qquad\qquad y_p = -\left(\dfrac{e^{-3x}}{10} + 2x^4 + 8x^3 + 48\right)$.

The roots of the characteristic equation of (a) are $1, i, -i$. Hence

(f) $$y_c = c_1 e^x + c_2 \sin x + c_3 \cos x.$$

The general solution of (a), is therefore the sum of (e) and (f).

EXERCISE 25

1. Evaluate (see Comment 25.22).
   - (a) $(D - 3)^{-1}(x^3 + 3x - 5)$.   (b) $(D - 1)^{-1}x^2$.
   - (c) $(D^2 - 3D + 2)^{-1} \sin 2x$.   (d) $(D^2 - 1)^{-1}(2x)$.
   - (e) $(4D^2 - 5D)^{-1}(x^2 e^{-x})$.

2. Evaluate (see Comment 25.24).
   - (a) $D^{-1}(2x + 3)$.   (b) $D^{-3}x$.   (c) $D^{-2}(3e^{3x})$.   (d) $D^{-2}(2 \sin 2x)$.

Find a particular solution of each of the following differential equations by means of the inverse operator; use the appropriate method outlined in numbers 1 to 7 of Lesson 25B.

3. $y'' + 3y' + 2y = 4$.
4. $y'' + y' + y = x^2$.
5. $y'' - y = 2x$.
6. $y'' - 3y' + 2y = x$.
7. $y' - y = 3x^2$.
8. $y'' + y' = x^2 + 2x$.
9. $y''' - y'' = 2x^3$.
10. $y^{(4)} - y''' + y'' = 6$.
11. $y'' + 3y' + 2y = 12e^x$.
12. $y'' + 3y' + 2y = e^{ix}$.
13. $y'' + y = 3e^{-2x}$.

14. $y'' - y = \sin x$. *Hint.* See 15.
15. $y'' - y = \cos x$.
    Hint. For 14 and 15, solve
    $y'' - y = e^{ix}$.
16. $y'' - 3y' + 2y = 3 \sin x$.
17. $y'' + a^2 y = \sin ax$. *Hint.* See 18.
18. $y'' + a^2 y = \cos ax$.
    Hint. For 17 and 18, solve $y'' + a^2 y = e^{iax}$, $a \neq 0$.
19. $4y'' - 5y' = x^2 e^{-x}$.
20. $y'' + y' + y = 3x^2 e^x$.

21. $y'' - 2y' - 8y = 9xe^x$.
22. $y''' + 3y'' + 3y' + y = e^{-x}(2 - x^2)$.
23. $y'' + 4y = 4x \sin 2x$. *Hint.* See 24.
24. $y'' + 4y = 4x \cos 2x$.
    Hint. For 23 and 24, solve $y'' + 4y = 4xe^{2ix}$.
25. $y'' - y = 2e^x$.
26. $y'' + y' - 2y = 3e^{-2x}$.
27. $y''' - 5y'' + 8y' - 4y = e^x$.
28. $y^{(4)} - 3y''' - 6y'' + 28y' - 24y = e^{2x}$.
29. $y''' - 11y'' + 39y' - 45y = e^{3x}$.
30. $y'' - 2y' + y = 7e^x$.
31. $y''' + y' = \sin x$. *Hint.* See 32.
32. $y''' + y' = \cos x$.
    Hint. For 31 and 32, solve $y''' + y' = e^{ix}$.
33. $y''' - 3y'' + 3y' - y = 2e^x$.
34. $y'' + 3y' + 2y = 8 + 6e^x + 2 \sin x$.
35. $y'' + 3y' + 2y = 2(e^{-2x} + x^2)$.
36. $y^{(5)} + 2y''' + y' = 2x + \sin x + \cos x$.
37. $y' - 3y = x^3 + 3x - 5$.
38. Solve Examples 25.36 and 25.37 by following the order suggested in Comment 25.35. Show that the results differ from the text answers, if they do differ, by a term which is a constant multiple of a term in y_c.

39. Solve Example 25.61, by interchanging the order of the operators in the denominator of (c). Show that your answer differs from the text answer by a term which is a constant multiple of a term in y_c.

ANSWERS 25

1. (a) $-\frac{1}{27}(9x^3 + 9x^2 + 33x - 34)$.   (b) $-(x^2 + 2x + 2)$.

   (c) $\frac{3}{20}\cos 2x - \frac{1}{20}\sin 2x$.   (d) $-2x$.   (e) $\dfrac{e^{-x}}{729}(81x^2 + 234x + 266)$.

2. (a) $x^2 + 3x$.   (b) $x^4/24$.   (c) $e^{3x}/3$.   (d) $-\frac{1}{2}\sin 2x$.

3. $y_p = 2$.

4. $x^2 - 2x$.

5. $-2x$.

6. $y_p = \dfrac{x}{2} + \dfrac{3}{4}$.

7. $y_p = -3(x^2 + 2x + 2)$.

8. $y_p = x^3/3$.

9. $y_p = -2\left(\dfrac{x^5}{20} + \dfrac{x^4}{4} + x^3 + 3x^2\right)$.

10. $y_p = 3x^2$.

11. $y_p = 2e^x$.

12. $y_p = \dfrac{1 - 3i}{10}\,e^{ix}$.

13. $y_p = \frac{3}{5}e^{-2x}$.

14. $y_p = -\frac{1}{2}\sin x$.

15. $y_p = -\frac{1}{2}\cos x$.

16. $y_p = \frac{1}{10}(\sin x + 3\cos x)$.

17. $y_p = -\dfrac{x}{2a}\cos ax$.

18. $y_p = \dfrac{x}{2a}\sin ax$.

19. See 1(e).

20. $y_p = \frac{1}{3}e^x(3x^2 - 6x + 4)$.

21. $y_p = -xe^x$.

22. $y_p = \dfrac{x^3 e^{-x}}{60}(20 - x^2)$.

23. $y_p = \dfrac{x}{4}(\sin 2x - 2x\cos 2x)$.

24. $y_p = \dfrac{x}{4}(2x\sin 2x + \cos 2x)$.

25. $y_p = xe^x$.

26. $y_p = -xe^{-2x}$.

27. xe^x.

28. $y_p = x^3 e^{2x}/30$.

29. $y_p = -x^2 e^{3x}/4$.

30. $y_p = 7x^2 e^x/2$.

31. $y_p = -\frac{1}{2}x\sin x$.

32. $y_p = -\frac{1}{2}x\cos x$.

33. $y_p = x^3 e^x/3$.

34. $y_p = 4 + e^x + \frac{1}{5}(\sin x - 3\cos x)$.

35. $y_p = x^2 - 3x + \frac{7}{2} - 2xe^{-2x}$.

36. $y_p = x^2 + \dfrac{x^2}{8}(\cos x - \sin x)$.

37. See 1(a).

LESSON 26.   Solution of a Linear Differential Equation by Means of the Partial Fraction Expansion of Inverse Operators.

LESSON 26A. Partial Fraction Expansion Theorem. In algebra, an expression of the type

(a) $$\frac{2}{x - 1} + \frac{3}{x + 1}$$

can be simplified by the usual method of finding the least common denominator. There results

(b)
$$\frac{5x-1}{x^2-1}.$$

Conversely, if we start with (b) we can expand it into the partial fractions (a) by means of the **partial fraction expansion theorem** of algebra. We do this by factoring the denominator of (b), and then determining A and B so that

(c)
$$\frac{5x-1}{x^2-1} = \frac{A}{x-1} + \frac{B}{x+1}.$$

Putting the right side of (c) under one common denominator, the equation becomes

(d)
$$\frac{5x-1}{x^2-1} = \frac{Ax+A+Bx-B}{x^2-1}.$$

Since the denominators on both sides of the equal sign are alike, the A's and B's must be chosen so that their respective numerators are also alike, i.e., so that

(e)
$$x(A+B) + (A-B) \equiv 5x-1.$$

Equation (e) will be an identity in x if the coefficients of like powers of x on both sides of the identity sign are equal. Hence we must have

(f)
$$A+B = 5,$$
$$A-B = -1.$$

Solving (f) simultaneously, we find $A = 2$ and $B = 3$. Substituting these values in (c) gives us the partial fraction expansion of (b), namely

(g)
$$\frac{5x-1}{x^2-1} = \frac{2}{x-1} + \frac{3}{x+1}.$$

We thus are able to go from (a) to (b) or from (b) to (a).

Comment 26.1. The general rule for determining the form of each numerator in a partial fraction expansion of a quotient $P(x)/Q(x)$, where $P(x)$ is of *degree less than* $Q(x)$, will be evident from the following example. Let

(h)
$$Q(x) = (x+a)(x^3+b)(x^2+c)^2(x+d)^3,$$

and let $P(x)$ be a polynomial of degree less than $Q(x)$. Then the partial fraction expansion of $P(x)/Q(x)$ will have the form

(i) $$\frac{P(x)}{Q(x)} = \frac{A}{x+a} + \frac{Bx^2 + Cx + D}{x^3 + b} + \frac{Ex + F}{x^2 + c} + \frac{Gx + H}{(x^2 + c)^2}$$

$$+ \frac{I}{x+d} + \frac{J}{(x+d)^2} + \frac{K}{(x+d)^3}.$$

Note that:

1. Each numerator is a polynomial of degree one less than the degree of the term **inside** the parenthesis of its denominator.
2. A term such as $(x^2 + c)^2$ has the exponent two **outside** the parenthesis. Hence $(x^2 + c)$ appears twice in the denominator, once as $(x^2 + c)$, the second time as $(x^2 + c)^2$.
3. A term such as $(x + d)^3$ has exponent three **outside** the parenthesis. Hence $(x + d)$ appears three times in the denominator, once as $(x + d)$, the second time as $(x + d)^2$, and the third time as $(x + d)^3$.

Comment 26.11. If the numerator of a fraction is a constant and the denominator has *distinct* zeros, then there is a neat method which will quickly give the numerators of a partial fraction expansion. Let

(a) $$f(x) = (x - r_1)(x - r_2) \cdots (x - r_k) \cdots (x - r_n),$$

where the zeros $r_1, r_2, \cdots, r_n$ are *distinct*. By (a)

(b) $$f(r_k) = (r_k - r_1)(r_k - r_2) \cdots 0 \cdots (r_k - r_n) = 0.$$

By comment 26.1, the partial fraction expansion of $1/f(x)$ will have the form

(c) $$\frac{1}{f(x)} = \frac{A_1}{x - r_1} + \frac{A_2}{x - r_2} + \cdots + \frac{A_k}{x - r_k} + \cdots + \frac{A_n}{x - r_n}.$$

Multiply (c) by $(x - r_k)$. Since by (b), $f(r_k) = 0$, there results

(d) $$\frac{x - r_k}{f(x) - f(r_k)} = \frac{A_1(x - r_k)}{x - r_1} + \cdots + A_k + \cdots + \frac{A_n(x - r_k)}{x - r_n}$$

Let $x \to r_k$. The left side of (d) will approach $1/f'(r_k)$ and its right side will approach A_k. Hence,

(e) $$A_k = \frac{1}{f'(r_k)}, \quad k = 1, 2, \cdots, n.$$

Substituting (e) in (c), we obtain

(26.12) $$\frac{1}{f(x)} = \frac{1}{f'(r_1)(x - r_1)} + \frac{1}{f'(r_2)(x - r_2)} + \cdots + \frac{1}{f'(r_n)(x - r_n)}.$$

Example 26.13. Find the partial fraction expansion of

(a)
$$\frac{3}{x^2 - 1}.$$

Solution. Comparing (a) with (26.12), we see that $f(x) = x^2 - 1 = (x - 1)(x + 1)$. Therefore, $r_1 = 1$, $r_2 = -1$, $f'(x) = 2x$, $f'(r_1) = f'(1) = 2$, $f'(r_2) = f'(-1) = -2$. Hence, by (26.12),

(b)
$$\frac{3}{x^2 - 1} = 3\left[\frac{1}{2(x - 1)} - \frac{1}{2(x + 1)}\right].$$

Example 26.131. Find the partial fraction expansion of

(a)
$$\frac{3}{(x - 2)(x^2 + x + 1)}.$$

Solution. We can if we wish factor

$$x^2 + x + 1 = \left[\left(x + \frac{1 + \sqrt{3}\,i}{2}\right)\left(x + \frac{1 - \sqrt{3}\,i}{2}\right)\right]$$

and then use (26.12). Or we can use the rules of partial fraction expansion given in Comment 26.1. Using this latter method, we have

(b)
$$\frac{3}{(x - 2)(x^2 + x + 1)} = \frac{A}{x - 2} + \frac{Bx + C}{x^2 + x + 1}$$

$$= \frac{Ax^2 + Ax + A + Bx^2 + Cx - 2Bx - 2C}{(x - 2)(x^2 + x + 1)}.$$

For (b) to be an identity in x, the numerator in the last fraction must equal 3. Hence we must choose A, B, C, so that

(c)
$$(A + B)x^2 + (A - 2B + C)x + (A - 2C) \equiv 3.$$

Equating coefficients of like powers of x on both sides of the identity sign, we obtain

(d)
$$A + B = 0, \qquad A - 2B + C = 0, \qquad A - 2C = 3.$$

Solving (d) simultaneously for A, B, C gives

(e)
$$A = \tfrac{3}{7}, \qquad B = -\tfrac{3}{7}, \qquad C = -\tfrac{9}{7}.$$

Substituting these values in (b), we have

(f)
$$\frac{3}{(x - 2)(x^2 + x + 1)} = \frac{3}{7(x - 2)} - \frac{3(x + 3)}{7(x^2 + x + 1)}.$$

Example 26.132. Find the partial fraction expansion of

(a)
$$\frac{1}{(x + 1)(x - 1)^2}.$$

Solution. Here $F(x) = (x + 1)(x - 1)^2$ has a repeated root. Hence we cannot use (26.12), but must fall back on the rules of partial fraction expansion given in Comment 26.1. Therefore,

(b)
$$\frac{1}{(x + 1)(x - 1)^2} = \frac{A}{x + 1} + \frac{B}{x - 1} + \frac{C}{(x - 1)^2}$$
$$= \frac{A(x - 1)^2 + B(x^2 - 1) + C(x + 1)}{(x - 1)^2(x + 1)}.$$

For (b) to be an identity in x, the numerator in the last fraction must equal one. Hence we must choose A, B, and C so that

(c)
$$A(x - 1)^2 + B(x^2 - 1) + C(x + 1) \equiv 1.$$

Instead of equating coefficients of like powers of x and then solving for A, B, C as we did in the previous example, an alternate simpler method in this case, is to let $x = 1$ in (c). There results

(d)
$$2C = 1, \qquad C = \tfrac{1}{2}.$$

If we let $x = -1$ in (c), we obtain

(e)
$$4A = 1, \qquad A = \tfrac{1}{4}.$$

If we let $x = 0$ in (c), we obtain

(f)
$$A - B + C = 1.$$

With $A = \tfrac{1}{4}$, $C = \tfrac{1}{2}$, we find $B = -\tfrac{1}{4}$. Substituting these values in (b), we obtain

(g)
$$\frac{1}{(x + 1)(x - 1)^2} = \frac{1}{4(x + 1)} - \frac{1}{4(x - 1)} + \frac{1}{2(x - 1)^2}.$$

Comment 26.14. Analogously it can be shown that:

1. If $y_p = \dfrac{1}{P(D)} Q(x)$, then the inverse operator can also be expanded into partial fractions just as if it were an ordinary polynomial.

2. Applying *each member* of a partial fraction expansion of $1/P(D)$ to $Q(x)$ and adding the results will give the same answer as will applying $1/P(D)$ to $Q(x)$.

3. If $P(D)$ has distinct factors, then it is permissible to take advantage of (26.12) to find its partial fraction expansion.

LESSON 26B. **First Method of Solving a Linear Equation by Means of the Partial Fraction Expansion of Inverse Operators.** We illustrate the method by means of examples.

Example 26.15. Find the general solution of

(a) $y''' - 5y'' + 8y' - 4y = 2e^{4x}$, $(D^3 - 5D^2 + 8D - 4)y = 2e^{4x}$.

Solution. Here

(b) $P(D) = D^3 - 5D^2 + 8D - 4 = (D - 2)^2(D - 1)$.

Therefore, by Definition 25.2,

(c) $y_p = \dfrac{1}{D^3 - 5D^2 + 8D - 4}(2e^{4x}) = \dfrac{1}{(D - 2)^2(D - 1)}2e^{4x}$.

Following the partial fraction expansion method outlined above, we find

(d) $\dfrac{1}{(D - 2)^2(D - 1)} = \dfrac{1}{D - 1} - \dfrac{1}{D - 2} + \dfrac{1}{(D - 2)^2}$.

Hence by (d) and Comment 26.14, (c) can be written as

(e) $y_p = \dfrac{1}{D - 1}(2e^{4x}) - \dfrac{1}{D - 2}(2e^{4x}) + \dfrac{1}{(D - 2)^2}(2e^{4x})$.

Applying (25.4) to each term on the right of (e), we obtain, with $b = 2$, $a = 4$, and $P(D)$ equal to the respective denominators,

(f) $y_p = \dfrac{2e^{4x}}{3} - \dfrac{2e^{4x}}{2} + \dfrac{2e^{4x}}{4} = \tfrac{1}{6}e^{4x}$,

which is a particular solution of (a). The roots of the characteristic equation of (a) are 2, 2, 1. Hence

(g) $y_c = (c_1 + c_2 x)e^{2x} + c_3 e^x$.

The general solution of (a) is therefore the sum of (f) and (g).

Example 26.16. Find a particular solution of

(a) $y'' - y = 2e^{3x}$, $(D^2 - 1)y = 2e^{3x}$.

Solution. Here $P(D) = D^2 - 1$. Therefore by Definition 25.2,

(b) $y_p = \dfrac{1}{D^2 - 1}(2e^{3x}) = \dfrac{1}{(D - 1)(D + 1)}(2e^{3x})$.

By Comment 26.14 and (26.12), with $f(x) \equiv P(D) = D^2 - 1$, $f'(x) \equiv$

$P'(D) = 2D$, $r_1 = 1$, $r_2 = -1$, $P'(r_1) = 2$, $P'(r_2) = -2$, we can write
(h) as

(c) $$y_p = \frac{1}{2(D-1)}(2e^{3x}) - \frac{1}{2(D+1)}(2e^{3x}).$$

Applying (25.4) to each term in the right of (c), we obtain

(d) $$y_p = \frac{e^{3x}}{2} - \frac{e^{3x}}{4} = \frac{e^{3x}}{4}.$$

Comment 26.2.   If an inverse operator is of the second order and has distinct factors, we can generalize its partial fraction expansion. By (26.12), if $f(x) = x^2 - (r_1 + r_2)x + r_1 r_2 = (x - r_1)(x - r_2)$, then

(a) $$\frac{1}{f(x)} = \frac{1}{(x - r_1)(x - r_2)} = \frac{1}{f'(r_1)(x - r_1)} + \frac{1}{f'(r_2)(x - r_2)},$$

where r_1 and r_2 are distinct. Here $f'(x) = 2x - (r_1 + r_2)$. Hence $f'(r_1) = 2r_1 - (r_1 + r_2) = r_1 - r_2$ and $f'(r_2) = 2r_2 - (r_1 + r_2) = -r_1 + r_2$. Therefore (a) becomes

(b) $$\frac{1}{(x - r_1)(x - r_2)} = \frac{1}{(r_1 - r_2)}\frac{1}{(x - r_1)} - \frac{1}{(r_1 - r_2)}\frac{1}{(x - r_2)}.$$

Analogously,

(26.21) $$\frac{1}{(D - r_1)(D - r_2)} = \frac{1}{r_1 - r_2}\frac{1}{D - r_1} - \frac{1}{r_1 - r_2}\frac{1}{D - r_2}$$
$$= \frac{1}{r_1 - r_2}\left[\frac{1}{D - r_1} - \frac{1}{D - r_2}\right],$$

which may be taken as a formula for the partial fraction expansion of a second order inverse operator whose zeros r_1 and r_2 are *distinct*.

Example 26.22.   Find a particular solution of

(a) $$y'' + 2y' + 2y = 3xe^x, \qquad (D^2 + 2D + 2)y = 3xe^x.$$

Solution.   Here

(b) $$P(D) = (D^2 + 2D + 2).$$

Therefore by Definition 25.2

(c) $$y_p = \frac{1}{D^2 + 2D + 2}(3xe^x).$$

The roots of $D^2 + 2D + 2$ are $r_1 = -1 + i$, $r_2 = -1 - i$. Hence,

(d) $$r_1 - r_2 = -1 + i + 1 + i = 2i.$$

Therefore, by (26.21) and Comment 26.14, we can write (c) as

(e) $$y_p = \frac{1}{2i}\left[\frac{1}{D+1-i}(3xe^x) - \frac{1}{D+1+i}(3xe^x)\right].$$

Applying (25.51) to each term on the right of (e), we obtain, with $u = 3x$, $a = 1$,

(f) $$y_p = \frac{1}{2i}\left[e^x \frac{1}{D+2-i}(3x) - e^x \frac{1}{D+2+i}(3x)\right].$$

By (25.3), the series expansions of the inverse operators in (f) are [write $1/(D+2-i) = 1/[(2-i)(1 + D/(2-i))]$ and then use ordinary division]

$$\frac{1}{2-i}\left(1 - \frac{D}{2-i} + \cdots\right) \quad \text{and} \quad \frac{1}{2+i}\left(1 - \frac{D}{2+i} + \cdots\right).$$

Therefore, by (25.3), (f) is equal to

(g) $$y_p = \frac{e^x}{2i}\left[\frac{3}{2-i}\left(1 - \frac{D}{2-i} + \cdots\right)x\right.$$
$$\left. - \frac{3}{2+i}\left(1 - \frac{D}{2+i} + \cdots\right)x\right]$$
$$= \frac{3e^x}{2i}\left[\frac{x}{2-i} - \frac{1}{(2-i)^2} - \frac{x}{2+i} + \frac{1}{(2+i)^2}\right]$$
$$= \frac{3e^x}{2i}\left(\frac{2i}{5}x - \frac{8i}{25}\right) = \frac{3}{5}xe^x - \frac{12}{25}e^x,$$

which is a particular solution of (a).

LESSON 26C. A Second Method of Solving a Linear Equation by Means of the Partial Fraction Expansion of Inverse Operators.

Example 26.3. Find a particular solution of

(a) $$y'' - 3y' + 2y = \sin x, \qquad (D^2 - 3D + 2)y = \sin x.$$

Solution. Here

(b) $$P(D) = D^2 - 3D + 2.$$

Therefore, by Definition 25.2,

(c) $$y_p = \frac{1}{D^2 - 3D + 2}\sin x,$$

where y_p is a particular solution of (a). The zeros of $D^2 - 3D + 2$ are

$r_1 = 2, r_2 = 1$. Hence,

$$r_1 - r_2 = 2 - 1 = 1.$$

By (26.21) and Comment 26.14, (c) becomes

(d) $$y_p = \frac{1}{D-2} \sin x - \frac{1}{D-1} \sin x.$$

Let

(e) $$y_{1p} = \frac{1}{D-2} \sin x, \quad y_{2p} = \frac{-1}{D-1} \sin x.$$

Therefore

(f) $$y_p = y_{1p} + y_{2p}.$$

By Definition 25.2, y_{1p} and y_{2p} are particular solutions respectively of

(g) $$(D - 2)y = \sin x, \quad (D - 1)y = -\sin x.$$

A particular solution of each equation in (g) is respectively

(h) $$y_{1p} = -\frac{2 \sin x + \cos x}{5}, \quad y_{2p} = \frac{\sin x + \cos x}{2}.$$

Substituting these values in (f), we obtain

(i) $$y_p = -\tfrac{2}{5} \sin x - \tfrac{1}{5} \cos x + \tfrac{1}{2} \sin x + \tfrac{1}{2} \cos x$$
$$= \tfrac{1}{10} \sin x + \tfrac{3}{10} \cos x,$$

which is a particular solution of (a).

Comment 26.31.   If we had solved (a) by the method of polynomial operators, as outlined in Lesson 24C, we would have written (a) as

$$(D - 2)(D - 1)y = Q(x),$$

and then, in effect, solved two linear equations in *succession*. By the above method, we solve two linear equations *independently*.

EXERCISE 26

1. Find the partial fraction expansion of each of the following.

(a) $\dfrac{2}{x^2 - 1}$.   (b) $\dfrac{3}{(x - 2)(x^2 - 1)}$.   (c) $\dfrac{2x + 1}{x^2 - 1}$.

(d) $\dfrac{2x^2 + 1}{x^2 - 1}$.   (e) $\dfrac{x}{(x - 1)^2}$.   (f) $\dfrac{x + 1}{(x^2 + 1)(x - 1)}$.

Find a particular solution of each of the following equations. Use methods of Lessons 26B or C.

2. $y'' + y = x^2 + e^{2x}$.
3. $y'' - 2y' - 3y = 3e^{ix}$.
4. $y'' - 2y' - 3y = 3 \sin x$.
5. $y'' - 2y' - 3y = 3 \cos x$.
6. $y'' + 4y = 2e^{-x}$.
7. $y'' - 3y' + 2y = xe^{-x}$.
8. $y'' - y' - 6y = x + e^{2x}$.

9. $y''' - 3y'' + 3y' - y = e^x$.
10. $y''' - y'' + y' - y = 4e^{-3x} + 2x^4$.
11. $y''' - y'' = 2x^3$.
12. $y^{(4)} - y''' + y'' = 6$.
13. $y^{(4)} - 3y''' - 6y'' + 28y' - 24y = e^{2x}$.
14. $y''' - 11y'' + 39y' - 45y = e^{3x}$.
15. $y^{(5)} + 2y''' + y' = 2x + \sin x + \cos x$.

ANSWERS 26

1. (a) $\dfrac{1}{x - 1} - \dfrac{1}{x + 1}$.

(b) $\dfrac{1}{x - 2} - \dfrac{3}{2(x - 1)} + \dfrac{1}{2(x + 1)}$.

(c) $\dfrac{3}{2(x - 1)} + \dfrac{1}{2(x + 1)}$.

(d) $2 + \dfrac{3}{2}\left(\dfrac{1}{x - 1} - \dfrac{1}{x + 1}\right)$.

(e) $\dfrac{1}{x - 1} + \dfrac{1}{(x - 1)^2}$.

(f) $\dfrac{-x}{x^2 + 1} + \dfrac{1}{x - 1}$.

2. $y_p = x^2 - 2 + \tfrac{1}{5}e^{2x}$.
3. $-\tfrac{3}{10}(2 \cos x + \sin x) - i\tfrac{3}{10}(2 \sin x - \cos x)$.
4., 5. See 3.

6. $y_p = \tfrac{2}{5}e^{-x}$.

7. $y_p = \tfrac{1}{36}(6xe^{-x} + 5e^{-x})$.

8. $y_p = -\tfrac{1}{4}e^{2x} - \tfrac{1}{36}(6x - 1)$.

9. $x^3e^x/6$.

10. $y_p = -\left(\dfrac{e^{-3x}}{10} + 2x^4 + 8x^3 + 48\right)$.

11. $y_p = -2\left(\dfrac{x^5}{20} + \dfrac{x^4}{4} + x^3 + 3x^2\right)$.

12. $y_p = 3x^2$.

13. $y_p = x^3e^{2x}/30$.

14. $y_p = -x^2e^{3x}/4$.

15. $y_p = x^2 + \dfrac{x^2}{8}(\cos x - \sin x)$.

LESSON 27. The Laplace Transform. Gamma Function.

LESSON 27A. Improper Integral. Definition of a Laplace Transform. For a clearer understanding of the material of this lesson, a knowledge of the meaning of the improper integral $\int_0^\infty f(x)\, dx$ is essential. We shall therefore briefly review this subject for you.

Let $f(x)$ be a continuous function on the interval $I: 0 \leqq x \leqq h$, $h > 0$. If, as $h \to \infty$, the definite integral $\int_0^h f(x)\, dx$ approaches a finite limit K, we say the improper integral $\int_0^\infty f(x)\, dx$ exists and **converges** to this value K. In that event we write

(27.1) $$\int_0^\infty f(x)\, dx = \lim_{h \to \infty} \int_0^h f(x)\, dx = K.$$

If the limit on the right does not exist, we say the improper integral on the left **diverges** and does not exist.

Let $f(x)$ be a continuous function on the interval $I: 0 < x \leqq h, h > 0$. If as $h \to \infty$ and $\epsilon \to 0$, the definite integral $\int_\epsilon^h f(x)\, dx$ approaches a finite limit L, we say the improper integral $\int_0^\infty f(x)\, dx$ exists and converges to this value L. In that event we write

$$(27.11) \qquad \int_0^\infty f(x)\, dx = \lim_{\substack{h \to \infty \\ \epsilon \to 0}} \int_\epsilon^h f(x)\, dx = L.$$

If the limit on the right side does not exist, we say the improper integral on the left diverges and does not exist.

Example 27.111. Determine whether the following integral exists.

(a) $$\int_0^\infty \frac{1}{x+1}\, dx.$$

Solution.

(b) $$\int_0^h \frac{1}{x+1}\, dx = \log\,(h+1).$$

As $h \to \infty$, $\log\,(h+1) \to \infty$. Hence the improper integral (a) diverges and does not exist.

Example 27.112. Evaluate

(a) $$\int_0^\infty \frac{1}{x^2+1}\, dx.$$

Solution.

(b) $$\lim_{h \to \infty} \int_0^h \frac{1}{x^2+1}\, dx = \lim_{h \to \infty} (\text{Arc tan } h - \text{Arc tan } 0) = \frac{\pi}{2}.$$

Hence, by (27.1), the integral (a) exists and converges to $\pi/2$. We can therefore write

$$\int_0^\infty \frac{1}{x^2+1}\, dx = \frac{\pi}{2}.$$

The following additional information will also be needed.

(27.113)

(a) $\lim_{x \to \infty} \dfrac{e^{-sx}}{s} = 0,\quad$ if $s > 0$.

(b) $\lim_{x \to \infty} \dfrac{e^{-sx}}{s^2} = 0,\quad$ if $s > 0$.

(c) $\lim_{x \to \infty} \dfrac{xe^{-sx}}{s} - 0,\quad$ if $s > 0$.

(d) $\lim_{x \to \infty} x^n e^{-sx} = 0,\quad$ if $s > 0$, n real.

Finally we shall need the following theorem, which we state without proof.

Theorem 27.12. *If the improper integral*

$$(27.121) \qquad \int_0^\infty e^{-sx} f(x)\, dx, \quad 0 \leqq x < \infty,$$

converges for a value of $s = s_0$, *then it converges for every* $s > s_0$.

The integral (27.121), if it exists, is a function of s. It is called the Laplace transform of $f(x)$ and is written as $L[f(x)]$. Hence the following definition:

Definition 27.13. Let $f(x)$ be defined on the interval $I: 0 \leqq x < \infty$. Then the **Laplace transform of** $f(x)$ is defined by

$$(27.14) \qquad L[f(x)] = F(s) = \int_0^\infty e^{-sx} f(x)\, dx,$$

where it is assumed that $f(x)$ is a function for which the integral on the right exists for some value of s.

Example 27.15. Find the Laplace transform of the function $f(x) = 1$, $x \geqq 0$.

Solution. By (27.14), with $f(x) = 1$,

$$(a) \qquad L[1] = F(s) = \int_0^\infty e^{-sx}\, dx$$

$$= \lim_{h \to \infty} \int_0^h e^{-sx}\, dx$$

$$= \lim_{h \to \infty} \left(-\frac{1}{s} e^{-sx} \right) \Big|_0^h$$

$$= \lim_{h \to \infty} \left(\frac{-e^{-sh}}{s} + \frac{1}{s} \right)$$

$$= \frac{1}{s}, \quad \text{if } s > 0 \text{ [by (27.113)(a)]}.$$

Example 27.151. Find the Laplace transform of the function $f(x) = \hat{x}$, $x \geqq 0$, where $\hat{x}$ is the function defined by $y = x$, see (6.15).

Solution. By (27.14), with $f(x) = \hat{x}$,

$$(a) \qquad L[\hat{x}] = F(s) = \int_0^\infty x e^{-sx}\, dx$$

$$= \lim_{h \to \infty} \int_0^h x e^{-sx}\, dx$$

$$= \lim_{h \to \infty} \left[e^{-sx} \left(-\frac{x}{s} - \frac{1}{s^2} \right) \right]_0^h$$

$$= \lim_{h \to \infty} \left(-\frac{he^{-sh}}{s} - \frac{e^{-sh}}{s^2} + \frac{1}{s^2} \right)$$

$$= \frac{1}{s^2}, \quad \text{if } s > 0 \text{ [by (27.113)(c) and (b)]}.$$

LESSON 27B.   Properties of the Laplace Transform.

Theorem 27.16. *If the Laplace transform of $f_1(x)$ converges for $s > s_1$ and the Laplace transform of $f_2(x)$ converges for $s > s_2$, then for s greater than the larger of s_1 and s_2,*

$$(27.17) \qquad L[c_1 f_1 + c_2 f_2] = c_1 L[f_1] + c_2 L[f_2],$$

where c_1 and c_2 are constants, i.e., the Laplace transformation is a linear operator.

Proof. By (27.14) and the hypothesis of the theorem,

(a) $$c_1 L[f_1] = c_1 \int_0^\infty e^{-sx} f_1(x) \, dx, \quad s > s_1,$$

$$c_2 L[f_2] = c_2 \int_0^\infty e^{-sx} f_2(x) \, dx, \quad s > s_2.$$

Hence by Theorem 27.12, both of the above integrals exist for all $s > s_1$ and s_2. Therefore for $s > s_1$ and s_2,

(b) $$c_1 L[f_1] + c_2 L[f_2] = \int_0^\infty e^{-sx} c_1 f_1(x) \, dx + \int_0^\infty e^{-sx} c_2 f_2(x) \, dx$$

$$= \int_0^\infty e^{-sx} [c_1 f_1 + c_2 f_2] \, dx = L[c_1 f_1 + c_2 f_2].$$

By repeated application of (27.17), it can be proved that

$$(27.171) \quad L[c_1 f_1 + c_2 f_2 + \cdots + c_n f_n]$$
$$= c_1 L[f_1] + c_2 L[f_2] + \cdots + c_n L[f_n].$$

We state the following theorem without proof.

Theorem 27.172. *If $f_1(x)$ and $f_2(x)$ are each continuous functions of x and $f_1 = f_2$, then $L[f_1] = L[f_2]$. Conversely if $L[f_1] = L[f_2]$, and f_1, f_2 are each continuous functions of x, then $f_1 = f_2$.*

Definition 27.18. If F is the Laplace transform of a *continuous* function f, i.e., if

$$L[f] = F,$$

then the **inverse Laplace transform** of F, written as $L^{-1}[F]$, is f, i.e.,

$$L^{-1}[F] = f.$$

The inverse Laplace transform, in other words, recovers the continuous function f when F is given.

In Example 27.151 we showed that $L[\hat{x}] = 1/s^2$. Hence the continuous function which is the inverse transform of $1/s^2$ is $\hat{x}$, i.e., $L^{-1}[1/s^2] = \hat{x}$.

Theorem 27.19. *The inverse Laplace transformation is a linear operator, i.e.,*

(27.191) $$L^{-1}[c_1 F_1 + c_2 F_2] = c_1 L^{-1}[F_1] + c_2 L^{-1}[F_2].$$

Proof. Let

(a) $$F_1 = L[f_1], \qquad F_2 = L[f_2],$$

where f_1 and f_2 are continuous functions. Therefore by Definition 27.18,

(b) $$L^{-1}[F_1] = f_1, \qquad L^{-1}[F_2] = f_2.$$

By (27.17)

(c) $$L[c_1 f_1 + c_2 f_2] = c_1 L[f_1] + c_2 L[f_2],$$

which, by (a), can be written as

(d) $$L[c_1 f_1 + c_2 f_2] = c F_1 + c_2 F_2.$$

By Definition 27.18, we obtain from (d),

(e) $$L^{-1}[c_1 F_1 + c_2 F_2] = c_1 f_1 + c_2 f_2.$$

Hence by (b), (e) becomes

(f) $$L^{-1}[c_1 F_1 + c_2 F_2] = c_1 L^{-1}[F_1] + c_2 L^{-1}[F_2].$$

LESSON 27C. Solution of a Linear Equation with Constant Coefficients by Means of a Laplace Transform. The method we are about to describe for solving the linear equation

(27.2) $$a_n y^{(n)}(x) + a_{n-1} y^{(n-1)}(x) + \cdots + a_1 y'(x) + a_0 y = f(x),$$

where $a_0, a_1, \cdots, a_n$ are constants and $a_n \neq 0$, is known by the name of the **Laplace transform method.** As the name suggests, it attains its objective by transforming one function into another. However, unlike the differential operator, the Laplace transform accomplishes the transformation by means of the integral in (27.14). We remark that the Laplace method has one advantage over the other methods thus far studied for solving (27.2): it will immediately give a particular solution of (27.2) *satisfying given initial conditions.*

By multiplying (27.2) by e^{-sx} and integrating the result from zero to infinity, we obtain

$$(27.21) \quad \int_0^\infty e^{-sx}[a_n y^{(n)} + a_{n-1} y^{(n-1)} + \cdots + a_1 y' + a_0 y]\, dx$$

$$= \int_0^\infty e^{-sx} f(x)\, dx, \quad s > s_0,$$

which is equivalent to

$$(27.22) \quad a_n \int_0^\infty e^{-sx} y^{(n)}\, dx + a_{n-1} \int_0^\infty e^{-sx} y^{(n-1)}\, dx + \cdots$$

$$+ a_1 \int_0^\infty e^{-sx} y'\, dx + a_0 \int_0^\infty e^{-sx} y\, dx = \int_0^\infty e^{-sx} f(x)\, dx, \quad s > s_0.$$

By (27.14), we note that each integral in (27.22) is a Laplace transform. Hence the equation can be written as

$$(27.23) \quad a_n L[y^{(n)}] + a_{n-1} L[y^{(n-1)}] + \cdots + a_1 L[y']$$
$$+ a_0 L[y] = L[f(x)], \quad s > s_0.$$

Comment 27.24. Equation (27.23) is easily obtained from (27.2). Insert an L after each constant coefficient in (27.2) and place brackets around $y(x)$, $y'(x)$, $\cdots$, $y^{(n)}(x)$ and $f(x)$.

Our next task is to evaluate $L[y^{(n)}]$. By (27.14),

$$(27.241) \qquad\qquad L[y^{(n)}] = \int_0^\infty e^{-sx} y^{(n)}\, dx.$$

If $n = 0$, (27.241) becomes (remember, n is an order, not an exponent)

$$(27.25) \qquad\qquad L[y] = \int_0^\infty e^{-sx} y\, dx, \quad s > s_0.$$

If $n = 1$, (27.241) becomes

$$(27.26) \qquad\qquad L[y'] = \int_0^\infty e^{-sx} y'\, dx, \quad s > s_0.$$

Integrating (27.26) by parts, we obtain with $u = e^{-sx}$, $dv = y'\, dx$,

$$(27.27) \quad L[y'] = \lim_{h \to \infty} [e^{-sx} y(x)]_0^h + s \int_0^\infty e^{-sx} y\, dx$$

$$= \lim_{h \to \infty} [e^{-sh} y(h) - y(0)] + s \int_0^\infty e^{-sx} y\, dx, \quad s > s_0.$$

We now make an additional assumption that $y(x)$, which is the solution we seek, is a function such that*

$$(27.28) \quad \lim_{x \to \infty} e^{-sx} y^{(k)}(x) = 0, \quad k = 0, 1, 2, \cdots, n - 1, \quad s > s_0.$$

Then (27.27) becomes, with the help of (27.25),

$$(27.29) \qquad\qquad L[y'] = -y(0) + sL[y].$$

If $n = 2$, (27.241) becomes

$$(27.3) \qquad\qquad L[y''] = \int_0^\infty e^{-sx} y'' \, dx, \quad s > s_0.$$

The integral in (27.3) can be evaluated by two successive integrations by parts. However an easier method is to make use of (27.29). If in it we replace y by y', we obtain

$$(27.301) \qquad\qquad L[y''] = -y'(0) + sL[y'].$$

By (27.29), we can write (27.301) as

$$(27.31) \qquad \begin{aligned} L[y''] &= -y'(0) + s(-y(0) + sL[y]) \\ &= s^2 L[y] - [y'(0) + sy(0)]. \end{aligned}$$

Similarly, by (27.31),

$$(27.311) \qquad\qquad L[y'''] = s^2 L[y'] - [y''(0) + sy'(0)].$$

By (27.29), (27.311) becomes

$$(27.32) \qquad \begin{aligned} L[y'''] &= s^2(-y(0) + sL[y]) - [y''(0) + sy'(0)] \\ &= s^3 L[y] - [y''(0) + sy'(0) + s^2 y(0)]. \end{aligned}$$

And in general it can be shown that

$$(27.33) \quad L[y^{(n)}] = s^n L[y] - [y^{(n-1)}(0) + sy^{(n-2)}(0) + \cdots \\ + s^{n-2} y'(0) + s^{n-1} y(0)].$$

Hence, by (27.33), we can now write (27.23) as

$$(27.4) \qquad a_n s^n L[y] - a_n[y^{(n-1)}(0) + sy^{(n-2)}(0) + \cdots \\ + s^{n-2} y'(0) + s^{n-1} y(0)] \\ + a_{n-1} s^{n-1} L[y] - a_{n-1}[y^{(n-2)}(0) + sy^{(n-3)}(0) + \cdots \\ + s^{n-3} y'(0) + s^{n-2} y(0)] \\ \cdots \cdots \cdots \cdots \cdots \cdots \cdots \cdots \cdots \cdots \cdots \cdots \cdots \\ + a_2 s^2 L[y] - a_2[y'(0) + sy(0)] \\ + a_1 s L[y] - a_1 y(0) \\ + a_0 L[y] = L[f(x)]$$

---

*See D. V. Widder, *Advanced Calculus*, Prentice-Hall (1961), for proof that the solution $y(x)$, obtained by the Laplace transform method, is a function which satisfies (27.28).

Collecting coefficients of like terms, (27.40) becomes

(27.41) $[a_n s^n + a_{n-1} s^{n-1} + \cdots + a_2 s^2 + a_1 s + a_0] L[y]$
$- [a_n s^{n-1} + a_{n-1} s^{n-2} + \cdots + a_2 s + a_1] y(0)$
$- [a_n s^{n-2} + a_{n-1} s^{n-3} + \cdots + a_3 s + a_2] y'(0)$
$\cdots \cdots \cdots \cdots \cdots \cdots \cdots \cdots \cdots \cdots \cdots \cdots$
$- [a_n s + a_{n-1}] y^{(n-2)}(0)$
$- a_n y^{(n-1)}(0) = L[f(x)].$

Examine (27.41) carefully. $L[y]$ is the Laplace transform of the solution $y(x)$ we seek, but which we do not know as yet; $y(0)$, $y'(0)$, $\cdots$, $y^{(n-1)}(0)$ are constants given by initial conditions; $L[f(x)]$ is the Laplace transform of the function $f(x)$ which appears in the given linear equation (27.2).

Just as there are integral tables, there are tables of Laplace transforms which will give $L[f(x)]$. By Definition 27.13, $L[f(x)]$ is a function of s. Now look again at (27.41). By solving it for $L[y]$, the right side of the resulting equation will be wholly a function of s. Let us call this function $G(s)$. The problem of finding the particular solution $y(x)$ is thus reduced to one of hunting in the tables for that continuous function y whose Laplace transform is $G(s)$, i.e., $L[y] = G(s)$; $y = L^{-1}[G(s)]$. In short, the Laplace transform method has changed the original differential equation involving derivatives, to an algebraic equation involving a function of s.

Note. If the initial conditions give the values of y, y', y'', $\cdots$, $y^{(n-1)}$ at $x = x_0 \neq 0$, it is always possible to translate the axes by letting $x = \bar{x} + x_0$, so that $\bar{x} = 0$ when $x = x_0$. The given differential equation can then be solved in terms of $\bar{x}$, and $\bar{x}$ replaced afterwards by $x - x_0$.

In solving linear equations by the Laplace transform method, we may use either the two equations (27.23) and (27.33), or the equivalent equation (27.41).

Comment 27.42. If $f(x) = 0$, then by (27.14)

(27.43) $L[0] = \int_0^\infty e^{-sx} 0 \, dx = 0.$

Example 27.431. Use the method of Laplace transforms to solve

(a) $y' + 2y = 0,$

for which $y(0) = 2$.

Solution. *Method 1. By use of (27.41).* Comparing (a) with (27.2), we see that $n = 1$, $a_1 = 1$, $a_0 = 2$. By (27.43), $L[0] = 0$. Hence (a) becomes, with the aid of the given initial condition $y(0) = 2$ and (27.41),

(b) $(s + 2) L[y] - (1)(2) = 0.$

Therefore

(c)
$$L[y] = \frac{2}{s+2}.$$

Referring to a table of Laplace transforms (there is a short one at the end of Lesson 27D) we find, see (27.82),

(d)
$$L[e^{-2x}] = \frac{1}{s+2}; \quad \text{therefore } L[2e^{-2x}] = \frac{2}{s+2}.$$

Hence by (c), (d) and Theorem 27.172,

(e)
$$y = 2e^{-2x},$$

which is the required solution.

Method 2. By use of (27.23) and (27.33). As noted in comment 27.24, we can write (a) as

(f)
$$L[y'] + 2L(y) = L[0].$$

By (27.33), or directly from (27.29), (f) becomes

(g)
$$sL[y] - y(0) + 2L[y] = L[0].$$

Solving (g) for $L[y]$ and noting from (27.43) that $L[0] = 0$, and from the initial conditions that $y(0) = 2$, we obtain

(h)
$$L[y] = \frac{2}{s+2},$$

which is the same as (c) above.

Example 27.44. Use the method of Laplace transforms to solve

(a)
$$y'' + 2y' + y = 1,$$

for which $y(0) = 2$, $y'(0) = -2$.

Solution. Comparing (a) with (27.2), we see that $n = 2$, $a_2 = 1$, $a_1 = 2$, $a_0 = 1$. Hence by (27.41), (a) can be written as

(b)
$$(s^2 + 2s + 1)L[y] - (s + 2)y(0) - y'(0) = L[1].$$

In Example 27.15, we found that $L[1] = 1/s$. Substituting this value and the initial conditions in (b), and then solving for $L[y]$, we obtain

(c)
$$L[y] = \frac{2s^2 + 2s + 1}{s(s+1)^2} = \frac{1}{s} + \frac{1}{s+1} - \frac{1}{(s+1)^2}.$$

Referring to a table of Laplace transforms we find, see (27.8), (27.82) and (27.83),

(d) $L[1] = \dfrac{1}{s}$, $L[e^{-x}] = \dfrac{1}{s+1}$, $L[xe^{-x}] = \dfrac{1}{(s+1)^2}$.

Hence by (27.171),

(e) $L[1 + e^{-x} - xe^{-x}] = \dfrac{1}{s} + \dfrac{1}{s+1} - \dfrac{1}{(s+1)^2}$.

Therefore by (c), (e) and Theorem 27.172,

(f) $y = 1 + e^{-x} - xe^{-x}$.

Alternate Method of Solution. As noted in Comment 27.24, we can write (a) as

(g) $L[y''] + 2L[y'] + L[y] = L[1]$.

By (27.33), or directly from (27.31) and (27.29), (g) becomes

(h) $s^2 L[y] - y'(0) - sy(0) - 2y(0) + 2sL[y] + L[y] = L[1]$,

which simplifies to (b) above.

Example 27.45.   Solve

(a) $y'' + 3y' + 2y = 12e^{2x}$,

for which $y(0) = 1; y'(0) = -1$.

Solution.   By (27.41), (a) can be written as

(b) $(s^2 + 3s + 2)L[y] - (s+3)y(0) - y'(0) = L[12e^{2x}] = 12L[e^{2x}]$.

From a table of Laplace transforms, we find [see (27.82)], $L[e^{2x}] = 1/(s-2)$. Substituting this value and the initial conditions in (b), we obtain

(c) $L[y] = \dfrac{s^2 + 8}{(s+2)(s+1)(s-2)} = \dfrac{3}{s+2} - \dfrac{3}{s+1} + \dfrac{1}{s-2}$.

Referring to a table of Laplace transforms we find, see (27.82),

(d) $L[e^{-2x}] = \dfrac{1}{s+2}$, $L[e^{-x}] = \dfrac{1}{s+1}$, $L[e^{2x}] = \dfrac{1}{s-2}$.

Hence by (27.171),

(e) $L[3e^{-2x} - 3e^{-x} + e^{2x}] = \dfrac{3}{s+2} - \dfrac{3}{s+1} + \dfrac{1}{s-2}$.

Therefore by (c) and (e)

(f) $y = 3e^{-2x} - 3e^{-x} + e^{2x}$.

Alternate Method of Solution. As noted in Comment 27.24, we can write (a) as

(g) $$L[y''] + 3L[y'] + 2L[y] = 12L[e^{2x}].$$

By (27.31) and (27.29), (g) becomes

(h) $s^2L[y] - y'(0) - sy(0) - 3y(0) + 3sL[y] + 2L[y] = 12L[e^{2x}],$

which reduces to (b) above.

LESSON 27D. Construction of a Table of Laplace Transforms.
In this lesson, we shall find the Laplace transforms of a few simple functions.

$$(27.5) \qquad L[k] = \int_0^\infty e^{-sx} k \, dx = k \lim_{h \to \infty} \int_0^h e^{-sx} \, dx$$

$$= k \lim_{h \to \infty} \left[\frac{-e^{-sx}}{s} \right]_0^h = k \lim_{h \to \infty} \left[\frac{-e^{-sh}}{s} + \frac{1}{s} \right]$$

$$= \frac{k}{s}, \quad \text{if } s > 0 \text{ [by (27.113)(a)].}$$

$$(27.51) \quad L[x^n] = \int_0^\infty e^{-sx} x^n \, dx = \lim_{h \to \infty} \int_0^h e^{-sx} x^n \, dx$$

$$= \lim_{h \to \infty} e^{-sx} \left(\frac{-x^n}{s} - \frac{nx^{n-1}}{s^2} - \frac{n(n-1)x^{n-2}}{s^3} \right.$$

$$\left. - \cdots - \frac{n!x}{s^n} - \frac{n!}{s^{n+1}} \right) \Big|_0^h$$

$$= \frac{n!}{s^{n+1}}, \quad s > 0, n = 1, 2, \cdots.$$

[If $s > 0$, then by (27.113), $e^{-sh} h^n \to 0$ as $h \to \infty$, and when $x = 0$, each term excepting the last equals zero.]

$$(27.52) \quad L[e^{ax}] = \int_0^\infty e^{-sx} e^{ax} \, dx = \int_0^\infty e^{(a-s)x} \, dx = \lim_{h \to \infty} \left[\frac{e^{(a-s)x}}{a-s} \right]_0^h$$

$$= \frac{1}{a-s} \lim_{h \to \infty} (e^{(a-s)h} - 1) = \frac{1}{s-a}, \quad \text{if } s > a.$$

By (18.84) and (27.52),

$$(27.53) \qquad L[\sin ax] = L \left[\frac{e^{iax} - e^{-iax}}{2i} \right]$$

$$= \frac{1}{2i} \left(\frac{1}{s - ia} - \frac{1}{s + ia} \right)$$

$$= \frac{1}{2i} \left(\frac{2ai}{s^2 + a^2} \right) = \frac{a}{s^2 + a^2}.$$

Many theorems exist which will aid in the computations of the Laplace transforms of more complicated functions. Unfortunately we cannot enter into a detailed study of them. However, to show you the power of these theorems, we shall state below a relatively simple one, and use it to find the Laplace transforms of functions which otherwise would be more difficult to obtain.

Theorem 27.6. *If*

$$(27.61) \qquad F(s) = L[f(x)] = \int_0^\infty e^{-sx} f(x) \, dx, \quad s > s_0,$$

then

$$(27.62) \quad F'(s) = -L[xf(x)] = -\int_0^\infty e^{-sx} x f(x) \, dx, \quad s > s_0,$$

$$F''(s) = L[x^2 f(x)] = \int_0^\infty e^{-sx} x^2 f(x) \, dx, \quad s > s_0,$$

$$\vdots$$

$$F^{(n)}(s) = (-1)^n L[x^n f(x)] = (-1)^n \int_0^\infty e^{-sx} x^n f(x) \, dx, \quad s > s_0.$$

Note. Each Laplace transform in (27.62) can be obtained by differentiating the previous function of s with respect to s. The theorem in effect states that if $F(s) = L[f(x)]$, then one can find the Laplace transform of $xf(x)$ by differentiating $-F(s)$; of $x^2 f(x)$ by differentiating $F(s)$ twice, etc.

Example 27.63. Compute

(a) \qquad\qquad 1. $L[x \sin ax]$, \qquad 2. $L[x^2 \sin ax]$.

Solution. By (27.53)

$$(b) \qquad\qquad L[\sin ax] = \frac{a}{s^2 + a^2} = F(s).$$

Hence by (b) and Theorem 27.6,

$$(c) \qquad\qquad L[x \sin ax] = -F'(s) = \frac{2as}{(s^2 + a^2)^2},$$

and

$$(d) \qquad\qquad L[x^2 \sin ax] = F''(s) = \frac{2a(3s^2 - a^2)}{(s^2 + a^2)^3}.$$

Another theorem, called the **Faltung theorem,** which is helpful in evaluating integrals and one which we shall prove, is the following.

Theorem 27.7. *If*

$$(27.71) \qquad F(s) = L[f(x)] \quad \text{and} \quad G(s) = L[g(x)],$$

then

$$(27.72) \quad L\left[\int_0^x f(x-t)g(t)\,dt\right] = L\left[\int_0^x f(t)g(x-t)\,dt\right]$$

$$= L[f(x)] \cdot L[g(x)] = F(s) \cdot G(s).$$

Proof. Let $u = x - t$. Therefore with x constant, $du = -dt$; $u = 0$ when $t = x$, and $u = x$ when $t = 0$. Making these substitutions in the first integral in (27.72), it becomes

$$(27.73) \quad L\left[\int_x^0 -f(u)g(x-u)\,du\right] = L\left[\int_0^x f(u)g(x-u)\,du\right].$$

In the second integral of (27.73), replace the dummy variable of integration u by t. The result is the second integral in (27.72). Hence we have proved the first equality in (27.72).

By (27.14),

$$(27.74) \quad L\left[\int_{t=0}^x f(x-t)g(t)\,dt\right] = \int_{x=0}^{\infty} e^{-sx}\left[\int_{t=0}^x f(x-t)g(t)\,dt\right]dx.$$

$$= \int_{x=0}^{\infty}\left[\int_{t=0}^x e^{-sx}f(x-t)g(t)\,dt\right]dx.$$

In Fig. 27.741, the shaded part indicates the region in the (x,t) plane

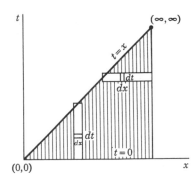

Figure 27.741

over which the integrations on the right of (27.74) take place. The first integration from $t = 0$ to $t = x$ will give the area of the vertical rectangle of width dx. The second integration from $x = 0$ to $x = \infty$ will give the area of the entire shaded region.

Let us now invert the order of integration, i.e., let us integrate first over dx and then over dt. A first integration, see Fig. 27.741, from $x = t$ to $x = \infty$, will give the area of the horizontal rectangle of width dt. A second integration from $t = 0$ to $t = \infty$ will give the area of the entire shaded region. Hence we can write (27.74) as

$$(27.75) \quad L\left[\int_{t=0}^x f(x-t)g(t)\,dt\right] = \int_{t=0}^{\infty}\left[\int_{x=t}^{\infty} e^{-sx}f(x-t)g(t)\,dx\right]dt$$

$$= \int_{t=0}^{\infty} g(t)\left[\int_{x=t}^{\infty} e^{-sx}f(x-t)\,dx\right]dt.$$

In the last integral of (27.75), let $w = x - t$. Then $dw = dx$ (remember t is a constant in this integration) $f(x \quad t) = f(w)$ and $e^{-sx} = e^{-s(w+t)}$. Also $w = 0$ when $x = t$, and $w \to \infty$ when $x \to \infty$. Making all these substitutions in this integral and then changing w, the dummy variable of integration, back to x, we obtain

$$(27.76) \quad L\left[\int_{t=0}^{x} f(x-t)g(t)\, dt\right] = \int_{t=0}^{\infty} g(t)\left[\int_{x=0}^{\infty} e^{-sx}e^{-st}f(x)\, dx\right] dt$$

$$= \int_{0}^{\infty} e^{-st}g(t)\, dt \int_{0}^{\infty} e^{-sx}f(x)\, dx$$

$$= \int_{0}^{\infty} e^{-sx}g(x)\, dx \int_{0}^{\infty} e^{-sx}f(x)\, dx$$

$$= L[g(x)] \cdot L[f(x)].$$

By (27.71), the last expression on the right of (27.76) is $G(s)F(s)$.

Example 27.77. Use Theorem 27.7 to show that

(a)

1. $\displaystyle\int_{0}^{x}(x-t)^a t^b\, dt = \frac{a!b!}{(a+b+1)!}\, x^{a+b+1},$

$$a > -1,\, b > -1,\, x > 0.$$

2. $\displaystyle\int_{0}^{1}(1-t)^a t^b\, dt = \frac{a!b!}{(a+b+1)!}.$

Solution. Let

(b) $$f(x) = x^a, \qquad g(x) = x^b.$$

Then

(c) $$f(x-t) = (x-t)^a, \qquad g(t) = t^b.$$

By (27.72)

(d) $$L\left[\int_{0}^{x} f(x-t)g(t)\, dt\right] = L[f(x)] \cdot L[g(x)].$$

Substituting (b) and (c) in (d), we obtain

(e) $$L\left[\int_{0}^{x}(x-t)^a t^b\, dt\right] = L(x^a) \cdot L(x^b).$$

From a table of Laplace transforms, we find [see (27.81)]

(f) $$L[x^a] \cdot L[x^b] = \frac{a!}{s^{a+1}} \cdot \frac{b!}{s^{b+1}} = \frac{a!b!}{s^{a+b+2}}$$

$$= L\left[\frac{a!b!}{(a+b+1)!}\, x^{a+b+1}\right].$$

Replacing the right side of (e) by its value in (f), we have by Theorem 27.172,

$$\text{(g)} \qquad \int_0^x (x - t)^a t^b \, dt = \frac{a!b!}{(a + b + 1)!} x^{a+b+1}.$$

To prove (a) 2, set $x = 1$ in (g).

Remark. The functions in (a) are known as **beta functions**.

Short Table of Laplace Transforms

	If $f(x) =$	Then $L[f(x)] = F(s) =$
(27.8)	k	$\dfrac{k}{s}, \quad s > 0$
(27.81)	x^n	$\dfrac{n!}{s^{n+1}}, \quad s > 0, n = 1, 2, \cdots$
(27.82)	e^{ax}	$\dfrac{1}{s - a}, \quad s > a$
(27.83)	$x^n e^{ax}$	$\dfrac{n!}{(s - a)^{n+1}}, \quad s > a, n = 1, 2, \cdots$
(27.84)	$\sin ax$	$\dfrac{a}{s^2 + a^2}$
(27.85)	$\cos ax$	$\dfrac{s}{s^2 + a^2}$
(27.86)	$x \sin ax$	$\dfrac{2as}{(s^2 + a^2)^2}$
(27.87)	$x \cos ax$	$\dfrac{s^2 - a^2}{(s^2 + a^2)^2}$
(27.88)	$e^{ax} \sin bx$	$\dfrac{b}{(s - a)^2 + b^2}$
(27.89)	$e^{ax} \cos bx$	$\dfrac{s - a}{(s - a)^2 + b^2}$
(27.9)	$\displaystyle\int_0^x f(x - t)g(t) \, dt$	$L[f(x)] \cdot L[g(x)] = F(s)G(s)$

LESSON 27E. The Gamma Function. By means of a function called the gamma function, it is possible to give meaning to the factorial function $n!$—ordinarily defined only for positive integers—when n is any number except a negative integer. Formulas such as (27.81) and (27.83) will then have meaning when $n = 0$ or $\frac{1}{2}$ or $-\frac{1}{2}$ or $2\frac{1}{4}$, etc. We digress momentarily, therefore, to give you those essentials of the gamma function which we shall need for our present and future purposes.

Definition 27.91. The **gamma function of** k, written as $\Gamma(k)$, is defined by the improper integral

$$(27.92) \qquad \Gamma(k) = \int_0^\infty x^{k-1} e^{-x} \, dx, \quad k > 0.$$

If $k = 1$, (27.92) becomes

$$(27.93) \qquad \Gamma(1) = \int_0^\infty e^{-x} \, dx = \lim_{h \to \infty} \left[-e^{-x}\right]_0^h = 1.$$

Integration of (27.92) by parts gives, with $u = e^{-x}$, $dv = x^{k-1} \, dx$,

$$(27.94) \qquad \Gamma(k) = \lim_{h \to \infty} \left[\frac{e^{-x} x^k}{k}\right]_0^h + \frac{1}{k} \int_0^\infty x^k e^{-x} \, dx, \quad k > 0.$$

By (27.113), the first term in the right of (27.94) approaches zero as $h \to \infty$, and is zero when $x = 0$. The second term by (27.92) equals $\Gamma(k + 1)/k$. Substituting these values in (27.94), we obtain

$$(27.95) \qquad \Gamma(k) = \frac{1}{k} \Gamma(k + 1), \quad k > 0.$$

Hence by (27.95)

$$(27.96) \qquad \Gamma(k + 1) = k\Gamma(k), \quad k > 0.$$

By (27.93), $\Gamma(1) = 1$. Therefore by (27.96), when

$$(27.961) \qquad \begin{array}{llll} k = 1: & \Gamma(2) = 1\Gamma(1) = 1 & = 1! \\ k = 2: & \Gamma(3) = 2\Gamma(2) = 2 \cdot 1 & = 2! \\ k = 3: & \Gamma(4) = 3\Gamma(3) = 3 \cdot 2 \cdot 1 & = 3! \\ k = 4: & \Gamma(5) = 4\Gamma(4) = 4 \cdot 3 \cdot 2 \cdot 1 & = 4! \end{array}$$

And in general, when $k = n$, where n is a positive integer,

$$(27.97) \qquad \Gamma(n + 1) = n!$$

By (27.95)

$$(27.971) \qquad \Gamma(k) = \frac{\Gamma(k + 1)}{k}, \quad k \neq 0.$$

In (27.971) replace k by $k + 1$. There results

$$(27.972) \qquad \Gamma(k + 1) = \frac{\Gamma(k + 2)}{k + 1}, \quad k \neq -1.$$

Now substitute in (27.971) the value of $\Gamma(k + 1)$ as given (27.972). We thus obtain

$$(27.973) \qquad \Gamma(k) = \frac{\Gamma(k + 2)}{k(k + 1)}, \quad k \neq 0, -1.$$

If in (27.972) we replace k by $k + 1$, there results

$$(27.974) \qquad \Gamma(k + 2) = \frac{\Gamma(k + 3)}{k + 2}, \quad k \neq -2.$$

Now substitute (27.974) in (27.973). There results

$$(27.975) \qquad \Gamma(k) = \frac{\Gamma(k + 3)}{k(k + 1)(k + 2)}, \quad k \neq 0, -1, -2.$$

In general it can be shown that

$$(27.98) \qquad \Gamma(k) = \frac{\Gamma(k + n)}{k(k + 1)(k + 2) \cdots (k + n - 1)},$$

$$k \neq 0, -1, -2, \cdots, -(n - 1).$$

By (27.971) and (27.98) we can extend the definition of $\Gamma(k)$, which, by (27.92), was defined only for $k > 0$, to include negative values of k, provided $k \neq 0, -1, -2, \cdots$. For example, if $k = -\frac{1}{2}$, then by (27.971), we can define

$$(27.981) \qquad \Gamma(-\tfrac{1}{2}) = -2\Gamma(\tfrac{1}{2}).$$

If, therefore, we know the value of $\Gamma(\frac{1}{2})$, we then also know the value of $\Gamma(-\frac{1}{2})$. And if we know the value of $\Gamma(-\frac{1}{2})$, we then also know the value of $\Gamma(-\frac{3}{2})$. For by (27.971), we can define

$$(27.982) \qquad \Gamma(-\tfrac{3}{2}) = -\tfrac{2}{3}\Gamma(-\tfrac{1}{2}), \text{ etc.}$$

Tables of values of the gamma function exist just as they do for $\sin x$, $\log x$, or e^x. From such tables we find, for example, $\Gamma(\frac{1}{2}) = \sqrt{\pi}$. Hence by (27.981) and (27.982),

$$(27.983) \qquad \Gamma(-\tfrac{1}{2}) = -2\sqrt{\pi}.$$
$$\Gamma(-\tfrac{3}{2}) = -\tfrac{2}{3}(-2\sqrt{\pi}) = \tfrac{4}{3}\sqrt{\pi}, \text{ etc.}$$

By means of the gamma function and its extended definition, we are thus able to give meaning to $n!$ when n is any number excepting a negative integer. For, by (27.97), tables of values of the gamma function, and (27.96),

$$(27.984) \qquad 0! = \Gamma(1) = 1.$$
$$(-\tfrac{1}{2})! = \Gamma(\tfrac{1}{2}) = \sqrt{\pi}.$$
$$(-\tfrac{3}{2})! = \Gamma(-\tfrac{1}{2}) = -2\sqrt{\pi}.$$
$$(\tfrac{1}{2})! = \Gamma(\tfrac{3}{2}) = \tfrac{1}{2}\Gamma(\tfrac{1}{2}) = \tfrac{1}{2}\sqrt{\pi}.$$

In Fig. 27.985, we have drawn a graph of the gamma function.

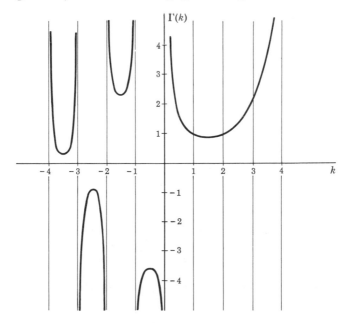

Figure 27.985

Example 27.986. Compute

(a) $$L[x^{-1/2}].$$

Solution. By (27.14)

(b) $$L[x^{-1/2}] = \int_0^\infty e^{-sx} x^{-1/2}\, dx.$$

You will find in a table of integrals, that the value of the improper integral on the right of (b) is $\Gamma(\tfrac{1}{2})/\sqrt{s}$. By (27.984), $\Gamma(\tfrac{1}{2}) = (-\tfrac{1}{2})! = \sqrt{\pi}$. Hence (b) becomes

(c) $$L[x^{-1/2}] = \frac{\sqrt{\pi}}{\sqrt{s}}.$$

Example 27.987. Compute

(a) $$L[x^{n-1/2}], \quad n = 1, 2, 3, \cdots.$$

Solution. By (c) of Example 27.986

(b) $$L[x^{-1/2}] = \sqrt{\pi}\, s^{-1/2} = F(s).$$

Therefore by (b) and (27.62) of Theorem 27.6,

(c) $\qquad L[x^{1/2}] = L[x(x^{-1/2})] = -F'(s) = \dfrac{\sqrt{\pi}}{2} s^{-3/2},$

$\qquad\qquad L[x^{3/2}] = L[x^2(x^{-1/2})] = F''(s) = \dfrac{3\sqrt{\pi}}{2^2} s^{-5/2},$

$\qquad\qquad L[x^{5/2}] = L[x^3(x^{-1/2})] = -F'''(s) = \dfrac{3 \cdot 5\sqrt{\pi}}{2^3} s^{-7/2}.$

And in general

(d) $\quad L[x^{n-1/2}] = L[x^n(x^{-1/2})] = (-1)^n F^{(n)}(s)$

$\qquad\qquad = \dfrac{1 \cdot 3 \cdot 5 \cdots (2n-1)\sqrt{\pi}}{2^n} s^{-(n+1/2)}, \quad n = 1, 2, 3, \cdots.$

Example 27.988. Compute

(a) $\qquad\qquad\qquad\qquad L[x^{-1/2}e^{ax}].$

Solution. By (27.14)

(b) $\qquad\qquad L[x^{-1/2}e^{ax}] = \displaystyle\int_0^\infty e^{-(s-a)x} x^{-1/2}\, dx.$

The integral on the right of (b) has the value $\Gamma(\frac{1}{2})/\sqrt{s-a}$. By (27.984) $\Gamma(\frac{1}{2}) = (-\frac{1}{2})! = \sqrt{\pi}$. Hence (b) becomes

(c) $\qquad\qquad\qquad L[x^{-1/2}e^{ax}] = \dfrac{\sqrt{\pi}}{\sqrt{s-a}}, \quad s > a.$

Additional Table of Laplace Transforms

	If $f(x) =$	Then $L[f(x)] = F(s) =$
(27.99)	$x^{-1/2}$	$\dfrac{\sqrt{\pi}}{\sqrt{s}}$
(27.991)	$x^{n-1/2}$	$\dfrac{1 \cdot 3 \cdot 5 \cdots (2n-1)\sqrt{\pi}}{2^n} s^{-(n+1/2)}, \; n = 1, 2 \cdots$
(27.992)	x^n	$\dfrac{n!}{s^{n+1}}, \; n > -1$
(27.993)	$x^{-1/2}e^{ax}$	$\dfrac{\sqrt{\pi}}{\sqrt{s-a}}$
(27.994)	$x^{n-1/2}e^{ax}$	$\dfrac{1 \cdot 3 \cdot 5 \cdots (2n-1)\sqrt{\pi}}{2^n} (s-a)^{-(n+1/2)}, \; n-1, 2, \cdots$
(27.995)	$x^n e^{ax}$	$\dfrac{n!}{(s-a)^{n+1}}, \; n > -1$

Following exactly the method outlined in Example 27.987, you will find

$$(27.989) \quad L[x^{n-1/2}e^{ax}] = \frac{1 \cdot 3 \cdot 5 \cdots (2n-1)\sqrt{\pi}}{2^n} (s-a)^{-(n+1/2)},$$

$$s > a, \, n = 1, 2, 3, \cdots.$$

Remark. We see, therefore, that with the gamma function, the Laplace transforms of x^n and $x^n e^{ax}$ as given in (27.81) and (27.83) now also have meaning when $n > -1$.

EXERCISE 27

1. Evaluate those of the following improper integrals which converge.

(a) $\int_0^\infty \frac{x \, dx}{\sqrt{x^2+1}}$.   (b) $\int_0^\infty \frac{2x \, dx}{(1+x^2)^2}$.   (c) $\int_0^\infty \frac{dx}{x^2}$.

(d) $\int_0^\infty \frac{dx}{(x+1)^3}$.

Find the Laplace transform of each of the functions 2–5.

2. $\sinh ax, \, x \geqq 0$.   3. $\cosh ax, \, x \geqq 0$.   4. $ax + b, \, x \geqq 0$.
5. $f(x) = k, \quad 0 \leqq x < 1$.
         $= 0, \quad x \geqq 1$.

With the aid of Theorem 27.6, find the Laplace transform of each of the following functions.

6. $x \sinh ax$.   7. $x \cosh ax$.   8. $x \cos ax$.
9. xe^{ax}.   10. $x^2 e^{ax}$.   11. $x^3 e^{ax}$.

Use the method of Laplace transforms to find a solution of each of the following differential equations satisfying the given initial conditions.

12. $y' - y = 0, \quad y(0) = 1$.
13. $y' - y = e^x, \quad y(0) = 1$.
14. $y' + y = e^{-x}, \quad y(0) = 1$.
15. $y'' + 4y' + 4y = 0, \quad y(0) = 1, \, y'(0) = 1$.
16. $y'' - 2y' + 5y = 0, \quad y(0) = 2, \, y'(0) = 4$.
17. $3y''' + 5y'' + y' - y = 0, \, y(0) = 0, \, y'(0) = 1, \, y''(0) = -1$.
18. $y'' - 5y' - 6y = e^{3x}, \quad y(0) = 2, \, y'(0) = 1$.
19. $y'' - y' - 2y = 5 \sin x, \quad y(0) = 1, \, y'(0) = -1$.
20. $y''' - 2y'' + y' = 2e^x + 2x, \quad y(0) = 0, \, y'(0) = 0, \, y''(0) = 0$.
21. $y'' + y' + y = x^2, \quad y(0) = 1, \, y'(0) = 1$.
22. Evaluate each of the following gamma functions.

(a) $\Gamma(6)$.   (b) $\Gamma(7)$.   (c) $\Gamma(-5/2)$.   (d) $\Gamma(5/2)$.   (e) $\Gamma(7/2)$.

23. Evaluate each of the following factorial functions.

(a) $(-\frac{5}{2})!$ (b) $(-\frac{7}{2})!$ (c) $(\frac{3}{2})!$ (d) $(\frac{5}{2})!$

24. Verify the correctness of (27.989). *Hint.* See the solution of Example 27.987.

25. In Example 27.986, we used the fact that

$$\int_0^\infty x^{-1/2} e^{-sx}\, dx = \Gamma(\tfrac{1}{2})/\sqrt{s}, \quad s > 0.$$

Prove it. *Hint.* Make the substitution $u = sx$. Then use (27.92).

26. Prove, in general, that

$$\int_0^\infty x^n e^{-sx}\, dx = \frac{\Gamma(n+1)}{s^{n+1}}, \quad n > -1,\ s > 0.$$

See hint in 25.

ANSWERS 27

1. (a) Diverges.   (b) 1.   (c) Diverges.   (d) $\tfrac{1}{2}$.

2. $\dfrac{a}{s^2 - a^2}$, $s > a$.   **3.** $\dfrac{s}{s^2 - a^2}$, $s > a$.   **4.** $\dfrac{a + bs}{s^2}$, $s > 0$.

5. $\dfrac{k(1 - e^{-s})}{s}$.   **6.** $\dfrac{2as}{(s^2 - a^2)^2}$.   **7.** $\dfrac{s^2 + a^2}{(s^2 - a^2)^2}$.

8. $\dfrac{s^2 - a^2}{(s^2 + a^2)^2}$.   **9.** $\dfrac{1}{(s - a)^2}$.   **10.** $\dfrac{2}{(s - a)^3}$.   **11.** $\dfrac{3!}{(s - a)^4}$.

12. $y = e^x$.   **13.** $y = (x + 1)e^x$.

14. $y = (x + 1)e^{-x}$.   **15.** $y = (1 + 3x)e^{-2x}$.

16. $y = (2 \cos 2x + \sin 2x)e^x$.

17. $y = \dfrac{9}{16} e^{x/3} + \left(\dfrac{x}{4} - \dfrac{9}{16}\right) e^{-x}$.

18. $y = \tfrac{10}{21}e^{6x} + \tfrac{45}{28}e^{-x} - \tfrac{1}{12}e^{3x}$.

19. $y = \tfrac{1}{3}e^{2x} + \tfrac{1}{6}e^{-x} - \tfrac{3}{2} \sin x + \tfrac{1}{2} \cos x$.

20. $y = x^2 + 4x + 4 + (x^2 - 4)e^x$.

21. $y = e^{-x/2}\left(\cos \dfrac{\sqrt{3}}{2} x + \dfrac{7\sqrt{3}}{3} \sin \dfrac{\sqrt{3}}{2} x\right) + x^2 - 2x$.

22. (a) 5!   (b) 6!   (c) $-\tfrac{8}{15}\sqrt{\pi}$ (d) $\tfrac{3}{4}\sqrt{\pi}$ (e) $\tfrac{15}{8}\sqrt{\pi}$.

23. (a) $\tfrac{4}{3}\sqrt{\pi}$ (b) $-\tfrac{8}{15}\sqrt{\pi}$ (c) $\tfrac{3}{4}\sqrt{\pi}$ (d) $\tfrac{15}{8}\sqrt{\pi}$.

Chapter 6

Problems Leading
to Linear Differential Equations
of Order Two

In this chapter we shall consider the motion of a particle whose equation of motion satisfies a differential equation of the form

(a) $$\frac{d^2x}{dt^2} + 2r\frac{dx}{dt} + \omega_0{}^2x = f(t),$$

where $f(t)$ is a continuous function of t defined on an interval I, and r and ω_0 are positive constants. If $f(t) \equiv 0$, then (a) simplifies to

(b) $$\frac{d^2x}{dt^2} + 2r\frac{dx}{dt} + \omega_0{}^2x = 0.$$

If $r = 0$, then (a) becomes

(c) $$\frac{d^2x}{dt^2} + \omega_0{}^2x = f(t).$$

Finally if both $r = 0$ and $f(t) \equiv 0$, then (a) becomes

(d) $$\frac{d^2x}{dt^2} + \omega_0{}^2x = 0.$$

In the lessons which follow we shall name and discuss each of these four important equations (a), (b), (c), and (d).

LESSON 28. Undamped Motion.

LESSON 28A. Free Undamped Motion. (Simple Harmonic Motion.) Many objects have a natural vibratory motion, oscillating back and forth about a fixed point of equilibrium. A particle oscillating in this manner in a medium in which the resistance or damping factor is negligible is said to execute *free undamped motion*, more commonly called *simple harmonic motion*. Two examples are a displaced helical spring and a pendulum. There are various ways of defining this motion. Ours will be the following.

Definition 28.1. A particle will be said to execute **simple harmonic motion** if its equation of motion satisfies a differential equation of the form

(28.11) $$\frac{d^2x}{dt^2} + \omega_0{}^2 x = 0,$$

where ω_0 is a positive constant, and x gives the position of the particle as a function of the time t.

By the method of Lesson 20D, you can verify that the solution of (28.11) is

(28.12) $$x = c_1 \cos \omega_0 t + c_2 \sin \omega_0 t.$$

By (20.57) we can also write the solution (28.12) in the form

(28.13) $$x = \sqrt{c_1{}^2 + c_2{}^2} \sin (\omega_0 t + \delta) = c \sin (\omega_0 t + \delta),$$

or by (20.58) with $+\delta$ replacing $-\delta$, in the form

(28.14) $$x = \sqrt{c_1{}^2 + c_2{}^2} \cos (\omega_0 t + \delta) = c \cos (\omega_0 t + \delta).$$

Hence an equivalent definition of simple harmonic motion is the following.

Definition 28.141. **Simple harmonic motion** is the motion of a particle whose position x as a function of the time t is given by any of the equations (28.12), (28.13), (28.14).

Example 28.15. A particle moving on a straight line is attracted to the origin by a force F. If the force of attraction is proportional to the distance x of the particle from the origin, show that the particle will execute simple harmonic motion. Describe the motion.

Solution. By hypothesis

(a) $$F = -kx,$$

where $k > 0$ is a proportionality constant. The negative sign is necessary because when the particle is at P_1, see Fig. 28.16, x is positive and F acts in a negative direction; when it is at P_2, x is negative and F acts in a positive direction. F and x, therefore, always have opposite signs.

Hence by (16.1) with s replaced by x, (a) becomes

(b) $$F = m\frac{d^2x}{dt^2} = -kx, \qquad \frac{d^2x}{dt^2} = -\frac{k}{m}x.$$

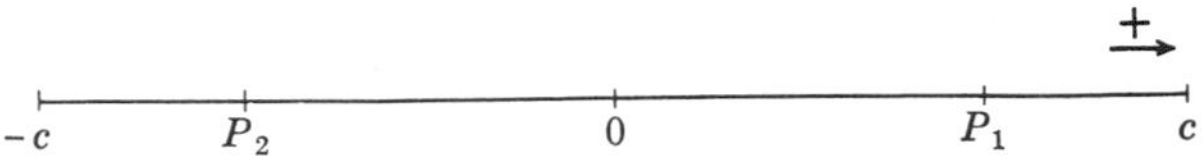

Figure 28.16

Since k and m are positive constants, we may in (b) replace k/m by a new constant ω_0^2. The differential equation of motion (b) is then

(c) $$\frac{d^2x}{dt^2} + \omega_0^2 x = 0,$$

which is the same as (28.11). By Definition 28.1, therefore, the particle executes simple harmonic motion.

Description of the Motion. The solution of (c), by (28.13), is

(d) $$x = c \sin(\omega_0 t + \delta).$$

Differentiation of (d) gives

(e) $$\frac{dx}{dt} = v = c\omega_0 \cos(\omega_0 t + \delta),$$

where v is the velocity of the particle. Since the value of the sine of an angle lies between -1 and 1, we see from (d) that $|x| \leqq |c|$. Hence the particle can never go beyond the points c and $-c$, Fig. 28.16. These points are thus the maximum displacements of the particle from the origin 0. When $|x| = |c|$, we have, by (d), $|\sin(\omega_0 t + \delta)| = 1$, which implies that $\cos(\omega_0 t + \delta) = 0$. Therefore, when $|x| = |c|$ the velocity v, by (e), is zero. We have thus shown that the velocity of the particle at the end points $\pm c$, is zero.

When $x = 0$, i.e., when the particle is at the origin, then, by (d), $\sin(\omega_0 t + \delta) = 0$, from which it follows that $|\cos(\omega_0 t + \delta)| = 1$. Since the value of the cosine of an angle lies between -1 and 1, we see, by (e), that the particle reaches its maximum speed $|v| = |c\omega_0|$, when it is at the origin. For values of x between 0 and $|c|$, you can verify, by means of (d) and (e), that the speed of the particle will be between 0 and its maximum value $|c\omega_0|$; its speed increasing as the particle goes from c, where its speed is zero, to the origin, where its speed is maximum. After crossing the origin, its speed decreases until its velocity is again zero at $-c$. The particle will now move in the other direction—remember the force which is directed toward the origin never ceases to act on the particle—its speed increasing until it reaches its maximum speed at 0, and then decreasing until it is zero again at c. *The particle thus oscillates back and forth, moving in an endless cycle from c to* −c *to* c.

Example 28.17. A particle P moves on the circumference of a circle of radius c with angular velocity ω_0 radians per second. Call Q the point of projection of P on a diameter of the circle. Show that the point Q executes simple harmonic motion.

Solution. See Fig. 28.18. Let

$P_0(x_0,y_0)$ be the position of the particle at time $t = 0$,
$P(x,y)$ be the position of the particle at any later time t,
δ be the central angle formed by a diameter, taken to be the x axis, and the radius OP_0.

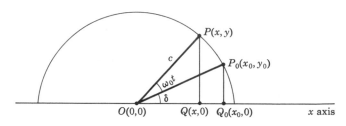

Figure 28.18

In time t the central angle through which the particle has rotated is $\omega_0 t$. (For example if $\omega_0 = 2$ radians/sec, then in $\frac{1}{2}$ second, P has swept out a central angle equal to 1 radian; at the end of 2 seconds the central angle swept out is 4 radians; at the end of t seconds, the central angle swept out is $2t$ radians.) The projections on the diameter of the positions of the particle at $t = 0$ and $t = t$, are respectively, $Q_0(x_0,0)$ and $Q(x,0)$. It is evident from Fig. 28.18 that

$$(28.2) \qquad x = c \cos (\omega_0 t + \delta),$$

an equation which expresses the position of the particle's projection Q on a diameter, as a function of the time t. A comparison of (28.2) with (28.14) shows that they are alike. Hence the point Q executes simple harmonic motion.

The alternate form (28.13) may be obtained by measuring the initial angle δ from the y axis instead of from the x axis, Fig. 28.22. From the figure we see that

$$(28.21) \qquad x = c \cos \left(\frac{\pi}{2} + \delta + \omega_0 t\right) = -c \sin (\omega_0 t + \delta)$$

$$\equiv C \sin (\omega_0 t + \delta).$$

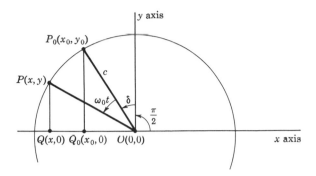

Figure 28.22

Comment 28.23. **Description of the Motion.** As P moves once around the circle, its projection Q (Fig. 28.18) moves to one extremity of the diameter, changes direction and goes to the other extremity, changes direction again and returns to its original position headed in the *same starting direction*. Note the similarity of Q's motion to the motion of the particle in Example 28.15.

LESSON 28B. Definitions in Connection with Simple Harmonic Motion. For convenience, we recopy the three solutions of (28.11). Each, by Definition 28.141, is the equation of motion of a particle executing simple harmonic motion.

(28.24) $$x = c_1 \cos \omega_0 t + c_2 \sin \omega_0 t.$$

(28.25) $$x = c \sin (\omega_0 t + \delta).$$

(28.26) $$x = c \cos (\omega_0 t + \delta).$$

If you will refer to Figs. 28.16 and 28.18, and reread the "description of the motion" paragraphs of Example 28.15 and Comment 28.23, the definitions which follow will be more meaningful to you.

Definition 28.3. The center 0 of the line segment on which the particle moves back and forth is called the **equilibrium position** of the particle's motion.

Definition 28.31. The absolute value of the constant c in (28.25) and in (28.26) is called the **amplitude** of the motion. It is the farthest displacement of the particle from its equilibrium position.

Comment 28.32. By (28.13), $|c| = \sqrt{c_1{}^2 + c_2{}^2}$. If therefore the solution of (28.11) is written in the form (28.24), then the amplitude of the motion is $\sqrt{c_1{}^2 + c_2{}^2}$.

Definition 28.33. The constant δ in (28.25) and in (28.26) is called the **phase** or the **phase angle** of x. [Note that when $t = 0$ in (28.26), $x = c \cos \delta$, and from Fig. 28.18, we observe that $c \cos \delta$ is the initial position $(x_0, 0)$ of the particle's projection.]

If, in (28.26), $t = 0, 2\pi/\omega_0, 4\pi/\omega_0, \cdots, 2n\pi/\omega_0$, n an integer, then for each t, $x = c \cos \delta$. This means that in time $2\pi/\omega_0$, the particle has made one complete revolution around the circle and is back at its starting position, headed in the *same starting direction*. Or equivalently, the particle has made one complete oscillation along a diameter or a line segment and is back at its starting position headed in the same starting direction. Hence the following definition.

Definition 28.34. The constant

$$(28.35) \qquad\qquad T = \frac{2\pi}{\omega_0},$$

where ω_0 is the constant in (28.24), (28.25), and (28.26), is called the **period** of the motion. It is the time it takes the particle to make one complete oscillation about its equilibrium position. If the constant ω_0 is the angular velocity of a particle moving on the circumference of a circle, then the period is the time required for the particle to make one complete revolution or equivalently the time it takes the particle's projection on the diameter to make one complete oscillation about the center of the circle.

For example, if a particle makes two complete revolutions, or equivalently two complete oscillations, in one second, i.e., if $\omega_0 = 4\pi$ rad/sec, then its period by (28.35) is $\frac{1}{2}$ second; if it makes $\frac{1}{4}$ of a revolution or of a complete oscillation in one second, i.e., if $\omega_0 = \pi/2$ rad/sec, then its period is 4 seconds. Note that the reciprocal of T gives the number of complete revolutions or oscillations made by the particle in one second. For example when the period T, which is the time required to make one complete revolution or oscillation, is $\frac{1}{2}$, the particle makes two complete oscillations in one second; when the period $T = 4$, the particle makes $\frac{1}{4}$ of a complete oscillation in one second. Hence the following definition.

Definition 28.36. The constant

$$(28.37) \qquad\qquad \nu = \frac{1}{T} = \frac{\omega_0}{2\pi}$$

is called the **natural (undamped) frequency** of the motion. It gives the number of complete revolutions or cycles made by the particle in a unit of time, or equivalently it gives the number of complete oscillations made by the particle on a line segment in a unit of time.

An alternative definition of natural undamped frequency that is often used is the following.

Definition 28.38.   The constant ω_0 in (28.24) to (28.26), (which can be thought of as the angular velocity of a particle moving on the circumference of a circle) is also called the **natural (undamped) frequency** of the motion.

Comment 28.39.   We shall use ν when we wish to express the frequency of the motion in cycles per unit of time and use ω_0 when we wish to express it in radians per unit of time. In the example after Definition 28.34 where $\omega_0 = 4\pi$ rad/sec, the frequency ν, by (28.37), is thus 2 cycles/sec; the frequency ω_0 is 4π radians/sec.

Comment 28.4.   **A Summary.** The differential equation of motion of a particle executing simple harmonic motion has the form

$$\frac{d^2x}{dt^2} + \omega_0{}^2 x = 0.$$

Its solutions may be written in any of the following forms.

$$x = c_1 \cos \omega_0 t + c_2 \sin \omega_0 t,$$
$$x = c \cos (\omega_0 t + \delta),$$
$$x = c \sin (\omega_0 t + \delta).$$

They are the equations of motion of a particle executing simple harmonic motion. The motion can be looked at as the motion of the projection on a diameter, of a particle moving with angular velocity ω_0 on the circumference of a circle of radius c. Or it may be looked at as the motion of a particle attracted to an origin by a force which is proportional to the distance of the particle from the origin. The particle moves back and forth forever across its equilibrium position. The farthest position reached by the particle from its equilibrium position is given by c. Its speed at c and $-c$ is zero; at the origin its speed is greatest. The time required to make a complete oscillation is $T = 2\pi/\omega_0$; the number of complete oscillations in a unit of time is $1/T$.

Example 28.5.   A particle executes simple harmonic motion. The natural (undamped) frequency of the motion is 4 rad/sec. If the object starts from the equilibrium position with a velocity of 4 ft/sec, find:

1. The equation of motion of the object.
2. The amplitude of the motion.
3. The phase angle.
4. The period of the motion.
5. The frequency of the motion in cycles per second.

Solution. Since the particle executes simple harmonic motion, its equation of motion, by (28.26), is

(a) $\qquad x = c \cos(\omega_0 t + \delta), \qquad v = \dfrac{dx}{dt} = -\omega_0 c \sin(\omega_0 t + \delta).$

The frequency of 4 rad/sec implies, by Definition 28.38, $\omega_0 = 4$. Hence (a) becomes

(b) $\qquad\qquad x = c \cos(4t + \delta), \qquad v = -4c \sin(4t + \delta).$

The initial conditions are $t = 0$, $x = 0$, $v = 4$. Substituting these values in the two equations in (b), we obtain

(c) $\qquad\qquad\qquad 0 = c \cos \delta, \qquad 4 = -4c \sin \delta.$

Since the amplitude $c \neq 0$, we find from the first equation in (c), $\delta = \pm \pi/2$. With this value of δ, the second equation in (c) gives $c = \mp 1$. Hence the equation of motion (b) becomes

(d) $\qquad\qquad x = \cos\left(4t - \dfrac{\pi}{2}\right) \text{ or } x = -\cos\left(4t + \dfrac{\pi}{2}\right),$

which is the answer to 1. The answers to the remaining questions are:

2. Amplitude of the motion, by Definition 28.31, $= 1$ foot.
3. The phase angle, by Definition 28.33, $= -\pi/2$ radians.
4. The period of the motion, by Definition 28.34, $= \pi/2$ seconds.
5. The frequency of the motion in cycles per second, by Definition 28.36, equals $2/\pi$ cps.

Example 28.51. A particle executes simple harmonic motion. The amplitude and period of the motion are 6 feet and $\pi/4$ seconds respectively. Find the velocity of the particle as it crosses the point $x = -3$ feet.

Solution. Since the particle executes simple harmonic motion, its equation of motion, by (28.26), is

(a) $\qquad\qquad\qquad x = c \cos(\omega_0 t + \delta).$

The amplitude of 6 feet implies, by Definition 28.31, that $c = 6$ (or -6). The period of $\pi/4$ seconds implies, by Definition 28.34, that

(b) $\qquad\qquad\qquad \dfrac{2\pi}{\omega_0} = \dfrac{\pi}{4}, \quad \omega_0 = 8.$

Hence the equation of motion (a) becomes, using $c = 6$,

(c) $\qquad\qquad x = 6 \cos(8t + \delta), \qquad \dfrac{dx}{dt} = -48 \sin(8t + \delta).$

When $x = -3$, we find from the first equation in (c),

(d) $\qquad \cos (8t + \delta) = -\frac{1}{2}, \qquad 8t + \delta = 120° \text{ or } 240°,$

and from the second equation in (c)

(e) $\qquad v = \dfrac{dx}{dt} = -48 \sin (120° \text{ or } 240°)$

$\qquad\qquad = -48(\pm\frac{1}{2}\sqrt{3}) = \mp24\sqrt{3} \text{ ft/sec.}$

The plus sign is to be used if the body is moving in the positive direction, the minus sign if the body is moving in the negative direction.

EXERCISE 28A AND B

1. Verify the accuracy of each of the solutions of (28.11) as given in (28.12), (28.13), and (28.14).
2. If a ball were dropped in a hole bored through the center of the earth, it would be attracted toward the center with a force directly proportional to the distance of the body from the center. Find: (a) the equation of motion, (b) the amplitude of the motion, (c) the period, and (d) the frequency of the motion. NOTE. This problem was originally included in Exercise 16A, 40.
3. A particle executes simple harmonic motion. Its period is 2π sec. If the particle starts from the position $x = -4$ ft with a velocity of 4 ft/sec, find: (a) the equation of motion, (b) the amplitude of the motion, (c) the frequency of the motion, (d) the phase angle, and (e) the time when the particle first crosses the equilibrium position.
4. A particle executes simple harmonic motion. At $t = 0$, its velocity is zero and it is 5 ft from the equilibrium position. At $t = \frac{1}{4}$, its velocity is again 0 and its position is again 5 ft.

   (a) Find its position and velocity as functions of time.
   (b) Find its frequency and amplitude.
   (c) When and with what velocity does it first cross the equilibrium position?

5. A particle executes simple harmonic motion. At the end of each $\frac{3}{4}$ sec, it passes through the equilibrium position with a velocity of ± 8 ft/sec.

   (a) Find its equation of motion.
   (b) Find the period, frequency and amplitude of the motion.

6. A particle executes simple harmonic motion. Its amplitude is 10 ft and its frequency 2 cps.

   (a) Find its equation of motion.
   (b) With what velocity does it pass the 5-ft mark?

7. A particle executes simple harmonic motion. Its period is $a\pi$ sec and its velocity at $t = 0$, when it crosses the point $x = x_1$, is $\pm v_1$. Find its equation of motion.
8. A particle executes simple harmonic motion. Its frequency is 3 cps. At $t = 1$ sec, it is 3 ft from the equilibrium position and moving with a velocity of 6 ft/sec. Finds its equation of motion.
9. A particle executes simple harmonic motion. When it is 2 ft from its equilibrium position, its velocity is 6 ft/sec; when it is 3 ft, its velocity is 4 ft/sec. Find the period of its motion, also its frequency.

10. A particle moves on the circumference of a circle of radius 6 ft with angular velocity $\pi/3$ rad/sec. At $t = 0$, it makes a central angle of 150° with a fixed diameter, taken as the x axis.

    (a) Find the equation of motion of its projection on the diameter.
    (b) With what velocity does its projection cross the center of the circle?
    (c) What are its period, amplitude, frequency, phase angle?
    (d) Where on the diameter is the particle's projection at $t = 0$?

11. A particle moves on the circumference of a circle of radius 4 ft. Its projection on a diameter, taken as the x axis, has a period of 2 sec. At $x = 4$, its velocity is zero. Find the equation of motion of the projection.

12. A particle weighing 8 lb moves on a straight line. It is attracted to the origin by a force F that is proportional to the particle's distance from the origin. If this force is 6 lb at a distance of -2 ft, find the natural frequency of the system.

13. A particle of mass m moving in a straight line is repelled from the origin 0 by a force F. If the force is proportional to the distance of the particle from 0, find the position of the particle as a function of time. Note. The particle no longer executes simple harmonic motion.

14. If at $t = 0$ the velocity of the particle of problem 13 is zero and it is 10 ft from the origin, find the position and velocity of the particle as functions of time.

15. At $t = 0$ the velocity of the particle of problem 13 is $-a\sqrt{k}$ ft/sec and it is a feet from the origin. If the repelling force has the value kx, find the position of the particle as a function of time. Show that if $m < 1$, the particle will never reach the origin; that if $m = 1$, the particle will approach the origin but never reach it.

16. A particle executes simple harmonic motion. At $t = 0$, it is -5 ft from its equilibrium position, its velocity is 6 ft/sec and its acceleration is 10 ft/sec². Find its equation of motion and its amplitude.

17. Let $|c|$ be the amplitude of a particle executing simple harmonic motion. We proved in the text, see description of the motion, Example 28.15, that the particle has zero speed at the points $\pm c$ and maximum speed at the equilibrium position $x = 0$.

    (a) Prove that the particle has maximum acceleration at $x = \pm c$ and has zero acceleration at $x = 0$. *Hint.* Take the second derivative of the equation of motion (28.25).
    (b) What is the magnitude of the maximum velocity and of the maximum acceleration?
    (c) If v_m is the maximum velocity and a_m is the maximum acceleration of a particle, show that its period is $T = 2\pi v_m/a_m$ and its amplitude is $A = v_m{}^2/a_m$.

18. Two points A and B on the surface of the earth are connected by a frictionless, straight tube. A particle placed at A is attracted toward the center of the earth with a force directly proportional to the distance of the particle from the center.

    (a) Show that the particle executes simple harmonic motion inside the tube. *Hint.* Take the origin at the center of the tube and let x be the distance of the particle from the center at any time t. Call b the distance of the center of the tube from the center of the earth and R the radius of the earth. Take a diameter of the earth parallel to the tube as a polar axis and let (r,θ) be the polar coordinates of the particle at any time t. Then show that the component of force moving the particle is $F \cos \theta$ and that $\cos \theta = x/r$. Initial conditions are $t = 0$, $x = \sqrt{R^2 - b^2}$, $dx/dt = 0$.

(b) Show that the equation of motion is

$$x = \sqrt{R^2 - b^2} \cos \sqrt{k/m}\, t,$$

where k is a proportionality constant and m is the mass of the particle.

(c) Show that for a given mass, the time to get from A to B is the same for any two points A and B on the surface of the earth. *Hint.* Show that the period of the motion is a constant.

ANSWERS 28A AND B

2. (a) $x = (4000)(5280) \cos (t/812)$ in ft.   (b) 4000 mi.   (c) 85 min.
(d) 0.7 oscillation/hr.

3. (a) $x = 4\sqrt{2} \sin \left(t - \dfrac{\pi}{4}\right)$.   (b) $4\sqrt{2}$ ft.   (c) 1 rad/sec or $\left(\dfrac{1}{2\pi}\right)$ cps.
(d) $-\pi/4$.   (e) $\pi/4$ sec.

4. (a) $x = 5 \cos (8\pi t)$; $v = -40\pi \sin (8\pi t)$.   (b) 8π rad/sec, 5 ft.
(c) $\frac{1}{16}$ sec, -40π ft/sec.

5. (a) $x = \dfrac{6}{\pi} \sin \left(\dfrac{4\pi}{3}\, t\right)$.   (b) $\frac{3}{2}$ sec, $4\pi/3$ rad/sec, or $\frac{2}{3}$ cps, $6/\pi$ ft.

6. (a) $x = 10 \sin (4\pi t + \delta)$.   (b) $\pm 20\pi\sqrt{3}$ ft/sec.

7. $x = \sqrt{x_1{}^2 + a^2 v_1{}^2/4} \sin (2t/a + \delta)$, where $\sin \delta = x_1/\sqrt{x_1{}^2 + a^2 v_1{}^2/4}$.

8. $x = \sqrt{9 + 1/\pi^2} \sin (6\pi t + \delta)$, where $\delta = \text{Arc} \sin (3\pi/\sqrt{9\pi^2 + 1})$.

9. π sec, 2 rad/sec or $1/\pi$ cps.

10. (a) $x = 6 \cos (\pi t/3 + 5\pi/6)$.   (b) $\pm 2\pi$ ft/sec.
(c) 6 sec, 6 ft, $\frac{1}{6}$ cps, $5\pi/6$ rad.   (d) $-3\sqrt{3}$ ft from center.

11. $x = 4 \cos (\pi t)$.

12. $\sqrt{12}$ rad/sec or $\sqrt{3}/\pi$ cps.

13. $x = c_1 e^{\sqrt{k/m}\, t} + c_2 e^{-\sqrt{k/m}\, t}$), where k is a proportionality constant.

14. $x = 5(e^{\sqrt{k/m}\, t} + e^{-\sqrt{k/m}\, t})$, $v = 5\sqrt{k/m}\ (e^{\sqrt{k/m}\, t} - e^{-\sqrt{k/m}\, t})$.

15. $x = \dfrac{a}{2} [(1 - \sqrt{m})e^{\sqrt{k/m}\, t} + (1 + \sqrt{m})e^{-\sqrt{k/m}\, t}]$.

16. $x = 3\sqrt{2} \sin \sqrt{2}\, t - 5 \cos \sqrt{2}\, t$; $\sqrt{43}$ ft.

17. (b) $v = c\omega_0$, $a = c\omega_0{}^2$, where c is the amplitude and ω_0 is the frequency of the motion in radians per unit of time.

LESSON 28C.   Examples of Particles Executing Simple Harmonic Motion.   Harmonic Oscillators.   A dynamical system which vibrates with simple harmonic motion is called a **harmonic oscillator**. Below we give two examples of harmonic oscillators.

Example A. The Motion of a Particle Attached to an Elastic Helical Spring. Hooke's Law. The unstretched natural length of an elastic helical spring is l_0 feet, Fig. 28.6(a). A weight w pounds is attached to it and brought to rest, Fig. 28(b). Because of the stretch due to the

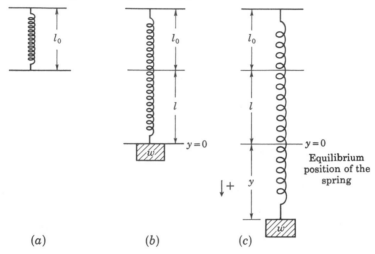

(a) (b) (c)

Figure 28.6

attached weight, a force or tension is created in the spring which tries to restore the spring to its original unstretched, natural length. By **Hooke's law**, this upward force of the spring is proportional to the distance l, by which the spring has been stretched. Hence,

(28.61)   The upward force of the spring $= kl$,

where $k > 0$ is a proportionality constant, called the **spring constant** or the **stiffness coefficient** of the spring. The downward force acting on the spring is the weight w pounds that is attached to it. If the spring is on the surface of the earth, then $w = mg$, where m is the mass in slugs of the attached weight and g is the acceleration due to the gravitational force of the earth in feet per second per second. Since the spring is in equilibrium, the upward force must equal the downward force. Hence by (28.61)

(28.62) $kl = mg.$

Let $y = 0$ be the equilibrium position of the spring with the weight w pounds attached to it. If the spring with this weight attached is now stretched an additional distance y, Fig. 28.6(c), then the following forces will be acting on the spring.

1. An upward force due to the tension of the spring which, by Hooke's law, is now $k(l + y)$.
2. A downward force due to the weight w pounds attached to the spring which is equal to mg.

By Newton's second law of motion, the net force acting on a system is equal to the mass of the system times its acceleration. Hence, with the positive direction taken as downward,

$$(28.621) \qquad m \frac{d^2y}{dt^2} = mg - k(l + y) = mg - kl - ky.$$

By (28.62), (28.621) simplifies to

$$(28.63) \qquad m \frac{d^2y}{dt^2} = -ky, \qquad \frac{d^2y}{dt^2} + \frac{k}{m} y = 0,$$

which is the differential equation of motion of an elastic helical spring. Since it has the same form as (28.11), we now know that a displaced helical spring with weight attached and with no resistance will execute simple harmonic motion about the equilibrium position $y = 0$. The solution of (28.63) is

$$(28.64) \qquad y = c \cos (\sqrt{k/m}\, t + \delta) \quad \text{or} \quad y = c \sin (\sqrt{k/m}\, t + \delta).$$

Example 28.65. A 5-pound body attached to an elastic helical spring stretches it 4 inches. After it comes to rest, it is stretched an additional 6 inches and released. Find its equation of motion, period, frequency, and amplitude. Take the second for a unit of time.

Solution. Since 5 pounds stretches the spring 4 inches $= \frac{1}{3}$ foot, we have by (28.62), with $mg = 5$, $l = \frac{1}{3}$,

$$(a) \qquad \frac{k}{3} = 5, \qquad k = 15,$$

which gives the stiffness coefficient of the spring. With $g = 32$, the mass m of the attached body is $\frac{5}{32}$. Therefore the differential equation of motion, by (28.63), is

$$(b) \qquad \frac{5}{32} \frac{d^2y}{dt^2} = -15y, \qquad \frac{d^2y}{dt^2} + 96y = 0.$$

Its solution, by (28.64), or by solving it independently, is

$$(c) \qquad y = c \cos (\sqrt{96}\, t + \delta), \qquad \frac{dy}{dt} = -\sqrt{96}\, c \sin (\sqrt{96}\, t + \delta).$$

The initial conditions are $t = 0$, $y = \frac{1}{2}$, $v = dy/dt = 0$. Inserting these values in both equations of (c), we obtain

(d) $$\tfrac{1}{2} = c \cos \delta,$$
$$0 = -\sqrt{96}\, c \sin \delta.$$

Since c, the amplitude, cannot equal 0, the second equation in (d) implies $\delta = 0$. With this value of δ, the first equation in (d) gives $c = \tfrac{1}{2}$. Substituting these values in the first equation of (c), we find for the equation of motion

(e) $$y = \tfrac{1}{2} \cos \sqrt{96}\, t.$$

The period of the motion, by Definition 28.34, is therefore $2\pi/\sqrt{96}$ seconds; the frequency, by Definition 28.36, is $\sqrt{96}/2\pi$ cps, or by Definition 28.38, $\sqrt{96}$ rad/sec; the amplitude, by Definition 28.31, is $\tfrac{1}{2}$ foot.

Example 28.66.   A body attached to an elastic helical spring executes simple harmonic motion. The natural (undamped) frequency of the motion is 2 cps and its amplitude is 1 foot. Find the velocity of the body as it passes the point $y = \tfrac{1}{2}$ foot.

Solution.   Since the body attached to the helical spring is executing simple harmonic motion, its equation of motion by (28.64) is

(a) $$y = c \cos (\sqrt{k/m}\, t + \delta).$$

The amplitude of 1 foot implies, by Definition 28.31, that $c = 1$. The frequency of 2 cps implies, by Definition 28.36, that $2 = \sqrt{k/m}/2\pi$, or $\sqrt{k/m} = 4\pi$. Substituting these values in (a) it becomes

(b) $$y = \cos (4\pi t + \delta), \qquad v = \frac{dy}{dt} = -4\pi \sin (4\pi t + \delta).$$

When $y = \tfrac{1}{2}$, we obtain from the first equation in (b)

(c) $$\tfrac{1}{2} = \cos (4\pi t + \delta), \qquad 4\pi t + \delta = 60°, 300°.$$

Hence, by (c), when $y = \tfrac{1}{2}$,

(d) $$\sin (4\pi t + \delta) = \pm \frac{\sqrt{3}}{2}.$$

Substituting (d) in the second equation of (b), we obtain

(e) $$v = \frac{dy}{dt} = \pm 2\sqrt{3}\,\pi,$$

which is the velocity of the body as it passes the point $y = \tfrac{1}{2}$. The plus sign is to be used if the body is moving downward; the minus sign if it is moving upward.

Example B.   The Motion of a Simple Pendulum.   A displaced simple pendulum of length l, with weight $w = mg$ attached (Fig. 28.7) will execute an oscillatory motion. If the angle of swing is very small, then as we shall show, the motion will closely approximate a simple harmonic motion.

The effective force F which moves the pendulum is the component of the weight acting in a direction tangent to the arc of swing. From Fig. 28.7,

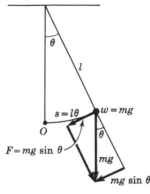

θ is positive if the weight w is to the right of O; negative if to the left of O.

$\omega = d\theta/dt$ is positive if the weight w moves counterclockwise; negative if it moves clockwise.

F is positive if it acts in a direction to move the pendulum counterclockwise; negative if the direction is clockwise.

Note that when θ is positive, F is negative; when θ is negative, F is positive.

Figure 28.7

we see that the value of this component is $mg \sin \theta$, where θ is the angle of swing and m and g have their usual meanings. (The tension in the string and the component of the weight in the direction of the string do not affect the motion.) Therefore by Newton's second law

(28.71) $$F = m\frac{dv}{dt} = -mg \sin \theta, \qquad \frac{dv}{dt} = -g \sin \theta.$$

The distance s over which the weight moves along the arc is

(28.72) $$s = l\theta, \quad \text{therefore } \frac{ds}{dt} = l\frac{d\theta}{dt}.$$

Since $v = ds/dt$, we obtain from the second equation in (28.72),

(28.73) $$v = l\frac{d\theta}{dt}; \qquad \frac{dv}{dt} = l\frac{d^2\theta}{dt^2}.$$

Substituting in (28.71) the value of dv/dt as given in (28.73), we have

(28.74) $$l\frac{d^2\theta}{dt^2} + g \sin \theta = 0.$$

Equation (28.74) does not as yet have the form (28.11) of simple harmonic motion. However, we know from the calculus that $\sin \theta = \theta - \theta^3/3! +$

$\theta^5/5! - \cdots$. We may therefore write (28.74) as

(28.75) $l\dfrac{d^2\theta}{dt^2} + g\theta - \dfrac{g\theta^3}{3!} + \dfrac{g\theta^5}{5!} - \cdots = 0.$

Hence, for a pendulum swinging through a small angle, it is not unreasonable to assume that the error made in linearizing the equation by dropping terms in θ^3 and higher powers of θ will also be small. Hence (28.75) becomes

(28.76) $l\dfrac{d^2\theta}{dt^2} + g\theta = 0, \qquad \dfrac{d^2\theta}{dt^2} + \dfrac{g}{l}\,\theta = 0.$

But this last equation is now the differential equation of a particle executing simple harmonic motion. The solution of (28.76) is

(28.77) $\theta = c \cos (\sqrt{g/l}\, t + \delta),$

where c is the amplitude of θ in radians.

Example 28.771.   A simple pendulum whose length is 5 ft swings with an amplitude of $\frac{1}{10}$ radian. Find the period of the pendulum and its velocity as it crosses the equilibrium position, i.e., the position when $\theta = 0$.

Solution.   The given amplitude of $\frac{1}{10}$ radian implies that $c = \frac{1}{10}$ in (28.77). Therefore, with $l = 5$ and $g = 32$, (28.77) becomes

(a) $\theta = \frac{1}{10} \cos (\sqrt{32/5}\, t + \delta),$

                 $\dfrac{d\theta}{dt} = -\frac{1}{10}\sqrt{32/5} \sin (\sqrt{32/5}\, t + \delta).$

The period of the pendulum, by Definition 28.34, is

(b) $T = 2\pi\sqrt{5/32} = \dfrac{\pi}{4}\sqrt{10}$ sec.

When $\theta = 0$, we have by (a),

(c) $0 = \cos (\sqrt{32/5}\, t + \delta),$

which implies that

(d) $\sin (\sqrt{32/5}\, t + \delta) = \pm 1.$

Substituting (d) in the second equation of (a), there results

(e) $\dfrac{d\theta}{dt} = \pm\frac{1}{10}\sqrt{32/5} = \pm\frac{2}{25}\sqrt{10}.$

By the first equation in (28.73) and (e), we obtain with $l = 5$,

(f) $v = 5(\pm\frac{2}{25}\sqrt{10}) = \pm\frac{2}{5}\sqrt{10}$ ft/sec, which is the velocity of the pendulum as it crosses the central position. The plus sign is to be used if the weight is moving counterclockwise; the minus sign if it is moving clockwise.

EXERCISE 28C

1. Verify the accuracy of the solution of (28.63) as given in (28.64).
2. Verify the accuracy of the solution of (28.76) as given in (28.77).
3. Solve (28.63) for y as a function of t if at $t = 0$, $y = y_0$, $v = v_0$.
4. Show that the sum $\dfrac{k}{2} y^2 + \dfrac{m}{2} \left(\dfrac{dy}{dt}\right)^2$, where y is the solution of (28.63) as given in (28.64), is a constant. NOTE. The first term gives the **potential energy** of a mass attached to a helical spring displaced a distance y from its equilibrium position. The second term gives its **kinetic energy**. Since the sum of the two energies is a constant, a mass oscillating on a helical spring with simple harmonic motion satisfies the **law of the conservation of energy**.
5. Prove the law of the conservation of energy for the undamped helical spring (see problem 4) by multiplying (28.63) by dy/dt and then integrating the resulting equation with respect to t. For hint, see answer.
6. A 12-lb body attached to a helical spring stretches it 6 in. After it comes to rest, it is stretched an additional 4 in. and released. Find the equation of motion of the body; also the period, frequency, and amplitude of the motion. With what velocity will it cross the equilibrium position?
7. A 10-lb body stretches a spring 3 in. After it comes to rest, it is stretched an additional 6 in. and released.

    (a) Find the position and velocity of the body at the end of 1 sec.
    (b) When and with what velocity will the object first pass the equilibrium position?
    (c) What is the velocity of the body when it is -3 in. from equilibrium?
    (d) When will its velocity be 2 ft/sec and where will it be at that instant?
    (e) What are the period, frequency, and amplitude of the motion?

8. A spring is stretched 8 in. by a 16-lb weight. The weight is then removed and a 24-lb weight attached. After the system is brought to rest, the spring is stretched an additional 10 in. and released with a downward velocity of 6 ft/sec.

    (a) Find the equation of motion, period, frequency, amplitude and phase angle.
    (b) Answer the same questions if the weight were given an upward velocity of 6 ft/sec instead of a downward one.

9. A heavy rubber band of natural length l_0 is stretched 2 ft when a weight W is attached to it. Find the equation of motion and the maximum stretch of the band, if at the point l_0 the weight is given a downward velocity of 6 ft/sec. Assume the rubber band obeys Hooke's law.
10. The spring constant of a helical spring is 27. When a weight of 96 lb is attached, it vibrates with an amplitude of 3/2 ft.

    (a) What are its period and frequency?
    (b) With what velocity does it pass the point $y = -1$ ft?

11. A piece of steel wire is stretched 8 in. when a 128-lb weight is attached to it. After it is brought to rest, it is displaced from its equilibrium position. Find the frequency of its motion. Assume the wire obeys Hooke's law.

12. A helical spring is stretched 2 in. when a 5-lb weight is attached to it. If the 5-lb weight is removed and a weight of W lb attached, the spring oscillates with a frequency of 6 cps. Find the weight W.

13. When two springs, suspended parallel to each other, are connected by a bar at their bottom, they act as if they were one spring. The spring constant of the system is then equal to the sum of the spring constant of each spring. Assume a weight of 8 lb will stretch one spring 2 in. and the other spring 3 in.

    (a) What is the spring constant of each spring?
    (b) If the two springs are suspended parallel to each other and connected by a bar at their bottom, what is the spring constant of the system?
    (c) If a weight of 8 lb is attached to the system, brought to rest, and then released after being displaced 6 in., find the amplitude, period, and frequency of the motion.

14. When two springs are connected in series, i.e., when one spring is attached to the end of the other, the spring constant k of the system is given by the formula

$$\frac{1}{k} = \frac{1}{k_1} + \frac{1}{k_2},$$

    where k_1 and k_2 are the respective spring constants of each spring. Assume the springs of problem 13 are attached in series.

    (a) What is the spring constant of the system?
    (b) If a weight of 8 lb is attached to the bottom spring, brought to rest, displaced 6 in, and then released, find the amplitude, period, and frequency of the motion.

The solutions to problems 15–17 will be facilitated if reference is made to the solution given in problem 3.

15. A spring is stretched l ft by a weight W and brought to rest. It is then given a downward velocity of v_0 ft/sec.

    (a) Find the lowest point reached by the weight and the elapsed time.
    (b) Find the amplitude, frequency, and period of the motion.

16. A spring is stretched l ft by a weight W and brought to rest. It is then stretched an additional a ft and given a downward velocity v_0.

    (a) Find the amplitude, frequency, and period of the motion.
    (b) Answer the same questions if the initial velocity is v_0 upward instead of downward.

17. A helical spring has a period of 4 sec when a 16-lb weight is attached. If the 16-lb weight is removed and a weight W attached, the spring oscillates with a period of 3 sec. Find the weight W.

In problems 18–28, it is assumed that θ, the angle a pendulum makes with the vertical, i.e., the angle the pendulum makes with its equilibrium position, is sufficiently small so that equations (28.76) and (28.77) are applicable. Remember that in these problems angular velocity $\omega = d\theta/dt$, and linear velocity $v = l \, d\theta/dt = l\omega$, where l is the length of the pendulum from point of support to the center of mass of the bob. For positive and negative directions, see Fig. 28.7.

18. A simple pendulum of length l ft is given an angular velocity of ω_0 rad/sec from the position $\theta = \theta_0$.

    (a) Find the position of the bob of the pendulum as a function of time.
    (b) Find amplitude, period, frequency, and phase angle.
    (c) With what velocity, angular and linear, does the pendulum cross the equilibrium position?

19. A simple pendulum of length l ft is released from the position $\theta = \theta_0$.

    (a) Find the position of the pendulum as a function of time.
    (b) Find amplitude, period, and frequency. Compare results with problem 18.
    (c) With what velocity, angular and linear, does the pendulum cross the equilibrium position?

20. A simple pendulum of length 2 ft is released from the position $\theta = \frac{1}{12}$ rad at time $t = 0$.

    (a) Find the value of θ when $t = \pi/16$.
    (b) When and with what velocity, angular and linear, does the pendulum cross the equilibrium position?

21. A simple pendulum of length 6 in. is given an angular velocity of magnitude $\frac{1}{2}$ rad/sec toward the vertical from the position $\theta = \frac{1}{10}$ rad. Find the equation of motion, amplitude, period, frequency, and phase angle. *Hint.* At $t = 0$, $\omega = -\frac{1}{2}$ rad/sec.

22. A simple pendulum of length 2 ft is given an angular velocity of $\frac{1}{2}$ rad/sec toward the vertical from the position $\theta = -\frac{1}{10}$ rad. Find the equation of motion and the amplitude.

23. Solve problem 22, if the pendulum is given an angular velocity of $\frac{1}{2}$ rad/sec away from the vertical from the position $\theta = \frac{1}{10}$ rad.

24. A simple pendulum of length l ft is started by giving it an angular velocity of ω_0 rad/sec from its equilibrium position.

    (a) Find the equation of motion.
    (b) Find the value of θ and the time when the pendulum reaches its maximum displacement.

25. A clock pendulum is regulated so that 1 sec elapses each time it passes its equilibrium position. (Its period is therefore 2 sec.) What is the length of the pendulum?

26. It is desired that a clock pendulum cross its equilibrium position each second and have an amplitude of $\frac{1}{15}$ rad. What should its angular velocity be as it crosses the equilibrium position? *Hint.* The length of the pendulum has already been found in 25, the equation of motion in problem 24.

27. The amplitude of a pendulum is 0.05 rad. What fraction of its period has elapsed when $\theta = 0.025$ rad? For hint, see answer.

28. A clock pendulum has a length of 6 in. Each time it crosses the equilibrium position, it makes a tick. How many ticks will it make in 1 hr?

29. In the pendulum example in the text and in the problems above, the weight of the wire supporting the mass was considered negligible and therefore ignored. If, however, this weight is not negligible, then, instead of (28.71), the differential equation of motion of the pendulum and bob is

(28.772) $$I \frac{d^2\theta}{dt^2} = -(mg \sin \theta)l,$$

where I is the moment of inertia* of the pendulum and bob about an axis which passes through the point of suspension and is perpendicular to the plane of rotation of the pendulum; m is the mass of pendulum and bob; l is the length of the pendulum from the point of support to the center of gravity of pendulum and bob. [For the case of a pendulum of negligible mass, the moment of inertia I of a particle of mass m attached at a distance l from the axis of rotation is ml^2. Substitution of this value of I in (28.772) will yield (28.74).]

With the aid of (28.772), find the equation of motion and the period of a pendulum consisting of a uniform rod of length l and swinging through a small angle. *Hint.* The moment of inertia I of a uniform rod of length l and mass m about an axis through one end is given by

$$I = \lim_{n\to\infty} \sum_{i=1}^{n} \delta x_i^2 \, \Delta x_i = \int_0^l \delta x^2 \, dx = \frac{ml^2}{3} \text{ with } \delta = m/l.$$

Since the rod is uniform, its mass may be considered as concentrated at its center, i.e., its center of gravity is at its geometric center.

In problems 30–34, we no longer assume that the angle of swing is small. Hence we cannot, in (28.74), replace $\sin \theta$ by θ.

30. Solve (28.74) for $d\theta/dt$ with the initial conditions $t = 0$, $\theta = \theta_0$, $\omega = 0$. *Hint.* Multiply the equation by $2d\theta/dt$ and then make use of the identity

$$2\left(\frac{d^2\theta}{dt^2}\right)\frac{d\theta}{dt} = \frac{d}{dt}\left(\frac{d\theta}{dt}\right)^2.$$

Answer is

(28.78) $$\omega = d\theta/dt = \pm\sqrt{2g/l}\,\sqrt{\cos\theta - \cos\theta_0}.$$

31. By integrating (28.78), remember the initial condition is $t = 0$, $\theta = \theta_0$, show that

(28.79) $$t(\theta) = \pm\sqrt{l/2g}\int_{\theta_0}^{\theta} \frac{du}{\sqrt{\cos u - \cos\theta_0}}.$$

Since at $t = 0$, $\theta = \theta_0$, $\omega = 0$, one-half of a period elapses when $\theta = -\theta_0$. Use this fact, (28.79), and the fact that $\cos u$ is an even function, i.e., $\cos(-u) = \cos u$, to show that the period T of the pendulum is given by

(28.791) $$\frac{T}{2} = \sqrt{\frac{l}{2g}}\int_{-\theta_0}^{\theta_0} \frac{du}{\sqrt{\cos u - \cos\theta_0}},$$

$$T = 2\sqrt{2}\,\sqrt{l/g}\int_0^{\theta_0} \frac{du}{\sqrt{\cos u - \cos\theta_0}}.$$

---

*For definition of moment of inertia, see Lesson 30M-B.

In (28.791), replace $\cos u$ by its equal $1 - 2\sin^2 \frac{u}{2}$, $\cos \theta_0$ by its equal $1 - 2\sin^2 \frac{\theta_0}{2}$. Show that the resulting equation is

(28.7911) $$T = 2\sqrt{l/g} \int_0^{\theta_0} \frac{du}{\sqrt{\sin^2 (\theta_0/2) - \sin^2 (u/2)}} .$$

Finally in (28.7911), make the substitutions

(28.7912) $$\sin \frac{u}{2} = \sin \frac{\theta_0}{2} \sin \phi, \qquad \tfrac{1}{2} \cos \frac{u}{2} \, du = \sin \frac{\theta_0}{2} \cos \phi \, d\phi.$$

Show that when $u = 0, \phi = 0$; when $u = \theta_0, \phi = \pi/2$ and that (28.7911) therefore becomes

(28.7913) $$T = 4\sqrt{l/g} \int_0^{\pi/2} \frac{d\phi}{\sqrt{1 - k^2 \sin^2 \phi}}, \qquad k = \sin \frac{\theta_0}{2}.$$

The integral in (28.7913) is known as the **complete elliptic integral of the first kind.** It cannot be expressed in terms of elementary functions, but it can be evaluated* by writing the integrand as a power series

(28.7914) $$(1 - k^2 \sin^2 \phi)^{-1/2} = 1 + \tfrac{1}{2}k^2 \sin^2 \phi + \frac{1 \cdot 3}{2 \cdot 4} k^4 \sin^4 \phi + \cdots.$$

Substituting the above series in (28.7913) and carrying out the integration, we obtain, since, when n is an *even* positive integer,

$$\int_0^{\pi/2} \sin^n \phi \, d\phi = \frac{1 \cdot 3 \cdot 5 \cdots (n - 1)}{2 \cdot 4 \cdot 6 \cdots n} \frac{\pi}{2},$$

(28.7915) $$T = 2\pi\sqrt{l/g} \left[1 + \frac{k^2}{2^2} + \left(\frac{1 \cdot 3}{2 \cdot 4} \right)^2 k^4 + \cdots \right],$$

which gives the **actual period of a pendulum.**

By (28.77), the period of a pendulum, approximated by replacing $\sin \theta$ by θ, is

(28.7916) $$T = 2\pi\sqrt{\frac{l}{g}} .$$

Assume the initial displacement of a pendulum at $t = 0$, is $\theta_0 = 4°$. Then by (28.7913), $k = \sin 2° = 0.0349$ and $k^2 = 0.00122$. Hence by (28.7915), the more exact period of the pendulum is

(28.7917) $$T \text{ (exact)} = 2\pi\sqrt{l/g} \left(1 + \frac{0.00122}{4} + \frac{9(0.00122)^2}{64} + \cdots \right)$$
$$= 2\pi\sqrt{l/g} \, (1.000305) \text{ sec},$$

and by (28.7916) the approximate period is

(28.7918) $$T \text{ (approx.)} = 2\pi\sqrt{l/g} \text{ sec.}$$

*A table of values of this integral can be found, just as one can find a table of values of $\sin x$; see, for example, Pierce's Tables.

By (28.7917) and (28.7918), we see, therefore, that

(28.7919) T (exact) $= 1.000305T$ (approx.).

Suppose now, we design a clock with pendulum of sufficient length l so that its period, using the approximation formula (28.7918) is 2 sec, and its amplitude is 4°. Therefore, by (28.7919), its more exact period is 2.00061 sec. This means that in each 2 sec the clock is incorrect by 0.00061 sec, its second hand has moved 2 sec whereas it should have moved 2.00061 sec; the clock is thus too slow by this much. Calculate the error in the clock at the end of each 24 hr. *Ans.* 26 sec slow.

32. Solve (28.74) for $d\theta/dt$ with the initial conditions $t = 0$, $\theta = 0$, $\omega = \omega_0$. See hint in 30. Answer is

(28.792) $$\left(\frac{d\theta}{dt}\right)^2 = \omega_0{}^2 - \frac{2g}{l}(1 - \cos\theta) = \omega_0{}^2\left[1 - \frac{2g}{l\omega_0{}^2}(1 - \cos\theta)\right].$$

33. In (28.792), make the substitution

(28.793) $$k^2 = \frac{4g}{l\omega_0{}^2} \quad \text{and} \quad \sin^2\frac{\theta}{2} = \tfrac{1}{2}(1 - \cos\theta).$$

Show that at time t, remember at $t = 0$, $\theta = 0$,

(28.7931) $$t(\theta) = \pm\int_0^\theta \frac{du}{\omega_0\sqrt{1 - k^2\sin^2(u/2)}} \equiv \pm\frac{2}{\omega_0}\int_0^{\theta/2}\frac{d\phi}{\sqrt{1 - k^2\sin^2\phi}}.$$

The integral in (28.7931) is called an **incomplete elliptic integral of the first kind**.

34. In the text, we derived the differential equation for a pendulum of negligible weight swinging along the arc of a circle. The resulting equation (28.74) was not simple harmonic. Assume now that the end of the pendulum moves along a curve which is given by the parametric equations

(a) $x = a(2\theta) + a\sin 2\theta,$
                       $y = a - a\cos 2\theta,$

where a is a positive constant and θ is the inclination of the tangent to the curve at a point P, Fig. 28.794. Hence, the differential equation of motion of the pendulum is

$$m\left(\frac{dv}{dt}\right) = -mg\sin\theta.$$

Replacing v by its equal ds/dt, we obtain

(b) $$\frac{d^2s}{dt^2} = -g\sin\theta,$$

where s is the distance along the curve measured from the lowest position 0. It can be shown that the distance s along an arc of the curve (a) from its lowest point is $s = 4a\sin\theta$. Solve for $\sin\theta$ and substitute this value in (b). Show that the resulting equation is now simple harmonic. Solve for s as a function of t, find the period of the motion and establish the fact that the period is a constant.

The parametric equations (a) above are the equations of an inverted cycloid traced by a point on a circle of radius a. (When a circle rolls along a straight line, the curve traced by a point P on its circumference, is called a **cycloid**.) In (a) above, 2θ is the central angle made by lines drawn from the center of the circle, one vertical, the other to P. It can then be proved that the inclination of the tangent to the cycloid at P is θ. If, therefore, you can devise a pendulum whose bob will move along the path of an inverted cycloid, the pendulum will swing in simple harmonic motion regardless of the angle of swing, and its period will be independent of its amplitude.

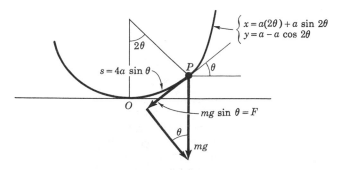

Figure 28.794

Archimedes' principle states that a body partially or totally submerged in a liquid is buoyed up by a force equal to the weight of the liquid displaced. For example, if a 5-pound body floats on water, then the weight of the water displaced is 5 pounds, i.e.,

V (in ft^3 of water displaced) $\times$ 62.5 (weight of a ft^3 of water) $= 5$.

The fact that the body is resting or floating on water implies that the downward force due to the weight $W = mg$ of the body must be the same as the upward buoyant force of the water, namely the weight of water displaced.

Call $y = 0$ the equilibrium position of the lower end of a body when it is floating in a liquid. If the body is now depressed a distance y from its equilibrium position, an additional upward force will act on the body due to the additional liquid displaced. If the body's cross-sectional area is A, then the volume of additional liquid displaced is Ay and the weight of this additional liquid displaced is $(Ay)\rho$, where ρ is the weight per unit volume of the liquid. Since this additional upward force is the net force acting on the depressed body, we have, by Newton's second law of motion, with positive direction downward,

(28.795) $$m\frac{d^2y}{dt^2} = -\rho Ay.$$

A floating body, therefore, when depressed from its equilibrium position executes simple harmonic motion.

With the aid of (28.795), solve the following problems:

35. A body of mass m has cross-sectional area A sq ft. It is depressed y_0 ft from its equilibrium position in a liquid whose weight per ft^3 is ρ and released. (a) Show that its equation of motion is

(28.796) $$y = y_0 \cos \sqrt{\rho A/m}\, t,$$

where y is the distance of the body from equilibrium. *Hint.* At $t = 0$, $y = y_0$, $v = 0$. (b) If the body is a right circular cylinder with vertical axis and radius r, show that its period is

(28.797) $$T = \frac{2}{r} \sqrt{\frac{m\pi}{\rho}}.$$

36. A cylindrical buoy of radius 2 ft and weighing 64 lb floats in water with its axis vertical. It is depressed $\frac{1}{2}$ ft and released.

    (a) Find its equation of motion, amplitude, and period.
    (b) How far below the water line is the equilibrium position.

37. A cubic block of wood weighing 96 lb has cross-sectional area 4 ft. It is depressed slightly in a liquid which weighs 50 lb/cu ft and released. What is its period of oscillation?

38. A cylindrical buoy of radius 1 ft floats in water with its axis vertical. Its period, when depressed and released, is 4 sec. What is its weight?

39. A cubical block of wood is 2 ft on a side. When depressed and released in water, it oscillates with a period of 1 sec. What is the specific gravity of the wood? *Hint.* Use (28.796) to find the mass m of the cubical block. Then find its weight per cubic foot. Then compare this figure with the weight of a cubic foot of water.

40. When a cylinder with vertical axis is depressed and then released in water, it vibrates with a period T_0 sec. When water is replaced by another liquid, it vibrates with a period of $2T_0$ sec. Find the weight of the liquid per cubic foot. *Hint.* Use (28.797) to find ρ.

41. A body whose cross-sectional area is A, displaces, when in equilibrium, l ft of a liquid. The body is then depressed and released. Show that its differential equation of motion is

(28.798) $$\frac{d^2 y}{dt^2} = -\frac{g}{l}\, y,$$

where y is the distance of the body from equilibrium at time t. *Hint.* In equilibrium, the weight of the body equals the weight of the water displaced. The weight of displaced water is $(Al)\rho$. Therefore $Al\rho = mg$. Solve for ρ and substitute in (28.795).

42. Solve (28.798). What is the period of vibration?

43. A cylinder with vertical axes vibrates in water with a period of 2 sec. How far from the surface is its equilibrium position? *Hint.* Make use of the solution of (28.798) as found in problem 42.

44. A spherical body of radius r is in equilibrium when half of it is submerged in a liquid. It is given a displacement and released. Find its differential equation of motion and, if the displacement is small in comparison with the radius r. also find its approximate period. For hints see answer.

ANSWERS 28C

3. $y = v_0\sqrt{m/k} \sin(\sqrt{k/m}\, t) + y_0 \cos(\sqrt{k/m}\, t)$

$\equiv \sqrt{y_0^2 + (m/k)v_0^2} \sin(\sqrt{k/m}\, t + \delta)$,

where $\delta = \text{Arc sin} \dfrac{y_0}{\sqrt{y_0^2 + (m/k)v_0^2}}$.

5. To integrate the first term make the substitution $u = dy/dt$, $du = (d^2y/dt^2)\, dt$.

6. $y = \frac{1}{3}\cos 8t$, $\pi/4$ sec, 8 rad/sec, or $4/\pi$ cps, $\frac{1}{3}$ ft, $\pm\frac{8}{3}$ ft/sec.

7. (a) 0.16 ft, -5.37 ft/sec.   (b) $\sqrt{2}\pi/32$ sec, -5.66 ft/sec,

(c) $\pm 2\sqrt{6}$ ft/sec,   (d) 0.31 sec, -0.47 ft,

(e) $\sqrt{2}\,\pi/8$ sec, $8\sqrt{2}$ rad/sec, $\frac{1}{2}$ ft.

8. (a) $y = \dfrac{3\sqrt{2}}{4}\sin(4\sqrt{2}\,t) + \dfrac{5}{6}\cos(4\sqrt{2}\,t) \equiv \dfrac{\sqrt{131}}{6\sqrt{2}}\sin(4\sqrt{2}\,t + 0.67)$,

$\dfrac{\pi\sqrt{2}}{4}$ sec, $4\sqrt{2}$ rad/sec, $\sqrt{131}/6\sqrt{2}$ ft, 0.67 rad.

(b) $y = -\dfrac{3\sqrt{2}}{4}\sin(4\sqrt{2}\,t) + \dfrac{5}{6}\cos(4\sqrt{2}\,t)$; remaining answers are the same as in (a).

9. $y = \frac{3}{2}\sin 4t - 2\cos 4t$. Amplitude is 5/2 ft. Maximum stretch is 9/2 ft.

10. (a) $2\pi/3$ sec, 3 rad/sec or $3/2\pi$ cps.   (b) $\pm\frac{3}{2}\sqrt{5}$ ft/sec.

11. $4\sqrt{3}$ rad/sec.

12. $\dfrac{20}{3\pi^2}$ lb.

13. (a) 48 and 32.   (b) 80.   (c) $\frac{1}{2}$ ft, $\sqrt{5}\,\pi/20$ sec, $4\sqrt{5}/\pi$ cps.

14. (a) 19.2.   (b) $\frac{1}{2}$ ft, $2\pi/\sqrt{76.8}$ sec, $\sqrt{76.8}/2\pi$ cps.

15. (a) $y = v_0\sqrt{m/k} = v_0\sqrt{l/g}$ ft, $t = \dfrac{\pi}{2}\sqrt{m/k} = \dfrac{\pi}{2}\sqrt{l/g}$ sec.

(b) $v_0\sqrt{l/g}$ ft, $\sqrt{k/m} = \sqrt{g/l}$ rad/sec, $2\pi\sqrt{l/g}$ sec.

16. (a) $A = \sqrt{a^2 + lv_0^2/g}$, period and frequency are the same as in 15.

(b) Same answers as in (a).

17. 9 lb.

18. (a) $\theta = \theta_0\cos\sqrt{g/l}\,t + \omega_0\sqrt{l/g}\sin\sqrt{g/l}\,t \equiv \sqrt{\theta_0^2 + (l/g)\omega_0^2}\sin(\sqrt{g/l}\,t + \delta)$,

where $\delta = \text{Arc sin}\left(\theta_0/\sqrt{\theta_0^2 + (l/g)\omega_0^2}\right)$.

(b) $\sqrt{\theta_0^2 + (l/g)\omega_0^2}$ ft, $2\pi\sqrt{l/g}$ sec, $(1/2\pi)\sqrt{g/l}$ cps.

δ is given in (a).

(c) $\omega = \pm\sqrt{(g/l)\theta_0^2 + \omega_0^2}$ rad/sec, $v = \pm\sqrt{lg\theta_0^2 + l^2\omega_0^2}$ ft/sec.

19. (a) $\theta = \theta_0\cos\sqrt{g/l}\,t$.

(b) $A = \theta_0$ rad, $T = 2\pi\sqrt{l/g}$ sec, $\nu = \dfrac{1}{2\pi}\sqrt{g/l}$ cps.

(c) $\omega = \pm\theta_0\sqrt{g/l}$ rad/sec, $v = \pm\theta_0\sqrt{gl}$ ft/sec.

20. (a) $\theta = \sqrt{2}/24$ rad.   (b) $t = \frac{1}{8}(2n + 1)\pi$, $n = 0, 1, \cdots$, $\pm\frac{1}{3}$ rad/sec, $\pm\frac{2}{3}$ ft/sec.

21. $\theta = \frac{1}{10}\cos 8t - \frac{1}{16}\sin 8t \equiv \sqrt{89}/80 \sin (8t - 4.15)$, $\sqrt{89}/80$ ft, $\pi/4$ sec, $4/\pi$ cps, 4.15 rad.

22. $\theta = -\frac{1}{10}\cos 4t + \frac{1}{8}\sin 4t$, $A = \sqrt{41}/40$ ft.

23. $\theta = \frac{1}{10}\cos 4t + \frac{1}{8}\sin 4t$, $A = \sqrt{41}/40$ ft.

24. (a) $\theta = \omega_0\sqrt{l/g}\sin (\sqrt{g/l}\,t)$.   (b) $\theta = \omega_0\sqrt{l/g}$ rad, $t = (\pi/2)\sqrt{l/g}$ sec.

25. $32/\pi^2$ ft.   **26.** 0.21 rad/sec.

27. $T/6$. *Hint.* In the solution given in problem 18, $\theta_0 = 0.05$ and take $\omega_0 = 0$. Find t when $\theta = 0.025$.

28. 9167.

29. $\theta = c_1 \cos \sqrt{3g/2l}\,t + c_2 \sin \sqrt{3g/2l}\,t$, $T = 2\pi\sqrt{2l/3g}$ sec.

34. $s = c_1 \sin (t/2)\sqrt{g/a} + c_2 \cos (t/2)\sqrt{g/a}$, $T = 4\pi\sqrt{a/g}$.

36. (a) $y = \frac{1}{2}\cos (5\sqrt{5\pi}\,t)$, $A = \frac{1}{2}$ ft, $T = \frac{2}{5}\sqrt{\pi/5}$ sec.   (b) 0.08 ft.

37. 0.77 sec.   **38.** 2546 lb.   **39.** 0.405.   **40.** One-fourth the weight of water.

42. $y = c \cos \sqrt{g/l}\,t$, $T = 2\pi\sqrt{l/g}$ sec.

43. g/π^2 ft $= 3\frac{1}{4}$ ft approx.

44. Let y be the displaced distance of the sphere from equilibrium. The buoyant force of the liquid, therefore, is equal to the weight of liquid displaced, i.e., it is the weight of a volume of liquid equal to one-half the volume of the sphere minus the volume of a segment of a sphere of height $r - y$. The volume of a segment of a sphere is $(\pi h^2/3)(3r - h)$, where r is the radius of the sphere and h is the height of the segment. And since the body is in equilibrium when only one-half of it is submerged, the weight of the liquid per unit volume must be 2ρ, where ρ is the weight per unit volume of the sphere. The differential equation motion is

$$\frac{d^2y}{dt^2} = -\frac{g}{2}\left[3\frac{y}{r} - \left(\frac{y}{r}\right)^3\right].$$

If the displacement y is small in comparison with r, then it is reasonable to assume that the error made in linearizing the equation by dropping $\frac{(y/r)^3}{}$ will also be small. The approximate period is therefore $T = 2\pi\sqrt{2r/3g}$.

LESSON 28D. Forced Undamped Motion. The motion of a particle of mass m that satisfies a differential equation of the form

(28.8) $$m\frac{d^2y}{dt^2} + m\omega_0^2 y = f(t), \qquad \frac{d^2y}{dt^2} + \omega_0^2 y = \frac{1}{m}f(t)$$

where $f(t)$ is a **forcing function** attached to the system and ω_0 is defined in 28.38, is called **forced undamped motion,** in contrast to the *free* undamped motion (i.e., simple harmonic motion) when $f(t) \equiv 0$. Let us assume that the forcing function $f(t) = mF \sin (\omega t + \beta)$ where F is a constant. Then (28.8) becomes

(28.81) $$\frac{d^2y}{dt^2} + \omega_0^2 y = F \sin (\omega t + \beta).$$

If we set the left side of (28.81) equal to zero, and solve the resulting homogenous equation, we obtain the complementary function

$$(28.82) \qquad\qquad y_c = c \sin (\omega_0 t + \delta).$$

A particular solution y_p of (28.81) will then depend on the relative values of the natural (undamped) frequency ω_0 of the system and the **impressed frequency** ω of the forcing function $mF \sin (\omega t + \beta)$. We shall treat each of the two possibilities in the two cases below.

Case 1. $\omega \neq \omega_0$. If $\omega \neq \omega_0$, then a particular solution of (28.81) is

$$(28.83) \qquad\qquad y_p = \frac{F}{\omega_0{}^2 - \omega^2} \sin (\omega t + \beta).$$

Hence, by (28.82) and (28.83), the general solution of (28.81) is

$$(28.84) \qquad y = c \sin (\omega_0 t + \delta) + \frac{F}{\omega_0{}^2 - \omega^2} \sin (\omega t + \beta).$$

The motion of the system is now the sum of two separate and distinct motions, each of which is simple harmonic. The displacement or departure of the particle from its equilibrium position is therefore the sum of two separate (harmonic) displacements with respective amplitudes of c and $F/(\omega_0{}^2 - \omega^2)$. The maximum value of the displacement or departure, however, cannot exceed $|c| + |F/(\omega_0{}^2 - \omega^2)|$. Since all these letters denote constants, the displacement or departure has finite magnitude. A motion in which the displacement or departure of a particle from its equilibrium position remains finite with time is called a **stable motion.** If, however, ω_0 and ω are nearly alike, then $(\omega_0{}^2 - \omega^2)$ will be small, and since this term appears in the denominator of (28.84), the departure or displacement of the particle from equilibrium will be large, i.e., the vibrations of the system will be big, and if sufficiently large, a breakdown of the system may result.

The behavior of the particle's motion is again influenced by two motions with different frequencies, the natural (undamped) frequency ω_0 and the forcing frequency ω. If ω_0/ω is a rational number, say 3, then the motion due to y_c will make three revolutions while the motion due to y_p is making only one. Therefore, during the time interval $2\pi/\omega$ (the period of one revolution due to the y_p motion), the motion of the system will be erratic. However, at the end of this time interval, the position of the particle due to the y_c and y_p motions will again be its starting one, and the system will again repeat its erratic behavior. The motion of the system will thus have an appearance somewhat like that shown in Fig.

28.85. If, however, ω_0/ω is an irrational number, then the motion will not have a repetitive pattern.

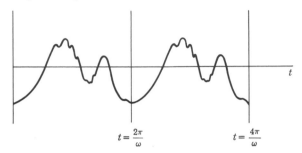

$$t = \frac{2\pi}{\omega} \qquad t = \frac{4\pi}{\omega}$$

Figure 28.85

Case 2. $\omega = \omega_0$. If $\omega = \omega_0$, the differential equation of motion (28.81) becomes

$$(28.9) \qquad \frac{d^2y}{dt^2} + \omega_0{}^2 y = F \sin(\omega_0 t + \beta).$$

Now, however, a term in the y_c part of the solution as given in (28.82), agrees, except for phase and constant coefficient, with the function on the right of (28.9). Hence the trial function y_p, by Lesson 21A, Case 2, must be of the form

$$(28.91) \qquad y_p = At \sin(\omega_0 t + \beta) + Bt \cos(\omega_0 t + \beta).$$

Following the method described in Lesson 21A, Case 2, we find

$$(28.92) \qquad y_p = -\frac{F}{2\omega_0} t \cos(\omega_0 t + \beta).$$

Hence the general solution of (28.9), by (28.82) and (28.92), is

$$(28.93) \qquad y = c \sin(\omega_0 t + \delta) - \frac{F}{2\omega_0} t \cos(\omega_0 t + \beta).$$

The maximum departure or displacement of the motion is $|c| + \left| \dfrac{F}{2\omega_0} t \right|$.

The presence of the variable t in the second term implies that the departure or displacement due to this part of the motion increases with time, see Fig. 28.94. A motion in which the departure or displacement increases beyond all bounds as time passes is called an **unstable motion**. In such cases, a mechanical breakdown of the system is bound to occur. This condition, where ω, the frequency of the forcing function, equals ω_0, the natural (undamped) frequency of the system, is known as **undamped resonance,** and ω_0 is called the **undamped resonant frequency.**

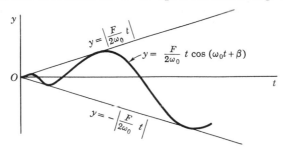

Figure 28.94

Comment 28.941. In engineering circles, the function $f(t)$ of (28.8) is referred to as the **input** of the system, the solution $y(t)$ of (28.8) as the **output** of the system.

Example 28.95. The differential equation of motion of a system is

(a) $$y'' + 4y = \cos 2t.$$

Find its equation of motion. Is the motion stable or unstable? What is the undamped resonant frequency?

Solution. Setting the left side of (a) equal to zero, and solving, we obtain the complementary function

(b) $$y_c = c \cos (2t + \delta).$$

Since the right side of (a) agrees, except for phase, with the complementary function (b), the trial function y_p, by Lesson 21A, Case 2, must be of the form

(c) $$y_p = At \sin 2t + Bt \cos 2t.$$

Following the method outlined in this lesson, we find

(d) $$y_p = \tfrac{1}{4}t \sin 2t.$$

Hence the general solution of (a) is

(e) $$y = c \cos (2t + \delta) + \tfrac{1}{4}t \sin 2t.$$

The presence of the factor $t/4$ in the amplitude of the second term implies that the displacement increases with time. The motion is therefore unstable. A comparison of (a) with (b) also shows that the condition of undamped resonance is present since the frequency of the forcing function and the natural (undamped) frequency of the system are both 2 radians per unit of time. Hence the undamped resonant frequency is 2 rad/unit of time.

Example 28.951. The differential equation of motion of a system is

(a) $$y'' + 4y = 6 \sin t.$$

Find its equation of motion. Is the motion stable or unstable?

Solution. The general solution of (a), by any method you wish to use, is

(b) $$y = c \cos (2t + \delta) + 2 \sin t.$$

The maximum displacement is $|c| + 2$, a finite quantity. Hence the motion is stable.

Example 28.96. A 16-lb weight stretches a spring 6 in. A forcing function $f(t) = 10 \sin 2t$ is attached to the system. Find the equation of motion. What is the maximum displacement of the weight? Is the motion stable or unstable? What frequency of the forcing function would produce resonance?

Solution Because of the presence of the forcing function, (28.63) must be modified to read

(a) $$m \frac{d^2 y}{dt^2} + ky = 10 \sin 2t.$$

Here $mg = 16$, $m = \frac{16}{32} = \frac{1}{2}$. Since 16 pounds stretches the spring 6 inches, we have, by (28.62),

(b) $$\tfrac{1}{2}k = 16, \qquad k = 32.$$

Hence (a) becomes

(c) $$\frac{1}{2} \frac{d^2 y}{dt^2} + 32y = 10 \sin 2t, \qquad \frac{d^2 y}{dt^2} + 64y = 20 \sin 2t.$$

The general solution of (c) is

(d) $$y = c \cos (8t + \delta) + \tfrac{1}{3} \sin 2t.$$

Its maximum displacement is $|c| + \frac{1}{3}$, a finite quantity. Therefore the motion is stable. To produce resonance, the frequency of the forcing function would have to be 8 rad/unit of time.

EXERCISE 28D

1. Verify the accuracy of the solution of (28.81) as given in (28.84).
2. Verify the accuracy of the particular solution of (28.9) as given in (28.92).
3. (a) Solve (28.81) with $\omega \neq \omega_0$ and initial conditions $t = 0$, $y = y_0$, $v = v_0$. Use for the y_c solution the form $y_c = c_1 \sin \omega_0 t + c_2 \cos \omega_0 t$.
   (b) Write the solution if $\beta = 0$.
4. In (28.81), replace sin $(\omega t + \beta)$ by cos ωt. Solve the equation, with $\omega \neq \omega_0$, and initial conditions $t = 0$, $y = y_0$, $v = v_0$.
5. Solve (28.9) with $\beta = 0$ and initial conditions $t = 0$, $y = y_0$, $v = v_0$.

When a forcing function $f(t)$ is attached to a helical spring, the differential equation of motion, as given in (28.63), must be modified to read

(28.961)
$$m \frac{d^2 y}{dt^2} + ky = f(t).$$

Use this equation to solve the helical spring problems below. Take positive direction downward.

6. A 4-lb body stretches a helical spring 1 in. A forcing function $f(t) = \frac{1}{8} \sin 8\sqrt{6}\, t$ is attached to the system. Find the equation of motion. Is the motion stable or unstable? What is the undamped resonant frequency?
7. A 16-lb body stretches a helical spring 4 in. A forcing function $f(t) = \sin 4\sqrt{6}\, t$ is attached to the system. After it is brought to rest, it is displaced 6 in. and given a downward velocity of 4 ft/sec. Find the equation of motion of the body. Is the motion stable or unstable? What is the undamped resonant frequency? *Hint.* At $t = 0$, $y = \frac{1}{2}$, $v = 4$.
8. An 8-lb body stretches a helical spring 2 ft. After it is brought to rest, a forcing function $f(t) = \sin 6t$ is attached to the system causing it to vibrate. Find:
   (a) Distance and velocity as functions of time (*Hint.* At $t = 0$, $y = 0$, $v = 0$).
   (b) The natural frequency of the system.
   (c) The forcing frequency.
   (d) The maximum possible displacement or departure of the body from its equilibrium position.
   (e) Whether the motion is stable or unstable.

9. Answer the same questions as in problem 8, if at $t = 0$, the body is held at rest 4 ft below the equilibrium position and then given a velocity of -3 ft/sec. In addition, write the complementary function in the form $y_c = c \cos (kt + \delta)$.
10. Solve (28.8) if $f(t) = Ft$ where F is a constant. Is the motion stable or unstable?
11. A particle weighing 16 lb and moving on a horizontal line, is attracted to an origin 0 by a force which is proportional to its distance from 0. When the particle is at $x = -2$, this force is 9 lb. In addition, a forcing function $f(t) = \sin 3t$ is impressed on the system. If at $t = 0$, $x = 2$, $v = 0$, find (a) the equation of motion of the particle and (b) the resonant frequency of the system.
12. In problem 11, change sin $3t$ to cos $2t$. Find: (a) the equation of motion, (b) the natural frequency of the system, (c) the forcing frequency, and (d) the maximum possible displacement of the particle from 0.

13. A body attached to a helical spring oscillates with a period of $\pi/8$ sec. A forcing function attached to the system produces resonance. What is the frequency of the forcing function?

14. A mass m is attached to a helical spring whose spring constant is k. At $t = 0$, it is brought to rest and a constant forcing function $f(t) = 1/b$ lb is impressed on the system. After b sec, the force is removed.

(a) Find the position y of the mass as a function of time. *Hint.* First solve with $f(t) = 1/b$. Find $y(b)$ and $y'(b)$. Now solve the equation with $f(t) = 0$ and initial conditions $t = b$, $y = y(b)$, $dy/dt = y'(b)$.

Remark. After the input or forcing function $1/b$ is removed, note that it still is possible to have an output $y(t)$. Note, too, that if b is small, say $1/50$, then $f(t) = 50$ lb, but that it acts for only $1/50$ sec. It is as if the mass were given a sudden blow by a force that was immediately removed. Finally note that the output or response function $y(t)$ is continuous for $t \geqq 0$, even though the input or forcing function $f(t)$ is discontinuous. The latter can be written as

$$f(t) = \begin{cases} \dfrac{1}{b}, & 0 \leqq t \leqq b, \\ 0, & t > b. \end{cases}$$

If $b = 0$, $f(t)$ does not exist. In engineering circles, however, the fictitious forcing function $f(t)$ which results when $b = 0$, is called a **unit impulse;** in physics it is called a **Dirac δ-function.**

(b) Show that as $b \to 0$, the solution $y(t)$—let us call it $y_0(t)$—approaches

$$y_0(t) = (1/k)\sqrt{k/m} \sin \sqrt{k/m}\, t.$$

Hint. Use the fact that $\lim\limits_{\theta \to 0} (\sin \theta/\theta) = 1$. Now prove that $y_0(t)$ satisfies the equation $m(d^2y/dt^2) + ky = 0$ with initial conditions $t = 0$, $y = 0$, $dy/dt = 1/m$. The function $y_0(t)$ is called the **impulsive response** or **the response of the system to a unit impulse.**

15. A 16-lb weight stretches a spring 8 ft. At $t = 0$, it is brought to rest and a forcing function $f(t)$, defined by

$$f(t) = \begin{cases} e^t, & 0 \leqq t \leqq 1, \\ 0, & t > 1 \end{cases}$$

is impressed on the system. Find the equation of motion. (See hint in 14.)

A forcing function which has a different formula for a different time interval is called an **intermittent force.**

16. (a) Show that the solution of

$$\frac{d^2y}{dt^2} + \omega_0{}^2 y = F \sin (\omega t), \quad \omega \neq \omega_0,$$

with initial conditions $t = 0$, $y = 0$, $dy/dt = 0$, is

(28.97) $$y = \frac{F}{\omega_0{}^2 - \omega^2} \left(\sin \omega t - \frac{\omega}{\omega_0} \sin \omega_0 t \right).$$

(b) Show that if $\omega = \omega_0 + \epsilon$, where $\epsilon > 0$ is assumed to be small, then the solution (28.97) becomes

(28.971) $y = \dfrac{F}{(2\omega_0 + \epsilon)\epsilon} [\sin \omega_0 t - \sin (\omega_0 + \epsilon)t] + \dfrac{F}{(2\omega_0 + \epsilon)\omega_0} \sin \omega_0 t.$

(c) Using the identity $\sin A - \sin B = 2 \cos \dfrac{A+B}{2} \sin \dfrac{A-B}{2}$, show that if we ignore the last term on the right of (28.971)—we shall refer to it again later—(28.971) can be written as

(28.972) $y = - \dfrac{F}{2\left(\omega_0 + \dfrac{\epsilon}{2}\right)\epsilon} \left[2 \cos \left(\omega_0 + \dfrac{\epsilon}{2}\right)t \sin \dfrac{\epsilon t}{2}\right].$

(d) As we pointed out in this lesson of the text, the phenomenon of undamped resonance occurs when $\omega = \omega_0$. The amplitude of the motion then increases with time so that an unstable motion results. In this problem, we have taken $\omega = \omega_0 + \epsilon$, $\epsilon \neq 0$, so that $\omega \neq \omega_0$. However, as $\epsilon \to 0$, ω approaches the resonant frequency ω_0. We now make the assumption that ϵ, although not zero, is very small in comparison with ω_0. Hence we commit a relatively small error if in (28.972) we replace $\omega_0 + \epsilon/2$ by ω_0. Show that then (28.972) becomes

(28.973) $y = \left(-\dfrac{F}{\epsilon \omega_0} \sin \dfrac{\epsilon t}{2}\right) \cos \omega_0 t.$

(e) We can get an idea of the appearance of the graph of the motion given by (28.973), if we look at the function defined by it, as a harmonic motion $\cos \omega_0 t$ with a time varying amplitude

(28.974) $A = -\dfrac{F}{\epsilon \omega_0} \sin \dfrac{\epsilon t}{2}.$

Equation (28.974) itself defines a simple harmonic motion whose period is $4\pi/\epsilon$. Since ϵ is assumed small, the period of A is large. This means that the amplitude of (28.973) is varying slowly. It is, therefore, called appropriately a **slowly varying amplitude,** and the function $\cos \omega_0 t$ is said to be **amplitude modulated.** The graph of A is given by the broken lines in Fig. 28.975.

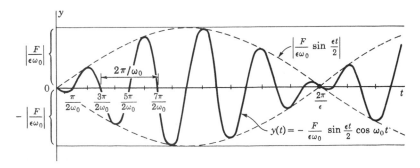

Figure 28.975

By (28.973), show that for each value of t such that $\cos \omega_0 t = \mp 1$, $y(t) = \pm (F/\epsilon\omega_0) \sin (\epsilon t/2)$. Hence show that the graph of the solution $y(t)$ will touch the upper part of the dotted curve in Fig. 28.975, for each value of t such that $\cos \omega_0 t = -1$ and touch the lower part of the dotted curve for each value of t such that $\cos \omega_0 t = 1$.

(f) Show that $y(t) = 0$ when

$$t = \frac{\pi}{2\omega_0}, \frac{3\pi}{2\omega_0}, \frac{5\pi}{2\omega_0}, \ldots, \frac{(2n+1)\pi}{2\omega_0},$$

and that the period of $\cos \omega_0 t$ is therefore $2\pi/\omega_0$. Since ω_0 is much greater than ϵ, this period $2\pi/\omega_0$ is much smaller than the period $4\pi/\epsilon$ of the slowly varying amplitude A of (28.974), whose graph is represented by the dotted lines in Fig. 28.975. Show, therefore, that the graph of $y(t)$ will thus resemble the solid lines shown in Fig. 28.975.

The variations in the amplitude of $y(t)$ are known as **beats**. When the amplitude is largest, the sound is loudest. This phenomenon of beats can be heard when two tuning forks with almost but not identical frequencies are set into vibration simultaneously. Note that the last term in (28.971), which was omitted in arriving at (28.973), represents a simple harmonic motion. It does not affect the phenomenon of beats.

Remark 1. Each musical note in an instrument has a definite frequency associated with it. When a standard note and its corresponding musical note are sounded at the same time, beats will result if their frequencies differ slightly, i.e., if they are not in tune. When the musical note of the instrument is adjusted so that beats disappear, the musical note is then in tune with the standard note. Can you see how this result can be used to tune an instrument?

Remark 2. We assumed in this problem $\omega = \omega_0 + \epsilon$. Therefore, as $\epsilon \to 0$, $\omega \to \omega_0$, the natural frequency of the system. Hence the undamped resonant case discussed in the text is the limit of the undamped modulated vibrations discussed in this problem.

ANSWERS 28D

3. (a) $y = \dfrac{1}{\omega_0} \left(v_0 - \dfrac{F\omega \cos \beta}{\omega_0{}^2 - \omega^2} \right) \sin \omega_0 t + \left(y_0 - \dfrac{F \sin \beta}{\omega_0{}^2 - \omega^2} \right) \cos \omega_0 t$

$\qquad + \dfrac{F}{\omega_0{}^2 - \omega^2} \sin (\omega t + \beta).$

(b) $y = \dfrac{1}{\omega_0} \left(v_0 - \dfrac{F\omega}{\omega_0{}^2 - \omega^2} \right) \sin \omega_0 t + y_0 \cos \omega_0 t + \dfrac{F}{\omega_0{}^2 - \omega^2} \sin \omega t.$

4. $y = \dfrac{v_0}{\omega_0} \sin \omega_0 t + \left(y_0 - \dfrac{F}{\omega_0{}^2 - \omega^2} \right) \cos \omega_0 t + \dfrac{F}{\omega_0{}^2 - \omega^2} \cos \omega t.$

5. $y = \dfrac{F + 2\omega_0 v_0}{2\omega_0{}^2} \sin \omega_0 t + y_0 \cos \omega_0 t - \dfrac{Ft \cos \omega_0 t}{2\omega_0}.$

6. $y = c \sin (8\sqrt{6}\, t + \delta) - \dfrac{\sqrt{6}}{96} t \cos (8\sqrt{6}\, t).$ Unstable. $\omega_0 = 8\sqrt{6}$ rad/sec.

7. $y = \dfrac{1 + 16\sqrt{6}}{96} \sin 4\sqrt{6}\,t + \tfrac{1}{2} \cos 4\sqrt{6}\,t - \dfrac{\sqrt{6}}{24}\,t \cos 4\sqrt{6}\,t$.

Unstable. $\omega_0 = 4\sqrt{6}$ rad/sec.

8. (a) $y = \tfrac{1}{10}(3 \sin 4t - 2 \sin 6t)$, $v = \tfrac{6}{5}(\cos 4t - \cos 6t)$.
   (b) $\omega_0 = 4$ rad/sec.   (c) $\omega = 6$ rad/sec.   (d) $\tfrac{1}{2}$ ft.   (e) Stable.

9. (a) $y = -\tfrac{9}{20} \sin 4t + 4 \cos 4t - \tfrac{1}{5} \sin 6t$
   $\equiv 4.03 \sin (4t + \delta) - \tfrac{1}{5} \sin 6t$, where $\delta = \pi - $ Arc sin 80/80.5.
   $v = -16.1 \cos (4t + \delta) - \tfrac{6}{5} \cos 6t$.
   (b), (c), (e) same as in 8.   (d) $4.03 + \tfrac{1}{5} = 4.23$ ft.

10. $y = c_1 \cos \omega_0 t + c_2 \sin \omega_0 t + Ft/\omega^2$. Unstable.

11. (a) $x = \tfrac{1}{9} \sin 3t + 2 \cos 3t - \tfrac{1}{3}t \cos 3t$.   (b) $\omega_0 = 3$ rad/sec.

12. (a) $x = \tfrac{2}{5} \cos 2t + \tfrac{8}{5} \cos 3t$.   (b) 3 rad/sec.   (c) 2 rad/sec.
   (d) $\tfrac{2}{5} + \tfrac{8}{5} = 2$.

13. 16 rad/sec.

14.
$$y(t) = \begin{cases} \dfrac{1}{bk}\,(1 - \cos \sqrt{k/m}\,t),\ 0 \leqq t \leqq b, \\[2mm] \dfrac{1}{bk}\,[\cos \sqrt{k/m}\,(t - b) - \cos \sqrt{k/m}\,t] \\[2mm] = \dfrac{2}{bk} \sin\left[\sqrt{k/m}\left(t - \dfrac{b}{2}\right)\right] \sin \sqrt{k/m}\,\dfrac{b}{2},\quad t > b. \end{cases}$$

15.
$$y(t) = \begin{cases} \dfrac{2e^t}{5} - \tfrac{1}{5} \sin 2t - \tfrac{2}{5} \cos 2t,\quad 0 \leqq t \leqq 1, \\[2mm] \tfrac{1}{5}(2e \sin 2 + e \cos 2 - 1) \sin 2t \\[2mm] + \tfrac{1}{5}(2e \cos 2 - e \sin 2 - 2) \cos 2t,\quad t > 1. \end{cases}$$

LESSON 29. Damped Motion.

In the previous lesson, we ignored the important factor of resistance or damping. In this lesson we shall discuss the more realistic motion of a particle that is subject to a resistance or damping force. We shall assume, for illustrative purposes, that the resisting force is proportional to the first power of the velocity. Frequently it will not be. In such cases more complicated methods, beyond the scope of this text, will be needed to solve the resulting differential equation.

LESSON 29A. Free Damped Motion. (Damped Harmonic Motion).

Definition 29.1. A particle will be said to execute free damped motion, more commonly called **damped harmonic motion**, if its equation of motion satisfies a differential equation of the form

$$(29.11)\quad m\,\frac{d^2 y}{dt^2} + 2mr\,\frac{dy}{dt} + m\omega_0{}^2 y = 0,\qquad \frac{d^2 y}{dt^2} + 2r\,\frac{dy}{dt} + \omega_0{}^2 y = 0,$$

where the coefficient $2mr > 0$ is called the **coefficient of resistance** of the system. As before ω_0 is the natural (undamped) frequency of the system and m is the mass of the particle.

The characteristic equation of (29.11) is $m^2 + 2rm + \omega_0^2 = 0$, whose roots are

$$(29.12) \qquad m = -r \pm \sqrt{r^2 - \omega_0^2}.$$

The solution of (29.11) will thus depend on the character of the roots of (29.12), i.e., whether they are real, imaginary, or multiple. We shall consider each case separately.

Case 1. $r^2 > \omega_0^2$. If $r^2 > \omega_0^2$, the roots in (29.12) are real and unequal. Hence the solution of (29.11) is

$$(29.13) \qquad y = c_1 e^{(-r + \sqrt{r^2 - \omega_0^2})t} + c_2 e^{(-r - \sqrt{r^2 - \omega_0^2})t}.$$

Since both exponents in (29.13) are negative quantities (verify it) we can write (29.13) as

$$(29.14) \qquad y = c_1 e^{At} + c_2 e^{Bt}, \quad A < 0, B < 0.$$

If $c_1 \neq 0$, $c_2 \neq 0$, and c_1, c_2 have the same sign, then because $e^z > 0$ for all z, there is no value of t for which $y = 0$. Hence, in this case, the graph of (29.14) cannot cross the t axis. If, however, $c_1 \neq 0$, $c_2 \neq 0$, and c_1, c_2 have opposite signs, then setting $y = 0$ in (29.14) and solving it for t will determine the t intercepts of its graph. Therefore setting $y = 0$ in (29.14), we obtain

$$(29.15) \qquad e^{(A-B)t} = -\frac{c_2}{c_1},$$

$$(A - B)t = \log\left(\frac{-c_2}{c_1}\right),$$

$$t = \frac{1}{A - B} \log\left(\frac{-c_2}{c_1}\right).$$

From (29.15), we deduce that there can be only one value of t for which $y = 0$. We have thus shown that the curve representing the motion given by (29.14) can cross the t axis once at most. Further, by (27.113), $y \to 0$ as $t \to \infty$. [Remember A and B in (29.14) are negative.]

Differentiation of (29.14) gives

$$(29.16) \qquad \frac{dy}{dt} = c_1 A e^{At} + c_2 B e^{Bt}, \quad A < 0, B < 0.$$

Since this equation has the same form as (29.14), it, too, can have, at most, only one value of t for which $dy/dt = 0$. Hence the curve determined by (29.14) can have at most only *one* maximum or minimum point.

The motion is therefore *nonoscillatory* and dies out with time. In Fig. 29.17 we have drawn graphs of a few possible motions.

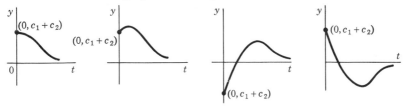

Figure 29.17

Comment 29.18. In this case, where $r^2 > \omega_0{}^2$, the resisting or **damping force** represented by r overpowers the restoring force represented by ω_0 and hence prevents oscillations. The system is called **overdamped.**

Example 29.19. A helical spring is stretched 32 inches by an object weighing 2 pounds, and brought to rest. It is then given an additional pull of 1 ft and released. If the spring is immersed in a medium whose coefficient of resistance is $\frac{1}{2}$, find the equation of motion of the object. Assume the resisting force is proportional to the first power of the velocity. Also draw a rough graph of the motion.

Solution. Because of the presence of a resisting factor, whose coefficient of resistance is $\frac{1}{2}$, the differential equation of motion (28.63) for the helical spring must be changed to read

(a) $$m\frac{d^2y}{dt^2} = -ky - \frac{1}{2}\frac{dy}{dt}.$$

In this example, since 2 pounds stretches the spring 32 inches $= \frac{8}{3}$ feet, we have, by (28.62),

(b) $$\tfrac{8}{3}k = 2, \quad k = \tfrac{3}{4}.$$

The mass m of the object is $\frac{2}{32} = \frac{1}{16}$. Hence (a) becomes

(c) $$\frac{1}{16}\frac{d^2y}{dt^2} + \frac{1}{2}\frac{dy}{dt} + \frac{3}{4}y = 0, \qquad \frac{d^2y}{dt^2} + 8\frac{dy}{dt} + 12y = 0,$$

whose general solution is

(d) $$y = c_1e^{-2t} + c_2e^{-6t}, \qquad y' = -2c_1e^{-2t} - 6c_2e^{-6t}.$$

The initial conditions are $t = 0$, $y = 1$, $dy/dt = 0$. Substituting these values in (c), we obtain

(e) $$1 = c_1 + c_2,$$
$$0 = -2c_1 - 6c_2.$$

The solution of (e) is $c_1 = \frac{3}{2}$, $c_2 = -\frac{1}{2}$. Hence (d) becomes

(f) $\qquad y = \frac{3}{2}e^{-2t} - \frac{1}{2}e^{-6t}, \qquad y' = -3e^{-2t} + 3e^{-6t}.$

Setting $y = 0$ in (f), we find $t = -\frac{1}{4}\log 3 = -0.27$. Setting $y' = 0$, we find $t = 0$ and by the first equation in (f), $y = 1$ when $t = 0$. Hence the curve has a maximum at $t = 0$, $y = 1$. Setting $y'' = 0$, we find the curve has an inflection point at $t = \frac{1}{4}\log 3 = 0.27$, $y = 0.77$. A rough graph of the motion is given in Fig. 29.191. The motion is non-oscillatory. The maximum displacement occurs at $t = 0$, i.e., at the beginning of its motion; the displacement then gradually dies out.

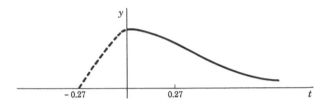

Figure 29.191

Case 2. $r^2 = \omega_0{}^2$. If $r^2 = \omega_0{}^2$, the roots of (29.12) are $-r$ twice. Hence the solution of (29.11), by Lesson 20C, is

(29.2) $\qquad y = c_1 e^{-rt} + c_2 t e^{-rt},$
$\qquad\qquad y' = -rc_1 e^{-rt} + c_2 e^{-rt} - rc_2 t e^{-rt}.$

Since $r > 0$, by (27.113), both e^{-rt} and $te^{-rt} \to 0$ as $t \to \infty$. And as in the previous case, there is only one value of t at most for which y and $y' = 0$. Therefore as in the previous case, the motion is nonoscillatory, and dies out with time. The graphs of some of its possible motions are similar to those shown in Fig. 29.17.

Comment 29.21. In this case, where $r = \omega_0$, the resisting or damping force represented by r is just as strong as the restoring force represented by ω_0 and hence prevents oscillations. For this reason the system is said to be **critically damped.**

Case 3. $r^2 < \omega_0{}^2$. If $r^2 < \omega_0{}^2$, the roots in (29.12) are imaginary and can be written as

(29.3) $\qquad m = -r \pm i\sqrt{\omega_0{}^2 - r^2}.$

The solution of (29.11), by Lesson 20D, is therefore

(29.31) $\qquad y = ce^{-rt}\sin(\sqrt{\omega_0{}^2 - r^2}\,t + \delta).$

Because of the sine term in the solution, the motion is oscillatory. The **damped amplitude** of the motion is ce^{-rt} and since $r > 0$, this factor decreases as t increases and approaches zero as t approaches ∞. Hence with time, the particle vibrates with smaller and smaller oscillations about its equilibrium position.

Each function defined in (29.31) is not periodic since its values do not repeat. However, because the motion is oscillatory, we say the function is **damped periodic** and define its **damped period** to be the time it takes the particle, starting at the equilibrium position, to make one complete oscillation. Hence its **damped period** is said to be

(29.32)
$$T = \frac{2\pi}{\sqrt{\omega_0^2 - r^2}}.$$

The **damped frequency** of the motion is $\sqrt{\omega_0^2 - r^2}$ radians per unit of time, or $\sqrt{\omega_0^2 - r^2}/2\pi$ cycles per unit of time.

The exponential term e^{-rt} is called appropriately the **damping factor.** Since this factor decreases with time, the motion eventually dies down. When $t = 1/r$, the damping factor is $1/e$. The time it takes the damping factor to reach this value $1/e$ is called the **time constant.** Hence the time constant $\tau = 1/r$.

A graph of the function defined by (29.31) is given in Fig. 29.33. It is *an oscillatory motion whose amplitude decreases with time.*

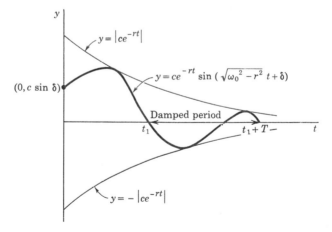

Figure 29.33

Comment 29.34. In this case, where $r^2 < \omega_0^2$, the damping force represented by r is weaker than the restoring force represented by ω_0 and thus cannot prevent oscillations. For this reason the system is called **underdamped.**

Example 29.35. If the coefficient of resistance in example 29.19 is $\frac{3}{8}$ instead of $\frac{1}{2}$, find:

1. The equation of motion of the system.
2. The damping factor.
3. The damped amplitude of the motion.
4. The damped period of the motion.
5. The damped frequency of the motion.
6. The time constant.

Solution. The differential equation of motion (c) in example 29.19 now becomes

(a) $$\frac{1}{16}\frac{d^2y}{dt^2} + \frac{3}{8}\frac{dy}{dt} + \frac{3}{4}y = 0, \qquad \frac{d^2y}{dt^2} + 6\frac{dy}{dt} + 12 = 0.$$

Its solution is

(b) $$y = ce^{-3t}\sin(\sqrt{3}\,t + \delta).$$

Differentiation of (b) gives

(c) $$y' = -3ce^{-3t}\sin(\sqrt{3}\,t + \delta) + c\sqrt{3}\,e^{-3t}\cos(\sqrt{3}\,t + \delta).$$

The initial conditions are $t = 0$, $y = 1$, $y' = 0$. Substituting these values in (b) and (c), we obtain

(d) $$1 = c\sin\delta,$$
$$0 = -3c\sin\delta + \sqrt{3}\,c\cos\delta.$$

Substituting in the second equation of (d), the value of c as given in the first equation, we obtain

(e) $$0 = -3 + \sqrt{3}\cot\delta, \qquad \cot\delta = \frac{3}{\sqrt{3}} = \sqrt{3},$$

$$\delta = \frac{\pi}{6} \text{ or } \frac{7\pi}{6}, \qquad \sin\delta = \pm\frac{1}{2}.$$

Hence, by the first equation in (d), choosing $\sin\delta = \frac{1}{2}$, we have,

(f) $$c = 2.$$

The equation of motion (b) therefore becomes

(g) $$y = 2e^{-3t}\sin\left(\sqrt{3}\,t + \frac{\pi}{6}\right),$$

which is the answer to 1. The answers to the remaining questions follow.

2. The damping factor is e^{-3t}.
3. The damped amplitude of the motion is $2e^{-3t}$ feet.

4. The damped period of the motion is $2\pi/\sqrt{3}$ seconds.

5. The damped frequency of the motion is $\sqrt{3}$ rad/sec $\equiv \dfrac{\sqrt{3}}{2\pi}$ cps.

6. The time constant $\tau = \frac{1}{3}$ sec.

EXERCISE 29A

1. Verify the accuracy of the solution of (29.11) with $r^2 > \omega_0{}^2$, as given in (29.13).
2. Verify that each of the exponents in (29.13) is a negative quantity.
3. Verify the accuracy of the solution (d) of Example 29.19.
4. Verify the accuracy of the solution of (29.11) with $r^2 = \omega_0{}^2$, as given in (29.2)
5. Verify the accuracy of the solution of (29.11) with $r^2 < \omega_0{}^2$, as given in (29.31).
6. Verify the accuracy of the solution (b) of Example 29.35.
7. Show that the system whose differential equation is

$$\frac{d^2y}{dt^2} + 2a\,\frac{dy}{dt} + b^2 y = 0,\; a > 0,$$

is: (a) overdamped and the motion not oscillatory if $a^2 > b^2$, (b) critically damped and the motion not oscillatory if $a^2 = b^2$, (c) underdamped and the motion oscillatory if $a^2 < b^2$.

8. (a) Solve the differential equation

(29.36)
$$\frac{d^2y}{dt^2} + 2a\,\frac{dy}{dt} + by = 0.$$

Note that here b is not squared as in 7.

(b) Show that the motion of the system is stable only if $a > 0$ and $b > 0$. (For definitions of stable and unstable, see Lesson 28D). *Hint.* Show that if $a > 0$, $b > 0$, each independent solution of (29.36) approaches zero as $t \to \infty$. Hence the distance y from equilibrium approaches zero. Consider each other possibility $a > 0$, $b < 0$; $a < 0$, $b > 0$; $a < 0$, $b < 0$, and show that in each case $y \to \infty$ as $t \to \infty$.

(c) Show that if $a < 0$, $b > 0$ and $a^2 < b$, the motion, although unstable, is oscillatory; if $a > 0$, $b < 0$, or if $a < 0$, $b < 0$, or if $a < 0$, $b > 0$, and in each case $a^2 > b$, the motion, although unstable, is not oscillatory.

9. With the help of the answers to problems 7 and 8, determine, without solving, whether the motion of the system, whose differential equation is:

(a) $\dfrac{d^2y}{dt^2} - \dfrac{dy}{dt} - 2y = 0.$

(b) $\dfrac{d^2y}{dt^2} - 3\dfrac{dy}{dt} + 2y = 0.$

(c) $\dfrac{d^2y}{dt^2} + 2\dfrac{dy}{dt} + 5y = 0.$

(d) $\dfrac{d^2y}{dt^2} + 4\dfrac{dy}{dt} + 4y = 0.$

(e) $\dfrac{d^2y}{dt^2} + 4\dfrac{dy}{dt} - 4y = 0.$

(f) $\dfrac{d^2y}{dt^2} - 2\dfrac{dy}{dt} + 5y = 0.$

(g) $\dfrac{d^2y}{dt^2} + 6\dfrac{dy}{dt} + 6y = 0.$

is stable or unstable; oscillatory or not oscillatory. Also determine whether the system is underdamped, critically damped, or overdamped. Check your answer by solving each equation. Draw a rough graph of each motion.

10. A particle moves on a straight line according to the law

$$\frac{d^2x}{dt^2} + 2r\frac{dx}{dt} + x = 0,$$

where r is a constant and x is the displacement of the particle from its equilibrium position.

(a) For what values of r will the motion be stable; unstable; oscillatory; not oscillatory. For what values of r will the system be underdamped; critically damped; overdamped.

(b) Check your answers by solving the equation with $r = \frac{1}{2}, r = 1, r = 2,$ $r = -\frac{1}{2}, r = -1$.

(c) For what value of r will the motion be oscillatory and have a damped period equal to 3π?

(d) Is there a value of r that will make the damped period less than 2π?

11. A particle moves on a straight line in accordance with the law

$$\frac{d^2x}{dt^2} + 4\frac{dx}{dt} + 13x = 0.$$

At $t = 0, x = 0, v = 12$ ft/sec.

(a) Solve the equation for x as a function of t.

(b) What is the damping factor, the damped amplitude, the damped period, the damped frequency, the time constant?

(c) Find the time required for the damped amplitude—and hence also for the damping factor—to decrease by 50 percent. *Hint.* The damped amplitude is $4e^{-2t}$. When $t = 0, 4e^{-2t} = 4$. You want t so that $4e^{-2t} = 2$.

(d) What percentage of its original value has the damping factor, and therefore the damped amplitude, after one half period has elapsed? *Hint.* The damped period is $2\pi/3$. Evaluate e^{-2t} when $t = \pi/3$. What is the damped amplitude at that instant?

(e) Where is the particle and with what velocity is it moving when $t = \pi/6$ sec?

(f) Draw a rough graph of the curve.

12. A particle moves in a straight line in accordance with the law

$$\frac{d^2x}{dt^2} + 10\frac{dx}{dt} + 16x = 0.$$

At $t = 0, x = 1$ ft, $v = 4$ ft/sec.

(a) Find the equation of motion.

(b) Is the motion oscillatory?

(c) What is the maximum value of x? When does x attain this maximum value?

(d) Draw a rough graph of the curve. Does the curve cross the t axis for $t > 0$?

13. A particle is executing damped harmonic motion. In 10 sec, the damping factor has decreased by 80 percent. Its damped period is 2 sec. Find the differential equation of motion.

14. A particle of mass m moves in a straight line. It is attracted toward the origin by a force equal to k times its distance from the origin. The resistance is $2R$ times the velocity. Find the maximum value of m so that the motion will not be oscillatory.

15. A particle moves in a straight line in accordance with the law

$$\frac{d^2x}{dt^2} + 6\,\frac{dx}{dt} - 16x = 0.$$

At $t = 0$, the particle is at $x = 2$ ft and moving to the left with a velocity of 10 ft/sec.

(a) When will the particle change direction and go to the right?
(b) Will it ever change direction again?

When the damping or resisting factor of a system is not negligible, the differential equation (28.63) for the helical spring must be modified to read, with downward direction positive,

(29.37) $$m\,\frac{d^2y}{dt^2} + r\,\frac{dy}{dt} + ky = 0,$$

where we have assumed that the force of resistance is proportional to the first power of the velocity and $r > 0$ is the coefficient of resistance of the system. Note that here r replaces $2mr$ of (29.11).

Use (29.37) to solve the following problems, 16–25.

16. A weight of 16 lb stretches a helical spring $1\frac{3}{5}$ ft. The coefficient of resistance of the spring is 2. After it is brought to rest, it is given a velocity of 12 ft/sec.

(a) Find the equation of motion. Draw a rough graph of the motion.
(b) Find damping factor, damped amplitude, damped period, damped frequency, time constant.
(c) When will the weight stop for the first time and change direction? How far from equilibrium will it then be?
(d) When will it stop for the second time? How far from equilibrium will it be?
(e) Write a formula which will give the times when the weight crosses the equilibrium position and for the times of its successive stops.

17. A 16-lb weight stretches a spring 6 in. The coefficient of resistance is 8. After the spring is brought to rest, it is stretched an additional 3 in. and released. Find the equation of motion. Draw a rough graph of the motion.

18. In problem 17, change the coefficient of resistance to 10. Find the equation of motion. Draw a rough graph of the motion.

19. (a) Solve (29.37) if $r^2 < 4km$ and the initial conditions are $t = 0$, $y = y_0$, $v = 0$.
(b) What is the damped period of the motion?
(c) When will the damping factor, and therefore the damped amplitude, be p percent of its initial value? *Hint.* The damping factor is $e^{-rt/2m}$. At $t = 0$, the damping factor is $e^{-rt/2m} = 1 = 100$ percent. Therefore want t such that $e^{-rt/2m} = p/100$. Hence $-rt/2m = \log(p/100)$, $t = -(2m/r)\log(p/100)$.

(d) Call the time obtained in (c) t_0 sec. Therefore the damping factor at the end of t_0 sec is $e^{-rt_0/2m}$ and this damping factor, and therefore also the damped amplitude, is p percent of its value at $t = 0$. Show that at the end of every period of t_0 sec, the new damping factor is p percent of its value at the beginning of the period. *Hint.* Show that $e^{-r(t_0+t_0)/2m} = p^2/10^4$, i.e., show that it is $p^2/10^4$ of the original damped period and hence is p percent of the damped period at the end of t_0 sec. Or you can let $e^{-rt_0/2m} = 100$ percent. Then want t_1 such that $e^{-rt_1/2m} = p/100$. Find $t_1 = -(2m/r) \log p/100$ as in (c).

(e) When $t = T$, the damped period of the motion, the damping factor is $e^{-rT/2m}$. It has a definite value, say q percent of the value of the damping factor at $t = 0$. Show that at the end of each period of T sec, the damping factor, and therefore the damped amplitude, is q percent of the damping factor at the beginning of the period. *Hint.* See (d) above. This constant percentage, therefore, gives the percentage decrease in the displacement of a particle from equilibrium at the end of a period as compared with its displacement at the beginning of a period. Hence, the damped amplitude at the end of a period of T sec = q percent of the damped amplitude at the beginning of that period. Therefore,

$$(29.38) \qquad \log \left(\frac{\text{Damped amplitude at the beginning of a period of } T \text{ sec}}{\text{Damped amplitude at the end of that period}} \right)$$
$$= \log (100/q) = \text{a constant } D.$$

The constant D is called the **logarithmic decrement**. It is, as equation (29.38) shows, the constant positive difference between the logarithm of the damped amplitude at the beginning of a period of T sec and the logarithm of the damped amplitude at the end of that period.

(f) Find the logarithmic decrement of this problem. *Hint.* In (29.38) substitute the damped amplitude when $t = 0$ and when $t = T$ as found in (b).

(g) When will the body first reach the equilibrium position?

20. A 20-lb weight stretches a spring 3 in. After it comes to rest, it is given an additional stretch of 2 in. and released. The internal resistance of the spring is negligible but the resistance due to the air is $1/50$ of its velocity.

(a) Find the equation of motion and draw a rough graph of its motion.

(b) Find the damped amplitude, damping factor, damped period, damped frequency, time constant.

(c) When will the damping factor have decreased by 50 percent?

(d) Over what time intervals will the damping factor at the end of an interval be 50 percent of its value at the beginning of the interval?

(e) What percentage of its original value does the damping factor have, and therefore also the amplitude, at the end of a period? Note by (e) of problem 19 that the damping factor at the end of any period is this same percentage of the damping factor at the beginning of that period.

(f) Find the logarithmic decrement.

(g) When does the particle first cross the equilibrium position?

21. The oscillatory motion of a spring is given by

$$\frac{d^2y}{dt^2} + 2a \frac{dy}{dt} + by = 0, \quad a^2 < b.$$

It is observed that the damping factor has decreased by 80 percent in 10 sec and that its damped period is 2 sec. Find the values of a and b.

22. The natural frequency of a spring is 1 cps. After the spring is immersed in a resisting medium, its frequency is reduced to $\frac{2}{3}$ cps.

   (a) What is the damping factor?
   (b) What is the differential equation of motion?

23. In problem 21, find a and b if the period of the motion is 2 sec and the logarithmic decrement is $\frac{1}{5}$.

24. The differential equation of motion of a body attached to a helical spring is given by (29.37).

   (a) Solve the equation if its mass $m = r^2/4k$ and at $t = 0$, $y = y_0$, $v = v_0$.
   (b) Is the motion oscillatory or not oscillatory?
   (c) When will it reach its maximum displacement from equilibrium?
   (d) Show that from its maximum displacement it will move toward the equilibrium position but never reach it. *Hint.* Show that $y \to 0$ as $t \to \infty$ [see (27.113)].
   (e) Show that if $rv_0 = -2ky_0$, the body will never change its direction but will move continually toward the equilibrium position.

25. The differential equation of motion of a body attached to a helical spring is given by (29.37).

   (a) Solve the equation if $r^2 > 4km$ and at $t = 0$, $y = 0$, $v = v_0$.
   (b) Is the motion oscillatory or not oscillatory?
   (c) Show that the solution can also be written in the form

$$y = \frac{2mv_0}{\sqrt{r^2 - 4km}} e^{-rt/2m} \sinh \frac{\sqrt{r^2 - 4km}}{2m} t.$$

   Hint. See (18.9).
   (d) When will the body reach its maximum displacement from equilibrium?
   (e) Show that from its maximum displacement, it will move toward the equilibrium position but never reach it.

We have included below only a few pendulum problems because of the similarity in form of the pendulum equation and the helical spring equation—compare (29.37) with (29.381) below. The same questions asked for the spring could be asked for the pendulum. All one need do to obtain a solution for the pendulum is to replace k in the previous answers by mg/l and y by θ. Remember, linear velocity $v = l \, d\theta/dt = l\omega$, where ω is angular velocity.

26. A simple pendulum of length l, with weight mg attached, swings in a medium which offers a resisting force proportional to the first power of the linear velocity. Show that the differential equation of motion is

(29.381) $$\frac{d^2\theta}{dt^2} + \frac{r}{m} \frac{d\theta}{dt} + \frac{g}{l} \theta = 0,$$

where r is the coefficient of resistance of the system. *Hint.* Adjust (28.71) to take into account the resisting force, and remember linear velocity $v = l \dfrac{d\theta}{dt}$.

27. (a) Show that the pendulum in problem 26 is overdamped and the motion not oscillatory if $r^2/4m^2 > g/l$; critically damped and the motion not oscillatory if $r^2/4m^2 = g/l$; underdamped and the motion oscillatory if $r^2/4m^2 < g/l$.

   (b) Solve (29.381) if at $t = 0$, $\theta = 0$, $\omega = \omega_0$ and $r^2/4m^2 < g/l$.

28. A weight of 2 lb is attached to a pendulum 16 ft long. Find the smallest positive value of the coefficient of resistance r for which the pendulum will not oscillate.

29. A weight of 4 lb is attached to a pendulum swinging in a medium which offers a resistance of one-eighth of the linear velocity. It is desired that the period of the pendulum be 2π. How long must the pendulum be?

30. (a) Solve (29.381) if the pendulum is released from the position $\theta = \theta_0$. Assume oscillatory motion.

   (b) When will the pendulum first reach the equilibrium position?

ANSWERS 29A

7. $y = c_1 e^{(-a+\sqrt{a^2-b^2})\,t} + c_2 e^{(-a-\sqrt{a^2-b^2})\,t}$, $b^2 < a^2$,
$y = (c_1 + c_2 t)e^{-at}$, $b^2 = a^2$,
$y = e^{-at}(c_1 \cos \sqrt{b^2 - a^2}\,t + c_2 \sin \sqrt{b^2 - a^2}\,t)$, $a^2 < b^2$.

8. $y = c_1 e^{(-a+\sqrt{a^2-b})\,t} + c_2 e^{(-a-\sqrt{a^2-b})\,t}$, $b < a^2$,
$y = (c_1 + c_2 t)e^{-at}$, $b = a^2$,
$y = e^{-at}(c_1 \cos \sqrt{b - a^2}\,t + c_2 \sin \sqrt{b - a^2}\,t)$, $a^2 < b$.

9. (a) Unstable, not oscillatory, overdamped, $y = c_1 e^{2t} + c_2 e^{-t}$.

   (b) Unstable, not oscillatory, overdamped, $y = c_1 e^{2t} + c_2 e^{t}$.

   (c) Stable, oscillatory, underdamped, $y = c e^{-t} \sin (2t + \delta)$.

   (d) Stable, not oscillatory, critically damped, $y = c_1 e^{-2t} + c_2 t e^{-2t}$.

   (e) Unstable, not oscillatory, overdamped, $y = c_1 e^{(-2+2\sqrt{2})t} + c_2 e^{(-2-\sqrt{2})t}$.

   (f) Unstable, oscillatory, underdamped, $y = c e^{t} \cos (2t + \delta)$.

   (g) Stable, not oscillatory, overdamped, $y = c e^{(-3+\sqrt{3})t} + c_2 e^{(-3-\sqrt{3})t}$.

10. (a) Stable only if $r \geq 0$; underdamped and oscillatory if $0 < r < 1$; critically damped and not oscillatory if $r = 1$; overdamped and not oscillatory if $r > 1$; unstable and oscillatory if $-1 < r < 0$; unstable and not oscillatory if $r \leq -1$.

   (b) $y = c e^{-t/2} \sin (\sqrt{3}\,t/2 + \delta)$,
   $y = (c_1 + c_2 t)e^{-t}$,
   $y = c_1 e^{(-2+\sqrt{3})t} + c_2 e^{(-2-\sqrt{3})t}$,
   $y = c_1 e^{t/2} \sin (\sqrt{3}\,t/2 + \delta)$,
   $y = (c_1 + c_2 t)e^{t}$.

   (c) $r = \sqrt{5}/3$.   (d) No.

11. (a) $x = 4e^{-2t} \sin 3t$.   (b) e^{-2t}, $4e^{-2t}$ ft, $2\pi/3$ sec, 3 rad/sec or $3/2\pi$ cps, $\frac{1}{2}$ sec.   (c) $t = \frac{1}{2} \log 2$ sec $= 0.35$ sec.   (d) 12.3 percent, 0.49 ft.   (e) 1.4 ft, -2.8 ft/sec.

12. (a) $x = 2e^{-2t} - e^{-8t}$.   (b) No.   (c) $x = 1.19$ ft, $t = 0.12$ sec.   (d) No.

13. $\dfrac{d^2y}{dt^2} + 0.322 \dfrac{dy}{dt} + 9.896y = 0$.   **14.** R^2/k.

15. (a) 0.223 sec. (b) No.

16. (a) $y = 3e^{-2t} \sin 4t$.   (b) e^{-2t}, $3e^{-2t}$ ft, $\pi/2$ sec, 4 rad/sec, $\frac{1}{2}$ sec.
   (c) 0.277 sec, 1.54 ft.   (d) $0.277 + \pi/4$ sec, -0.32 ft.
   (e) $n\pi/4$ sec; $0.277 + n\pi/4$ sec, $n = 0, 1, 2, \cdots$.

17. $y = e^{-8t}(\frac{1}{4} + 2t)$.   **18.** $y = \frac{1}{3}e^{-4t} - \frac{1}{12}e^{-16t}$.

19. (a) $y = y_0 \sqrt{\dfrac{4km}{4km - r^2}}\, e^{-rt/2m} \cos\left(\dfrac{\sqrt{4km - r^2}}{2m}\, t + \delta\right)$,

    where $\delta = \text{Arc tan }(-r/\sqrt{4km - r^2})$.

   (b) $T = 4\pi m/\sqrt{4km - r^2}$ sec.

   (f) log decrement $= 2\pi r/\sqrt{4km - r^2}$.

   (g) $t = m(\pi - 2\delta)/\sqrt{4km - r^2}$ sec.

20. (a) $y = \frac{1}{6}e^{-0.016t} \cos(8\sqrt{2}\, t + \delta)$, approximately, where $\tan \delta = -0.0014$,
    $\delta = -0.0014$ radian,
   (b) $\frac{1}{6}e^{-0.016t}$ ft, $e^{-0.016t}$, $\pi\sqrt{2}/8$ sec, $8\sqrt{2}$ rad/sec, or $4\sqrt{2}/\pi$ cps, $62\frac{1}{2}$ sec.
   (c) and (d) 43.3 sec.   (e) 99 percent.   (f) 0.009.   (g) 0.14 sec.

21. $a = 0.161$, $b = 9.896$.

22. (a) $e^{-4.68t}$.   (b) $y'' + 9.37y' + 39.5y = 0$.

23. $a = 0.1$, $b = 9.88$.

24. (a) $y = \left[y_0 + \left(v_0 + \dfrac{2ky_0}{r}\right)t\right]e^{-2kt/r}$.

   (b) Not oscillatory.   (c) $t = v_0 r^2/2k(rv_0 + 2ky_0)$ sec.

25. (a) $y = \dfrac{mv_0}{\sqrt{r^2 - 4km}}\, e^{-rt/2m}\left(e^{t\sqrt{r^2-4km}/2m} - e^{-t\sqrt{r^2-4km}/2m}\right)$

   (b) Not oscillatory

   (d) $\tanh \dfrac{\sqrt{r^2 - 4km}}{2m}\, t = \dfrac{\sqrt{r^2 - 4km}}{r}$,

     $t = \dfrac{2m}{\sqrt{r^2 - 4km}} \tanh^{-1} \dfrac{\sqrt{r^2 - 4km}}{r}$.

27. (b) $\theta = 2\omega_0 m \sqrt{\dfrac{l}{4m^2 g - lr^2}}\, e^{-rt/2m} \sin \sqrt{\dfrac{g}{l} - \dfrac{r^2}{4m^2}}\, t$.

28. $r = \dfrac{1}{4\sqrt{2}}$ lb-sec/ft.   **29.** $l = 25.6$ ft.

30. In answers to 19(a) and (g), replace k by mg/l and y by θ.

LESSON 29B. Forced Motion with Damping. The motion of a
particle that satisfies the differential equation

(29.4) $$m\frac{d^2y}{dt^2} + 2mr\frac{dy}{dt} + m\omega_0^2 y = f(t),$$

$$\frac{d^2y}{dt^2} + 2r\frac{dy}{dt} + \omega_0^2 y = \frac{1}{m}f(t),$$

where, as before, $2mr$ is the coefficient of resistance of the system, ω_0 is
the natural (undamped) frequency of the system, m is the mass of the

particle and $f(t)$ is a forcing function attached to the system, is called **forced damped motion** in contrast to the *free* damped motion (i.e., damped harmonic motion) when $f(t) \equiv 0$. In engineering circles, $f(t)$ is called the **input** of the system and the solution $y(t)$ of (29.4) the **output** of the system. Let us assume the forcing function $f(t) = mF \sin (\omega t + \beta)$ where F is a constant. Then (29.4) becomes

(29.41) $$\frac{d^2 y}{dt^2} + 2r \frac{dy}{dt} + \omega_0^2 y = F \sin (\omega t + \beta).$$

The different possible complementary functions y_c obtained by setting the left side of (29.41) equal to zero and solving it will be the same as those given in the three cases of Lesson 29A. The trial solution y_p for all such solutions y_c is

(29.42) $$y_p = A \sin (\omega t + \beta) + B \cos (\omega t + \beta).$$

Following the method outlined in Lesson 21A, we find that

(29.43) $$A = \frac{F(\omega_0^2 - \omega^2)}{[(\omega_0^2 - \omega^2)^2 + (2r\omega)^2]},$$

$$B = \frac{-F(2r\omega)}{[(\omega_0^2 - \omega^2)^2 + (2r\omega)^2]}.$$

Let (see Fig. 29.44),

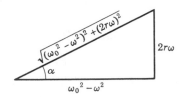

Figure 29.44

(29.45) $$\cos \alpha = \frac{\omega_0^2 - \omega^2}{\sqrt{(\omega_0^2 - \omega^2)^2 + (2r\omega)^2}},$$

$$\sin \alpha = \frac{2r\omega}{\sqrt{(\omega_0^2 - \omega^2)^2 + (2r\omega)^2}}.$$

Substituting these values in (29.43) and the resulting expressions for A and B in (29.42), we obtain

(29.46) $$y_p = \frac{F}{\sqrt{(\omega_0^2 - \omega^2)^2 + (2r\omega)^2}}$$

$$\times [\cos \alpha \sin (\omega t + \beta) - \sin \alpha \cos (\omega t + \beta)].$$

Hence the general solution of (29.41) is

$$(29.47) \quad y = y_c + \frac{F}{\sqrt{(\omega_0{}^2 - \omega^2)^2 + (2r\omega)^2}} \sin (\omega t + \beta - \alpha),$$

where y_c is any one of the functions given in Lesson 29A. As we saw there, the motion due to the y_c part of the solution (29.47), in all cases, whether oscillatory or nonoscillatory, dies out with time. For this reason this part of the motion has been called appropriately the **transient motion.** The equation of motion (29.47) is thus a complicated one only for the time in which the transient motion is effective. Thereafter the motion will be due entirely to the y_p part of the solution as given by the second term on the right of (29.47). This part of the motion has therefore been appropriately named the **steady state motion.**

Comment 29.48. In many physical problems, the transient motion is the least important part of the motion. However, there are cases where it is of major importance.

By (29.41) and (29.47), we see that the steady state motion has the same frequency as the forcing function $f(t)$, namely ω rad/sec, but is out of phase with it and that the *amplitude* of the steady state motion is

$$(29.5) \qquad A = \frac{F}{\sqrt{(\omega_0{}^2 - \omega^2)^2 + (2r\omega)^2}} .$$

If $\omega = \omega_0$ [the condition for (undamped) resonance], the amplitude reduces to the interesting form

$$(29.51) \qquad A = \frac{F}{2r\omega_0} .$$

If $\omega \neq \omega_0$, then by differentiating (29.5) with respect to ω and setting the resulting expression for $dA/d\omega$ equal to zero, we obtain

$$(29.52) \qquad 2(\omega_0{}^2 - \omega^2)(-2\omega) + 8r^2\omega = 0,$$

from which we find

$$(29.53) \quad \omega^2 = \omega_0{}^2 - 2r^2, \qquad \omega = \sqrt{\omega_0{}^2 - 2r^2}, \quad \omega_0{}^2 > 2r^2.$$

Hence if a resisting force is present, and if ω, the frequency of the forcing function, is not equal to ω_0, the natural (undamped) frequency of a system, then, for fixed F, the amplitude A of the steady state motion will be a maximum if ω has the value given in (29.53). A forcing function $f(t)$, having this frequency ω, is then said to be in **resonance** with the system. Substituting this value of ω in (29.5), we find that the maximum amplitude is

$$(29.531) \qquad A_{\max} = \frac{F}{2r\sqrt{\omega_0{}^2 - r^2}} .$$

Assume now that $2r$, the coefficient of resistance of a system per unit mass, is **small**. Hence we commit a small error if we omit the r^2 term in (29.531). We thus obtain

$$(29.532) \qquad A_{\max} \approx \frac{F}{2r\omega_0},$$

the same amplitude obtained in (29.51) when $\omega = \omega_0$. Further, we showed in Lesson 29A, Case 3, that the natural (damped) frequency of a system is $\sqrt{\omega_0^2 - r^2}$, which, for small r, is close to the resonant frequency $\sqrt{\omega_0^2 - 2r^2}$, i.e., it is close to the frequency which will produce the maximum amplitude.

We infer from all the above remarks that if a resisting force is present and ω, the *frequency of the forcing function* $f(t)$, *equals* ω_0, *the natural (undamped) frequency of a system, or is close to* $\sqrt{\omega_0^2 - r^2}$, *the natural (damped) frequency of the system, then the amplitude of the system is inversely proportional to the damping or resisting factor* $2r$. Hence if $2r$ is small, A will be large, and tremendous vibrations may be produced. That is why soldiers crossing a bridge may be ordered to break step (although the chances are that this precaution is unnecessary), for it is feared that if the frequency which they create with their footbeat is the same as the natural (undamped) frequency of the bridge, or near its damped frequency, and if in addition the internal resistance of the bridge is small, the vibrations may become so large as to cause a breakage. The walls of Jericho, so some assert, came tumbling down because the sound the trumpeteers made with their trumpets caused a wave motion whose frequency equaled the natural (undamped) frequency of the walls. Students at Cornell University used to find it amusing either to create a wave motion in the old suspension bridge over the gorge or to get it to swing violently from side to side. They would march across it in a straight line with a rhythmic beat or walk with a sailor's gait, first emphasizing one side, then the other. To timid souls, however, it was never very amusing—terrifying would be a more descriptive word. On such occasions, it was impossible to walk across the bridge with an even step or in a straight line, depending on whether the bridge was waving or swinging.

We cite two more examples of this phenomenon and ones which you can easily experience or may have already experienced.

1. A swing, with a child seated on it, when displaced from its equilibrium position, will move back and forth across the equilibrium position with a natural (damped) frequency. If you now apply a force to the swing with a frequency close to this natural (damped) frequency, then for a fixed F and small r, the maximum amplitude will equal, approximately, $F/2r\omega_0$. Hence, if r is small, the amplitude of swing will be large. If you want a still larger amplitude, you must increase F.

2. When you jump off a diving board, the end of the board will vibrate about its equilibrium position with a natural (damped) frequency. If instead of jumping off, you now jump up and down above the end of the board with a frequency near this natural (damped) frequency, you will be able to make the magnitude of the oscillation large. If r is small, the maximum amplitude, for a given F, will equal, approximately, $F/2r\omega_0$.

The ratio

$$(29.54) \qquad M = \frac{\text{Amplitude of } y_p}{F/\omega_0{}^2},$$

where F and $\omega_0{}^2$ are given in (29.41), is called the **magnification ratio of the system** or the **amplification ratio of the system.** By (29.54) and (29.47), this magnification ratio is

$$(29.55) \qquad M = \frac{\omega_0{}^2}{\sqrt{(\omega_0{}^2 - \omega^2)^2 + (2r\omega)^2}}$$

$$= \frac{1}{\sqrt{\left[1 - \left(\dfrac{\omega}{\omega_0}\right)^2\right]^2 + 4\left(\dfrac{r}{\omega_0}\right)^2\left(\dfrac{\omega}{\omega_0}\right)^2}}$$

Since ω_0 is fixed, the amplification ratio of a system depends on the frequency ω of the forcing function $f(t)$ and the coefficient of resistance per unit mass $2r$. In practical applications where ω is also fixed, the resistance $2r$ is made large if one wishes the magnifying response to be small as, for example, in vibrations of machinery and in shock absorbers; the resistance $2r$ is made small, if one wishes the response to be large, as, for example, in a radio receiver.

If in (29.55), we let

$$(29.56) \qquad \mu = \frac{\omega}{\omega_0} \quad \text{and} \quad \nu = \frac{r}{\omega_0},$$

the equation becomes

$$(29.561) \qquad M = \frac{1}{\sqrt{(1 - \mu^2)^2 + 4\nu^2\mu^2}}.$$

The quantity μ, by (29.56), is thus the ratio of the impressed or input frequency ω to the natural (undamped) frequency ω_0. The quantity ν may be looked at as measuring the amount of damping present for a fixed ω_0. For each fixed value of ν, M is a function of μ. Hence it is possible to draw a graph of the magnification M for each such fixed value of ν. For example, if ν is $\frac{1}{2}$, then, by (29.561),

$$(29.562) \qquad M(\mu) = \frac{1}{\sqrt{(1 - \mu^2)^2 + \mu^2}}.$$

If $\nu = 0$, which implies by (29.56) that $r = 0$, then, by (29.561),

$$(29.563) \qquad M(\mu) = \frac{1}{1 - \mu^2}.$$

By (29.563), we see that as $\mu \to 1$, which implies by (29.56) that $\omega \to \omega_0$, $M \to \infty$.

Example 29.564. A forcing function $f(t) = \frac{5}{2}\cos 2t$ is applied to the motion given in Example 29.19. Find the steady state motion and the amplification ratio of the system. Is resonance possible?

Solution. With $f(t) = \frac{5}{2}\cos 2t$, the differential equation of motion (c) in Example 29.19 becomes

(a) $$\frac{1}{16}\frac{d^2y}{dt^2} + \frac{1}{2}\frac{dy}{dt} + \frac{3}{4}y = \frac{5}{2}\cos 2t,$$

(b) $$\frac{d^2y}{dt^2} + 8\frac{dy}{dt} + 12y = 40\cos 2t.$$

A particular solution of (b) is

(c) $$y_p = 2\sin 2t + \cos 2t,$$

which is the steady state motion. By Comment 28.32, the amplitude of the motion defined by (c) is $\sqrt{2^2 + 1^2} = \sqrt{5}$. Comparing (b) with (29.41), we see that $\omega^2 = 12$, $2r = 8$, $F = 40$. Therefore by (29.54), the magnification ratio of the system is

(d) $$M = \frac{\sqrt{5}}{40/12} = \frac{3\sqrt{5}}{10}.$$

And since $\omega_0{}^2 = 12 < 2r^2 = 32$, resonance is not possible. See (29.53).

Comment 29.6. For easy reference, we have listed in the table on page 365 the different differential equations discussed thus far in this chapter, and the pertinent information related to each.

EXERCISE 29B

1. Verify the values of A and B as given in (29.43).
2. Verify the solution (29.46).
3. Verify (29.53).
4. Verify the accuracy of the solution (c) of Example 29.564.
5. For what value of ω will the magnification ratio as given in (29.55) be a maximum? Find this maximum value.
6. A particle moves according to the law

$$\frac{d^2y}{dt^2} + 2\frac{dy}{dt} + 9y = 5\sin 2t.$$

Differential Equations Discussed in Lessons 28 and 29

Equation	Solution	Name of Motion	Amplitude	Type of Motion	Frequency
$\dfrac{d^2y}{dt^2} + \omega_0^2 y = 0$	$y_c = c_1 \cos \omega_0 t + c_2 \sin \omega_0 t$ $y_c = c \sin(\omega_0 t + \delta)$ $y_c = c \cos(\omega_0 t + \delta)$	Simple harmonic or free undamped motion.	$\sqrt{c_1^2 + c_2^2}$ c c	Perpetual motion. Stable.	ω_0 rad or $\dfrac{\omega_0}{2\pi}$ cycles per unit time.
$\dfrac{d^2y}{dt^2} + \omega_0^2 y$ $= F \sin(\omega t + \beta)$	$y = y_c$ as given above $+$ $\dfrac{F}{\omega_0^2 - \omega^2} \sin(\omega t + \beta)$, $\omega_0 \neq \omega$	Forced undamped motion.	$c + \dfrac{F}{\omega_0^2 - \omega^2}$	Oscillatory. Stable.	
	$y = y_c - \dfrac{F}{2\omega_0} t \cos(\omega t + \beta)$, $\omega_0 = \omega$		$c - \dfrac{F}{2\omega_0} t$	(Resonance) Unstable.	
$\dfrac{d^2y}{dt^2} + 2r \dfrac{dy}{dt}$ $+ \omega_0^2 = 0$	$y_c = c_1 e^{(-r+\sqrt{r^2-\omega_0^2})\,t}$ $+ c_2 e^{(-r-\sqrt{r^2-\omega_0^2})\,t}$, $r > \omega_0$.	Damped harmonic motion or free damped motion.	None	Non-oscillatory. Stable.	
	$y_c = c_1 e^{-rt} + c_2 t e^{-rt}$, $r = \omega_0$				
	$y_c = c e^{-rt} \sin(\sqrt{\omega_0^2 - r^2}\,t + \delta)$, $r < \omega_0$		$c e^{-rt}$	Oscillatory. Stable.	$\sqrt{\omega_0^2 - r^2}$ rad or $\dfrac{\sqrt{\omega_0^2 - r^2}}{2\pi}$ cycles per unit of time.
$\dfrac{d^2y}{dt^2} + 2r \dfrac{dy}{dt} + \omega_0^2 y$ $= F \sin(\omega t + \beta)$	$y = y_c + \dfrac{F \sin(\omega t + \beta - \alpha)}{\sqrt{(\omega_0^2 - \omega^2)^2 + (2r\omega)^2}}$	Forced damped motion.	$c + \dfrac{F}{\sqrt{(\omega_0^2 - \omega^2)^2 + (2r\omega)^2}}$	Oscillatory. Stable.	Of steady state motion, ω rad or $\omega/2\pi$ cycles per unit time.

(a) Find the steady state motion; also the amplitude, period, and frequency of the steady state motion.

(b) What is the magnification ratio of the system?

(c) What frequency of the forcing function will produce resonance?

7. A particle moves according to the law

$$m\frac{d^2y}{dt^2} + 2mr\frac{dy}{dt} + m\omega_0^2 y = f(t).$$

(a) Find the equation of motion if $f(t) = mF\cos\omega t$. Assume $r^2 < \omega_0^2$.

(b) What is the transient motion; the steady state motion?

(c) What is the amplification ratio of the system?

8. A particle moves in accordance with the law

$$\frac{d^2y}{dt^2} + 4\frac{dy}{dt} + 16y = f(t).$$

(a) What frequency of the function $f(t)$ will make the period of the steady state motion $\pi/3$?

(b) What frequency of the function $f(t)$ will produce resonance?

9. A particle moves according to the law

$$\frac{d^2y}{dt^2} + 5\frac{dy}{dt} + 6y = e^{-t}\sin 2t.$$

(a) Solve for y as a function of t.

(b) What is the input; the output?

(c) Describe the motion.

10. In Exercise 29A, 8, we asked you to show that the motion of a particle whose differential equation is $\frac{d^2y}{dt^2} + 2a\frac{dy}{dt} + by = 0$, is stable only if $a > 0$, $b > 0$. Since the addition to the equation of a function $f(t)$ does not affect the complementary function y_c, it follows that $a > 0$, $b > 0$ is also a necessary condition for the stability of the motion of a particle whose differential equation is

$$\frac{d^2y}{dt^2} + 2a\frac{dy}{dt} + by = f(t).$$

Prove that it is not a sufficient condition by solving the equation

$$\frac{d^2y}{dt^2} + 5\frac{dy}{dt} + 6y = 12e^t,$$

and then showing that the solution $y(t) \to \infty$ as $t \to \infty$.

When the damping or resisting factor of a system is not negligible and a forcing function $f(t)$ is attached to it, the differential equation (28.63) for the hélical spring must be modified to read [See also (29.37).]

(29.7) $$m\frac{d^2y}{dt^2} + r\frac{dy}{dt} + ky = f(t),$$

where r is the coefficient of resistance of the system. Use (29.7) to solve the next two problems.

11. A 16-lb weight stretches a spring 1 ft. The spring is immersed in a medium whose coefficient of resistance is 4. After the spring is brought to rest, a forcing function $10 \sin 2t$ is applied to the system.

   (a) Find the equation of motion.
   (b) What is the transient motion; the steady state motion?
   (c) Find the amplitude, period, and frequency of the steady state motion.
   (d) What is the magnification ratio of the system?

12. A 16-lb weight stretches a spring 6 in. Its coefficient of resistance is 2. The 16-lb weight is removed, replaced by a 64-lb weight and brought to rest. At $t = 0$, a forcing function $8 \cos 4t$ is applied to the system. Find the steady state motion and the amplification ratio of the system.

13. In (29.4), let

$$f(t) = m(A_1 \sin \omega_1 t + A_2 \sin \omega_2 t + \cdots + A_n \sin \omega_n t),$$

so that n different oscillations are impressed on the system.

   (a) Find the steady state motion. *Hint.* Use the superposition principle, see Comment 24.25; also Exercise 19,6.
   (b) What is the magnification ratio due to the input $mA_1 \sin \omega_1 t$, to $mA_2 \sin \omega_2 t$, $\cdots$, to $mA_n \sin \omega_n t$? A glance at the denominator of each magnification ratio term will show that those terms with frequencies close to ω_0 will be magnified to a much larger extent than those with frequencies farther away. A system of this kind thus acts as a filter. It responds to those vibrations with frequencies near ω_0 and ignores those vibrations with frequencies not near ω_0.

14. In Exercise 28D, 14, we introduced the discontinuous unit impulse function

$$f(t) = \begin{cases} \dfrac{1}{b}, & 0 \leq t \leq b, \\ 0, & t > b. \end{cases}$$

Solve the equation

$$\frac{d^2y}{dt^2} + 2\frac{dy}{dt} + 2y = f(t),$$

for y as a function of t, where $f(t)$ is the above function and initial conditions are $t = 0$, $y = 0$, $y' = 0$. *Hint.* First solve with $f(t) = 1/b$. Find $y(b)$ and $y'(b)$. Then solve the equation with $f(t) = 0$ and initial conditions $t = b$, $y = y(b)$, $dy/dt = y'(b)$.

15. Solve problem 14 if

$$f(t) = \begin{cases} e^{-t}, & 0 \leq t \leq 1. \\ 0, & t > 1. \end{cases}$$

Hint. See suggestions given in 14.

ANSWERS 29B

5. $\omega = \sqrt{\omega_0^2 - 2r^2}$, the same value of ω that makes the amplitude a maximum, see (29.53); $M(\omega) = 1/2r\sqrt{\omega_0^2 - r^2}$.

6. (a) $y_p = \dfrac{5}{\sqrt{41}}\sin(2t - \alpha)$, where $\alpha = \text{Arc tan }\tfrac{4}{5}$,

   $5\sqrt{41}$, π, 2 rad/sec.   (b) $1\sqrt{41}$.   (c) $\sqrt{7}$.

7. (a) $y = ce^{-rt}\cos(\sqrt{\omega_0^2 - r^2}\,t + \delta) + \dfrac{F\cos(\omega t - \alpha)}{\sqrt{(\omega_0^2 - \omega^2)^2 + (2r\omega)^2}}$, with

   $r^2 < \omega_0^2$ and α given by (29.45). (b) First term on right of (a); second term on right of (a). (c) Same as (29.55).

8. (a) $\omega = 6$.   (b) $\omega = \sqrt{8}$.

9. (a) $y = c_1 e^{-2t} + c_2 e^{-3t} - \dfrac{e^{-t}}{20}(\sin 2t + 3\cos 2t)$.

   (b) $e^{-t}\sin 2t$; solution $y(t)$ as given in (a).
   (c) Each term in $y(t)$ approaches zero as $t \to \infty$. The complementary function is not oscillatory; the particular solution, however, is damped oscillatory since $y \to 0$ as $t \to \infty$.

10. $y = c_1 e^{-2t} + c_2 e^{-3t} + e^t \to \infty$ as $t \to \infty$.

11. (a) $y = \dfrac{e^{-4t}}{26}(\sin 4t + 8\cos 4t) + \tfrac{1}{13}(7\sin 2t - 4\cos 2t)$.

   (b) First term in (a); second term in (a).   (c) $\sqrt{65}/13$, π sec, $1/\pi$ cps.
   (d) $8\sqrt{65}/65$.

12. $y_p = \sin 4t$; 4.

13. (a) $y_p = \dfrac{A_1\sin(\omega_1 t - \alpha_1)}{\sqrt{(\omega_0^2 - \omega_1^2)^2 + (2r\omega_1)^2}} + \cdots + \dfrac{A_n\sin(\omega_n t - \alpha_n)}{\sqrt{(\omega_0^2 - \omega_n^2)^2 + (2r\omega_n)^2}}$,

   where α_i, $i = 1, \cdots, n$, is defined as in (29.45).

   (b) $\dfrac{\omega_0^2}{\sqrt{(\omega_0^2 - \omega_1^2)^2 + (2r\omega_1)^2}}, \cdots, \dfrac{\omega_0^2}{\sqrt{(\omega_0^2 - \omega_n^2)^2 + (2r\omega_n)^2}}$.

14.
$$y = \begin{cases} \dfrac{1}{2b}[1 - e^{-t}(\sin t + \cos t)], & 0 \le t \le b. \\[2mm] \dfrac{e^{-t}}{2b}[\{e^b(\sin b + \cos b) - 1\}\sin t \\[1mm] \quad + \{e^b(\cos b - \sin b) - 1\}\cos t], & t > b. \end{cases}$$

15.
$$y = \begin{cases} e^{-t}(1 - \cos t), & 0 \le t \le 1. \\[1mm] e^{-t}[\sin 1 \sin t + (\cos 1 - 1)\cos t] \\[1mm] \quad = e^{-t}[\cos(t - 1) - \cos t], & t > 1. \end{cases}$$

LESSON 30. Electric Circuits. Analog Computation.

By Newton's laws of motion we were able to set up a relationship among active forces in a mechanical system. Analogous laws, known as Kirchhoff's (1824–1887) laws, make it likewise possible for us to set up a relationship among those forces which supply and use energy in an electrical system. In Lesson 30A below, we state one of these laws and apply it to a simple electric circuit.

LESSON 30A. Simple Electric Circuit. In the simple electric circuit which we have diagrammed in Fig. 30.1, the source of energy in the circuit is marked E. It may be a cell, battery, or generator. It supplies the energy in the form of an electrical flow of charged particles. The velocity of the particles is called a **current.** However, the energy source will produce this flow only when the key at A is moved to B. The circuit is then said to be **closed.** The **electromotive force** of the battery or other source of energy, usually written as **emf,** is defined as numerically equal to the energy supplied by the battery or source when one unit charge is carried around the complete circuit. For example, if three units of energy are supplied by a source when one unit charge is carried around the complete circuit, then its emf is three units. There

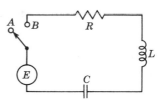

Figure 30.1

are, for the electrical system, as in the mechanical one, different systems of units in use. In the one we shall adopt, the unit of emf is called a **volt.** The other three elements in the circuit labeled R, L, and C are users of energy. In nontechnical terms, this means that a certain amount of energy is needed to move the electrical flow of charged particles across these barriers. We express the energy each uses by giving the **voltage drop** across it.*

From the physicist, we learn that:

(30.11)   the voltage drop across a **resistor** (R in figure) $= Ri$,

the voltage drop across an **inductor** (L in figure) $= L\dfrac{di}{dt}$,

the voltage drop across a **capacitor** (C in figure) $= \dfrac{1}{C}\,q$,

---

*The voltage drop across each element is easily measured by means of an instrument called a voltmeter. All one need do is to connect one wire of the voltmeter to one side of the element, another wire to the other side, and then read how far a pointer moves.

provided:

the **resistance** R of the resistor is measured in ohms,

the **coefficient of inductance** L of the inductor is measured in henrys,

the **capacitance** C of the capacitor is measured in farads,

the **charge** q in the circuit is measured in coulombs,

the **current** i in the circuit, which is defined to be the rate of change of the charge q, or the velocity of q, i.e.,

$$(30.12) \qquad i = \frac{dq}{dt},$$

is measured in amperes.

The resistor, as the name implies, resists the flow of the charged particles, and thus energy is needed to move the particles across it. The inductor's job is to keep the rate of flow of the charged particles as near constant as possible. It thus opposes an increase or a decrease in the current. The capacitor stores charged particles and thus interrupts the electrical flow. When the accumulated charges become too numerous for its capacity, the charged particles leap across the gap (that is when the spark occurs) and the particles then continue their course in the circuit.

Kirchhoff's second law states that the sum of the voltage drops in a closed circuit is equal to the electromotive force of the source of energy $E(t)$. Hence, by (30.11),

$$(30.13) \qquad Ri + L\frac{di}{dt} + \frac{1}{C}q = E(t).$$

By (30.12), we can write (30.13) as

$$(30.14) \qquad L\frac{d^2q}{dt^2} + R\frac{dq}{dt} + \frac{1}{C}q = E(t),$$

which is the differential equation of motion of the charge q in the circuit as a function of the time t.

To find the current i in the circuit as a function of the time t, we can either solve (30.14) for q and take its derivative, or we can differentiate (30.13) to obtain, with the help of (30.12), the differential equation

$$(30.15) \qquad L\frac{d^2i}{dt^2} + R\frac{di}{dt} + \frac{1}{C}i = \frac{d}{dt}E(t),$$

and then solve (30.15) for i.

Assume

$$(30.16) \quad E(t) = F\sin(\omega t + \beta); \text{ therefore } \frac{d}{dt}E(t) = F\omega\cos(\omega t + \beta).$$

Then (30.15) becomes

$$(30.17) \qquad L\frac{d^2i}{dt^2} + R\frac{di}{dt} + \frac{1}{C}i = F\omega \cos(\omega t + \beta),$$

$$\frac{d^2i}{dt^2} + \frac{R}{L}\frac{di}{dt} + \frac{1}{CL}i = \frac{F\omega}{L}\cos(\omega t + \beta).$$

Its solution, by any method you wish to use, is, assuming the roots of the characteristic equation are imaginary,

$$(30.18) \quad i = Ae^{-(R/2L)t}\sin\left(\frac{\sqrt{4CL - R^2C^2}}{2CL}t + \delta\right)$$

$$\overbrace{}^{i_c}$$

$$+ F\omega C\left[\frac{R\omega C\sin(\omega t + \beta) + (1 - CL\omega^2)\cos(\omega t + \beta)}{(R\omega C)^2 + (1 - CL\omega^2)^2}\right].$$

$$\underbrace{}_{i_p}$$

Let (Fig. 30.211)

$$(30.19) \qquad \sin\alpha = \frac{1 - CL\omega^2}{\sqrt{(R\omega C)^2 + (1 - CL\omega^2)^2}}.$$

$$\cos\alpha = \frac{R\omega C}{\sqrt{(R\omega C)^2 + (1 - CL\omega^2)^2}}.$$

Then the i_p part of (30.18) can be written as

$$(30.2) \qquad i_p = \frac{F\omega C}{\sqrt{(R\omega C)^2 + (1 - CL\omega^2)^2}}$$

$$\times\; [\sin(\omega t + \beta)\cos\alpha + \cos(\omega t + \beta)\sin\alpha]$$

$$= \frac{F\omega C}{\sqrt{(R\omega C)^2 + (1 - CL\omega^2)^2}}\;[\sin(\omega t + \beta + \alpha)].$$

Hence the solution (30.18) becomes

$$(30.21) \quad i = Ae^{-(R/2L)t}\sin\left(\frac{\sqrt{4CL - R^2C^2}}{2CL}t + \delta\right)$$

$$\overbrace{}^{i_c}$$

$$+ \frac{F\omega C}{\sqrt{(R\omega C)^2 + (1 - CL\omega^2)^2}}\sin(\omega t + \beta + \alpha)$$

$$\underbrace{}_{i_p}.$$

The current in the circuit, therefore consists of two parts, a damped harmonic motion due to the i_c part of the solution and a simple harmonic motion due to the i_p part. As in the mechanical case, the presence of the *damping factor* $e^{-(R/2L)t}$ causes the current due to the i_c part of the solution to die out in time. [If we had assumed real roots of the characteristic equation of (30.17), instead of imaginary ones, the current due to the i_c part of the solution would still die out in time. See the solutions for each of the different cases in the corresponding mechanical case, Lesson 29A.] The i_c part of the solution is therefore called appropriately the **transient current.** The current equation (30.21) will thus be a complicated one only for the time in which the transient current is effective. Thereafter the current will be determined entirely by the i_p part of the solution. The i_p current has therefore been named, also appropriately, the **steady state** current.

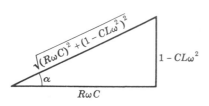

Figure 30.211

Comment 30.212. As in the mechanical case, the function $E(t)$ of (30.14) or $\dfrac{d}{dt} E(t)$ of (30.15) is called the **input** of the system; the solution of each equation the **output** of the respective system.

By (30.21) and (30.16), we see that the steady state current has the same frequency as that of the energy source $E(t)$, namely ω rad/sec, but is out of phase with it.

The *amplitude* of the steady state current is, by (30.21),

$$(30.22) \quad A = \frac{F\omega C}{\sqrt{(R\omega C)^2 + (1 - CL\omega^2)^2}} = \frac{F}{\sqrt{\left(R^2 + \left(\dfrac{1}{\omega C} - L\omega\right)^2\right)}}.$$

The denominator of the last expression in (30.22) is called the **impedance Z** of the circuit. When its value is a minimum, the amplitude A is a maximum. To find the value of ω that will make Z a minimum, for fixed R, C, and L, we differentiate the impedance equation

$$(30.23) \qquad Z = \sqrt{R^2 + \left(\frac{1}{\omega C} - L\omega\right)^2}$$

with respect to ω and set $dZ/d\omega$ equal to zero. The result is [square Z in (30.23) and then differentiate with respect to ω]

$$(30.24) \qquad 0 = 2\left(\frac{1}{\omega C} - L\omega\right)\left(-\frac{1}{C\omega^2} - L\right),$$

from which we obtain

$$(30.25) \qquad \omega^2 = \frac{1}{CL}, \quad \omega = \sqrt{1/CL}.$$

For this value of ω, the impedance Z of the current will be a minimum, the amplitude A will be a maximum, and as in the mechanical system, we say the electromotive force is in **resonance** with the circuit.

Substituting in the first equation of (30.22), this resonant value of $\omega^2 = 1/CL$ as given in (30.25), we obtain for the maximum value of the amplitude,

$$(30.26) \qquad A = \frac{F}{R}.$$

From (30.26) we observe that when resonance occurs, the maximum value of the amplitude A is inversely proportional to the resistance R. Hence when R is small, the maximum value of A is large, and when R is large, the maximum amplitude is small. The condition of resonance therefore is always dangerous unless the resistance R is sufficiently large to prevent a breakdown of the circuit. And if $R = 0$, a breakdown is bound to occur.

If we fix F, R, L, and ω, then by (30.22), the amplitude A of the steady state current is a function of the capacitance C. If $C = 0$, $A = 0$, and if C is adjusted so that $\sqrt{1/CL}$ is equal to the frequency ω of $E(t)$, then A will be largest. Hence by adjusting C, we can make the amplitude of the steady state current small or large. In a public address system when we want a large amplification ratio [this means, by (29.54), that we want the amplitude A of y_p to be relatively large], or in a home radio set when we want a lower amplification ratio, we adjust the capacitance C accordingly by turning a dial.

Example 30.27. A capacitor whose capacitance is 2/1010 farad, an inductor whose coefficient of inductance is $\frac{1}{20}$ henry, and a resistor whose resistance is 1 ohm are connected in series. If at $t = 0$, $i = 0$ and the charge on the capacitor is 1 coulomb, find the charge and the current in the circuit due to the discharge of the capacitor when $t = 0.01$ second.

Solution. Here $E(t) = 0$, $C = \frac{2}{1010}$, $L = \frac{1}{20}$, and $R = 1$. Hence (30.14) becomes

$$(a) \qquad \frac{1}{20}\frac{d^2q}{dt^2} + 1\frac{dq}{dt} + 505q = 0, \qquad \frac{d^2q}{dt^2} + 20\frac{dq}{dt} + 10{,}100q = 0.$$

Its solution is

$$(b) \qquad q = e^{-10t}(c_1 \sin 100t + c_2 \cos 100t).$$

Therefore

(c) $i = \dfrac{dq}{dt} = -10e^{-10t}(c_1 \sin 100t + c_2 \cos 100t)$

$\qquad\qquad + e^{-10t}(100c_1 \cos 100t - 100c_2 \sin 100t).$

The initial conditions are $t = 0$, $q = 1$, $i = 0$. Substituting these values in (b) and (c), we obtain

(d) $\qquad\qquad\qquad 1 = c_2,$

$\qquad\qquad 0 = -10c_2 + 100c_1, \quad c_1 = 0.1.$

In (b) and (c) replace c_1 and c_2 by these values. Then when $t = 0.01$ there results

(e) $q(0.01) = e^{-0.1}(0.1 \sin 1 + \cos 1) = 0.57$ coulomb,

$\quad i(0.01) = -10e^{-0.1}(0.1 \sin 1 + \cos 1) + e^{-0.1}(10 \cos 1 - 100 \sin 1)$

$\qquad\qquad\; = -76.9$ amperes.

The negative current indicates that the condenser is discharging, i.e., the charged particles are moving in a direction opposite to the one in which they moved when the capacitor was being charged.

Example 30.3. To the circuit of the previous problem is added a source of energy whose electromotive force $E = 50 \sin 120t$. Change the capacitance of the capacitor to 2×10^{-3} farad. At $t = 0$ seconds, the switch is closed. If at that instant there is no charge on the capacitor and no current in the circuit, find:

1. The equation of motion of the steady state current after the switch is closed.
2. The amplitude of the steady state current.
3. The frequency of the steady state current.
4. The value of the capacitance which will make the amplitude of the steady state current a maximum.

Solution. Here $\dfrac{d}{dt} E(t) = \dfrac{d}{dt}(50 \sin 120t) = 6000 \cos 120t$. Using the figures for L and R as given in Example 30.27 and of C as given above, the differential equation of motion (30.15) becomes

(a) $\qquad\qquad \dfrac{1}{20}\dfrac{d^2i}{dt^2} + \dfrac{di}{dt} + \dfrac{10^3}{2}i = 6000 \cos 120t,$

$\qquad\qquad \dfrac{d^2i}{dt^2} + 20\dfrac{di}{dt} + 10^4 i = 120{,}000 \cos 120t.$

Its steady state solution is

(b) $\qquad\qquad i_p = 11.5 \sin 120t - 21 \cos 120t.$

By Comment 28.32, the amplitude of the steady state current is

(c) $A = \sqrt{11.5^2 + 21^2} = 23.9.$

The frequency ω of the steady state current is 120 rad/sec, the same as the frequency of the source of energy $E(t)$.

By (30.25), the amplitude of the steady state current will be a maximum if C has a value such that $\sqrt{1/CL} = \omega$, i.e., when

(d) $C - \frac{1}{L\omega^2} = \frac{20}{120^2} = \frac{1}{720}$ farad.

LESSON 30B. Analog Computation. We recopy below the differential equation of motion (28.63) of a mechanical system with the coefficient of resistance and forcing function terms added, and the differential equations (30.14) and (30.15) of an electrical system.

(30.4) $m \frac{d^2y}{dt^2} + r \frac{dy}{dt} + ky = F \sin \omega t.$

(30.41) $L \frac{d^2q}{dt^2} + R \frac{dq}{dt} + \frac{1}{C} q = E(t).$

(30.42) $L \frac{d^2i}{dt^2} + R \frac{di}{dt} + \frac{1}{C} i = \frac{d}{dt} [E(t)].$

When placed underneath each other in this manner, the similarity in form of the two systems is striking. It should be evident to you that if in an electric circuit, Fig. 30.43(a), we insert a resistor $R = r$, an inductor $L = m$, a capacitor $C = 1/k$, and a source of energy $E = F \sin \omega t$ (or

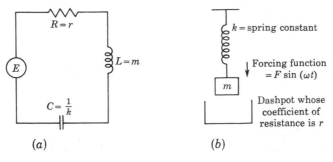

Figure 30.43

$-F\omega \cos \omega t$), the solution q of (30.41) [or i of (30.42)] will be the same as the solution y of (30.4). By solving the electrical system, it is then possible to determine the motion of a corresponding mechanical system, such as the one pictured in Fig. 30.43(b). Since it is usually less expensive and easier to set up a simple electric circuit than it is to construct a mechanical system, the importance of this fortunate coincidence should be evident to you. This method, which is now well developed, of computing the motion of a mechanical system from a simple electric circuit is known as **analog computation.**

However, because of the current accessibility to high-speed digital computers, the most accurate and least expensive method at present of computing the motion of a mechanical system is to use such a computer.

EXERCISE 30

1. Verify the accuracy of the solution of (30.17) as given in (30.18).

2. Verify the accuracy of the solution (b) of Example 30.27.

3. Verify the accuracy of the solution (b) of Example 30.3.

In the problems below, it is assumed, when not explicitly stated, that the coefficient of inductance L of the inductor is measured in henrys, the resistance R of the resistor is measured in ohms, the capacitance C of the capacitor is measured in farads, the charge q is in coulombs, the current i is in amperes and the emf of the source of energy is in volts.

4. If the emf i.e., if the source of energy, is missing from the circuit, then the differential equations (30.14) and (30.15) become respectively

$$(30.5) \qquad L\frac{d^2q}{dt^2} + R\frac{dq}{dt} + \frac{1}{C}q = 0,$$

$$(30.51) \qquad L\frac{d^2i}{dt^2} + R\frac{di}{dt} + \frac{1}{C}i = 0.$$

(a) What is the natural (undamped) frequency of vibrations of current and charge? *Hint.* Set $R = 0$.

(b) For what values or R will the charge and current subside to zero without oscillating; for what values of R will they oscillate before subsiding to zero?

(c) Find the general solutions for q and i as functions of time if $R^2 = 4L/C$. To what mechanical case is this situation comparable?

(d) Find q and i as functions of time if at $t = 0$, $q = q_0$ and $i = 0$. Assume $R^2 < 4L/C$.

5. For a certain LRC electric circuit, $L = \frac{1}{2}$, $C = \frac{1}{800}$.

(a) For what values of R will the current subside to zero without oscillating after the emf is removed from the circuit; for what value of R will it subside to zero with oscillations?

(b) What is the natural (undamped) frequency of the system?

6. A capacitor whose capacitance is 10^{-5} farad, an inductor whose coefficient of inductance is 10 henrys, and a resistor whose resistance is 3 ohms are connected in series. At $t = 0$, $i = 0$ and the charge on the capacitor is 0.5 coulomb. Find the charge and current in the circuit as functions of time due to the discharge of the capacitor.

7. If a resistance is missing from the circuit, then by (30.14)

(30.52)
$$L \frac{d^2 q}{dt^2} + \frac{q}{C} = E(t).$$

Equation (30.52) is the differential equation of the **harmonic oscillator** for the electric current and corresponds to the forced undamped motion of the mechanical system, see Lesson 28D.

(a) Solve for q and i as functions of time if $E(t) = 0$ and $t = 0$, $q = q_0$, $i = 0$. What is the natural (undamped) frequency of the system?
(b) Solve for q and i as functions of time if $E(t) = $ a constant emf E and $t = 0$, $q = 0$, $i = 0$.
(c) Solve for q and i as functions of time if $E(t) = E \sin \omega t$ and $t = 0$, $q = 0$, $i = 0$ (two cases). What value of ω will produce (undamped) resonance?

8. (a) Find q and i as functions of time if in (30.52) $C = 10^{-4}$, $L = 1$, $E(t) = 100$, and at $t = 0$, $q = 0$, $i = 0$.
(b) What is the natural (undamped) frequency of the system?
(c) What is the value of the current when $t = 0.02$ sec?

9. (a) Find q and i as functions of time if in (30.52) $C = 10^{-4}$, $L - 1$, $E(t) = 100 \sin 50t$ and at $t = 0$, $q = 0$, $i = 0$.
(b) What is the value of the current when $t = 0.02$ sec?
(c) What is the maximum value of the current?
(d) What is the natural (undamped) frequency of the system?

10. If the capacitance is missing from the circuit, then by (30.13),

(30.53)
$$L \frac{di}{dt} + Ri = E(t).$$

(a) Find i as a function of t if $E(t)$ is a constant emf E and at $t = 0$, $i = 0$. What is the transient current, the steady state current?
(b) Find i as a function of t if $E(t) = E \sin \omega t$ and at $t = 0$, $i = 0$. What is the transient current, the steady state current?

11. Find i as a function of t if in (30.53) $R = 20$, $L = 0.1$, and

(a) $E(t) = 10$,   (b) $E(t) = 100 \sin 50t$.

12. An inductor of L henries, a resistor of R ohms, and a capacitor of C farads are connected in series to a battery whose emf is E volts.

(a) Find q and i as functions of time. Assume $R^2 < 4L/C$.
(b) What is the frequency of the transient charge and current?
(c) Is there a steady state charge, a steady state current?

Hint. By (30.14), the differential equation is $L \frac{d^2 q}{dt^2} + R \frac{dq}{dt} + \frac{1}{C} q = E$.

13. (a) Find q and i as functions of time if in (30.14) and (30.15), $L = 1$, $R = 5$, $C = 10^{-4}$, $E(t) = 50$, and at $t = 0$, when the switch is closed, $q = 0$, $i = 0$.
(b) What is the frequency of the transient charge and current?
(c) What is the steady state charge?

14. (a) Find the steady state current if, in (30.15), $L = \frac{1}{20}$, $R = 5$, $C = 4 \times 10^{-4}$, $dE/dt = 200 \cos 100t$, and if at $t = 0$, when the switch is closed, $q = 0$, $i = 0$.

(b) What is the amplitude and frequency of the steady state current?

(c) For what value of the capacitance will the amplitude be a maximum?

(d) What should the frequency of the input $E(t)$ be in order that it be in resonance with the system?

(e) What is the maximum value of the amplitude for this resonant frequency?

(f) What is the impedance of the system?

15. Find the steady state charge and the steady state current if, in (30.14) and (30.15), $L = \frac{1}{20}$, $R = 20$, $C = 10^{-4}$, $E = 100 \cos 200t$. In regard to the steady state current, answer all questions (b) to (f) of 14.

16. In (30.15), let

$$E(t) = E_1 \sin \omega_1 t + E_2 \sin \omega_2 t + \cdots + E_n \sin \omega_n t,$$

so that n different frequencies are impressed on an electric system.

(a) Show that the steady state current is

(30.54) $\quad i_s = \dfrac{E_1 \omega_1 C}{\sqrt{(R\omega_1 C)^2 + (1 - CL\omega_1{}^2)^2}} \sin (\omega_1 t + \alpha_1) + \cdots$

$\quad + \dfrac{E_n \omega_n C}{\sqrt{(R\omega_n C)^2 + (1 - CL\omega_n{}^2)^2}} \sin (\omega_n t + \alpha_n),$

where α_i, $i = 1, \cdots, n$, is defined as in (30.19). *Hint.* Use the superposition principle, see Comment 24.25; also Exercise 19,6.

(b) Show that the amplitude of the steady state current due to the input $E_k \sin \omega_k t$ is

$$A_k = \dfrac{E_k}{\sqrt{R^2 + \left(\dfrac{1}{\omega_k C} - L\omega_k\right)^2}}.$$

We proved in the text that A_k will be largest when $\sqrt{1/CL}$ is equal to the frequency ω_k. Hence by adjusting C until $\sqrt{1/CL} = \omega_k$, we can make the amplitude of the response or output due to the input $E_k \sin \omega_k t$ larger than the amplitudes due to the other inputs. The electrical system will thus act as a filter, responding to those inputs whose frequencies are near $\sqrt{1/CL}$ and ignoring those inputs whose frequencies are farther away. If the inputs, for example, are coming from different radio stations which are broadcasting at different frequencies, you tune your radio to one of them by turning a dial and adjusting the capacitance until the amplitude of the output is greatest for that station's input. The amplitude A_k also has E_k in the numerator. Hence for good reception from station k, you would want its E_k to be larger, i.e., more powerful, than the E of other stations and the frequencies of the other stations to be not too close to ω_k. Compare this problem with Exercise 29B, 13.

ANSWERS 30

4. (a) $\sqrt{1/CL}$ rad/sec. (b) No oscillations if $R^2 \geqq 4L/C$, oscillations if $R^2 < 4L/C$. (c) $y = e^{-Rt/2L}(C_1 + C_2 t)$, critically damped case. NOTE. Here $y = q$ or i.

(d) $q = 2q_0 \sqrt{\dfrac{L}{4L - CR^2}}\, e^{-Rt/2L} \sin\left(\sqrt{\dfrac{1}{CL} - \dfrac{R^2}{4L^2}}\, t + \delta\right),$

$\delta = \text{Arc tan} \sqrt{\dfrac{4L}{CR^2} - 1}; \; i = dq/dt.$

5. (a) $R \geqq 40$, no oscillations; $R < 40$ oscillations.   (b) 40 rad/sec.

6. $q = \frac{1}{2}e^{-3t/20} \sin(100t + \delta)$ approximately, where $\delta = \text{Arc tan}\,(2000/3)$ approximately; $i = -\frac{3}{40}e^{-3t/20} \sin(100t + \delta) + 50e^{-3t/20} \cos(100t + \delta)$.

7. (a) $q = q_0 \cos \sqrt{1/CL}\, t$; $i = -\dfrac{q_0}{\sqrt{CL}} \sin \sqrt{1/CL}\, t$, $\sqrt{1/CL}$ rad/sec.

(b) $q = CE(1 - \cos \sqrt{1/CL}\, t)$, $i = dq/dt$.

(c) $q = \dfrac{CE}{1 - CL\omega^2}\,(\sin \omega t - \omega\sqrt{CL} \sin \sqrt{1/CL}\, t)$, $\omega \neq 1/\sqrt{CL}$;

$i = dq/dt$,

$q = \dfrac{EC}{2} \sin \dfrac{1}{\sqrt{CL}}\, t - \dfrac{E}{2} \sqrt{C/L}\, t \cos \dfrac{1}{\sqrt{CL}}\, t$, $\omega = 1/\sqrt{CL}$;

$i = dq/dt$; $\omega = 1/\sqrt{CL}$ rad/sec.

8. (a) $q = \frac{1}{100}(1 - \cos 100t)$, $i = \sin 100t$.   (b) 100 rad/sec.
(c) 0.909 amp.

9. (a) $q = \frac{1}{75}(\sin 50t - \frac{1}{2} \sin 100t)$; $i = \frac{2}{3}(\cos 50t - \cos 100t)$.
(b) $i(0.02) = 0.638$ amp.   (c) max $|i| = \frac{4}{3}$ amp.   (d) 100 rad/sec.

10. (a) $i = \dfrac{E}{R}(1 - e^{-Rt/L})$; $i_t = -\dfrac{E}{R} e^{-Rt/L}$; $i_s = \dfrac{E}{R}.$

(b) $i = \dfrac{E}{R^2 + L^2\omega^2}\,(R \sin \omega t - L\omega \cos \omega t + L\omega e^{-Rt/L})$,

$i_t = \dfrac{EL\omega}{R^2 + L^2\omega^2}\, e^{-Rt/L},$

$i_s = \dfrac{E}{R^2 + L^2\omega^2}\,(R \sin \omega t - L\omega \cos \omega t).$

11. (a) $i = \frac{1}{2}(1 - e^{-200t})$.   (b) $i = \frac{20}{17}(4 \sin 50t - \cos 50t + e^{-200t})$.

12. (a) $q = q_c + EC$, where q_c is the same as i_c in (30.18),

$i = \dfrac{dq}{dt}.$

(b) $\sqrt{4CL - R^2C^2}/4\pi CL$ cps.
(c) $q_s = EC$. There is no steady state current. In taking the derivative of q, the constant EC vanishes.

13. (a) $q = -10^{-3}e^{-2.5t}(0.125 \sin 99.97t + 5 \cos 99.97t) + 0.005$,

$i = dq/dt = 0.500e^{-2.5t} \sin 99.97t$.

(b) $99.97/2\pi = 15.9$ cps.
(c) 0.005. For a short time the charge on the capacitor will oscillate about this figure and approach this figure as $t \to \infty$.

14. (a) $i_s = \frac{2}{85}(\sin 100t + 4 \cos 100t)$.   (b) $\frac{2}{85}\sqrt{17}$, $100/2\pi$ cps.
(c) $C = 2 \times 10^{-3}$ farad.   (d) $100\sqrt{5}$ rad/sec.   (e) 2/5 amp.
(f) $5\sqrt{17}$ ohms.

15. (a) $q_s = 5 \times 10^{-3}(\sin 200t + 2 \cos 200t)$, $i_s = \cos 200t - 2 \sin 200t$.
(b) $\sqrt{5}$, $200/2\pi$ cps.   (c) 5×10^{-4}.   (d) $200\sqrt{5}$.   (e) 5 amp.
(f) $20\sqrt{5}$.

LESSON 30M.

MISCELLANEOUS TYPES OF PROBLEMS LEADING TO LINEAR EQUATIONS OF THE SECOND ORDER

A. Problems Involving a Centrifugal Force. When a body is whirled in a circle at the end of string, a force, directed toward the center of the circle, must be exerted to prevent the body from flying off; the faster the rotation, the more powerful the force. Since this force is directed toward the center of the path, it has been called the **centripetal force** or **central force**. And since the body remains in its path, there must be an outward force in the opposite direction equal to the central force. This force is called the **centrifugal force**. It has been proved that the centripetal force required to hold a mass m in a circular path of radius r, moving with a linear velocity v is

$$(30.6) \qquad \text{C.F.} = \frac{mv^2}{r}.$$

Hence this formula must also give the centrifugal force of the mass m. The linear velocity v of the particle is $v = r\, d\theta/dt$, where θ is the central angle measured in radians through which the particle is rotated. Substituting this value of v in (30.6), we obtain

$$(30.61) \qquad \text{C.F.} = \frac{m}{r}\left(r\frac{d\theta}{dt}\right)^2 = mr\omega^2,$$

where ω is the angular velocity of the particle.

With the help of (30.61), solve the following problems.

1. A smooth straight tube rotates in a vertical plane about its mid-point with constant angular velocity ω. A particle of mass m inside the tube is free to slide without friction.

   (a) Find the differential equation of motion of the particle. *Hint.* There are two forces acting on the particle at time t, see Fig. 30.62.
   (b) Solve the equation with $t = 0$, $r = r_0$, $dr/dt = v_0$.
   (c) From the introductory remarks, it is clear that if the particle is too far from 0 or if dr/dt is too great, the particle will fly off from an end of the tube; for certain values of r and dr/dt, it will not. Find values of the initial conditions r_0 and $v_0 = dr/dt$ so that the particle will execute simple harmonic motion. Write the resulting equation of motion for these values. Can you identify it? Draw the figure.

2. Solve problem 1, if the tube rotates in a horizontal plane about a vertical axis. Assume at $t = 0$, $r = 0$ and $dr/dt = v_0$.

3. Solve problem 1, if at $t = 0$, $r = 0$, $dr/dt = 0$.

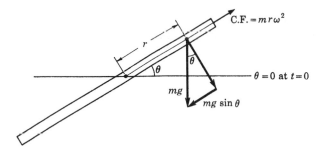

Figure 30.62

B. Rolling Bodies. Newton's first law of motion states that a body at rest or moving with uniform velocity will remain in the respective state of rest or motion unless a force acts on it. We say the body has **inertia,** i.e., it resists having its status changed. Similarly, a body at rest or rotating about an axis with a constant angular velocity will remain in the respective state of rest or rotation, unless a **torque** or a **moment of force** acts on it; for definition of torque, see (30.63) below. In this case we say the body has **rotational inertia,** also called **moment of inertia.** By definition, the moment of inertia I of a particle is

$$(30.621) \qquad\qquad I = mx^2,$$

where m is the mass of the particle located x units from the axis of rotation. In the calculus, you were taught how to calculate the moment of inertia of different bodies. For example, for a solid cylinder of radius r and mass m rotating about an axis coinciding with the axis of the cylinder, $I = mr^2/2$. It is as if the entire mass of the cylinder were concentrated at a distance $r^2/2$ units from the axis.

We define **the torque** or **moment of force** L as follows, see Fig. 30.631.

(30.63) $L = x$ times the component of the force F acting at right angles to the line joining the axis of rotation and the point P where F is being applied; x is the distance between the axis and P.

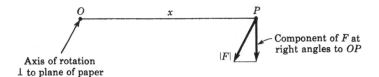

Figure 30.631

Finally, it has been proved that corresponding to the law $F = $ mass $\times$ acceleration governing the linear motion of a body, the law governing the rotational motion of a body is given by

$$(30.64) \qquad L = I\alpha \equiv I\frac{d^2\theta}{dt^2} \equiv I\frac{d\omega}{dt},$$

where α is the angular acceleration of the body, ω is its angular velocity, and θ is the central angle through which the body has rotated from $\theta = 0$.

With the help of equations (30.621) to (30.64), solve the following problems.

4. A cord is wound a few turns around a solid cylindrical spool of mass m and radius r. One end of the cord is attached to the ceiling. See Fig. 30.65. At $t = 0$, the spool, which is being held against the ceiling with axis horizontal, is released.

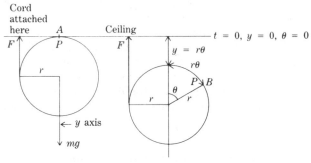

Figure 30.65

(a) If gravity is the only acting force, find the differential equation of motion of the spool. *Hint.* By Newton's law, mass $\times$ acceleration of body must equal the net force acting on the body. These forces are F and mg as shown in Fig. 30.65. By (30.63), (30.64) and the fact that $I = mr^2/2$ for a solid cylinder rotating about its axis, we have

$$Fr = I\alpha = \frac{mr^2}{2}\frac{d^2\theta}{dt^2}.$$

When the cylinder has rolled through a central angle θ so that the point of the spool initially at A is now at B, the distance y from the ceiling is $r\theta$. Therefore,

$$y = r\theta, \quad \frac{d^2y}{dt^2} = r\frac{d^2\theta}{dt^2}, \quad \frac{d^2\theta}{dt^2} = \frac{1}{r}\frac{d^2y}{dt^2}.$$

Hence $F = (m/2)(d^2y/dt^2)$.

(b) Solve the differential equation for y as a function of t. Remember at $t = 0$, $y = 0$, $dy/dt = 0$.

5. Answer questions (a) and (b) of problem 4 if there is a resisting force due to friction and air of $(m/80)(dy/dt)$. (c) What is the limiting velocity?

6. At $t = 0$, a solid cylinder of radius r and mass m is placed at the top of an incline and released, Fig. 30.66. Assume it rolls without slipping and that a frictional force F acts to oppose the motion.

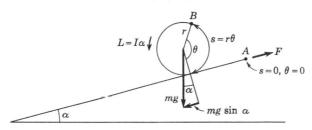

Figure 30.66

(a) Find the differential equation of motion. *Hint.* The only difference between this problem and 4 is that mg is replaced by $mg \sin \alpha$.

(b) Solve the differential equation. Remember at $t = 0$, $0 = 0$, $ds/dt = 0$.

C. Twisting Bodies. When a spring is stretched, a force results, proportional to the amount of stretch, that tries to restore the spring to its original natural length. Similarly, when a hanging wire is twisted by rotating a bob about it as an axis, where the bob is rigidly attached to it at one end, a torque or moment of force results that tries to restore the wire to its original position. This torque L is, in many cases, proportional to the angle θ through which the bob is turned. By (30.64), therefore,

$$(30.67) \qquad\qquad I \frac{d^2\theta}{dt^2} = -k\theta,$$

where k is called the **torsional stiffness constant.** The negative sign is necessary, because when θ is turning clockwise, the torque acts counterclockwise; hence torque and θ have opposite signs.

7. (a) Solve (30.67) for θ as a function of time if the torque is equal in magnitude to the angle θ, i.e., $k = 1$.

(b) If the bob returns to its equilibrium position at the end of each $\frac{1}{8}$ second, find the moment of inertia of the bob with respect to the wire as an axis. Assume the mass of the wire is negligible.

D. Bending of Beams. We consider a beam with the following properties.

1. It is relatively long in comparison with its width and thickness.
2. Every cross section is uniform.
3. The center of gravity of each cross section lies on a straight line, called the **axis of the beam.** It is the line joining $(0,0)$ to $(L,0)$ in Fig. 30.7.

If a beam merely rests on supports at its two ends, it is called a **simple beam** and it is said to be **simply supported** at the ends. If a beam is supported only at one end, as for example, when it is embedded in masonry at one end and hangs freely at the other end, it is called a **cantilever beam.**

In Fig. 30.7, we have drawn a simple beam of length L ft with rectangular cross sections whose centers of gravity lie in the geometrical center of

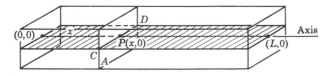

Figure 30.7

the rectangle. (However, beams may have other shaped cross sections, as long as all cross sections are uniform and the center of gravity of each lies on a straight line.) We may look on such a beam as composed of fibers parallel to the axis of the beam, each of whose length is L ft. When a load is distributed along a simple beam, a sag develops so that the fibers on one side of the beam are compressed and fibers on the other side are

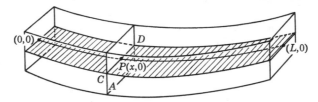

Figure 30.71

stretched, Fig. 30.71. It follows, therefore, that somewhere between the two sides, a **neutral surface** exists that is neither stretched nor compressed (shaded area in Fig. 30.71), i.e., it retains its original length L. The intersection of this neutral surface with a vertical plane through the axis of the beam is called the **elastic curve** of the beam. It is the curve joining $(0,0)$ to $(L,0)$ in Fig. 30.71.

It has been proved in mechanics that

$$(30.72) \qquad M(x) = \frac{EI}{R},$$

where:

$M(x)$ is the **bending moment** at any cross section A, x units from one end of the beam. The bending moment at A is defined as the algebraic sum of all the moments of force

acting on only one side of A about an axis through the
center of the cross section A, marked CD in the figure.
(For definition of a moment of force, see (30.63) above.)
I is the moment of inertia of the cross section A about its
center axis CD (for definition of moment of inertia, see
(30.621) above).
R is the radius of curvature of the elastic curve of the beam.
E is a proportionality constant, called **Young's modulus** or
modulus of elasticity. It is dependent only on the material
of which the beam is made.

The radius of curvature is given by the formula

$$R = [1 + (y')^2]^{3/2}/y''.$$

Its substitution in (30.72) gives

(30.73) $\qquad M(x) = EIy''[1 + (y')^2]^{-3/2}$
$\qquad\qquad\qquad = EIy''[1 - \tfrac{3}{2}(y')^2 + \tfrac{15}{8}(y')^4 - \cdots].$

Since the bending is usually slight, y' is very small. Hence it is not unreasonable to assume that we commit a small error if in (30.73) we neglect
$(y')^2$ and higher powers of y'. Equation (30.73) thus simplifies to

(30.74) $\qquad\qquad\qquad M(x) = EIy'',$

which is the differential equation of the elastic curve of the beam.

We shall arbitrarily assume that an *upward force gives a positive moment*
and that a *downward force gives a negative moment.*

With the help of (30.74), solve the following problems.

8. A horizontal beam of length $2L$ ft is simply supported at its ends. The
weight of the beam is evenly distributed and equals w lb/ft.

(a) Find the equation of the elastic curve. *Hint.* See Fig. 30.75. The total
weight of the beam is $2Lw$ lb. Therefore the upward force at each end

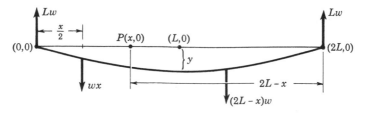

Figure 30.75

is Lw lb. Since the beam is uniform, we can consider the weight of the
beam from $(0,0)$ to $P(x,0)$ as concentrated at its mid-point $(x/2, 0)$.
Hence the downward force at this mid-point is wx lb. The bending

moment $M(x)$ at P is therefore—remember the bending moment is the algebraic sum of all moments of force acting on *one* side of the cross section A whose axis goes through P—

$$M(x) = (Lw)x - (wx)\left(\frac{x}{2}\right) = Lwx - \frac{wx^2}{2}.$$

Substitute this value of $M(x)$ in (30.74) and solve. The initial conditions are $x = 0$, $y = 0$; $x = L$, $y' = 0$. (Note there is also a third initial condition $x = 2L$, $y = 0$. However, the three are not mutually independent. Use of any two of the three will result in a solution which satisfies the third condition. Verify this statement.)

(b) What is the maximum sag? *Hint.* The maximum sag occurs when $x = L$.

(c) Show that the same bending moment at P results, if the forces to the right of P were used. *Hint.* The bending moment at P due to the forces on the right is

$$M(x) = Lw(2L - x) - [(2L - x)w]\left(\frac{2L - x}{2}\right).$$

Simplify the right side.

9. A horizontal beam of length $2L$ ft is simply supported at its ends and carries a weight W lb at its center. If the weight of the beam is negligible compared to W, find the equation of the elastic curve and the sag at the center. See Fig. 30.76. Two cases must be considered.

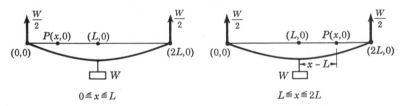

Figure 30.76

Case 1. If P is to the left of the mid-point, the only active force to the left of P is $W/2$. Hence the bending moment at P is

(a) $M(x) = \dfrac{W}{2} x$ which can be written as $\frac{1}{2}WL - \frac{1}{2}W(L - x)$, $0 \leqq x < L$.

Case 2. If P is to the right of the mid-point, then the active forces to the left of P are $W/2$ upward and W downward. Hence the bending moment at P is

(b) $M(x) = \dfrac{W}{2} x - W(x - L)$ which can be written as $\frac{1}{2}WL + \frac{1}{2}W(L - x)$,
$$L < x \leqq 2L.$$

Cases 1 and 2 can therefore be treated as one if we write

(c) $$M(x) = \tfrac{1}{2}WL \mp \tfrac{1}{2}W(L - x),$$

where it is understood that the minus sign is to be used when $0 \leqq x < L$ and the plus sign when $L < x \leqq 2L$. When $x = L$, (a), (b), and (c) are the same. Hence the solution obtained by using (c) is also valid when $x = L$. Initial conditions are $x = 0$, $y = 0$; $x = 2L$, $y = 0$. (A third initial condition $x = L$, $y' = 0$ is not independent of the other two. Verify that it satisfies the derivative of the solution.)

10. Solve problem 9, if the weight of the beam is not negligible and is w lb/ft. *Hint.* Follow all the instructions given in 8 and 9. As in 9 there will be two cases, one when P is to left of center, the other when P is to right of center. The bending moment at P due to the forces to the left of P are

$$M(x) = \left(wL + \frac{W}{2}\right)x - \frac{wx^2}{2}$$

$$= wLx - \tfrac{1}{2}wx^2 - \tfrac{1}{2}W(L - x) + \tfrac{1}{2}WL, \quad 0 \leqq x < L,$$

if P is to left of the center, and

$$M(x) = \left(wL + \frac{W}{2}\right)x - \frac{wx^2}{2} - W(x - L)$$

$$= wLx - \tfrac{1}{2}wx^2 + \tfrac{1}{2}W(L - x) + \tfrac{1}{2}WL, \quad L < x \leqq 2L,$$

if P is to right of the center.

When $x = L$, both bending moments are the same. Thus both cases can be combined if you take

$$M(x) = wLx - \tfrac{1}{2}wx^2 \mp \tfrac{1}{2}W(L - x) + \tfrac{1}{2}WL,$$

where it is understood that the minus sign is to be used when $0 \leqq x < L$ and the plus sign when $L < x \leqq 2L$.

11. A horizontal beam of length 30 ft is simply supported at its ends and carries a weight of 360 lb at its center. If the weight of the beam is negligible, find the equation of the elastic curve for each half beam. What is its sag? Solve independently. Check your results with solutions given in problem 9.

12. A simply supported horizontal beam of length $2L$ carries a weight W lb attached to it at a distance $2L/3$ from one end. Assume the weight of the beam is negligible. Find the equation of the elastic curve. *Hint.* The end of the beam closer to the weight now supports $2W/3$ lb; the other end supports only $W/3$ lb. Two cases will be needed as in 9 and 10, one if P is to the left of W, the other if P is to the right of W. The bending moment at P using forces to the left of P are

$$M(x) = \frac{2W}{3}\,x, \quad 0 \leqq x < \frac{2L}{3},$$

if P is to the left of W, and

$$M(x) = \frac{2W}{3}\,x - W\left(x - \frac{2L}{3}\right), \quad \frac{2L}{3} < x \leqq 2L,$$

if P is to the right of W.

Initial conditions are $x = 0$, $y = 0$; $x = 2L$, $y = 0$. Note also, since the elastic curve is continuous at $x = 2L/3$ and has a tangent there, that when $x = 2L/3$, the value of y and the value of the derivative y' for each of the two curves must be the same. These conditions are known respectively as the **condition of continuity of the curve** and the **condition of continuity**

of the slope. You will need to use these facts in order to evaluate some of the constants of integration.

13. A horizontal beam of length $2L$ ft and of uniform weight w lb/ft is embedded in concrete at both ends. Find the equation of the elastic curve and the maximum sag. Take the origin at one end of the beam. Here in addition to the usual moments found in problem 8, there is an additional moment of force at each end acting to keep the beam horizontal, i.e., the masonry at each end prevents the beam in its immediate neighborhood from sagging. Call this unknown moment of force M. The initial conditions are $x = 0$, $y = 0$; $x = 0$, $y' = 0$; $x = L$, $y' = 0$; $x = 2L$, $y = 0$; $x = 2L$, $y' = 0$. There are five sets of initial conditions. Use of three, say the first three, will enable you to evaluate M and the constants of integration. Verify that the resulting equation satisfies the other two initial conditions.

14. Solve problem 13, if the beam also supports a weight W at its center. *Hint.* Here, in addition to the usual moments found in problem 10, there is a moment M at each end. As in problem 10, there are two cases to be considered, one when P is to the left of center, the other when P is to the right of center. It will be easier to treat each case separately instead of combining them as we did in 9 and 10. Initial conditions are $x = 0$, $y = 0$; $x = 0$, $y' = 0$; $x = L$, $y' = 0$; $x = 2L$, $y = 0$; $x = 2L$, $y' = 0$. Note that the condition $x = L$, $y' = 0$, applies to each case, since the curve is continuous at $x = L$. There is one more condition than you need. However, it is not independent of the others. Verify that the one you omit satisfies the solution.

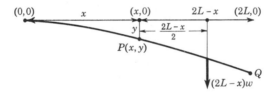

Figure 30.77

15. A cantilever beam of length $2L$ and of uniform weight w lb/ft is embedded in concrete at one end. Find the equation of its elastic curve and the maximum deflection. See Fig. 30.77. In this case, it will be easier to consider moments of force to the right of P. Since the beam is uniform, there is a downward force at the center of PQ. And since this is the only acting force to the right of P, the bending moment at P is

$$M(x) \;=\; -w(2L - x)\left(\frac{2L - x}{2}\right) \;=\; -\frac{w}{2}\,(2L - x)^2.$$

Substitute this value in (30.74). Initial conditions are $x = 0$, $y = 0$; $x = 0$, $y' = 0$.

16. A cantilever beam of length $2L$ and of negligible weight supports a load of W lb at its center.

(a) Find the equation of its elastic curve, the deflection at its center, and its maximum deflection. *Hint.* See Fig. 30.78. There are two cases to be considered. Take moments to right of P. Initial conditions are $x = 0$, $y = 0$; $x = 0$, $y' = 0$. When P is to the right of center, there are no forces to the right of P. Hence the bending moment $M(x) = 0$ at P. You will also need to use the fact that when $x = L$, the solution y and the slope y' for the case $0 \leqq x < L$ must agree respectively with the solution y and the slope y' for the case $L < x \leqq 2L$.

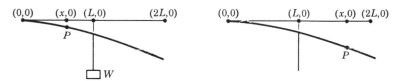

Figure 30.78

(b) Find the maximum deflection and the equation of the elastic curve if the weight W were placed at the end of the beam. *Hint.* Here there is only one case to consider, namely, P to the left of W. The only force to the right of P contributing to the bending moment $M(x)$ at P is the weight W.

17. Solve problem 16(a) if the weight of the beam is not negligible and is w lb/ft. *Hint.* There will be two cases as in 16. The bending moment at P will be the sum of the bending moments given in 15 and 16. And remember when $x = L$, the solution y and the slope y' must agree for the two cases.

18. A horizontal beam of length $2L$ is embedded in concrete at one end and is simply supported at the other end with both ends at the same level. A weight W is suspended at its mid-point and the beam itself weighs w lb/ft. Find the equation of the elastic curve. Take the origin at the embedded end. *Hint.* We need two cases, Case 1 when P is to the left of the mid-point; Case 2 when P is to the right of the mid-point. See Fig. 30.79. Take moments to the right of P. Call F the unknown upward force at $(2L,0)$. Initial conditions are $x = 0$, $y = 0$; $x = 0$, $y' = 0$; $x = 2L$, $y = 0$. And remember, when $x = L$, the solution y and the slope y' must agree for both cases.

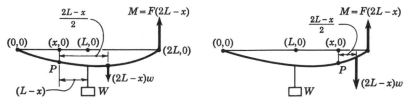

Figure 30.79

19. A spring board, fixed at one end only, may be considered as a cantilever beam. It is desired that its maximum deflection be 1 ft when a 240-lb man steps on the end. If the board is 20 ft long and weighs 5 lb/ft, find the value of the constant EI. *Hint.* The formula for maximum deflection is the sum of the maximum deflections given in 15 and 16(b). And remember in these formulas L is one-half the length of the board.

ANSWERS 30M.

1. (a) $m \dfrac{d^2 r}{dt^2} = mr\omega^2 - mg \sin \omega t$.

(b) $r = r_0 \dfrac{e^{\omega t} + e^{-\omega t}}{2} + \left(\dfrac{2\omega v_0 - g}{2\omega^2} \right) \dfrac{e^{\omega t} - e^{-\omega t}}{2} + \dfrac{g}{2\omega^2} \sin \omega t$

$\qquad = r_0 \cosh \omega t + \dfrac{2\omega v_0 - g}{2\omega^2} \sinh \omega t + \dfrac{g}{2\omega^2} \sin \omega t$.

(c) $r_0 = 0$, $v_0 = \dfrac{g}{2\omega}$. Resulting equation is

$\qquad r = \dfrac{g}{2\omega^2} \sin \omega t = \dfrac{g}{2\omega^2} \sin \theta$.

In polar coordinates it is the equation of a circle with center at $(g/4\omega^2, \pi/2)$ and radius $g/4\omega^2$.

2. $r = \dfrac{v_0}{2\omega} (e^{\omega t} - e^{-\omega t})$.

3. $r = \dfrac{g}{4\omega^2} (2 \sin \omega t - e^{\omega t} + e^{-\omega t})$.

4. (a) $m \dfrac{d^2 y}{dt^2} = mg - F$, $\dfrac{d^2 y}{dt^2} = \tfrac{2}{3}g$. (b) $y = gt^2/3$.

5. (a) $\dfrac{3}{2} \dfrac{d^2 y}{dt^2} + \dfrac{1}{80} \dfrac{dy}{dt} = g$.

(b) $y = 9600g(e^{-t/120} - 1) + 80gt$, $v = -80ge^{-t/120} + 80g$. (c) $80g$.

6. (a) $m \dfrac{d^2 s}{dt^2} = mg \sin \alpha - F$, $F = \dfrac{m}{2} \dfrac{d^2 s}{dt^2}$, $\dfrac{3}{2} \dfrac{d^2 s}{dt^2} = g \sin \alpha$.

(b) $s = (g \sin \alpha)t^2/3$.

7. (a) $\theta = c_1 \cos (\sqrt{1/I}\, t + \delta)$. (b) $I = 1/(64\pi^2)$.

8. (a) $EIy = \dfrac{wx}{24} (4Lx^2 - x^3 - 8L^3)$, $0 \leqq x \leqq 2L$.

(b) sag $= 5wL^4/24EI$ ft.

9. $EIy = \dfrac{W}{12} [3Lx^2 \mp (L - x)^3 - 6L^2 x + L^3]$, $0 \leqq x \leqq 2L$;

$EIy = \dfrac{Wx}{12} (x^2 - 3L^2)$, $0 \leqq x \leqq L$;

$EIy = \dfrac{Wx}{12} (x^2 - 3L^2) + \dfrac{W}{6} (L - x)^3$, $L \leqq x \leqq 2L$;

sag $= WL^3/6EI$ ft.

10. $EIy = $ sum of the results obtained in 8 and 9; sag $= $ sum of the results obtained in 8 and 9.

11. $EIy = 30x(x^2 - 675)$, $0 \leqq x \leqq 15$,

$\qquad = 30x(x^2 - 675) + 60(15 - x)^3$, $15 \leqq x \leqq 30$;

sag $= 202{,}500/EI$ ft.

12. $EIy = \dfrac{Wx}{81}(9x^2 - 20L^2)$, $0 \leqq x \leqq 2L/3$;

$EIy = \dfrac{Wx}{81}(9x^2 - 20L^2) - \dfrac{W}{6}\left(x - \dfrac{2L}{3}\right)^3$, $\dfrac{2L}{3} \leqq x \leqq 2L$.

13. $EIy'' = M + Lwx - \dfrac{wx^2}{2}$, $M = -L^2w/3$;

$EIy = \dfrac{wx^2}{24}(4Lx - x^2 - 4L^2) = -\dfrac{w}{24}x^2(2L - x)^2$;

sag $= wL^4/24EI$ ft.

14. Moment $M = -\tfrac{1}{3}wL^2 - \dfrac{WL}{4}$;

$EIy'' = \left(wL + \dfrac{W}{2}\right)x - \dfrac{wx^2}{2} + M$, $0 \leqq x \leqq L$;

$EIy = \dfrac{w}{24}(4Lx^3 - x^4 - 4L^2x^2) + \dfrac{W}{24}(2x^3 - 3Lx^2)$, $0 \leqq x \leqq L$;

$EIy'' = \left(wL + \dfrac{W}{2}\right)x - \dfrac{wx^2}{2} - W(x - L) + M$, $L \leqq x \leqq 2L$,

$Ely = \dfrac{w}{24}(4Lx^3 - x^4 - 4L^2x^2) + \dfrac{W}{24}[2x^3 - 4(x - L)^3 - 3Lx^2]$,
$$L \leqq x \leqq 2L,$$

sag $= (wL^4 + WL^3)/24EI$.

15. $EIy = \dfrac{w}{24}[16L^4 - 32L^3x - (2L - x)^4]$

$= \dfrac{w}{24}(8Lx^3 - 24L^2x^2 - x^4)$.

Maximum deflection $= 2wL^4/EI$.

16. (a) $EIy = \dfrac{W}{6}[L^3 - 3L^2x - (L - x)^3]$

$= \dfrac{W}{6}(x^3 - 3Lx^2)$, $0 \leqq x \leqq L$;

$EIy = \dfrac{W}{6}(L^3 - 3L^2x)$, $L \leqq x \leqq 2L$.

Deflection at mid-point $WL^3/3\,EI$; maximum deflection $5WL^3/6EI$.

(b) $EIy = \dfrac{W}{6}(x^3 - 6Lx^2)$, maximum deflection $8WL^3/3EI$.

17. $EIy = \dfrac{w}{24}(8Lx^3 - 24L^2x^2 - x^4) + \dfrac{W}{6}(x^3 - 3Lx^2)$, $0 \leqq x \leqq L$;

$EIy = \dfrac{w}{24}(8Lx^3 - 24L^2x^2 - x^4) + \dfrac{W}{6}(L^3 - 3L^2x)$, $L \leqq x \leqq 2L$.

Deflection at midpoint: $(17wL^4 + 8WL^3)/24EI$; deflection at end point: $(12wL^4 + 5WL^3)/6EI$.

18. $F = \frac{3}{4}wL + \frac{5W}{16}$.

$$EIy'' = -W(L - x) - \frac{w}{2}(2L - x)^2 + F(2L - x);$$

$$EIy = \frac{w}{48}(10Lx^3 - 12L^2x^2 - 2x^4) + \frac{W}{96}(11x^3 - 18Lx^2),$$

$$0 \leqq x \leqq L;$$

$$EIy'' = -\frac{w}{2}(2L - x)^2 + F(2L - x),$$

$$EIy = \frac{w}{48}(10Lx^3 - 12L^2x^2 - 2x^4)$$

$$+ \frac{W}{96}(16L^3 - 48L^2x + 30Lx^2 - 5x^3), \quad L \leqq x \leqq 2L.$$

19. 740,000.

Chapter 7

Systems of Differential Equations.
Linearization of First Order Systems

LESSON 31. Solution of a System of Differential Equations.

LESSON 31A. Meaning of a Solution of a System of Differential Equations. In algebra it is frequently necessary to solve a system of simultaneous equations of the type

(a) $2x + 3y = 5$, (b) $x^2 - 3y = 2$, (c) $x + 3y - 2z = 15$,
$\quad x - y = 15$; $\quad\quad x + y^2 = 5$; $\quad\quad x - y + z = 7$,
$\quad\quad\quad\quad\quad\quad\quad\quad\quad\quad\quad\quad\quad\quad\quad 3x + 2y - z = 12.$

Similarly, it is frequently necessary to solve a system of differential equations of the type

(d)
$$x \frac{dx}{dt} + y^2 \frac{dy}{dt} + 2x - 3y = 3t,$$
$$\frac{d^2x}{dt^2} + 2y \frac{dy}{dt} + 3 \frac{dy}{dt} - 3x = e^t,$$

where x and y are dependent variables and t is an independent variable. In Lessons 33 and 34, we discuss and solve numerous physical problems which give rise to such systems.

A solution of an algebraic system of two equations is a pair of values of x and y such that this pair satisfies both equations. For example, $x = 10$, $y = -5$ is a solution of (a), since this pair of values satisfies both equations. Analogously we say that the pair of functions $x(t)$, $y(t)$, each defined on a common interval I, is a solution of the system (d), if this pair satisfies both equations identically on I, i.e., if in *each* equation of (d) an identity results when x is replaced by $x(t)$, y by $y(t)$, and their respective derivatives by $x'(t)$, $y'(t)$, etc.

The extension of the meaning of a solution of a system of three or more equations should be apparent.

LESSON 31B.   Definition and Solution of a System of First Order Equations.

Definition 31.1.   The pair of equations

(31.11) $$\frac{dx}{dt} = f_1(x,y,t), \qquad \frac{dy}{dt} = f_2(x,y,t),$$

where f_1 and f_2 are functions of x, y, t, defined on a common set S, is called a **system of two first order equations**. A **solution** of (31.11) will then be a *pair* of functions $x(t)$, $y(t)$, each defined on a common interval I contained in S, satisfying both equations of (31.11) identically.

A generalization of this type of system is given in the following definition.

Definition 31.12.   The system of n equations

(31.13) $$\frac{dy_1}{dt} = f_1(y_1, y_2, \cdots, y_n, t),$$

$$\frac{dy_2}{dt} = f_2(y_1, y_2, \cdots, y_n, t),$$

$$\cdots \cdots \cdots \cdots \cdots \cdots$$

$$\frac{dy_n}{dt} = f_n(y_1, y_2, \cdots, y_n, t),$$

where $f_1, \cdots, f_n$ are each functions of y_1, y_2, $\cdots$, y_n, t, defined on a common set S, is called a **system of n first order equations**.

Definition 31.14.   A **solution** of the system (31.13) is a set of functions $y_1(t)$, $y_2(t)$, $\cdots$, $y_n(t)$, each defined on a common interval I contained in S, satisfying all equations of (31.13) identically.

Comment 31.141.   In Lesson 62, you will find a criterion, Theorem 62.12, which gives a sufficient condition for the *existence* and *uniqueness* of a solution of the system (31.13) satisfying the n initial conditions,

(31.15) $$y_1(t_0) = a_1, \qquad y_2(t_0) = a_2, \cdots, y_n(t_0) = a_n.$$

Comment 31.16.   In Lesson 62B, we show how a nonlinear differential equation of order greater than one can be reduced to a system of first order equations. If, therefore, we can devise methods for finding solutions of the system (31.13), then theoretically any nonlinear differential equation can be solved. In Lesson 34, in fact, we solve certain *special* types of second order nonlinear differential equations by reducing them to a system of two first order equations.

Comment 31.17.   Solutions of systems of first order equations will not in general be expressible explicitly or implicitly in terms of elementary functions. Only a few very special first order systems will have such solutions. Even this simple looking pair of first order equations,

$$\frac{dx}{dt} = e^{t^2}, \qquad \frac{dy}{dt} = x$$

cannot be solved in terms of elementary functions. In later lessons, we shall show you how a pair of first order equations may sometimes be solved by means of series methods, numerical methods, and by a method known as Picard's method of successive approximations. We wish to impress on you that the examples of first order systems which we have solved below are of a very special kind, *artificially* designed to enable us to obtain solutions in terms of elementary functions.

Example 31.18. Solve the first order system

(a) $$\frac{dx}{dt} = \frac{t}{x^2}, \qquad \frac{dy}{dt} = \frac{y}{t^2}, \quad x \neq 0, t \neq 0.$$

(Note that each equation is of the separable type.)

Solution. From the first equation, we obtain

(b) $$\frac{x^3}{3} = \frac{t^2}{2} + c_1', \qquad x^3 = \tfrac{3}{2}t^2 + c_1,$$

and from the second, provided $y \neq 0$,

(c) $$\log y = -\frac{1}{t} + c_2', \qquad y = c_2 e^{-1/t}.$$

The pair of functions defined by (b) and (c) is a solution of the system (a).

Example 31.19. Solve the first order system

(a) $$\frac{dx}{dt} = 2e^{2t}, \qquad \frac{dy}{dt} = \frac{x^2 - y}{t}, \quad t \neq 0.$$

(Note that the first equation has no x or y in it.)

Solution. Solving the first equation in (a), we obtain

(b) $$x = e^{2t} + c_1.$$

Substituting (b) in the second equation of (a), there results

(c) $$\frac{dy}{dt} + \frac{y}{t} = \frac{e^{4t} + 2c_1 e^{2t} + c_1{}^2}{t},$$

which is a first order linear equation in y. Its solution, by any method you wish to choose, is

(d) $yt = \int (e^{4t} + 2c_1 e^{2t} + c_1{}^2)\, dt = \frac{1}{4} e^{4t} + c_1 e^{2t} + c_1{}^2 t + c_2,$

$y = (\frac{1}{4} e^{4t} + c_1 e^{2t} + c_1{}^2 t + c_2) t^{-1}, \; t \neq 0.$

Your can verify that the pair of functions defined by (b) and (d) satisfies both equations of (a) and is, therefore, a solution of the system.

LESSON 31C. Definition and Solution of a System of Linear First Order Equations. A special type of first order system is one in which the functions $f_1(x,y,t)$ and $f_2(x,y,t)$ of (31.11) are *linear* in x and y. This means that each equation of the system has the form

(31.2) $$\frac{dx}{dt} = f_1(t)x + g_1(t)y + h_1(t),$$

$$\frac{dy}{dt} = f_2(t)x + g_2(t)y + h_2(t).$$

A pair of equations of this type is called a **system of two linear first order equations.**. Note that x and y both have the exponent *one*, but that no such restriction is placed on the independent variable t. A generalization of this type of system is given in the following definition.

Definition 31.21. The system of n equations

(31.22) $\dfrac{dy_1}{dt} = f_{11}(t)y_1 + f_{12}(t)y_2 + \cdots + f_{1n}(t)y_n + Q_1(t),$

$\dfrac{dy_2}{dt} = f_{21}(t)y_1 + f_{22}(t)y_2 + \cdots + f_{2n}(t)y_n + Q_2(t),$

. . . .

$\dfrac{dy_n}{dt} = f_{n1}(t)y_1 + f_{n2}(t)y_2 + \cdots + f_{nn}(t)y_n + Q_n(t).$

is called a **system of n linear first order equations.**

Comment 31.23. In Lesson 62C you will find a criterion, Theorem 62.3, which gives a sufficient condition for the *existence* and *uniqueness* of a solution of the system (31.22) satisfying the initial conditions,

(31.24) $y_1(t_0) = a_1, \qquad y_2(t_0) = a_2, \cdots, y_n(t_0) = a_n.$

Comment 31.24. No standard method is known of finding a solution in terms of elementary functions, if one exists, of a general linear first order system (31.22). If, however, all coefficients $f_{ij}(t)$, $i = 1, \cdots, n$; $j = 1, \cdots, n$, are *constants*, then standard methods of solution are available. These methods are discussed in Lesson 31D, where the system (31.22) with constant coefficients, is included in the larger class of systems of

linear equations with constant coefficients of order greater than or equal
to one. In the examples solved below, where the coefficients are not constants, we again impress upon you the fact that they have been artificially
selected to yield elementary functions for solutions.

Example 31.25.   Solve the linear first order system

(a) $$\frac{dx}{dt} = 2xt - x, \qquad \frac{dy}{dt} = 2yt + x.$$

(Note that the first equation has no y in it.)

Solution.   The first equation in (a) is the separable type discussed in
Lesson 6C. By the method outlined there, we obtain the general solution

(b) $$x = c_1 e^{t^2 - t}.$$

Substituting (b) in the second equation of (a), there results

(c) $$\frac{dy}{dt} = 2ty + c_1 e^{t^2 - t},$$

which is a first order equation, linear in y. Its solution, by the method of
Lesson 11B, is

(d) $$ye^{-t^2} = c_1 \int e^{-t}\, dt = -c_1 e^{-t} + c_2,$$
$$y = e^{t^2}(c_2 - c_1 e^{-t}).$$

You can verify that the pair of functions defined by (b) and (d) satisfies
both equations of (a) and is therefore a solution of the system (a).

Example 31.26.   Solve the linear first order system

(a) $$\frac{dx}{dt} = 2e^{2t}, \qquad \frac{dy}{dt} = \frac{x - y}{t}.$$

Solution.   The solution of the first equation is

(b) $$x = e^{2t} + c_1.$$

Substituting (b) in the second equation of (a), there results

(c) $$\frac{dy}{dt} = \frac{e^{2t} - y + c_1}{t}, \qquad \frac{dy}{dt} + \frac{y}{t} = \frac{e^{2t} + c_1}{t}.$$

The solution of (c), by the method of Lesson 11B, is

(d) $$yt = \int (e^{2t} + c_1)\, dt, \qquad y = (\tfrac{1}{2}e^{2t} + c_1 t + c_2)t^{-1}.$$

The pair of functions defined by (b) and (d) is a solution of the system (a).

LESSON 31D. Solution of a System of Linear Equations with Constant Coefficients by the Use of Operators. Nondegenerate Case. The pair of equations

(31.3)　　　　　$f_1(D)x + g_1(D)y = h_1(t)$,

　　　　　　　　$f_2(D)x + g_2(D)y = h_2(t)$,

where D is the operator d/dt and the coefficients of x and y are polynomial operators as defined in Lesson 24A, is called a **system of two linear differential equations.** A generalization of this type of system is given in the following definition.

Definition 31.31. The system of equations

(31.32)　　$P_{11}(D)y_1 + P_{12}(D)y_2 + \cdots + P_{1n}(D)y_n = h_1(t)$,

　　　　　　$P_{21}(D)y_1 + P_{22}(D)y_2 + \cdots + P_{2n}(D)y_n = h_2(t)$,

　　　　　$\cdots\cdots\cdots\cdots\cdots\cdots\cdots\cdots\cdots\cdots\cdots\cdots\cdots$

　　　　　　$P_{n1}(D)y_1 + P_{n2}(D)y_2 + \cdots + P_{nn}(D)y_n = h_n(t)$,

where D is the operator d/dt and the coefficients of $y_1, y_2, \cdots, y_n$ are polynomial operators, is called a **system of n linear differential equations.**

Definition 31.321. A **solution** of the linear system (31.32) is a set of functions $y_1(t), y_2(t), \cdots, y_n(t)$, each defined on a common interval I, satisfying all equations of the system (31.32) identically; the solution is a **general** one if, in addition, the set of functions $y_1(t), \cdots, y_n(t)$ contains the correct number of arbitrary constants; see Theorem 31.33 below.

Examples of systems of linear equations are

(a)　　　　　$(2D + 3)x + (5D - 1)y = e^t$,

　　　　　　　$(D - 1)x + (3D + 1)y = \sin t;$

(b)　　　　　$(D^2 + 3D - 1)x + y = 6 + t^2$,

　　　　　　　$(D + 2)x - (D^2 + D)y = t;$

(c)　　　　　$(3D^2 + 1)x + D^3y - (D + 1)z = t^2 + 2$,

　　　　　　　$(D - 1)x + (D^2 + 1)y + (D^2 - 2)z = e^t$,

　　　　　　　$Dx - Dy - (D^3 + 5)z = 2t + 5.$

Systems of linear equations with constant coefficients lend themselves readily to solutions by means of operators and, if $t \geqq 0$, by Laplace transforms. Although we shall confine our attention primarily to a system of two linear equations and touch briefly on a system of three linear equations, the extension of the method of solution to a larger system of linear differential equations with constant coefficients will be made apparent. In this lesson we shall solve a linear system by the use of operators; in Lesson 31H by means of the Laplace transform.

Since polynomial operators with constant coefficients obey all the rules of algebra summarized in 24.523, the method we shall be able to use for solving a system (31.32) will be similar to that used in solving an algebraic system of simultaneous equations. There are, however, two important differences between the two systems.

1. The operator symbol D does not represent a numerical quantity such as 2, $\sqrt{3}$, etc. It represents a differential operator, operating on a function. Hence the order in which operators are written is important.
2. Solutions of algebraic systems do not usually have arbitrary constants; general solutions of systems of linear differential equations usually do. By Definition 31.321, a general solution, in addition to satisfying the system, must contain the correct number of such constants. The relevant theorem needed in this connection for the two equation system (31.3) is the following.

Theorem 31.33. *The number of arbitrary constants in the general solution $x(t)$, $y(t)$ of the linear system (31.3) is equal to the order of*

$$(31.34) \qquad f_1(D)g_2(D) - g_1(D)f_2(D),$$

provided $f_1(D)g_2(D) - g_1(D)f_2(D) \neq 0$.

The proof of the theorem has been deferred to Lesson 31E.

Comment 31.341. If the difference in (31.34) is zero, the system is called **degenerate**. We shall discuss the degenerate case in Lesson 31F.

Comment 31.35. The quantity $a_1b_2 - a_2b_1$, where a_1, a_2, b_1, b_2 are constants, appears so frequently in mathematical literature that it has been given a special name. It is called a **determinant**.* A determinant is also usually written as

$$\begin{vmatrix} a_1 & a_2 \\ b_1 & b_2 \end{vmatrix} ; \quad \text{hence,} \quad \begin{vmatrix} a_1 & b_1 \\ b_1 & b_2 \end{vmatrix} \equiv a_1b_2 - a_2b_1.$$

Since (31.34) has the same form as a determinant, we shall also refer to it as a determinant, and write

$$(31.36) \qquad \begin{vmatrix} f_1(D) & g_1(D) \\ f_2(D) & g_2(D) \end{vmatrix} \equiv f_1(D)g_2(D) - g_1(D)f_2(D).$$

---

*The solution of the pair of equations $a_1x + b_1y = c_1$, $a_2x + b_2y = c_2$ is $x = (c_1b_2 - c_2b_1)/(a_1b_2 - a_2b_1)$, $y = (a_1c_2 - a_2c_1)/(a_1b_2 - a_2b_1)$. Note that the quantity $(a_1b_2 - a_2b_1)$ appears in the denominator of both equations and thus determines whether a solution exists; see Lesson 63A. Hence the name determinant.

Observe that the members of the left side of (31.36) are the coefficients of (31.3) written in the same relative position in which they appear there and that each term on the right side is a cross product of these coefficients in a definite order with a minus sign between them. We shall therefore refer to (31.36) as the **determinant of the system** (31.3). Using this terminology, we can say, by Comment 31.341, that the system (31.3) is degenerate if its determinant (31.36) is zero.

For the remainder of this Lesson 31D, we consider only nondegenerate systems.

The usual procedure followed to solve the algebraic system

$$(a) \qquad \begin{aligned} 2x + 3y &= 7, \\ 3x - 2y &= 4, \end{aligned}$$

is to multiply the first by 3, the second by -2, and then add the two. In this way we eliminate x, thus obtaining

$$13y = 13, \qquad y = 1.$$

To find x, we can again start with (a) and eliminate y, or we can substitute $y = 1$ in either of the two equations. By either method, we find $x = 2$. This pair of values $x = 2$, $y = 1$ satisfies both equations and is therefore a solution of (a).

We follow an identical procedure in solving the system (31.3). Multiplying the first by $f_2(D)$, the second by $-f_1(D)$ and adding the two resulting equations will eliminate x and yield a linear differential equation in y which can be solved by previous methods. Substituting this value of y in either of the two given equations will enable us to find $x(t)$. [Or we can eliminate y in the given equations and solve for $x(t)$.] The unfortunate feature of this standard method is that in multiplying each equation by a polynomial operator, we usually raise the order of the given equations and thus introduce superfluous constants in the pair of functions $x(t)$, $y(t)$. It then becomes necessary, if the pair is to be a general solution of (31.3), to determine how these constants are related: usually a tedious task. We shall show by examples how to eliminate such superfluous constants and thus obtain the general solution of (31.3). Later we shall describe a method which will immediately give the correct number of constants in the pair of functions $x(t)$, $y(t)$, and hence will immediately yield a general solution of (31.3)—see Lesson 31E.

Example 31.361. Solve the system

$$(a) \qquad \begin{aligned} 2\frac{dx}{dt} - x + \frac{dy}{dt} + 4y &= 1, \\ \frac{dx}{dt} - \frac{dy}{dt} &= t - 1. \end{aligned}$$

Solution. In operator notation, with $D = d/dt$, we can write (a) as

(b)
$$(2D - 1)x + (D + 4)y = 1,$$
$$Dx - \qquad Dy = t - 1.$$

Following the procedure outlined above, we multiply the first equation by D, the second by $(D + 4)$ and add the two. There results

(c) $$(3D^2 + 3D)x = D(1) + (D + 4)(t - 1) = 4t - 3.$$

The general solution of (c), by any of the previous methods discussed, is

(d) $$x(t) = c_1 + c_2 e^{-t} + \tfrac{2}{3}t^2 - \tfrac{7}{3}t.$$

Substituting (d) in the second equation of (b), we obtain

(e) $$D(c_1 + c_2 e^{-t} + \tfrac{2}{3}t^2 - \tfrac{7}{3}t) - Dy = t - 1,$$

which simplifies to

(f) $$Dy = -c_2 e^{-t} + \frac{t}{3} - \frac{4}{3}.$$

Integration of (f) gives its general solution,

(g) $$y(t) = c_2 e^{-t} + \frac{t^2}{6} - \tfrac{4}{3}t + c_3.$$

By Comment 31.35, the determinant of (b) is $(2D - 1)(-D) - D(D + 4) = -3D^2 - 3D$ which is of order two. Hence, by Theorem 31.33, the general solution of (b) must contain only two arbitrary constants. But the pair of functions $x(t)$, $y(t)$ as given in (d) and (g) respectively has three. To find a relationship among the three constants, we use the fact, see Definition 31.321, that the solution of a system of equations is a set of functions which satisfies each equation of the system identically. Since $y(t)$ was obtained by substituting $x(t)$ in the second equation of (b), we know that the pair of functions $x(t)$, $y(t)$ will satisfy this equation identically. Hence we substitute (d) and (g) in the first equation. Making these substitutions, there results

(h) $$(2D - 1)(c_1 + c_2 e^{-t} + \tfrac{2}{3}t^2 - \tfrac{7}{3}t) + (D + 4)\left(c_2 e^{-t} + \frac{t^2}{6} - \tfrac{4}{3}t + c_3\right) = 1.$$

Performing the indicated operations in (h), we obtain

(i) $$-6 - c_1 + 4c_3 = 1, \qquad 4c_3 = c_1 + 7, \qquad c_3 = \frac{c_1 + 7}{4}.$$

Hence if $c_3 = (c_1 + 7)/4$, (h) will be an identity in t. Substituting this value in (g), we obtain the corrected function

(j) $$y(t) = c_2 e^{-t} + \frac{t^2}{6} - \frac{4}{3}t + \frac{c_1 + 7}{4}.$$

The pair of functions $x(t)$, $y(t)$ defined by (d) and (j) now contains the correct number of two arbitrary constants. You can verify that this pair of functions satisfies both equations of (a) and is, therefore, by Definition 31.321, its general solution.

Example 31.37. Solve the system

(a)
$$\frac{dx}{dt} - x + \frac{dy}{dt} + y = 0,$$

$$2\frac{dx}{dt} + 2x + 2\frac{dy}{dt} - 2y = t.$$

Solution. In operator notation, we can write (a) as

(b)
$$(D - 1)x + (D + 1)y = 0,$$
$$(2D + 2)x + (2D - 2)y = t.$$

Multiplying the first equation in (b) by $(2D + 2)$, the second by $-(D - 1)$ and adding the two, there results

(c)
$$[(2D + 2)(D + 1) - (D - 1)(2D - 2)]y = -(D - 1)t,$$
$$8Dy = t - 1,$$
$$Dy = \frac{t}{8} - \frac{1}{8}, \qquad y(t) = \frac{t^2}{16} - \frac{t}{8} + c_1.$$

Substituting this value of $y(t)$ in the first equation of (b), we obtain

(d)
$$(D - 1)x + (D + 1)\left[\frac{t^2}{16} - \frac{t}{8} + c_1\right] = 0,$$

which simplifies to the first order linear equation

(e)
$$(D - 1)x = -\frac{t^2}{16} + \frac{1}{8} - c_1.$$

The general solution of (e) is

(f)
$$x(t) = \frac{t^2}{16} + \frac{t}{8} + c_1 + c_2 e^t.$$

The determinant of (b), by Comment 31.35, is $(D - 1)(2D - 2) - (D + 1)(2D + 2) = -8D$ which is of order one. Hence by Theorem 31.33, the pair of functions $x(t)$, $y(t)$ as given in (f) and (c) respectively should have only one arbitrary constant. But the pair has two. To find the relationship between the constants we proceed as we did in the previous example. Substituting (f) and (c) in the second equation of (b) (we have already used the first), there results

(g)
$$2(D + 1)\left[\frac{t^2}{16} + \frac{t}{8} + c_1 + c_2 e^t\right] + 2(D - 1)\left[\frac{t^2}{16} - \frac{t}{8} + c_1\right] = t,$$

which simplifies to

(h) $$2\left[\frac{t}{2} + 2c_2 e^t\right] = t, \quad c_2 e^t = 0.$$

Equation (h) will be an identity in t only if $c_2 = 0$. Substituting this value in (f) we obtain the corrected function

(i) $$x(t) = \frac{t^2}{16} + \frac{t}{8} + c_1.$$

The pair of functions $x(t)$, $y(t)$ defined in (i) and (c) respectively is, by Definition 31.321, the general solution of (a).

Example 31.38. Solve the system

(a) $$(D + 3)x + (D + 1)y = e^t,$$
$$(D + 1)x + (D - 1)y = t.$$

Solution. Multiplying the first by $-(D - 1)$, the second by $(D + 1)$ and adding the two, we obtain

(b) $$[-(D^2 + 2D - 3) + (D^2 + 2D + 1)]x = 1 + t,$$
$$4x = 1 + t, \quad x(t) = \tfrac{1}{4}(t + 1).$$

Substituting this value of $x(t)$ in the first equation of (a), there results

(c) $$(D + 3)\left(\frac{t}{4} + \frac{1}{4}\right) + (D + 1)y = e^t, \quad (D + 1)y = e^t - 1 - \tfrac{3}{4}t.$$

The general solution of (c) by any method you wish to choose is

(d) $$y(t) = c_1 e^{-t} + \frac{e^t}{2} - \tfrac{1}{4} - \tfrac{3}{4}t.$$

The determinant of (a) is $(D + 3)(D - 1) - (D + 1)^2 = -4$ which is of order zero. Hence, by Theorem 31.33, the pair of functions $x(t)$, $y(t)$ should have no arbitrary constant. But the pair as given in (b) and (d) has one. We know the pair satisfies the first equation of (a) identically, since we used it to find $y(t)$. We therefore substitute $x(t)$ and $y(t)$ in the second equation of (a). There results

(e) $$(D + 1)\left(\frac{t}{4} + \frac{1}{4}\right) + (D - 1)\left(c_1 e^{-t} + \frac{e^t}{2} - \tfrac{1}{4} - \tfrac{3}{4}t\right) = t,$$
$$\frac{1}{4} + \frac{t}{4} + \frac{1}{4} - c_1 e^{-t} + \frac{e^t}{2} - \frac{3}{4} - c_1 e^{-t} - \frac{e^t}{2} + \frac{1}{4} + \tfrac{3}{4}t = t,$$

which simplifies to

(f) $$t - 2c_1 e^{-t} = t.$$

Hence (f) will be an identity in t, only if $c_1 = 0$. Substituting $c_1 = 0$ in (d) gives the corrected function

(g) $$y(t) = \frac{e^t}{2} - \frac{1}{4} - \frac{3}{4}t.$$

By Definition 31.321, the pair of functions $x(t)$, $y(t)$ defined by (b) and (g) is the general solution of (a).

Example 31.39. Solve the system

(a)
$$\frac{d^2x}{dt^2} - 4x + \frac{dy}{dt} = 0,$$

$$-4\frac{dx}{dt} + \frac{d^2y}{dt^2} + 2y = 0.$$

Solution. In operator notation we can write (a) as

(b)
$$(D^2 - 4)x + \qquad Dy = 0,$$
$$(-4D)x + (D^2 + 2)y = 0.$$

Multiply the first equation by $(4D)$, the second by $(D^2 - 4)$ and add the two. There results

(c) $\quad [4D^2 + (D^2 + 2)(D^2 - 4)]y = 0, \qquad (D^4 + 2D^2 - 8)y = 0.$

The general solution of (c) is

(d) $$y(t) = c_1 e^{\sqrt{2}t} + c_2 e^{-\sqrt{2}t} + c_3 \cos 2t + c_4 \sin 2t.$$

Substituting (d) in the second equation of (b) gives

(e) $\quad -4Dx + 2c_1 e^{\sqrt{2}t} + 2c_2 e^{-\sqrt{2}t} - 4c_3 \cos 2t - 4c_4 \sin 2t$
$$+ 2c_1 e^{\sqrt{2}t} + 2c_2 e^{-\sqrt{2}t} + 2c_3 \cos 2t + 2c_4 \sin 2t = 0,$$

$$Dx = c_1 e^{\sqrt{2}t} + c_2 e^{-\sqrt{2}t} - \tfrac{1}{2}c_3 \cos 2t - \tfrac{1}{2}c_4 \sin 2t,$$

$$x(t) = \frac{\sqrt{2}}{2} c_1 e^{\sqrt{2}t} - \frac{\sqrt{2}}{2} c_2 e^{-\sqrt{2}t} - \tfrac{1}{4}c_3 \sin 2t + \tfrac{1}{4}c_4 \cos 2t + c_5.$$

The determinant of (b), by Comment 31.35, is $(D^2 - 4)(D^2 + 2) - D(-4D)$ which is of order four. Hence by Theorem 31.33, the pair of functions $x(t)$, $y(t)$ should have four arbitrary constants. But the pair as given in (d) and (e) has five. You can verify that the substitution of $x(t)$, $y(t)$ in the first equation of (b)—we have already used the second—will be an identity if $c_5 = 0$. Hence this pair of functions, with $c_5 = 0$, is the general solution of (a).

LESSON 31E.　An Equivalent Triangular System.　The standard method outlined above for solving the system (31.3) can be tedious and time consuming, if by its use the pair of functions thus obtained has more than the required number of constants.　A method therefore which would immediately give the correct number of constants in the pair of functions should indeed be welcomed.　We shall now develop such a method.

For convenience we recopy the system (31.3),

$$(31.4) \qquad f_1(D)x + g_1(D)y = h_1(t),$$
$$f_2(D)x + g_2(D)y = h_2(t).$$

From it, we obtain a new system in the following manner.　We *retain either one of the equations* in (31.4).　Let us say we retain the first.　The second is then changed as follows.　Multiply the one retained, here the first, by any arbitrary operator $k(D)$ and add it to the second.　The new system thus becomes

$$(31.41) \qquad f_1(D)x + \qquad\qquad g_1(D)y = h_1(t),$$
$$[f_1(D)k(D) + f_2(D)]x + [g_1(D)k(D) + g_2(D)]y = k(D)h_1(t) + h_2(t).$$

The new system (31.41) is equivalent to the first system (31.4) in the sense that a pair of functions $x(t)$, $y(t)$ which satisfies the first system will also satisfy the second system, and conversely, a solution of the system (31.41) will satisfy the system (31.4).

By Comment 31.35, the determinant of (31.4) is

$$(31.42) \qquad f_1g_2 - g_1f_2,$$

and the determinant of (31.41) is

$$(31.43) \qquad f_1g_1k + f_1g_2 - f_1g_1k - f_2g_1 = f_1g_2 - g_1f_2.$$

Hence we see that the determinants of both systems are the same.　Therefore the order of the determinant of our original system (31.4) must be the same as the order of the determinant of the equivalent system (31.41).

Let us assume momentarily—we shall show you later how to do it—that by always *retaining one equation in a system* and changing the other in the manner described above, we can obtain a new system, equivalent to (31.4), in the form

$$(31.44) \qquad F_1(D)x \qquad\qquad = H_1(t),$$
$$F_2(D)x + G_2(D)y = H_2(t).$$

Hence the determinant of the system (31.44) is the same as the determinant of the system (31.4) and the orders of their determinants are also the same.

We call such an equivalent system, i.e., one in which a coefficient of x or of y is zero, an **equivalent triangular system.** But now note. If we solve the system (31.44) for x, using the first equation in (31.44), its general solution $x(t)$ will have as many arbitrary constants as the order of $F_1(D)$. Substituting this solution in the second equation of (31.44) will result in a linear equation in y whose order is equal to that of $G_2(D)$. Hence, in solving this equation for y, we shall introduce additional constants equal to the order of $G_2(D)$. Therefore the total number of arbitrary constants in the pair of functions $x(t)$, $y(t)$ of (31.44), if solved first for x and then for y, will equal the sum of the orders of $F_1(D)$ and $G_2(D)$. But, by Comment 31.35, the determinant of (31.44) is $F_1(D)G_2(D)$ whose order is exactly the sum of the orders of F_1 and G_2. (Remember in multiplying differential operators, the orders are added just as if they were exponents, for example $D^2 D^3 = D^5$.) We have thus shown that for the system (31.44) the pair of functions $x(t)$, $y(t)$, if the system is solved first for $x(t)$ and then for $y(t)$, will contain the correct number of constants (since the pair satisfies each equation identically), and that this number equals the order of the determinant of (31.44). But the order of the determinant of (31.44) is the same as the order of the determinant of the original system (31.4) from which it came. We have therefore not only proved that a solution of (31.44) will also be a solution of (31.4) and contain the correct number of arbitrary constants, but have at the same time **proved Theorem 31.33.**

If the original system is already in the special form

(31.45) (a)
$$f_1(D)x \qquad\quad = h_1(t),$$
$$g_2(D)y = h_2(t),$$

or in the triangular form

(31.45) (b)
$$f_1(D)x \qquad\qquad = h_1(t),$$
$$f_2(D)x + g_2(D)y = h_2(t),$$

then the pair of functions $x(t)$, $y(t)$ obtained by solving the system will contain the correct number of constants. In solving (b), it is essential to obtain $x(t)$ first and then $y(t)$. If you solve first for y by eliminating x, superfluous constants may be introduced.

Example 31.46. Solve the system

(a)
$$(D+1)x \quad\;\; = t,$$
$$D^2 y = e^{2t}.$$

Solution. The general solution of the first equation in (a), by the method of Lesson 11B, is

(b)
$$x(t) = t - 1 + c_1 e^{-t}.$$

The general solution of the second equation in (a) is

(c) $$y(t) = \tfrac{1}{4}e^{2t} + c_2 t + c_3.$$

You can verify, by Theorem 31.33, that the pair of functions in (b) and (c) has the correct number of three constants and is therefore the general solution of (a).

Example 31.47. Solve the system

(a)
$$\begin{aligned}
(3D^2 + 3D)x &= 4t - 3, \\
(D - 1)x - D^2 y &= t^2.
\end{aligned}$$

Solution. The general solution of the first equation in (a), by any of the previous methods discussed, is

(b) $$x(t) = c_1 + c_2 e^{-t} + \tfrac{2}{3}t^2 - \tfrac{7}{3}t.$$

Hence

(c)
$$\begin{aligned}
(D - 1)x &= -c_2 e^{-t} + \tfrac{4}{3}t - \tfrac{7}{3} - c_1 - c_2 e^{-t} - \tfrac{2}{3}t^2 + \tfrac{7}{3}t \\
&= -\tfrac{7}{3} - c_1 - 2c_2 e^{-t} + \tfrac{11}{3}t - \tfrac{2}{3}t^2.
\end{aligned}$$

Substituting (c) in the second equation of (a), there results

(d) $$D^2 y = -\tfrac{7}{3} - c_1 - 2c_2 e^{-t} + \tfrac{11}{3}t - \tfrac{5}{3}t^2.$$

Integrating (d) twice, we obtain the general solution

(e) $$y(t) = c_4 + c_3 t - \left(\frac{7}{6} + \frac{c_1}{2}\right)t^2 + \tfrac{11}{18}t^3 - \tfrac{5}{36}t^4 - 2c_2 e^{-t}.$$

The pair of functions defined in (b) and (e) is the general solution of (a). You can verify, by Theorem 31.33, that this pair of functions contains the correct number of four arbitrary constants.

Before proceeding to the general system, we consider one more special type, namely one in which a coefficient of x or y is a constant. Hence the system is of the form

(31.48)
$$\begin{aligned}
f_1(D)x + \quad ky &= h_1(t), \\
f_2(D)x + g_2(D)y &= h_2(t).
\end{aligned}$$

In this case it will be possible to obtain an equivalent triangular system in one step by the following procedure. *Retain the equation which contains the constant coefficient*—in (31.48) it is the first—and change the second by multiplying the first by $-g_2(D)/k$ and adding it to the second.

Example 31.49. Solve the system

(a)
$$\begin{aligned}
(3D - 1)x + 4y &= t, \\
Dx - Dy &= t - 1.
\end{aligned}$$

Solution. Following the procedure outlined above we *copy the first equation,* and obtain a second equation by multiplying the first by $D/4$ and adding it to the second. We thus obtain the equivalent triangular system

(b)
$$(3D - 1)x + 4y = t,$$

$$\tfrac{1}{4}(3D^2 + 3D)x = \frac{D}{4}(t) + t - 1 = t - \tfrac{3}{4}.$$

The general solution of the second equation in (b), by any of the methods previously discussed, is

(c)
$$x(t) = c_1 + c_2 e^{-t} + \tfrac{2}{3}t^2 - \tfrac{7}{3}t.$$

Substituting this value of x in the first equation in (b), we obtain

(d)
$$(3D - 1)(c_1 + c_2 e^{-t} + \tfrac{2}{3}t^2 - \tfrac{7}{3}t) + 4y = t,$$

which simplifies to

(e)
$$y(t) = \frac{c_1 + 7}{4} + c_2 e^{-t} + \frac{1}{6}t^2 - \frac{4}{3}t.$$

The determinant of (a), by (31.36), is

$$(3D - 1)(-D) - 4D = -3D^2 - 3D,$$

whose order is two. Note that the pair of functions $x(t)$, $y(t)$ contains the correct number of two constants and is therefore the general solution of (a).

Example 31.5. Solve the system

(a)
$$(D + 4)x + Dy = 1,$$
$$(D - 2)x + y = t^2.$$

Solution. We *copy the second equation* and obtain a new first equation by multiplying the second by $-D$ and adding it to the first. We thus obtain the equivalent triangular system

(b)
$$(-D^2 + 3D + 4)x = -2t + 1,$$
$$(D - 2)x + y = t^2.$$

The general solution of the first equation in (b) by any of the methods previously discussed is

(c)
$$x(t) = c_1 e^{4t} + c_2 e^{-t} - \frac{t}{2} + \frac{5}{8}.$$

Substituting (c) in the second equation of (b) gives

(d)
$$(D - 2)\left(c_1 e^{4t} + c_2 e^{-t} - \frac{t}{2} + \frac{5}{8}\right) + y = t^2,$$

$$y(t) = -2c_1 e^{4t} + 3c_2 e^{-t} + t^2 - t + \tfrac{7}{4}.$$

The determinant of (a) is $(D + 4) - D(D - 2) = -D^2 + 3D + 4$, which is of order two. Note that the pair of functions $x(t)$, $y(t)$ contains the correct number of two constants, and is therefore the general solution of (a).

Example 31.51. Solve the system

(a)
$$(D + 1)x + y = e^t,$$
$$(D^2 + 1)x + (D - 1)y = t.$$

Solution. We *copy the first equation*, and change the second by multiplying the first by $-(D - 1)$ and adding it to the second equation. We thus obtain the equivalent system

(b)
$$(D + 1)x + y = e^t,$$
$$2x = -(D - 1)e^t + t = t, \quad x(t) = \frac{t}{2}.$$

Substituting in the first equation of (b), the value of x as found in the second equation, there results

(c)
$$(D + 1)\frac{t}{2} + y = e^t, \quad y(t) = e^t - \frac{1}{2} - \frac{t}{2}.$$

The determinant of (a), by (31.36), is $(D + 1)(D - 1) - (D^2 + 1) = D^2 - 1 - D^2 - 1 = -2$, which is of order zero. Note that the pair of functions $x(t)$, $y(t)$ contains this correct number of zero constants, and is therefore the general solution of (a).

We shall now show you a method by which you can reduce a general system to an equivalent triangular one. We demonstrate the method by an example. Consider the system

(31.52)
$$(D^3 - D^2 + 2D + 1)x + (D^4 + 3D^2 + 1)y = 0,$$
$$(D - 2)x + (D^3 - 3)y = 0.$$

Of the four polynomial operators in (31.52), concentrate on the one of lowest order. In the above example it is $D - 2$. Retain the equation in which it appears, multiply the equation by $-D^2$ and add it to the first. The resulting equivalent system is

$$(D^2 + 2D + 1)x + (-D^5 + D^4 + 6D^2 + 1)y = 0,$$
$$(D - 2)x + (D^3 - 3)y = 0.$$

Note that by this process, we were able to reduce the order of the coefficient of x in the first equation from three to two. The polynomial operator of lowest order in this new system is again $(D - 2)$. We therefore again retain the equation in which it appears, multiply it by $-D$ and add it to the first. We thus obtain the equivalent system

$$(4D + 1)x + (-D^5 + 6D^2 + 3D + 1)y = 0,$$
$$(D - 2)x + \qquad\qquad\qquad (D^3 - 3)y = 0.$$

Note that the order of the coefficient of x in the first equation has been reduced from two to one. There are now two polynomial operators with the same lowest order. We can retain either one. Because the coefficient of D in $(D - 2)$ is one, it will be found easier to retain this equation. Multiply it by -4 and add it to the first equation. We thus obtain the equivalent system

$$9x + (-D^5 - 4D^3 + 6D^2 + 3D + 13)y = 0,$$
$$(D - 2)x + \qquad\qquad\qquad (D^3 - 3)y = 0.$$

The system is now of the form (31.48). Therefore, retaining the first equation, multiplying it by $-(D - 2)/9$ and adding it to the second will finally give the equivalent triangular system

$$9x + (-D^5 - 4D^3 + 6D^2 + 3D + 13)y = 0,$$
$$\left[(D^5 + 4D^3 - 6D^2 - 3D - 13)\,\frac{D - 2}{9} + D^3 - 3\right]y = 0.$$

In the manner described above, it is always possible to reduce the *order* of the coefficient of one polynomial operator in the system to zero. When that point has been reached, the system will be in the form (31.48). One more step will then give the required equivalent triangular system.

Comment 31.53. Use of the above method will always enable you to obtain an equivalent triangular system. Use of a little ingenuity may at times enable you to obtain it sooner. For example, if the given system is, or if in the course of your work, it becomes,

$$D^3x + (D^2 - 2)y = e^t,$$
$$-2D^3x + (D + 3)y = 4t + 2,$$

then retaining the first, multiplying it by two and adding it to the second will give you an equivalent triangular system immediately, even though D^3 is not the polynomial operator of lowest order.

Example 31.531. Solve the system

(a)
$$(D + 1)x + (D + 1)y = 1,$$
$$D^2x - \qquad Dy = t - 1.$$

Solution. Here there are three polynomial operators of the same lowest order. It will be found easiest to *retain the second* and change the first by adding the two equations. There results the equivalent system

(b)
$$(D^2 + D + 1)x + \quad y = t,$$
$$D^2x - Dy = t - 1.$$

The system is now of the form (31.48). We therefore copy the *first* equation and change the second by multiplying the first by D and adding it to the second. We thus obtain the equivalent triangular system

(c)
$$(D^2 + D + 1)x + y = t,$$
$$(D^3 + 2D^2 + D)x \quad = t.$$

A general solution of the second equation is

(d)
$$x(t) = c_1 + c_2e^{-t} + c_3te^{-t} + \frac{t^2}{2} - 2t.$$

Substituting this value of $x(t)$ in the first equation of (c), we obtain

(e)
$$y(t) = t - (D^2 + D + 1)\left(c_1 + c_2e^{-t} + c_3te^{-t} + \frac{t^2}{2} - 2t\right)$$
$$= t - \left(c_2e^{-t} - 2c_3e^{-t} + c_3te^{-t} + 1 - c_2e^{-t} + c_3e^{-t}\right.$$
$$\left. - c_3te^{-t} + t - 2 + c_1 + c_2e^{-t} + c_3te^{-t} + \frac{t^2}{2} - 2t\right),$$
$$y(t) = -\left[c_1 + c_2e^{-t} + c_3(-e^{-t} + te^{-t}) + \frac{t^2}{2} - 2t - 1\right].$$

The determinant of (a), by (31.36), is $(D + 1)(-D) + (D + 1)(D^2)$ which is of order three. Note that our pair of functions $x(t)$, $y(t)$ contains the correct number of three constants, and is therefore the general solution of (a).

Example 31.54. Solve the system

(a)
$$(2D - 1)x + (D + 4)y = 1,$$
$$Dx - \quad Dy = t - 1.$$

Solution. Here all four polynomial operators in (a) are of the same order one. We can therefore retain either equation of (a). It will, however, be found easier to *retain the second* and change the first by adding the second to it. We thus obtain the equivalent system

(b)
$$(3D - 1)x + 4y = t,$$
$$Dx - Dy = t - 1.$$

The system is now of the form (31.48). We therefore *retain the first equation* and change the second by multiplying the first by $D/4$ and adding it to the second. There results the equivalent triangular system

(c)
$$(3D - 1)x + 4y = t,$$
$$(3D^2 + 3D)x = 4t - 3.$$

The general solution of the second equation in (c) is

(d)
$$x(t) = c_1 + c_2 e^{-t} + \tfrac{2}{3}t^2 - \tfrac{7}{3}t.$$

Substituting this value of x in the first equation of (c), we obtain

(e)
$$(3D - 1)(c_1 + c_2 e^{-t} + \tfrac{2}{3}t^2 - \tfrac{7}{3}t) - t = -4y,$$

which simplifies to

(f)
$$y(t) = c_2 e^{-t} + \frac{t^2}{6} - \frac{4}{3}t + \frac{c_1 + 7}{4}.$$

The pair of functions (d) and (f) are the same as those obtained in Example 31.361 by using the standard method of elimination. Note how much easier the above method is. Note too that we obtained the correct number of two constants immediately.

Example 31.55. Solve the system

(a)
$$(D - 1)x + (D + 1)y = 0,$$
$$(D + 1)x + (D - 1)y = \frac{t}{2}.$$

Solution. We *retain the first equation* and change the second by multiplying the first by -1 and adding it to the second. We thus obtain the equivalent system

(b)
$$(D - 1)x + (D + 1)y = 0,$$
$$2x - \qquad 2y = \frac{t}{2}.$$

Retain the second equation and change the first by multiplying the second by $(D + 1)/2$ and adding to the first. There results the equivalent triangular system

(c)
$$2Dx \qquad = \frac{(D + 1)}{4} t = \tfrac{1}{4}(1 + t),$$
$$2x - 2y = \frac{t}{2}.$$

Integrating the first equation in (c), we obtain

(d)
$$2x = \frac{t^2}{8} + \frac{t}{4} + 2c, \qquad x(t) = \frac{t^2}{16} + \frac{t}{8} + c.$$

Substituting (d) in the second equation of (c), we obtain

(e) $\qquad 2y = \dfrac{t^2}{8} + \dfrac{t}{4} - \dfrac{t}{2} + 2c, \qquad y(t) = \dfrac{t^2}{16} - \dfrac{t}{8} + c.$

These are the same functions obtained in Example 31.37. Note how much easier the above method is over the previous one. Note, too, that we obtained the correct number of one constant immediately.

Example 31.56.   Solve the system

(a) $\qquad\qquad\qquad (D + 3)x + (D + 1)y = e^t,$
$\qquad\qquad\qquad\quad (D + 1)x + (D - 1)y = t.$

Solution.   We *retain the second equation* and change the first by multiplying the second by -1 and adding it to the first. We thus obtain the equivalent system

(b) $\qquad\qquad\qquad\quad 2x \qquad\quad + 2y = e^t - t,$
$\qquad\qquad\quad (D + 1)x + (D - 1)y = t.$

Retain the first equation and change the second by multiplying the first by $-(D - 1)/2$ and adding it to the second. There results the equivalent triangular system

(c) $\qquad 2x + 2y = e^t - t,$
$\qquad\ 2x \qquad = -\tfrac{1}{2}(D - 1)(e^t - t) + t = \tfrac{1}{2}(t + 1).$

From the second equation of (c), we obtain

(d) $\qquad\qquad\qquad\qquad x(t) = \tfrac{1}{4}(t + 1).$

Subtracting the second equation in (c) from the first, we obtain

(e) $\qquad 2y = e^t - \dfrac{3t}{2} - \dfrac{1}{2}, \qquad y(t) = \dfrac{e^t}{2} - \dfrac{3t}{4} - \dfrac{1}{4}.$

These are the same functions $x(t)$, $y(t)$ obtained in Example 31.38. Note that by this method we obtained the correct number of zero constants immediately.

LESSON 31F.   Degenerate Case. $f_1(D)g_2(D) - g_1(D)f_2(D) = 0.$ An algebraic system of equations may be degenerate. In such cases the system will have no solutions or infinitely many solutions. For example, the systems

(a) $2x + 3y = 5,$ $\qquad$ (b) $2x + 3y = 5,$
$\quad\ 2x + 3y = 7;$ $\qquad\qquad 4x + 6y = 7,$

have no solutions. On the other hand, each of the systems

(c) $2x + 3y = 5$,   (d) $2x + 3y = 0$,

    $4x + 6y = 10$; $4x + 6y = 0$,

has infinitely many solutions. Note that in all four examples, the determinant formed by the coefficients of x and y is zero. Note also that there are no solutions, when in trying to eliminate x or y, the right side does not also reduce to zero. There are infinitely many solutions when the right side reduces to zero.

Similarly, we call the system of linear differential equations (31.4) degenerate whenever its determinant

$$\begin{vmatrix} f_1(D) & g_1(D) \\ f_2(D) & g_2(D) \end{vmatrix} = f_1(D)g_2(D) - g_1(D)f_2(D)$$

is zero. As in the algebraic system, there will be no solutions if, in trying to eliminate x or y, the right side of the system is not zero; there will be infinitely many if the right side is zero.

Example 31.6. Show that the system

(a) $$Dx - Dy = t,$$
$$Dx - Dy = t^2,$$

is degenerate. Find the number of solutions it has.

Solution. By (31.36), the determinant of (a) is $-D^2 - (-D^2) = 0$. Hence the system is degenerate. Since the right side does not reduce to zero when we eliminate x or y, it has no solutions. This example corresponds to (a) at the bottom of page 413.

Example 31.61. Show that the system

(a) $$Dx - Dy = t,$$
$$4Dx - 4Dy = 4t,$$

is degenerate. Find the number of solutions it has.

Solution. The determinant of (a) is $-4D^2 - (-4D^2) = 0$. Hence the system is degenerate. Its right side, however, reduces to zero when we eliminate x or y. In this case there are infinitely many solutions of the system. For example, the pair, $x = \dfrac{t^2}{2} + c$, $y = 5c$ is a solution; the pair $x = \dfrac{t}{2} + c_1$, $y = \dfrac{t}{2} - \dfrac{t^2}{2} + c_2$ is a solution. [In either equation, define $x(t)$ arbitrarily, and solve for $y(t)$. This pair of functions will also satisfy the other equation.] This example corresponds to (c) at top of page.

Example 31.7.   Show that the system

(a)
$$(D + 1)x + (D + 1)y = 0,$$
$$(D - 1)x + (D - 1)y = 0,$$

is degenerate. Find its solutions.

Solution.   The determinant of (a) is

(b) $\begin{vmatrix} D+1 & D+1 \\ D-1 & D-1 \end{vmatrix} = (D+1)(D-1) - (D+1)(D-1) = 0.$

Hence the system (a) is degenerate. By (24.21), we can write (a) as

(c)
$$(D + 1)(x + y) = 0,$$
$$(D - 1)(x + y) = 0.$$

Let $u = x + y$. Therefore (c) becomes $(D + 1)u = 0$, and $(D - 1)u = 0$. The solution of the first equation is $u = c_1 e^{-t}$; the solution of the second equation is $u = c_2 e^t$. Hence the solution of the system (c) is

(d) $\qquad x + y = c_1 e^{-t} \quad \text{and} \quad x + y = c_2 e^t.$

However, if $c_1 \neq 0$ and $c_2 \neq 0$, then the pair of solutions of (d) is inconsistent, i.e., there are no functions $x(t)$ and $y(t)$ that will satisfy them simultaneously. If however, $c_1 = c_2 = 0$, then the infinitely many functions, such that

(e) $\qquad x + y = 0 \quad \text{or} \quad x(t) = -y(t)$

will satisfy the system (c).

LESSON 31G.   Systems of Three Linear Equations.   We discuss briefly the system of three linear equations:

(31.71)
$$f_1(D)x + g_1(D)y + h_1(D)z = k_1(t),$$
$$f_2(D)x + g_2(D)y + h_2(D)z = k_2(t),$$
$$f_3(D)x + g_3(D)y + h_3(D)z = k_3(t).$$

Its determinant is defined as

(31.72) $\begin{vmatrix} f_1(D) & g_1(D) & h_1(D) \\ f_2(D) & g_2(D) & h_2(D) \\ f_3(D) & g_3(D) & h_3(D) \end{vmatrix} = \begin{aligned} & f_1 g_2 h_3 + f_2 g_3 h_1 + f_3 g_1 h_2 - f_1 g_3 h_2 \\ & - f_2 g_1 h_3 - f_3 g_2 h_1. \end{aligned}$

The method of finding a general solution of the system (31.71) follows the same rules outlined previously in solving a two-equation system. The number of constants in the general solution of (31.71), i.e., in the set of functions $x(t)$, $y(t)$, $z(t)$ satisfying (31.71), as in the case of a two-equation system, must equal the order of the determinant (31.72), provided this determinant is not zero.

The usual standard method for solving the system (31.71) is to eliminate one of the variables, say z, in the same manner we eliminate this variable in an algebraic system of three equations. The system (31.71) can thus be reduced to the system

$$(31.73) \qquad \begin{aligned} F_1(D)x + G_1(D)y &= l_1(t), \\ F_2(D)x + G_2(D)y &= l_2(t). \end{aligned}$$

Finally by eliminating, in the usual manner, one variable from the system (31.73), say y, we obtain the single linear equation

$$(31.74) \qquad F(D)x = k(t),$$

which we can solve for x. Substituting this value of x in either equation of (31.73) will enable us to solve for y. Substituting both values x and y in any one of the equations in (31.71) will enable us to solve for z. Since in this process, we usually must multiply by an operator in order to effect the desired eliminations, the number of constants in the set of functions $x(t)$, $y(t)$, $z(t)$ will generally be more than required. We then have to go through the tedious process of finding what relationship exists among the constants by substituting the three functions $x(t)$, $y(t)$, $z(t)$ in either of the two equations of (31.71) which were not used to find $z(t)$. Since the three functions must satisfy each equation in (31.71) identically, we can thus determine from the equation a relationship among the constants, just as we did in the two-equation system.

Fortunately it is always possible, in exactly the same manner described for the two-equation system, to reduce the system (31.71) to an equivalent triangular system of the form

$$(31.75) \qquad \begin{aligned} F_1(D)x &= k_1(t), \\ F_2(D)x + G_2(D)y &= k_2(t), \\ F_3(D)x + G_3(D)y + H_3(D)(z) &= k_3(t). \end{aligned}$$

If solutions are then obtained in the order $x(t)$, $y(t)$, $z(t)$ starting with the first equation in (31.75), this set of functions will contain the correct number of constants required in the general solution of the system (31.71).

Example 31.76. Reduce the following system to an equivalent triangular one and determine the number of arbitrary constants needed in the general solution.

$$(a) \qquad \begin{aligned} (D-2)x + \quad y - \quad z &= t, \\ -x + (2D+1)y + \quad 2z &= 1, \\ 2x + \quad 6y + Dz &= 0. \end{aligned}$$

Solution. Here we can retain any one of the three equations. We decide to *retain the first equation* and change the second equation by multiplying the first one by 2 and adding it to the second. We change the third equation by multiplying the first by D and adding it to the third. There results the equivalent system

(b)
$$(D - 2)x + \qquad\qquad y - z = t,$$
$$(2D - 5)x + (2D + 3)y \qquad = 2t + 1,$$
$$(D^2 - 2D + 2)x + (D + 6)y \qquad = 1.$$

We *retain the first and third equations* and change the second by multiplying the third by -2 and adding it to the second. We thus obtain the equivalent system

(c)
$$(D - 2)x + \qquad\qquad y - z = t,$$
$$(-2D^2 + 6D - 9)x - \qquad 9y \qquad = 2t - 1,$$
$$(D^2 - 2D + 2)x + (D + 6)y \qquad = 1.$$

We *retain the first and second equations* and change the third by multiplying the second by $(D + 6)/9$ and adding it to the third. We thus finally obtain the equivalent triangular system

(d)
$$(D - 2)x + y - z = t,$$
$$(-2D^2 + 6D - 9)x - 9y \qquad = 2t - 1,$$
$$(-2D^3 + 3D^2 + 9D - 36)x \qquad = 12t + 5.$$

Solving the third equation for x will give three constants. Substituting this value of x in the second equation will give an equation in y of order zero. Hence the solution $y(t)$ will contain no new constants. Substituting the solutions $x(t)$, $y(t)$ in the first equation will give an equation in z that is also of order zero and hence the solution $z(t)$ will not contain any new constants. We will thus have three arbitrary constants in the set of functions $x(t)$, $y(t)$, $z(t)$. The determinant of (a), by (31.72), is also of order three, and hence only three constants are needed in the general solution of (a). We see, therefore, that by using this method the set of functions obtained will contain the correct number of constants.

The extension of the standard method of solution to systems of linear equations of order higher than three should now be apparent to you. To determine the number of arbitrary constants in the general solution, you will need to know the order of the determinant of the system. For this information we refer you to any book on determinants and matrices. If, however, you solve the system by first reducing it to an equivalent triangular one, your solution will always contain the correct number of constants.

LESSON 31H. Solution of a System of Linear Differential Equations with Constant Coefficients by Means of Laplace Transforms. A system of linear differential equations with constant coefficients can also be solved by means of Laplace transforms. Unlike the operational method, however, the Laplace transform method can be used only if initial conditions are given and the interval over which the solutions are valid is $0 \leqq t < \infty$. On the other hand, the Laplace method has an advantage over the preceding operational one in that it will immediately yield a particular solution satisfying given initial conditions without the necessity of evaluating arbitrary constants.

Although we have confined our attention to a system of two differential equations with two dependent variables, the extension of the method to a larger system will be apparent.

Example 31.8. Solve the system

(a)
$$\frac{d^2x}{dt^2} - 4x + \frac{dy}{dt} = 0,$$

$$-4\frac{dx}{dt} + \frac{d^2y}{dt^2} + 2y = 0,$$

for which $x(0) = 0$, $x'(0) = 1$, $y(0) = -1$, $y'(0) = 2$.

Solution. In Laplace notation, we can write (a) as—see Comment 27.24—

(b) $$L[x''(t)] - 4L[x(t)] + L[y'(t)] = 0,$$
$$-4L[x'(t)] + L[y''(t)] + 2L[y(t)] = 0.$$

By (27.33), or directly from (27.29) and (27.31), (b) becomes

(c) $$s^2L[x(t)] - x'(0) - sx(0) - 4L[x(t)] + sL[y(t)] - y(0) = 0,$$
$$-4sL[x(t)] + 4x(0) + s^2L[y(t)] - y'(0) - sy(0) + 2L[y(t)] = 0.$$

Inserting the initial conditions in (c) and simplifying the resulting equations, we obtain

(d) $$(s^2 - 4)L[x] + \qquad sL[y] = 1 - 1 = 0,$$
$$-4sL[x] + (s^2 + 2)L[y] = 2 - s.$$

Just as with operators, we now solve (d) as if they were ordinary algebraic expressions in $L[x]$ and $L[y]$. Multiply the first equation in' (d) by $4s$, the second by $s^2 - 4$, and add the two. The result is

(e) $$(s^4 + 2s^2 - 8)L[y] = -s^3 + 2s^2 + 4s - 8.$$

Hence

(f) $$L[y] = \frac{-s^3 + 2s^2 + 4s - 8}{(s^2 + 4)(s^2 - 2)} = \frac{1}{6}\left[\frac{1+\sqrt{2}}{s+\sqrt{2}} + \frac{1-\sqrt{2}}{s-\sqrt{2}} - \frac{8(s-2)}{s^2+4}\right].$$

Referring to a table of Laplace transforms, we find that

(g) $$L[(1 + \sqrt{2})e^{-\sqrt{2}t}] = \frac{1 + \sqrt{2}}{s + \sqrt{2}},$$

$$L[(1 - \sqrt{2})e^{\sqrt{2}t}] = \frac{1 - \sqrt{2}}{s - \sqrt{2}},$$

$$L[-8\cos 2t] = \frac{-8s}{s^2 + 2^2}, \qquad L[8\sin 2t] = 8\frac{2}{s^2 + 2^2}.$$

Therefore, by (f) and (g),

(h) $y(t) = \frac{1}{6}[(1 + \sqrt{2})e^{-\sqrt{2}t} + (1 - \sqrt{2})e^{\sqrt{2}t} - 8\cos 2t + 8\sin 2t].$

Again as with operators, two methods are available for finding $x(t)$. We can start with (d) again and eliminate $y(t)$ or we can substitute (f) in either equation of (d) and solve for $L[x(t)]$. Using this latter method, we obtain by (f) and the first equation in (d)

(i) $$(s^2 - 4)L[x] + s\left(\frac{-s^3 + 2s^2 + 4s - 8}{(s^2 + 4)(s^2 - 2)}\right) = 0.$$

(j) $$L[x] = \frac{s(s - 2)^2(s + 2)}{(s - 2)(s + 2)(s^2 + 4)(s^2 - 2)} = \frac{s(s - 2)}{(s^2 + 4)(s^2 - 2)}$$

$$= \frac{-1}{12}\left[\frac{2 - \sqrt{2}}{s - \sqrt{2}} + \frac{2 + \sqrt{2}}{s + \sqrt{2}} - 4\left(\frac{s + 2}{s^2 + 4}\right)\right].$$

Referring to a table of Laplace transforms, we find that

(k) $\frac{-1}{12}L[(2 - \sqrt{2})e^{\sqrt{2}t} + (2 + \sqrt{2})e^{-\sqrt{2}t} - 4\cos 2t - 4\sin 2t]$

equals the expression on the extreme right of (j). Therefore, by (j) and (k),

(l) $x(t) = \frac{-1}{12}[(2 - \sqrt{2})e^{\sqrt{2}t} + (2 + \sqrt{2})e^{-\sqrt{2}t} - 4\cos 2t - 4\sin 2t].$

Example 31.81. Solve the system (see Example 31.5)

(a) $$\frac{dx}{dt} + 4x + \frac{dy}{dt} = 1,$$

$$\frac{dx}{dt} - 2x + y = t^2,$$

for which $x(0) = 2$ and $y(0) = -1$.

Solution. In Laplace notation (a) can be written as

(b) $$L[x'] + 4L[x] + L[y'] = L[1],$$
$$L[x'] - 2L[x] + L[y] = L[t^2].$$

By (27.33) or directly from (27.29), (b) becomes

(c) $sL[x(t)] - x(0) + 4L[x(t)] + sL[y(t)] - y(0) = L(1),$
       $sL[x(t)] - x(0) - 2L[x(t)] + L[y(t)] = L(t^2).$

Referring to a table of Laplace transforms, we find that

(d) $L(1) = \dfrac{1}{s}, \qquad L(t^2) = \dfrac{2}{s^3}.$

Substituting these values and the initial conditions in (c), and simplifying the resulting equations, we obtain

(e) $(s + 4)L[x(t)] + sL[y(t)] = \dfrac{1}{s} + 1,$

                 $(s - 2)L[x(t)] + L[y(t)] = \dfrac{2}{s^3} + 2.$

Multiplying the second equation in (e) by $-s$ and adding it to the first, there results

(f) $(s + 4)L(x) + (-s^2 + 2s)L(x) = \dfrac{1}{s} + 1 - \dfrac{2}{s^2} - 2s.$

Solving (f) for $L[x(t)]$ gives

(g) $L[x(t)] = \dfrac{2s^3 - s^2 - s + 2}{s^2(s^2 - 3s - 4)} = \dfrac{(s + 1)(2s^2 - 3s + 2)}{s^2(s - 4)(s + 1)}$

        $= \dfrac{2s^2 - 3s + 2}{s^2(s - 4)} = \dfrac{5}{8s} - \dfrac{1}{2s^2} + \dfrac{11}{8(s - 4)}.$

Referring to a table of Laplace transforms, we find that

(h) $L\left[\dfrac{5}{8}\right] = \dfrac{5}{8s}, \qquad L\left[-\dfrac{1}{2}t\right] = -\dfrac{1}{2s^2}, \qquad L\left[\dfrac{11}{8}e^{4t}\right] = \dfrac{11}{8(s - 4)}.$

Hence,

(i) $L\left[\dfrac{5}{8} - \dfrac{t}{2} + \dfrac{11}{8}e^{4t}\right] = \dfrac{5}{8s} - \dfrac{1}{2s^2} + \dfrac{11}{8(s - 4)}.$

Therefore by (g) and (i)

(j) $x(t) = \dfrac{5}{8} - \dfrac{t}{2} + \dfrac{11}{8}e^{4t}.$

To find $y(t)$, we substitute (g) in either equation of (e). You can verify that

(k) $y(t) = t^2 - t + \tfrac{7}{4} - \tfrac{11}{4}e^{4t}.$

Note. These are the same results you would have obtained if you had inserted the initial conditions in (c) and (d) of Example 31.5 where we solved this same problem by means of operators. Verify it.

EXERCISE 31

Solve each of the following systems of equations.

1. $\dfrac{dx}{dt} = -x^2, \quad \dfrac{dy}{dt} = -y.$

2. $\dfrac{dx}{dt} = 3e^{-t}, \quad \dfrac{dy}{dt} = x + y.$

3. $\dfrac{dx}{dt} = x^2 t, \quad \dfrac{dy}{dt} = yt^2.$

4. $\dfrac{dx}{dt} = 2t, \quad \dfrac{dy}{dt} = 3x + 2t, \quad \dfrac{dz}{dt} = x + 4y + t.$

5. $\dfrac{dx}{dt} = e^t, \quad \dfrac{dy}{dt} = \dfrac{x - y}{t}.$

6. $\dfrac{dx}{dt} = x + \sin t, \quad \dfrac{dy}{dt} = t - y.$

7. $\dfrac{dx}{dt} = 3x + 2e^{3t}, \quad \dfrac{dx}{dt} + \dfrac{dy}{dt} - 3y = \sin 2t.$

8. $\dfrac{dx}{dt} = y, \quad \dfrac{dy}{dt} = -x + 2y.$

9. $3\dfrac{dx}{dt} + 3x + 2y = e^t, \quad 4x - 3\dfrac{dy}{dt} + 3y = 3t.$

10. $\dfrac{d^2x}{dt^2} + 4x = 3\sin t, \quad \dfrac{dx}{dt} - \dfrac{d^2y}{dt^2} + y = 2\cos t.$

11. $\dfrac{d^2x}{dt^2} - 4x - 2\dfrac{dy}{dt} + y = t, \quad 2\dfrac{dx}{dt} + x + \dfrac{d^2y}{dt^2} = 0.$

12. $2\dfrac{dx}{dt} + \dfrac{dy}{dt} - x = e^t, \quad 3\dfrac{dx}{dt} + 2\dfrac{dy}{dt} + y = t.$

13. $\dfrac{d^2x}{dt^2} + x - \dfrac{d^2y}{dt^2} - y = -\cos 2t, \quad 2\dfrac{dx}{dt} - \dfrac{dy}{dt} - y = 0.$

14. $\dfrac{d^2x}{dt^2} - \dfrac{dy}{dt} = 1 - t, \quad \dfrac{dx}{dt} + 2\dfrac{dy}{dt} = 4e^t + x.$

15. $\dfrac{d^2x}{dt^2} - x + \dfrac{d^2y}{dt^2} + y = 0, \quad \dfrac{dx}{dt} + 2x + \dfrac{dy}{dt} + 2y = 0.$

16. $\dfrac{dx}{dt} + \dfrac{dy}{dt} + y = t, \quad \dfrac{d^2x}{dt^2} + \dfrac{d^2y}{dt^2} + \dfrac{dy}{dt} + x + y = t^2.$

Verify that each of the following systems 17–20 is degenerate $(D = d/dt)$. Find solutions if they exist.

17. $Dx + 2Dy = e^t,$
    $Dx + 2Dy = t.$

18. $Dx - Dy = e^t,$
    $3Dx - 3Dy = 3e^t.$

19. $(D^2 - 1)x + (D^2 - 1)y = 0.$
    $(D^2 + 4)x + (D^2 + 4)y = 0.$

20. $(D - 2)x + (D - 2)y = t$,
$(D + 3)x + (D + 3)y = t$.

21. Reduce the following system to an equivalent triangular one. How many arbitrary constants should the general solution contain?

$$
\begin{aligned}
(D - 2)x + && 3y - && z &= t, \\
-x + (D + 3)y + && && 2z &= 1, \\
2x + && 39y + (D - 4)z &= 0.
\end{aligned}
$$

22. Solve each of the following systems of equations.

(a) $(D - 1)x$ $= 0$,
$-x + (D - 3)y$ $= 0$,
$-x + \quad y + (D - 2)z = 0$.

(b) $(D - 1)x$ $= 0$,
$3x + 2(D + 1)y$ $= 0$,
$2y + (2D - 3)z = 0$.

(c) $(D - 2)x$ $= 0$,
$-3x + (D + 2)y$ $= 0$,
$-2y + (D - 3)z = 0$.

(d) $Dx - \quad y + z = 0$,
$-x + (D - 1)y$ $= 0$,
$-x + \quad (D - 1)z = 0$.

Solve each of the following systems of equations by the method of the Laplace transform.

23. $\dfrac{dx}{dt} - y = t$,

$x - \dfrac{dy}{dt} = 1$, $x(0) = 2$, $y(0) = 1$.

24. $3\dfrac{dx}{dt} + 3x + 2y = e^t$,

$4x - 3\dfrac{dy}{dt} + 3y = 3t$, $x(0) = 1$, $y(0) = -1$.

25. $\dfrac{dx}{dt} + \dfrac{dy}{dt} - 4y = 1$,

$x + \dfrac{dy}{dt} - 3y = t^2$, $x(0) = 2$, $y(0) = -2$.

26. $\dfrac{d^2x}{dt^2} - \dfrac{dy}{dt} = 1 - t$,

$\dfrac{dx}{dt} + 2\dfrac{dy}{dt} = 4e^t + x$, $x(0) = 0$, $y(0) = 0$, $x'(0) = 1$.

ANSWERS 31

1. $x^{-1} = t + c_1$, $y = c_2 e^{-t}$.

2. $x = -3e^{-t} + c_1$, $y = \frac{3}{2}e^{-t} - c_1 + c_2 e^t$.

3. $x^{-1} = -\frac{1}{2}(t^2 + c_1)$, $y = c_2 e^{t^3/3}$.

4. $x = t^2 + c_1$, $\quad y = t^3 + t^2 + 3c_1 t + c_2$,

$\quad z = t^4 + \frac{5}{3}t^3 + (6c_1 + \frac{1}{2})t^2 + (c_1 + 4c_2)t + c_3$.

5. $x = e^t + c_1$, $\quad y = t^{-1}e^t + c_1 + c_2 t^{-1}$.

6. $x = c_1 e^t - \frac{1}{2}(\sin t + \cos t)$, $\quad y = t - 1 + c_2 e^{-t}$.

7. $x = (2t + c_1)e^{3t}$, $\quad y = -\dfrac{3.\sin 2t + 2\cos 2t}{13} - e^{3t}(3t^2 + 3c_1 t + 2t + c_2)$.

8. $x = c_1 t e^t + c_2 e^t$, $\quad y = c_1 t e^t + (c_1 + c_2)e^t$.

9. $x = c_1 e^{t/3} + c_2 e^{-t/3} - 6t$, $\quad y = -2c_1 e^{t/3} - c_2 e^{-t/3} + \frac{1}{2}e^t + 9t + 9$.

10. $x = \sin t + c_1 \sin 2t + c_2 \cos 2t$,

$\quad y = \frac{1}{2}\cos t - \frac{2}{5}c_1 \cos 2t + \frac{2}{5}c_2 \sin 2t + c_3 e^t + c_4 e^{-t}$.

11. $x = c_1 \sin t + c_2 \cos t + c_3 e^t + c_4 e^{-t}$,

$\quad y = (c_1 - 2c_2)\sin t + (2c_1 + c_2)\cos t - 3c_3 e^t + c_4 e^{-t} + t + 2$.

12. $x = c_1 e^t + c_2 e^{-t} + \frac{3}{2}t e^t + 1$, $\quad y = t - 2 - (\frac{1}{2} + c_1)e^t - 3c_2 e^{-t} - \frac{3}{2}t e^t$.

13. $x = c_1 e^t + c_2 \sin t + c_3 \cos t - \frac{1}{15}(3\cos 2t + 4\sin 2t)$,

$\quad y = c_1 e^t + (c_2 - c_3)\sin t + (c_2 + c_3)\cos t - \frac{4}{15}(2\cos 2t + \sin 2t)$.

14. $x = c_1 e^{-t} + c_2 e^{t/2} + 2e^t + 2t$,

$\quad y = -c_1 e^{-t} + \dfrac{c_2}{2} e^{t/2} + 2e^t + \dfrac{t^2}{2} - t + c_3$.

15. $x = 5ce^{-2t}$, $\quad y = -3ce^{-2t}$.

16. $x = t^2 + t - 1$, $\quad y = -t$.

17. No solutions.

18. Infinitely many. Define $x(t)$ arbitrarily and solve for $y(t)$.

19. Infinitely many; $x(t) = -y(t)$.

20. No solutions.

21. One such equivalent triangular system is

$$(D - 2)x + \qquad 3y - z = t,$$
$$(2D - 5)x + (D + 9)y \qquad = 2t + 1,$$
$$(D^2 - 12D + 25)x \qquad\qquad = -10t - 2.$$

General solution should contain three arbitrary constants.

22. (a) $x = c_1 e^t$, $\quad y = c_2 e^{3t} - \dfrac{c_1}{2} e^t$, $\quad z = c_3 e^{2t} - \dfrac{3c_1}{2} e^t - c_2 e^{3t}$.

(b) $x = c_1 e^t$, $\quad y = \dfrac{-3c_1}{4} e^t + c_2 e^{-t}$, $\quad z = -\dfrac{3c_1}{2} e^t + \dfrac{2c_2}{5} e^{-t} + c_3 e^{3t/2}$.

(c) $x = 4c_1 e^{2t}$, $\quad y = 3c_1 e^{2t} + 5c_2 e^{-2t}$, $\quad z = -6c_1 e^{2t} - 2c_2 e^{-2t} + c_3 e^{3t}$.

(d) $x = c_1 + c_2 e^t$, $y = -c_1 + (c_2 t + c_3)e^t$, $z = -c_1 + (c_2 t - c_2 + c_3)e^t$.

23. $x = \frac{3}{2}e^t + \frac{1}{2}e^{-t}$, $\quad y = \frac{3}{2}e^t - \frac{1}{2}e^{-t} - t$.

24. $x = \frac{19}{2}e^{t/3} - \frac{17}{2}e^{-t/3} - 6t$, $\quad y = -19e^{t/3} + \frac{17}{2}e^{-t/3} + 9t + 9 + \frac{1}{2}e^t$.

25. $x = \dfrac{7}{4}e^{2t} - 4te^{2t} + t^2 + \dfrac{3t}{2} + \dfrac{1}{4}$, $\quad y = -\dfrac{9}{4}e^{2t} - 4te^{2t} + \dfrac{t}{2} + \dfrac{1}{4}$.

26. $x = \frac{4}{3}e^{-t} - \frac{10}{3}e^{t/2} + 2e^t + 2t$,

$\quad y = -\dfrac{4}{3}e^{-t} - \dfrac{5}{3}e^{t/2} + 2e^t + \dfrac{t^2}{2} - t + 1$.

LESSON 32. Linearization of First Order Systems.

Let the x and y components of the velocity of a particle be given by

$$(32.1) \qquad \frac{dx}{dt} = f(x,y), \qquad \frac{dy}{dt} = g(x,y),$$

where the functions $f(x,y)$ and $g(x,y)$ are assumed to have continuous partial derivatives for all (x,y). This system of two first order equations may, in many problems, be difficult to solve. What we do in such cases is to find the first three terms of the Taylor series expansion of $f(x,y)$ and $g(x,y)$, and then eliminate the parameter t between them. The Taylor series expansions of these functions about the point $(0,0)$ are, see Lesson 38,

$$(32.11) \quad \frac{dx}{dt} = f(x,y) = a_0 + a_1 x + a_2 y + a_3 x^2 + a_4 xy + a_5 y^2 + \cdots,$$

$$\frac{dy}{dt} = g(x,y) = b_0 + b_1 x + b_2 y + b_3 x^2 + b_4 xy + b_5 y^2 + \cdots.$$

Using only the first three terms in each expansion, and dividing the second by the first, we obtain

$$(32.12) \qquad \frac{dy}{dx} = \frac{b_0 + b_1 x + b_2 y}{a_0 + a_1 x + a_2 y}.$$

The solution of this equation will give what is called a *first approximation* to the path of the particle.* The equation (32.12) is now of the type with linear coefficients discussed in Lesson 8. As commented in 8.26, the resulting implicit solution of (32.12) will frequently be so complicated as to be of little practical use in determining the path of the particle. We shall now show how important information as to the character of the particle's motion can often be obtained more easily from the differential equation (32.12) itself than from its complicated implicit solution.

In Lesson 8B, we showed how by a translation of axes, it is always possible to transform a differential equation of the form (32.12) with linear coefficients (assuming these represent nonparallel lines) to one of the form $(a_1 \bar{x} + b_1 \bar{y})\, dx + (a_2 \bar{x} + b_2 \bar{y})\, dy = 0$ with homogeneous coefficients. All we need do is to translate the origin to the point of intersection of the two nonparallel lines represented by these linear coefficients. Hence we shall assume in what follows that this shift of origin has been made. We therefore need consider only equations of the form

$$(32.13) \quad \frac{dy}{dx} = \frac{a_1 x + b_1 y}{a_2 x + b_2 y}, \qquad \frac{a_1}{a_2} \neq \frac{b_1}{b_2} \quad \text{and} \quad a_2 x + b_2 y \neq 0.$$

---

*It is possible for the first approximation to be very wide of the actual path of the particle because the omitted terms are relevant and important.

To save writing, we define

$$(32.14) \qquad\qquad D = a_1 b_2 - b_1 a_2.$$

Note that D is the determinant

$$\begin{vmatrix} a_1 & b_1 \\ a_2 & b_2 \end{vmatrix}$$

whose elements are the coefficients of x and y in (32.13).

We consider separately each of the cases resulting from different values of a_1, a_2, b_1, b_2. In all cases, dy/dx is meaningless at the origin $(0,0)$. Hence, in all cases, no solution will lie on this point. However, every other point in the plane is an ordinary point, including points in a neighborhood of the origin, no matter how small. Hence, by Definition 5.41, *the origin is, in all cases, a singular point.* For convenience, however, we shall say *"solutions through the origin." This expression is to be interpreted to mean solutions through points in a neighborhood of the origin, but not through the origin itself.*

Case 1. $a_1 = b_2 = 0$ and $b_1 = a_2$. In this case (32.13) simplifies to

$$(32.15) \qquad\qquad \frac{dy}{dx} = \frac{y}{x}, \quad x \neq 0,$$

whose solutions are

$$(32.16) \qquad\qquad y = cx, \quad x \neq 0.$$

It is a family of straight lines through the origin $(0,0)$. Thus for this special case, the family of integral curves of (32.13) is easily drawn.

Case 2. $b_1 = -a_2$. In this case, (32.13) becomes

$$(32.17) \qquad\qquad \frac{dy}{dx} = \frac{a_1 x - a_2 y}{a_2 x + b_2 y}.$$

Its solution by the method of Lesson 7B, is

$$(32.18) \qquad\qquad a_1 x^2 - 2a_2 xy - b_2 y^2 = c.$$

Since $b_1 = -a_2$, we obtain by (32.14),

$$(32.181) \qquad\qquad D = a_1 b_2 + a_2{}^2,$$

which is also the discriminant of the quadratic equation (32.18).

From analytic geometry we know the following.

1. If $D < 0$ and $c \neq 0$, (32.18) is a family of ellipses with center at the origin; if $c = 0$, only the point $(0,0)$ satisfies the equation. But the

point $(0,0)$ is excluded since, by (32.17), dy/dx is meaningless there. Hence the family of integral curves of (32.13) will resemble those shown in Fig. 32.19(a).

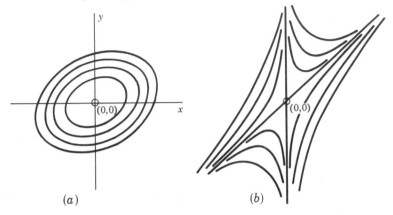

Figure 32.19

2. If $D > 0$ and $c \neq 0$, (32.18) is a family of hyperbolas with center at the origin; if $c = 0$, (32.18) degenerates into two straight lines through the origin which are the asymptotes of the family of hyperbolas. Hence the family of integral curves of (32.13) will resemble those shown in Fig. 32.19(b).

For this special Case 2, therefore, the family of integral curves of (32.13) can also be easily drawn.

Case 3. *General case, $a_1b_2 \neq a_2b_1$, with no other restrictions placed on* a_1, a_2, b_1, b_2. Let $P(x,y)$ be a point on an integral curve of (32.13),

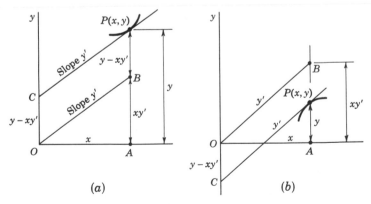

Figure 32.2

Fig. 32.2(a). The slope of the tangent line CP to the curve is therefore y'. Assume this line does not go through the origin O. Draw OB parallel to CP. Its slope is also y'. From Fig. 32.2(a), we see that (tan $\sphericalangle AOB = y' = AB/x$; therefore $AB = xy'$)

$$(32.21) \qquad\qquad OC = AP - AB = y - xy'.$$

If therefore OC lies above the x axis, then $y - xy' > 0$. Further we know from the calculus that if $y'' > 0$ at a point P of a curve, then the curve in a neighborhood of P is concave upward and a tangent at P lies below the curve. We have thus proved that if at a point $P(x,y)$ on an integral curve, $y'' > 0$ and $y - xy' > 0$, then the tangent line at P separates curve and origin in a neighborhood of P.

A similar conclusion can be proved if $y'' < 0$ and $y - xy' < 0$, see Fig. 32.2(b).

If, however, y'' and $y - xy'$ have different signs at $P(x,y)$, then origin and an integral curve near P will lie on the same side of the tangent line at P. For example, in Fig. 32.2(a), draw at P an integral curve which is concave downward so that $y'' < 0$.

By differentiation of our original differential equation (32.13), we obtain after simplification,

$$(32.22) \qquad\qquad y'' = \frac{a_1b_2 - a_2b_1}{(a_2x + b_2y)^2}\,(y - xy').$$

By (32.13), the denominator of the fraction in (32.22) is not equal to zero. Hence it is always positive since it is squared. Its numerator by (32.14) is D. If therefore

$(32.221) \qquad D > 0$, y'' and $(y - xy')$ have the same sign,

$\qquad\qquad\qquad D < 0$, y'' and $(y - xy')$ have opposite signs.

Because of (32.221) and the remarks after (32.21), we can now assert the following.

Comment 32.23. If $y - xy' \neq 0$, i.e., if the tangent line at a point P of an integral curve of (32.13) does not go through the origin, and if D of (32.14) > 0, then curve and origin, in a neighborhood of P, are separated by the tangent at P; if $D < 0$, then curve and origin, in a neighborhood of P, lie on the same side of the tangent at P.

If the tangent at a point P of an integral curve of (32.13) goes through the origin, then it is evident from Fig. 32.2(a) or (b), that $OC = 0$ and therefore by (32.21), that $y - xy' = 0$. This means that if (32.13) has rectilinear solutions (i.e., straight line solutions) through the origin, these solutions must satisfy the equation $y - xy' = 0$. For example, the straight line solutions through the origin which we found in Case 1, namely $y = cx$, satisfy this requirement. For then $y' = c$ and $y - xy' = cx - xc = 0$.

The question we now ask is the following. Are there, for the *general case*, rectilinear solutions of (32.13) which go through the origin? The answer is: all those solutions which satisfy the equation $y - xy' = 0$. From (32.13) we find, after simplification,

$$(32.231) \qquad y - xy' = \frac{b_2 y^2 + (a_2 - b_1)xy - a_1 x^2}{a_2 x + b_2 y}.$$

Hence the locus of those points (x,y) which will make the numerator of (32.231) zero will make $y - xy' = 0$. Such loci will then be rectilinear solutions of (32.13). To find them, we set

$$(32.24) \qquad b_2 y^2 + (a_2 - b_1)xy - a_1 x^2 = 0.$$

Let us call Δ the discriminant of (32.24). Then

$$(32.25) \qquad \begin{aligned} \Delta &= (a_2 - b_1)^2 + 4a_1 b_2 = a_2{}^2 - 2a_2 b_1 + b_1{}^2 + 4a_1 b_2 \\ &= a_2{}^2 + 2a_2 b_1 + b_1{}^2 - 4a_2 b_1 + 4a_1 b_2 \\ &= (a_2 + b_1)^2 + 4(a_1 b_2 - a_2 b_1). \end{aligned}$$

By (32.14), we can write (32.25) as

$$(32.26) \qquad \Delta = (a_2 + b_1)^2 + 4D.$$

From algebra, we know that if

$(32.27) \qquad$ (a) $\Delta > 0$, then (32.24) will have two real, distinct factors, say $(ax + by)(cx + dy) = 0$,

$\qquad$ (b) $\Delta = 0$, then (32.24) will have one real repeated factor,

$\qquad$ (c) $\Delta < 0$, then (32.24) has only imaginary factors.

In case (a), there will thus be two rectilinear integral curves of (32.13) through the origin. These two lines will separate the plane into four regions in which all other integral curves of (32.13), for which $y - xy' \neq 0$, will lie.

In case (b), there will be only one rectilinear integral curve through the origin. This line will divide the plane into two regions in which all other integral curves of (32.13), for which $y - xy' \neq 0$, will lie.

In case (c), there will be no real rectilinear solution of (32.13).

By (32.26), we note further that when $D > 0$, Δ is also > 0. But when $D < 0$, Δ may take on any of the three values in (32.27). We have then four possibilities to consider in the general Case 3: one when $D > 0$; three when $D < 0$.

Case 3-1. *D of (32.14) > 0, [and therefore Δ of (32.26) > 0]*. [Note. If $D > 0$ and also $b_1 = -a_2$, then the special Case 2 applies; see 2 after (32.181) of that case.] For this Case 3-1 we know from the remarks above, that (32.13) has two rectilinear solutions, each of which is a factor of

(32.24). All points on these two lines will make $y - xy' = 0$. All other integral curves for which $y - xy' \neq 0$, will lie in the four regions made by the two rectilinear solutions. And since $D > 0$, we know by Comment 32.23, that a tangent line drawn at any point of an integral curve will separate curve and origin. The integral curves will thus have the general appearance of those shown in Fig. 32.281. The rectilinear solutions are asymptotes of the integral curves. The origin itself, however, is a singular point. The curves, while not hyperbolas, will have their general appearance.

Example 32.28. Discuss, without solving the equation, the character of the solutions of

(a) $$y' = \frac{4x - y}{2x + y}.$$

Solution. Comparing (a) with (32.13), we see that $a_1 = 4, b_1 = -1$, $a_2 = 2, b_2 = 1$. By (32.14), $D = 4 + 2 = 6 > 0$. Hence this Case 3-1 applies. There should therefore be, and as we shall now show, there are two factors of (32.24) each of which is a rectilinear solution of (a). Substituting in (32.24) the above values of a_1, a_2, b_1, b_2, we obtain

(b) $$y^2 + 3xy - 4x^2 = 0, \qquad (y + 4x)(y - x) = 0.$$

Hence the two rectilinear solutions of (a) are

(c) $$y + 4x = 0, \qquad y - x = 0.$$

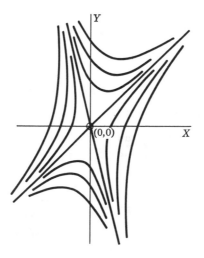

Figure 32.281

[Verify that the functions defined in (c) are solutions of (a).] They are shown in heavy lines in Fig. 32.281. Note that a tangent drawn at a point of a nonrectilinear integral curve separates curve and origin.

Case 3-2. *D of (32.14) < 0.* [Note. If $D < 0$ and also $b_1 = -a_2$, then the special Case 2 applies; see 1 after (32.181) of that case.] For this Case 3–2, as remarked earlier, Δ of (32.26) may take on any of the three values in (32.27). We shall therefore need to consider each of these three possibilities separately. Before doing so, however, it will be necessary for us to have additional information. We therefore digress momentarily in order to obtain this information.

The family of solutions of (32.13) has slope y'. Let y_1' be the slope of an isogonal trajectory family which cuts this given family in a positive $\sphericalangle \alpha$, measured counter clockwise from y' to y_1', Fig. 32.3.

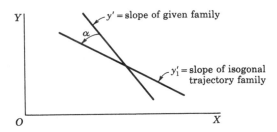

Figure 32.3

By a formula in analytic geometry, therefore,

$$(32.31) \qquad \tan \alpha = \frac{y_1' - y'}{1 + y_1'y'}.$$

Solving (32.31) for y', we obtain

$$(32.32) \qquad y' = \frac{y_1' - \tan \alpha}{1 + (\tan \alpha)y_1'}.$$

Replacing y' in (32.32) by its value as given in (32.13), there results

$$(32.33) \qquad \frac{y_1' - \tan \alpha}{1 + (\tan \alpha)y_1'} = \frac{a_1 x + b_1 y}{a_2 x + b_2 y}.$$

The solution of (32.33) for y_1' is

$$(32.34) \qquad y_1' = \frac{(a_2 \tan \alpha + a_1)x + (b_2 \tan \alpha + b_1)y}{(a_2 - a_1 \tan \alpha)x + (b_2 - b_1 \tan \alpha)y},$$

which has the same form as (32.13). We have thus proved that the differential equation of a family of isogonal trajectories, making an $\sphericalangle \alpha$ with the given family of solutions of (32.13), has the same form as (32.13).

The determinant D_I, whose elements are the coefficients of x and y in (32.34), is

(32.35)

$$D_I = (a_2 \tan \alpha + a_1)(b_2 - b_1 \tan \alpha) - (b_2 \tan \alpha + b_1)(a_2 - a_1 \tan \alpha)$$
$$= (a_1 b_2 - a_2 b_1)(\tan^2 \alpha + 1) = D(\tan^2 \alpha + 1) = D \sec^2 \alpha,$$

where D is given by (32.14). Since $\sec^2 \alpha > 0$, we see from (32.35) that D_I and D have the same sign.

By Case 2, the family of solutions of (32.34), which remember is an isogonal family of trajectories of the family of solutions of (32.13), will be ellipses or hyperbolas with center at the origin if

(32.36) $$b_2 \tan \alpha + b_1 = -(a_2 - a_1 \tan \alpha),$$

i.e., if

(32.37) $$\tan \alpha = \frac{a_2 + b_1}{a_1 - b_2}.$$

They are, by Case 2, ellipses if $D_I < 0$ or equivalently, as noted above, if $D < 0$.

Comment 32.371. We have thus proved that if $D < 0$ and α is an angle which satisfies (32.37), then the family of isogonal trajectories which cuts the family of integral curves of (32.13) in this angle α, measured from the integral family counterclockwise to the isogonal trajectory family, is a family of ellipses with center at the origin. If $\alpha \neq 0$, each integral curve will therefore approach the origin in one direction and recede infinitely in the other direction. The origin itself is a singular point. If $\alpha = 0$, then the isogonal trajectory family coincides with the integral family and the integral family is thus also a family of ellipses as in Case 2. [Note that if $\alpha = 0$, then $\tan \alpha = 0$ and by (32.37), $a_2 = -b_1$ as in Case 2.]

We are now ready to examine each of the three possibilities under Case 3-2.

Case 3-2(a). D of $(32.14) < 0$ and Δ of $(32.26) > 0$. In this case, there will be, by (32.27)(a), two rectilinear solutions of (32.13) through the origin. These lines will separate the plane into four regions. For a non-rectilinear integral curve of (32.13), we now know the following facts.

1. It cannot cross either of the rectilinear solutions.
2. By Comment 32.23, origin and integral curve, in a neighborhood of a point P on it, lie on the same side of the tangent at P.
3. If α is an angle which satisfies (32.37), then the family of isogonal trajectories cutting the integral family of (32.13) in this angle α, measured

from the integral family counterclockwise to the isogonal trajectory family, is a family of ellipses, with center at the origin, satisfying (32.34).

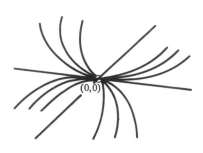

In addition, it has been proved that

4. An integral curve will approach the origin in a direction tangent to one of the rectilinear solutions; see Exercise 32,12.
5. A radius vector to a point P moving along an integral curve away from the origin will approach coincidence with the second rectilinear solution.

Figure 32.38

The graphs of the integral curves will therefore resemble those shown in Fig. 32.38. The two heavy lines are the rectilinear solutions.

Example 32.39.   Discuss, without solving, the character of the solutions of

(a) $$y' = \frac{2x + 4y}{x - y}.$$

Solution.   Comparing (a) with (32.13), we see that $a_1 = 2$, $b_1 = 4$, $a_2 = 1$, $b_2 = -1$. By (32.14), $D = -2 - 4 = -6 < 0$ and by (32.26), $\Delta = (1 + 4)^2 + 4D = 25 - 24 = 1 > 0$. Hence this Case 3–2(a) applies. Since $\Delta > 0$, there should be, and as we shall now show, there are two rectilinear solutions of (a), each of which is a factor of (32.24). Substituting in (32.24) the values of a_1, b_1, a_2, b_2, we obtain

(b) $$y^2 + 3xy + 2x^2 = 0, \qquad (y + 2x)(y + x) = 0.$$

Each of the lines $y + 2x = 0$ and $y + x = 0$ are solutions of (a). (Verify it.) They are shown in heavy lines in Fig. 32.391. All other integral curves must lie inside the four regions formed by these two lines.

From (32.37), we find $\tan \alpha = 5/3$, so that α is approximately 59°. With this value of $\tan \alpha$, (32.34) becomes

$$y_1' = \frac{(\frac{5}{3} + 2)x + (-\frac{5}{3} + 4)y}{(1 - \frac{10}{3})x + (-1 - \frac{20}{3})y} = \frac{\frac{11}{3}x + \frac{7}{3}y}{-\frac{7}{3}x - \frac{23}{3}y} = \frac{11x + 7y}{-7x - 23y},$$

which has, as it should, the same form as (32.17). Hence by (32.18), its solution is

(c) $$11x^2 + 14xy + 23y^2 = c.$$

A rotation of the axes through an angle of approximately 65°18′ will eliminate the xy term and (c) will become, approximately,

(d)
$$26\bar{x}^2 + 8\bar{y}^2 = c,$$

which represents, as it also should, a family of ellipses. Some of these ellipses are shown in Fig. 32.391. Each nonrectilinear integral curve of

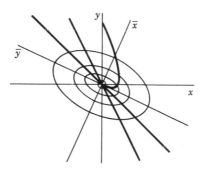

Figure 32.391

(a) will cut this family at the constant angle $\alpha = 59°$, measured counterclockwise from the integral curve to the ellipse. We have shown a part of one such integral curve in the figure.

Case 3-2(b). *D of (32.14) < 0 and Δ of (32.26) = 0.* In this case, there will be, by 32.27(b), only one rectilinear integral curve through the origin. This line will separate the plane into two regions. The same comments we made in the previous lesson in regard to a nonrectilinear integral curve will apply here. These are:

1. It cannot cross the rectilinear solution.
2. By Comment 32.23, the origin and each integral curve in a neighborhood of a point P on it, lie on the same side of the tangent at P.
3. If α is an angle which satisfies (32.37), then the family of isogonal trajectories cutting the given family of solutions of (32.13) in this angle α, measured from the integral family counterclockwise to the trajectory family, is a family of ellipses, with center at the origin, satisfying (32.34).
4. An integral curve will approach the origin in a direction tangent to the rectilinear solution; see Exercise 32,13.
5. A radius vector drawn to a point P moving along the integral curve away from the origin will approach coincidence with the rectilinear solution.

The graphs of the integral curves will thus resemble those shown in Fig. 32.4.

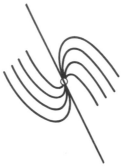

Figure 32.4

Example 32.41. Discuss, without solving, the character of the solutions of

(a) $$y' = \frac{x + 3y}{x - y}.$$

Solution. Comparing (a) with (32.13), we see that $a_1 = 1$, $b_1 = 3$, $a_2 = 1$, $b_2 = -1$. By (32.14), $D = -1 - 3 < 0$. By (32.26) $\Delta = (4)^2 + 4(-4) = 0$. Hence this Case 3-2(b) applies. There should be, and as we shall now show, there is, only one rectilinear solution of (a). Substituting in (32.24) the values of a_1, a_2, b_1, b_2 give above, we obtain

(b) $$y^2 + 2xy + x^2 = 0, \qquad (y + x)^2 = 0.$$

Hence the line $y = -x$ is the only rectilinear solution of (a). We have left it to you as an exercise, to find the trajectory family of ellipses as we did in Example 32.39, and the angle α of (32.37) which each integral curve makes with these ellipses. Draw some of these ellipses and an integral curve. Its graph should resemble a curve of Fig. 32.4.

Case 3-2(c). *D* of *(32.14)* < 0, Δ of *(32.26)* < 0. In this case by (32.27)(c), there are no rectilinear solutions of (32.13). And if α is an angle satisfying (32.37), then by Comment 32.371, the family of isogonal trajectories which cuts the given family of solutions of (32.13) in this angle α measured from the integral family counterclockwise to the trajectory family, is a family of ellipses with center at the origin. If therefore $\alpha = 0$, the family of solutions of the original equation (32.13) coincides with the trajectory family and thus are also ellipses with center at the origin. Hence if $\alpha = 0$, every integral curve encircles the origin.

If, however, $\alpha \neq 0$, then each integral curve of (32.13) will cut the isogonal trajectory family of ellipses, which are integral curves of (32.34), in this constant angle α. We shall now prove for this Case 3-2(c), that when $\alpha \neq 0$, an integral curve will also encircle the origin.

The equation of the family of straight lines through the origin is

(a) $$y = mx; \qquad y' = m.$$

Replacing m in the second equation of (a) by its value y/x obtained from the first equation, we have

(b) $$y' = m = \frac{y}{x}.$$

Let θ be the angle of intersection of a line of (a) and an integral curve of (32.13), measured from the integral curve to the line, Fig. 32.42. Then, by a formula in analytic geometry,

(c) $$\tan \theta = \frac{\dfrac{y}{x} - \dfrac{a_1 x + b_1 y}{a_2 x + b_2 y}}{1 + \dfrac{y}{x} \dfrac{a_1 x + b_1 y}{a_2 x + b_2 y}} = \frac{b_2 \left(\dfrac{y}{x}\right)^2 + (a_2 - b_1) \dfrac{y}{x} - a_1}{b_1 \left(\dfrac{y}{x}\right)^2 + (a_1 + b_2) \dfrac{y}{x} + a_2}.$$

By (b) we can write (c) as

(d) $$\tan \theta = \frac{b_2 m^2 + (a_2 - b_1)m - a_1}{b_1 m^2 + (a_1 + b_2)m + a_2}.$$

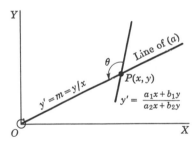

Figure 32.42

An integral curve of (32.13), therefore, cuts each line of (a) at the angle θ given by (d). [For each line of (a), there is a distinct m and therefore by (d), a definite angle θ.] The numerator of (d) is

(e) $$b_2 m^2 + (a_2 - b_1)m - a_1.$$

Its discriminant is

(f) $$(a_2 - b_1)^2 + 4 a_1 b_2,$$

which is exactly the same as the discriminant Δ of (32.24) [see (32.25)]. Since by our assumption for this Case 3-2(c), $\Delta < 0$, it follows that the expression in (f) is less than zero. Hence, as noted in (32.27)(c), there are no real values of m which will make (e) zero. Since (e) is the numerator of (d), this means that the angle θ in (d) is never equal to zero. An integral curve of (32.13), since it cuts every straight line through the origin at an angle $\theta \neq 0$, must therefore encircle the origin.

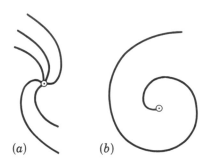

(a)   (b)

Figure 32.43

Hence in both cases, $\alpha = 0$ and $\alpha \neq 0$, an integral curve of (32.13) will go completely around the origin. In the first case, it will enclose the origin in the form of an ellipse as in Fig. 32.19(a); in the second case in the form of spirals as shown in Fig. 32.43(a) and (b). In both cases the origin is a singular point.

Example 32.44. Discuss, without solving, the character of the solutions of

(a) $$\frac{dy}{dx} = -\frac{2x - y + 1}{x + y}.$$

(NOTE. This equation was solved in Example 8.25. Its implicit solution can be found there.)

Solution. A translation of the axes so that the origin is at $(-\frac{1}{3}, \frac{1}{3})$, see Example 8.25, transforms (a) to the form

(b) $$\frac{dy}{dx} = \frac{-2x + y}{x + y}.$$

Comparing (b) with (32.13), we see that $a_1 = -2$, $b_1 = 1$, $a_2 = 1$, $b_2 = 1$. You can verify that D of (32.14) < 0 and Δ of (32.26) < 0. Hence this Case 3–2(c) applies. By (32.37)

(c) $$\tan \alpha = -\tfrac{2}{3}, \ \alpha = -33°41'.$$

With this value of $\tan \alpha$, (32.34) becomes

(d) $$y_1' = \frac{(-\frac{2}{3} - 2)x + (-\frac{2}{3} + 1)y}{(1 - \frac{4}{3})x + (1 + \frac{2}{3})y} = \frac{-8x + y}{-x + 5y}.$$

Its solution by (32.18) is

(e) $$8x^2 - 2xy + 5y^2 = c.$$

You can verify that (e) has, as it should, only imaginary factors. A rotation of the axes through an angle of approximately $73°10'$ will transform (e) to

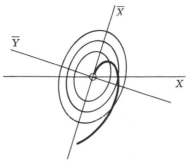

(f) $4.7\bar{x}^2 + 8.3\bar{y}^2 = c,$

which represents a family of ellipses. A few of these ellipses are shown in Fig. 32.45. Each integral curve must cut each ellipse at the constant angle $-33°41'$, measured from integral curve to ellipse. We have also shown in this figure part of one such integral curve.

Figure 32.45

Comment 32.46. **Position of the Particle as a Function of Time.** In the above discussions of the path of a particle, we eliminated the important element of time. Hence, although we have shown the general appearance of the path, in the absence of an equation connecting the position of the particle with time, we cannot know where the particle is at any instant. To obtain this information, we must change equation (32.13) back to the system from which it came, namely

(32.47) $\dfrac{dx}{dt} = a_2 x + b_2 y, \qquad \dfrac{dy}{dt} = a_1 x + b_1 y,$

and then solve it by the methods of Lesson 31E. In operator notation, we can write (32.47) as

(32.48) $(D - a_2)x - b_2 y = 0,$
                $-a_1 x + (D - b_1 y) = 0.$

Following the method outlined in Lesson 31E, we *retain the first equation* and change the second by multiplying the first by $(D - b_1)/b_2$ and adding it to the second. There results the equivalent triangular system

(32.49) $(D - a_2)x - b_2 y = 0,$
                $\left[\dfrac{(D - b_1)(D - a_2)}{b_2} - a_1 \right] x = 0.$

The second equation in (32.49) reduces to

(32.5) $[D^2 - (a_2 + b_1)D + a_2 b_1 - a_1 b_2]x = 0.$

The characteristic equation of (32.5) is

(32.51) $m^2 - (a_2 + b_1)m + a_2 b_1 - a_1 b_2 = 0,$

whose roots are

$$(32.52) \qquad m_1 = \frac{a_2 + b_1 + \sqrt{(a_2 - b_1)^2 + 4a_1b_2}}{2},$$

$$m_2 = \frac{a_2 + b_1 - \sqrt{(a_2 - b_1)^2 + 4a_1b_2}}{2}$$

Hence the solution of (32.5) is

$$(32.53) \qquad x = c_1 e^{m_1 t} + c_2 e^{m_2 t},$$

where m_1 and m_2 have the values given in (32.52). Note that the character of the roots m_1 and m_2, and therefore the solution x, depends on whether the discriminant $(a_2 - b_1)^2 + 4a_1b_2$ is positive, zero, or negative.

Substituting this value of x in the first equation in (32.48), we obtain

$$(32.54) \qquad b_2 y = (D - a_2)(c_1 e^{m_1 t} + c_2 e^{m_2 t})$$
$$= m_1 c_1 e^{m_1 t} + m_2 c_2 e^{m_2 t} - a_2 c_1 e^{m_1 t} - a_2 c_2 e^{m_2 t}$$
$$= c_1(m_1 - a_2)e^{m_1 t} + c_2(m_2 - a_2)e^{m_2 t}.$$

Hence

$$(32.55) \qquad y = \frac{1}{b_2} [c_1(m_1 - a_2)e^{m_1 t} + c_2(m_2 - a_2)e^{m_2 t}].$$

By Theorem 31.33, the pair of functions $x(t)$, $y(t)$ as given in (32.53) and (32.55), have the correct number of two arbitrary constants. The value of the constants c_1 and c_2 will depend on the position of the particle at $t = 0$. The solutions $x(t)$, $y(t)$ of (32.53) and (32.55) will then give the position of the particle at later times t, see, for example, Exercise 32,11.

EXERCISE 32

Discuss without solving the character of the solution of each of the following differential equations. (First determine to which of the different cases each belongs.) Draw rectilinear solutions where these exist. Draw a rough graph of an integral curve.

1. $\dfrac{dy}{dx} = \dfrac{y + 1}{x - 1}$.

2. $\dfrac{dy}{dx} = \dfrac{x - 3y}{3x + y}$.

3. $\dfrac{dy}{dx} = \dfrac{x - 3y}{3x - 10y}$.

4. $\dfrac{dy}{dx} = \dfrac{4x + y}{-2x + y}$.

5. $\dfrac{dy}{dx} = \dfrac{2x + y}{3x + y}$.

6. $\dfrac{dy}{dx} = \dfrac{3x + y}{2x + y}$.

7. $\dfrac{dy}{dx} = \dfrac{-2x + y}{x + y}$.

8. $\dfrac{dy}{dx} = \dfrac{x + y}{3x - y}$.

9. $\dfrac{dy}{dx} = \dfrac{-x - y}{x + 2y}$.

10. $\dfrac{dy}{dx} = \dfrac{x - y}{x + y}$.

11. Assume the original system from which the equation of problem 4 above came is

$$\frac{dx}{dt} = -2x + y, \qquad \frac{dy}{dx} = 4x + y,$$

and that the initial conditions are $x(0) = 1$, $y(0) = 2$. Find the parametric equations of the path and the values of x and y when $t = 1$.

12. Prove statement 4 of Case 3–2(a). *Hint.* Follow the proof given in Case 3–2(c).

13. Prove statement 4 of Case 3–2(b). *Hint.* Follow the proof given in Case 3–2(c).

ANSWERS 32

1. Case 1.   2. Case 2; family of hyperbolas.
3. Case 2; family of ellipses.   4. Case 3–1; $y = 4x$, $y = -x$.
5. Case 3–2(a); $y = 0.73x$, $y = -2.73x$.
6. Case 3–1; $y = 1.3x$, $y = -2.3x$.
7. Case 3–2(c).   8. Case 3–2(b); $y = x$.   9. Case 2; family of ellipses.
10. Case 2; family of hyperbolas.
11. $x(t) = \frac{2}{5}e^{-3t} + \frac{3}{5}e^{2t}$, $y(t) = -\frac{2}{5}e^{-3t} + \frac{12}{5}e^{2t}$; $x = 4.45$, $y = 17.71$.

Chapter 8

Problems Giving Rise to Systems of Equations. Special Types of Second Order Linear and Nonlinear Equations Solvable by Reduction to Systems

LESSON 33. **Mechanical, Biological, Electrical Problems Giving Rise to Systems of Equations.**

LESSON 33A. **A Mechanical Problem—Coupled Springs.** Two masses m_1 and m_2 rest on a frictionless, horizontal table. They are connected to each other and to two fixed supports by three unstretched springs S_1, S_2, S_3 with respective spring constants k_1, k_2, k_3, Fig. 33.1(a).

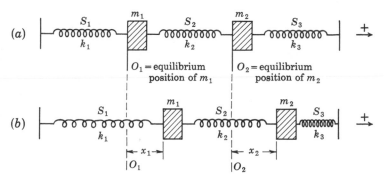

Figure 33.1

[For the meaning of a spring constant, see (28.61).] The masses can be displaced from their equilibrium position by holding, for example, m_1, moving m_2 to the right and then releasing both. Figures 33.1(a) and (b) show the position of the two masses before and after a displacement. We wish to find the equation of motion of each mass.

Denote by O_1 the position of m_1 and by O_2 the position of m_2, when the system is at rest. Let x_1 be the displacement of m_1 from its equilibrium or rest position O_1, and x_2 the displacement of m_2 from its equilibrium position O_2, at time t. Hence when m_1 is at x_1 and m_2 is at x_2, Fig. 33.1(b), spring S_1 has been elongated a distance x_1 from its equilibrium position and spring S_2 has been elongated a distance $x_2 - x_1$ from its equilibrium position. For example, if $x_1 = 1$ foot and $x_2 = \frac{3}{2}$ feet, then spring S_2 has been elongated only $\frac{1}{2}$ foot, i.e., a distance $x_2 - x_1$. Therefore the restoring forces acting on m_1, with the positive direction to the right, are two in number.

1. A spring force due to S_1 acting to the left. By Hooke's law, which you will find in Example A of Lesson 28C, this force is $-k_1x_1$.
2. A spring force due to S_2 acting to the right. By Hooke's law, this force is $k_2(x_2 - x_1)$. (Remember the spring S_2 wants to restore itself to its original size and will therefore pull on m_1 to the right and on m_2 to the left.)

Hence the differential equation of motion of m_1 is

$$(33.11) \qquad m_1 \frac{d^2 x_1}{dt^2} = -k_1 x_1 + k_2(x_2 - x_1).$$

A similar analysis shows that the restoring forces acting on m_2 are two in number.

1. A spring force due to S_2 acting to the left. By Hooke's law, this force is $-k_2(x_2 - x_1)$.
2. A spring force due to S_3 acting to the left. Remember S_3 has been squeezed a distance x_2 and it wants to restore itself to its original size. Therefore, by Hooke's law, this force is $-k_3x_2$.

Hence the differential equation of motion of m_2 is

$$(33.111) \qquad m_2 \frac{d^2 x_2}{dt^2} = -k_2(x_2 - x_1) - k_3 x_2.$$

The solution of the system of linear differential equations (33.11) and (33.111) will give the equations of motion $x_1(t)$ and $x_2(t)$ of the **coupled springs** diagrammed in Fig. 33.1.

If, in addition to the spring forces, a forcing function $F_1(t)$ is attached to m_1 and a forcing function $F_2(t)$ is attached to m_2 as shown in Fig. 33.121, then the pair of equations (33.11) and (33.111) becomes

$$(33.12) \qquad m_1 \frac{d^2 x_1}{dt^2} + k_1 x_1 + k_2(x_1 - x_2) = F_1(t),$$

$$m_2 \frac{d^2 x_2}{dt^2} + k_3 x_2 + k_2(x_2 - x_1) = F_2(t).$$

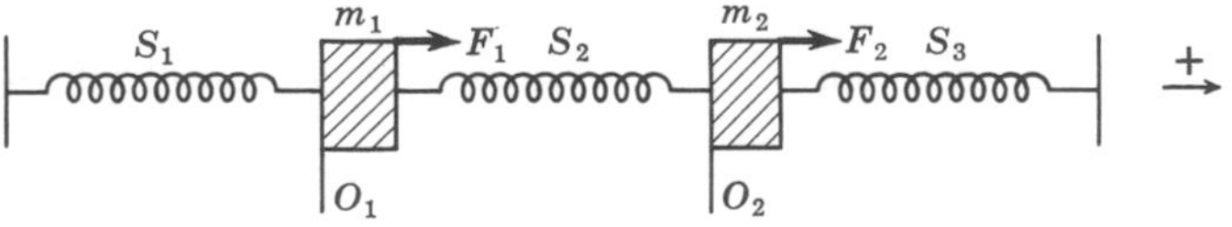

Figure 33.121

And if we assume that because of friction or because the system is immersed in a medium, there is a damping force $r_1 \dfrac{dx_1}{dt}$ acting on m_1 and a damping force $r_2 \dfrac{dx_2}{dt}$ acting on m_2 then the system (33.12) becomes

(33.122) $m_1 \dfrac{d^2 x_1}{dt^2} + r_1 \dfrac{dx_1}{dt} + k_1 x_1 + k_2(x_1 - x_2) = F_1(t),$

$m_2 \dfrac{d^2 x_2}{dt^2} + r_2 \dfrac{dx_2}{dt} + k_3 x_2 + k_2(x_2 - x_1) = F_2(t).$

If all spring constants have the same value k, then (33.122) simplifies to the linear system

(33.13) $m_1 \dfrac{d^2 x_1}{dt^2} + r_1 \dfrac{dx_1}{dt} + 2k x_1 - k x_2 = F_1(t),$

$m_2 \dfrac{d^2 x_2}{dt^2} + r_2 \dfrac{dx_2}{dt} + 2k x_2 - k x_1 = F_2(t).$

In operator notation, we can write (33.13) as

(33.14) $(m_1 D^2 + r_1 D + 2k)x_1 - k x_2 = F_1(t),$

$-k x_1 + (m_2 D^2 + r_2 D + 2k)x_2 = F_2(t).$

We can eliminate x_2 by multiplying the first equation of (33.14) by $(m_2 D^2 + r_2 D + 2k)$, the second by k and adding the two. We thus obtain

(33.15) $\{m_1 m_2 D^4 + (m_1 r_2 + m_2 r_1)D^3 + [r_1 r_2 + 2k(m_1 + m_2)]D^2$
$ + [2k(r_1 + r_2)]D + 3k^2\}x_1 = (m_2 D^2 + r_2 D + 2k)F_1(t) + kF_2(t),$

an equation which is now linear in x_1. Its general solution can be obtained by the usual methods outlined in previous lessons. When x_1 has been determined, the general solution for x_2 can be obtained from the first equation in (33.14). Note that the general solution of (33.15) for $x_1(t)$ will contain four constants. Substituting this value of $x_1(t)$ in the first equation of (33.14) and solving it for $x_2(t)$ will add no new constants. By Theorem 31.33, the correct number of constants required in the general solution of the system (33.15) is four. Hence the pair of functions $x_1(t)$, $x_2(t)$ will contain the correct number of constants.

Comment 33.16. The characteristic equation of the left side of (33.15) set equal to zero is of the fourth degree. Hence, as we remarked in Comment 18.26, these roots may not be easy to find, especially if they are imaginary. Remember these roots are needed in order to write the complementary function y_c of the general solution of (33.15). If however $m_1 = m_2$, i.e., if both masses are the same, and $r_1 = r_2$, then the elimination of x_2 in (33.14) will result in

$$(33.17) \quad [(m_1D^2 + r_1D + 2k)^2 - k^2]x_1$$
$$= (m_1D^2 + r_1D + 2k)F_1(t) + kF_2(t).$$

But now if we set the left side equal to zero, its characteristic equation is $(m_1m^2 + r_1m + 2k)^2 - k^2 = 0$ which factors into

$$(33.18) \qquad (m_1m^2 + r_1m + k)(m_1m^2 + r_1m + 3k) = 0.$$

Its roots can thus be readily obtained by setting each factor equal to zero.

Comment 33.19. If the forcing functions $F_1(t)$ and $F_2(t)$ are identically zero and if $r_1 \geqq 0$, $r_2 \geqq 0$, then as we showed in Lessons 28A and 29A, the motion of the system is stable. If the forcing functions are not identically zero, then the motion is much more complicated and a breakdown of the system may occur if the phenomenon of resonance is present, see Exercise 33A, 1(b). (If you have forgotten the meaning of resonance, reread Lessons 28D and 29B.)

EXERCISE 33A

1. (a) Solve the system (33.13) if $m_1 = m_2 = m$; $r_1 = r_2 = 0$; $F_1(t) = F_2(t) = 0$.
   (b) The system in (a) has two frequencies. These are called **normal frequencies.** (In simple harmonic motion, a system had only one natural frequency.) Find the normal frequencies of (a). **Resonance** occurs if a forcing function, impressed on the system, has either of these normal frequencies.
   (c) Find the solution if at $t = 0$, m_2 is held fixed and m_1 is moved to the right A units and released. *Hint.* Initial conditions are $t = 0$, $x_1 = A$, $x_2 = 0$, $dx_1/dt = 0$, $dx_2/dt = 0$.
2. (a) Solve the system given by equations (33.11) and (33.111) if $m_1 = 1$, $m_2 = 2$, $k_1 = 1$, $k_2 = 2$, $k_3 = 3$.
   (b) What are its normal frequencies?

3. If you write the system of problem 1(a) in operator notation, you will obtain

(a) $\qquad (mD^2 + 2k)x_1 - kx_2 = 0, \qquad -kx_1 + (mD^2 + 2k)x_2 = 0.$

Since each differential equation of the system is linear, it is reasonable to assume that its solutions will have the form

(b) $$x_1 = c_1 e^{\omega t}, \quad x_2 = c_2 e^{\omega t}.$$

Show that the substitution of (b) in (a) yields the system

(c) $$(m\omega^2 + 2k)c_1 - kc_2 = 0, \quad -kc_1 + (m\omega^2 + 2k)c_2 = 0.$$

Note that (c) is obtainable from (a) by replacing D by ω, x_1 by c_1, x_2 by c_2. The system (c) will have nontrivial solutions for c_1 and c_2 (i.e., solutions other than $c_1 = 0$, $c_2 = 0$) only if the determinant

(d) $$\begin{vmatrix} m\omega^2 + 2k & -k \\ -k & m\omega^2 + 2k \end{vmatrix} = 0,$$

see Theorem 63.42. Expand the determinant and solve the resulting equation, called the **characteristic equation**, for ω. Note that the imaginary part of the solution for ω is the same as the normal frequencies found in 1(b). You will have thus proved that the *imaginary part* of the solution of ω, where ω is obtained by the process outlined above, will give the normal frequencies of the system.

4. Use the method outlined in problem 3 to find the normal frequencies of the system of problem 2. What is the determinant? Compare with answer to 2(b).

5. (a) Show that when $x_1 = x_2$, the system of coupled springs solved in problem 1(a) has only the one normal frequency $\sqrt{k/m}$ rad/unit time; that when $x_1 = -x_2$, it has only the one normal frequency $\sqrt{3k/m}$ rad/unit time.

   (b) Show that if the initial conditions are so chosen that $c_3 = c_4 = 0$ in the answer to problem 1(a), then $x_1 = x_2$ and the system will vibrate with only the one normal frequency $\sqrt{k/m}$; if chosen so that $c_1 = c_2 = 0$, then $x_1 = -x_2$ and the system will vibrate with only the one normal frequency $\sqrt{3k/m}$.

   (c) Let $y_1 = x_1 + x_2$ and $y_2 = x_1 - x_2$. Show that the solution of 1(a) becomes respectively

   $$y_1 = C_1 \sin \sqrt{k/m}\, t + C_2 \cos \sqrt{k/m}\, t,$$

   $$y_2 = C_3 \sin \sqrt{3k/m}\, t + C_4 \cos \sqrt{3k/m}\, t.$$

   Note that each equation has only one normal frequency. The first equation (where $y_1 = x_1 + x_2$) corresponds to a motion where the two masses move in coordination to the left and to the right with the same amplitude; the second equation (where $y_1 = x_1 - x_2$) corresponds to a motion where the two masses move in opposite directions with the same amplitude. These new coordinates y_1 and y_2 which replace x_1 and x_2 are called **normal coordinates.** They are useful in simplifying certain problems in the theory of vibrations.

6. (a) Solve the system (33.13) if $m_1 = m_2 = 2$, $k = 32$, $r_1 = r_2 = 0$ and $F_1(t) = F_2(t) = 4 \cos t$.

   (b) In (a), replace $F_1(t)$ and $F_2(t)$ by $4 \cos \omega t$. What values of ω will produce resonance?

7. Two masses m_1 and m_2 are connected by two springs S_1 and S_2 with respective spring constants k_1 and k_2 as shown in Fig. 33.191. After the system is brought to rest, the masses are displaced from their equilibrium position.

Assume no damping forces and no forcing functions.

(a) Find the differential equations of motion of the system.

(b) Find the equations of motion of the system if $k_1 = k_2 = k$ and $m_1 = m_2 = m$.

(c) What are the normal frequencies of the system?

8. Find the normal frequencies of the system of problem 7 by means of the method outlined in problem 3. What is the determinant? Compare with answer to 7(c).

9. (a) Use the method outlined in problem 3 to find the normal frequencies of the system shown in Fig. 33.191, i.e., with $m_1 \neq m_2$ and $k_1 \neq k_2$. See problem 7.

(b) Write the normal frequencies if $m_1 = m_2 = m$.

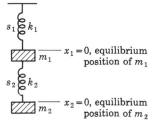

Figure 33.191

10. Solve the system of Fig. 33.191 if $k_1 = 1$, $k_2 = 2$, $m_1 = 1$, $m_2 = 2$, and if, after the system is brought to rest, the mass m_2 is held fixed, the mass m_1 is displaced A units downward and both masses released. *Hint.* At $t = 0$, $x_1 = A$, $x_2 = 0$, $dx_1/dt = 0$, $dx_2/dt = 0$.

11. In the system illustrated in Fig. 33.191, assume that the mass m_1 is damped by a shock absorber whose coefficient of resistance is r and that a forcing function $F = F_0 \sin \omega t$ is attached to m_1. (a) Show that the differential equations of motion are

$$(m_1 D^2 + rD + k_1 + k_2)x_1 - k_2 x_2 = F_0 \sin \omega t,$$
$$-k_2 x_1 + (m_2 D^2 + k_2)x_2 = 0.$$

Hint. See answer to problem 7(a). Modify it to include the damping factor and the forcing function. (b) By eliminating x_2, show that

$$[(m_1 D^2 + rD + k_1 + k_2)(m_2 D^2 + k_2) - k_2{}^2]x_1 = F_0(k_2 - m_2 \omega^2) \sin \omega t.$$

Remark. The general solution of this last equation for x_1 will, as usual, be the sum of a complementary function x_{1c} and a particular solution x_{1p}. It can be shown that the motion due to the x_{1c} part of the solution is a transient one and subsides with time. If, therefore, k_2 and m_2 are so chosen that $k_2 = m_2 \omega^2$, then the right side of the equation is zero. Hence $x_{1p} = 0$, and the motion due to the x_{1p} part of the solution is also zero. This means that if $k_2 = m_2 \omega^2$, the mass m_1 will in time approach a position x_1 and remain there. Hence, the mass m_1 will not vibrate. This result has practical use when it is necessary that the vibrations of an instrument be negligible.

ANSWERS 33A

1. (a) $x_1 = c_1 \sin \sqrt{k/m}\, t + c_2 \cos \sqrt{k/m}\, t + c_3 \sin \sqrt{3k/m}\, t$
$+ c_4 \cos \sqrt{3k/m}\, t$,

$x_2 = c_1 \sin \sqrt{k/m}\, t + c_2 \cos \sqrt{k/m}\, t - c_3 \sin \sqrt{3k/m}\, t$
$- c_4 \cos \sqrt{3k/m}\, t$.

(b) $\sqrt{k/m}$ rad/unit time; $\sqrt{3k/m}$ rad/unit time.

(c) $x_1 = \dfrac{A}{2}(\cos\sqrt{k/m}\,t + \cos\sqrt{3k/m}\,t)$,

$\quad x_2 = \dfrac{A}{2}(\cos\sqrt{k/m}\,t - \cos\sqrt{3k/m}\,t)$.

2. (a) $x_1 = c_1\cos(\sqrt{1.31}\,t + \delta_1) + c_2\cos(\sqrt{4.19}\,t + \delta_2)$,

$\quad x_2 = \frac{1}{2}[1.69c_1\cos(\sqrt{1.31}\,t + \delta_1) - 1.19c_2\cos(\sqrt{4.19}\,t + \delta_2)]$.

(b) $\sqrt{1.31}$ rad/unit time, $\sqrt{4.19}$ rad/unit time.

4. $\begin{vmatrix} \omega^2 + 3 & -2 \\ -2 & 2\omega^2 + 5 \end{vmatrix}$.

6. (a) $x_1 = c_1\sin 4t + c_2\cos 4t + c_3\sin 4\sqrt{3}\,t + c_4\cos 4\sqrt{3}\,t + \frac{2}{15}\cos t$,

$\quad x_2 = c_1\sin 4t + c_2\cos 4t - c_3\sin 4\sqrt{3}\,t - c_4\cos 4\sqrt{3}\,t + \frac{2}{15}\cos t$.

(b) $\omega = 4$ rad/unit time and $\omega = 4\sqrt{3}$ rad/unit time.

7. (a) $m_1\dfrac{d^2x_1}{dt^2} = -k_1x_1 + k_2(x_2 - x_1)$,

$\quad m_2\dfrac{d^2x_2}{dt^2} = -k_2(x_2 - x_1)$.

(b) $x_1 = c_1\sin\sqrt{0.38k/m}\,t + c_2\cos\sqrt{0.38k/m}\,t + c_3\sin\sqrt{2.62k/m}\,t$
$\quad + c_4\cos\sqrt{2.62k/m}\,t$.

$\quad x_2 = 1.62c_1\sin\sqrt{0.38k/m}\,t + 1.62c_2\cos\sqrt{0.38k/m}\,t$
$\quad\quad - 0.62c_3\sin\sqrt{2.62k/m}\,t - 0.62c_4\sin\sqrt{2.62k/m}\,t$.

(c) $\omega = \sqrt{0.38k/m}$ and $\omega = \sqrt{2.62k/m}$.

8. $\begin{vmatrix} m\omega^2 + 2k & -k \\ -k & m\omega^2 + k \end{vmatrix}$.

9. (a) $\begin{vmatrix} m_1\omega^2 + k_1 + k_2 & -k_2 \\ -k_2 & m_2\omega^2 + k_2 \end{vmatrix}$.

$\quad \omega = \sqrt{-\dfrac{1}{2}\left(\dfrac{k_1 + k_2}{m_1} + \dfrac{k_2}{m_2}\right) \pm \dfrac{1}{2}\sqrt{\left(\dfrac{k_1 + k_2}{m_1} + \dfrac{k_2}{m_2}\right)^2 - \dfrac{4k_1k_2}{m_1m_2}}}$.

(b) $\omega = \sqrt{-\dfrac{k_1 - 2k_2 \pm \sqrt{k_1{}^2 + 4k_2{}^2}}{2m}}$.

10. $x_1 = \dfrac{A}{6}\left[(3 - \sqrt{3})\cos\left(\dfrac{\sqrt{6} - \sqrt{2}}{2}\right)t + (3 + \sqrt{3})\cos\left(\dfrac{\sqrt{6} + \sqrt{2}}{2}\right)t\right]$,

$x_2 = \dfrac{A\sqrt{3}}{6}\left[\cos\left(\dfrac{\sqrt{6} - \sqrt{2}}{2}\right)t - \cos\left(\dfrac{\sqrt{6} + \sqrt{2}}{2}\right)t\right]$, or

$x_1 = A(0.211\cos 0.518t + 0.789\cos 1.932t)$,

$x_2 = A(0.289\cos 0.518t - 0.289\cos 1.932t)$.

LESSON 33B. A Biological Problem. There is a constant struggle for survival among different species. One species survives by eating members of another species; a second by preventing itself from being eaten. Certain birds live on fish. When, for example, there is an abundant supply of fish, the population of this bird species flourishes and grows. When these birds become too numerous and consume too many fish, thus reducing the fish population, then their own bird population begins to diminish. As the number of birds decreases, the fish population increases. And as the fish population increases, the bird population starts increasing, etc., in an endless cycle of periodic increases and decreases in the respective populations of the two species.

Problems concerning the rise and fall of the populations of interacting species have been studied extensively by biologists and mathematicians. The theoretical mathematical results which we shall develop in this lesson agree fairly accurately with actual population trends of certain interacting species.[*]

We shall consider the problem of determining the *populations of two interacting species:* a parasitic species P which hatches its eggs in a host species H. Unfortunately for the H species, the deposit of an egg in a member of H causes this member's death. For each 100 of population:

Let H_b denote the number of births per year of the H species,
$\quad H_d$ denote the number of deaths per year of the H species, if no P
$\qquad$ species were present, i.e., H_d denotes H's natural death rate,
$\quad P_d$ denote the number of natural deaths per year of the P species.

We now make the following assumptions:

1. That H_b, H_d, and P_d are constants. Hence if at time t, x is the population of the H species, y the population of the P species, then in time Δt

(a) $H_b \dfrac{x}{100} \Delta t$ is the approximate number of births of the host H,

(b) $-H_d \dfrac{x}{100} \Delta t$ is the approximate number of natural deaths of the
$\qquad\qquad$ host H,

(c) $-P_d \dfrac{y}{100} \Delta t$ is the approximate number of natural deaths of the
$\qquad\qquad$ parasite P.

2. That the number of eggs per year deposited by the P species resulting in the death of the H species is proportional to the probability that the members of the two species meet. Since this probability depends on the product xy of the two populations, we assume that in time Δt,

(d) $$-kxy\Delta t$$

---

[*]V. A. Kostitzin, *Mathematical Biology*, London, G. G. Harrap & Co. Ltd. (1939); Vito Volterra, *Les Associations Biologiques au Point de Vue Mathématique*, Paris, Hermann & Co. (1935).

is the approximate number of deaths of the host H due to the presence of the parasite P, where k is a proportionality constant. Hence, also, since each such death of H implies the laying of an egg by P,

(e) $kxy\Delta t$

is the approximate number of births of P.

Therefore in time Δt, the approximate changes in the H and P species are respectively

(33.192) $\Delta x = H_b \dfrac{x}{100}\Delta t - H_d \dfrac{x}{100}\Delta t - kxy\Delta t,$

$\Delta y = kxy\Delta t - P_d \dfrac{y}{100}\Delta t.$

In the first equation of (33.192), let $h = (H_b - H_d)/100$, assumed >0, i.e., h is the net natural annual percentage increase of the population of H. In the second equation of (33.192), let $p = P_d/100$, i.e., p is the natural annual percentage death rate of P. Then (33.192) simplifies to

(33.2) $\Delta x = hx\Delta t - kxy\Delta t,$

$\Delta y = kxy\Delta t - py\Delta t.$

Note that we have used the words "approximate changes" in connection with the system (33.2). It is conceivable that in some time intervals

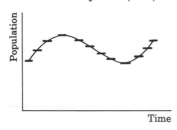

Figure 33.21

Δt, no births or deaths take place; in other time intervals, one or more births or deaths occur. A graph of the population of a species as a function of time is therefore not a continuous curve. It will show sudden jumps or drops as in Fig. 33.21. However, if the population is large, it is convenient to assume that there is a continuous curve which will approximate the population trend. Further, one need not stretch one's credulity too far to believe that the system of equations

(33.22) $\dfrac{dx}{dt} = hx - kxy,$

$\dfrac{dy}{dt} = kxy - py,$

which result, when in (33.2), we divide each equation by Δt and let $\Delta t \to 0$, will give good approximations for large populations and long periods of time.

The equations in (33.22) form a system of two nonlinear first-order equations which cannot be solved by the methods thus far discussed. In later chapters we shall describe other methods by which an approximate solution of the system (33.22) may be obtained, if the numerical values of h, k, and p are known, see Exercise 39,5.

It is possible, however, to solve the system (33.22) as an implicit function of x, if we divide the second equation by the first. Making this division, we obtain

$$(33.23) \quad \frac{dy}{dx} = \frac{y(kx - p)}{x(h - ky)}, \quad \left(\frac{h}{y} - k\right) dy = \left(k - \frac{p}{x}\right) dx.$$

Integration of the second equation in (33.23) gives

$$(33.24) \qquad \log y^h - ky = kx - \log x^p - \log c,$$

from which we obtain

$$(33.25) \quad \log cy^h x^p = ky + kx, \quad cy^h x^p = e^{ky} e^{kx}, \quad cy^h e^{-ky} = x^{-p} e^{kx}.$$

If at $t = 0$, the population of the host species is x_0 and the population of the parasitic species is y_0, then the last equation in (33.25) becomes

$$(33.26) \qquad c = x_0^{-p} y_0^{-h} e^{k(x_0 + y_0)},$$

which is the value of the arbitrary constant in the solution.

As we have frequently mentioned, implicit solutions are usually of little value. Here it would be difficult indeed to plot a graph of (33.25) showing the population y as a function of x.

Although we cannot at this stage solve the system (33.22) for x and y as functions of time, nevertheless certain important useful information can be obtained from it. Suppose, for example, that at a certain moment of time, the populations of our two interacting species are

$$(33.27) \qquad\qquad x = \frac{p}{k}, \qquad y = \frac{h}{k}.$$

Then by (33.22) $dy/dt = 0$ and $dx/dt = 0$. Since the rate of change of the two populations at that moment is zero, there are no increases or decreases. Let us call these populations the *equilibrium populations* of x and y respectively. Let

$$(33.28) \qquad\qquad x = \frac{p}{k} + X, \qquad y = \frac{h}{k} + Y,$$

where X and Y represent the respective deviations of the equilibrium populations of x and y. Substituting (33.28) in (33.22), we obtain

$$(33.29) \qquad \frac{d}{dt}\left(\frac{p}{k} + X\right) = h\left(\frac{p}{k} + X\right) - k\left(\frac{p}{k} + X\right)\left(\frac{h}{k} + Y\right),$$

$$\frac{d}{dt}\left(\frac{h}{k} + Y\right) = k\left(\frac{p}{k} + X\right)\left(\frac{h}{k} + Y\right) - p\left(\frac{h}{k} + Y\right).$$

These equations simplify to

$$(33.291) \qquad \frac{dX}{dt} = -pY - kXY, \qquad \frac{dY}{dt} = hX + kXY.$$

If we assume that the deviations X and Y are small, we may, without serious error, discard the XY terms in (33.291). These equations then become

$$(33.292) \qquad \frac{dX}{dt} = -pY, \qquad \frac{dY}{dt} = hX.$$

But now we recognize that the system of equations is exactly the same as those discussed in Lesson 32 under the heading "Linearization of First Order Systems." Dividing the second equation in (33.292) by the first, we have

$$(33.293) \qquad \frac{dY}{dX} = -\frac{hX}{pY}.$$

Its solution, obtained by solving it directly, or by recognizing that this differential equation comes under Case 2 of Lesson 32 where its solution is given in (32.18), is

$$(33.294) \qquad hX^2 + pY^2 = c.$$

Both p and h are positive quantities. (Remember h is the net natural percentage increase of the host population, assumed > 0, and p is the natural percentage death rate of the parasite population.) If therefore $h = p$, (33.294) is the equation of a family of circles; if $h \neq p$, it is a family of ellipses. We have thus shown that if the populations of two interacting species remain stationary over long periods of time, as they do in many cases, then the graph of their population deviations from equilibrium is a circle or an ellipse, Fig. 33.295.

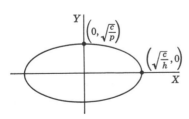

Figure 33.295

Note how the mathematical equation (33.294) supports our opening remarks. When the deviation X of the host species equals $\sqrt{c/h}$ and is thus largest, the deviation Y of the parasite species is zero, i.e., the parasite population is normal. But now there are more hosts available in which the parasites can lay their eggs. The population of the parasite species thus begins to increase, while that of the host species decreases. But as the parasites get too numerous, they kill more hosts, until the parasite deviation is maximum and the host deviation is zero. When this point is reached, there are relatively so many parasites around, that they begin killing off some of the host's normal population. As the host's population decreases below normal, the parasitic population, because there are now insufficient hosts, must also decrease, until finally a point is reached when its deviation is zero and the host's deviation has its largest negative value $-\sqrt{c/h}$. The host's population is now too low to support a normal parasitic population. The parasitic deviation therefore begins to decrease below zero, thus allowing the host's deviation to increase, etc., in an endless cycle of periodic increases and decreases.

LESSON 33C. An Electrical Problem. More Complex Circuits.
In Lesson 30, we discussed a simple electric circuit in which the charged particles moved in one path. A circuit in which the charged particles can move in different paths will, as we shall show below, give rise to a system of differential equations. To enable us to write differential equations with correct signs, we shall adopt the following convention. If a branch of the circuit contains a source of energy E, we shall indicate by an arrowhead that the direction of the current is from the side which we label $+$, to the side which we label $-$. In every other branch of the circuit, we put in arrowheads arbitrarily. It is only essential that the arrowheads remain fixed in a problem.

Example 33.3. Set up a system of differential equations for the current in the electric circuit shown in Fig. 33.31.

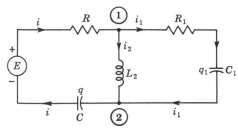

Figure 33.31

Solution. By **Kirchhoff's first law**, the algebraic sum of the currents at any junction point is zero. If we call the current whose arrowhead is directed toward a junction point positive and the current whose arrowhead is directed away from a junction point negative, then at the junction point marked 1 in Fig. 33.31,

(a) $\qquad\qquad i - i_1 - i_2 = 0, \qquad i = i_1 + i_2.$

And at junction point 2,

(b) $$i_1 + i_2 - i = 0, \qquad i = i_1 + i_2.$$

Applying Kirchhoff's second law to the left circuit, we obtain

(c) $$Ri + L_2 \frac{di_2}{dt} + \frac{1}{C} q = E(t).$$

If no source of energy is present in a closed circuit, then his second law states that the algebraic sum of the voltage drops is zero. Applying this law to the right circuit, we obtain

(d) $$R_1 i_1 + \frac{1}{C_1} q_1 - L_2 \frac{di_2}{dt} = 0.$$

Differentiating (c) and (d) and replacing $i(= dq/dt)$, wherever it appears by its value in (a), we have

(e) $$L_2 \frac{d^2 i_2}{dt^2} + R\left(\frac{di_1}{dt} + \frac{di_2}{dt}\right) + \frac{1}{C}(i_1 + i_2) = \frac{dE}{dt},$$

$$L_2 \frac{d^2 i_2}{dt} - R_1 \frac{di_1}{dt} - \frac{1}{C_1} i_1 = 0.$$

The linear system (e) can be solved for i_1 and i_2 by the method of Lesson 31D or E. The current i will then be the sum of these two currents.

Comment 33.311. We obtained two equations in the above problem by using the left and right circuits shown in Fig. 33.31. It is possible to obtain additional equations by using other circuits. Each of these, however, will not be independent of the two obtained in (c) and (d). For example, we could, if we had wished, taken the circuit going completely around the border and have thus obtained

$$Ri + R_1 i_1 + \frac{1}{C_1} q_1 + \frac{1}{C} q = E.$$

This equation, however, is the sum of (c) and (d).

Example 33.32. Set up a system of differential equations for the currents in the electric circuit shown in Fig. 33.33.

Solution. By Kirchhoff's first law, the algebraic sum of the currents at any junction point is zero. Hence at the junction points correspondingly marked in Fig. 33.33, we have

(a)
$$\begin{aligned}
&①: \quad i_1 - i_2 - i_4 = 0, \qquad i_4 = i_1 - i_2; \\
&②: \quad i_2 - i_5 - i_3 = 0, \qquad i_5 = i_2 - i_3; \\
&③: \quad i_3 + i_6 - i_1 = 0, \qquad i_6 = i_1 - i_3.
\end{aligned}$$

At junction point ④, $i_4 + i_5 - i_6 = 0$. And we see from the sums of ①, ②, and ③, that this condition is satisfied.

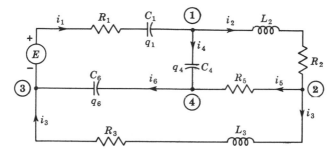

Figure 33.33

Applying Kirchhoff's second law to the upper left, the upper right and the lower circuit respectively, we obtain the linear system

(b)
$$R_1 i_1 + \frac{1}{C_1} q_1 + \frac{1}{C_4} q_4 + \frac{1}{C_6} q_6 = E,$$

$$L_2 \frac{di_2}{dt} + R_2 i_2 + R_5 i_5 - \frac{1}{C_4} q_4 = 0,$$

$$-\frac{1}{C_6} q_6 - R_5 i_5 + L_3 \frac{di_3}{dt} + R_3 i_3 = 0.$$

Differentiating (b) and replacing d/dt by D and dq_n/dt by i_n, there results

(c)
$$R_1 D i_1 + \frac{1}{C_1} i_1 + \frac{1}{C_4} i_4 + \frac{1}{C_6} i_6 = \frac{dE}{dt},$$

$$L_2 D^2 i_2 + R_2 D i_2 + R_5 D i_5 - \frac{1}{C_4} i_4 = 0,$$

$$L_3 D^2 i_3 - \frac{1}{C_6} i_6 - R_5 D i_5 + R_3 D i_3 = 0.$$

Substituting in (c) the values of i_4, i_5, i_6 as given in (a), we obtain

(d)
$$\left(R_1 D + \frac{1}{C_1} + \frac{1}{C_4} + \frac{1}{C_6} \right) i_1 - \frac{1}{C_4} i_2 - \frac{1}{C_6} i_3 = \frac{dE}{dt},$$

$$-\frac{1}{C_4} i_1 + \left(L_2 D^2 + R_2 D + R_5 D + \frac{1}{C_4} \right) i_2 - R_5 D i_3 = 0,$$

$$-\frac{1}{C_6} i_1 - R_5 D i_2 + \left(L_3 D^2 + R_5 D + R_3 D + \frac{1}{C_6} \right) i_3 = 0.$$

The system (d) can be solved for i_1, i_2, and i_3 by the method of Lesson 31G.

Comment 33.34. As remarked in Comment 33.311, other equations in addition to those found in (b) can be obtained by considering different circuits. These, however, will not be independent of those in (b). For example, we could have taken the circuit going completely around the border. This equation, however, can be obtained from (b). Verify it. Can you find other possible circuits?

EXERCISE 33C

1. A transformer consists of a primary coil and a secondary coil. The primary coil has an emf of $E(t)$ volts, a resistance of R_1 ohms, and an inductance of L_1 henrys. The secondary coil has a resistance of R_2 ohms and an inductance of L_2 henrys. Let i_1 and i_2 be the respective currents in each coil. It has been shown that the differential equations for the currents in the coils are

(33.4)
$$L_1 \frac{di_1}{dt} + R_1 i_1 + M \frac{di_2}{dt} = E(t),$$

$$L_2 \frac{di_2}{dt} + R_2 i_2 + M \frac{di_1}{dt} = 0,$$

where M is a constant called the **mutual inductance**.
(a) Solve the system (33.4) if $E(t) = 0$ and $M^2 < L_1 L_2$.
(b) Show that the exponents a and b, as given in the answer section, are negative quantities and therefore that both i_1 and i_2 approach zero as $t \to \infty$.
(c) Solve the system (33.4) if $E(t)$ is a constant E_0 and $M^2 < L_1 L_2$. Show that as $t \to 0$, the current i_1 of the primary coil approaches the steady state current E_0/R_1 and the current i_2 of the secondary coil approaches zero as $t \to \infty$.
2. (a) Set up a system of differential equations for the current in the electric circuit shown in Fig. 33.41.

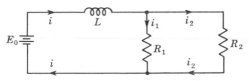

Figure 33.41

(b) Solve the system with initial conditions $t = 0$, $i = 0$ when the switch is closed, and show that the circuit is equivalent to one in which the resistors R_1, R_2 are replaced by one resistor R, whose coefficient of resistance is given by

$$\frac{1}{R} = \frac{1}{R_1} + \frac{1}{R_2}, \qquad R = \frac{R_1 R_2}{R_1 + R_2}.$$

Remark. If more resistors $R_3, \cdots, R_n$ were inserted in the circuit of Fig. 33.41, parallel to R_1 and R_2, it follows from (b) that this circuit is equivalent to one in which the resistors $R_1, R_2, \cdots, R_n$ are replaced by one resistor R whose coefficient of resistance is given by

$$\frac{1}{R} = \frac{1}{R_1} + \frac{1}{R_2} + \cdots + \frac{1}{R_n}.$$

3. (a) Set up a system of differential equations for the charges in the electric circuit shown in Fig. 33.42.
   (b) Solve the system for q and i as functions of time with initial conditions $t = 0$, $q_1 = q_2 = 0$ when the switch is closed, and show that the circuit is equivalent to one in which the capacitors C_1, C_2 are replaced by one capacitor whose capacitance is given by $C = C_1 + C_2$.

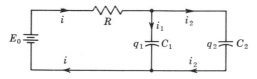

Figure 33.42

Remark. If more capacitors $C_3, \cdots, C_n$ were inserted in the circuit of Fig. 33.42, parallel to C_1 and C_2, it follows from (b) that this circuit is equivalent to one in which the capacitors $C_1, C_2, \cdots, C_n$ are replaced by one capacitor whose capacitance is given by $C = C_1 + C_2 + \cdots + C_n$.

In the problems below, set up a system of differential equations for the current in each of the circuits diagrammed. Solve as many as you wish.

4. See Fig. 33.43. At $t = 0$, $i = 0$, $q = q_1 = q_2 = 0$.

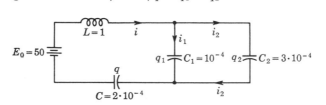

Figure 33.43

5. In problem 4, replace $E_0 = 50$ by $E(t) = 100 \sin 60t$.
6. See Fig. 33.44. At $t = 0$, $i = 0$, $q = q_2 = 0$.

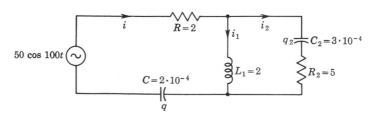

Figure 33.44

7. See Fig. 33.45. At $t = 0$, $i = 0$, $q_2 = 0$.

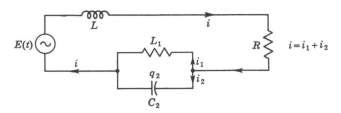

Figure 33.45

8. See Fig. 33.46. Initial conditions are $t = 0$, $i = 0$, $q_1 = q_2 = 0$.

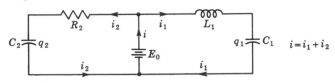

Figure 33.46

9. See Fig. 33.47.

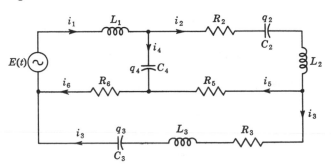

Figure 33.47

10. See Fig. 33.48.

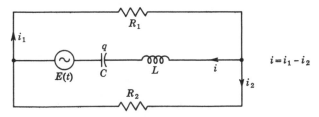

Figure 33.48

11. See Fig. 33.49.

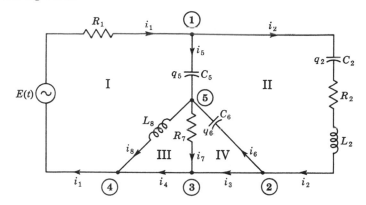

Figure 33.49

ANSWERS 33C

1. (a) $i_1 = c_1e^{at} + c_2e^{bt}$, where

$$a, b = \frac{-(L_2R_1 + L_1R_2) \pm \sqrt{(L_2R_1 + L_1R_2)^2 - 4R_1R_2(L_1L_2 - M^2)}}{2(L_1L_2 - M^2)}$$

$$= \frac{-(L_2R_1 + L_1R_2) \pm \sqrt{(L_2R_1 - L_1R_2)^2 + 4M^2R_1R_2}}{2(L_1L_2 - M^2)},$$

$$i_2 = \frac{1}{MR_2}[(L_1L_2 - M^2)(ac_1e^{at} + bc_2e^{bt}) + L_2R_1(c_1e^{at} + c_2e^{bt})].$$

(c) $i_1 = c_1e^{at} + c_2e^{bt} + (E_0/R_1)$; i_2 is the same as in 1(a).

2. (a) $(LD + R_1)i_1 + LDi_2 = E_0$, $\quad R_1i_1 - R_2i_2 = 0$.

(b) $i = i_1 + i_2 = \dfrac{E_0(R_1 + R_2)}{R_1R_2}[1 - e^{-R_1R_2t/L(R_1+R_2)}]$

$\equiv \dfrac{E_0}{R}(1 - e^{-Rt/L})$.

3. (a) $\left(RD + \dfrac{1}{C_1}\right)q_1 + RDq_2 = E$, $\quad \dfrac{1}{C_1}q_1 - \dfrac{1}{C_2}q_2 = 0$.

(b) $q_1 = C_1E_0(1 - e^{-t/R(C_1+C_2)})$, $\quad q_2 = C_2E_0(1 - e^{-t/R(C_1+C_2)})$,

$i = i_1 + i_2 = \dfrac{dq_1}{dt} + \dfrac{dq_2}{dt} = \dfrac{E_0}{R}e^{-t/R(C_1+C_2)} \equiv \dfrac{E_0}{R}e^{-t/RC}$.

4. $Di + \dfrac{q_1}{C_1} + \dfrac{q}{C} = E_0$, $\quad \dfrac{1}{C_2}q_2 - \dfrac{1}{C_1}q_1 = 0$, $i = i_1 + i_2$.

$$\begin{cases} D^2i_1 + D^2i_2 + \dfrac{1}{C_1}i_1 + \dfrac{1}{C}(i_1 + i_2) = 0, \\ \dfrac{1}{C_2}i_2 - \dfrac{1}{C_1}i_1 = 0. \end{cases}$$

5. In 4 replace E_0 by 100 sin 60t. Therefore $\dfrac{d}{dt} E(t) = 6000 \cos 60t$.

6. $\left(RD + LD^2 + \dfrac{1}{C} \right) i_1 + \left(RD + \dfrac{1}{C} \right) i_2 = -5000 \sin 100t,$

$L_1 D^2 i_1 - \left(R_2 D + \dfrac{1}{C_2} \right) i_2 = 0.$

7. $Ri + L \dfrac{di}{dt} + L_1 \dfrac{di_1}{dt} = E(t),$

$Ri + L \dfrac{di}{dt} + \dfrac{1}{C_2} q_2 = E(t),$

$-L_1 \dfrac{di_1}{dt} + \dfrac{1}{C_2} q_2 = 0.$

The three equations are not independent. The third can be obtained by subtracting the second from the first. To solve the system, it is easier to use the first and third equations.

8. $L_1 \dfrac{di_1}{dt} + \dfrac{1}{C_1} q_1 = E_0, \quad R_2 i_2 + \dfrac{1}{C_2} q_2 = E_0.$

$L_1 \dfrac{d^2 q_1}{dt^2} + \dfrac{1}{C_1} q_1 = E_0, \quad R_2 \dfrac{dq_2}{dt} + \dfrac{1}{C_2} q_2 = E_0.$

$q_1 = C_1 E_0 \left(1 - \cos \dfrac{1}{\sqrt{C_1 L_1}} t \right) \dfrac{-E_0 \sqrt{C_1 L_1}}{R_2} \sin \dfrac{t}{\sqrt{C_1 L_1}},$

$i_1 = \dfrac{C_1 E_0}{\sqrt{C_1 L_1}} \sin \dfrac{1}{\sqrt{C_1 L_1}} t - \dfrac{E_0}{R_2} \cos \dfrac{t}{\sqrt{C_1 L_1}}.$

$q_2 = C_2 E_0 (1 - e^{-t/R_2 C_2}), \quad i_2 = \dfrac{E_0}{R_2} e^{-t/R_2 C_2}.$

9. $L_1 \dfrac{di_1}{dt} + \dfrac{1}{C_4} q_4 + R_6 i_6 = E(t),$

$R_2 i_2 + \dfrac{1}{C_2} q_2 + L_2 \dfrac{di_2}{dt} + R_5 i_5 - \dfrac{1}{C_4} q_4 = 0,$

$L_3 \dfrac{di_3}{dt} + R_3 i_3 + \dfrac{1}{C_3} q_3 - R_6 i_6 - R_5 i_5 = 0.$

Substitute $i_n = \dfrac{dq_n}{dt}$ and $i_4 = i_1 - i_2, \ i_5 = i_2 - i_3, \ i_6 = i_1 - i_3.$

10. $R_1 \dfrac{di_1}{dt} + L \dfrac{d^2 i_1}{dt^2} - L \dfrac{d^2 i_2}{dt^2} + \dfrac{1}{C} (i_1 - i_2) = \dfrac{d}{dt} E(t), \quad R_1 i_1 + R_2 i_2 = 0.$

11. Junction point ①: $i_1 - i_5 - i_2 = 0, \quad i_5 = i_1 - i_2.$
Junction point ②: $i_2 - i_6 - i_3 = 0, \quad i_6 = i_2 - i_3.$
Junction point ③: $i_3 + i_7 - i_4 = 0, \quad i_7 = i_4 - i_3.$
Junction point ④: $i_4 + i_8 - i_1 = 0, \quad i_8 = i_1 - i_4.$

At junction point ⑤: $i_5 + i_6 - i_7 - i_8 = 0$, which satisfies the four equations above. Set up, in the usual manner, the differential equation for each circuit marked I, II, III, and IV in diagram. For I, it is

$$R_1 i_1 + \frac{1}{C_5} q_5 + L_8 \frac{di_8}{dt} = E(t).$$

Although there are other choices of circuits, their equations will not be independent of the chosen four.

LESSON 34. Plane Motions Giving Rise to Systems of Equations.

In Lesson 16, we discussed the motion of a particle constrained to move along a straight line. In this lesson, we consider the motion of a particle free to move in a plane.

LESSON 34A. Derivation of Velocity and Acceleration Formulas.
If a particle is free to move in a plane, then a change in the direction of its velocity will be equally as important as a change in the magnitude of its velocity. As mentioned in Lesson 16C, quantities in which both magnitude and direction play a role are called vectors.

Since Newton's second law of motion is also applicable to particles which move in a plane, we have, by (16.1),

(34.1) $$F = ma = m\frac{dv}{dt} \equiv m\dot{v}.$$

Remark. A dot over a variable means its derivative with respect to time; two dots over the variable means its second derivative with respect to time. Hence $\dot{x} \equiv dx/dt$, $\ddot{x} \equiv d^2x/dt^2$, $\dot{v} \equiv dv/dt$, etc.

The mass m is not a vector quantity. We therefore see, by (34.1), that the acceleration of a particle acted on by a force, not only has magnitude F/m, but also has the *same direction* as F.

In Fig. 34.11, the vector **F** represents the magnitude and direction of a force F. It is convenient to break up this vector force into two components, one, F_x, to represent that part of the force which accelerates the particle in the x direction, the other, F_y, to represent that part of the force which accelerates the particle in the

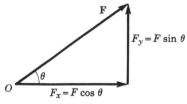

Figure 34.11

y direction. If the inclination of the force F is θ, we see from Fig. 34.11, that

(34.12) $$F_x = F \cos \theta, \qquad F_y = F \sin \theta.$$

Since $F_x =$ mass times a_x, the acceleration of a particle in the x direction

and F_y = mass times a_y, the acceleration of a particle in the y direction, we obtain from (34.12), the system of equations

$$(34.13) \qquad F \cos \theta = F_x = ma_x = m \frac{d^2x}{dt^2} \equiv m\ddot{x},$$

$$F \sin \theta = F_y = ma_y = m \frac{d^2y}{dt^2} \equiv m\ddot{y}.$$

In a similar manner we can break up any vector quantity into its x and y components. In Fig. (34.14) we have shown, for example, the x and y components of a velocity vector **v**.

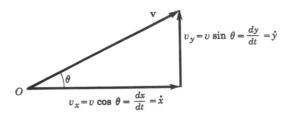

Figure 34.14

For certain problems, it is often convenient to use polar coordinates instead of rectangular coordinates. The vector quantity is then broken up into two components: one along the radial r direction, the other in a direction perpendicular to it. In Fig. 34.15, we have broken up the vector **v** into these two components v_r and v_θ.

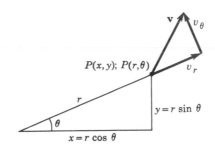

Figure 34.15

Let (x,y) be the coordinates of a point P in a rectangular system and (r,θ) its coordinates in a polar system, Fig. 34.15. Then we see from the figure, that

$$(34.16) \qquad x = r \cos \theta, \qquad y = r \sin \theta.$$

Differentiation of (34.16) gives

(34.17)
$$\frac{dx}{dt} = \cos\theta\,\frac{dr}{dt} - r\sin\theta\,\frac{d\theta}{dt},$$

$$\frac{dy}{dt} = \sin\theta\,\frac{dr}{dt} + r\cos\theta\,\frac{d\theta}{dt}.$$

A second differentiation gives

(34.18)
$$\frac{d^2x}{dt^2} - \cos\theta\,\frac{d^2r}{dt^2} - 2\sin\theta\,\frac{dr}{dt}\,\frac{d\theta}{dt} - r\cos\theta\left(\frac{d\theta}{dt}\right)^2 - r\sin\theta\,\frac{d^2\theta}{dt^2},$$

$$\frac{d^2y}{dt^2} = \sin\theta\,\frac{d^2r}{dt^2} + 2\cos\theta\,\frac{dr}{dt}\,\frac{d\theta}{dt} - r\sin\theta\left(\frac{d\theta}{dt}\right)^2 + r\cos\theta\,\frac{d^2\theta}{dt^2}.$$

These formulas, (34.17) and (34.18), are valid for every value of θ. Hence they must hold in particular when $\theta = 0$. But when $\theta = 0$, $x = r$ and the direction perpendicular to r is the y direction. Hence the components of velocity and acceleration in a radial direction and in a direction perpendicular to it are

(34.19)
$$v_r = \frac{dx}{dt}, \qquad v_\theta = \frac{dy}{dt}, \qquad a_r = \frac{d^2x}{dt^2}, \qquad a_\theta = \frac{d^2y}{dt^2}.$$

Substituting the first two equations of (34.19) in (34.17), the second two in (34.18), we obtain with $\theta = 0$,

(34.2)
$$v_r = \frac{dr}{dt} \equiv \dot{r}, \qquad v_\theta = r\,\frac{d\theta}{dt} \equiv r\dot{\theta}.$$

(34.21)
$$a_r = \frac{d^2r}{dt^2} - r\left(\frac{d\theta}{dt}\right)^2 \equiv \ddot{r} - r\dot{\theta}^2,$$

$$a_\theta = 2\,\frac{dr}{dt}\,\frac{d\theta}{dt} + r\,\frac{d^2\theta}{dt^2} \equiv 2\dot{r}\dot{\theta} + r\ddot{\theta}.$$

Formulas (34.2) and (34.21) give respectively the components of the velocity and acceleration vectors of a particle along the radial axis and in a direction perpendicular to it, at the point P where the curve crosses the x axis. Since the x axis can be chosen in any direction, these equations are valid for every point P of the particle's path.

EXERCISE 34A

1. In Fig. 34.22, we have shown the x and y components of a velocity vector $\mathbf{v}$ as well as its components in a radial direction and in a direction perpendicular to it. With the aid of this diagram, prove (34.2). *Hint.* First show that

(a)
$$v_x = \frac{dx}{dt} = v\cos\alpha, \qquad v_y = \frac{dy}{dt} = v\sin\alpha.$$

Then show that

(b) $\qquad v_r = v \cos (\alpha - \theta) = v \cos \alpha \cos \theta + v \sin \alpha \sin \theta,$

$$= \frac{dx}{dt} \cos \theta + \frac{dy}{dt} \sin \theta.$$

In (b), replace dx/dt and dy/dt by their values as given in (34.17).

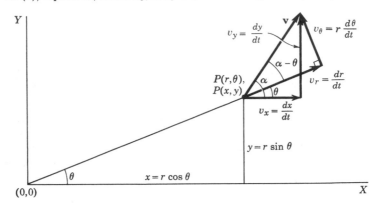

Figure 34.22

Similarly, show that

(c) $\qquad v_\theta = v \sin (\alpha - \theta) = v \sin \alpha \cos \theta - v \cos \alpha \sin \theta$

$$= \frac{dy}{dt} \cos \theta - \frac{dx}{dt} \sin \theta.$$

In (c), replace dy/dt and dx/dt by their values as given in (34.17).

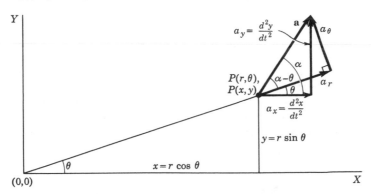

Figure 34.23

2. In Fig. 34.23, we have shown the x and y components of an acceleration vector **a** as well as its components in a radial direction and in a direction per-

pendicular to it. With the aid of this diagram, prove (34.21). *Hint.* First
show that

(a) $$a_x = \frac{d^2 x}{dt^2} = a \cos \alpha, \qquad a_y = \frac{d^2 y}{dt^2} = a \sin \alpha.$$

Then show that

(b) $$a_r = a \cos (\alpha - \theta) = a \cos \alpha \cos \theta + a \sin \alpha \sin \theta$$

$$= \frac{d^2 x}{dt^2} \cos \theta + \frac{d^2 y}{dt^2} \sin \theta.$$

In (b), replace $d^2 x/dt^2$ and $d^2 y/dt^2$ by their values as given in (34.18).
Similarly, show that

(c) $$a_\theta = a \sin (\alpha - \theta) = a \sin \alpha \cos \theta - a \cos \alpha \sin \theta$$

$$= \frac{d^2 y}{dt^2} \cos \theta - \frac{d^2 x}{dt^2} \sin \theta.$$

In (c) replace $d^2 y/dt^2$ and $d^2 x/dt^2$ by their values as given in (34.18).

3. A particle of mass m is attracted to a fixed point O by a force F. The particle
moves in a plane with a constant speed but not in a straight line. Show that
the particle moves in a circle with center at O. *Hint.* The component a_t of
the acceleration vector a in a direction tangent to the path of the particle
measures the change in the speed of the particle. Since this speed is constant, $a_t = 0$. Hence the acceleration acts only in a direction perpendicular
to the path of the particle. And since F and a are in the same direction, F
and the constant speed v_0 are perpendicular to each other. Then show
$dy/dx = -x/y$.

LESSON 34B. The Plane Motion of a Projectile.

Example 34.3.   A particle of mass m is projected from the earth with
a velocity v_0 at an angle α with the horizontal. The only force acting on
it is that of gravity. Assuming a level terrain, find:

1. The equation of the particle's path.
2. Its horizontal range.
3. The maximum height it will reach.
4. The value of α for which the range will be a maximum.
5. When the particle will reach the ground.

Solution.   Refer to Fig. 34.31. We take the x and y axes in a plane
which is perpendicular to the ground, and which contains the given
velocity vector $\mathbf{v_0}$. The z axis is, as usual, perpendicular to both x and y
axes. Since by assumption the only active force F is that of gravity, the
components of F in the x, y, z directions are respectively $F_x = 0$, $F_y =$
$-mg$, $F_z = 0$. Hence

(a) $$m\ddot{x} = 0, \qquad m\ddot{y} = -mg, \qquad m\ddot{z} = 0.$$

Integration of (a) gives

(b) $v_x = \dot{x} = c_1,$ $v_y = \dot{y} = -gt + c_2,$ $v_z = \dot{z} = c_3.$

Since the velocity vector v_0 lies in the xy plane, it follows that at $t = 0$, i.e., at the moment of projection, see Fig. 34.31,

(c) $v_x = v_0 \cos \alpha,$ $v_y = v_0 \sin \alpha,$ $v_z = 0,$

where α is the angle which v_0 makes with the horizontal x axis. Substituting these values in (b), we find

(d) $c_1 = v_0 \cos \alpha,$ $c_2 = v_0 \sin \alpha,$ $c_3 = 0.$

Hence (b) becomes

(e) $v_x = \dot{x} = v_0 \cos \alpha,$ $v_y = \dot{y} = -gt + v_0 \sin \alpha,$ $v_z = \dot{z} = 0.$

Integration of (e) gives

(f) $x = (v_0 \cos \alpha)t + c_4,$ $y = -g\dfrac{t^2}{2} + (v_0 \sin \alpha)t + c_5,$ $z = c_6.$

If we now choose our origin at the point where the particle is projected, then at $t = 0$, $x = 0$, $y = 0$, $z = 0$. Substituting these values in (f), we

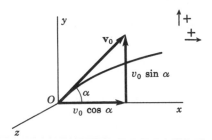

Figure 34.31

find $c_4 = 0$, $c_5 = 0$, $c_6 = 0$. The parametric equations of the path of the particle are therefore

(g) $x = (v_0 \cos \alpha)t,$ $y = (v_0 \sin \alpha)t - \dfrac{gt^2}{2},$ $z = 0.$

Since $z = 0$, it follows that a *projectile subject only to a gravitational force moves in a plane* containing the vector v_0.

   To find the equation of the path in rectangular coordinates, we eliminate t by solving the first equation in (g) for t and substituting this value in the second equation. There results

(h) $y = (\tan \alpha)x - \left(\dfrac{g \sec^2 \alpha}{2v_0{}^2} \right) x^2,$

which is the equation of a parabola through the origin. Since the coefficient of x^2 is negative, the curve is concave downward.

The horizontal range of the projectile, i.e., the distance from the origin to the point where the particle strikes the ground, is obtained by setting $y = 0$ in (h), for when $y = 0$ the particle is at ground level. Equation (h) then becomes, if $\alpha \neq \pi/2$,

(i) $\qquad 0 = (\sin \alpha)x - \dfrac{g}{2} \dfrac{x^2}{v_0^2 \cos \alpha}, \qquad x = \dfrac{v_0^2}{g}(2 \sin \alpha \cos \alpha),$

$\qquad x = \dfrac{v_0^2}{g} \sin 2\alpha,$

which gives the horizontal range of the particle.

The maximum height is reached when $y' = 0$. Therefore, differentiating (h) with respect to x and setting $y' = 0$, we obtain

(j) $\qquad 0 = \tan \alpha - \dfrac{g \sec^2 \alpha}{v_0^2} x, \qquad x = \dfrac{v_0^2}{g} \sin \alpha \cos \alpha.$

When x has the value in (j), y by (h) has the value

(k) $\qquad y = \dfrac{v_0^2}{g} \sin^2 \alpha - \dfrac{v_0^2}{2g} \sin^2 \alpha = \dfrac{v_0^2}{2g} \sin^2 \alpha,$

which gives the maximum height reached by the particle.

The range will be a maximum when, in (i), α is so chosen that x is a maximum. Hence differentiating the last equation in (i) with respect to α, and setting $dx/d\alpha = 0$, we have

(l) $\qquad 0 = \dfrac{2v_0^2}{g} \cos 2\alpha, \qquad \alpha = \dfrac{\pi}{4}.$

The range therefore will be a maximum if the projectile is fired at an angle of 45°. By (i), this maximum range is v_0^2/g.

The particle will reach the ground when $y = 0$. Setting $y = 0$ in (g), we obtain

(m) $\qquad (v_0 \sin \alpha)t - \dfrac{gt^2}{2} = 0, \qquad t = \dfrac{2v_0}{g} \sin \alpha,$

which is the time it takes the particle to reach the ground.

EXERCISE 34B

In problems 1–14, assume the air resistance is negligible.

1. A projectile is fired from the earth with a velocity of 1600 ft/sec at an angle of 45°. Find the equation of motion, the maximum height reached and the range of the projectile.

2. Start with the equation of motion (h) of Example 34.3.
   (a) Show that the coordinates of the vertex of the parabola are

$$\left(\frac{v_0^2 \sin \alpha \cos \alpha}{g}, \quad \frac{v_0^2 \sin^2 \alpha}{2g}\right).$$

   (b) Show that the distance of the vertex to the focus is $v_0^2 \cos^2 \alpha/2g$, and therefore that the equation of the directrix is $y = v_0^2/2g$. Note that the equation has no α in it. Hence the parabolic orbits of all projectiles fired with a given velocity have the same directrix regardless of the angle at which they are fired.
   (c) Finally show that this constant height of the directrix above the horizontal is the distance a projectile reaches when fired straight up with an initial velocity of v_0 ft/sec.

   If you have succeeded in answering the above questions, you have proved that every parabolic orbit of a projectile fired with the same velocity v_0 has the same constant directrix whose height is the distance the projectile would reach if fired vertically.

3. We showed in Example 34.3 that if a projectile is fired at an angle of $45°$, its range will be a maximum and will equal v_0^2/g. An artillery piece, whose muzzle velocity is v_0 ft/sec, is located at a distance $D < v_0^2/g$ from an object at the same level as itself. Show that there are two angles at which the artillery piece can be fired and hit the object—one as much greater than $45°$ as the other is less. Find these angles. *Hint.* In (h) of Example 34.3, you want α such that when $y = 0$, $x = D$. Make use of the identity

$$2 \sin \alpha \cos \alpha = \sin 2\alpha = \cos\left(2\alpha - \frac{\pi}{2}\right).$$

4. The muzzle velocity of an artillery piece is 800 ft/sec. Assuming a level terrain, answer the following questions.
   (a) An object is 3.8 mi away. Can it be hit?
   (b) An object is 15,000 ft away. At what angles must the artillery piece be fired in order to hit the object?
   (c) What is the maximum height reached by the shell of (b)?
   (d) When did the shell reach the object?
   (e) If a mountain of height 6000 ft is 4000 ft from the artillery piece, is it still possible to hit the object?

5. A projectile, fired with a velocity of 96 ft/sec, reaches its maximum height in 2 sec. Assume a level terrain.
   (a) Find the angle of projection of the particle. *Hint.* $\dfrac{dy}{dx} = \dfrac{dy/dt}{dx/dt}$. Therefore $dy/dx = 0$ if $dy/dt = 0$.
   (b) Find the maximum height reached by the particle.
   (c) What is the range of the projectile from the point fired?

6. A projectile is fired from a height of y_0 ft above a level terrain, with a velocity of v_0 ft/sec and at an angle α with the horizontal. Find:
   (a) The equation of the particle's path.
   (b) Its horizontal range—take the x axis on ground level.
   (c) Its maximum height.
   (d) When it will reach the ground.
   (e) At what angle and with what velocity it will strike the ground. (*Hint.* If θ is the angle, $\tan \theta = (dy/dt)/(dx/dt)$ and $|v| = \sqrt{(dx/dt)^2 + (dy/dt)^2}$.)
   (f) The value of α that will make the range a maximum.

7. A projectile is fired from a height of 50 ft above a level terrain with a velocity of 64 ft/sec at an angle of 45°. Answer the questions asked for in 6, with the exception of (f). In regard to (f), find the angle of projection that will make the horizontal range a maximum and the value of this maximum range.

8. The maximum range of a projectile when fired on a level terrain is 1000 ft.
   (a) What is its muzzle velocity?
   (b) What is the maximum horizontal distance it can travel if the projectile is fired from a 40-ft-high platform? *Hint.* First find the firing angle for maximum horizontal range, see problem 6.

9. The maximum distance a boy can throw a ball on level ground is 100 ft. Neglecting the height of the boy, find
   (a) The velocity with which the ball leaves his hand,
   (b) The maximum horizontal distance he can throw the ball if he stands on a roof which is 40 ft above the ground. *Hint.* First find the throwing angle for maximum horizontal range, see problem 6.

10. Two athletes, one $6\frac{1}{2}$ ft tall, the other $5\frac{1}{2}$ ft tall, can each put a shot with the same velocity of 36 ft/sec. At what angle should the shot leave each athlete's hand in order to get the maximum horizontal range? Assume the shot leaves from heights of 6 ft and 5 ft respectively. How much farther will the taller athlete's throw go?

11. Answer the questions in problem 6, excepting (f), if $\alpha = 0$, i.e., if the projectile is fired horizontally from a distance y_0 ft above the horizontal.

12. A projectile is fired with a velocity v_0 at an angle α with the horizontal. The terrain makes an angle β with the horizontal.
   (a) Find the range of the projectile. *Hint.* Call R the range of the projectile. Then the projectile will hit the terrain when $x = R\cos\beta$ and $y = R\sin\beta$. Substitute in (h) of Example 34.3.
   (b) Find the value of α which will make the range a maximum. *Hint.* Make use of double angle formulas.
   (c) What is the maximum range?

13. In problem 3, we gave two angles at which a projectile could be fired in order to hit an object located within range and on the same level as the firing weapon.
   (a) Solve this same problem if the object to be hit is on the top of a hill whose angle of inclination is β and whose distance D from the firing point is less than or equal to the range $v_0^2/g(1 + \sin\beta)$, as given in (c) of problem 12. *Hint.* Call (X, Y) the coordinates of the object. Then $D = \sqrt{X^2 + Y^2}$, $\sin\beta = Y/D$, $\cos\beta = X/D$. Replace R by D in answer to 12(a). Solve for α. Use the fact that

$$\cos A \sin B = \tfrac{1}{2}[\sin(A + B) - \sin(A - B)]$$

and

$$\sin(2\alpha - \beta) = \cos\left(2\alpha - \beta - \frac{\pi}{2}\right).$$

   (b) Show, by means of the solution found in (a), that it is possible to hit an object only if its X and Y coordinates satisfy the inequality

$$\sqrt{X^2 + Y^2} + Y \leqq v_0^2/g.$$

   Hint. Use the fact—see answer to (a)—that

$$(gX^2 + Yv_0^2)/(v_0^2\sqrt{X^2 + Y^2})$$

must be $\leqq 1$. Note that if $Y = 0$, so that object is on a level terrain, $X \leqq v_0{}^2/g$ as we saw previously.

14. A man is hunting with a gun whose muzzle velocity is 224 ft/sec. He aims for a bird on the top of a tree 150 ft high and 1500 ft away. Is the bird in danger of being hit?

In problems 1–14, we ignored air resistance. In the problems below, we shall assume the projectile is fired from the earth and is subject not only to a gravitational force but also to an air resistance which is proportional to the first power of the velocity. We shall also assume that the force of the air resistance acts in a direction opposite to that of the velocity, i.e., that it acts along a tangent to the projectile's path and in a direction to oppose the motion. Call R the proportionality factor of the air resistance.

15. A projectile is fired on a level terrain at an angle α with the horizontal and with a velocity of v_0 ft/sec.
    (a) Find the parametric equations of the particle's path. *Hint.* Modify equation (a) of Example 34.3 to take into account the components of the force of the air resistance in the x and y directions.
    (b) What is the maximum height reached by the particle? *Hint.* Set $\dfrac{dy}{dx} = \dfrac{dy/dt}{dx/dt}$ equal to zero.

16. A projectile is fired in a horizontal direction with a velocity of v_0 ft/sec from a height of y_0 ft. Find the parametric equations of its path.
17. An anti-aircraft gun fires a shell almost vertically with an initial velocity of v_0 ft/sec. The horizontal component of the air resistance is therefore negligible. Assume the gun makes an angle α with the horizontal, and the vertical component of resistance is $R\, dy/dt$.
    (a) Find the parametric equations of the path of the shell.
    (b) Assume the shell weighs 60 lb, the muzzle velocity is 2000 ft/sec, the angle of elevation is 80°, and the vertical component of the air resistance is $1/20\, dy/dt$. Find the parametric equations of the shell's path, the maximum height attained by it, and the time required to reach this maximum height.

ANSWERS 34B

1. $y = x - x^2/80{,}000$, 20,000 ft, 80,000 ft.

3. $\alpha = \dfrac{\pi}{4} \pm \dfrac{1}{2}\,\text{Arc cos }\dfrac{Dg}{v_0{}^2}$.

4. (a) No. The maximum range is 3.7879 mi.   (b) $\left(\dfrac{\pi}{4} \pm 0.36\right)$ rad.

   (c) 1693 ft or 8307 ft, approx.   (d) 20.6 sec or 45.6 sec, approx.
   (e) Yes. When $x = 4000$ ft, the height y of the projectile is approximately 6498 ft, if the larger angle of elevation is used.

5. (a) $\alpha = \text{Arc sin }(\tfrac{2}{3})$.   (b) $y = 64$ ft.   (c) $x = 143$ ft.

6. (a) $x = (v_0 \cos \alpha)t, \quad y = y_0 + (v_0 \sin \alpha)t - \dfrac{gt^2}{2}$;

   $y = y_0 + (\tan \alpha)x - \dfrac{g \sec^2 \alpha}{2v_0{}^2}\,x^2$.

(b) Range is given by the positive value of x for which

$$\frac{g \sec^2 \alpha}{2v_0{}^2} x^2 - (\tan \alpha)x - y_0 = 0.$$

(c) $y = y_0 + \dfrac{v_0{}^2 \sin^2 \alpha}{2g}$.

(d) $t = x/(v_0 \cos \alpha)$, where x is given by (b),

(e) $\tan \theta = \dfrac{v_0 \sin \alpha - gt}{v_0 \cos \alpha}$,

$\quad |v| = \sqrt{v_0{}^2 - 2v_0(\sin \alpha)gt + (gt)^2}$, where t is given by (d).

(f) $\sin \alpha = v_0/\sqrt{2(v_0{}^2 + gy_0)}$.

7. (a) $y = 50 + x - \dfrac{x^2}{128}$.   (b) $x = 166.4$ ft.   (c) 82 ft.

(d) $t = 3.7$ sec.   (e) $122°00'$, 85.6 ft/sec.

(f) $\sin \alpha = 0.6$, $\alpha = 37°$, approx., 171 ft, approx.

8. (a) $v = 80 \sqrt{5}$ ft/sec.   (b) 1039 ft, approx. ($\alpha = 44°$, approx.).

9. (a) $40\sqrt{2}$ ft/sec.   (b) 134 ft ($\alpha = 37°$, approx.).

10. Use the answer given in 6(f) to find the angle α for each athlete. Then use the range equation as given in 6(b).

11. (a) $x = v_0 t$, $y = y_0 - \dfrac{gt^2}{2}$, $y = y_0 - \dfrac{g}{2v_0{}^2} x^2$.

(b) $x = v_0 \sqrt{2y_0/g}$ ft.

(c) $y = y_0$ ft.

(d) $t = x/v_0$, where x is given by (b).

(e) $\tan \theta = -gt/v_0$, $|v| = \sqrt{v_0{}^2 + (gt)^2}$.

12. (a) Range $R = \dfrac{2v_0{}^2 \cos \alpha \sin (\alpha - \beta)}{g \cos^2 \beta}$.

(b) $\alpha = \dfrac{\pi}{4} + \dfrac{\beta}{2}$.   (c) Maximum range $R = \dfrac{v_0{}^2}{g(1 + \sin \beta)}$.

13. (a) $\alpha = \dfrac{\pi}{4} + \dfrac{\beta}{2} \pm \tfrac{1}{2} \text{Arc} \cos \left(\dfrac{gX^2 + Yv_0{}^2}{v_0{}^2 \sqrt{X^2 + Y^2}} \right)$.

14. No. See 13(b).

15. (a) $x = \dfrac{m}{R} v_0 \cos \alpha (1 - e^{-Rt/m})$,

$\quad y = \dfrac{m}{R} \left(\dfrac{mg}{R} + v_0 \sin \alpha \right) (1 - e^{-Rt/m}) - \dfrac{mg}{R} t$,

where R is the proportionality factor of the air resistance.

(b) $t = \dfrac{m}{R} \log \left(1 + \dfrac{Rv_0}{mg} \sin \alpha \right)$,

$\quad y = \dfrac{mv_0}{R} \sin \alpha - \dfrac{m^2 g}{R^2} \log \left(1 + \dfrac{Rv_0}{mg} \sin \alpha \right)$.

16. $x = \dfrac{mv_0}{R} (1 - e^{-Rt/m})$, $y = y_0 + \dfrac{m^2 g}{R^2} (1 - e^{-Rt/m}) - \dfrac{mg}{R} t$.

17. (a) $x = (v_0 \cos \alpha)t$, $y = $ as in 15(a).

(b) $x = 347.3t$, $y = 118,861(1 - e^{-0.0267t}) - 1200t$, $30,153$ ft, 36.4 sec.

LESSON 34C. Definition of a Central Force. Properties of the Motion of a Particle Subject to a Central Force. Assume a particle in motion is *attracted* to a fixed point O, by a force F. In such cases we say the particle moves subject to a **central force,** and call the fixed point O to which the particle is attracted, the **center of attraction.**

In Example 28.15, we discussed a special central force problem where the particle moved on a line. In the remainder of this lesson we consider the motion of a particle subject to a central force where the particle is free to move in space. We prove below certain properties which are common to the motions of all particles subject to a central force.

Property A. A Particle in Motion Subject to a Central Force Moves in a Plane Which Contains the Fixed Point O. We assume a particle, moving in space, is subject to a central force F. Let the fixed point O, toward which the force is directed, be the origin of a coordinate system. Let $P(x,y,z)$ be the rectangular coordinates of the particle, let r be the distance of the particle from O at time t and let α, β, γ be the direction angles of the force F, Fig. 34.4.

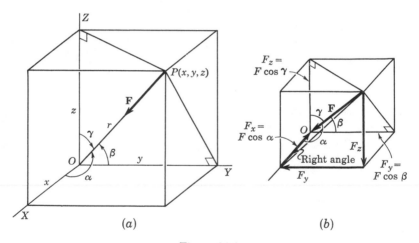

(a) (b)

Figure 34.4

The components of F in the x,y,z directions are respectively, see Figs. 34.4(a) and (b),

$$(34.41) \quad F_x = F \cos \alpha = F \frac{x}{r}, \quad F_y = F \cos \beta = F \frac{y}{r}, \quad F_z = F \cos \gamma = F \frac{z}{r}.$$

Since $F_x = m\ddot{x}$, $F_y = m\ddot{y}$, $F_z = F\ddot{z}$, we obtain, by (34.41), the system

$$(34.42) \qquad m\ddot{x} = F \frac{x}{r}, \qquad m\ddot{y} = F \frac{y}{r}, \qquad m\ddot{z} = F \frac{z}{r}.$$

Multiply the first equation in (34.42) by $-y$, the second by x and add the two. There results

(34.43) $$m x\ddot{y} - m y\ddot{x} = 0, \qquad x\ddot{y} - y\ddot{x} = 0.$$

In an analogous manner, we can obtain, respectively, from the second and third equations in (34.42), and from its first and third equations,

(34.44) $$y\ddot{z} - z\ddot{y} = 0, \qquad z\ddot{x} - x\ddot{z} = 0.$$

Integrating each of the equations in (34.43) and (34.44) with respect to time, we obtain

(34.45) $$x\dot{y} - y\dot{x} = c_1, \qquad y\dot{z} - z\dot{y} = c_2, \qquad z\dot{x} - x\dot{z} = c_3.$$

[Verify that the derivative of each equation in (34.45) gives the respective equation in (34.43) or (34.44).] We now multiply the first equation in (34.45) by z, the second by x, the third by y, and add all three. There results

(34.46) $$c_2 x + c_3 y + c_1 z = 0,$$

which is the equation of a plane through the origin, i.e., through the fixed point O.

Property B. A Particle in Motion Subject to a Central Force Satisfies the Law of the Conservation of Angular Momentum. We assume that a particle in motion is subject to a central force F. By property A, the particle moves in a plane. By the definition of a central force, F is always directed toward a fixed point O, which we take as the origin of a polar coordinate system. Let $P(r,\theta)$ be the coordinates of the particle's position at time t. Call F_r the component of F acting along the radial axis r, and F_θ the component of F acting in a direction perpendicular to r. Since F always acts toward O along a radius vector, its component F_θ is zero. Therefore, by (34.21),

(a) $$F_\theta = m a_\theta = m(2\dot{r}\dot{\theta} + r\ddot{\theta}) = 0,$$
$$2\dot{r}\dot{\theta} + r\ddot{\theta} = 0.$$

If we multiply the last equation in (a) by r, it becomes $2 r\dot{r}\dot{\theta} + r^2\ddot{\theta} = 0$, which is equivalent to

(b) $$\frac{d}{dt}(r^2\dot{\theta}) = 0.$$

Integration of (b) gives

(34.47) $$r^2\dot{\theta} = h,$$

where h is a constant. By definition, the **angular momentum** of a particle of mass m rotating about an axis perpendicular to the plane of its motion is $mr^2\dot{\theta}$, where r is the distance of the particle from the axis of rotation and $\dot{\theta}$ is its angular velocity about this axis. We see, therefore, by (34.47), that h *is the angular momentum of a particle per unit mass.* And since h is a constant, the equation tells us that the angular momentum of the particle is conserved. We have thus proved that a particle in motion subject to a central force satisfies the **law of the conservation of angular momentum.**

Property C. A Particle in Motion Subject to a Central Force Sweeps out Equal Areas in Equal Intervals of Time. (Note. This property is essentially a restatement of property B.) The area of a *circular*

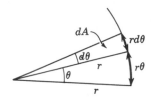

Figure 34.48

sector of radius r and central angle θ is $r^2\theta/2$, Fig. 34.48. Hence when a radius vector r turns through an infinitesimal angle $d\theta$, it sweeps out an area equal to

(34.481) $dA = \tfrac{1}{2}r^2\,d\theta, \qquad \dfrac{dA}{dt} = \tfrac{1}{2}r^2\dfrac{d\theta}{dt}.$

Substituting (34.47) in the second equation of (34.481), we obtain, with initial conditions $A = 0$, $t = 0$,

(34.49) $\dfrac{dA}{dt} = \dfrac{h}{2}, \qquad \displaystyle\int_{A=0}^{A} dA = \int_{t=0}^{t} \dfrac{h}{2}\,dt, \qquad A = \dfrac{ht}{2}.$

In words the first equation in (34.49) says that the rate of change of the area A is a constant. (Remember h is a constant.) The last equation says that equal areas are swept out in equal times. We have thus proved that a particle in motion subject to a central force sweeps out equal areas in equal time intervals.

EXERCISE 34C

1. (a) By (34.47), $r^2\dot{\theta} = h$, where (r,θ) are the polar coordinates of a particle moving subject to a central force. Show that in rectangular coordinates $h = x\dot{y} - y\dot{x}$. *Hint.* $\tan\theta = y/x$. Differentiate with respect to time and solve for $\dot{\theta}$.

(b) Hence show that the areal velocity dA/dt, in rectangular coordinates, is given by $dA/dt = \frac{1}{2}(x\dot{y} - y\dot{x})$. *Hint.* See (34.49).

LESSON 34D. Definitions of *Force Field, Potential, Conservative Field.* Conservation of Energy in a Conservative Field. Assume a force **F** acts on a unit mass placed at each point (x,y,z) of a region of space. Hence at each point of the region, we can represent the magnitude and direction of **F** by drawing a vector **F** there. We call **F** a **vector point-function** throughout this region, since it is a vector whose components along the X, Y, Z axes are functions of the space coordinates x,y,z of the unit mass. A region of this type is an example of a force field. Its formal definition follows.

Definition 34.5. A region in space, having the property that at every one of its points a vector point-function **F** exists that gives the magnitude and direction of the force acting on a particle of unit mass placed there, is called a **field of force** or a **force field.**

A region in the neighborhood of the solar system is a field of force. At each point of the region, a vector point-function exists due to the members of the solar system. The region in the neighborhood of a current bearing wire is a force field. It is called an electromagnetic field.

By Definition 9.23, $F_x\, dx + F_y\, dy + F_z\, dz$ is called an exact differential if there exists a function $U(x,y,z)$ such that

(34.51) $$dU = F_x\, dx + F_y\, dy + F_z\, dz,$$

or equivalently, such that

(34.52) $$\frac{\partial U}{\partial x} = F_x, \qquad \frac{\partial U}{\partial y} = F_y, \qquad \frac{\partial U}{\partial z} = F_z.$$

Definition 34.53. A force field or a field of force is called **conservative** if there exists a function $U(x,y,z)$ such that its differential dU satisfies (34.51), or equivalently if its partial derivatives with respect to x,y,z respectively satisfy (34.52), where F_x, F_y, and F_z are the x,y,z components of a force F acting in the field. The function $U(x,y,z)$ itself is called a **force function** and the *negative* of $U(x,y,z)$ is called the **potential** or the **potential energy** of the force field.

Comment 34.54. Not every force field has a potential $-U(x,y,z)$.

Comment 34.6. The potential $(-U)$ may be looked upon as a function whose partial derivatives with respect to x,y,z give respectively the components of a force F in the negative x, y, and z directions.

Example 34.61. A particle moving in a force field is subject to a central force F whose magnitude is proportional to its distance r from a fixed point O. Show that the force field is conservative.

Solution. By hypothesis

(a) $$F = -kr,$$

where $k > 0$ is a proportionality constant. The magnitude of the components of F in the x,y,z directions are given in (34.41). Hence substituting (a) in (34.41), we obtain

(b) $$F_x = \frac{x}{r}(-kr) = -kx, \qquad F_y = \frac{y}{r}(-kr) = -ky,$$

$$F_z = \frac{z}{r}(-kr) = -kz.$$

Let us take for the force function U of Definition 34.53,

(c) $$U(x,y,z) = -\frac{kr^2}{2} + C \equiv -\frac{k}{2}(x^2 + y^2 + z^2) + C.$$

Therefore

(d) $$\frac{\partial U}{\partial x} = -\frac{k}{2}(2x) = -kx, \qquad \frac{\partial U}{\partial y} = -ky, \qquad \frac{\partial U}{\partial z} = -kz.$$

A comparison of (d) with (b) shows that the values on the right of the equations in (d) are respectively F_x, F_y, F_z. Hence by Definition 34.53, the force field is conservative.

Example 34.62. A particle moving in a force field is subject to a central force F whose magnitude is inversely proportional to the square of its distance r from a fixed point O. Show that the force field is conservative.

Solution. By hypothesis

(a) $$F = -\frac{k}{r^2},$$

where $k > 0$ is a proportionality constant. Substituting this value of F in (34.41), we obtain

(b) $$F_x = \frac{x}{r}\left(-\frac{k}{r^2}\right) = -\frac{kx}{r^3}, \qquad F_y = -\frac{ky}{r^3}, \qquad F_z = -\frac{kz}{r^3}.$$

Let us take for the force function U of Definition 34.53,

(c) $$U(x,y,z) = \frac{k}{r} + C \equiv \frac{k}{\sqrt{x^2 + y^2 + z^2}} + C.$$

Therefore

(d) $\dfrac{\partial U}{\partial x} = -\dfrac{kx}{\sqrt{(x^2+y^2+z^2)^3}} = -\dfrac{kx}{r^3},\quad \dfrac{\partial U}{\partial y} = -\dfrac{ky}{r^3},\quad \dfrac{\partial U}{\partial z} = -\dfrac{kz}{r^3}.$

A comparison of (d) with (b) shows that the values on the right of the equations in (d) are respectively F_x, F_y, F_z. Hence by Definition 34.53, the force field is conservative.

Property D. Conservation of Energy in a Conservative Field.
Let F be a force acting on a particle moving in a conservative field. Therefore by Definition 34.53, there exists a function $U(x,y,z)$ such that

(34.63) $$dU = \frac{\partial U}{\partial x}\,dx + \frac{\partial U}{\partial y}\,dy + \frac{\partial U}{\partial z}\,dz,$$

where

$$\frac{\partial U}{\partial x} = F_x,\qquad \frac{\partial U}{\partial y} = F_y,\qquad \frac{\partial U}{\partial z} = F_z.$$

Since $F_x = m\ddot{x}$, $F_y = m\ddot{y}$, $F_z = m\ddot{z}$, we have

(34.64) $$\frac{\partial U}{\partial x} = m\ddot{x},\qquad \frac{\partial U}{\partial y} = m\ddot{y},\qquad \frac{\partial U}{\partial z} = m\ddot{z}.$$

Multiplying the first equation in (34.64) by dx, the second by dy, the third by dz, and adding all three, we obtain, with the help of (34.63),

(34.65) $$m\ddot{x}\,dx + m\ddot{y}\,dy + m\ddot{z}\,dz = \frac{\partial U}{\partial x}\,dx + \frac{\partial U}{\partial y}\,dy + \frac{\partial U}{\partial z}\,dz = dU.$$

Integration of (34.65) with respect to time gives

(34.66) $$\frac{m\dot{x}^2}{2} + \frac{m\dot{y}^2}{2} + \frac{m\dot{z}^2}{2} = U + C.$$

The left side of (34.66) is defined as the **kinetic energy** of a particle. By Definition 34.53, $-U$ is its potential energy. Hence (34.66) tells us that the sum of the kinetic and potential energies of a particle in a conservative field is a constant. This fact, namely that the sum of kinetic and potential energies of a particle is a constant, is known as the **law of the conservation of energy.** We have thus proved the law of the conservation of energy for a particle moving in a conservative field.

EXERCISE 34D

1. A particle moving in a force field is subject to a central force F whose magnitude is proportional to its distance r^2 from a fixed point O. Show that the force field is conservative.

2. A particle moving in a force field is subject to a central force F whose magnitude is proportional to its distance r^3 from a fixed point O. Show that the force field is conservative.

3. A particle moving in a force field is subject to a central force F whose magnitude is proportional to its distance r^n from a fixed point O, where n is a positive number. Show that the force field is conservative.

4. A particle moving in a force field is subject to a central force F whose magnitude is inversely proportional to its distance r from a fixed point O. Show that the force field is conservative.

5. A particle moving in a force field is subject to a central force F whose magnitude is inversely proportional to its distance r^n from a fixed point O, where n is a positive number greater than 1. Show that the force field is conservative.

6. Can you think of a force field which is not conservative?

ANSWERS 34D

1. Force function $U = -\dfrac{k}{3} r^3 + C.$

2. Force function $U = -\dfrac{k}{4} r^4 + C.$

3. Force function $U = -\dfrac{k}{n+1} r^{n+1} + C.$

4. Force function $U = -\dfrac{k}{2} \log r^2.$

5. Force function $U = \dfrac{k}{(n-1)r^{n-1}}.$

6. A field in which energy is being dissipated as the particle moves. For example, a field in which a resisting force, proportional to velocity, is present cannot be a conservative field.

LESSON 34E. Path of a Particle in Motion Subject to a Central Force Whose Magnitude Is Proportional to Its Distance from a Fixed Point O.

We assume that a particle in motion of mass m is subject to a *central* force F whose magnitude is proportional to its distance r from a fixed point O. We already know many facts about the particle. By properties A, B, C of Lesson 34C, we know that it moves in a plane, that it satisfies the law of the conservation of angular momentum and that it sweeps out equal areas in equal times. By Definition 34.5, the region in which the particle moves is a force field. By Example 34.61, this field is conservative. Hence by property D following Example 34.62, we also know that the particle satisfies the law of the conservation of energy.

To find the equation of its path, we take the x,y axes in the plane of the particle's motion with the origin at the fixed point O toward which the force acts. Let $P(x,y)$ be the coordinates of the position of the particle at time t in a rectangular system and (r,θ) its coordinates in a polar system. The components of F in the x and y directions are, by (34.42) (remember the force is central so that $F_z = 0$),

(34.7) $$F_x = m\ddot{x} = \frac{Fx}{r}, \qquad F_y - m\ddot{y} - \frac{Fy}{r}.$$

By hypothesis

(34.71) $$F = -k^2 mr,$$

where for convenience we have used $k^2 m$ for the proportionality constant. The minus sign is necessary because the force is acting toward O and the positive direction is outward from O. Substituting (34.71) in (34.7), we obtain the linear system of equations

(34.72) $$\ddot{x} = -k^2 x, \qquad \ddot{y} = -k^2 y.$$

Their respective solutions, obtained by any of the methods previously discussed, are

(34.73) $\quad x = c_1 \cos kt + c_2 \sin kt, \qquad y = c_3 \cos kt + c_4 \sin kt.$

These are the parametric equations of the path. You can verify by referring to Theorem 31.33 that the pair of functions in (34.73) contains the correct number of four arbitrary constants. Hence in a specific problem four initial conditions will be needed, $x(0)$, $x'(0)$, $y(0)$, $y'(0)$. The **period of the motion** of the particle, by Definition 28.34, is $2\pi/k$. It is the time it takes the particle to return to its initial starting position, headed in the same starting direction.

To find the equation of the path in rectangular coordinates, we must eliminate the parameter t between the two equations in (34.73). The easiest way to do this is to first solve them simultaneously for $\sin kt$ and $\cos kt$ in terms of x and y. The result is

(34.74) $\quad \sin kt = \dfrac{c_3 x - c_1 y}{c_2 c_3 - c_1 c_4}, \quad \cos kt = \dfrac{c_4 x - c_2 y}{c_1 c_4 - c_2 c_3}, \quad c_1 c_4 - c_2 c_3 \neq 0.$

Squaring both equations in (34.74) and then adding them, we obtain for the path of the particle in rectangular coordinates,

(34.75) $\quad 1 = \dfrac{(c_3 x - c_1 y)^2 + (c_4 x - c_2 y)^2}{(c_1 c_4 - c_2 c_3)^2}, \quad c_1 c_4 - c_2 c_3 \neq 0,$

which can be written as

(34.76) $\quad (c_3{}^2 + c_4{}^2) x^2 - 2(c_1 c_3 + c_2 c_4) xy$
$\qquad\qquad + (c_1{}^2 + c_2{}^2) y^2 - (c_1 c_4 - c_2 c_3)^2 = 0, \quad c_1 c_4 - c_2 c_3 \neq 0.$

From analytic geometry we know that if the constants in

(34.77) $$Ax^2 + 2Bxy + Cy^2 + D = 0$$

are such that $B^2 - AC < 0$, and

Case 1. $D \neq 0$, $AD < 0$, then the equation represents an ellipse with center at the origin.

Case 2. $D \neq 0$, $AD > 0$, the equation has no locus.

Case 3. $D = 0$, the equation represents a single point.

A comparison of (34.76) with (34.77) shows that

(a) $$B^2 - AC = (c_1 c_3 + c_2 c_4)^2 - (c_3^2 + c_4^2)(c_1^2 + c_2^2)$$
$$= -(c_1 c_4 - c_2 c_3)^2,$$

and

(b) $$AD = -(c_3^2 + c_4^2)(c_1 c_4 - c_2 c_3)^2.$$

Both of the above expressions are less than zero if $c_1 c_4 - c_2 c_3 \neq 0$. Hence if, in (34.76), $c_1 c_4 - c_2 c_3$, which corresponds to D of (34.77), is not zero, Case 1 above applies and (34.76) is the equation of an ellipse with *center at the origin*.

We have thus proved that the orbit of a particle, attracted to an origin O by a central force F whose magnitude is proportional to its distance from O, is an ellipse with center at the fixed point O. Hence we have also proved for this case that the force is directed toward the **center of the ellipse**.

Comment 34.78.   If the particle is constrained to move toward the fixed point O along a radius vector so that θ is constant, then $d\theta/dt = 0$, $F = F_r$, and by (34.21) and (34.71),

$$F_r = ma_r = m\ddot{r} = -k^2 mr, \quad \ddot{r} = -k^2 r.$$

This last equation, as we showed in Example 28.15, is, as it should be, the differential equation of motion of a particle executing simple harmonic motion.

Example 34.79.   A particle weighing 16 pounds is 10 feet from a fixed point O and is given an initial velocity of 15 ft/sec in a direction perpendicular to the x axis. If a *central* force F acts on the particle with a magnitude which is one-eighth of the distance of the particle from the fixed point O, find the equation of its path and the period of the motion.

Solution.   We take the origin at the fixed point O, and the x,y axes in the plane of the particle's motion. By hypothesis

(a) $$F = -\tfrac{1}{8}r,$$

and by (34.42),

(b) $$F_x = m\ddot{x} = \frac{Fx}{r}, \qquad F_y = m\ddot{y} = \frac{Fy}{r},$$

where r is the distance of the particle from O. Substituting (a) in (b), we obtain

(c) $$m\ddot{x} = -\frac{x}{8}, \qquad m\ddot{y} = -\frac{y}{8}.$$

By hypothesis $m = \frac{16}{32} = \frac{1}{2}$. Hence (c) becomes

(d) $$\ddot{x} + \frac{x}{4} = 0, \qquad \ddot{y} + \frac{y}{4} = 0,$$

whose solutions, by any method you wish to choose, are

(e) $$x = c_1 \cos\frac{t}{2} + c_2 \sin\frac{t}{2}, \qquad y = c_3 \cos\frac{t}{2} + c_4 \sin\frac{t}{2}.$$

Therefore

(f) $$\frac{dx}{dt} = -\frac{c_1}{2}\sin\frac{t}{2} + \frac{c_2}{2}\cos\frac{t}{2}, \qquad \frac{dy}{dt} = -\frac{c_3}{2}\sin\frac{t}{2} + \frac{c_4}{2}\cos\frac{t}{2}.$$

The initial conditions are $t = 0$, $x = 10$, $y = 0$, $dx/dt = 0$, $dy/dt = 15$. Substituting these values in (e) and (f), we obtain

(g) $$c_1 = 10, \qquad c_2 = 0, \qquad c_3 = 0, \qquad c_4 = 30.$$

Hence (e) becomes

(h) $$x = 10 \cos\frac{t}{2}, \qquad y = 30 \sin\frac{t}{2}.$$

By eliminating the parameter t, we obtain

(i) $$\frac{x^2}{10^2} + \frac{y^2}{30^2} = 1,$$

which is the equation of an ellipse with *center at the origin or fixed point O*. Its graph is shown in Fig. 34.791. The period of the motion, obtained

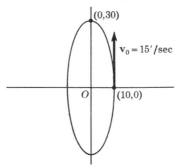

Figure 34.791

from (h), is 4π seconds. It is the time it will take the particle to make a complete circuit of the ellipse.

EXERCISE 34E

1. Verify the accuracy of the solution of (34.72) as given in (34.73).
2. Verify the accuracy of (34.74), (34.75), and (34.76).
3. A particle in motion of mass m is subject to a central force F whose magnitude is proportional to its distance from a fixed point O. Initially it is x_0 ft from the origin and is given a velocity of v_0 ft/sec in a direction perpendicular to the x axis.
   (a) Find the parametric equations of its path; also the equation of its path in rectangular coordinates. Take k for the proportionality constant instead of k^2m as we did in the text.
   (b) For what relative values of x_0, v_0 will the path be a circle?
4. A body weighing 16 lb is attracted to a fixed point O by a force whose magnitude is one-eighth the distance of the particle from O. Initially it is 12 ft from O and given a velocity of v_0 ft/sec in a direction perpendicular to the x axis.
   (a) Find the parametric equations of motion; also the equation of motion in rectangular coordinates.
   (b) What initial velocity v_0 will make the eccentricity of the orbit $\frac{1}{2}$?
5. Solve problem 3(a), if initially the particle is x_0 ft from the origin and is given a velocity of v_0 ft/sec in a direction making an angle θ with the x axis.
6. A body weighing 16 lb is 12 ft from a fixed point O. It is given an initial velocity of 20 ft/sec in a direction making an angle of 45° with the x axis. A central force acts on the particle with a magnitude equal to one-eighteenth of the distance of the body from O. Find the equation of its path and the period of its motion. Verify that the equation satisfies Case 1 after (34.77) and is therefore the equation of an ellipse.
7. A particle in motion of mass m is repelled from a fixed point O with a force proportional to its distance from O. Initially it is x_0 ft from the origin and is given a velocity of v_0 ft/sec in a direction perpendicular to the x axis. (a) Find the equation of motion. (b) What type conic is it? (c) Show that properties A, B, C, D of Lessons 34C and D are also valid when the central force is repelling instead of attracting.
8. Set up the system of differential equations of motion for the particle of problem 3 if in addition there is a force of resistance proportional to the velocity.

ANSWERS 34E

3. (a) $x = x_0 \cos \sqrt{k/m}\, t$, $y = v_0\sqrt{m/k} \sin \sqrt{k/m}\, t$; $\dfrac{x^2}{x_0^2} + \dfrac{ky^2}{mv_0^2} = 1$.
   (b) $x_0^2 = mv_0^2/k$.

4. (a) $x = 12 \cos \frac{1}{2}t$, $y = 2v_0 \sin \frac{1}{2}t$, $\dfrac{x^2}{144} + \dfrac{y^2}{4v_0^2} = 1$.
   (b) $v_0 = 3\sqrt{3}$ ft/sec, or $4\sqrt{3}$ ft/sec.

5. $x = x_0 \cos \sqrt{k/m}\, t + v_0\sqrt{m/k} \cos \theta \sin \sqrt{k/m}\, t$,

   $y = v_0\sqrt{m/k} \sin \theta \sin \sqrt{k/m}\, t$;

   $(\sin^2 \theta)x^2 - (2 \sin \theta \cos \theta)xy + \left(\dfrac{kx_0^2}{mv_0^2} + \cos^2 \theta\right) y^2 - x_0^2 \sin^2 \theta = 0$.

6. $x = 12 \cos \frac{1}{3}t + 30\sqrt{2} \sin \frac{1}{3}t$, $y = 30\sqrt{2} \sin \frac{1}{3}t$; 6π sec;
   $x^2 - 2xy + 1.08y^2 - 144 = 0$.

7. (a) $x = x_0 \cosh \sqrt{k/m}\, t,\ y = v_0 \sqrt{m/k} \sinh \sqrt{k/m}\, t,$

$$\frac{x^2}{x_0{}^2} - \frac{ky^2}{mv_0{}^2} = 1.$$

(b) Hyperbola.

8. $m\ddot{x} = -kx - r\dot{x},\ m\ddot{y} = -ky - r\dot{y}$, where k and r are the proportionality constants respectively for the force and the resistance.

LESSON 34F. Path of a Particle in Motion Subject to a Central Force Whose Magnitude Is Inversely Proportional to the Square of Its Distance from a Fixed Point O. We assume that a particle in motion, of mass m, is subject to a central force F whose magnitude is inversely proportional to the square of its distance r from a fixed point O. By properties A, B, C of Lesson 34C, we know that:

1. The particle moves in a plane.
2. It satisfies the law of the conservation of angular momentum.
3. It sweeps out equal areas in equal times.

By Definition 34.5, the region in which the particle moves is a force field. Hence, by Example 34.62 and property D following it, we also know that:

4. This field is conservative and the particle therefore satisfies the law of the conservation of energy.

To find the equation of the path of the particle, we take the x,y axes in the plane of the particle's motion and the origin at the fixed point O toward which the force F is directed. Let $P(x,y)$ be the coordinates of the position of the particle at time t in a rectangular system and (r,θ) its coordinates in a polar system. By (34.42),

(34.8) $$F_x = m\ddot{x} = \frac{Fx}{r}, \qquad F_y = m\ddot{y} = \frac{Fy}{r}.$$

By hypothesis

(34.81) $$F = -\frac{Km}{r^2},$$

where for convenience we have taken Km, $K > 0$, for the proportionality constant. The minus sign is necessary because the force acts toward O and the positive direction is outward from O. Substituting (34.81) in (34.8), we obtain the system of equations

(34.82) $$\ddot{x} = -\frac{K}{r^3}x, \qquad \ddot{y} = -\frac{K}{r^3}y.$$

If in (34.82), we substitute for r its equal $\sqrt{x^2 + y^2}$, the resulting equations form a nonlinear system which is difficult to solve. It turns out that the path of the particle can be found more easily by using polar coordinates. Call F_r the component of F in the radial r direction and F_θ

its component in a direction perpendicular to F_r. Then by (34.21),

$$(34.83) \quad F_r = ma_r = m(\ddot{r} - r\dot{\theta}^2), \quad F_\theta = ma_\theta = m(2\dot{r}\dot{\theta} + r\ddot{\theta}).$$

Since F is a central force, $F_\theta = 0$. Setting the second equation in (34.83) equal to zero, and then multiplying it by r/m, we find $2r\dot{r}\dot{\theta} + r^2\ddot{\theta} = 0$, which is equivalent to

$$(34.84) \quad \frac{d}{dt}(r^2\dot{\theta}) = 0, \quad r^2\dot{\theta} = h, \quad \dot{\theta} = h/r^2,$$

where h is the same constant we introduced in (34.47), i.e., h is the angular momentum of the particle per unit mass. In the first equation of (34.83), substitute for F_r its value as given in (34.81) (remember here $F_r \equiv F$ since the force acts only along r) and for $\dot{\theta}$ its value h/r^2 as given in (34.84). We thus obtain

$$(34.85) \quad -\frac{Km}{r^2} = m\left(\ddot{r} - r\frac{h^2}{r^4}\right), \quad \ddot{r} - \frac{h^2}{r^3} = -\frac{K}{r^2}.$$

Although methods of solving the nonlinear equation (34.85) are given in both Lessons 35A and 35C which follow—see also Exercise 35,11—use of either of these methods will give a solution of t as a function of r. It turns out to be easier to analyze the path of the particle if we solve the second equation in (34.85) for r as a function of θ. To accomplish this end, we use the substitution

$$(34.86) \quad u = \frac{1}{r}, \quad r = \frac{1}{u}.$$

[Note that u as defined in (34.86) is the force function U of Example 34.62.] Substituting (34.86) in the last equation of (34.84), we obtain

$$(34.87) \quad \dot{\theta} = hu^2.$$

Two differentiations of the second equation in (34.86) give, with the help of (34.87),

$$(34.88) \quad \dot{r} = -\frac{1}{u^2}\frac{du}{dt} = -\frac{1}{u^2}\frac{du}{d\theta}\frac{d\theta}{dt} = -\frac{1}{u^2}\frac{du}{d\theta}hu^2 = -h\frac{du}{d\theta},$$

$$\ddot{r} = -h\frac{d^2u}{d\theta^2}\dot{\theta} = -h\frac{d^2u}{d\theta^2}hu^2 = -h^2u^2\frac{d^2u}{d\theta^2}.$$

Substituting the last value of $\ddot{r}$ of (34.88) and the value of r of (34.86) in the second equation of (34.85), we obtain

$$(34.89) \quad -h^2u^2\frac{d^2u}{d\theta^2} - h^2u^3 = -Ku^2,$$

which simplifies to the linear equation

$$(34.891) \quad \frac{d^2u}{d\theta^2} + u = \frac{K}{h^2}, \quad h \neq 0.$$

Its solution, by any method you wish to choose, is

(34.892) $$u = \frac{K}{h^2} + c \cos (\theta - \theta_0), \quad h \neq 0,$$

where c and θ_0 are arbitrary constants. Since K, h, and c are constants, we can write (34.892) in a more useful form by replacing c by a new constant Ke/h^2. We thus obtain

(34.893) $$u = \frac{K}{h^2} [1 + e \cos (\theta - \theta_0)], \quad h \neq 0.$$

By (34.86), $u = 1/r$. If we now make this substitution in (34.893) and choose our axes so that $\theta_0 = 0$, the equation simplifies to

(34.894) $$r = \frac{h^2}{K(1 + e \cos \theta)}, \quad h \neq 0,$$

which is the equation, in polar coordinates, of the path of a particle moving subject to a central force whose magnitude varies inversely as the square of its distance from a fixed point O.

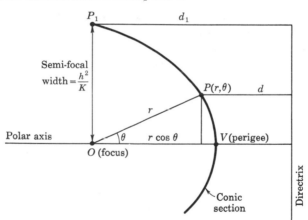

Figure 34.895

We digress momentarily to review for you the proof that (34.894) is the equation of a conic section whose eccentricity is e, whose semifocal width is h^2/K and which has one focus at the origin. In Fig. 34.895, we have illustrated such a conic. The point on a particle's path that is nearest the point O to which the particle is attracted is called the **perigee** of the path. By definition, the ratio r/d for every point P on a conic is equal to its eccentricity e. Hence, for the two points P and P_1 on the conic we have respectively,

(a) $\quad e = \dfrac{r}{d} \quad$ and $\quad e = \dfrac{h^2}{Kd_1}; \quad d = \dfrac{r}{e} \quad$ and $\quad d_1 = \dfrac{h^2}{eK}, \quad h \neq 0.$

From the figure, we see that

(b) $$d_1 = d + r \cos \theta.$$

In (b), replace d and d_1 by their values as given in the last two equations of (a). There results

(c) $$\frac{h^2}{eK} = \frac{r}{e} + r \cos \theta, \qquad r = \frac{h^2}{K(1 + e \cos \theta)}, \quad h \neq 0,$$

which is the same as (34.894). Additionally, we know from analytic geometry that if

(d) $e < 1$, the conic is an ellipse,

            $e = 1$, the conic is a parabola,

            $e > 1$, the conic is an hyperbola.

If $e = 0$, (34.894) becomes $r = h^2/K$, which, in polar coordinates, is the equation of a circle whose radius is h^2/K.

Comment 34.896. We have thus proved that, if $h \neq 0$, the orbit of a particle, attracted to a fixed point O by a central force which satisfies the inverse square law (34.81), is a conic section with one focus at the fixed point O. If $e < 1$, the conic section is an ellipse. Since one focus is at the fixed point O, which we also took to be the origin of our coordinate system, O cannot be the center of the ellipse. If therefore a particle, subject to a force which satisfies the inverse square law (34.81), moves in an elliptical orbit, the central force is directed toward a **focus** of the ellipse and not toward its center. Contrast this result with that obtained in Lesson 34E. We found there that if F varies *directly* as the distance r from O, then the force is directed toward the **center** of the ellipse.

Comment 34.8961. If $h = 0$, then by (34.84), $\dot{\theta} = 0$. Therefore θ is a constant. Hence the particle must move on a line. By (34.85), with $h = 0$, the differential equation of motion simplifies to

$$\ddot{r} = -\frac{K}{r^2}.$$

For possible methods of solving it, see Lesson 35, also Exercise 35,7 and 22.

Comment 34.897. Determining the Constants of Integration h, e, θ_0 of (34.893). The path of the particle, by (34.893), with u replaced by its equal $1/r$, is

(a) $$\frac{1}{r} = \frac{K}{h^2} [1 + e \cos (\theta - \theta_0)],$$

where h, e, and θ_0 are constants which were introduced by integrations. Assume that when the particle is at the point P_0 of its path, its distance from O is r_0 and that it is moving with a velocity v_0 in a direction making

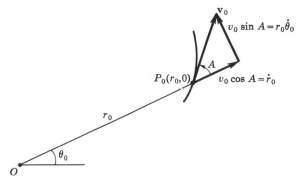

Figure 34.898

an angle A with the line joining O to P_0, Fig. 34.898. For convenience, we measure subsequent values of the angle θ from this line OP_0. The initial conditions are, therefore, [for v_r, v_θ values see Fig. 34.898 and (34.2)]

(b) $r = r_0,$ $\theta = 0,$ $v = v_0,$ $A = $ angle shown in Fig. 34.898,

$$v_r = v_0 \cos A = \dot{r}_0, \qquad v_\theta = v_0 \sin A = r_0 \dot{\theta}_0.$$

The substitution of (b) in (a) gives

(c) $$\frac{1}{r_0} = \frac{K}{h^2}[1 + e \cos \theta_0], \qquad e \cos \theta_0 = \frac{h^2}{r_0 K} - 1.$$

Let α be the angle measured from the radius vector to a tangent to a curve at $P(r,\theta)$, Fig. 34.899. Let **v** be the velocity of the particle at P.

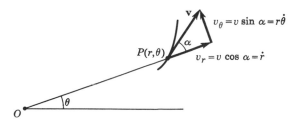

Figure 34.899

Therefore the components of **v** in the radial direction and in a direction perpendicular to it are respectively

(d) $$v_r = v \cos \alpha, \qquad v_\theta = v \sin \alpha.$$

By (34.2), $v_r = \dot{r}$ and $v_\theta = r\dot{\theta}$. Hence (d) becomes

(e) $$\dot{r} = v \cos \alpha, \qquad r\dot{\theta} = v \sin \alpha.$$

By (34.84), $r\dot{\theta} = h/r$. Therefore the second equation in (e) can be written as

(f) $$h = rv \sin \alpha.$$

Hence when $r = r_0$, $v = v_0$ and $\alpha = A$, we obtain, by (f) and (b),

(g) $$h = r_0 v_0 \sin A = r_0{}^2 \dot{\theta}_0.$$

Differentiation of (a) with respect to θ gives

(h) $$-\frac{1}{r^2}\frac{dr}{d\theta} = -\frac{Ke}{h^2} \sin (\theta - \theta_0).$$

Also

$$
\begin{aligned}
-\frac{1}{r^2}\frac{dr}{d\theta} &= -\frac{1}{r^2}\frac{dr}{dt}\frac{dt}{d\theta} \\
&= -\frac{1}{r^2}\frac{dr}{dt}\frac{r^2}{h} \quad \text{[by (34.84)]} \\
&= -\frac{1}{h}\frac{dr}{dt} \\
&= -\frac{1}{h}v \cos \alpha \quad \text{[by (e) above]}.
\end{aligned}
$$

Hence, by equating this last expression with the right side of (h), we obtain

(i) $$\frac{Ke}{h} \sin (\theta - \theta_0) = v \cos \alpha.$$

Inserting in (i) the initial conditions $\theta = 0$, $v = v_0$, $\alpha = A$, $v_0 \cos A = \dot{r}_0$, there results

(j) $$\frac{Ke}{h} \sin (-\theta_0) = v_0 \cos A, \qquad -e \sin \theta_0 = \frac{hv_0}{K} \cos A = \frac{h}{K}\dot{r}_0.$$

Squaring the second equations in (c) and (j) and adding them, we have

(k) $$e^2 = \left(\frac{h^2}{r_0 K} - 1\right)^2 + \left(\frac{h}{K}\dot{r}_0\right)^2 = \frac{h^4}{r_0{}^2 K^2} - \frac{2h^2}{r_0 K} + 1 + \frac{h^2 v_0{}^2}{K^2} \cos^2 A.$$

We can simplify (k) somewhat, by noting from (g) that

(l) $$h^2 = r_0{}^2 v_0{}^2 \sin^2 A, \qquad \frac{h^2}{r_0{}^2 v_0{}^2} = 1 - \cos^2 A,$$

$$\cos^2 A = 1 - \frac{h^2}{r_0{}^2 v_0{}^2}.$$

Substituting the last equation of (l) in (k), we find

(m)
$$e^2 = \frac{h^4}{r_0{}^2K^2} - \frac{2h^2}{r_0K} + 1 + \frac{h^2v_0{}^2}{K^2} - \frac{h^4}{r_0{}^2K^2}$$
$$= 1 - \frac{2h^2}{r_0K} + \frac{h^2v_0{}^2}{K^2}.$$

By (g), we can determine h. By (k) or (m), we can then determine e; remember K is a proportionality constant and not a constant of integration. With c and h known, we can determine θ_0 by the second equations in (c) and (j). After θ_0 is known, we can then choose our axes to make $\theta_0 = 0$, and thus obtain (34.894).

Comment 34.9.   Energy Considerations Related to the Inverse Square Law. In property D following Example 34.62, we showed that the sum of the kinetic and potential energies of a particle moving in a conservative field is a constant, see (34.66). Therefore, by (34.66),

(a)
$$\tfrac{1}{2}mv^2 - U = E,$$

where we have replaced the constant c by the energy constant E, and the velocity components, $\dot{x}, \dot{y}, \dot{z}$, by v. In Example 34.62, we showed that if $F = -k/r^2$, then the force field is conservative and the potential energy function $-U = -k/r$. In the example of this Lesson 34F, $F = -Km/r^2$, see (34.81). Hence the potential energy function $-U = -Km/r$. Substituting this value of U in (a), it becomes

(b)
$$\tfrac{1}{2}mv^2 - \frac{Km}{r} = E.$$

Inserting in (b), the initial values $v = v_0$ and $r = r_0$, we obtain— remember E is constant for all v and r—

(c)
$$\tfrac{1}{2}mv_0{}^2 - \frac{Km}{r_0} = E.$$

By (m) of Comment 34.897,

(d)
$$1 - e^2 = \frac{2h^2}{r_0K} - \frac{h^2v_0{}^2}{K^2}.$$

Multiply (c) by $-2h^2/K^2m$. There results·

(e)
$$-\frac{2h^2}{K^2m} E = -\frac{2h^2}{K^2m}\left(\tfrac{1}{2}mv_0{}^2 - \frac{Km}{r_0}\right) = -\frac{h^2v_0{}^2}{K^2} + \frac{2h^2}{r_0K}.$$

Since the right sides of (d) and (e) are the same, we can equate their left sides. We have thus shown that

(f)
$$1 - e^2 = -\frac{2h^2}{K^2m} E = -\frac{2}{m}\left(\frac{h}{K}\right)^2 E.$$

The mass m is a positive quantity; so also is $(h/K)^2$. Hence we conclude from (f), that if $E < 0$,

$$\text{(g)} \qquad\qquad 1 - e^2 > 0, \qquad e^2 < 1, \qquad e < 1,$$

and the particle, therefore, must move in an elliptic orbit. If $E = 0$, then $1 - e^2 = 0$, $e = 1$, and the particle must move in a parabolic orbit; if $E > 0$, then $e > 1$ and the particle must move in a hyperbolic orbit. We infer from all this that a particle in motion in a conservative field, whose orbit is elliptic, must initially have had negative energy. Conversely, if a particle with negative energy is projected into a conservative field, it will move in an elliptic orbit. Analogous remarks can be made for the other two types of orbits.

Remark. A particle will have negative energy initially, if its kinetic energy, which is the first term of (a), is less than the force function $U = Km/r$.

Comment 34.91. Equation (34.894) gives the position r of a particle as a function of θ, $h \neq 0$. It would be desirable to express r as a function of t so that we can know where the particle is at any moment. By (34.85) $\ddot{r} = h^2/r^3 - K/r^2$. As mentioned previously, two methods of solving this equation will be given in Lesson 35. (Also see Exercise 35,11.) Unfortunately, use of either method gives t as a function of r. The problem of solving the resulting equation for r as a function of t turns out to be exceedingly difficult.

EXERCISE 34F

1. Verify the accuracy of the solution of (34.891) as given in (34.892).
2. A body weighing 16 lb is 12 ft from a fixed point O. It is given an initial velocity of 6 ft/sec in a direction perpendicular to the x axis. Find its equation of motion if it is subject to a central force whose magnitude is equal to $120/r^2$, where r is the distance of the particle from O. *Hint.* Follow the method of the text. Initial conditions in rectangular coordinates are $t = 0$, $x = x_0 = 12$, $y = y_0 = 0$, $\dot{x}_0 = 0$, $\dot{y}_0 = v_0 = 6$. In polar coordinates initial conditions are $t = 0$, $r = r_0 = 12$, $\theta = \theta_0 = 0$, $\dot{r}_0 = 0$, $\dot{\theta} = \dot{\theta}_0 = v_0/r_0 = \frac{6}{12} = \frac{1}{2}$. What is the eccentricity of the path?
3. A body weighing 16 lb is 10 ft from a fixed point O. It is subject to a central force F whose magnitude is $100/r^2$, where r is the distance of the particle from O. What initial velocity should be given the particle, in a direction perpendicular to the x axis, in order that the particle may (a) move in an elliptic orbit of eccentricity $\frac{1}{2}$, (b) move in a circular orbit (*hint*, orbit is circular if $e = 0$), (c) move in a parabolic orbit, (d) move in a hyperbolic orbit of eccentricity 2?

In the problems below, we shall consider the motion of a satellite of the earth, where the satellite has been set in motion by being ejected from a rocket. These problems are central force problems, obeying the inverse square law (34.81). The satellite will be attracted toward the center of the earth with a force inversely proportional to the square of the distance r of the satellite from this center. Call R the radius of the earth.

4. In (34.81), replace F by ma, where a is the acceleration of the particle.
   (a) Show that

$$(34.911) \qquad K = gR^2.$$

    Hint. When $r = R$, $a = -g$.
   (b) Show that the equation of motion (34.892) becomes

$$(34.912) \qquad u = \frac{gR^2}{h^2} + C \cos(\theta - \theta_0),$$

and that (34.894) becomes

$$(34.913) \qquad r = \frac{h^2}{gR^2[1 + e(\cos\theta - \theta_0)]},$$

where

$$(34.914) \qquad h = r^2\dot{\theta}.$$

5. Assume in (34.913) that at $t = 0$, $\theta = 0$, that the last rocket is fired at a distance $r = r_0$ from the center of the earth and that it has ejected the satellite with a velocity v_0 in a direction making an angle A with the radius vector joining the rocket to the center of the earth. See Fig. 34.898. Show that the constants h, e, and θ_0 in (34.913) are given respectively by

$$(34.915) \qquad h = r_0^2\dot{\theta}_0,$$

$$(34.916) \qquad e^2 = \left(\frac{r_0^3\dot{\theta}_0^2}{gR^2} - 1\right)^2 + \left(\frac{r_0^2\dot{\theta}_0}{gR^2}\dot{r}_0\right)^2,$$

$$(34.917) \qquad \cos\theta_0 = \frac{1}{e}\left(\frac{r_0^3\dot{\theta}_0^2}{gR^2} - 1\right), \qquad \sin\theta_0 = -\frac{1}{e}\left(\frac{r_0^2\dot{\theta}_0\dot{r}_0}{gR^2}\right),$$

where $\dot{r}_0$ and $r_0\dot{\theta}_0$ are the initial velocity components of v_0 in the radial direction and in a direction perpendicular to the radial axis. *Hint.* With $K = gR^2$ as given in (34.911), equation (34.913) is the same as (a) of Comment 34.897. Make use of (g), (k), (c), and (j) of this comment.

6. Show that the orbit of a satellite of the earth will be a circle if

$$(34.918) \qquad r_0^3\dot{\theta}_0^2 = gR^2 \quad \text{and} \quad \dot{r}_0 = 0.$$

Hint. By (34.913), the orbit is circular if $e = 0$; by (34.916) $e = 0$ if (34.918) holds.

7. A satellite is ejected by a rocket into a *circular* orbit 300 mi above the earth's surface. Find its period of rotation. *Hint.* The period of the satellite, by Definition 28.34, is $T = 2\pi/\omega$, where $\omega = \dot{\theta}$ is its angular velocity. Therefore by (34.918), $T = (2\pi r_0/R)\sqrt{r_0/g}$. Here $R = 4000$ mi, $r_0 = 4300$ mi.

8. If the satellite is very close to the surface of the earth so that r_0 is very close to R, then (34.918) can be written as

$$(34.919) \qquad \dot{\theta}_0 = \sqrt{g/R}.$$

Show that the period in this case, for a circular orbit, is approximately 85 min. See hint in problem 7 for period formula. [Compare with answer to Exercise 28A and B, 2(c).]

9. At a distance r_0 from the center of the earth, a rocket propels a satellite in a direction perpendicular to the radius vector joining the rocket to the center. The velocity of the satellite is v_0. Hence $\dot{r}_0 = 0$, $\dot{\theta}_0 = v_0/r_0$ and the angle A in Fig. 34.898 is 90°.

(a) Show that if $\dot{\theta}_0^2 < gR^2/r_0^3$, then $\theta_0 = \pi$. *Hint.* In (34.916), $\dot{r}_0 = 0$. Solve for e and, since the eccentricity is always positive, choose the proper sign to make $e > 0$. Substitute this value of e in (34.917).

(b) Show that the orbit of the satellite is then

$$r = \frac{r_0^4 \dot{\theta}_0^2}{gR^2(1 - e\cos\theta)}, \text{ where } e = 1 - \frac{r_0^3 \dot{\theta}_0^2}{gR^2}, \text{ or}$$

$$r = \frac{r_0^4 \dot{\theta}_0^2}{gR^2 - (gR^2 - r_0^3 \dot{\theta}_0^2)\cos\theta}.$$

Hint. Use (34.913), (34.915), and (34.916).

(c) Show that the satellite is farthest from the center of the earth, called the **apogee** of the orbit, when $r = r_0$, $\theta = 0$, i.e., the apogee is at the point where the satellite is released. *Hint.* The distance r is largest when the denominator in equation (b) above is smallest. The denominator is smallest when the negative term in it is largest. This negative term is largest when $\theta = 0$, $\cos\theta = 1$. Solve for r with $\cos\theta = 1$.

(d) Show that the satellite's **perigee**, i.e., the point of the satellite's orbit nearest the center of the earth, occurs when $\theta = \pi$, and that its distance from the center of the earth is then $r_0^4 \dot{\theta}_0^2/(2gR^2 - r_0^3 \dot{\theta}_0^2)$. See hint in (c) above.

(e) Show that the satellite will make a complete orbit without hitting the earth if $\dot{\theta}_0^2 > 2gR^3/[r_0^3(r_0 + R)]$. *Hint.* The perigee of the orbit, as given in (d) above, must be greater than the radius R of the earth.

(f) Show that if $\dot{\theta}_0^2 > gR^2/r_0^3$, then $\theta_0 = 0$. See hint in (a) of this problem. Show that the orbit of the satellite is then

$$r = \frac{r_0^4 \dot{\theta}_0^2}{gR^3(1 + e\cos\theta)}, \text{ where } e = \frac{r_0^3 \dot{\theta}^2}{gR^2} - 1, \text{ or}$$

$$r = \frac{r_0^4 \dot{\theta}_0^2}{gR^2 + (r_0^3 \dot{\theta}_0^2 - gR^2)\cos\theta}.$$

See hint in (b) above. Show that the perigee of the orbit occurs when $r = r_0$, $\theta = 0$, i.e., at the point where the satellite is released; that the apogee of the orbit occurs when $\theta = \pi$ and that its distance from the center of the earth is then $r_0^4 \dot{\theta}_0^2/(2gR^2 - r_0^3 \dot{\theta}_0^2)$, provided $2gR^2 > r_0^3 \dot{\theta}_0^2$. See hints in (c) and (d) above. Finally show that if $2gR^2 \leqq r_0^3 \dot{\theta}_0^2$, the orbit will not have an apogee. *Hint.* The denominator of the apogee's distance formula above will then be negative or zero.

If $2gR^2 = r_0^3 \dot{\theta}_0^2$, then $r_0^2 \dot{\theta}_0^2 = 2gR^2/r_0$. If the satellite is ejected at or near the surface of the earth so that $r_0 = R$, then the last equation becomes $r_0 \dot{\theta}_0 = \sqrt{2gR}$. But $v_0 = r_0 \dot{\theta}_0$ so that $v_0 = \sqrt{2gR}$ which is the escape velocity of a body fired from the surface of the earth, see (i) of Example (16.36).

ANSWERS 34F

2. $h = r_0 v_0 = 72, m = \frac{1}{2}, mk = 120, k = 240,$
$e = 0.8, r = 21.6/(1 + 0.8 \cos \theta).$

3. (a) 5.48 ft/sec.   (b) 4.47 ft/sec.   (c) 6.32 ft/sec.   (d) 7.75 ft/sec.

7. 5700 sec, approximately, or 95 min. The first satellite put into orbit in 1957 by the U.S.S.R., known as the Sputnik, had a nearly circular orbit of 300 miles above the earth's surface and a period of 96 min.

LESSON 34G. Planetary Motion. Newton's law of universal gravitation states that every two bodies in the universe attract each other with a force proportional to the product of their masses and inversely proportional to the square of the distance separating them. Let M be the mass of the sun and m the mass of a planet. It can be proved that we do not commit a serious error if we consider the sun as fixed, its mass M as concentrated at its center, the planet as a particle, and sun and planet as isolated bodies. Then by Newton's law of universal gravitation,

$$(34.92) \qquad\qquad F = -\frac{GMm}{r^2},$$

where r is the distance of a planet from the sun's center, and G is a proportionality constant called the **gravitational constant.** Replacing the constant GM in (34.92) by a new constant K, it becomes $F = -Km/r^2$, which is the same equation as (34.81) of Lesson 34F. Since this force F is directed toward a fixed point O, namely the sun's center, it is a central force. Hence planetary motion is exactly the same as the motion of the particle discussed in Lesson 34F. We can therefore assert that:

1. A planet moves in a plane.
2. The orbit of a planet is a conic section whose equation, by (34.894), is

$$r = \frac{h^2}{K(1 + e \cos \theta)}, \quad h \neq 0,$$

where h is the angular momentum of the planet per unit mass, e is the eccentricity of its orbit and K is the product of the gravitational constant G and the mass M of the sun.

3. The planets satisfy the law of the conservation of angular momentum.
4. The planets sweep out equal areas in equal intervals of time.
5. The force field in which the planets move is conservative; hence the planets satisfy the law of the conservation of energy.
6. The sun is at one focus of the planet's orbit. [See Comment 34.896.]

Comment 34.93. *The Orbits of the Earth and the Other Planets of Our Solar System Are Ellipses with the Sun at One Focus.* Hence for the planets of our solar system $e < 1$. This means, as we showed in Comment 34.9, that each planet, at the beginning of its existence, had negative energy.

The orbits of comets* which appear after long intervals of time are extremely elongated ellipses whose eccentricity is near 1, almost close to parabolas. Those bodies for which $e \geq 1$ have parabolic or hyperbolic orbits. They leave the solar system and never return.

EXERCISE 34G

1. Find the equation of motion of a planet of mass m if its distance at perigee, i.e., its distance nearest the sun, is r_0 and its velocity there is v_0. *Hint.* Take the axis of the ellipse through the perigee. Then at $t = 0, r = r_0, \theta = \theta_0 = 0$, $\dot{r} = 0, \dot{\theta} = v_0/r_0$. See Fig. 34.898.
2. Find the approximate equation of Halley's comet. *Hint.* See footnote at the bottom of page: $e = 0.967$, $a - c = 0.587$, where a is the semimajor axis and c is the distance of the focus from the center of the elliptic orbit.
3. A comet at rest at an infinite distance away from the sun is attracted toward the sun in accordance with the inverse square law. If its distance at perigee is r_0, find the equation of its path and show that its orbit is parabolic. *Hint.* Take axis so that $\theta_0 = 0$ in (34.892). At $t = 0, \theta = \pi, u = 1/r = 0$. When $\theta = 0, u = 1/r_0$.

ANSWERS 34G

1. $r = \dfrac{r_0{}^2 v_0{}^2}{K + (r_0 v_0{}^2 - K) \cos \theta}$.

2. $\dfrac{x^2}{316.41} + \dfrac{y^2}{20.54} = 1$. Figures in astronomical units.

3. $r = \dfrac{2r_0}{1 + \cos \theta}$. Orbit is parabolic since e, the coefficient of $\cos \theta$, is one.

LESSON 34H. Kepler's (1571–1630) Laws of Planetary Motion. Proof of Newton's Inverse Square Law. Kepler's three laws of planetary motion are:

1. Each planet moves in an elliptical orbit with the sun at one focus.
2. The radius vector connecting sun and planet sweeps out equal areas in equal times.
3. The square of the period of a planet is proportional to the cube of the semimajor axis of its orbit.

We have already proved 1 and 2: see numbers 6 and 4 of Lesson 34G. We shall now prove 3.

Proof of 3. By (34.894), the orbit of a planet is given by

(a) $$r = \frac{h^2}{K(1 + e \cos \theta)},$$

---

*The famous Halley's comet has an elliptical orbit whose eccentricity is 0.967. Its period is 76 years. Its perigee is 0.587 astronomical units (an astronomical unit is the distance of the earth to the sun, approximately 92,900,000 mi). Since it last visited us in 1910, it will be again visible in 1986.

where one focus is at the origin of a coordinate system, called $(\overline{0},\overline{0})$ in Fig. 34.94, and the semifocal width is h^2/K, called L in the figure. Let $(0,0)$ be the center of the ellipse. With respect to the center of the ellipse,

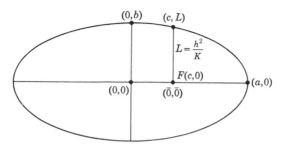

Figure 34.94

let $(c,0)$ be the coordinates of the focus, $(a,0)$, $(0,b)$, be the coordinates of the ends of the semimajor and semiminor axes respectively. The equation of the ellipse with respect to its center as an origin is, therefore,

(b)
$$\frac{x^2}{a^2} + \frac{y^2}{b^2} = 1,$$

where $c^2 = a^2 - b^2$. If $x^2 = c^2 = a^2 - b^2$, then $y^2 = L^2$, and therefore, by (b),

(c)
$$\frac{a^2 - b^2}{a^2} + \frac{L^2}{b^2} = 1, \qquad \frac{L^2}{b^2} = \frac{b^2}{a^2}, \qquad L = \frac{b^2}{a}.$$

The semifocal width L also equals h^2/K. Substituting this value of L in the last equation of (c), we obtain

(d)
$$\frac{h^2}{K} = \frac{b^2}{a}, \qquad b^2 = \frac{ah^2}{K}.$$

The first equation in (34.49) holds for every particle subject to a central force. It therefore holds for the planets. The last equation in (34.49) resulted when we took for our initial conditions $A = 0$, $t = 0$. Hence if $A = 0$, $t = 0$,

(e)
$$A = \tfrac{1}{2}ht,$$

gives the area A swept out by a planet in time t. Call T the period of a planet's orbit, i.e., the time it takes a planet to make a complete circuit of its orbit. When the planet has made a complete circuit, it has swept out the area of the ellipse, namely πab. Hence when $A = \pi ab$ and $t = T$, we obtain by (e)

(f) $$T = \frac{2\pi ab}{h}, \qquad T^2 = \frac{4\pi^2 a^2 b^2}{h^2}.$$

In (f), replace b^2 by its value as given in (d). There results

(34.95) $$T^2 = \frac{4\pi^2 a^2}{h^2} \cdot \frac{ah^2}{K} = \frac{4\pi^2 a^3}{K}.$$

Since $K(= GM)$ is a constant, and a is the semimajor axis of the ellipse, (34.95) says that the square of the period of a planet is proportional to the cube of the semimajor axis of its orbit.

Proof of Newton's Inverse Square Law from Kepler's Laws. We have proved Kepler's three laws from Newton's universal law of gravitation. Historically, however, Kepler preceded Newton and hence the former knew nothing of the law of gravitation. It is indeed remarkable that Kepler was able to deduce his three laws from an intensive study of the recordings of the positions of the planets made by direct observations. It was Newton who used Kepler's laws as a hypothesis to develop his own universal law of gravitation. Part of his problem was thus the inverse of the one we solved. He assumed that a planet moves in an elliptical orbit with the sun at one focus, that it sweeps out equal areas in equal times, and then set out to prove that the planet must therefore be subject to a central force directed toward a focus, whose magnitude varies inversely as the square of the distance of the planet from the sun. The proof follows.

Proof. By Kepler's second law, dA/dt is a constant. Therefore by the second equation in (34.481),

(a) $$\tfrac{1}{2} r^2 \dot{\theta} = \frac{c}{2}, \qquad r^2 \dot{\theta} = c,$$

where c is a constant. Differentiation of the second equation in (a) and multiplying the result by a constant mass m, gives

(b) $$m(2r\dot{r}\dot{\theta} + r^2\ddot{\theta}) = 0, \qquad m(2\dot{r}\dot{\theta} + r\ddot{\theta}) = 0.$$

The second equation of (b), by (34.21), is $ma_\theta = 0$. But ma_θ is the component of force acting on a particle in a direction perpendicular to the radius vector. Since this component is zero, the force acting on a planet must be a central one; i.e., the force always acts along a radius vector toward or away from the sun.

By Kepler's first law, a planet moves in an elliptical orbit with the sun at one focus. We showed in Lesson 34F, that the equation of an ellipse in polar coordinates with one focus at the origin is

(c) $$r = \frac{A}{1 + e \cos \theta},$$

where $e < 1$ is its eccentricity and A is its semifocal width. Differentiation of (c) gives

(d) $$\dot{r} = \frac{Ae \sin \theta}{(1 + e \cos \theta)^2} \dot{\theta} = \frac{A^2}{(1 + e \cos \theta)^2} \frac{e \sin \theta}{A} \dot{\theta}.$$

Making use of (c) and (a), we can write (d) as

(e) $$\dot{r} = r^2 \frac{e \sin \theta}{A} \frac{c}{r^2} = \frac{ce}{A} \sin \theta.$$

Differentiating (e) and then using (a), we obtain

(f) $$\ddot{r} = \frac{c}{A} (e \cos \theta)\dot{\theta} = \frac{c^2}{Ar^2} (e \cos \theta).$$

Solving (c) for $e \cos \theta$, there results

(g) $$e \cos \theta = \frac{A - r}{r}.$$

Substituting this value in (f), we obtain

(h) $$\ddot{r} = \frac{c^2}{Ar^2} \left(\frac{A - r}{r} \right) = \frac{c^2}{r^3} - \frac{c^2}{Ar^2}.$$

The component of a force in a radial direction is, by the first equation in (34.21),

(i) $$F_r = ma_r = m(\ddot{r} - r\dot{\theta}^2).$$

In (i), replace $\ddot{r}$ by its value in (h) and $\dot{\theta}$ by its value in (a). There results

(j) $$F_r = m\left[\frac{c^2}{r^3} - \frac{c^2}{Ar^2} - \frac{c^2}{r^3} \right] = -\frac{mc^2}{Ar^2}.$$

We showed above that the force acting on a planet is toward or away from the sun. Since m, c^2, and A are positive constants, (j) tells us that the force acting on a planet is directed toward the sun, and its magnitude is inversely proportional to the square of its distance from the sun.

Comment 34.951. The inverse square law just proved is only part of the universal law of gravitation. Newton's studies of the gravitational force of the earth plus his observations of the moon's orbit about the earth, plus his own genius, enabled him to formulate his famous law of universal gravitation as stated at the beginning of Lesson 34G.

Comment 34.96. In Lesson 34C, we proved that every particle subject to a central force obeys Kepler's second law, i.e., it sweeps out equal areas in equal times. Hence Newton's inverse square law is only a *sufficient* condition for Kepler's equal area law, not a *necessary* one. However, the inverse square law is a necessary and sufficient condition for Kepler's first law: the orbit of a particle subject to a central force is an ellipse with the force directed toward a **focus**. Note that if F is proportional to r as in Lesson 34E, the orbit is also elliptical, but the force is directed toward the **center** of the ellipse.

EXERCISE 34H

1. (a) Take 240,000 miles as the semimajor axis of the moon's orbit and its period as 27.3 days. Use Kepler's third law to find the value of the proportionality constant k for the earth, where k replaces $4\pi^2/K$ in (34.95). Use for units 1000 mi and hour.
   (b) In Exercise 34F,7, we found $T = 95$ min for the period of the circular orbit of a satellite of the earth 300 mi above its surface. Take 4300 mi for the semimajor axis of the satellite's orbit and calculate k. Use for units 1000 mi and hour.
   (c) In Exercise 34F,8, we found $T = 85$ min for the period of the orbit of a satellite close to the earth's surface. Take 4000 mi for the semimajor axis of the orbit and calculate k. Use for units 1000 mi and hour.
   Compare results in (a), (b), and (c). *Ans.* $k = 0.031$.

   Remark. Next time you read of an earth satellite which has been successfully orbited, and its perigee and apogee are given, use this value of k and Kepler's third law to calculate the period of the orbit and see if it agrees with the observed period.

2. (a) Use Kepler's third law to calculate the value of the proportionality constant k for the sun. Use for units 1,000,000 mi and day. Take the period of the earth as 365 days and the semimajor axis of its orbit as 93,000,000 mi. *Ans.* $k = 0.165$.
   (b) Use this value of k to calculate the period of one of the other planets from its known distance from the sun, or calculate its mean distance from the sun from its known period.

By (34.891),

(a) $$\frac{d^2u}{d\theta^2} + u = \frac{K}{h^2}, \quad h \neq 0.$$

By (34.81) and (34.86),

(b) $$F = -\frac{Km}{r^2}, \quad K = -\frac{Fr^2}{m} = -\frac{F}{mu^2}.$$

Hence (a) becomes

(34.97) $$\frac{d^2u}{d\theta^2} + u = -\frac{Fr^2}{mh^2} = -\frac{F}{mh^2u^2},$$

which is the differential equation, in polar coordinates, of the orbit of a particle subject to a central force. For different values of the force F, there will be different orbits. Conversely, if the equation $r = r(\theta)$ in polar coordinates of the orbit of a particle is known, (34.97) will give the central force F which causes the particle to move in this orbit. All one need do to find F is to substitute in (34.97) the values of u and $d^2u/d\theta^2$ and solve for F. (Remember $u = 1/r$.)

Use the above facts and (34.97) to solve the following problems. Assume in all cases that F is a central force and that the fixed point O toward which the force acts is at the origin.

3. The orbit of a particle is an ellipse with one focus at the origin. Show that the force F obeys the inverse square law. (Note. This assertion has already been proved in this lesson; see proof of Newton's inverse square law from Kepler's laws.) *Hint.* The equation of the orbit in polar coordinates is

$$r = \frac{A}{1 + e \cos \theta}.$$

Therefore

$$u = \frac{1 + e \cos \theta}{A}, \qquad \frac{d^2u}{d\theta^2} = -\frac{e}{A} \cos \theta.$$

Substitute the last two values in (34.97). Solve for F. Remember that m, h, and A are constants. *Ans.* $F = -mh^2/Ar^2$.

4. The orbit of a particle is a circle, with the origin a point of the circumference. Show that the force F is inversely proportional to the fifth power of the distance of the particle from the origin. *Hint.* The equation of the orbit in polar coordinates is $r = 2a \cos \theta$, where a is the radius of the circle. Therefore $\sec \theta = 2a/r = 2au$. Also make use of the fact that $\sec^2 \theta = 1 + \tan^2 \theta$. *Ans.* $F = -8a^2h^2m/r^5$.

5. The orbit of a particle is an ellipse with the origin at the center of the ellipse. Show that the force F is proportional to the distance of the particle from the center, see Lesson 34E. *Hint.* The equation of the orbit in polar coordinates is

$$r^2 = \frac{A^2}{1 - e^2 \cos^2 \theta}.$$ Follow suggestions in problem 3.

Ans. $F = -mh^2(1 - e^2)r/A^4$.

6. The orbit of a particle is the spiral $r = e^\theta$. Find the force F.
Ans. $F = -2mh^2/r^3$.

7. The orbit of a particle is the lemniscate $r^2 = a^2 \cos 2\theta$. Find the force F.
Ans. $F = -3mh^2a^4/r^7$.

8. The orbit of a particle is the cardioid $r = a(1 + \cos \theta)$. Find the force F.
Ans. $F = -3mh^2a/r^4$.

9. The orbit of a particle is a circle with center at the origin. Find the force F.
Hint. The equation of the orbit is $r = a$. *Ans.* $F = -mh^2/a^3$.

EXERCISE 34M

MISCELLANEOUS TYPES OF PROBLEMS LEADING TO SYSTEMS OF EQUATIONS

1. A particle moves in a plane. If the x and y components of its velocity are equal respectively to the y and x coordinates of its position, find the equation of its path.
2. (a) Solve problem 1, if the word velocity is changed to acceleration. (b) Find the equation of its path if initially the particle is at the origin and has a velocity of 15 ft/sec in a direction whose slope is $\frac{3}{4}$.

3. Solve the following system of differential equations. They are used in certain problems of electron motion.

$$m\frac{d^2x}{dt^2} + eH\frac{dy}{dt} \doteq eE, \qquad m\frac{d^2y}{dt^2} - eH\frac{dx}{dt} = 0,$$

where m = the mass of the electron,
   e = the charge of the electron,
   H = the intensity of the magnetic field,
   E = the intensity of the electric field.
Assume that initially the electron is at the origin and its velocity is zero.

In Lesson 30M-C, we discussed the problem of a wire twisted by rotating a bob at one end, where the resulting torque or moment of force was proportional to the angle of twist, see (30.63), (30.64), and (30.67). For convenience we recopy (30.67).

(34.971) $$I\frac{d^2\theta}{dt^2} = -k\theta,$$

where k is a proportionality constant, called the **torsional stiffness constant,** and θ is the angle through which the wire has twisted from an equilibrium position. Make use of (34.971) to solve problems 4-6.

4. Three disks are connected by shafts. The moment of inertia of the two end disks is I, of the middle disk $2I$, Fig. 34.98. The torsional stiffness constant of each of the two shafts connecting the three disks is k. If a torque $2T_0 \sin \omega t$

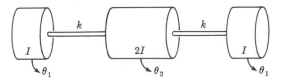

Figure 34.98

is applied to the center disk, find the angular motion of the disks. Assume no resistance and that initially the disks are at rest and the shafts are in their untwisted equilibrium position. *Hint.* Call θ_1 the angular displacement from equilibrium at time t of an end disk and θ_2 the angular displacement from equilibrium of the middle disk. If the end disks are considered fixed at that instant, then at time t, the shafts connecting them to the middle disk have twisted through an angle, $\theta_2 - \theta_1$. Hence the restoring torque acting on the middle disk is $2k(\theta_2 - \theta_1)$. If the middle disk is considered as fixed, then the shaft connecting it to an end disk has twisted through an angle, $-(\theta_2 - \theta_1)$. Hence the restoring torque acting on an end disk is $k(\theta_1 - \theta_2)$. Now make use of (34.971) taking into account the applied torque acting on the center disk. The differential equations can be found in the answer section.

5. Three disks and a driving wheel are connected by shafts, Fig. 34.99. The right end disk is free. Each disk has the same moment of inertia I, and the three shafts have the same torsional stiffness constant, k. Set up the system of

differential equations for the angular motion $\theta_1(t)$, $\theta_2(t)$, $\theta_3(t)$ of the three disks from their equilibrium positions due to an angular motion θ of the driving wheel. Assume that initially the disks are at rest and the shafts are in their untwisted equilibrium position. (See hint given in problem 4.)

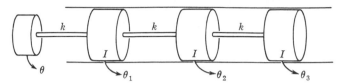

Figure 34.99

6. Solve problem 5 if there is also a resisting torque operating on the disks proportional to the first power of their angular velocities. This proportionality constant is called the **torsional resistance constant.** Assume the torsional resistance constant for each disk is R.

ANSWERS 34M

1. $x = c_1 e^t + c_2 e^{-t}$, $y = c_1 e^t - c_2 e^{-t}$; $x^2 - y^2 = c$.

2. (a) $x = c_1 e^t + c_2 e^{-t} + c_3 \cos t + c_4 \sin t$,
   $y = c_1 e^t + c_2 e^{-t} - c_3 \cos t - c_4 \sin t$.
   (b) $x = \frac{21}{4}(e^t - e^{-t}) + \frac{3}{2} \sin t$, $y = \frac{21}{4}(e^t - e^{-t}) - \frac{3}{2} \sin t$.

3. $x = \dfrac{Em}{eH^2}\left(1 - \cos \dfrac{He}{m} t\right)$, $y = \dfrac{E}{H} t - \dfrac{Em}{eH^2} \sin \dfrac{eH}{m} t$.
   They are the parametric equations of a cycloid. For the definition of a cycloid, see Exercise 28C,34.

4. $I \dfrac{d^2\theta_1}{dt^2} = -k(\theta_1 - \theta_2)$, for each end disk,

   $2I \dfrac{d^2\theta_2}{dt^2} = -2k(\theta_2 - \theta_1) + 2T_0 \sin \omega t$, for the middle disk.

   Solutions are,

   $\theta_1 = c_1 + c_2 t + c_3 \sin \sqrt{2k/I}\, t + c_4 \cos \sqrt{2k/I}\, t + \dfrac{kT_0 \sin \omega t}{I\omega^2(I\omega^2 - 2k)}$,

   $\theta_2 = c_1 + c_2 t - c_3 \sin \sqrt{2k/I}\, t - c_4 \cos \sqrt{2k/I}\, t + \dfrac{T_0(k - \omega^2 I) \sin \omega t}{I\omega^2(I\omega^2 - 2k)}$.

   Initial conditions are $t = 0$, $\theta_1 = 0$, $\theta_2 = 0$, $\dot{\theta}_1 = 0$, $\dot{\theta}_2 = 0$.

5. $I \dfrac{d^2\theta_1}{dt^2} = -k(\theta_1 - \theta) - k(\theta_1 - \theta_2)$,

   $I \dfrac{d^2\theta_2}{dt^2} = -k(\theta_2 - \theta_1) - k(\theta_2 - \theta_3)$,

   $I \dfrac{d^2\theta_3}{dt^2} = -k(\theta_3 - \theta_2)$;

or

$$(ID^2 + 2k)\theta_1 - k\theta_2 = k\theta(t),$$
$$-k\theta_1 + (ID^2 + 2k)\theta_2 - k\theta_3 = 0,$$
$$-k\theta_2 + (ID^2 + k)\theta_3 = 0.$$

Initial conditions at $t = 0$ are $\theta_1 = \theta_2 = \theta_3 = \dot{\theta}_1 = \dot{\theta}_2 = \dot{\theta}_3 = 0$.

6. Add the term $-R \, d\theta_1/dt$ to the right side of the first equation of 5; $-R \, d\theta_2/dt$ to the right side of the second equation; $-R \, d\theta_3/dt$ to the right side of the third equation.

LESSON 35.   Special Types of Second Order Linear and Nonlinear Differential Equations Solvable by Reduction to a System of Two First Order Equations.

In previous lessons, we outlined standard methods by which solutions in terms of elementary functions could be obtained for certain types of first order differential equations and for *linear* differential equations with *constant* coefficients of order $n > 1$. For the *nonlinear* differential equation of order $n > 1$ and for *linear equations with nonconstant coefficients* of order $n > 1$, such standard methods are available only if the equation belongs to one of several special kinds. In Lesson 23 and Exercise 23, 18 and 22, we discussed such special kinds of linear equations with nonconstant coefficients. In this lesson we discuss three special types of nonlinear equations of order two for which a standard method of solution is available. (See also Exercise 35, 23 for an additional type.) These same methods can, of course, also be used if the equation is linear.

LESSON 35A.   Solution of a Second Order Nonlinear Differential Equation in Which y' and the Independent Variable x Are Absent. Equations of this type that we shall consider will be those which can be written in the form

$$(35.1) \qquad\qquad y'' = f(y),$$

where $f(y)$ is defined on an interval $I: a \leq y \leq b$. Note that y' and x are missing. The substitution $u = y'$, $u' = y''$ will change (35.1) into the equivalent first order system

$$(35.11) \qquad\qquad \frac{dy}{dx} = u, \qquad \frac{du}{dx} = f(y).$$

The second equation in (35.11) can be solved for u as follows. Multiply it by $2u$ to obtain $2uu' = 2uf(y)$. Replace $2uu'$ by its equal $(d/dx)(u^2)$, and u by dy/dx. Hence

$$(35.12) \qquad\qquad \frac{d}{dx}(u^2) = 2\frac{dy}{dx}f(y).$$

Therefore

(35.13) $$d(u^2) = 2f(y)\,dy.$$

Integration of (35.13) gives

(35.14) $$u^2 = 2\int f(y)\,dy = F(y) + c_1.$$

Substituting (35.14) in the first equation of (35.11), we have

(35.15) $$\frac{dy}{dx} = \pm\sqrt{F(y) + c_1}.$$

If $\sqrt{F(y) + c_1} \neq 0$, we obtain from (35.15)

(35.16) $$\pm\int \frac{dy}{\sqrt{F(y) + c_1}} = \int dx + c_2.$$

Remark. We do not wish to imply that (35.16) will always be integrable in terms of elementary functions. In the example below, $f(y)$ has been chosen carefully so that it will be. In general, it will not be.

Example 35.17. Solve the nonlinear equation

(a) $$y'' = 4y^{-3}, \quad y \neq 0.$$

Solution. Following the method outlined above, we substitute $u = y'$, $u' = y''$ in (a) to obtain the equivalent first order system.

(b) $$\frac{dy}{dx} = u, \quad \frac{du}{dx} = 4y^{-3}.$$

Multiplying the second equation in (b) by $2u$, there results

(c) $$2u\frac{du}{dx} = 8y^{-3}u, \quad \frac{d}{dx}(u^2) = 8y^{-3}\frac{dy}{dx},$$

$$d(u^2) = 8y^{-3}\,dy, \quad u^2 = \int 8y^{-3}\,dy = -4y^{-2} + c_1, \quad c_1 > 0,$$

$$u = \pm\sqrt{c_1 - 4y^{-2}} = \pm\frac{\sqrt{c_1 y^2 - 4}}{y}, \quad y > \frac{2}{\sqrt{c_1}} \text{ or } y < \frac{-2}{\sqrt{c_1}}.$$

Substituting this last value of u in the first equation of (b), we have

(d) $$\frac{y\,dy}{\pm\sqrt{c_1 y^2 - 4}} = dx.$$

Integration of (d) now gives

(e) $$\pm\frac{1}{c_1}\sqrt{c_1 y^2 - 4} = x + c_2, \quad \pm\sqrt{c_1 y^2 - 4} = c_1 x + c_1 c_2,$$

$$c_1 y^2 = (c_1 x + c_1 c_2)^2 + 4, \quad y > \frac{2}{\sqrt{c_1}}, \ y < \frac{-2}{\sqrt{c_1}}, \ c_1 > 0.$$

LESSON 35B.   Solution of a Second Order Nonlinear Differential Equation in Which the Dependent Variable y Is Absent. Equations of this type that we shall consider will be those which can be written in the form

(35.2) $$y'' = f(x,y').$$

Note that y is missing. The substitution $u = y'$, $u' = y''$ will change (35.2) into the equivalent first order system

(35.21) $$\frac{dy}{dx} = u, \qquad \frac{du}{dx} = f(x,u).$$

The second equation is now a first order equation in u and hence may be solvable by the methods of Chapter 2. If it is and its solution is $u = u(x) + c_1$, then by the first equation in (35.21),

(35.22) $$y = \int [u(x) + c_1]\, dx + c_2.$$

Example 35.23.   Solve the nonlinear equation

(a) $$y'' = x(y')^2.$$

Solution.   Following the method outlined above, we substitute $u = y'$, $u' = y''$ in (a) to obtain the equivalent first order system

(b) $$\frac{dy}{dx} = u, \qquad \frac{du}{dx} = xu^2.$$

If $u \neq 0$, we can write the second equation in (b) as

(c) $$\frac{du}{u^2} = x\, dx.$$

Its solution is $u = -2/(x^2 + c)$, which we write as

(d) $$u = \frac{-2}{x^2 \pm c_1{}^2},$$

with the understanding that the plus sign is to be used if $c > 0$; the minus sign if $c < 0$. Substituting (d) in the first equation of (b), we have

(e) $$\frac{dy}{dx} = -\frac{2}{x^2 \pm c_1{}^2}, \qquad dy = \frac{-2dx}{x^2 \pm c_1{}^2},$$

By integrating (e), we obtain

$$y = -2\left[\frac{1}{c_1} \text{Arc tan}\left(\frac{x}{c_1}\right) + c_2\right] \quad \text{and} \quad y = -\frac{1}{c_1}\log\left(\frac{x - c_1}{x + c_1}\right) + c_2.$$

If $c_1 = 0$, then by (e),

(f) $$dy = -\frac{2}{x^2}\,dx, \qquad y = \frac{2}{x} + C, \quad x \neq 0.$$

LESSON 35C.   Solution of a Second Order Nonlinear Equation in Which the Independent Variable x Is Absent.   Equations of this type that we shall consider will be those which can be written in the form

(35.3) $$y'' = f(y,y').$$

Note that x is missing. If we let $u = y'$, then

(35.31) $$y'' = \frac{du}{dx} = \frac{du}{dy}\frac{dy}{dx} = \frac{du}{dy}u.$$

Substituting these values in (35.3) will change it into the equivalent first order system

(35.32) $$\frac{dy}{dx} = u, \qquad u\frac{du}{dy} = f(y,u).$$

The second equation in (35.32) is now a first order equation in u and hence may be solvable by the methods of Chapter 2. If it is and its solution is $u = u(y) + c_1$, then by the first equation in (35.32),

(35.33) $$\int \frac{dy}{u(y) + c_1} = \int dx = x + c_2.$$

Example 35.34.   Solve the nonlinear equation

(a) $$yy'' = y^3 + (y')^2.$$

Solution.   Following the method outlined above, we substitute $u = y'$, $y'' = u\dfrac{du}{dy}$ in (a) to obtain the first order system

(b) $$\frac{dy}{dx} = u, \qquad uy\frac{du}{dy} = y^3 + u^2.$$

The second equation in (b) can be written as

(c) $$\frac{du}{dy} - \frac{1}{y}u = y^2 u^{-1}, \quad u \neq 0, y \neq 0,$$

which is a Bernoulli equation. Hence, following the method outlined in Lesson 11D, we multiply (c) by $2u$ to obtain

(d) $$2u\frac{du}{dy} - \frac{2}{y}u^2 = 2y^2,$$

which can be written as

(e) $$\frac{d}{dy}(u^2) - \frac{2}{y}(u^2) = 2y^2, \quad u \neq 0, \ y \neq 0,$$

an equation linear in u^2. Its solution by the method of Lesson 11B is (the integrating factor is $e^{\int -2dy/y} = e^{-2 \log y} = 1/y^2$)

(f) $$\frac{u^2}{y^2} = \int 2dy, \quad u = \pm y\sqrt{2y + c_1}, \quad u \neq 0, \ y \neq 0, \ 2y + c_1 > 0.$$

Substituting these values of u in the first equation of (b), we obtain

(g) $$\frac{dy}{dx} = \pm y\sqrt{2y + c_1}; \quad \frac{dy}{y\sqrt{2y + c_1}} = \pm dx, \quad y \neq 0, \ 2y + c_1 > 0.$$

The solution of the second equation in (g), if $y \neq 0$, $2y + c_1 > 0$, is

(h) $$\frac{2}{\sqrt{c_1}} \log \frac{\sqrt{2y + c_1} - \sqrt{c_1}}{\sqrt{y}} = \pm x + c_2, \quad c_1 > 0,$$

$$\frac{2}{\sqrt{-c_1}} \text{Arc} \tan \frac{\sqrt{2y + c_1}}{\sqrt{-c_1}} = \pm x + c_2, \quad c_1 < 0.$$

If $c_1 = 0$, then (f) becomes $u = \pm\sqrt{2} \, y^{3/2}$. Therefore by (b)

(i) $$\frac{dy}{dx} = \pm\sqrt{2} \, y^{3/2}, \quad \pm y^{-3/2} \, dy = \sqrt{2} \, dx, \quad \mp 2y^{-1/2} = \sqrt{2} \, x + c_2$$

$$(\sqrt{2} \, x - c_2)^2 y = 4, \quad y \neq 0.$$

Note. The function $y = 0$ which we had to discard in order to obtain (h) and (i) also satisfies (a). It is a particular solution of (a) not obtainable from the families (h) or (i).

EXERCISE 35

Solve each of the following differential equations.

1. $y'' = 2yy'$.
2. $y^3 y'' = k$.
3. $yy'' = (y')^2 - 1$.
4. $x^2 y'' + xy' = 1$.
   (Note that the equation is linear.)
5. $xy'' - y' = x^2$.
   (Note that the equation is linear.)

6. $(y + 1)y'' = 3(y')^2$.
7. $\ddot{r} = -k/r^2$.
   (See Comment 34.8961.)
8. $y'' = \frac{3}{2}ky^2$.
9. $y'' = 2ky^3$.
10. $yy'' + (y')^2 - y' = 0$.

11. $\ddot{r} = \dfrac{h^2}{r^3} - \dfrac{k}{r^2}$. (See Comment 34.91.)

12. $yy'' + (y')^3 - (y')^2 = 0$.
13. $yy'' - 3(y')^2 = 0$.
14. $(1 + x^2)y'' + (y')^2 + 1 = 0$.
15. $(1 + x^2)y'' + 2x(y' + 1) = 0$. (Note that the equation is linear.)

Find a particular solution of each of the following differential equations satisfying the given initial conditions.

16. $(y + 1)y'' = 3(y')^2$, $y(1) = 0$, $y'(1) = -\frac{1}{2}$.
17. $y'' = y'e^y$, $y(3) = 0$, $y'(3) = 1$.
18. $y'' = 2yy'$, $y(0) = 1$, $y'(0) = 2$.
19. $2y'' = e^y$, $y(0) = 0$, $y'(0) = 1$.
20. $x^2y'' + xy' = 1$, $y(1) = 1$, $y'(1) = 2$. (Note that the equation is linear.)
21. $xy'' - y' = x^2$, $y(1) = 0$, $y'(1) = -1$. (Note that the equation is linear.)
22. $\ddot{r} = -\dfrac{k}{r^2}$, $r(0) = 1$, $\dot{r}(0) = 0$, $\ddot{r}(0) = -\frac{1}{2}$. See problem 7 above. *Hint.*
    After finding t as a function of r in integral form, substitute either $r = \cos^2 u$ or $r = u^2$.

23. A differential equation is said to be **homogeneous in x**, if by arbitrarily assigning degree n to x^n and degree $-n$ to $y^{(n)}$, each nonzero term of the equation is of the same degree. In computing the degree of a term, the degree of each of its members is added. For example, x^2 is of degree 2; y is of degree zero; y'' is of degree -2; x^2yy'' is of degree zero. Equations homogeneous in x may be solvable by the substitutions, see also Exercise 23,18,

$$x = e^u, \qquad u = \log x,$$

$$\frac{dy}{dx} = \frac{1}{x}\frac{dy}{du}, \qquad \frac{d^2y}{dx^2} = \frac{1}{x^2}\left(\frac{d^2y}{du^2} - \frac{dy}{du}\right),$$

etc., and then making use of one of the methods of Lesson 35. Verify that each of the following nonlinear differential equations is homogeneous in x and solve.
(a) $xyy'' - 2x(y')^2 + yy' = 0$.
(b) $xyy'' + x(y')^2 - yy' = 0$.
(c) $xyy'' - 2x(y')^2 + (y + 1)y' = 0$.

ANSWERS 35

1. $y = c_1 \tan [c_1(x + c_2)]$.
2. $c_1y^2 = (c_1x + c_1c_2)^2 + k$.
3. $c_1y = \sinh (c_1x + c_2)$.

4. $y = \dfrac{(\log x)^2}{2} + c_1 \log x + c_2$.

5. $y = \dfrac{x^3}{3} + c_1x^2 + c_2$.

6. $(y + 1)^{-2} = c_1x + c_2$.

7. $t = \displaystyle\int \sqrt{\dfrac{r}{2k + c_1r}}\, dr + c_2$. To evaluate the integral, let $r = u^2$. See 22.

8. $x = \pm \displaystyle\int \dfrac{dy}{\sqrt{ky^3 + c_1}} + c_2$.

9. $x = \pm \displaystyle\int \dfrac{dy}{\sqrt{ky^4 + c_1}} + c_2$.

10. $x = y + c_1 \log (y - c_1) + c_2; y = c.$

11. $t = c_2 \pm \dfrac{1}{c_1} \sqrt{c_1 r^2 + 2kr - h^2} \pm \dfrac{k}{c_1^{3/2}} \log \left[r + \dfrac{k}{c_1} + \sqrt{\dfrac{c_1 r^2 + 2kr - h^2}{c_1}} \right].$

12. $y = x + c_1 \log (c_2 y).$

13. $y^{-2} = c_1 x + c_2.$

14. $c_1^2 y = (c_1^2 + 1) \log (c_1 x + 1) - c_1 x + c_2.$

15. $y = c_1 \text{ Arc tan } x - x + c_2.$

16. $(y + 1)^{-2} = x.$

17. $y = -\log (4 - x).$

18. $y = \tan \left(x + \dfrac{\pi}{4} \right).$

19. $e^{y/2} = \dfrac{2}{2 - x}.$

20. $y = \dfrac{(\log x)^2}{2} + 2 \log x + 1.$

21. $y = \dfrac{x^3}{3} - x^2 + \dfrac{2}{3}.$

22. $k = \frac{1}{2}, c_1 = -1, t = \displaystyle\int_1^r \pm \sqrt{\dfrac{r}{1 - r}}\, dr = \mp (\text{Arc cos } \sqrt{r} + \sqrt{r(1 - r)}).$

NOTE. Since the limits of integration are from 1 to r and $r < 1$, use of the plus sign in the integrand will give a negative time; use of the minus sign will give a positive time.

23. (a) $y^{-1} = c_1 \log x + c_2.$ (b) $y^2 = c_1 x^2 + c_2.$
(c) $2 \text{ Arc tan } (c_1 y) = c_1 \log x + c_2.$

LESSON 36.   Problems Giving Rise to Special Types of Second Order Nonlinear Equations.

LESSON 36A.   The Suspension Cable. A cable, chain, string, or similar object supported at two ends is called a **suspension cable**. It may support a load attached to it as in the case of a bridge, or it may hang under its own weight. We consider the latter possibility first. We shall determine for it the shape of the curve that the cable assumes.

In Fig. 36.1, we have drawn a cable, assumed to be perfectly flexible and inextensible, supported at two ends, A, B, and hanging under its own weight. Whether the cable will have the appearance shown in Fig. 36.1(a) or in Fig. 36.1(b) will depend on the length s of the cable relative to the length AB between the points of support. We consider the more general case shown in Fig. 36.1(a), where the cable at its lowest point does not necessarily have a horizontal tangent as it does in Fig. 36.1(b).

Let $P(x,y)$ and $P_1(x + \Delta x, y + \Delta y)$ be two neighboring points on the curve AB, and call Δs the length of the arc between them. The forces acting on this piece of cable, considered as isolated from the rest of the system are [refer to Fig. 36.1(a)]:

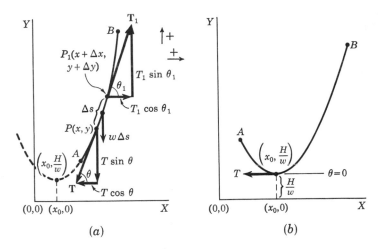

Figure 36.1

1. A tension **T** at P directed along the tangent to the curve.
2. A tension $\mathbf{T}_1$ at P_1 also directed along the tangent to the curve but in a direction opposite to that of **T**.
3. A force due to the weight of the cable of length Δs, directed downward. If we assume a homogeneous cable whose weight is uniformly distributed and is w pounds per foot, then the weight of the cable of length Δs is $w\,\Delta s$.

By hypothesis, the cable is perfectly flexible and inextensible. And since it is in equilibrium, the portion PP_1 of the cable will retain its shape under the action of the three forces 1, 2, 3 above, even if it were cut at P and P_1, just as if it were a rigid body. But the condition for static equilibrium of the piece of cable PP_1 is that the algebraic sum of the components of the system of forces acting along an arbitrary direction be zero. Hence equating to zero, the algebraic sum of the horizontal components of the three forces 1, 2, 3 above, we obtain [keep referring to Fig. 36.1(a)].

$$(36.11) \qquad\qquad T_1 \cos \theta_1 - T \cos \theta = 0,$$

where θ and θ_1 are the angles shown in Fig. 36.1(a).

Equating to zero the vertical components of these forces, we obtain

$$(36.12) \qquad\qquad T_1 \sin \theta_1 - T \sin \theta - w\Delta s = 0.$$

The change in the horizontal tension $(T \cos \theta)$, is

$$(36.13) \qquad\qquad \Delta(T \cos \theta) = T_1 \cos \theta_1 - T \cos \theta;$$

the change in the vertical tension $T \sin \theta$, is

$$(36.14) \qquad \Delta(T \sin \theta) = T_1 \sin \theta_1 - T \sin \theta.$$

Substituting (36.13) in (36.11) and (36.14) in (36.12), we obtain respectively

$$(36.15) \qquad \Delta(T \cos \theta) = 0, \qquad \Delta(T \sin \theta) = w\Delta s.$$

Dividing each of the equations in (36.15) by Δx and letting $\Delta x \to 0$, there results the system of equations

$$(36.16) \qquad \frac{d}{dx}(T \cos \theta) = 0, \qquad \frac{d}{dx}(T \sin \theta) = w\frac{ds}{dx}.$$

From the first equation in (36.16), we find that

$$(36.17) \qquad T \cos \theta = H,$$

where H is a constant. Since $T \cos \theta$ is a constant and equal to H, we may write the second equation in (36.16) as

$$(36.18) \qquad \frac{d}{dx}\left(\frac{T \sin \theta}{T \cos \theta}\right) = \frac{w}{H}\frac{ds}{dx},$$

which simplifies to

$$(36.19) \qquad \frac{d}{dx}(\tan \theta) = k\frac{ds}{dx},$$

where

$$(36.2) \qquad k = \frac{w}{H}.$$

But $\tan \theta$ is the slope y' of the curve of the cable at P. Hence

$$(36.21) \qquad \tan \theta = \frac{dy}{dx}.$$

Substituting (36.21) in (36.19) and recalling that the differential ds of the arc PP_1 of the cable is given by the formula $ds = \sqrt{dx^2 + dy^2}$ so that

$$\frac{ds}{dx} = \sqrt{1 + (dy/dx)^2},$$

we obtain

$$(36.22) \qquad \frac{d^2y}{dx^2} = k\sqrt{1 + (dy/dx)^2}.$$

This is a nonlinear second order equation of the type discussed in Lesson 35B with y missing. Following the method outlined there, we let $u = y'$,

$u' = y''$ and thus obtain from (36.22), the system

$$(36.23) \qquad \frac{dy}{dx} = u, \qquad \frac{du}{dx} = k\sqrt{1 + u^2},$$

By the method of Lesson 6C, and with the constant of integration taken to be kx_0, the solution of the second equation in (36.23) is

$$(36.24) \qquad \log(u + \sqrt{1 + u^2}) = kx - kx_0,$$
$$u + \sqrt{1 + u^2} = e^{k(x - x_0)}.$$

The second equation in (36.24) can be solved for u by the usual algebraic means. A simpler method, however, is the following. Take the reciprocal of each side of the second equation in (36.24) and rationalize the denominator of the resulting left side. There results

$$(36.25) \qquad u - \sqrt{1 + u^2} = -e^{-k(x - x_0)}.$$

Adding (36.25) and the second equation in (36.24), we obtain

$$(36.26) \qquad u = \tfrac{1}{2}[e^{k(x - x_0)} - e^{-k(x - x_0)}].$$

By (18.9), we can write (36.26) as

$$(36.261) \qquad u = \sinh[k(x - x_0)].$$

Replacing u in (36.261) by its value as given in (36.23) and k by its value as given in (36.2), we obtain

$$(36.27) \qquad y' = \sinh\left[\frac{w}{H}(x - x_0)\right].$$

Integration of (36.27) gives

$$(36.28) \qquad y = \frac{H}{w}\cosh\left[\frac{w}{H}(x - x_0)\right] + c,$$

which is the equation of the curve of a hanging cable supported at two points P and P_1. Since the same form of equation (36.28) results if the points P and P_1 are taken anywhere on AB, (36.28) is the equation of a hanging cable supported at two ends A,B. Keep in mind that as P and P_1 shift, the values of H and x in (36.28) change.

There are three constants in our solution (36.28), H, x_0, and c. From (36.27), we see that $y' = 0$ when $x = x_0$, (remember $\sinh 0 = 0$). This means that x_0 is the x coordinate of that point on the curve (36.28) where its slope is zero. This point is, therefore, a minimum point of the curve. If when $x = x_0$, we now choose our x axis so that y has the value H/w, then by (36.28), $c = 0$, (remember $\cosh 0 = 1$). In Fig. 36.1(a), we have shown the point $(x_0, H/w)$. If we now choose our y axis so that

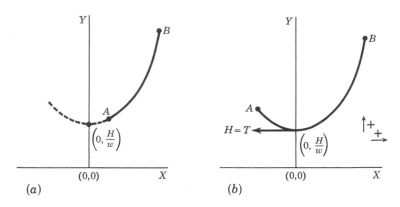

Figure 36.29

it goes through the point $(x_0, H/w)$, then $x_0 = 0$. With reference to these new axes, i.e., where the y axis goes through the lowest point of the curve (36.28), and the x axis is H/w units below this lowest point, see Fig. 36.29(a), (36.28) simplifies to

$$(36.3) \qquad y = \frac{H}{w} \cosh\left(\frac{w}{H} x\right),$$

and (36.27) simplifies to

$$(36.31) \qquad y' = \sinh\left(\frac{w}{H} x\right).$$

A curve whose equation has the form (36.3) is called a **catenary.**

Finally to find H we make use of the fact that the length AB of the cable is s. Let a be the x coordinate of the point A; b the x coordinate of the point B. Then from the calculus, we know

$$(36.32) \qquad s = \int_a^b \sqrt{1 + (y')^2}\, dx.$$

By (36.31), we can write (36.32) as

$$(36.321) \qquad s = \int_a^b \sqrt{1 + \sinh^2\left(\frac{w}{H} x\right)}\, dx$$

$$= \int_a^b \cosh\frac{wx}{H}\, dx$$

$$= \frac{H}{w} \sinh\frac{w}{H} x \Big|_a^b.$$

Therefore

(36.33) $$s = \frac{H}{w}\left[\sinh\left(\frac{w}{H}b\right) - \sinh\left(\frac{w}{H}a\right)\right].$$

In (36.33), s, w, a, b are given constants. This equation therefore determines H.

Comment 36.34. Let us assume that s is sufficiently long so that the cable hangs as in Fig. 36.1(b). Hence the point $(x_0, H/w)$ at which the curve has a horizontal tangent is a point of the cable. The tension T of the cable at that point is therefore also horizontal. But when the tension T is horizontal, the angle θ in Fig. 36.1(a) is zero. And when $\theta = 0$, we see from (36.17) that $T = H$. Hence for this *special case, H is the tension of the cable at its lowest point*, see Fig. 36.29(b). And if $x_0 = 0$, *the length s of a catenary from its lowest point $(0, H/w)$ to any point $P(x,y)$ on it*, for this special case, is therefore [in (36.33) take $a = 0$, $b = x$],

(36.35) $$s = \frac{H}{w}\sinh\frac{w}{H}x.$$

Example 36.36. A cable 50 feet long, weighing 5 lb/ft, hangs under its own weight between two supports 30 feet apart. Find:

1. The equation of the curve.
2. The tension at its lowest point.
3. The sag.

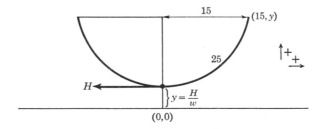

Figure 36.37

Solution (Fig. 36.37). If we choose the origin so that the y axis goes through the lowest point $(0, H/w)$ of the cable, then by (36.3) the equation of the curve formed by the cable is

(a) $$y = \frac{H}{w}\cosh\frac{w}{H}x;$$

and by (36.35), the length of the cable from $x = 0$ to $x = x$ is

(b) $$s = \frac{H}{w}\sinh\frac{w}{H}x.$$

Inserting in (b) the given conditions, $w = 5$, $x = 15$, $s = 25$, we find

(c) $$25 = \frac{H}{5} \sinh \frac{5}{H} 15, \qquad \frac{125}{H} = \sinh \frac{75}{H}.$$

Using a table of values of $\sinh x$, the approximate value of H which will satisfy (c) is

(d) $$H = 40.8 \text{ lb}.$$

This is the tension of the cable at its lowest point. When $H = 40.8$ and $w = 5$, we obtain from (a)

(e) $$y = 8.16 \cosh \frac{x}{8.16},$$

which is the equation of the curve of the cable. From it we find that when $x = 15$,

(f) $$y = 8.16 \cosh \frac{15}{8.16} = 8.16(3.22) = 26.28 \text{ ft},$$

and that when $x = 0$, $y = 8.16$. Hence the sag of the cable is $26.28 - 8.16 = 18.12$ feet.

Comment 36.38. If the specific weight of a body is not uniform but is a function of x, then its total weight W from $x = 0$ to $x = x$ is given by

(36.4) $$W = \int_0^x f(x) \, dx,$$

where $f(x)$ is the **specific weight** of the body, i.e., its weight per unit length. For example if $f(x) = 50 + x$ pounds, then when $x = 10$ feet, the specific weight of the body at that point is 60 lb/ft; when $x = 10.1$ feet the specific weight at that point is 60.1 lb/ft. By (36.4), its total weight W from $x = 0$ to say $x = 20$ feet is

$$W = \int_0^{20} (50 + x) \, dx = \left[50x + \frac{x^2}{2} \right]_0^{20} = 1200 \text{ lb}.$$

Assume that a perfectly flexible cable of negligible weight supports, by means of vertical rods, a horizontal load such as a bridge, whose specific weight is $f(x)$. The rods are equally spaced and close enough to form a continuous system. The weights of the rods are also negligible. We wish to find the equation of the curve formed by the cable.

Following the steps used to arrive at the equation of the hanging cable with no load attached, we find that no changes occur until we reach (36.12). In this equation, $w\Delta s$ must be replaced, see Fig. 36.41, by $f(x)\Delta x$, where $f(x)$ is the specific weight of the horizontal load measured from the origin, which is taken to be the lowest point of the curve. Continuing from there on, and replacing $w\Delta s$ by $f(x)\Delta x$, we obtain eventually, in

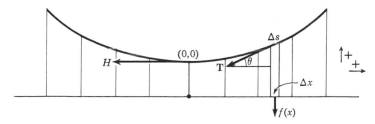

Figure 36.41

place of (36.18),

(36.42)
$$\frac{d}{dx}\left(\frac{T \sin \theta}{T \cos \theta}\right) = \frac{f(x)}{H},$$

$$d(\tan \theta) = \frac{f(x)}{H}\,dx,$$

$$d\left(\frac{dy}{dx}\right) = \frac{f(x)}{H}\,dx.$$

Integrating the last equation and making use of the fact that when $x = 0$, $dy/dx = 0$, we have

(36.43)
$$\frac{dy}{dx} = \frac{1}{H}\int_0^x f(x)\,dx.$$

A second initial condition is $x = 0$, $y = 0$. By Comment 36.34, the H in this equation is the tension at the lowest point of the cable.

Example 36.44. A perfectly flexible suspension cable, attached to two towers at the same level, 100 feet apart, supports a bridge (assumed rigid) by means of vertical rods connecting cable and bridge. The rods are equally spaced and close enough to form an approximately continuous system. The weights of cable and rods are negligible in comparison with the weight of the bridge whose specific weight is given by $f(x) = 10 + x^2/50$ lb/ft. If the length of the cable is such that its sag is 25 feet when the bridge is horizontal, find the equation of the curve in which the cable hangs and the tension at its lowest point. Assume that the origin is taken at the lowest point of the cable.

Solution. See Fig. 36.441. By (36.43) and the given specific weight $f(x) = 10 + x^2/50$, we obtain

(a)
$$H\frac{dy}{dx} = \int_0^x \left(10 + \frac{x^2}{50}\right)dx = 10x + \frac{x^3}{150}.$$

Integration of (a) gives

(b)
$$Hy = 5x^2 + \frac{x^4}{600} + c_1.$$

The substitution in (b) of the initial condition $x = 0$, $y = 0$ gives $c_1 = 0$; the substitution of the initial condition $x = 50$, $y = 25$, gives

(c) $$25H = 5(50)^2 + \frac{50^4}{600}, \qquad H = 917,$$

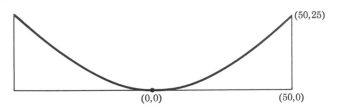

(50,25)

(0,0)   (50,0)

Figure 36.441

which is the tension of the cable at its lowest point. Hence (b) becomes

(d) $$y = \frac{1}{917}\left(5x^2 + \frac{x^4}{600}\right),$$

which is the equation of the curve of the cable.

EXERCISE 36A

1. A cable of length $2s$ and uniform weight w lb/ft hangs from two supports on the same level, Fig. 36.45. The supports are $2L$ ft apart with $L < s$. The tension at the lowest point of the cable is H.

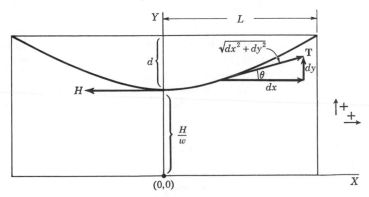

Figure 36.45

(a) Show that the sag d is given by

(36.46) $$d = \frac{H}{w}\left(\cosh\frac{wL}{H} - 1\right).$$

Hint. Use (36.3). The sag is the difference in values when $x = L$ and $x = 0$.

(b) Show that the length s of the cable from $x = 0$ to $x = L$ is

(36.47)
$$s = \frac{H}{w} \sinh \frac{wL}{H}.$$

Hint. Use (36.35) with $x = L$.

(c) Show that the tension T at any point of the cable is given by

(36.48)
$$T = H \cosh \frac{w}{H} x = wy.$$

Hint. By (36.17) (see also Fig. 36.45) $T = H \sec \theta = H \dfrac{\sqrt{dx^2 + dy^2}}{dx} =$
$H\sqrt{1 + (dy/dx)^2}$. Replace y' by its value as given in (36.31). *Remark.* Since $\cosh x$ is an increasing function of x, the tension is a maximum when x is largest, i.e., at a point of support.

(d) Show that the total weight of the cable is given by

(36.49)
$$W = 2H \sinh (wL/H).$$

Hint. Use (36.47) which gives the length of cable from $x = 0$, to $x = L$.

(e) Show that the horizontal tension H is given by

(36.491)
$$e^{wL/H} = (s + d)/(s - d),$$

$$H = \frac{wL}{\log \left[(s + d)/(s - d) \right]} \text{ lb.}$$

Hint. In (36.46) and (36.47), change the hyperbolic functions to their exponential forms, see (18.9) and (18.91), and then show that the right side of the first equation in (36.491) simplifies to $e^{wL/H}$.

(f) Show that the horizontal tension H is also given by

(36.492)
$$H = w(s^2 - d^2)/2d.$$

Hint. Use (36.46), (36.47), and the fact that $\cosh^2 x - \sinh^2 x = 1$.

2. A telephone wire, weighing 0.04 lb/ft, is attached to poles at intervals of 300 ft. The sag at the center is 7.5 ft.

   (a) Find the tension at the lowest point of the curve. *Hint.* Use (36.46). Remember in this formula, the distance between supports is $2L$.
   (b) What is the length of the wire between poles? *Hint.* Use (36.47). Remember in this formula the length of wire is $2s$. Use (36.491) or (36.492) to check your result.
   (c) What is the tension at the supports? *Hint.* Use (36.48) with $x = 150$.
   (d) Find the equation of the curve. *Hint.* Use (36.3).

3. A cable 100 ft long, weighing 4 lb/ft, hangs under its own weight between two supports on the same level and 80 ft apart. Find:

   (a) The equation of the curve.
   (b) The tension of the cable at its lowest point; at a point of the cable midway between the lowest point and a point of support; at a point midway horizontally between the lowest point and a point of support; at a point of support.
   (c) The sag.
   (d) The slope of the curve at a point of support.

   For hints, see answer section.

4. A cable hangs under its own weight between two supports on the same level. The slope of the curve formed by the cable at a point of support is 0.2013. Its sag at the center of the cable is 12 ft. Find:

   (a) The distance between supports.
   (b) The length of the cable.
   (c) The equation of the curve.

   Hint. Use (36.31) to find wL/H and solve for H/w in terms of L. Then use (36.46), (36.47), (36.3) in that order.

5. A chain 117.5 ft long hangs under its own weight between two supports on the same level 100 ft apart.

   (a) Find the sag of the chain. *Hint.* Use (36.47) to find H/w. Then use (36.46).
   (b) Find the equation of the curve.
   (c) Find the maximum tension if the chain weighs 2 lb/ft. *Hint.* See remark after (36.48). Use (36.48).

6. A chain hangs under its own weight between two supports on the same level 100 ft apart. Its sag is 7.55 ft.

   (a) Find the length of the chain. *Hint.* Solve (36.46) for H/w. Then use (36.47).
   (b) Find the equation of the curve.

7. A cable hangs under its own weight between two supports on the same level 50 ft apart. The slope of the curve formed by the cable at a point of support is 0.5211. Find:

   (a) The length of the cable.
   (b) The sag.
   (c) The equation of the curve.
   (d) The tension of the cable at its lowest point if it weighs 0.5 lb/ft, i.e., find H.

   Hint. Use (36.31) to find w/H. Then use (36.47), (36.46), (36.3) in that order.

8. A cable, 100 ft long, hangs under its own weight between two supports on the same level. Its sag at the center is 10 ft.

   (a) Find the distance between the supports.
   (b) Find the equation of the curve.

   Hint. Use (36.491) to find wL/H. Then use (36.46) or (36.47).

9. A cable 200 ft long hangs under its own weight between two supports on the same level. Its sag at the center is 20 ft. Its maximum tension, see remark after (36.48), is 60 lb. Find its weight w per foot. For hints, see answer section.

10. Start with the general equation of the hanging cable as given in (36.28). It contains three constants, x_0, c, H. By choosing our origin in a special way, we were able to make $x_0 = 0$, $c = 0$. H was then determined by using the given length s of the cable. This time let us take the origin at the lower point A of the hanging cable and call (x_1, y_1) the coordinates of the higher point B, see Fig. 36.493. Show that the values of the three constants can be

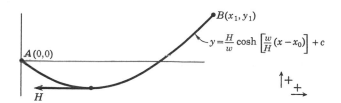

Figure 36.493

obtained from the equations

(A)
$$0 = \frac{H}{w} \cosh \frac{w}{H} x_0 + c,$$

(36.494)
$$y_1 = \frac{H}{w} \cosh \left[\frac{w}{H} (x_1 - x_0) \right] + c,$$

(B)
$$s = \int_0^{x_1} \sqrt{1 + (y')^2}\, dx = \frac{H}{w} \left[\sinh \frac{w}{H} (x_1 - x_0) + \sinh \frac{w}{H} x_0 \right]$$
$$= \frac{2H}{w} \left[\sinh \frac{wx_1}{2H} \cosh \frac{w(x_1 - 2x_0)}{2H} \right].$$

By subtracting (A) from (36.494), show that

(36.495)
$$y_1 = \frac{H}{w} \left[\cosh \frac{w(x_1 - x_0)}{H} - \cosh \frac{wx_0}{H} \right]$$
$$= \frac{2H}{w} \left[\sinh \frac{wx_1}{2H} \sinh \frac{w(x_1 - 2x_0)}{2H} \right].$$

Finally show by (B) and (36.495)—square both equations and then subtract the second from the first—

$$s^2 - y_1{}^2 = \frac{4H^2}{w^2} \sinh^2 \left(\frac{wx_1}{2H} \right),$$

(36.496)
$$\sinh \left(\frac{wx_1}{2H} \right) = \frac{w}{2H} \sqrt{s^2 - y_1{}^2}.$$

Following the hint given after (36.48), show that the tension at any point of the cable is given by

(36.497)
$$T = H \cosh \frac{w}{H} (x - x_0) = w(y - c),$$

where c is a constant of integration whose value is given by (A) and (36.494).

11. Using the equations found in 10, solve the following problem. See Fig. 36.493. A cable of length 100 ft and uniform weight of 2 lb/ft hangs from two supports A and B, whose horizontal distance is 50 ft apart. Support B is 20 ft higher than A. Find:

  (a) The equation of the hanging cable. [*Hint.* Use (36.496), (36.495), and (36.494) in that order.]
  (b) The coordinates of its lowest point, i.e., the point where $y' = 0$.
  (c) The tension at both supports and at its lowest point.

In problems 12–16, assume the weights of the cable and rods are negligible in comparison with the weight of a *horizontal* supported load, that the cable is perfectly flexible, that the rods are equally spaced and close enough to form a continuous system, and that the *origin* is at the lowest point of the cable.

12. A suspension bridge is $2x$ ft long and is supported by vertical rods. The weight of the bridge is uniformly distributed and is w lb per unit horizontal distance. Find the equation of the curve formed by the cable. *Hint.* In (36.43), $f(x) = w$.

13. Show that the tension T at any point of the cable of problem 12 is given by

(36.498) $$T = \sqrt{H^2 + w^2 x^2}.$$

See (36.48) and hint following; here y' is obtained from problem 12.

14. A suspension bridge is 200 ft long. The supporting ends of the cable are 50 ft above the bridge and the center of the bridge is 10 ft below the center of the cable. The weight of the bridge is uniformly distributed and is w lb per unit horizontal distance.

    (a) Find the tension at the lowest point of the cable, i.e., find H. *Hint.* In the solution to 12, use the fact that $x = 100$, $y = 40$.
    (b) Find the equation of the curve formed by the cable.
    (c) What is the tension at the supports? *Hint.* See (36.498).
    (d) What is the slope of the cable at the supports?

15. A suspension bridge is $2L$ ft long. The weight of the bridge is uniformly distributed and is w lb/ft. The supports of the cable holding the bridge are a ft above the origin.

    (a) Find the tension at its lowest point, i.e., find H. *Hint.* In the solution to problem 12, use the fact that when $x = L$, $y = a$.
    (b) Find the equation of the curve.
    (c) What is the tension at the supports? *Hint.* See (36.498).

16. A suspension bridge is 200 ft long. The sag at its center is 50 ft. The specific weight of the bridge is given by $f(x) = 100 + x^2$. Find:

    (a) The tension at the lowest point, i.e., find H.
    (b) The equation of the curve formed by the cable.

17. A uniform, flexible cable weighing w_1 lb/ft supports a horizontal bridge. The weight of the bridge is uniformly distributed and is w_2 lb/ft. The weight of the supporting rods is negligible. Show that the differential equation of the curve of the cable is given by

(36.499) $$H \frac{d^2 y}{dx^2} = w_1 \sqrt{1 + (dy/dx)^2} + w_2.$$

Hint. Combine equations (36.22) and (36.42) with $k = w_1/H$ and $f(x) = w_2$.

18. If the rods in problem 17 are not of negligible weight and weigh w_3 lb/ft, show that the differential equation of the curve of the cable is given by

(36.499i) $$H \frac{d^2 y}{dx^2} = w_1 \sqrt{1 + (dy/dx)^2} + w_2 + w_3 y,$$

where y is the length of a rod. Assume that the rods are so close together as to form an approximate continuous system. *Hint.* The weight due to a rod of width Δx is $w_3 y\,\Delta x$.

19. Holes are bored in the ends of slender uniform rods of varying lengths and the rods are then strung on a cord of negligible weight. Assume that the rods just touch each other, that each weighs w lb/sq ft and that their other ends

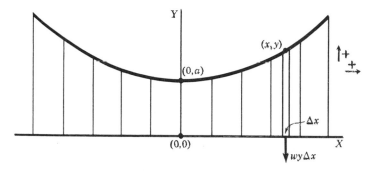

Figure 36.4992

lie on a horizontal line, Fig. 36.4992. Find the equation of the curve formed by the cord. Take the origin at a distance a feet below the lowest point of the curve. *Hint.* In (36.15), replace $w\,\Delta s$ by $wy\,\Delta x$. Initial conditions are $x = 0$, $y = a$, $dy/dx = 0$.

20. For a nonhomogeneous hanging cable whose weight is a function of its distance from its lowest point, show that the differential equation of the curve formed by this hanging cable is

(36.4993)
$$H\frac{d^2 y}{dx^2} = \rho(s)\sqrt{1 + (dy/dx)^2},$$

where $\rho(s)$ is the specific weight of the cable, i.e., its weight per unit length at a distance s units from the lowest point of the cable. *Hint.* In (36.15), replace w by $\rho(s)$.

21. If in (36.4993), $\rho(s) = aH/(a^2 - x^2)$, and the origin is taken at the lowest point of the cable so that $x = 0$, $y = 0$, $dy/dx = 0$, show that the curve formed by the hanging cable will be an arc of the circle.

$$x^2 + (y - a)^2 = a^2.$$

22. An arched bridge is to be built of stone of uniform density. The weight of the stone is w lb for each square foot of facing. The bridge is so constructed that at each point of the arch the resultant tension due to the weight of the stones above it acts in a direction tangent to the arch, see Fig. 36.4994. Find the equation of the arch. Take the x axis at the top of the bridge, the y axis through the highest point of the arch, the origin as indicated in the figure and the positive direction downward. *Hint.* Since the tension at each point acts in a direction tangent to the curve, equations (36.15) apply with $w\,\Delta s$ replaced by $wy\,\Delta x$. Initial conditions are $x = 0$, $y = h$, $dy/dx = 0$.

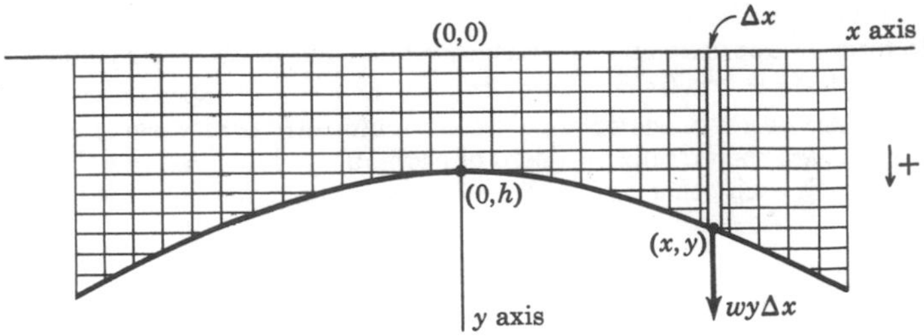

Figure 36.4994

23. Assume a road is built on top of the masonry of problem 22. The weight of the road is distributed uniformly and weighs kw lb per linear foot. Find the equation of the curve. *Hint.* In (36.15) replace $w \, \Delta s$ by $wy\Delta x + kw \, \Delta x$.

ANSWERS 36A

2. (a) 60 lb.   (b) 300.5 ft.   (c) 60.3 lb.   (d) $y = 1500 \cosh (x/1500)$.
3. (a) Use (36.47) to find H. Then use (36.3).   (b) Use (36.35) to find x when $s = 25$. Then use (36.48) with x equal successively to 0; its value when $s = 25; 20; 40$.   (c) Use (36.46).   (d) Use (36.31) with $x = 40$.
4. (a) 239.3 ft.   (b) 240.9 ft.   (c) $y = 598 \cosh (x/598)$.
5. (a) 27.2 ft.   (b) $y = 50 \cosh (x/50)$.   (c) 154.3 lb.
6. (a) 101.5 ft.   (b) $y = (500/3) \cosh (3x/500)$.
7. (a) 52.11 ft.   (b) 6.38 ft.   (c) $y = 50 \cosh (x/50)$.   (d) 25 lb.
8. (a) 97.3.   (b) $y = 120 \cosh 0.0083x$.
9. Use (36.492) to find H/w. Use (36.47) to find L. Use (36.48) with $x = L$ (write $H = w(H/w)$ and solve for w); $w = 3/13$ lb/ft.
11. (a) $H = 23.35$ approx., $x_0 = 22.6$ approx., $c = -41.3$ approx.,

$$y = 11.7 \cosh \left(\frac{x - 22.6}{11.7} \right) - 41.3.$$

(b) $(22.4, -29.7)$.   (c) 82.7 lb at A, 122.7 lb at B, 23.35 lb at its lowest point.

12. $y = \dfrac{w}{H} \dfrac{x^2}{2}$, a parabola.

14. (a) $H = 125w$.   (b) $y = x^2/250$.   (c) $T = 25w\sqrt{41}$.
    (d) Slope $= \pm 0.8$.

15. (a) $H = wL^2/2a$.   (b) $y = ax^2/L^2$.   (c) $(wL/2a)\sqrt{L^2 + 4a^2}$.

16. (a) $H = 530,000/3$.   (b) $y = (3/530,000) \left(50x^2 + \dfrac{x^4}{12} \right)$.

19. $y = \dfrac{a}{2} (e^{\sqrt{w/H}\,x} + e^{-\sqrt{w/H}\,x}) = a \cosh \sqrt{w/H}\,x$.

22. Same as in 19 with a replaced by h. Since the tension at each point of the bridge is tangent to the curve, it follows that no mortar will be needed if the bridge has the shape of a catenary.

23. $y = (h + k) \cosh \sqrt{w/H}\,x - k$.

LESSON 36B.   A Special Central Force Problem.

Example 36.5.   A particle weighing 16 pounds and 10 feet from a fixed point is given a velocity of 15 ft/sec in a direction perpendicular to the x axis. The particle is attracted to the fixed point by a force F whose magnitude is inversely proportional to the cube of its distance from the point. If the proportionality constant is 8, find the distance of the particle from the fixed point as a function of the time.

Solution.   Let us take the origin of a coordinate system at the fixed point. We have already proved in Lesson 34C that a particle subject to a central force moves in a plane. Since the force F acts only in a direction along the radius vector, the components of F are, with the proportionality constant equal to 8,

(a) $$F_r = -\frac{8}{r^3} = ma_r, \qquad F_\theta = 0 = ma_\theta.$$

Here the mass $m = \frac{16}{32} = \frac{1}{2}$. Hence by (34.21), (a) becomes

(b) $$\frac{1}{2}(\ddot{r} - r\dot{\theta}^2) = -\frac{8}{r^3}, \qquad \frac{1}{2}(2\dot{r}\dot{\theta} + r\ddot{\theta}) = 0.$$

After multiplication by $2r$, the second equation in (b) is equivalent to

(c) $$\frac{d}{dt}(r^2\dot{\theta}) = 0,$$

from which we obtain

(d) $$r^2\dot{\theta} = c_1.$$

The initial conditions are $t = 0$, $r = 10$, $v_\theta = 15$. Therefore, by the second equation of (34.2), $r\,d\theta/dt = 15$. Substituting these values in (d), we find $c_1 = 150$. Hence (d) becomes

(e) $$\dot{\theta} = \frac{150}{r^2}.$$

In the first equation of (b) replace $\dot{\theta}$ by its value as given in (e). There results

(f) $$\frac{1}{2}\left(\ddot{r} - \frac{150^2}{r^3}\right) = -\frac{8}{r^3}, \qquad \ddot{r} = \frac{150^2 - 16}{r^3} = \frac{22{,}484}{r^3},$$

an equation which is of the type discussed in Lesson 35A. As suggested there, we let $u = dr/dt$ and obtain the system

(g) $$\dot{r} = u, \qquad \dot{u} = \frac{22{,}484}{r^3}.$$

Following the procedure outlined in Lesson 35A, we multiply the second equation of (g) by $2u$. The remaining steps are given without further comment.

(h)
$$2u\dot{u} = 2\,\frac{22{,}484}{r^3}\,u = 2\,\frac{22{,}484}{r^3}\,\dot{r},$$

$$d(u^2) = 2\,\frac{22{,}484}{r^3}\,dr, \qquad u^2 = -\,\frac{22{,}484}{r^2} + c_2.$$

At $t = 0$, $r = 10$, $u = \dot{r} = 0$. Substituting these values in the last equation of (h), we find $c_2 = 22{,}484/100$. Hence by (h) and (g),

(i)
$$u^2 = \dot{r}^2 = -22{,}484\left(\frac{1}{r^2} - \frac{1}{100}\right) = 22{,}484\left(\frac{r^2 - 100}{100r^2}\right),$$

$$\left(\frac{dt}{dr}\right)^2 = \frac{100r^2}{22{,}484(r^2 - 100)},$$

$$dt = \frac{10r}{149.9\sqrt{r^2 - 100}}\,dr.$$

Integration of (i) gives

(j)
$$t = \frac{1}{14.99}\,\sqrt{r^2 - 100} + c_3.$$

At $t = 0$, $r = 10$. Therefore $c_3 = 0$. Hence (j) becomes

(k)
$$(14.99t)^2 = r^2 - 100$$

$$r^2 = 100 + 224.84t^2$$

$$r = \sqrt{100 + 224.84t^2}.$$

Comment 36.51. Note that for this special central force problem, we were able to find r as a function of t relatively easily. In Comment 34.91, we remarked on the difficulty in finding r as a function of t in the case of a particle moving subject to the inverse square law.

EXERCISE 36B

1. A particle of mass m is attracted to a fixed point O by a force F that varies inversely as the cube of the distance r of the particle from O. Initially the particle is a units from O and is given a velocity v_0 in a direction perpendicular to the x axis. Find:

   (a) t as a function of r.
   (b) r as a function of t.
   (c) r as a function of θ, where θ is the polar angle.

   Take mk for the proportionality constant. *Hint.* In (c), after you obtain an equation for $\dot{r}^2$ divide the equation by $\dot{\theta}^2 = a^2v_0^2/r^4$ and note that $\dot{r}/\dot{\theta} = dr/d\theta$. Initial conditions are $t = 0$, $r = a$, $\theta = 0$, $\dot{r} = 0$, $v_0 = a\dot{\theta}_0$.

2. Solve problem 1 if the particle is repelled from instead of being attracted to O.

3. A particle of mass m is attracted to a fixed point O by a force F that varies inversely as the fifth power of the distance r of the particle from O. Take $k^2 m$ for the proportionality constant. Initially the particle is a units from O and is given a velocity $v_0 = k/(a^2\sqrt{2})$ in a direction perpendicular to the x axis. Find r as a function θ. See also Exercise 34H,4. *Hint.* You should obtain $\dot{r}^2 = -k^2/(2a^2r^2) + k^2/(2r^4)$ and $\dot{\theta}^2 = k^2/(2a^2r^4)$. Divide $\dot{r}^2$ by $\dot{\theta}^2$ and note that $\dot{r}/\dot{\theta} = dr/d\theta$. Initial conditions are $t = 0, r = a, \theta = 0, \dot{r} = 0$, $v_0 = a\dot{\theta}_0 = k/(a^2\sqrt{2})$.

ANSWERS 36B

1. (a) $t^2 = \dfrac{a^2}{a^2v_0{}^2 - k}(r^2 - a^2)$,

(b) $r^2 = a^2 + \dfrac{a^2v_0{}^2 - k}{a^2}t^2$,

(c) $r = a \sec \dfrac{\sqrt{a^2v_0{}^2 - k}}{av_0}\theta$.

Note that there are three different possible orbits depending on whether k is less than, equal to or greater than $a^2v_0{}^2$.

2. $t^2 = \dfrac{a^2}{a^2v_0{}^2 + k}(r^2 - a^2)$, $r^2 = a^2 + \dfrac{a^2v_0{}^2 + k}{a^2}t^2$,

$r = a \sec \dfrac{\sqrt{a^2v_0{}^2 + k}}{av_0}\theta$.

3. $r = a \cos \theta$, which is the equation of a circle through the origin.

LESSON 36C.   A Pursuit Problem Leading to a Second Order Nonlinear Differential Equation.

In Lesson 17, we solved pursuit problems leading to a first order differential equation. We shall now solve a pursuit problem which leads to a second order nonlinear differential equation.

Example 36.6. A fighter pilot sights an enemy plane at a distance and starts in pursuit, always keeping the nose of his plane in the direction of the enemy plane. Assume the enemy plane is flying in a straight line at V_E miles/hour, and the fighter plane is flying at V_F miles/hour. Find the equation of the path of the fighter plane as seen by an observer on the ground. NOTE. This is the same type of problem as the relative pursuit problem of Lesson 17B. Here, however, the origin is fixed in space and does not move with the enemy plane.

Solution (Fig. 36.61). Let the origin be the position of the enemy plane, and $P_0(x_0, y_0)$ be the position of the fighter plane, at time $t = 0$. Let the x axis represent the line of flight of the enemy plane. Then at the end of t seconds, the enemy plane has gone a distance equal to $V_E t$ miles along the x axis, and the fighter plane, which is now at $P(x, y)$, a distance

$V_F t$ miles along the arc of his path. Hence the fighter plane's distance in time t is given by

(a) $\quad V_F t = -\int_{y_0}^{y} \sqrt{1 + (dx/dy)^2}\, dy, \quad t = -\frac{1}{V_F}\int_{y_0}^{y} \sqrt{1 + (dx/dy)^2}\, dy.$

(The minus sign is necessary since $y_0 > y$.) At P the fighter's direction is toward the bomber's position $(V_E t, 0)$ and is tangent to the curve of pur-

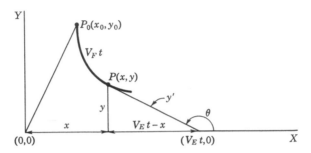

Figure 36.61

suit there. Hence the slope of the curve at P is $y' = \tan\theta = -y/(V_E t - x)$. Therefore at P

(b) $\quad \dfrac{dy}{dx} = \dfrac{y}{x - V_E t}; \quad \dfrac{dx}{dy} = \dfrac{x - V_E t}{y}; \quad t = \dfrac{1}{V_E}\left(x - y\,\dfrac{dx}{dy}\right).$

Equating the values of t in (a) and (b), we obtain

(c) $\quad \dfrac{1}{V_E}\left(x - y\,\dfrac{dx}{dy}\right) = -\dfrac{1}{V_F}\int_{y_0}^{y}\sqrt{1 + (dx/dy)^2}\, dy.$

Differentiation of (c) with respect to y gives, by (9.15),

(d) $\quad \dfrac{1}{V_E}\left(\dfrac{dx}{dy} - y\,\dfrac{d^2x}{dy^2} - \dfrac{dx}{dy}\right) = -\dfrac{1}{V_F}\sqrt{1 + (dx/dy)^2},$

which simplifies to, with V_E/V_F replaced by a new constant k,

(e) $\quad y\,\dfrac{d^2x}{dy^2} = k\sqrt{1 + (dx/dy)^2}, \quad k = \dfrac{V_E}{V_F}.$

This equation is of the type discussed in Lesson 35B, with x and y interchanged. Following the procedure outlined there, we let

(f) $\quad\quad\quad u = \dfrac{dx}{dy}, \quad \dfrac{du}{dy} = \dfrac{d^2x}{dy^2}.$

Substituting these values in (e), we obtain the system of first order equations

(g) $$\frac{dx}{dy} = u, \qquad y\frac{du}{dy} = k\sqrt{1 + u^2}.$$

The solution of the second equation in (g) is

(h) $$\log(u + \sqrt{1 + u^2}) = k\log y + \log c_1,$$

which we can write as

(i) $$\sqrt{1 + u^2} = c_1 y^k - u.$$

Squaring (i) and simplifying the result, we obtain

(j) $$1 = c_1{}^2 y^{2k} - 2c_1 y^k u.$$

Replacing u by its value in (f), we have

(k) $$\frac{dx}{dy} = \frac{c_1{}^2 y^{2k} - 1}{2c_1 y^k} = \frac{1}{2}\left[c_1 y^k - \frac{y^{-k}}{c_1}\right].$$

If $k = V_E/V_F$, is not equal to one, i.e., if the velocities of fighter and bomber are not the same, then integration of (k) gives

(l) $$x = \frac{1}{2}\left[\frac{c_1}{k+1}y^{k+1} + \frac{y^{1-k}}{c_1(k-1)}\right] + c_2, \quad k = \frac{V_E}{V_F} \neq 1.$$

At $t = 0$, $x = x_0$, $y = y_0$, $dy/dx = y_0/x_0$, which implies, by (f), that $u = dx/dy = x_0/y_0$. Substituting these values in (l) and (k), we obtain respectively

(m) $$x_0 = \frac{1}{2}\left[\frac{c_1}{k+1}y_0{}^{k+1} + \frac{y_0{}^{1-k}}{c_1(k-1)}\right] + c_2,$$

(n) $$\frac{x_0}{y_0} = \frac{c_1{}^2 y_0{}^{2k} - 1}{2c_1 y_0{}^k}.$$

With the aid of equations (m) and (n), we can obtain, in a particular problem, the values of c_1 and c_2. The curve of pursuit is given by (l).

If $k = 1$, then by (k)

(o) $$\frac{dx}{dy} = \frac{1}{2}\left(c_1 y - \frac{1}{c_1 y}\right),$$

$$x = \frac{1}{2}\left(c_1\frac{y^2}{2} - \frac{1}{c_1}\log y + c_2\right).$$

EXERCISE 36C

1. A dog sees a cat and starts in pursuit always running in the direction of the cat. The dog's speed is v_d ft/sec, the cat's v_c ft/sec. Assume that initially the dog is at $(0,a)$, the cat at $(0,0)$ and that the cat runs straight for a tree located on the x axis.

(a) Let $k = v_c/v_d$. Determine the equation of the dog's path if $k \neq 1$ and if $k = 1$. *Hint.* Follow the method of Example 36.6. Initial conditions are $t = 0$, $x = 0$, $y = a$, $u = dx/dy = 0$.

(b) Show that if $k < 1$, i.e., if the cat's speed is less than the dog's speed, the cat will reach the tree in safety provided the distance of the tree from the origin is $< ak/(1 - k^2)$. *Hint.* Show that when $y = 0$, $x = ak/(1 - k^2)$ and therefore that dog and cat would meet at this distance from the origin.

(c) When will the dog reach the cat, assuming the tree's distance is $> ak/(1 - k^2)$?

(d) Show that if $k > 1$, $x \to \infty$ as $y \to 0$ and if $k = 1$, $x \to \infty$ as $y \to 0$. In both cases, it is evident that the dog cannot reach the cat.

2. A fighter plane, located at $(0,0)$, sights a bomber at $(a,0)$ and starts in pursuit always keeping the nose of his plane in the direction of the bomber. Assume the bomber is flying parallel to the y axis at v_B mi/hr, and the fighter plane's speed is at v_F mi/hr.

(a) Let $k = v_B/v_F$. Find the equation of the fighter's path if $k \neq 1$, and if $k = 1$. *Hint.* Follow method of Example 36.6. Initial conditions are $t = 0$, $x = 0$, $y = 0$, $dy/dx = 0$.

(b) Show that if $k = v_B/v_F < 1$, i.e., if the bomber's speed is less than the fighter's speed, the fighter will reach the bomber at $x = a$, $y = ak/(1 - k^2)$. Hence show that the fighter will reach the bomber in time

$$t = \frac{ak}{v_B(1 - k^2)} = \frac{av_F}{v_F^2 - v_B^2}.$$

(c) Assume $a = 2$ mi, $v_F = 300$ mi/hr, $v_B = 200$ mi/hr. When will the fighter reach the bomber?

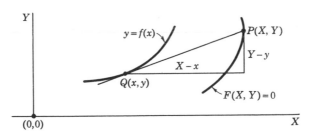

Figure 36.7

3. Assume that the pursued object does not move along one of the axes, but along a given curve $F(X,Y) = 0$, and that at time t, its position is $P(X,Y)$, Fig. 36.7. The pursuer, as usual, always moves toward the position P of the pursued object. If the pursuer's position at time t is $Q(x,y)$ and the equation

of his path is $y = f(x)$, we have, for the equation of the line PQ,

(36.71) $$(X - x)y' = Y - y.$$

With $Q(x,y)$ considered as fixed, the differential of (36.71) is

(36.72) $$(X - x)\, dy' + y'\, dX = dY.$$

The differential of

(36.73) $$F(X,Y) = 0$$

is

(36.74) $$\frac{\partial F}{\partial X}\, dX + \frac{\partial F}{\partial Y}\, dY = 0.$$

If the speed of Q is k times the speed of P, we have [speed $= |v| = |ds/dt| = \sqrt{(dx/dt)^2 + (dy/dt)^2}$, $|ds| = \sqrt{dx^2 + dy^2}$]

(36.75) $$\sqrt{dx^2 + dy^2} = k\sqrt{dX^2 + dY^2}.$$

With the help of equations (36.71) to (36.75), solve the following problem. A pursued plane P flies in a straight line making an angle of 45° with the x axis. Find the equation of the path of the pursuing plane Q if its speed is twice that of the pursued plane P. Choose the origin on P's path. *Hint.* Here (36.73) is $X - Y = 0$, (36.74) is $dX - dY = 0$, (36.75) is

(36.76) $\sqrt{1 + (y')^2}\, dx = 2\sqrt{1 + (dY/dX)^2}\, dX = 2\sqrt{1 + 1}\, dX = 2\sqrt{2}\, dX,$

(36.71) is $(X - x)y' = X - y$, (36.72) is $(X - x)\, dy' = (1 - y')\, dX$. Now show that $dX = \dfrac{(x - X)\, dy'}{y' - 1}$ and $y' - 1 = \dfrac{X - y}{X - x} - 1 = \dfrac{y - x}{x - X}$, $x - X = \dfrac{y - x}{y' - 1}$. Hence $dX = \dfrac{(y - x)\, dy'}{(y' - 1)^2}$. Substitute this value of dX in (36.76) to obtain

(36.77) $$(y' - 1)^2\sqrt{1 + (y')^2} = 2\sqrt{2}\,(y - x)\frac{d^2y}{dx^2}.$$

To solve (36.77), make the substitution

(36.78) $$u = y - x, \quad \frac{du}{dx} = \frac{dy}{dx} - 1, \quad \frac{dy}{dx} = 1 + \frac{du}{dx}, \quad \frac{d^2y}{dx^2} = \frac{d^2u}{dx^2}.$$

There should result

(36.79) $$\left(\frac{du}{dx}\right)^2 \sqrt{1 + \left(1 + \frac{du}{dx}\right)^2} = 2\sqrt{2}\, u\, \frac{d^2u}{dx^2}.$$

Since (36.79) does not contain the independent variable x, the method of Lesson 35C is applicable.

ANSWERS 36C

1. (a) $x = \dfrac{a}{2}\left[\dfrac{1}{k+1}\left(\dfrac{y}{a}\right)^{k+1} + \dfrac{1}{k-1}\left(\dfrac{y}{a}\right)^{1-k}\right] - \dfrac{ak}{k^2-1}$, $k \neq 1$.

$$x = \frac{1}{2}\left[\frac{y^2 - a^2}{2a} - a\log\frac{y}{a}\right], \quad k = 1.$$

(c) $t = \dfrac{a}{v_d(1 - k^2)} = \dfrac{av_d}{v_d{}^2 - v_c{}^2}$ sec.

2. (a) $y = \dfrac{1}{2}\left[\dfrac{(a - x)^{1+k}}{a^k(1 + k)} - \dfrac{a^k(a - x)^{1-k}}{1 - k}\right] + \dfrac{ak}{1 - k^2}, \quad k \neq 1.$

$$y = \frac{1}{2}\left[a\log\frac{a}{a - x} + \frac{x(x - 2a)}{2a}\right].$$

(c) $t = 43.2$ sec.

3. $2\sqrt{2}\, x = 2c_1\sqrt{y - x} - \sqrt{2}\,(y - x) - \dfrac{(y - x)^{3/2}}{3c_1} + c_2.$

LESSON 36D. Geometric Problems. In Lesson 13, we solved geometric problems which gave rise to a first order differential equation. In this lesson, we shall solve a geometric problem giving rise to one of the specific types of second order equations discussed in Lesson 35.

In rectangular coordinates the **radius of curvature** R of a curve $y = f(x)$ at a point $P(x,y)$ on it is given by the formula

(36.8) $$R = \frac{[1 + (y')^2]^{3/2}}{y''}.$$

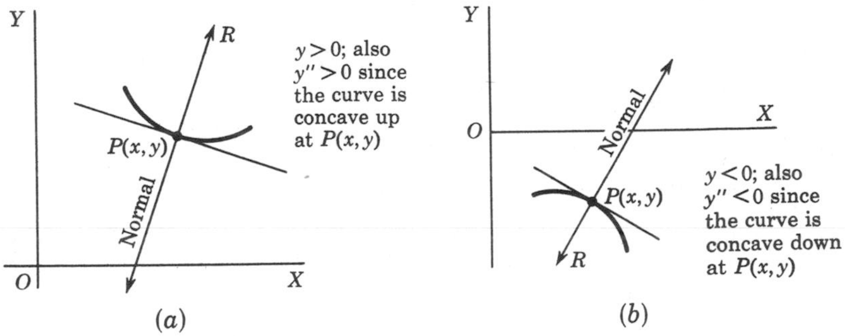

Figure 36.81

If a normal to the curve is drawn from the point P to the x axis, then it and the radius of curvature have:

1. Opposite directions when y and y'' have the same signs, see Fig. 36.81(a) and (b).
2. The same directions when y and y'' have opposite signs, see Fig. 36.82(a) and (b).

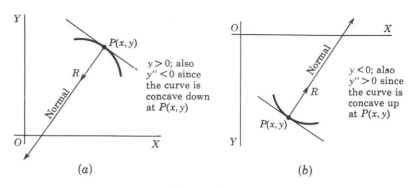

$y > 0$; also
$y'' < 0$ since
the curve is
concave down
at $P(x, y)$

$y < 0$; also
$y'' > 0$ since
the curve is
concave up
at $P(x, y)$

(a)   (b)

Figure 36.82

Example 36.83.   Find the family of curves whose radius of curvature is twice the length of the normal segment from a point on the curve to the x axis, (a) if normal and radius of curvature have opposite directions, (b) if the two have the same directions.

Solution.   The length of the normal segment from a point on the curve to the x axis is [see (i) of Exercise 13,1] $|y\sqrt{1 + (y')^2}|$. Hence by (36.8) and the hypothesis of the problem,

(a) $$\frac{[1 + (y')^2]^{3/2}}{y''} = \pm 2y\sqrt{1 + (y')^2},$$

where the plus sign is to be used when normal and radius of curvature have opposite directions, the minus sign when they have the same direction.   Division of (a) by $[1 + (y')^2]^{1/2}$ gives

(b) $$1 + (y')^2 = \pm 2yy'',$$

an equation which is of the type described in Lesson 35C.   In (b), we therefore make the substitutions

(c) $$y' = u, \qquad y'' = u\frac{du}{dy},$$

thus obtaining

(d) $$1 + u^2 = \pm 2yu\frac{du}{dy}, \qquad \frac{2u\,du}{1 + u^2} = \pm\frac{dy}{y}.$$

The solution of (d) is

(e) $$\log(1 + u^2) = \pm\log y + \log c_1.$$

From (e) and (c), we obtain the equation

(f) $1 + u^2 = c_1 y, \qquad \dfrac{dy}{dx} = \pm\sqrt{c_1 y - 1}, \qquad \dfrac{dy}{\sqrt{c_1 y - 1}} = \pm dx,$

when normal and radius of curvature have opposite directions, and the equation

(g) $1 + u^2 = \dfrac{c_1}{y}$, $\dfrac{dy}{dx} = \pm\sqrt{\dfrac{c_1 - y}{y}}$, $\dfrac{\sqrt{y}\,dy}{\sqrt{c_1 - y}} = \pm dx$,

when they have the same directions. The solution of the last equation in (f) is

(h) $\dfrac{2}{c_1}\sqrt{c_1 y - 1} = \pm(x + c_2)$, $4(c_1 y - 1) = c_1{}^2(x + c_2)^2$,

which is a family of parabolas with vertices at $(-c_2, 1/c_1)$ and axes parallel to the y axis. The solution of the last equation in (g) is (substitute $u = \sqrt{y}$)

(i) $c_1 \operatorname{Arc\,sin} \sqrt{y/c_1} - \sqrt{y(c_1 - y)} = \pm x + c_2$,

which is a family of cycloids. (For definition of a cycloid, see Exercise 28c, 34.)

EXERCISE 36D

1. Find the family of curves whose radius of curvature is equal to the length of the normal segment from a point on the curve to the x axis and has the same direction as the normal. Identify the family.
2. Solve problem 1, if the radius of curvature and normal have opposite directions. Identify the family.
3. Find the family of curves whose radius of curvature has a constant value k.

ANSWERS 36D

1. $(x - c_1)^2 + y^2 = c_2$. A family of circles with centers at $(c_1, 0)$ and radius $\sqrt{c_2}$.
2. $c_1 y = \cosh(\pm c_1 x + c_2)$. A family of catenaries.
3. $(x - c_1)^2 + (y - c_2)^2 = k^2$.

Chapter 9

Series Methods

Introductory Remarks. As we have repeatedly emphasized, few differential equations have solutions which can be expressed explicitly or implicitly in terms of elementary functions. When a solution cannot be expressed in this way, the problem of finding a solution of a differential equation is not entirely hopeless. There are available graphical methods, some of which were described in earlier lessons, numerical methods which will be discussed in the next chapter and series methods which we shall consider in this chapter. Furthermore, it is frequently true that an implicit solution in terms of elementary functions is less useful than a series solution or a numerical one. Implicit solutions are usually such complicated expressions that it is extremely difficult to find values of the dependent variable for given values of the independent variable.

For a clearer understanding of the subject matter of this chapter, it will be necessary to have a knowledge of Taylor series and to know certain definitions and theorems from analysis. Hence, before beginning a discussion of series methods for solving differential equations, we shall first review this needed material for you.

LESSON 37. Power Series Solutions of Linear Differential Equations.

LESSON 37A. Review of Taylor Series and Related Matters. A series of the form

$$(37.1) \quad a_0 + a_1(x - x_0) + a_2(x - x_0)^2 + a_3(x - x_0)^3 + \cdots,$$

where $a_0, a_1, a_2, \cdots, x_0$ are constants and x is a variable is called a **power series.** A power series may:

1. Converge only for the single value $x = x_0$.
2. Converge absolutely for values of x in a neighborhood of x_0, i.e., converge for $|x - x_0| < h$; diverge for $|x - x_0| > h$. At the end points $x_0 \pm h$, it may either converge or diverge.
3. Converge absolutely for all values of x, i.e., for $-\infty < x < \infty$.

In cases 2 and 3, the set of values of x for which the power series converges is called the **interval of convergence** of the series. In case 2, for example, if a series also converges for $x = x_0 \pm h$, then its interval of convergence is $x_0 - h \leqq x \leqq x_0 + h$; if it converges only for $x = x_0 + h$, but not for $x = x_0 - h$, then the interval of convergence is $x_0 - h < x \leqq x_0 + h$; etc. In case 3, the interval of convergence is the entire real axis.

Comment 37.11. *Interval of Convergence.* In the calculus you were taught certain tests by which you could determine an interval of convergence of a power series. A simple one and one which is frequently used is known as the "ratio test." It states that the series

$$u_1 + u_2 + u_3 + \cdots + u_n + \cdots$$

converges *absolutely* if

(37.12)
$$\lim_{n \to \infty} \left| \frac{u_{n+1}}{u_n} \right| = k < 1.$$

We give below examples of each of the three types of series. For convenience we have taken $x_0 = 0$.

Example 37.13. Determine the interval of convergence of the power series

(a)
$$1 + 1!x + 2!x^2 + 3!x^3 + \cdots + n!x^n + \cdots.$$

Solution. Here $u_n = n!x^n$, $u_{n+1} = (n + 1)!x^{n+1}$. Therefore

(b)
$$\left| \frac{u_{n+1}}{u_n} \right| = \left| \frac{(n + 1)!}{n!} \frac{x^{n+1}}{x^n} \right| = |(n + 1)x|.$$

For each $x \neq 0$, $|(n + 1)x| \to \infty$ as $n \to \infty$. Since this limit $\neq k < 1$, the series (a), by Comment 37.11, converges only for $x = 0$.

Example 37.14. Determine the interval of convergence of the power series

(a)
$$1 + x + \frac{1}{2}x^2 + \frac{1}{3}x^3 + \cdots + \frac{1}{n}x^n + \cdots.$$

Solution. Here $u_n = x^n$, $u_{n+1} = x^{n+1}$. Therefore

(b)
$$\lim_{n \to \infty} \left| \frac{u_{n+1}}{u_n} \right| = \lim_{n \to \infty} \left| \frac{(n + 1)}{n} \frac{x^{n+1}}{x^n} \right| = |x|.$$

Hence, by Comment 37.11, the series (a) converges absolutely for each x whose absolute value is less than one. The interval of convergence, however, is $-1 \leqq x < 1$, since the series converges for $x = -1$, but diverges for $x = 1$.

Example 37.15. Determine the interval of convergence of the power series

(a) $$1 - \frac{x^2}{2!} + \frac{x^4}{4!} - \frac{x^6}{6!} + \cdots + \frac{(-1)^{n-1}x^{2n-2}}{(2n-2)!} + \cdots .$$

Solution. Here $|u_n| = \dfrac{x^{2(n-1)}}{(2n-2)!}$, $|u_{n+1}| = \dfrac{x^{2n}}{(2n)!}$. Therefore,

(b) $$\lim_{n\to\infty} \left| \frac{u_{n+1}}{u_n} \right| = \lim_{n\to\infty} \frac{x^{2n}}{(2n)!} \frac{(2n-2)!}{x^{2n-2}} = \lim_{n\to\infty} \frac{x^2}{2n(2n-1)} = 0,$$

for each x. Hence, by Comment 37.11, the series (a) converges absolutely for all x. Its interval of convergence is therefore the entire real axis.

We now state a number of relevant theorems in connection with power series.

Theorem 37.16. *If a power series (37.1) converges on an interval I:* $|x - x_0| < R$, *where R is a positive constant, then the power series defines a function* $f(x)$ *which is continuous for each x in I.*

Comment 37.17. If one writes a power series, say

(a) $$1 + x + x^2 + x^3 + \cdots ,$$

which is convergent on $I\colon -1 < x < 1$, then by Theorem 37.16, the series (a) defines a function which is continuous on this interval. The question naturally arises: which function? This question is, in general, not easy to answer although it is for (a), because the series is a geometric one. This series, for each x for which $|x| < 1$, converges to $1/(1-x)$. Hence,

(b) $$f(x) = \frac{1}{1-x} = 1 + x + x^2 + x^3 + \cdots , \quad |x| < 1.$$

For example, when $x = \frac{1}{2}$, then $f(\frac{1}{2}) = \dfrac{1}{1-\frac{1}{2}} = 2$ and the geometric series on the right converges to 2. But if $x = 2$, then $f(x) = 1/(1-2) = -1$, and the series on the right of (b) certainly does not converge to -1. Now consider the power series

(c) $$x - \frac{x^3}{3} + \frac{x^5}{5} - \frac{x^7}{7} + \cdots + (-1)^{n-1}\frac{x^{2n-1}}{2n-1} + \cdots ,$$

whose interval of convergence is also $-1 < x < 1$. Hence by Theorem 37.16, it defines a function $f(x)$ which is continuous on this interval. In this case only one familiar with series might recognize that the series (c) defines the function Arc tan x. It is a fact, however, that many convergent power series cannot be linked to elementary functions for the very good reason that many convergent power series do not define elementary functions.

The converse of the question raised above is somewhat easier to answer, i.e., given a continuous function on an interval I, is there a power series which defines it? We shall state certain theorems below which not only will give us the answer to this question, but also will show us at the same time how to find the defining power series, if the function has one.

Theorem 37.2.   *If $f(x)$ is defined by a power series, i.e., if*

$$(37.21) \quad f(x) = a_0 + a_1(x - x_0) + a_2(x - x_0)^2$$
$$+ a_3(x - x_0)^3 + \cdots, \quad I: |x - x_0| < R,$$

then

$$(37.22) \quad f'(x) = a_1 + 2a_2(x - x_0) + 3a_3(x - x_0)^2 + \cdots,$$
$$I: |x - x_0| < R,$$

i.e., the power series obtained by differentiating each term of (37.21) defines (or converges to) the derivative of $f(x)$ on the same interval I.

Theorem 37.23.   *If $f(x)$ and $g(x)$ are defined by power series, i.e., if*

$$(37.231) \quad f(x) = a_0 + a_1(x - x_0) + a_2(x - x_0)^2 + \cdots, \quad |x - x_0| < R$$

and

$$(37.232) \quad g(x) = b_0 + b_1(x - x_0) + b_2(x - x_0)^2 + \cdots, \quad |x - x_0| < R,$$

then $f(x) = g(x)$ if and only if

$$(37.233) \qquad a_0 = b_0, \quad a_1 = b_1, \quad a_2 = b_2, \quad \cdots.$$

Theorem 37.24.   *If $f(x)$ is defined by a power series, i.e., if*

$$(37.25) \quad f(x) = a_0 + a_1(x - x_0) + a_2(x - x_0)^2 + \cdots$$
$$+ a_n(x - x_0)^n + \cdots, \quad |x - x_0| < R,$$

then

$$(37.26) \quad a_0 = f(x_0), \quad a_1 = f'(x_0),$$
$$a_2 = \frac{f''(x_0)}{2!}, \cdots, \quad a_n = \frac{f^n(x_0)}{n!}, \cdots.$$

Proof.   By Theorem 37.2, the successive derivatives of (37.25) are

$$(a) \quad f'(x) = a_1 + 2a_2(x - x_0) + 3a_3(x - x_0)^2 + \cdots$$
$$+ na_n(x - x_0)^{n-1} + \cdots,$$

$$f''(x) = 2a_2 + 3!a_3(x - x_0) + \cdots + n(n - 1)a_n(x - x_0)^{n-2} + \cdots,$$

$$f'''(x) = 3!a_3 + \cdots + n(n - 1)(n - 2)a_n(x - x_0)^{n-3} + \cdots,$$

$$\cdots \cdots \cdots \cdots \cdots \cdots \cdots \cdots \cdots \cdots \cdots \cdots \cdots \cdots \cdots \cdots$$

$$f^{(n)}(x) = n!a_n + (n + 1)!a_{n+1}(x - x_0) + \cdots.$$

In (37.25) and (a), set $x = x_0$ in each equation. There results

(b) $\qquad\qquad f(x_0) = a_0, \qquad f'(x_0) = a_1, \qquad f''(x_0) = 2!a_2,$

$\qquad\qquad\qquad f'''(x_0) = 3!a_3, \cdots, \qquad f^n(x_0) = n!a_n.$

Hence by (b),

(c) $\qquad\qquad\qquad a_0 = f(x_0), \qquad a_1 = f'(x_0),$

$$a_2 = \frac{f''(x_0)}{2!}, \qquad a_3 = \frac{f'''(x_0)}{3!}, \cdots, \qquad a_n = \frac{f^{(n)}(x_0)}{n!}.$$

If $f(x)$ is defined by a power series, then by (37.25) and (37.26),

$$(37.27) \quad f(x) = f(x_0) + f'(x_0)(x - x_0) + \frac{f''(x_0)}{2!}(x - x_0)^2$$

$$+ \frac{f'''(x_0)}{3!}(x - x_0)^3 + \cdots$$

$$+ \frac{f^{(n)}(x_0)}{n!}(x - x_0)^n + \cdots, \quad |x - x_0| < R.$$

If $x_0 = 0$, then (37.27) becomes

$$(37.28) \quad f(x) = f(0) + f'(0)x + \frac{f''(0)}{2!}x^2 + \frac{f'''(0)}{3!}x^3 + \cdots$$

$$+ \frac{f^{(n)}(0)}{n!}x^n + \cdots, \quad |x| < R.$$

Definition 37.3. The series on the right of (37.27) is called the **Taylor series** expansion of $f(x)$ in powers of $(x - x_0)$, or in a neighborhood of x_0.

Definition 37.31. The series on the right of (37.28) is called the **Maclaurin series** expansion of $f(x)$ in powers of x, or in a neighborhood of zero.

Example 37.32. Assume the function $f(x) = \sin x$ is defined by a Maclaurin series (37.28) on some interval. Find the series and its interval of convergence.

Solution. Taking successive derivatives of $f(x) = \sin x$ and evaluating them at $x = 0$, we obtain

(a) $\qquad\qquad\quad \begin{aligned} f(x) &= \sin x, & f(0) &= 0, \\ f'(x) &= \cos x, & f'(0) &= 1, \\ f''(x) &= -\sin x, & f''(0) &= 0, \\ f'''(x) &= -\cos x, & f'''(0) &= -1, \\ f^{(4)}(x) &= \sin x, & f^{(4)}(0) &= 0, \\ f^{(5)}(x) &= \cos x, & f^{(5)}(0) &= 1. \end{aligned}$

Substituting the right-hand values of (a) in (37.28), there results

(b) $\sin x = x - \dfrac{x^3}{3!} + \dfrac{x^5}{5!} - \cdots + (-1)^{n-1} \dfrac{x^{2n-1}}{(2n-1)!} + \cdots .$

Here $|u_n| = x^{2n-1}/(2n-1)!$. Therefore

(c) $\lim\limits_{n \to \infty} \left| \dfrac{u_{n+1}}{u_n} \right| = \lim\limits_{n \to \infty} \left| \dfrac{x^{2n+1}}{(2n+1)!} \dfrac{(2n-1)!}{x^{2n-1}} \right| = \lim\limits_{n \to \infty} \left| \dfrac{x^2}{(2n)(2n+1)} \right| = 0,$

for each x. Hence by Comment 37.11, the series (b) converges for all x.

Comment 37.33.   Theorem 37.24 says that *if* $f(x)$ is defined by a power series, then the coefficients in the series are given by (37.26). But it does not tell us whether $f(x)$ can be defined by a power series. Perhaps it cannot. Here is an example of a function which cannot be thus defined. Let

(a) $f(x) = e^{-1/x^2}, \quad x \neq 0$

$\qquad\qquad = 0, \quad x = 0.$

The function $f(x)$ is continuous at $x = 0$. Its derivatives at $x = 0$ are $\left[\text{use the definition of the derivative } f'(0) = \lim\limits_{h \to 0} \dfrac{f(h) - f(0)}{h}\right]$

(b) $f'(0) = 0,\ f''(0) = 0,\ f'''(0) = 0,\ \cdots .$

Substituting these values in (37.28), we obtain

(c) $f(x) = 0 + 0x + \dfrac{0}{2!}\,x^2 + \dfrac{0}{3!} + \cdots = 0 + 0 + 0 + \cdots,$

$\qquad\qquad -\infty < x < \infty.$

For an $x \neq 0$, the right side of (c) certainly does not converge to the function $f(x)$ defined by (a). For example, if $x = 0.1$, then by (a), $f(0.1) = e^{-1/0.01}$. The series on the right side of (c), however, converges to zero for all x.

If therefore we start with a continuous function $f(x)$ and obtain a power series whose coefficients are given by (37.26), we still need a theorem which will tell us whether the power series thus obtained actually defines or converges to the given function $f(x)$. For this purpose we introduce the following theorem.

Theorem 37.34.   *Taylor's Theorem.   If a function $f(x)$ has derivatives of all orders on an interval I: $|x - x_0| < h$, then*

(37.35) $f(x) = f(x_0) + f'(x_0)(x - x_0) + \dfrac{f''(x_0)}{2!}\,(x - x_0)^2 + \cdots$

$\qquad\qquad + \dfrac{f^n(x_0)}{n!}\,(x - x_0)^n + R_n(x),$

where $R_n(x)$, *the remainder term or the sum of all terms after the* $(n + 1)$
term, is given by

$$(37.36) \qquad R_n(x) = \frac{f^{(n+1)}(X)(x - x_0)^{n+1}}{(n + 1)!},$$

and X *is between* x_0 *and* x.

If $R_n(x) \to 0$ as $n \to \infty$, *then and only then is*

$$(37.37) \quad f(x) = f(x_0) + f'(x_0)(x - x_0) + \frac{f''(x_0)}{2!} (x - x_0)^2 + \cdots,$$

$$|x - x_0| < h,$$

i.e., the infinite series on the right actually converges to and defines the function $f(x)$ *on* I: $|x - x_0| < h$.

Remark. The series on the right of (37.35) is called a **Taylor series with remainder.**

Definition 37.4. A function $f(x)$ is said to be **analytic** at a point $x = x_0$ if it has a Taylor series expansion in powers of $(x - x_0)$ valid for each x in a neighborhood of x_0.

Definition 37.41. A function $f(x)$ is said to be **analytic on an interval** if it is analytic at each point of the interval.

Examples of Analytic Functions. It has been proved that each of the functions listed below is analytic on the interval indicated. Its Taylor series expansion is given by the series on the right.

$$(37.42) \qquad e^x = 1 + x + \frac{x^2}{2!} + \frac{x^3}{3!} + \cdots, \quad -\infty < x < \infty;$$

$$\cos x = 1 - \frac{x^2}{2!} + \frac{x^4}{4!} - \frac{x^6}{6!} + \cdots, \quad -\infty < x < \infty;$$

$$\sin x = x - \frac{x^3}{3!} + \frac{x^5}{5!} - \frac{x^7}{7!} + \cdots, \quad -\infty < x < \infty;$$

$$\text{Arc tan } x = x - \frac{x^3}{3} + \frac{x^5}{5} - \frac{x^7}{7} + \cdots, \quad -1 \leq x \leq 1.$$

Comment 37.43. Because of (37.26), an analytic function has one and only one Taylor series expansion. This fact implies that the same series results no matter what method is used to obtain it. Verify, for example, that the series $1/(1 - x) = 1 + x + x^2 + \cdots$, $|x| < 1$, can be obtained by Theorem 37.24 as well as by ordinary division.

LESSON 37B. Solution of Linear Differential Equations by Series Methods.   The series methods we shall describe in this lesson are especially well suited for finding a solution of the linear differential equation with nonconstant coefficients,

$$(37.5) \quad y^{(n)} + f_{n-1}(x)y^{(n-1)} + \cdots + f_1(x)y' + f_0(x)y = Q(x).$$

We shall therefore consider this class of equations first, and then proceed to a discussion of other types of differential equations and to systems of differential equations.

In Theorem 65.2, we state and prove a sufficient condition for the existence and uniqueness of a solution of (37.5) satisfying n initial conditions. Here we merely state, without proof, a sufficient condition for the existence of a *power series solution* of (37.5).

Theorem 37.51. *If each function* $f_0(x)$, $f_1(x)$, $\cdots$, $f_{n-1}(x)$, $Q(x)$ *in (37.5) is analytic at* $x = x_0$, *i.e., if each function has a Taylor series expansion in powers of* $(x - x_0)$ *valid* for* $|x - x_0| < r$, *then there is a unique solution* $y(x)$ *of (37.5) which is also analytic at* $x = x_0$, *satisfying the* n *initial conditions*

$$(37.52) \quad y(x_0) = a_0, \qquad y'(x_0) = a_1, \cdots, y^{(n-1)}(x_0) = a_{n-1},$$

i.e., the solution has a Taylor series expansion in powers of $(x - x_0)$ *also valid for* $|x - x_0| < r$.

Comment 37.53. A polynomial is a finite series. Hence the series is valid for all x. If therefore the functions $f_0(x), f_1(x), \cdots, f_{n-1}(x), Q(x)$ of (37.5) are each polynomials, then, by Theorem 37.51, every solution of (37.5) has a Taylor series expansion valid for all x.

Comment 37.54. An existence theorem tells you only whether a solution of a differential equation exists. It does not tell you how to find the solution.

First Series Method. By Successive Differentiations. We shall illustrate by examples the first method of finding a power series solution of a linear differential equation.

Example 37.541. Find by series methods, a particular solution of the linear equation

$$(a) \qquad y'' - (x + 1)y' + x^2 y = x$$

for which $y(0) = 1$, $y'(0) = 1$.

Solution. Comparing (a) with (37.5), we see that $f_0(x) = x^2, f_1(x) = -x - 1, Q(x) = x$. Since all these functions are polynomials, the series solution we shall obtain, by Comment 37.53, is valid for all x. Because

---

*We shall use the word "valid" to mean that for each x in a neighborhood of x_0 the series expansion converges to the value of the function.

we have been given values of the solution and its derivative when $x = 0$, we seek a solution in the form of the Maclaurin series (37.28). With $f(x)$ replaced by $y(x)$ it becomes

(b) $$y(x) = y(0) + y'(0)x + \frac{y''(0)}{2!} x^2 + \frac{y'''(0)}{3!} x^3$$

$$+ \frac{y^{(4)}(0)}{4!} x^4 + \cdots .$$

By the initial conditions, $x = 0$, $y = 1$, $y' = 1$. Substituting these values in (a) gives,

(c) $$y''(0) - 1 = 0, \qquad y''(0) = 1.$$

In (b), we now know, by the initial conditions and (c), the values of $y(0)$, $y'(0)$, $y''(0)$. To find the values of succeeding coefficients, we take successive derivatives of (a) and evaluate them at (0,1). The next two derivatives are

(d) $$y''' - (x + 1)y'' - y' + x^2 y' + 2xy = 1,$$
$$y^{(4)} - (x + 1)y''' - 2y'' + x^2 y'' + 4xy' + 2y = 0.$$

Hence when $x = 0$, $y = 1$, $y' = 1$, $y'' = 1$, we obtain from (d)

(e) $$y'''(0) = 3, \qquad y^{(4)}(0) = 3.$$

Substituting in (b), the initial conditions, (c) and (e), we have

(f) $$y(x) = 1 + x + \frac{x^2}{2} + \frac{x^3}{2} + \frac{x^4}{8} + \cdots, \qquad -\infty < x < \infty,$$

which gives the first five terms of a series solution of (a) satisfying the given initial conditions.

Example 37.55.   Find by power series methods, a particular solution of the linear equation

(a) $$y'' + \frac{x}{1 - x^2} y' - \frac{1}{1 - x^2} y = 0, \qquad |x| \neq 1,$$

for which $y(0) = 1$, $y'(0) = 1$.

Solution.   Since the initial conditions have been given in terms of $x = 0$, we seek a series solution in powers of x. Comparing (a) with (37.5), we see that $f_0(x) = -1/(1 - x^2)$, $f_1(x) = x/(1 - x^2)$, $Q(x) = 0$. The Maclaurin series expansion of $-1/(1 - x^2)$ is

(b) $$-\frac{1}{1 - x^2} = -(1 + x^2 + x^4 + x^6 + \cdots), \qquad |x| < 1.$$

The Maclaurin series expansion of $x/(1 - x^2)$ is also valid for $|x| < 1$. Hence by Theorem 37.51, each solution of (a) has a power series expansion which is valid for $|x| < 1$. By the initial conditions, $x = 0$, $y = 1$, $y' = 1$. Substituting these values in (a) gives

(c) $$y''(0) = 1.$$

In the Maclaurin series (37.28), namely

(d) $$y(x) = y(0) + y'(0)x + \frac{y''(0)}{2!} + \frac{y'''(0)}{3!} + \frac{y^{(4)}(0)}{4!}\,x^4 + \cdots,$$

we now know, by (c) and the initial conditions, the values of $y(0)$, $y'(0)$, $y''(0)$. To find the values of succeeding coefficients, we multiply (a) by $(1 - x^2)$, then take its successive derivatives and evaluate them at $(0,1)$. The next two derivatives are

(e) $$(1 - x^2)y''' - xy'' = 0, \qquad (1 - x^2)y^{(4)} - 3xy''' - y'' = 0.$$

Hence when $x = 0$, $y = 1$, $y' = 1$, $y'' = 1$, we find from (e)

(f) $$y'''(0) = 0, \qquad y^{(4)}(0) = 1.$$

Substituting in (d) the initial conditions, (c) and (f), we obtain

(g) $$y(x) = 1 + x + \frac{x^2}{2} + \frac{x^4}{24} + \cdots, \qquad -1 < x < 1,$$

which gives the first five terms of a series solution of (a) satisfying the given initial conditions.

Example 37.56. Find by power series methods a particular solution of the linear equation

(a) $$y''' + \frac{1}{x}\,y' - \frac{1}{x^2}\,y = 0, \qquad x \neq 0,$$

for which $y(1) = 1$, $y'(1) = 0$, $y''(1) = 1$.

Solution. Since the initial conditions have been given in terms of $x = 1$, we seek a series solution in powers of $x - 1$. Comparing (a) with (37.5), we see that $f_0(x) = -1/x^2$, $f_1(x) = 1/x$, $Q(x) = 0$. The series representation of $1/x^2$ in powers of $x - 1$ is $1 - 2(x - 1) + 3(x - 1)^2 - 4(x - 1)^3 + \cdots$, which is valid for $0 < x < 2$. The series representation of $1/x$ in powers of $(x - 1)$ is $1 - (x - 1) + (x - 1)^2 - (x - 1)^3 + \cdots$, which is also valid for $0 < x < 2$. Hence, by Theorem 37.51, each solution of (a) has a Taylor series expansion in powers of $x - 1$, valid for $0 < x < 2$. By the initial conditions, $x = 1$, $y = 1$, $y' = 0$, $y'' = 1$.

Substituting these values in (a) gives

(b) $\qquad\qquad y'''(1) + 0 - 1 = 0, \qquad y'''(1) = 1.$

In the Taylor series (37.27), namely

(c) $\quad y(x) = y(1) + y'(1)(x - 1) + \dfrac{y''(1)}{2!}(x - 1)^2 + \dfrac{y'''(1)}{3!}(x - 1)^3$

$$+ \frac{y^{(4)}(1)}{4!}(x - 1)^4 + \frac{y^{(5)}(1)}{5!}(x - 1)^5 + \cdots,$$

we now know the values of $y(1), y'(1), y''(1), y'''(1)$. To find the values of succeeding coefficients, we multiply (a) by x^2, then take its successive derivatives, and evaluate them at $(1,1)$. The next two derivatives are

(d) $\qquad\qquad x^2 y^{(4)} + 2xy''' + xy'' + y' - y' = 0,$

$$xy^{(4)} + 2y''' + y'' = 0, \qquad xy^{(5)} + 3y^{(4)} + y''' = 0.$$

When $x = 1, y = 1, y' = 0, y'' = 1, y''' = 1$, we find from (d)

(e) $\qquad\qquad y^{(4)}(1) = -3, \qquad y^{(5)}(1) = 8.$

Substituting in (c), the initial conditions, (b) and (e), we obtain

(f) $\quad y(x) = 1 + \dfrac{(x - 1)^2}{2} + \dfrac{(x - 1)^3}{6} - \dfrac{(x - 1)^4}{8} + \dfrac{(x - 1)^5}{15} - \cdots,$

$$0 < x < 2,$$

which gives the first six terms of a series solution of (a) satisfying the given initial conditions.

Second Series Method. Undetermined Coefficients. We outline a second method of obtaining a series solution, one which does not depend on taking derivatives. This method will therefore be more useful than the preceding one whenever it becomes too difficult to obtain successive derivatives.

Example 37.6. Find by power series methods a particular solution of the linear equation

(a) $\qquad\qquad y'' - (x + 1)y' + x^2 y = x$

for which $y(0) = 1, y'(0) = 1$.

Solution. This example is the same as 37.541. Hence we know that (a) has a series solution in powers of x valid for all x. A series solution in powers of x has the form

(b) $\qquad\qquad y(x) = a_0 + a_1 x + a_2 x^2 + a_3 x^3 + a_4 x^4 + \cdots.$

By Theorem 37.2, its two successive derivatives, also valid for all x, are

(c)
$$y'(x) = a_1 + 2a_2 x + 3a_3 x^2 + 4a_4 x^3 + \cdots,$$
$$y''(x) = 2a_2 + 6a_3 x + 12a_4 x^2 + \cdots.$$

Substituting (b) and (c) in (a), we see that $y(x)$ will be a solution of (a) if

(d) $2a_2 + 6a_3 x + 12a_4 x^2 + \cdots$
$$- (x + 1)(a_1 + 2a_2 x + 3a_3 x^2 + 4a_4 x^3 + \cdots)$$
$$+ x^2(a_0 + a_1 x + a_2 x^2 + a_3 x^3 + a_4 x^4 + \cdots) = x.$$

Carrying out the indicated operations in (d) and simplifying the result, we obtain

(e) $(2a_2 - a_1) + (6a_3 - 2a_2 - a_1 - 1)x$
$$+ (12a_4 - 3a_3 - 2a_2 + a_0)x^2 + \cdots = 0.$$

Because of Theorem 37.23 [take $g(x) = 0 + 0x + 0x^2 + \cdots$], equation (e) will be an identity in x if and only if each coefficient is zero. Hence we must have

(f)
$$2a_2 - a_1 = 0, \qquad a_2 = \frac{a_1}{2}.$$

$$6a_3 - 2a_2 - a_1 - 1 = 0, \qquad a_3 = \frac{2a_2 + a_1 + 1}{6}.$$

$$12a_4 - 3a_3 - 2a_2 + a_0 = 0, \qquad a_4 = \frac{3a_3 + 2a_2 - a_0}{12}.$$

By (37.26) and the initial conditions $y(0) = 1$, $y'(0) = 1$, we know, with $x_0 = 0$, that

(g)
$$a_0 = y(0) = 1, \qquad a_1 = y'(0) = 1.$$

Hence by (f),

(h)
$$a_2 = \tfrac{1}{2}, \qquad a_3 = \tfrac{1}{2}, \qquad a_4 = \frac{\tfrac{3}{2} + 1 - 1}{12} = \frac{1}{8}.$$

Substituting (g) and (h) in (b), we obtain

(i)
$$y(x) = 1 + x + \frac{x^2}{2} + \frac{x^3}{2} + \frac{x^4}{8} + \cdots, \qquad -\infty < x < \infty,$$

which agrees with (f) of Example 37.541.

Example 37.61. Find by power series methods a particular solution of the linear equation

(a)
$$y'' + \frac{x}{1 - x^2}\, y' - \frac{1}{1 - x^2}\, y = 0,$$

for which $y(0) = 1$, $y'(0) = 1$.

Solution. This example is the same as 37.55. Hence we know that (a) has a series solution in powers of x valid for $|x| < 1$. A series solution in powers of x has the form

(b) $$y = a_0 + a_1x + a_2x^2 + a_3x^3 + a_4x^4 + \cdots.$$

By Theorem 37.2, its two successive derivatives, also valid for $|x| < 1$, are

(c) $$y'(x) = a_1 + 2a_2x + 3a_3x^2 + 4a_4x^3 + 5a_5x^4 + \cdots,$$
$$y''(x) = 2a_2 + 6a_3x + 12a_4x^2 + 20a_5x^3 + \cdots.$$

The function $y(x)$ of (b) will be a solution of (a) if the substitution of (b) and (c) in (a) yields an identity in x. Multiplying (a) by $(1 - x^2)$, then making these substitutions and simplifying the result, we obtain

(d) $$(2a_2 - a_0) + 6a_3x + (12a_4 - a_2)x^2 + (20a_5 - 4a_3)x^3 + \cdots = 0.$$

Because of Theorem 37.23, equation (d) will be an identity in x, if and only if each of its coefficients is zero. Hence we must have

(e) $$a_2 = \frac{1}{2}a_0, \quad a_3 = 0, \quad a_4 = \frac{a_2}{12}, \quad a_5 = \frac{a_3}{5}, \cdots.$$

By (37.26) and the initial conditions $y(0) = 1$, $y'(0) = 1$, we know, with $x_0 = 0$, that

(f) $$a_0 = y(0) = 1, \quad a_1 = y'(0) = 1.$$

Hence by (e) and (f)

(g) $$a_2 = \tfrac{1}{2}, \quad a_3 = 0, \quad a_4 = \tfrac{1}{24}, \quad a_5 = 0, \cdots.$$

Substituting (f) and (g) in (b), we have

(h) $$y = 1 + x + \frac{x^2}{2} + \frac{1}{24}x^4 + \cdots, \quad |x| < 1,$$

which agrees with (g) of Example 37.55.

Example 37.62. Find by power series methods a particular solution of

(a) $$y''' + \frac{1}{x}y' - \frac{1}{x^2}y = 0,$$

for which $y(1) = 1$, $y'(1) = 0$, $y''(1) = 1$.

Solution. This example is the same as 37.56. Hence we know that (a) has a series solution in powers of $(x - 1)$ valid for $0 < x < 2$. A series solution in powers of $(x - 1)$ has the form

(b) $$y = a_0 + a_1(x - 1) + a_2(x - 1)^2 + a_3(x - 1)^3 + a_4(x - 1)^4 + \cdots.$$

By Theorem 37.2, its three successive derivatives, also valid for $0 < x < 2$, are

(c) $y' = a_1 + 2a_2(x - 1) + 3a_3(x - 1)^2$
$$+ 4a_4(x - 1)^3 + 5a_5(x - 1)^4 + \cdots,$$

$y'' = 2a_2 + 3!a_3(x - 1) + 3 \cdot 4a_4(x - 1)^2$
$$+ 4 \cdot 5a_5(x - 1)^3 + \cdots,$$

$y''' = 3!a_3 + 4!a_4(x - 1) + 60a_5(x - 1)^2 + \cdots.$

Substituting (b) and (c) in (a), we obtain, after multiplication by x^2,

(d) $x^2[3!a_3 + 4!a_4(x - 1) + 60a_5(x - 1)^2 + \cdots]$
$$+ x[a_1 + 2a_2(x - 1) + 3a_3(x - 1)^2$$
$$+ 4a_4(x - 1)^3 + 5a_5(x - 1)^4 + \cdots]$$
$$- [a_0 + a_1(x - 1) + a_2(x - 1)^2 + \cdots] = 0.$$

It will be easier to equate coefficients of like powers of x to zero, if we express x^2 and x in powers of $(x - 1)$. Their respective series are, see (38.25),

(e) $\qquad x^2 = 1 + 2(x - 1) + (x - 1)^2,$

$\qquad x = 1 + (x - 1).$

Substituting (e) in (d), and simplifying the resulting expression, we have

(f) $(6a_3 + a_1 - a_0) + (24a_4 + 12a_3 + 2a_2)(x - 1)$
$$+ (60a_5 + 48a_4 + 9a_3 + a_2)(x - 1)^2 + \cdots = 0.$$

Equating each coefficient in (f) to zero, we obtain

(g) $\qquad a_3 = \dfrac{a_0 - a_1}{6}, \qquad a_4 = \dfrac{1}{24}(-12a_3 - 2a_2),$

$a_5 = \tfrac{1}{60}(-48a_4 - 9a_3 - a_2).$

By (37.26) and the initial conditions, we know, with $x_0 = 1$, that

(h) $\quad a_0 = y(1) = 1, \qquad a_1 = y'(1) = 0, \qquad a_2 = \tfrac{1}{2}y''(1) = \tfrac{1}{2}.$

Hence by (g) and (h)

(i) $a_3 = \tfrac{1}{6}, \qquad a_4 = \tfrac{1}{24}(-2 - 1) = -\tfrac{1}{8}, \qquad a_5 = \tfrac{1}{60}(6 - \tfrac{3}{2} - \tfrac{1}{2}) = \tfrac{1}{15}.$

Substituting (h) and (i) in (b), we have

(j) $y = 1 + \dfrac{(x - 1)^2}{2} + \dfrac{(x - 1)^3}{6} - \dfrac{(x - 1)^4}{8} + \dfrac{(x - 1)^5}{15} + \cdots,$
$$0 < x < 2,$$

which agrees with (f) of Example 37.56.

Example 37.621. Find by power series methods, a general solution of the linear equation

(a) $$y'' + (\sin x)y' + e^x y = 0.$$

Solution. All coefficients in (a) are analytic at $x = 0$. We shall therefore seek a series solution of the form

(b) $$y = a_0 + a_1 x + a_2 x^2 + a_3 x^3 + a_4 x^4 + a_5 x^5 + \cdots.$$

Since, by (37.42),

(c) $$\sin x = x - \frac{x^3}{3!} + \frac{x^5}{5!} - \cdots, \quad -\infty < x < \infty,$$

$$e^x = 1 + x + \frac{x^2}{2!} + \frac{x^3}{3!} + \cdots, \quad -\infty < x < \infty,$$

each series solution of (a), by Theorem 37.51, is valid for all x. By Theorem 37.2, the two successive derivatives of (b) are

(d) $$y' = a_1 + 2a_2 x + 3a_3 x^2 + 4a_4 x^3 + 5a_5 x^4 + \cdots,$$

$$y'' = 2a_2 + 6a_3 x + 12a_4 x^2 + 20a_5 x^3 + \cdots.$$

Making the substitutions (b), (c), and (d) in (a), we obtain

(e) $(2a_2 + 6a_3 x + 12a_4 x^2 + 20a_5 x^3 + \cdots)$

$$+ \left(x - \frac{x^3}{3!} + \frac{x^5}{5!} - \cdots\right)(a_1 + 2a_2 x + 3a_3 x^2 + 4a_4 x^3 + 5a_5 x^4 + \cdots)$$

$$+ \left(1 + x + \frac{x^2}{2!} + \frac{x^3}{3!} + \cdots\right)$$

$$\times (a_0 + a_1 x + a_2 x^2 + a_3 x^3 + a_4 x^4 + a_5 x^5 + \cdots) = 0.$$

Performing the indicated operations in (e) and simplifying the result, we have

(f) $$(2a_2 + a_0) + (6a_3 + 2a_1 + a_0)x + \left(12a_4 + 3a_2 + a_1 + \frac{a_0}{2}\right) x^2$$

$$+ \left(20a_5 + 4a_3 + a_2 + \frac{a_1}{3} + \frac{a_0}{6}\right) x^3 + \cdots = 0.$$

Equating each coefficient in (f) to zero, we obtain

(g) $$a_2 = -\frac{a_0}{2}, \qquad a_3 = -\frac{2a_1 + a_0}{6} = -\frac{a_1}{3} - \frac{a_0}{6},$$

$$a_4 = -\frac{a_2}{4} - \frac{a_1}{12} - \frac{a_0}{24} = \frac{a_0}{8} - \frac{a_1}{12} - \frac{a_0}{24} = \frac{a_0}{12} - \frac{a_1}{12},$$

$$a_5 = -\frac{a_3}{5} - \frac{a_2}{20} - \frac{a_1}{60} - \frac{a_0}{120}$$

$$= -\frac{1}{5}\left(-\frac{a_1}{3} - \frac{a_0}{6}\right) - \frac{1}{20}\left(-\frac{a_0}{2}\right) - \frac{a_1}{60} - \frac{a_0}{120}$$

$$= \frac{a_0}{20} + \frac{a_1}{20}.$$

Substituting these values of the a's in (b), there results

(h) $y = a_0(1 - \frac{1}{2}x^2 - \frac{1}{6}x^3 + \frac{1}{12}x^4 + \frac{1}{20}x^5 + \cdots)$
              $+ a_1(x - \frac{1}{3}x^3 - \frac{1}{12}x^4 + \frac{1}{20}x^5 + \cdots)$, $-\infty < x < \infty$,

which gives terms to order five of a general solution of (a). Here a_0 and a_1 are arbitrary constants.

Comment 37.63. In finding a series solution of a differential equation, it will usually not be easy to write the general term of the series. In fact, it will, in most cases, be very difficult if not impossible. However, in each of the above examples where we stopped with a finite number of terms, it should be evident to you that for each x in the interval of convergence of the series, $y(x)$ can be computed to a desired degree of accuracy by using sufficient terms, just as e^x or $\sin x$ can be computed to a desired degree of accuracy from their respective series.

EXERCISE 37

1. Find the interval of convergence of each of the following series.

(a) $x^2 + \dfrac{x^4}{2!} + \dfrac{x^6}{3!} + \cdots + \dfrac{x^{2n}}{n!} + \cdots$.

(b) $1 + 8x + 8^2x^2 + \cdots + 8^{n-1}x^{n-1} + \cdots$.

(c) $1 + \dfrac{x+3}{2^2} + \dfrac{(x+3)^2}{3^2} + \cdots + \dfrac{(x+3)^{n-1}}{n^2} + \cdots$.

2. Obtain the first three nonzero terms of the Maclaurin series for each of the following functions. (a) $\cos x$. (b) e^x. (c) $\tan x$.

Obtain terms to order k of the particular solution of each of the following linear equations, where k is the number shown alongside each equation. Use *both* methods of this lesson. Also find an interval of convergence of each series solution.

3. $y' - xy + x^2 = 0$, $y(0) = 2, k = 5$.
4. $x^2y'' = x + 1$, $y(1) = 1, y'(1) = 0, k = 4$.
5. $xy'' + x^2y' - 2y = 0$, $y(1) = 0, y'(1) = 1, k = 4$.
6. $y'' + 3xy' + e^x y = 2x$, $y(0) = 1, y'(0) = -1, k = 4$.
7. $x^2y'' - 2xy' + (\log x)y = 0$, $y(1) = 0, y'(1) = \frac{1}{2}, k = 5$.
8. $(1 - x)y''' - 2xy' + 3y = 0$, $y(0) = 1, y'(0) = -1, y''(0) = 2, k = 6$.

Obtain terms to order k, in powers of $(x - x_0)$ of the general solution of each of the following equations, where k and x_0 are shown alongside each equation. Also find an interval of convergence of each series solution.

9. $y'' - xy' \mid 2y = 0$, $k - 7$, $x_0 = 0$.

10. $2(x^2 + 8)y'' + 2xy' + (x + 2)y = 0$, $k = 4$, $x_0 = 0$.

11. $xy'' + x^2y' - 2y = 0$, $k = 4$, $x_0 = 1$.

12. $y'' - xy' - y = \sin x$, $k = 5$, $x_0 = 0$.

ANSWERS 37

1. (a) $-\infty < x < \infty$. (b) $|x| < \frac{1}{8}$. (c) $-4 < x < -2$.

2. (a) $1 - \dfrac{x^2}{2!} + \dfrac{x^4}{4!} - \cdots$. (b) $1 + x + \dfrac{x^2}{2!} + \cdots$.

(c) $x + \dfrac{x^3}{3} + \dfrac{2x^5}{15} + \cdots$.

3. $y = 2 + x^2 - \dfrac{x^3}{3} + \dfrac{x^4}{4} - \dfrac{x^5}{15} + \cdots$, $-\infty < x < \infty$.

4. $y = 1 + (x - 1)^2 - \dfrac{(x - 1)^3}{2} + \dfrac{(x - 1)^4}{3} - \cdots$, $0 < x < 2$.

5. $y = x - 1 - \frac{1}{2}(x - 1)^2 + \frac{1}{3}(x - 1)^3 - \frac{1}{4}(x - 1)^4 + \cdots$, $0 < x < 2$.

6. $y = 1 - x - \dfrac{x^2}{2} + \dfrac{5x^3}{6} + \dfrac{x^4}{3} - \cdots$, $-\infty < x < \infty$.

7. $y = \frac{1}{2}(x - 1) + \frac{1}{2}(x - 1)^2 + \frac{1}{6}(x - 1)^3 - \frac{1}{24}(x - 1)^4 + \frac{1}{48}(x - 1)^5 - \cdots$,
$0 < x < 2$.

8. $y = 1 - x + x^2 - \dfrac{x^3}{2} - \dfrac{x^4}{12} - \dfrac{x^5}{60} - \dfrac{x^6}{48} + \cdots$, $-1 < x < 1$.

9. $y = a_0(1 - x^2) + a_1 \left(x - \dfrac{x^3}{6} - \dfrac{x^5}{120} - \dfrac{x^7}{1680} + \cdots \right)$, $-\infty < x < \infty$.

10. $y = a_0 \left(1 - \dfrac{x^2}{16} - \dfrac{x^3}{96} + \dfrac{5x^4}{1536} + \cdots \right) + a_1 \left(x - \dfrac{x^3}{24} - \dfrac{x^4}{192} + \cdots \right)$,
$-2\sqrt{2} < x < 2\sqrt{2}$.

11. $y = a_0 \left(1 + (x - 1)^2 - \dfrac{2(x - 1)^3}{3} + \dfrac{(x - 1)^4}{3} + \cdots \right)$

$+ a_1 \left((x - 1) - \dfrac{(x - 1)^2}{2} + \dfrac{(x - 1)^3}{3} - \dfrac{(x - 1)^4}{4} + \cdots \right)$, $0 < x < 2$.

12. $y = a_0 \left(1 + \dfrac{x^2}{2} + \dfrac{x^4}{8} + \cdots \right) + a_1 \left(x + \dfrac{x^3}{3} + \dfrac{x^5}{15} + \cdots \right)$

$+ \dfrac{x^3}{6} + \dfrac{x^5}{40} + \cdots$, $-\infty < x < \infty$.

LESSON 38.   Series Solution of $y' = f(x,y)$.

The two methods outlined in the previous lesson for finding a series solution of a linear equation carry over without change to the finding of a series solution of a general first order equation

$$(38.1) \qquad\qquad y' = f(x,y).$$

However, the determination of the interval of convergence for which the series solution is valid is a more difficult task. We shall first give a definition and then state without proof the relevant theorem we shall need in this connection.

Definition 38.11. A function $f(x,y)$ is **analytic** at a point (x_0,y_0) if it has a Taylor series expansion in powers of $(x - x_0)$ and $(y - y_0)$, valid in some rectangle $|x - x_0| < b$, $|y - y_0| < c$, i.e., $f(x,y)$ is analytic at (x_0,y_0) if

$$(38.12) \quad f(x,y) = a_{00} + [a_{10}(x - x_0) + a_{01}(y - y_0)]$$
$$+ [a_{20}(x - x_0)^2 + a_{11}(x - x_0)(y - y_0) + a_{02}(y - y_0)^2] + \cdots,$$

is valid for each (x,y) in the rectangle $|x - x_0| < b$, $|y - y_0| < c$ which has (x_0,y_0) at its center.

Comment 38.121. The method for finding the coefficients a_{ij} in (38.12) is essentially the same as that used in Theorem 37.24 to find the coefficients a_i in (37.25). These values of a_{ij} are

$$(38.13) \quad a_{00} = f(x_0,y_0);$$

$$a_{10} = \frac{\partial f(x_0,y_0)}{\partial x}, \qquad a_{01} = \frac{\partial f(x_0,y_0)}{\partial y};$$

$$a_{20} = \frac{1}{2!}\frac{\partial^2 f(x_0,y_0)}{\partial x^2}, \qquad a_{11} = \frac{2}{2!}\frac{\partial^2 f(x_0,y_0)}{\partial x\,\partial y},$$

$$a_{02} = \frac{1}{2!}\frac{\partial^2 f(x_0,y_0)}{\partial y^2};$$

$$a_{30} = \frac{1}{3!}\frac{\partial^3 f(x_0,y_0)}{\partial x^3}, \qquad a_{21} = \frac{3}{3!}\frac{\partial^3 f(x_0,y_0)}{\partial x^2\,\partial y},$$

$$a_{12} = \frac{3}{3!}\frac{\partial^3 f(x_0,y_0)}{\partial x\,\partial y^2}, \qquad a_{03} = \frac{1}{3!}\frac{\partial^3 f(x_0,y_0)}{\partial y^3};$$

$$\cdots\cdots\cdots\cdots\cdots\cdots\cdots\cdots\cdots\cdots\cdots.$$

The symbol $\partial^n f(x_0,y_0)/\partial x^n$ means evaluate $\partial^n f(x,y)/\partial x^n$ when $x = x_0$, $y = y_0$. Similarly, the symbol $\partial^n f(x_0,y_0)/\partial y^n$ means evaluate $\partial^n f(x,y)/\partial y^n$ when $x = x_0$, $y = y_0$.

In Theorem 58.5, we state and prove a sufficient condition for the existence and uniqueness of a particular solution of (38.1) satisfying an initial condition. Here we merely state without proof a sufficient condition for the existence of a *power series solution* of (38.1).

Theorem 38.14.   *If the function $f(x,y)$ of (38.1) is analytic at (x_0,y_0), i.e., if $f(x,y)$ has a Taylor series expansion in powers of $(x - x_0)$ and $(y - y_0)$, valid for $|x - x_0| < r$, $|y - y_0| < r$, and if for every (x,y) in this $2r \times 2r$ rectangle which has (x_0,y_0) at its center,*

$$(38.15) \qquad\qquad |f(x,y)| \leqq M,$$

where M is a positive number, then there is a unique particular solution $y(x)$ of (38.1), analytic at $x = x_0$, satisfying the initial condition $y(x_0) = y_0$, i.e., the solution has a Taylor series expansion

$$(38.16) \quad y(x) = y(x_0) + y'(x_0)(x - x_0) + \frac{y''(x_0)}{2!}(x - x_0)^2$$
$$+ \frac{y'''(x_0)}{3!}(x - x_0)^3 + \cdots,$$

valid in an interval about x_0. This interval is at least equal to

$$(38.17) \qquad\qquad I: |x - x_0| < \min\left(r, \frac{r}{3M}\right),$$

where r is given above and M is given in (38.15).

Remark.   This theorem is a special case of the more general theorem on systems stated in Lesson 39. See Theorem 39.12.

Comment 38.18.   Formula (38.17) gives a minimum interval on which the series in (38.16) converges to the solution $y(x)$ satisfying $y(x_0) = y_0$. The actual interval of convergence may be larger.

Example 38.2.   Find by series methods a particular solution of the nonlinear first order equation

$$(a) \qquad\qquad\qquad y' = x^2 + y^2,$$

for which

$$(b) \qquad\qquad\qquad y(0) = 1.$$

Solution.   Here $f(x,y)$ of Theorem 38.14 is $x^2 + y^2$ and $x_0 = 0$, $y_0 = 1$. In Comment 38.21 below, we show how to obtain the Taylor series expansion of $x^2 + y^2$ in powers of x and $(y - 1)$. This series is

$$(c) \qquad f(x,y) = x^2 + y^2 = 1 + 2(y - 1) + x^2 + (y - 1)^2.$$

Since the series is finite, it is valid for all x and y. We may therefore choose for r of Theorem 38.14 any value we please. For an arbitrary r and with

$|x| < r$, $|y - 1| < r$, (c) becomes

(d) $\qquad |x^2 + y^2| < 1 + 2r + r^2 + r^2 = 2r^2 + 2r + 1,$

which is the maximum value of $x^2 + y^2$ for all (x,y) in a $2r \times 2r$ rectangle which has $(0,1)$ at its center. It is therefore the M of (38.15). Hence, by the theorem there is a unique particular solution $y(x)$ of (a), analytic at $x = x_0 = 0$, satisfying (b), i.e., the solution has a series representation of the form (38.16), namely

(e) $\qquad y(x) = y(0) + y'(0)x + \dfrac{y''(0)}{2!} x^2 + \dfrac{y'''(0)}{3!} x^3 + \cdots,$

valid in an interval about $x = 0$. By (38.17), this interval is at least equal to [here r is arbitrary, M is given by (d)]

(f) $\qquad\qquad I: |x| < \min\left(r, \dfrac{r}{3(2r^2 + 2r + 1)}\right).$

To maximize I, we let $u = r/[3(2r^2 + 2r + 1)]$. Taking its derivative with respect to r, and setting the result equal to zero, we obtain

(g) $\quad 0 = 2r^2 + 2r + 1 - r(4r + 2), \quad -2r^2 + 1 = 0, \quad r = \dfrac{1}{\sqrt{2}}.$

For this value of r, the interval (f) becomes

(h) $\qquad I: |x| < \dfrac{1}{3\sqrt{2}(1 + \sqrt{2} + 1)} = \dfrac{1}{6(1 + \sqrt{2})} < 0.069.$

By Theorem 38.14, we now know that the series solution we shall obtain is valid for at least $|x| < 0.069$. We shall find this series solution by the two methods of Lesson 37B.

First Method. By Successive Differentiations. By (b), the initial conditions are $x = 0$, $y = 1$. Equation (a) and its next three successive derivatives, evaluated at $(0,1)$, are

(i) $\quad \begin{aligned} y' &= x^2 + y^2, & y'(0) &= 0 + 1 = 1; \\ y'' &= 2x + 2yy', & y''(0) &= 0 + 2 \cdot 1 \cdot 1 = 2; \\ y''' &= 2 + 2yy'' + 2(y')^2, & y'''(0) &= 2 + 2 \cdot 1 \cdot 2 + 2 \cdot 1 = 8; \\ y^{(4)} &= 2yy''' + 6y'y'', & y^4(0) &= 2 \cdot 1 \cdot 8 + 6 \cdot 1 \cdot 2 = 28. \end{aligned}$

Substituting (i) in (e), we obtain

(j) $\qquad\qquad y(x) = 1 + x + x^2 + \tfrac{4}{3}x^3 + \tfrac{7}{6}x^4 + \cdots,$

which gives the first five terms of the series solution of (a) satisfying (b), valid at least in the interval (h).

Second Method. Undetermined Coefficients. Since $x_0 = 0$, we seek a power series solution of the form

(k) $$y = a_0 + a_1x + a_2x^2 + a_3x^3 + a_4x^4 + \cdots.$$

Its derivative, by Theorem 37.2, is

(l) $$y' = a_1 + 2a_2x + 3a_3x^2 + 4a_4x^3 + 5a_5x^4 + \cdots.$$

Substituting (k) and (l) in (a), it becomes

(m) $$a_1 + 2a_2x + 3a_3x^2 + 4a_4x^3 + 5a_5x^4 + \cdots$$
$$= x^2 + (a_0 + a_1x + a_2x^2 + a_3x^3 + \cdots)^2$$
$$= x^2 + a_0{}^2 + 2a_0a_1x + (2a_0a_2 + a_1{}^2)x^2$$
$$+ (2a_0a_3 + 2a_1a_2)x^3 + \cdots.$$

By Theorem 37.23, (m) will be an identity in x, if the coefficient of each like power of x is zero. Hence we must have

(n) $$a_1 - a_0{}^2 = 0,$$
$$2a_2 - 2a_0a_1 = 0,$$
$$3a_3 - 1 - 2a_0a_2 - a_1{}^2 = 0,$$
$$4a_4 - 2a_0a_3 - 2a_1a_2 = 0.$$

The initial condition is $y(0) = 1$. Therefore by (37.26), with $x_0 = 0$,

(o) $$a_0 = y(0) = 1.$$

By (n) and (o)

(p) $$a_1 = 1, \qquad a_2 = 1,$$
$$a_3 = \frac{1 + 2 + 1}{3} = \frac{4}{3}, \qquad a_4 = \frac{1}{4}\left(\frac{8}{3} + 2\right) = \frac{7}{6}.$$

Substituting (o) and (p) in (k), there results

(q) $$y = 1 + x + x^2 + \tfrac{4}{3}x^3 + \tfrac{7}{6}x^4 + \cdots,$$

which is the same as (j) above.

Comment 38.21. We shall obtain the series representation of $f(x,y) = x^2 + y^2$, as given in (c) above, in two ways. First we shall obtain it by use of (38.13). This is a standard method. Starting with $f(x,y) = x^2 + y^2$, we take the successive partial derivatives called for in (38.13). These are

(38.22) $f(x,y) = x^2 + y^2$

$$\frac{\partial f}{\partial x} = 2x, \qquad \frac{\partial f}{\partial y} = 2y,$$

$$\frac{\partial^2 f}{\partial x^2} = 2, \qquad \frac{\partial^2 f}{\partial x\, \partial y} = 0, \qquad \frac{\partial^2 f}{\partial y^2} = 2.$$

All additional derivatives are zero. Hence by (38.13) and (38.22), with $x_0 = 0$, $y_0 = 1$, we find

(38.23) $a_{00} = 0 + 1 = 1,$

$a_{10} = 0, \quad a_{01} = 2,$

$a_{20} = \dfrac{1}{2!}\,(2) = 1, \quad a_{11} = 0, \quad a_{02} = \dfrac{1}{2!}\,(2) = 1.$

Substituting these values in (38.12), we obtain with $x_0 = 0$, $y_0 = 1$,

(38.24) $f(x,y) = x^2 + y^2 = 1 + 2(y - 1) + x^2 + (y - 1)^2,$

which is the same as (c) above.

A second method of obtaining (38.24) for this particular function $x^2 + y^2$, one which is much quicker and simpler, is to observe that x^2 is already in powers of x and that

(38.25) $y^2 = (y - 1)^2 + 2y - 1 = (y - 1)^2 + 2(y - 1) + 1.$

Comment 38.26. An easy way to find M in the above example, without the need of (c), is to observe that

(38.27) $|x^2 + y^2| \leqq |x^2| + |y^2|.$

If therefore $|x| < r$ and $|y - 1| < r$ so that $|y| < r + 1$, then by (38.27),

(38.28) $|x^2 + y^2| < r^2 + (r + 1)^2 = 2r^2 + 2r + 1,$

which is the same as (d) above.

Example 38.29. Find by power series methods a general solution, in powers of x, of the nonlinear equation

(a) $y' = x^2 - y^2.$

Find an interval of convergence of the series.

Solution. We shall seek a series solution of the form

(b) $y = a_0 + a_1 x + a_2 x^2 + a_3 x^3 + a_4 x^4 + \cdots.$

The derivative of (b), in its interval of convergence is, by Theorem 37.2,

(c) $$y' = a_1 + 2a_2x + 3a_3x^2 + 4a_4x^3 + \cdots.$$

Substituting (b) and (c) in (a), we obtain

(d) $a_1 + 2a_2x + 3a_3x^2 + 4a_4x^3 + \cdots = x^2 - a_0^2$
$$- 2a_0a_1x - (2a_0a_2 + a_1^2)x^2 - (2a_0a_3 + 2a_1a_2)x^3 - \cdots.$$

Equating coefficients of like powers of x, we have

(e) $a_1 = -a_0^2,$ $a_1 - -a_0^2;$
  $2a_2 = -2a_0a_1,$ $a_2 = -a_0a_1 = a_0^3;$
  $3a_3 = 1 - 2a_0a_2 - a_1^2,$ $a_3 = \frac{1}{3}(1 - 2a_0^4 - a_0^4) = \frac{1}{3} - a_0^4;$
  $4a_4 = -2a_0a_3 - 2a_1a_2,$ $a_4 = \frac{1}{4}[-2a_0(\frac{1}{3} - a_0^4) - 2(-a_0^2)(a_0^3)]$
  $\qquad\qquad\qquad\qquad\qquad\quad = -\frac{1}{6}a_0 + a_0^5.$

Substituting (e) in (b), there results

(f) $y' = a_0 - a_0^2x + a_0^3x^2 + (\frac{1}{3} - a_0^4)x^3 + (-\frac{1}{6}a_0 + a_0^5)x^4 + \cdots,$

which gives the first five terms of a series solution of (a). Here a_0 is the arbitrary constant. By (37.26), it is, for a particular solution, $y(0)$. To find an interval of convergence of the series solution, we proceed just as we did above in Example 38.2. Here $f(x,y) = x^2 - y^2$, $x_0 = 0$, $y_0 \geq 0$ is arbitrary. The Taylor series expansion of $x^2 - y^2$ in powers of x and $(y - y_0)$ is, see (38.25),

(g) $f(x,y) = x^2 - y^2 = x^2 - [(y - y_0)^2 + 2y_0(y - y_0) + y_0^2].$

Since the series is finite it is valid for all x and y. We may therefore choose for the r of Theorem 38.14, any value we please. For an arbitrary r and with $|x| < r$, $|y - y_0| < r$, (g) becomes

(h) $|f(x,y)| = |x^2 - y^2| \leq |x^2| + |(y - y_0)^2|$
  $\qquad\qquad + 2y_0|y - y_0| + y_0^2 < r^2 + r^2 + 2y_0r + y_0^2$
  $\qquad\qquad\qquad\qquad\qquad\qquad = 2(r^2 + y_0r) + y_0^2,$

which is an upper bound of $x^2 - y^2$ for all (x,y) in a $2r \times 2r$ rectangle which has $(0,y_0)$ at its center. It is therefore the M of Theorem 38.14. Hence, by (38.17), there is an interval I, at least equal to

(i) $$|x| < \min\left(r, \frac{r}{6(r^2 + y_0r) + 3y_0^2}\right),$$

on which the series (f) converges.

EXERCISE 38

1. Find a series representation of $f(x,y)$ in powers of x and y if (a) $f(x,y) = x \sin y$, (b) $f(x,y) = ye^x$.

2. Find a series representation of $f(x,y)$ in powers of $(x - 1)$, $(y - 1)$ if (a) $f(x,y) = xy$, (b) $f(x,y) = y \log x$.

Obtain terms to order k of the particular solution of each of the following differential equations, where k is the number shown alongside each equation. Use two methods. Also find an interval of convergence of each series solution.

3. $y' = x^2 - y^2$, $y(0) = 1, k = 4$.
4. $y' = x^2 - y^2$, $y(1) = 1, k = 4$.
5. $y' = y^2 - xy$, $y(0) = 2, k = 4$.
6. $y' = x^2 + \sin y$, $y(0) = \pi/2$, $k = 3$. *Hint.* Replace $\sin y$ by its series expansion in powers of y (for undetermined coefficient method).
7. $y' = x + e^y$, $y(0) = 0$, $k = 4$. *Hint.* Replace e^y by its series expansion in powers of y (for undetermined coefficient method).
8. $(1 - x - y)y' = 1$, $y(1) = -2, k = 4$. *Hint.* Write $(1 - x - y)y'$ as $[-(x - 1) - y]y'$ (for undetermined coefficient method).
9. $y' = \sin(xy) + x^2$, $y(0) = 3, k = 4$.
10. $y' = \log(xy)$, $y(1) = 1, k = 5$.
11. $y' = \cos(x + y)$, $y(0) = \pi/2, k = 3$.
12. $y' = y^2 + e^{x-1}$, $y(1) = 1, k = 3$.
13. $y' = \sqrt{1 + xy}$, $y(0) = 1, k = 4$.
14. $y' = 1 + xy^2$, $y(0) = 1, k = 4$.
15. $y' = \cos x + \sin y$, $y(0) = 0, k = 5$. *Hint.* Replace $\cos x$ and $\sin y$ by their series expansions.

Obtain terms to order k, in powers of x, of the general solution of each of the following differential equations, where k is shown alongside each equation.

16. $y' = x + y^2$, $k = 3$.
17. $y' = x + \dfrac{1}{y}$, $k = 4$. *Hint.* Write the equation as $yy' = xy + 1$.

ANSWERS 38

1. (a) $x\left(y - \dfrac{y^3}{3!} + \dfrac{y^5}{5!} - \cdots\right)$.   (b) $y\left(1 + x + \dfrac{x^2}{2!} + \dfrac{x^3}{3!} + \cdots\right)$.

2. (a) $(x - 1)(y - 1) + (x - 1) + (y - 1) + 1$.

     (b) $[1 + (y - 1)]\left[(x - 1) - \dfrac{(x - 1)^2}{2} + \dfrac{(x - 1)^3}{3} - \cdots\right]$.

3. $y = 1 - x + x^2 - \frac{2}{3}x^3 + \frac{5}{6}x^4 - \cdots$, $|x| < 0.069$.

4. $y = 1 + (x - 1)^2 - \dfrac{(x - 1)^3}{3} + \dfrac{(x - 1)^4}{6} - \cdots$, $|x - 1| < 0.04$.

5. $y = 2 + 4x + 7x^2 + \dfrac{40}{3}x^3 + \dfrac{307x^4}{12} + \cdots$, $|x| < 0.029$.

6. $y = \dfrac{\pi}{2} + x + \dfrac{1}{6}x^3 + \cdots$,　$|x| < \dfrac{r}{3(r^2 + 1)}$, r arbitrary, $|x| < \dfrac{1}{6}$.

7. $y = x + x^2 + \dfrac{x^3}{2} + \dfrac{5x^4}{12} + \cdots$,　$|x| = \dfrac{r}{3(r + e^r)}$, r arbitrary.

8. $y = -2 + \dfrac{x - 1}{2} + \dfrac{3}{16}(x - 1)^2 + \dfrac{7}{64}(x - 1)^3 + \dfrac{79}{1024}(x - 1)^4 + \cdots$.

9. $y = 3 + \frac{3}{2}x^2 + \frac{1}{3}x^3 - \frac{3}{4}x^4 \pm \cdots$.

10. $y = 1 + \frac{1}{2}(x - 1)^2 + \frac{1}{12}(x - 1)^4 - \frac{7}{120}(x - 1)^5 + \cdots$.

11. $y = \dfrac{\pi}{2} - \dfrac{x^2}{2} + \dfrac{x^3}{6} + \cdots$.

12. $y = 1 + 2(x - 1) + \frac{5}{2}(x - 1)^2 + \frac{19}{6}(x - 1)^3 + \cdots$.

13. $y = 1 + x + \dfrac{x^2}{4} + \dfrac{x^3}{8} - \dfrac{x^4}{64} + \cdots$.

14. $y = 1 + x + \dfrac{x^2}{2} + \dfrac{2x^3}{3} + \dfrac{x^4}{2} + \cdots$.

15. $y = x + \dfrac{x^2}{2} - \dfrac{x^4}{24} - \dfrac{x^5}{20} + \cdots$, $|x| < \dfrac{r}{6}$, r arbitrary.

16. $y = a_0 + a_0^2 x + (\frac{1}{2} + a_0^3)x^2 + (\frac{1}{3}a_0 + a_0^4)x^3 + \cdots$,

$\qquad\qquad |x| < \dfrac{r}{3[r^2 + (1 + 2y_0)r + y_0^2]}$, r arbitrary.

17. $y = a_0 + \dfrac{1}{a_0}x + \dfrac{a_0^3 - 1}{2a_0^3}x^2 + \dfrac{3 - a_0^3}{6a_0^5}x^3 + \dfrac{7a_0^3 - 15}{24a_0^7}x^4 + \cdots$.

LESSON 39. Series Solution of a Nonlinear Differential Equation of Order Greater Than One and of a System of First Order Differential Equations.

LESSON 39A. Series Solution of a System of First Order Differential Equations. In Theorem 62.12, we state a sufficient condition for the existence and uniqueness of a solution of a system of first order equations

$$(39.1) \qquad \frac{dy_1}{dt} = f_1(t, y_1, y_2, \cdots, y_n),$$

$$\frac{dy_2}{dt} = f_2(t, y_1, y_2, \cdots, y_n),$$

$$\cdots\cdots\cdots\cdots\cdots\cdots\cdots$$

$$\frac{dy_n}{dt} = f_n(t, y_1, y_2, \cdots, y_n),$$

satisfying the initial conditions

$$(39.11) \qquad y_1(t_0) = a_1, \qquad y_2(t_0) = a_2, \cdots, y_n(t_0) = a_n.$$

We now state without proof a sufficient condition for the existence of a *power series solution* of (39.1) satisfying (39.11).

Theorem 39.12. *If each function* $f_1, f_2, \cdots, f_n$ *of the system (39.1) is analytic at a point* $(t_0, a_1, a_2, \cdots, a_n)$, *i.e., if each function has a Taylor series expansion in powers of* $(t - t_0)$, $(y_1 - a_1)$, $\cdots$, $(y_n - a_n)$, *valid for* $|t - t_0| < r$, $|y_1 - a_1| < r$, $\cdots$, $|y_n - a_n| < r$, *and if for every point* $(t, y_1, y_2, \cdots, y_n)$ *in this* $(2r)^{n+1}$*-dimensional rectangle which has the point* $(t_0, a_1, a_2, \cdots, a_n)$ *at its center,*

$$(39.13) \qquad |f_i(t, y_1, y_2, \cdots, y_n)| < M, \quad i = 1, 2, \cdots, n,$$

where M is a positive constant, then an interval I exists on which there is one and only one set of functions, $y_1(x), y_2(x), \cdots, y_n(x)$, *each analytic at* $t = t_0$, *satisfying the given system (39.1) and the initial conditions (39.11), i.e., each function has a Taylor series expansion in powers of* $(t - t_0)$, *namely*

$$(39.14) \quad y_1(t) = a_1 + a_{11}(t - t_0) + a_{12}(t - t_0)^2 + a_{13}(t - t_0)^3 + \cdots,$$
$$y_2(t) = a_2 + a_{21}(t - t_0) + a_{22}(t - t_0)^2 + a_{23}(t - t_0)^3 + \cdots,$$
$$\cdots \cdots \cdots \cdots \cdots \cdots \cdots \cdots \cdots \cdots \cdots \cdots \cdots \cdots \cdots$$
$$y_n(t) = a_n + a_{n1}(t - t_0) + a_{n2}(t - t_0)^2 + a_{n3}(t - t_0)^3 + \cdots,$$

valid in an interval about t_0. *This interval is at least equal to*

$$(39.15) \qquad I \colon |t - t_0| < \min\left(r, \frac{r}{(n+2)M}\right),$$

where r is given above, n is the number of equations in the system (39.1) and M is given in (39.13).

The coefficients in (39.14) are given by

$$(39.16) \qquad a_{ij} = \frac{y_i^{(j)}(t_0)}{j!}.$$

For example, the coefficient a_{12} is, by (39.16), $y_1''(t_0)/2!$. Note that it agrees with the coefficient of $(x - x_0)^2$ in Taylor series (37.27). The coefficient of a_{23} is, by (39.16), $y_2'''(t_0)/3!$ which agrees with the coefficient of $(x - x_0)^3$ in (37.27).

Remark. If we look at the first order equation $dy/dx = f(x,y)$ as if it were a system of one equation $dy_1/dt = f_1(t,y_1)$, we see that Theorem 38.14 is only a special case of Theorem 39.12.

Example 39.17. Find by power series methods a particular solution of the first order system

(a) $$\frac{dx}{dt} = yt, \qquad \frac{dy}{dt} = xy,$$

for which

(b) $x(0) = 1, \qquad y(0) = 1.$

Solution. Comparing (a) with (39.1) and (b) with (39.11), we see, with x and y taking the places of y_1 and y_2, that

$$f_1 = yt, \qquad f_2 = xy, \qquad t_0 = 0, \qquad a_1 = 1, \qquad a_2 = 1.$$

By Theorem 39.12, we must find series expansions of f_1 and f_2 in powers of t, $(x - 1)$, $(y - 1)$. These are respectively

(c) $f_1 = yt = t(y - 1) + t,$

$\qquad f_2 = xy = (x - 1)(y - 1) + x + y - 1$
$\qquad\qquad = (x - 1)(y - 1) + (x - 1) + (y - 1) + 1.$

Since these series are finite, they are valid for all t, x, and y. We may therefore choose the r of Theorem 39.12 arbitrarily. Hence when $|t| < r$, $|x - 1| < r$, $|y - 1| < r$, we have by (c),

(d) $|f_1| < r^2 + r,$

$\qquad\qquad\qquad |f_2| < r^2 + r + r + 1 = r^2 + 2r + 1.$

For $r > 0$, $(r^2 + 2r + 1) > r^2 + r$. Therefore $r^2 + 2r + 1$ is the M of (39.13). Hence, by Theorem 39.12, there is a unique pair of functions $x(t)$, $y(t)$, each analytic at $t = 0$, satisfying the system (a) and the initial conditions (b), i.e., each function has a series expansion of the form (39.14), valid in an interval about $t = 0$. By (39.15), this interval is at least equal to [here $n = 2$, M is given by the second equation in (d), and r is arbitrary]

(e) $I: |t| < \min\left(r, \dfrac{r}{4(r^2 + 2r + 1)}\right).$

We find, in the usual manner, that the maximum value of I occurs when $r = 1$. Hence I will be a maximum if

(f) $I: |t| < \frac{1}{16} = 0.0625.$

We shall find a series solution of (a) by the usual two methods of Lesson 37B.

First Method. By Successive Differentiations. Here the independent variable is t, the dependent variables x and y. Since the initial conditions are given at $t = 0$, we seek series expansions of $x(t)$ and $y(t)$ of the form (37.28), namely

(g) $x(t) = x(0) + x'(0)t + \dfrac{x''(0)}{2}\,t^2 + \dfrac{x'''(0)}{3!}\,t^3 + \dfrac{x^{(4)}(0)}{4!}\,t^4 + \cdots,$

$\qquad y(t) = y(0) + y'(0)t + \dfrac{y''(0)}{2}\,t^2 + \dfrac{y'''(0)}{3!}\,t^3 + \dfrac{y^{(4)}(0)}{4!}\,t^4 + \cdots.$

Starting with each equation in (a) and taking its successive derivatives with respect to t, we obtain

(h) $\qquad x' = yt, \qquad\qquad y' = xy;$

$\qquad\qquad x'' = y + ty', \qquad\quad y'' = xy' + yx';$

$\qquad\qquad x''' = 2y' + ty'', \qquad y''' = xy'' + 2y'x' + yx'';$

$\qquad\qquad x^{(4)} = 3y'' + ty''', \quad y^{(4)} = xy''' + 3y''x' + 3y'x'' + yx''';$

$\qquad\qquad \cdots\cdots\cdots\cdots, \quad \cdots\cdots\cdots\cdots\cdots\cdots\cdots\cdots .$

We know from the initial conditions that

(i) $\qquad\qquad\qquad\qquad x(0) = 1, \qquad y(0) = 1.$

To find the values of succeeding coefficients in (g), we evaluate the derivatives in (h) in the order x', y', x'', y'', etc. These are with $t = 0$, $x = 1$, $y = 1$,

(j) $\qquad x'(0) = 1 \cdot 0 = 0, \qquad y'(0) = 1 \cdot 1 = 1;$

$\qquad\qquad x''(0) = 1 + 0 = 1, \qquad y''(0) = 1 \cdot 1 + 1 \cdot 0 = 1;$

$\qquad\qquad x'''(0) = 2 + 0 = 2, \qquad y'''(0) = 1 + 0 + 1 = 2;$

$\qquad\qquad x^{(4)}(0) = 3 + 0 = 3, \qquad y^{(4)}(0) = 2 + 0 + 3 + 2 = 7;$

$\qquad\qquad \cdots\cdots\cdots\cdots\cdots, \quad \cdots\cdots\cdots\cdots\cdots\cdots .$

Substituting (i) and (j) in (g), there results

(k) $\qquad\qquad x(t) = 1 + \dfrac{t^2}{2} + \dfrac{t^3}{3} + \dfrac{t^4}{8} + \cdots,$

$\qquad\qquad y(t) = 1 + t + \dfrac{t^2}{2} + \dfrac{t^3}{3} + \dfrac{7t^4}{24} + \cdots .$

which give terms to order four of a series solution of (a) satisfying (b), valid at least in the interval I of (f).

Second Method. By Undetermined Coefficients. Since the initial conditions are given at $t = 0$, we seek series expansions of the form

(l) $\qquad x(t) = a_0 + a_1 t + a_2 t^2 + a_3 t^3 + a_4 t^4 + \cdots,$

$\qquad\qquad y(t) = b_0 + b_1 t + b_2 t^2 + b_3 t^3 + b_4 t^4 + \cdots .$

Differentiating (l) with respect to t and setting each resulting equation equal to the respective right side of (a), we obtain

(m) $\qquad a_1 + 2a_2 t + 3a_3 t^2 + 4a_4 t^3 + \cdots = yt,$

$\qquad\qquad b_1 + 2b_2 t + 3b_3 t^2 + 4b_4 t^3 + \cdots = xy.$

In (m), replace x and y by their values in (l). Hence (m) becomes

(n) $a_1 + 2a_2t + 3a_3t^2 + 4a_4t^3 + \cdots = t(b_0 + b_1t + b_2t^2 + b_3t^3 + \cdots)$,

 $b_1 + 2b_2t + 3b_3t^2 + 4b_4t^3 + \cdots$
$$= (a_0 + a_1t + a_2t^2 + a_3t^3 + \cdots)(b_0 + b_1t + b_2t^2 + \cdots).$$

The pair of functions in (l) will be a solution of (a) if we choose the a's and b's so that each equation in (n) is an identity in t. Performing the indicated expansions in (n) and equating for each equation separately, coefficients of like powers of t, we obtain

(o) $\qquad a_1 = 0, \qquad b_1 = a_0b_0;$

$\qquad\qquad 2a_2 = b_0, \qquad 2b_2 = a_0b_1 + a_1b_0;$

$\qquad\qquad 3a_3 = b_1, \qquad 3b_3 = a_0b_2 + a_1b_1 + a_2b_0;$

$\qquad\qquad 4a_4 = b_2, \qquad 4b_4 = a_0b_3 + a_1b_2 + a_2b_1 + a_3b_0.$

By (b) and Theorem 37.24,

(p) $\qquad\qquad a_0[= x(0)] = 1, \qquad b_0\ [= y(0)] = 1.$

Hence from (o) and (p), we obtain, calculating the coefficients in the order a_1, b_1, a_2, b_2, etc.

(q) $a_1 = 0, \quad b_1 = 1; \quad a_2 = \frac{1}{2}b_0 = \frac{1}{2}, \quad b_2 = \frac{1}{2}(1 + 0) = \frac{1}{2};$

$\quad a_3 = \frac{1}{3}b_1 = \frac{1}{3}, \quad b_3 = \frac{1}{3}(\frac{1}{2} + 0 + \frac{1}{2}) = \frac{1}{3};$

$\quad a_4 = \frac{1}{4}b_2 = \frac{1}{8}, \quad b_4 = \frac{1}{4}(\frac{1}{3} + 0 + \frac{1}{2} + \frac{1}{3}) = \frac{7}{24}.$

Substituting (p) and (q) in (l), we find

(r) $$x(t) = 1 + \frac{t^2}{2} + \frac{t^3}{3} + \frac{t^4}{8} + \cdots,$$

$$y(t) = 1 + t + \frac{t^2}{2} + \frac{t^3}{3} + \frac{7t^4}{24} + \cdots,$$

which is the same as the solution (k) obtained previously.

LESSON 39B. Series Solution of a System of Linear First Order Equations. In Theorem 62.3 we state and prove a sufficient condition for the existence and uniqueness of a solution of a system of **linear** first order equations

(39.2) $\qquad \dfrac{dy_1}{dt} = f_{11}(t)y_1 + f_{12}(t)y_2 + \cdots + f_{1n}(t)y_n + Q_1(t),$

$\qquad \dfrac{dy_2}{dt} = f_{21}(t)y_1 + f_{22}(t)y_2 + \cdots + f_{2n}(t)y_n + Q_2(t),$

$\qquad \cdots\cdots\cdots\cdots\cdots\cdots\cdots\cdots\cdots\cdots\cdots\cdots\cdots$

$\qquad \dfrac{dy_n}{dt} = f_{n1}(t)y_1 + f_{n2}(t)y_2 + \cdots + f_{nn}(t)y_n + Q_n(t),$

satisfying the initial conditions

(39.21) $y_1(t_0) = a_1, \quad y_2(t_0) = a_2, \cdots, y_n(t_0) = a_n.$

We now state without proof a sufficient condition for the existence of a *power series solution* of (39.2) satisfying (39.21).

Theorem 39.22.   *If, in (39.2), each function f_{ij}, $i = 1, \cdots, n$, $j = 1$, $\cdots$, n and each function Q_i, $i = 1, \cdots, n$, is analytic at $t = t_0$, i.e., if each function has a Taylor series expansion in powers of $(t - t_0)$, valid for $|t - t_0| < r$, then there is a unique set of functions, $y_1(t), y_2(t), \cdots, y_n(t)$, each analytic at $t = t_0$, satisfying the given system (39.2) and the initial conditions (39.21), i.e., each function has a Taylor series expansion in powers of $(t - t_0)$, namely*

(39.23) $y_1(t) = a_1 + a_{11}(t - t_0) + a_{12}(t - t_0)^2 + a_{13}(t - t_0)^3 + \cdots,$

$\qquad\quad y_2(t) = a_2 + a_{21}(t - t_0) + a_{22}(t - t_0)^2 + a_{23}(t - t_0)^3 + \cdots,$

$$\cdots\cdots\cdots\cdots\cdots\cdots\cdots\cdots\cdots\cdots\cdots\cdots\cdots\cdots$$

$\qquad\quad y_n(t) = a_n + a_{n1}(t - t_0) + a_{n2}(t - t_0)^2 + a_{n3}(t - t_0)^3 + \cdots,$

valid for $|t - t_0| < r$. The coefficients in (39.23) are given by

(39.24) $a_{ij} = \dfrac{y_i^{(j)}(t_0)}{j!}.$

$$\left[\text{For example, } a_{13} = \frac{y_1'''(t_0)}{3!}, \quad a_{21} = \frac{y_2'(t_0)}{1!}, \text{ etc.} \right]$$

Example 39.25.   Find, by power series methods, a particular solution of the first order *linear* system

(a) $\dfrac{dx}{dt} = x \cos t, \quad \dfrac{dy}{dt} = t^2 y,$

for which

(b) $x(0) = 1, \quad y(0) = -1.$

Solution.   Comparing (a) and (b) with (39.2) and (39.21), we see, with x and y taking the place of y_1 and y_2, that

(c) $f_{11} = \cos t, \quad f_{22} = t^2, \quad t_0 = 0, \quad a_1 = 1, \quad a_2 = -1.$

Since f_{11} and f_{22} have Maclaurin series expansions, valid for all t, it follows by Theorem 39.22, that there is a unique pair of functions $x(t)$, $y(t)$, each analytic at $t = 0$, satisfying the system (a) and the initial conditions (b), i.e., $x(t)$ and $y(t)$ each has a series expansion in powers of t valid for all t.

There are the usual two methods available for finding a series solution of (a) satisfying (b). These have been described in Example 39.17. We shall use the more difficult one, in this case, of undetermined coefficients. By Theorem 39.22, the pair of functions $x(t)$, $y(t)$ have series expansions of the form (39.23). With $t_0 = 0$, these become

(d)
$$x(t) = a_0 + a_1 t + a_2 t^2 + a_3 t^3 + \cdots,$$
$$y(t) = b_0 + b_1 t + b_2 t^2 + b_3 t^3 + \cdots.$$

Differentiating (d) with respect to t, and setting each resulting right side equal to the respective right side of (a), we obtain

(e)
$$a_1 + 2a_2 t + 3a_3 t^2 + 4a_4 t^3 + \cdots = x \cos t,$$
$$b_1 + 2b_2 t + 3b_3 t^2 + 4b_4 t^3 + \cdots = t^2 y.$$

In (e) replace x and y by their values in (d), and replace $\cos t$, see (37.42), by its Maclaurin series expansion. There results

(f) $a_1 + 2a_2 t + 3a_3 t^2 + 4a_4 t^3 + \cdots$
$$= (a_0 + a_1 t + a_2 t^2 + \cdots)\left(1 - \frac{t^2}{2!} + \frac{t^4}{4!} - \cdots\right),$$
$$b_1 + 2b_2 t + 3b_3 t^2 + 4b_4 t^3 + \cdots = t^2(b_0 + b_1 t + b_2 t^2 + b_3 t^3 + \cdots).$$

The pair of functions defined in (d) will be a solution of (a) if we choose the a's and b's in (f) so that each equation is an identity in t. Performing the indicated operations in (f) and equating for each equation separately, coefficients of like powers of t, we obtain

(g)
$$a_1 = a_0, \qquad\qquad b_1 = 0,$$
$$2a_2 = a_1, \qquad\qquad 2b_2 = 0,$$
$$3a_3 = a_2 - \frac{a_0}{2}, \qquad 3b_3 = b_0,$$
$$4a_4 = a_3 - \frac{a_1}{2}, \quad 4b_4 = b_1.$$

By (b) and Theorem 37.24,

(h)
$$a_0 = x(0) = 1, \quad b_0 = y(0) = -1.$$

Hence we obtain from (h) and (g), calculating the coefficients in the order a_1, b_1, a_2, b_2, etc.,

(i) $a_1 = 1, \qquad b_1 = 0, \qquad a_2 = \frac{1}{2}, \qquad b_2 = 0, \qquad a_3 = \frac{1}{3}(\frac{1}{2} - \frac{1}{2}) = 0,$
$b_3 = \frac{1}{3}(-1) = -\frac{1}{3}, \qquad a_4 = \frac{1}{4}(-\frac{1}{2}) = -\frac{1}{8}, \qquad b_4 = \frac{1}{4}(0) = 0.$

Substituting (h) and (i) in (d), we find

(j)
$$x(t) = 1 + t + \frac{t^2}{2} - \frac{t^4}{8} + \cdots,$$

$$y(t) = -1 - \frac{t^3}{3} - \cdots,$$

which give terms to order four of a series solution of (a) satisfying (b), valid for all t.

LESSON 39C. Series Solution of a Nonlinear Differential Equation of Order Greater Than One. In Theorem 62.22, we state and prove a sufficient condition for the existence and uniqueness of a particular solution of an nth order differential equation

(39.3)
$$\frac{d^n y}{dx^n} = f(x,y,y',y'', \cdots, y^{(n-1)}),$$

satisfying a set of initial conditions

(39.31)
$$y(x_0) = y_0, \qquad y'(x_0) = y_1,$$
$$y''(x_0) = y_2, \cdots, y^{(n-1)}(x_0) = y_{n-1}.$$

The proof consists in showing how every nth order differential equation with given initial conditions can be changed to an equivalent first order system with equivalent initial conditions. Hence Theorem 62.22 is a by-product of Theorem 62.12 relating to first order systems. Analogously, the sufficient condition we now state for the existence of a power series solution of (39.3) satisfying (39.31) stems from Theorem 39.12 for a first order system.

Theorem 39.32. *If the function f of (39.3) is analytic at a point $(x_0,y_0,y_1, \cdots, y_{n-1})$, i.e., if f has a Taylor series expansion in powers of $(x - x_0)$, $(y - y_0)$, $(y' - y_1)$, $\cdots$, $(y^{(n-1)} - y_{n-1})$, valid for $|x - x_0| < r$, $|y - y_0| < r$, $|y' - y_1| < r$, $\cdots$, $|y^{(n-1)} - y_{n-1}| < r$, and if for every point $(x,y,y', \cdots, y^{(n-1)})$ in this $(2r)^{n+1}$-dimensional rectangle which has this point $(x_0,y_0,y_1, \cdots, y_{n-1})$ at its center,*

(39.33)
$$|f(x,y,y', \cdots, y^{(n-1)})| < M,$$

where M is a positive constant, then there is a unique particular solution of (39.3) analytic at $x = x_0$ satisfying (39.31), i.e., the solution has a Taylor series expansion in powers of $(x - x_0)$, namely

(39.34)
$$y(x) = y(x_0) + y'(x_0)(x - x_0) + \frac{y''(x_0)}{2!}(x - x_0)^2$$
$$+ \frac{y'''(x_0)}{3!}(x - x_0)^3 + \cdots,$$

valid in an interval about $x = x_0$. This interval is at least equal to

(39.35) $\qquad I: |x - x_0| < \min\left(r, \dfrac{r}{(n+2)M}\right),$

where r is given above, n is the order of the equation (39.3) and M is given in (39.33).

Example 39.36. Find by series methods a particular solution of the nonlinear equation

(a) $\qquad\qquad\qquad y''' = x^2 + y^2$

for which $y(1) = 1$, $y'(1) = 0$, $y''(1) = 2$.

Solution. Comparing (a) with (39.3) we see that $n = 3$ and $f = x^2 + y^2$. A comparison of the initial conditions with (39.31) shows that $x_0 = 1$, $y_0 = 1$, $y_1 = 0$, and $y_2 = 2$. The series expansion of f in powers of $(x - 1)$, $(y - 1)$, y', $(y'' - 2)$, by (38.25), is

(b) $x^2 + y^2 = 2 + 2(x - 1) + 2(y - 1) + (x - 1)^2 + (y - 1)^2.$

Since the series is finite, it is valid for all x and y. We may therefore choose any value we please for r of Theorem 39.32. For an arbitrary r and with $|x - 1| < r$, $|y - 1| < r$, (b) becomes

(c) $\qquad |x^2 + y^2| < 2 + 2r + 2r + r^2 + r^2 = 2(r^2 + 2r + 1),$

which is the M of (39.33).

Hence by Theorem 39.32, there is a unique particular solution $y(x)$ of (a), analytic at $x = 1$, satisfying the given initial conditions, i.e., the solution has a Taylor series expansion of the form (39.34), namely

(d) $\quad y(x) = y(1) + y'(1)(x - 1)$
$$+ \frac{y''(1)}{2!}(x - 1)^2 + \frac{y'''(1)}{3!}(x - 1)^3 + \cdots,$$

valid in an interval about $x = 1$. By (39.35), this interval is at least equal to [here $n = 3$, M is given by (c) and r is arbitrary]

(e) $\qquad\qquad I: |x - 1| < \dfrac{r}{5[2(r^2 + 2r + 1)]}.$

To maximize I, we let

(f) $\qquad\qquad\qquad u = \dfrac{r}{10(r^2 + 2r + 1)}.$

Taking its derivative with respect to r and setting the result equal to zero, we obtain

(g) $\qquad 0 = (r^2 + 2r + 1) - r(2r + 2) = -r^2 + 1, \quad r = 1.$

For this value of r, the interval I of (e) becomes

(h) $$|x - 1| < \tfrac{1}{40} = 0.025.$$

By Theorem 39.32, we now know that the series solution we shall obtain is valid for at least $|x - 1| < 0.025$. We shall use the usual two methods to find a series solution.

First Method. By Successive Differentiations. In (d), we know by the initial conditions that

(i) $$y(1) = 1, \qquad y'(1) = 0, \qquad y''(1) = 2.$$

To find the values of succeeding coefficients in (d), we evaluate (a) and its successive derivatives at $(1,1)$. These are, with $x = 1$, $y = 1$, $y' = 0$, $y'' = 2$,

(j) $\begin{aligned}
y''' &= x^2 + y^2, & y'''(1) &= 1 + 1 = 2; \\
y^{(4)} &= 2x + 2yy', & y^{(4)}(1) &= 2 + 0 = 2; \\
y^{(5)} &= 2 + 2yy'' + 2(y')^2, & y^{(5)}(1) &= 2 + 2\cdot 2 + 0 = 6; \\
y^{(6)} &= 2yy''' + 6y'y'', & y^{(6)}(1) &= 2\cdot 2 + 6\cdot 0\cdot 2 = 4; \\
\cdots\cdots\cdots\cdots\cdots&, & \cdots\cdots\cdots\cdots&\cdots\cdots
\end{aligned}$

Substituting (i) and (j) in (d), we obtain

(k) $$y = 1 + (x - 1)^2 + \tfrac{1}{3}(x - 1)^3 + \tfrac{1}{12}(x - 1)^4$$
$$+ \tfrac{1}{20}(x - 1)^5 + \tfrac{1}{180}(x - 1)^6 + \cdots,$$

which gives the first seven terms of a series solution of (a) satisfying the initial conditions, valid at least in the interval (h).

Second Method. By Undetermined Coefficients. Since the initial conditions have been given in terms of $x = 1$, we seek a series solution of the form

(l) $$y = a_0 + a_1(x - 1) + a_2(x - 1)^2 + a_3(x - 1)^3 + \cdots.$$

Its next three successive derivatives are

(m) $\begin{aligned}
y' &= a_1 + 2a_2(x - 1) + 3a_3(x - 1)^2 + 4a_4(x - 1)^3 \\
&\quad + 5a_5(x - 1)^4 + \cdots, \\
y'' &= 2a_2 + 6a_3(x - 1) + 12a_4(x - 1)^2 + 20a_5(x - 1)^3 + \cdots, \\
y''' &= 6a_3 + 24a_4(x - 1) + 60a_5(x - 1)^2 + \cdots.
\end{aligned}$

Substituting (l) and the last equation of (m) in (a), we obtain

(n) $$6a_3 + 24a_4(x - 1) + 60a_5(x - 1)^2 + \cdots$$
$$= x^2 + [a_0 + a_1(x - 1) + a_2(x - 1)^2 + \cdots]^2.$$

It will be easier to equate coefficients of like powers of x, if we find the series expansion of x^2 in powers of $(x - 1)$. By (38.25),

(o) $$x^2 = 1 + 2(x - 1) + (x - 1)^2.$$

In (n) replace x^2 by its value in (o), perform the indicated multiplication on the right, and then equate the coefficients of like powers of $(x - 1)$. There results

(p)
$$6a_3 = a_0{}^2 + 1,$$
$$24a_4 = 2 + 2a_0a_1,$$
$$60a_5 = 1 + 2a_0a_2 + a_1{}^2.$$

By Theorem 37.24 and the initial conditions,

(q) $$a_0 = 1, \quad a_1 = 0, \quad a_2 = \tfrac{2}{2} = 1.$$

Therefore by (p) and (q)

(r) $a_3 = \tfrac{2}{6} = \tfrac{1}{3}, \quad a_4 = \tfrac{1}{24}(2) = \tfrac{1}{12}, \quad a_5 = \tfrac{1}{60}(1 + 2) = \tfrac{1}{20}.$

Substituting (q) and (r) in (l), we obtain

(s) $y = 1 + (x - 1)^2 + \tfrac{1}{3}(x - 1)^3 + \tfrac{1}{12}(x - 1)^4 + \tfrac{1}{20}(x - 1)^5 + \cdots,$

which agrees with (k) to terms of order five.

Comment 39.37. As explained in Comment 38.26, we can find M without the need of (b), by observing that $|x^2 + y^2| \leq |x^2| + |y^2|$. Hence if $|x - 1| < r$, so that $|x| < r + 1$ and if $|y - 1| < r$, so that $|y| < r + 1$, then $|x^2 + y^2| < (r + 1)^2 + (r + 1)^2 = 2(r^2 + 2r + 1)$ as in (c) above.

Example 39.38. Find by series methods, a particular solution of the nonlinear equation

(a) $$y''' = (x - 1)^2 + y^2 + y' - 2$$

for which

(b) $$y(1) = 1, \quad y'(1) = 0, \quad y''(1) = 2.$$

Solution. Comparing (a) with (39.3) we see that $n = 3$ and $f = (x - 1)^2 + y^2 + y' - 2$. Comparing (b) with (39.31), we see that $x_0 = 1, y_0 = 1, y_1 = 0, y_2 = 2$. The series representation of f in powers of $(x - 1), (y - 1), (y' - 0), (y'' - 2)$, by (38.25), is

(c) $f(x,y,y') = (x - 1)^2 + 1 + 2(y - 1) + (y - 1)^2 + y' - 2,$

which is a finite series, and therefore valid for all x, y, y'. We may therefore choose r of Theorem 39.32 arbitrarily. Hence when $|x - 1| < r$, $|y - 1| < r$, $|y'| < r$, we have by (c)

(d) $\qquad |f(x,y,y')| < r^2 + |-1| + 2r + r^2 + r = 2r^2 + 3r + 1,$

which is the M of (39.33).

Hence, by Theorem 39.32, there is a unique particular solution $y(x)$ of (a) analytic at $x = 1$ satisfying (b), i.e., the solution has a Taylor series expansion of the form (39.34), namely

(e) $\qquad y(x) = y(1) + y'(1)(x - 1)$

$$+ \frac{y''(1)}{2!} (x - 1)^2 + \frac{y'''(1)}{3!} (x - 1)^3 + \cdots,$$

valid in an interval about $x = 1$. By (39.35), this interval is at least equal to [here $n = 3$, M is given by (d) and r is arbitrary]

(f) $\qquad\qquad\qquad I: |x - 1| < \dfrac{r}{5(2r^2 + 3r + 1)} \, .$

To maximize I, we let $u = r/5(2r^2 + 3r + 1)$, take its derivative with respect to r and set the result equal to zero. We thus obtain

(g) $\quad 0 = (2r^2 + 3r + 1) - r(4r + 3) = -2r^2 + 1, \quad r = 1/\sqrt{2}.$

For this value of r, the interval I of (f) becomes

(h) $\qquad\qquad\qquad I: |x - 1| < 0.034.$

We shall use the usual two methods of finding a series solution of (a) satisfying (b).

First Method. By Successive Differentiations. In (e), we know by (b)

(i) $\qquad\qquad y(1) = 1, \qquad y'(1) = 0, \qquad y''(1) = 2.$

To find the values of succeeding coefficients in (e), we evaluate (a) and its successive derivatives at $(1,1)$. These are, with $x = 1$, $y = 1$, $y' = 0$, $y'' = 2$,

(j) $\quad y''' = (x - 1)^2 + y^2 + y' - 2, \quad y'''(1) = 1 - 2 = -1.$
$\qquad y^{(4)} = 2(x - 1) + 2yy' + y'', \qquad y^{(4)}(1) = 2.$
$\qquad y^{(5)} = 2 + 2yy'' + 2(y')^2 + y''', \quad y^{(5)}(1) = 2 + 4 - 1 = 5.$
$\qquad \cdots \cdots \cdots \cdots \cdots \cdots \cdots \cdots , \quad \cdots \cdots \cdots \cdots \cdots \cdots$

Substituting (i) and (j) in (e), we obtain

(k) $\quad y = 1 + (x - 1)^2 - \tfrac{1}{6}(x - 1)^3 + \tfrac{1}{12}(x - 1)^4 + \tfrac{1}{24}(x - 1)^5 + \cdots,$

which are the first six terms of a series solution of (a) satisfying (b), valid at least in the interval I of (h).

Second Method. By Undetermined Coefficients. Since the initial conditions have been given in terms of $x = 1$, we seek a series solution of the form

(l) $y = a_0 + a_1(x - 1) + a_2(x - 1)^2 + a_3(x - 1)^3 + \cdots.$

Its next three successive derivatives are

(m) $y'(x) = a_1 + 2a_2(x - 1) + 3a_3(x - 1)^2 + 4a_4(x - 1)^3$
$$+ 5a_5(x - 1)^4 + \cdots,$$
$$y''(x) = 2a_2 + 6a_3(x - 1) + 12a_4(x - 1)^2 + 20a_5(x - 1)^3 + \cdots,$$
$$y'''(x) = 6a_3 + 24a_4(x - 1) + 60a_5(x - 1)^2 + \cdots.$$

Substituting (l) and the first and last equations of (m) in (a), we obtain

(n) $6a_3 + 24a_4(x - 1) + 60a_5(x - 1)^2 + \cdots$
$$= (x - 1)^2 + [a_0 + a_1(x - 1) + a_2(x - 1)^2 + \cdots]^2$$
$$+ a_1 + 2a_2(x - 1) + 3a_3(x - 1)^2 + \cdots - 2.$$

Equating coefficients of like powers of $(x - 1)$ in (n), we have

(o) $6a_3 = a_0{}^2 + a_1 - 2,$
$$24a_4 = 2a_0a_1 + 2a_2,$$
$$60a_5 = 1 + 2a_0a_2 + a_1{}^2 + 3a_3.$$

By (b) and Theorem 37.24, we know that

(p) $a_0 = 1, \qquad a_1 = 0, \qquad a_2 = 1.$

Therefore by (o) and (p),

(q) $a_3 = \tfrac{1}{6}(1 - 2) = -\tfrac{1}{6}, \qquad a_4 = \tfrac{1}{24}(2) = \tfrac{1}{12},$
$$a_5 = \tfrac{1}{60}(1 + 2 - \tfrac{1}{2}) = \tfrac{1}{24}.$$

Substituting (p) and (q) in (l), we obtain

(r) $y(x) = 1 + (x - 1)^2 - \tfrac{1}{6}(x - 1)^3 + \tfrac{1}{12}(x - 1)^4 + \tfrac{1}{24}(x - 1)^5 + \cdots,$

which is the same as that obtained previously in (k).

EXERCISE 39

Obtain terms to order k of the particular solution of each of the following systems of differential equations, where k is the number shown alongside each equation. Use two methods. Also find an interval of convergence of each series solution.

1. $\dfrac{dx}{dt} = e^t + y,\qquad \dfrac{dy}{dt} = e^{-t} + x,\quad x(0) = 0,\ y(0) = 0,\ k = 4.$

2. $\dfrac{dx}{dt} = y \sin t,\qquad \dfrac{dy}{dt} = t^2 + x,\quad x\left(\dfrac{\pi}{2}\right) = 2,\ y\left(\dfrac{\pi}{2}\right) = 1,\ k = 4.$

3. $\dfrac{dx}{dt} = t + x^2,\qquad \dfrac{dy}{dt} = t - y^2,\quad x(0) = 0,\ y(0) = 1,\ k = 4.$

4. $\dfrac{dx}{dt} = y + \sin t,\qquad \dfrac{dy}{dt} = x + \cos t,\quad x(0) = 0,\ y(0) = \dfrac{\pi}{2},\ k = 4.$

5. See (33.22), population problem.

$\dfrac{dx}{dt} = hx - kxy,\qquad \dfrac{dy}{dt} = kxy - py,$ where h, k, and p are constants, and $x(0) = 1,\ y(0) = 1,\ k = 3.$

6. $\dfrac{dx}{dt} = t^2 + x^2,\qquad \dfrac{dy}{dt} = yt,\quad x(0) = -1,\ y(0) = 1,\ k = 4.$

7. $\dfrac{dx}{dt} = xyt,\qquad \dfrac{dy}{dt} = x + t,\quad x(0) = 1,\ y(0) = -1,\ k = 4.$

Obtain terms to order k of the particular solution of each of the following differential equations, where k is shown alongside each equation. Use two methods. Find an interval of convergence.

8. $y'' = x^2 - y^2,\quad y(0) = 1,\ y'(0) = 0,\ k = 6.$

9. $y'' = xy - (y')^2,\quad y(0) = 2,\ y'(0) = 1,\ k = 4.$

10. $y'' = x^2 + \sin y,\quad y(0) = \dfrac{\pi}{2},\ y'(0) = 1,\ k = 5.$

11. $y'' = \cos x + \sin y,\quad y(0) = 0,\ y'(0) = 1,\ k = 4.$

12. $y''' = yy' + xy,\quad y(0) = 0,\ y'(0) = 1,\ y''(0) = 2,\ k = 5.$

13. $y'' = y' \log y + x,\quad y(0) = 1,\ y'(0) = 3,\ k = 4.$

14. $y''' = y^2 \log x + y',\quad y(1) = 1,\ y'(1) = 0,\ y''(1) = 1,\ k = 5.$

ANSWERS 39

1. $x = t + t^2 + \dfrac{t^3}{6} + \dfrac{t^4}{6} + \cdots,\ y = t + \dfrac{t^3}{2} + \cdots,\ |t| < \infty.$

2. $x = 1 + \left(t - \dfrac{\pi}{2}\right) + \dfrac{\pi^2 + 4}{8}\left(t - \dfrac{\pi}{2}\right)^2$

$\qquad\qquad + \dfrac{\pi}{6}\left(t - \dfrac{\pi}{2}\right)^3 - \dfrac{\pi^2}{48}\left(t - \dfrac{\pi}{2}\right)^4 + \cdots,$

$$y = 1 + \frac{\pi^2 + 4}{4}\left(t - \frac{\pi}{2}\right) + \frac{(\pi + 1)}{2}\left(t - \frac{\pi}{2}\right)^2$$

$$+ \frac{\pi^2 + 12}{24}\left(t - \frac{\pi}{2}\right)^3 + \frac{\pi}{24}\left(t - \frac{\pi}{2}\right)^4 + \cdots, \quad \left|t - \frac{\pi}{2}\right| < \infty.$$

3. $x = \dfrac{t^2}{2} + \dfrac{t^5}{20} + \cdots, \qquad y = 1 - t + \frac{3}{2}t^2 - \frac{4}{3}t^3 + \frac{17}{12}t^4 + \cdots,$

$$|t| < \frac{r}{4(r^2 + 3r + 1)}.$$

4. $x = \dfrac{\pi}{2}t + t^2 + \dfrac{\pi}{12}t^3 + \cdots, \quad y = \dfrac{\pi}{2} + t + \dfrac{\pi}{4}t^2 + \dfrac{1}{6}t^3 + \dfrac{\pi}{48}t^4 + \cdots,$

$$|t| < \infty$$

5. $x = 1 + (h - k)t + \dfrac{h^2 - 2hk + kp}{2}t^2$

$$+ \frac{h^3 - 3h^2k + 3hkp - hk^2 - kp^2 - k^2p + 2k^3}{3!}t^3 + \cdots,$$

$$y = 1 + (k - p)t + \frac{p^2 - 2kp + hk}{2}t^2$$

$$+ \frac{hk^2 + 3kp^2 - 2k^3 - 3hkp + h^2k + k^2p - p^3}{3!}t^3 + \cdots.$$

6. $x = -1 + t - t^2 + \frac{4}{3}t^3 - \frac{7}{6}t^4 + \cdots, \qquad y = 1 + \frac{1}{2}t^2 + \frac{1}{8}t^4 + \cdots.$

7. $x = 1 - \frac{1}{2}t^2 + \frac{1}{6}t^3 + \frac{1}{4}t^4 + \cdots, \qquad y = -1 + t + \dfrac{t^2}{2} - \dfrac{t^3}{6} + \dfrac{t^4}{12} + \cdots.$

8. $y = 1 - \frac{1}{2}x^2 + \frac{1}{6}x^4 - \frac{7}{360}x^6 + \cdots, \quad |x| < \dfrac{r}{4(2r^2 + 2r + 1)}, r$ arbitrary.

9. $y = 2 + x - \frac{1}{2}x^2 + \frac{2}{3}x^3 - \frac{1}{3}x^4 + \cdots, \quad |x| < \dfrac{r}{4(2r^2 + 4r + 1)}, r$ arbitrary.

10. $y = \dfrac{\pi}{2} + x + \dfrac{x^2}{2} + \dfrac{x^4}{24} - \dfrac{x^5}{40} + \cdots, \quad |x| < \dfrac{r}{4(r^2 + 1)}, r$ arbitrary.

11. $y = x + \dfrac{x^2}{2} + \dfrac{x^3}{6} + \cdots, \quad |x| < \dfrac{r}{8}, r$ arbitrary.

12. $y = x + x^2 + \dfrac{x^4}{24} + \dfrac{x^5}{15} + \cdots, \quad |x| < \dfrac{1}{5(2r + 1)}, r$ arbitrary.

13. $y = 1 + 3x + \frac{5}{3}x^3 - \frac{9}{8}x^4 + \cdots,$

$$|x| < \frac{r}{4[(r + 3)\log(r + 1) + r]}, r < 1.$$

14. $y = 1 + \dfrac{(x - 1)^2}{2} + \dfrac{(x - 1)^4}{12} - \dfrac{(x - 1)^5}{120} + \cdots,$

$$|x - 1| < \frac{r}{5[(r + 1)^2\log(r + 1) + r]}, r < 1.$$

LESSON 40.   Ordinary Points and Singularities of a Linear Differential Equation. Method of Frobenius.

LESSON 40A.   Ordinary Points and Singularities of a Linear Differential Equation. As remarked previously, power series methods are especially well suited for finding solutions of linear differential equations with nonconstant coefficients. These methods, however, cannot be applied indiscriminately. For example, if we tried to find a power series solution of

$$(40.1) \qquad x^2 y'' + xy' + (x^2 - \tfrac{1}{4})y = 0$$

in the form

$$(40.11) \qquad y(x) = a_0 + a_1 x + a_2 x^2 + a_3 x^3 + \cdots,$$

by the method of successive differentiations, we would run into trouble. By (37.26), $a_2 = y''(0)/2!$, and when $x = 0$, we see from (40.1) that y'' and, therefore, a_2 do not exist. If we tried to use the method of undetermined coefficients, we would find $a_0 = 0$, $a_1 = 0$, $a_2 = 0$, $\cdots$.

The trouble arises because after division by x^2 in (40.1), the coefficient of y' which becomes $1/x$ is not analytic at $x = 0$, i.e., it does not have a Maclaurin series expansion in powers of x. Hence the hypothesis of Theorem 37.51 is not satisfied. We distinguish between points x_0 which satisfy the hypothesis of Theorem 37.51 and those which do not by means of Definitions 40.2 and 40.22 which follow.

Definition 40.2.   A point $x = x_0$ is called an **ordinary** point of the linear differential equation

$$(40.21) \quad y^{(n)} + F_{n-1}(x)y^{(n-1)} + \cdots + F_1(x)y' + F_0(x)y = Q(x),$$

if each function $F_0, F_1, \cdots, F_{n-1}$, and Q is analytic at $x = x_0$. (Remember this means each function has a Taylor series expansion in powers of $x - x_0$ valid in a neighborhood of x_0.)

By Theorem 37.51, if $x = x_0$ is an ordinary point, then (40.21) has a solution which is also analytic at $x = x_0$, i.e., the solution has a Taylor series representation in powers of $(x - x_0)$ valid in a neighborhood of x_0.

Definition 40.22.   A point $x = x_0$ is called a **singularity** of (40.21), if one or more of the functions $F_0(x), \cdots, F_{n-1}(x), Q(x)$ is not analytic at $x = x_0$.

By Definition 40.22, the point $x = 0$ is therefore a singularity of (40.1).

For the remainder of this lesson, we shall confine our attention to a second order linear equation

$$(40.23) \qquad y'' + F_1(x)y' + F_2(x)y = 0,$$

where F_1 and F_2 are continuous functions of x on a common interval I. Its singularities, if there are any, have been divided into two kinds, regular singularities and irregular singularities.

Definition 40.24. If $x = x_0$ is a singularity of (40.23) and if the multiplication of $F_1(x)$ by $(x - x_0)$ and of $F_2(x)$ by $(x - x_0)^2$ result in functions, *each* of which is analytic at $x = x_0$, then the point $x = x_0$ is called a **regular singularity** of (40.23).

Example 40.25. Show that $x = 0$ and $x = 1$ are regular singularities of

(a) $$(x - 1)y'' + \frac{1}{x}\, y' - 2y = 0.$$

Solution. Dividing (a) by $(x - 1)$, we obtain

(b) $$y'' + \frac{1}{x(x - 1)}\, y' - \frac{2}{x - 1}\, y = 0.$$

Comparing (b) with (40.23), we see that

(c) $$F_1(x) = \frac{1}{x(x - 1)}, \qquad F_2(x) = -\frac{2}{x - 1}.$$

By Definition 40.22, $x = 0$ and $x = 1$ are singularities of (b).

Consider first the point $x = 0$. Following the instructions in Definition 40.24, we multiply F_1 by $(x - 0)$ and F_2 by $(x - 0)^2$. There results respectively new functions $1/(x - 1)$ and $-2x^2/(x - 1)$, each of which is analytic at $x = 0$. Their Taylor series expansions are in fact

(d) $$\frac{1}{x - 1} = -(1 + x + x^2 + \cdots),$$

$$-\frac{2x^2}{x - 1} = 2(x^2 + x^3 + x^4 + \cdots), \quad |x| < 1.$$

Hence by Definition 40.24, $x = 0$ is a regular singularity of (b) and therefore of (a).

Second we consider the point $x = 1$. Multiplication of F_1 by $(x - 1)$ and of F_2 by $(x - 1)^2$ give respectively

(e) $$\frac{1}{x} \quad \text{and} \quad -2(x - 1),$$

both of which are analytic at $x = 1$. Their Taylor series expansions are respectively $1 - (x - 1) + (x - 1)^2 - (x - 1)^3 + \cdots$, and $-2(x - 1)$. Hence by Definition 40.24, $x = 1$ is a regular singularity of (b) and therefore of (a).

Definition 40.26. If $x = x_0$ is a singularity of (40.23) and if the multiplication of $F_1(x)$ by $(x - x_0)$ and $F_2(x)$ by $(x - x_0)^2$ result in functions one or both of which are not analytic at $x = x_0$, then the point $x = x_0$ is called an **irregular singularity** of (40.23).

Example 40.27. Show that $x = 0$ and $x = 1$ are irregular singularities of

(a) $$(x - 1)^2 y'' + \frac{1}{x^2} y' + 2y = 0.$$

Solution. Division of (a) by $(x - 1)^2$ gives

(b) $$y'' + \frac{1}{x^2(x - 1)^2} y' + \frac{2}{(x - 1)^2} y = 0.$$

By Definition (40.22), $x = 0$ and $x = 1$ are singularities of (b). Consider first the point $x = 0$. Here $F_1 = \dfrac{1}{x^2(x - 1)^2}$. Multiplication of F_1 by $(x - 0)$ gives $1/x(x - 1)^2$, which is not analytic at $x = 0$. Hence by Definition 40.26, $x = 0$ is an irregular singularity of (b) and therefore of (a). Second, consider the point $x = 1$. Multiplication of F_1 by $(x - 1)$ gives $1/x^2(x - 1)$ which is not analytic at $x = 1$. Hence $x = 1$ is an irregular singularity of (b) and therefore of (a).

LESSON 40B. Solution of a Homogeneous Linear Differential Equation About a Regular Singularity. Method of Frobenius. If the linear equation

(40.3) $$y'' + F_1(x)y' + F_2(x)y = 0$$

has an *irregular* singularity at $x = x_0$, then the problem of finding a series solution is too difficult for discussion here. If, however, (40.3) has a *regular* singularity at $x = x_0$, then we shall describe a method for finding a series solution, valid in a neighborhood of x_0. It is known as the **method of Frobenius.** The series solution which Frobenius obtained, namely

(40.31) $$y = (x - x_0)^m [a_0 + a_1(x - x_0) + a_2(x - x_0)^2 + a_3(x - x_0)^3 + \cdots], \quad a_0 \neq 0,$$

is known as a **Frobenius series.**

Note that when $m = 0$ or a positive integer, the series becomes the usual Taylor series. However, for negative values of m or for nonintegral positive values of m, (40.31) is not a Taylor series. A Frobenius series, therefore, includes the Taylor series as a special case.

Assume x_0 is a regular singularity of (40.3). It therefore follows by Definitions 40.22 and 40.24, that F_1 or F_2 or both are not analytic at $x = x_0$, but that $(x - x_0)F_1$ and $(x - x_0)^2 F_2$ are. This means that F_1 has $(x - x_0)$ in its denominator and/or F_2 has $(x - x_0)^2$ in its denominator. Hence either $F_1(x) = f_1(x)/(x - x_0)$ or $F_2(x) = f_2(x)/(x - x_0)^2$ or both F_1 and F_2 have these respective forms. In any event, multiplication of (40.3) by $(x - x_0)^2$ will transform it into an equation of the form

$$(40.311) \qquad (x - x_0)^2 y'' + (x - x_0)f_1(x)y' + f_2(x)y = 0,$$

in which both $f_1(x)$ and $f_2(x)$ are now analytic at $x = x_0$.

The relevant theorem which will assure the existence of a Frobenius series solution (40.31) of (40.3) is the following.

Theorem 40.32. *Let x_0 be a regular singularity of (40.311). Then (40.311) has at least one Frobenius series solution of the form (40.31). It is valid in the common interval of convergence of $f_1(x)$ and $f_2(x)$ of (40.311), except perhaps for $x = x_0$, i.e., if each Taylor series expansion of $f_1(x)$ and $f_2(x)$ is valid in the interval $I: |x - x_0| < r$, then at least one Frobenius series solution is also valid in $I: |x - x_0| < r$ except perhaps for $x = x_0$.*

No loss in generality results if, in (40.311), we take $x_0 = 0$ since by a translation of axes we can always replace an expansion in powers of $(x - x_0)$ by one in powers of x. With this understanding, we rewrite (40.311) as

$$(40.33) \qquad x^2 y'' + xf_1(x)y' + f_2(x)y = 0,$$

where the functions $f_1(x)$ and $f_2(x)$ are analytic at $x = 0$. Hence each has a Taylor series expansion in powers of x valid in a neighborhood of $x = 0$. Let

$$(40.34) \qquad \begin{aligned} f_1(x) &= b_0 + b_1 x + b_2 x^2 + \cdots, \\ f_2(x) &= c_0 + c_1 x + c_2 x^2 + \cdots, \end{aligned}$$

be their respective series expansions.

The Frobenius series (40.31) and its next two derivatives are, with $x_0 = 0$,

$$(40.35) \quad \begin{aligned} y &= x^m(a_0 + a_1 x + a_2 x^2 + \cdots + a_n x^n + \cdots) \\ &= a_0 x^m + a_1 x^{m+1} + a_2 x^{m+2} + \cdots + a_n x^{m+n} + \cdots, \quad a_0 \neq 0, \\ y' &= a_0 m x^{m-1} + a_1(m + 1)x^m + a_2(m + 2)x^{m+1} + \cdots \\ &\quad + a_n(m + n)x^{m+n-1} + \cdots, \\ y'' &= a_0 m(m - 1)x^{m-2} + a_1 m(m + 1)x^{m-1} \\ &\quad + a_2(m + 1)(m + 2)x^m + \cdots \\ &\quad + a_n(m + n - 1)(m + n)x^{m+n-2} + \cdots. \end{aligned}$$

The function $y(x)$ will be a solution of (40.33) if it satisfies the equation. Hence substituting (40.35) in (40.33) and replacing at the same time the functions $f_1(x)$ and $f_2(x)$ by their series expansions in (40.34), we obtain

(40.36) $\quad x^2[a_0 m(m-1)x^{m-2} + a_1 m(m+1)x^{m-1} + \cdots$
$$+ a_n(m+n-1)(m+n)x^{m+n-2} + \cdots]$$
$$+ x(b_0 + b_1 x + b_2 x^2 + \cdots)$$
$$\times [a_0 m x^{m-1} + a_1(m+1)x^m + \cdots + a_n(m+n)x^{m+n-1} + \cdots]$$
$$+ (c_0 + c_1 x + c_2 x^2 + \cdots)$$
$$\times (a_0 x^m + a_1 x^{m+1} + \cdots + a_n x^{m+n} + \cdots) \equiv 0.$$

Expanding (40.36) and collecting coefficients of like powers of x, there results

(40.37) $\quad a_0[m(m-1) + b_0 m + c_0]x^m$
$$+ \{a_1[(m+1)m + b_0(m+1) + c_0] + a_0[b_1 m + c_1]\}x^{m+1}$$
$$+ \{a_2[(m+2)(m+1) + b_0(m+2) + c_0] + a_1[b_1(m+1) + c_1]$$
$$+ a_0[b_2 m + c_2]\}x^{m+2}$$
$$+ \{a_3[(m+3)(m+2) + b_0(m+3) + c_0] + a_2[b_1(m+2) + c_1]$$
$$+ a_1[b_2(m+1) + c_2] + a_0[b_3 m + c_3]\}x^{m+3}$$
$$\cdots\cdots\cdots\cdots\cdots\cdots\cdots\cdots\cdots\cdots\cdots\cdots\cdots\cdots$$
$$+ \{a_n[(m+n)(m+n-1) + b_0(m+n) + c_0]$$
$$+ a_{n-1}[b_1(m+n-1) + c_1]$$
$$+ a_{n-2}[b_2(m+n-2) + c_2] + \cdots$$
$$+ a_0[b_n m + c_n]\}x^{m+n} + \cdots \equiv 0.$$

Equation (40.37) will be an identity in x, if each of the coefficients of x^k, $k = m, \cdots, m+n$ is zero. Since we have assumed $a_0 \neq 0$, the first coefficient in (40.37) will be zero only if

(40.38) $\qquad\qquad m(m-1) + b_0 m + c_0 = 0.$

This equation has been given a special name. It is called the **indicial equation.** Since it is a quadratic equation in m, it has two roots. Let us call these roots m_1 and m_2. These roots may be:

1. Distinct, and their difference not equal to an integer.
2. Distinct, and their difference equal to an integer.
3. The same.

We shall consider each of these possibilities separately.

Case 1. The roots m_1 and m_2 of the indicial equation (40.38) are distinct and their difference is not an integer. Each of the roots $m = m_1$, and $m = m_2$ of the indicial equation (40.38) will make the first coefficient in (40.37) zero. We concentrate on the root m_1. Substituting it for m in the remaining coefficients in (40.37) and setting each equal to zero, will enable us to solve each of these equations respectively for a_1, a_2, $\cdots$, a_n, $\cdots$ in terms of a_0. [Remember the b's and c's are constants given by (40.34).] Since for this m_1 and for this set of values of a_1, $\cdots$, a_n, each coefficient in (40.37) is zero, (40.37) is an identity in x. Hence the substitution of this m_1 and this set of a's in the first equation of (40.35) will make $y(x)$ a solution of (40.33).

Following the same procedure outlined above, using the second root m_2 we obtain a second set of values of a_1, a_2, $\cdots$, a_n, $\cdots$ in terms of a_0 which with m_2 will make each coefficient in (40.37) zero. Hence the substitution of this second set of values of the a's and m_2 in the first equation of (40.35) will make y a second solution of (40.33).

Comment 40.381. Not every equation of the form (40.33) has two independent Frobenius series solutions. Some, as we shall show later, have only one. If, however, (40.33) has two Frobenius series solutions, then the following relevant theorem, stated without proof, supplements Theorem 40.32.

Theorem 40.39. *The two Frobenius series solutions of (40.33) are linearly independent. Each solution is valid for every x in the common interval of convergence of $f_1(x)$ and $f_2(x)$ except perhaps for $x = 0$.*

Comment 40.391. Theorem 40.39 can also be stated as follows. Each Frobenius series solution will converge for every x, except perhaps for $x = 0$, in a circle in the complex x plane, whose center is at 0, and whose radius extends at least to the next nearest singularity of (40.33), i.e., each solution is valid at least for $0 < |x| < a$, where a is the nearest singularity to 0.

Example 40.392. Find the interval of convergence of the series solution of

(a) $$x^2 y'' + \frac{x}{1 + x^2}\, y' - 3y = 0.$$

Solution. Division of (a) by x^2 and application of Definitions 40.22 and 40.24 shows that $x = 0$ is a regular singularity of (a). Comparing (a) with (40.33), we see that

(b) $$f_1(x) = \frac{1}{1 + x^2} = 1 - x^2 + x^4 - x^6 + \cdots, \quad |x| < 1,$$
$$f_2(x) = -3, \quad -\infty < x < \infty.$$

Hence, by Theorem (40.39), each Frobenius series solution of (a) is valid for $|x| < 1$, except perhaps for $x = 0$.

Remark. You can verify that $\pm i$ are also regular singularities of (a). Since $\pm i$ are singularities, it follows by Comment 40.391, that each

Frobenius series solution will converge for every x, except perhaps for $x = 0$, in a circle in the complex x plane of at least unit radius. This fact may help make clear why the interval of convergence of the series representation of $1/(1 + x^2)$ is $|x| < 1$. Since the function is continuous for all real x, it would seem that it ought to have a series representation valid for all real x.

Example 40.393. Find a Frobenius series solution of

(a) $\qquad x^2 y'' + x(x + \frac{1}{2})y' - (x^2 + \frac{1}{2})y = 0.$

Solution. Division by x^2 and application of Definitions 40.22 and 40.24, shows that $x = 0$ is a regular singularity of (a). We therefore seek a Frobenius series solution of the form (40.31), namely

(b) $\qquad y = x^m(a_0 + a_1 x + a_2 x^2 + \cdots), \quad a_0 \neq 0.$

If we wish, we can differentiate (b) twice to obtain y' and y'', substitute these values in (a) and then equate the coefficient of each like power of x to zero. But since we already did this work in obtaining (40.37), we may as well make use of this equation. To use it, we need to know the values of the b's and c's in it. By (40.34), these are the coefficients in the series representation of $f_1(x)$ and $f_2(x)$. Comparing (a) with (40.33) we see that

(c) $\qquad f_1(x) = \frac{1}{2} + x \quad \text{and} \quad f_2(x) = -\frac{1}{2} - x^2,$

both of which are already in series form. Hence comparing (c) with (40.34), we have

(d) $\qquad b_0 = \frac{1}{2}, \qquad b_1 = 1, \qquad c_0 = -\frac{1}{2}, \qquad c_1 = 0, \qquad c_2 = -1.$

All remaining b's and c's are zero. With these values of b_0 and c_0, the indicial equation (40.38) becomes

(e) $\qquad m(m - 1) + \frac{1}{2}m - \frac{1}{2} = 0, \qquad 2m^2 - m - 1 = 0,$

whose roots are $m = 1$, $m = -\frac{1}{2}$. Since these roots are distinct and do not differ by an integer, the method of this case is applicable. Following the method outlined above, we substitute the root $m = 1$ in the remaining coefficients in (40.37) and solve each for $a_1, a_2, \cdots$, in terms of a_0. The most effective way to use (40.37) is to set the coefficient of x^{m+n} equal to zero. When you do this, keep in mind that $m = 1$, b_0, b_1, c_0, c_1, c_2 have the values in (d) and that $b_2, b_3, \cdots, c_3, c_4, \cdots$ are zero. Substituting these values in the coefficient of x^{m+n}, we obtain

(f) $\qquad a_n[(1 + n)(1 + n - 1) + \frac{1}{2}(1 + n) - \frac{1}{2}]$
$\qquad\qquad + a_{n-1}[1(1 + n - 1) + 0] + a_{n-2}[0 - 1] = 0.$

Solving for a_n, there results

(g) $$\frac{2n^2 + 3n}{2} a_n = -na_{n-1} + a_{n-2}.$$

Formula (g) is a **recursion formula.** It will give the value of a_n for each $n \geq 2$. Before we can use it, therefore, we must find a_1. Setting the second coefficient in (40.37) equal to zero, we have, with $m = 1$ and the values of b_0, c_0, b_1, c_1 as given in (d),

(h) $$a_1[(2)(1) + \tfrac{1}{2}(1 + 1) - \tfrac{1}{2}] + a_0[1 + 0] = 0, \qquad a_1 = -\tfrac{2}{5}a_0.$$

We can now use (g) and (h) to obtain, when

(i)

$n = 2$:

$$7a_2 = -2a_1 + a_0 = \frac{4}{5}a_0 + a_0 = \frac{9}{5}a_0, \qquad a_2 = \frac{9}{35}a_0;$$

$n = 3$:

$$\frac{27}{2}a_3 = -3a_2 + a_1 = -\frac{27}{35}a_0 - \frac{2}{5}a_0 = -\frac{41}{35}a_0, \quad a_3 = -\frac{82}{945}a_0;$$

$n = 4$:

$$22a_4 = -4a_3 + a_2 = \frac{328}{945}a_0 + \frac{9}{35}a_0 = \frac{571}{945}a_0, \qquad a_4 = \frac{571}{20{,}790}a_0.$$

Substituting the value $m = 1$, and the above values of the a's in (b), we have

(j) $$y_1 = a_0 x \left(1 - \frac{2}{5}x + \frac{9}{35}x^2 - \frac{82}{945}x^3 + \frac{571}{20{,}790}x^4 - \cdots\right),$$

which are the first five terms of one series solution of (a). To obtain a second solution, we use the root $m = -\tfrac{1}{2}$ and proceed as above. The recursion formula, obtained from the coefficient of x^{m+n} in (40.37) becomes, with $m = -\tfrac{1}{2}$ and the values given in (d),

(k) $$a_n[(n - \tfrac{1}{2})(n - \tfrac{3}{2}) + \tfrac{1}{2}(n - \tfrac{1}{2}) + (-\tfrac{1}{2})]$$
$$+ a_{n-1}[n - \tfrac{3}{2}] + a_{n-2}[0 - 1] = 0,$$

which simplifies to

(l) $$\frac{2n^2 - 3n}{2} a_n = -\left(n - \frac{3}{2}\right)a_{n-1} + a_{n-2}, \quad n \geq 2.$$

As before, we find a_1 by setting the second coefficient in (40.37) equal to zero. There results

(m) $$a_1[\tfrac{1}{2}(-\tfrac{1}{2}) + (\tfrac{1}{2})(\tfrac{1}{2}) - \tfrac{1}{2}] + a_0(-\tfrac{1}{2}) = 0, \quad a_1 = -a_0.$$

Hence from (l) and (m), we obtain, when

(n) $n = 2$: $a_2 = -\frac{1}{2}a_1 + a_0 = \frac{1}{2}a_0 + a_0 = \frac{3}{2}a_0$;

$n = 3$: $\frac{9}{2}a_3 = -\frac{3}{2}a_2 + a_1 = -\frac{9}{4}a_0 - a_0 = -\frac{13}{4}a_0$, $a_3 = -\frac{13}{18}a_0$;

$n = 4$: $10a_4 = -\frac{5}{2}a_3 + a_2 = \frac{65}{36}a_0 + \frac{3}{2}a_0 = \frac{119}{36}a_0$, $a_4 = \frac{119}{360}a_0$.

Substituting $m = -\frac{1}{2}$ and the values of the above set of a's in (b), we obtain for the first five terms of a second series solution of (a)

(o) $\qquad y_2 = a_0 x^{-1/2}(1 - x + \frac{3}{2}x^2 - \frac{13}{18}x^3 + \frac{119}{360}x^4 - \cdots)$.

By Theorem 40.39, the two series solutions are linearly independent. Hence by Theorem 19.3 and Comment 19.41, a general solution of (a) is a linear combination of (j) and (o), namely

(p) $\quad y = c_1 x \left(1 - \dfrac{2}{5}x + \dfrac{9}{35}x^2 - \dfrac{82}{945}x^3 + \dfrac{571}{20{,}790}x^4 - \cdots\right)$

$\qquad\qquad + c_2 x^{-1/2}\left(1 - x + \dfrac{3}{2}x^2 - \dfrac{13}{18}x^3 + \dfrac{119}{360}x^4 - \cdots\right)$,

where a_0 has been incorporated into the arbitrary constants c_1 and c_2.

We observe from (c) that $f_1(x)$ and $f_2(x)$ are polynomials. Hence, by Comment 37.53, their series representations are valid for all x. Therefore, by Theorem 40.39, each Frobenius series solution is valid for all x except perhaps for $x = 0$. In this example the solution (j) is valid for all x. The second solution (o), because of the presence of $x^{-1/2}$, is valid for all x except $x = 0$. Hence the general solution (p) is valid for $0 < |x| < \infty$.

Note that for $x < 0$, the second series in (p) is imaginary. We can make it real by choosing $c_2 = ci$, where c is real.

Case 2. The roots m_1, m_2 of the indicial equation (40.38) differ by an integer. If the two roots of the indicial equation (40.38) differ by a nonzero integer, we can write them as m and $m + N$, where N is a positive integer. Since $m + N$ is a root of (40.38), it satisfies this equation. Hence,

(40.4) $\qquad (m + N)(m + N - 1) + b_0(m + N) + c_0 = 0$.

Now compare the left side of (40.4) with the coefficient of a_n in the last term of (40.37). They both will be exactly the same if n is replaced by N. This means that if we use the smaller root m in (40.37) to find a set of values for the a's which will make the coefficient of each x^k zero, we will be stopped when we reach the term in which a_N appears, since its coefficient will be zero. Hence we cannot solve this equation for a_N in terms of previous a's unless by accident, the remaining terms in the equation also add to zero. In this case, the equation will be satisfied for any arbitrary value of a_N. We can then continue to determine values of succeeding a's, i.e., of $a_{N+1}, a_{N+2}, \cdots$ in terms of a_0 and a_N.

If therefore the roots of the indicial equation (40.38) differ by a positive integer N, two possibilities may occur. Each is considered separately in Cases 2A and 2B below.

Case 2A. The coefficient of a_N in (40.37) is zero and the remaining terms in the coefficient of x^{m+N} also add to zero. In this case the larger root $m + N$ will determine, by (40.37), a set of values of the a's in terms of a_0; the smaller root m will determine two sets of values of the a's, one in terms of a_0 and the other in terms of a_N. However, the Frobenius series solution obtained by the larger root in terms of a_0 will not be linearly independent of the one in terms of a_0 obtained by the smaller root. Hence the smaller root alone will give, in this case, two independent solutions, whose linear combination will be a general solution of (40.33). It will have two arbitrary constants a_0 and a_N. These solutions will be valid in the same intervals given in Theorem 40.39.

Example 40.41. Find a general solution of

(a) $$x^2 y'' + xy' + \left(x^2 - \frac{1}{2^2} \right) y = 0.$$

[NOTE. This equation is known as the Bessel equation of index $\frac{1}{2}$. A discussion of the general Bessel equation of index n will be found in Lesson 42.]

Solution. By Definitions 40.22 and 40.24, it can be verified that $x = 0$ is a regular singularity of (a). Hence we seek a Frobenius series solution of the form

(b) $$y = x^m (a_0 + a_1 x + a_2 x^2 + \cdots).$$

Comparing (a) with (40.33), we see that

(c) $$f_1(x) = 1 \quad \text{and} \quad f_2(x) = -\tfrac{1}{4} + x^2,$$

both of which are already in series form. Hence comparing (c) with (40.34), we have

(d) $$b_0 = 1.$$
$$c_0 = -\tfrac{1}{4}, \quad c_1 = 0, \quad c_2 = 1.$$

All remaining b's and c's are zero. Therefore the indicial equation (40.38) becomes

(e) $$m(m - 1) + m - \tfrac{1}{4} = 0, \quad m^2 - \tfrac{1}{4} = 0,$$

whose roots are $m = \frac{1}{2}$ and $m = -\frac{1}{2}$. These roots differ by the integer 1. Hence for this example $N = 1$. When $m = -\frac{1}{2}$, which is the smaller root, we obtain, by setting the second coefficient in (40.37) equal to zero

(f) $a_1[(\frac{1}{2})(-\frac{1}{2}) + \frac{1}{2} - \frac{1}{4}] + a_0(0) = 0, \quad 0a_1 + 0a_0 = 0.$

The coefficient of a_1 is zero, but the other term in its equation is also zero. Hence this Case 2A is applicable. (Note. Since $N = 1$, $a_N = a_1$. We therefore could have anticipated from what we said above, that the coefficient of a_1 would be zero. We could not of course have anticipated that the coefficient of a_0 would also be zero.) Any values of a_0 and a_1 will satisfy the last equation in (f). The remaining a's, namely a_2, a_3, $\cdots$ obtained by equating each subsequent coefficient in (40.37) to zero, will now be solvable in terms of an arbitrary a_0 and an arbitrary a_1.

With $m = -\frac{1}{2}$, the recursion formula obtained by setting the coefficient of the general term x^{m+n} in (40.37) equal to zero, becomes, with the help of (d),

(g) $a_n[(-\frac{1}{2}+n)(-\frac{3}{2}+n) + (-\frac{1}{2}+n) - \frac{1}{4}] + a_{n-1}(0) + a_{n-2}(1) = 0,$

which simplifies to

(h) $(n^2 - n)a_n = -a_{n-2}, \quad n \geqq 2.$

From it we find (remember a_0 and a_1 are now arbitrary) when

(i) $n = 2$: $a_2 = -\frac{1}{2}a_0,$ $n = 3$: $a_3 = -\frac{1}{6}a_1,$

    $n = 4$: $a_4 = -\frac{1}{12}a_2 = \frac{1}{24}a_0,$ $n = 5$: $a_5 = -\frac{1}{20}a_3 = \frac{1}{120}a_1,$

    $n = 6$: $a_6 = -\frac{1}{30}a_4 = -\frac{1}{720}a_0,$ $n = 7$: $a_7 = -\frac{1}{42}a_5 = -\frac{1}{5040}a_1.$

Substituting $m = -\frac{1}{2}$ and the values of the above a's in (b), it becomes

(j) $y = a_0 x^{-1/2}\left(1 - \frac{1}{2}x^2 + \frac{1}{24}x^4 - \frac{1}{720}x^6 + \cdots\right)$

$+ a_1 x^{1/2}\left(1 - \frac{x^2}{6} + \frac{x^4}{120} - \frac{x^6}{5040} + \cdots\right),$

which are the first seven terms of a general series solution of (b). We observe from (c) that $f_1(x)$ and $f_2(x)$ are polynomials. Hence, by Comment 37.53, their series representations are valid for all x. Therefore, by Theorem 40.39, each Frobenius series solution converges for all x, except perhaps for $x = 0$. Here the first series converges for all x except $x = 0$, the second for all x. The general solution (j) converges for $0 < |x| < \infty$. Note that for $x < 0$, the first series is imaginary. We can make it real by choosing $a_0 = ci$ where c is real.

Case 2B. The coefficient of a_N in (40.37) is zero, but the remaining terms in the coefficient of x^{m+N} do not add to zero. In this case only the *larger* root $m + N$ of the indicial equation (40.38) will determine a set of values of the a's in terms of a_0. There will therefore be only one Frobenius series solution of (40.33).

Comment 40.5. A second independent solution of (40.33), it has been proved, will be of the form

$$(40.51) \qquad y_2(x) = u(x) - b_N y_1(x) \log x, \quad x > 0,$$

where N is the positive integral difference between the roots of the indicial equation (40.38), y_1 is a Frobenius series solution of (40.33) obtained with the larger root $m + N$, and $u(x)$ is a Frobenius series of the form

$$(40.52) \qquad u(x) = x^m(b_0 + b_1 x + b_2 x^2 + \cdots).$$

In (40.52), m is the *smaller* of the two roots of (40.38). By substituting (40.51), (40.52), and the necessary derivatives in (40.33), you will find that y_2 will be a solution of (40.33), if (see Exercise 40, 14)

$$(40.53) \qquad x^2 u'' + x f_1 u' + f_2 u = b_N[2x y_1' + (f_1 - 1)y_1].$$

If, in (40.53), you now substitute u and its derivatives as determined by (40.52), the Frobenius series solution y_1 and its derivative, then the resulting equation will enable you to find the values of the b's in (40.52). We shall, however, not discuss this matter further since **logarithmic solutions,** as these are called, have limited applications. See Exercise 40, 14–17.

Example 40.531. Find a solution of

$$(a) \qquad x^2 y'' - x(2 - x)y' + (2 + x^2)y = 0.$$

Solution. Division by x^2 and application of Definitions 40.22 and 40.24 shows that $x = 0$ is a regular singularity of (a). Hence we seek a Frobenius series of the form

$$(b) \qquad y = x^m(a_0 + a_1 x + a_2 x^2 + \cdots).$$

Comparing (a) with (40.33), we see that

$$(c) \qquad f_1(x) = -2 + x \quad \text{and} \quad f_2(x) = 2 + x^2,$$

both of which already are in series form. Hence, comparing (c) with (40.34), we find

$$(d) \qquad \begin{aligned} b_0 &= -2, & b_1 &= 1. \\ c_0 &= 2, & c_1 &= 0, & c_2 &= 1. \end{aligned}$$

All the remaining b's and c's are zero. The indicial equation (40.38) becomes

$$(e) \qquad m(m - 1) - 2m + 2 = 0, \qquad m^2 - 3m + 2 = 0,$$

whose roots are $m = 1$ and $m = 2$. These roots differ by an integer $N = 1$.

Using the smaller root $m = 1$ and the values in (d), we obtain by setting the second coefficient in (40.37) equal to zero,

(f) $\qquad a_1(2 - 4 + 2) + a_0(1 + 0) = 0, \qquad 0a_1 + a_0 = 0.$

The coefficient of $a_N = a_1$ is zero, but the coefficient of a_0 is not. Hence this root $m = 1$ will not lead to a solution. As remarked at the outset of this lesson, only the larger root in this case will give a solution.

Therefore using the larger root $m = 2$, we obtain by setting the second coefficient in (40.37) equal to zero,

(g) $\qquad a_1(3 \cdot 2 - 2 \cdot 3 + 2) + 2a_0 = 0, \qquad a_1 = -a_0.$

The recursion formula, obtained by setting the coefficient of x^{m+n} in (40.37) equal to zero becomes, with the help of (d),

(h) $\qquad a_n[(2 + n)(1 + n) - 2(2 + n) + 2] + (1 + n)a_{n-1} + a_{n-2} = 0,$

which simplifies to

(i) $\qquad (n^2 + n)a_n = -(n + 1)a_{n-1} - a_{n-2}, \quad n \geqq 2.$

By (g) and (i), when

(j) $\quad n = 2: \quad 6a_2 = -3a_1 - a_0 = 3a_0 - a_0, \quad a_2 = \dfrac{a_0}{3},$

$\qquad n = 3: \quad 12a_3 = -4a_2 - a_1 = -\dfrac{4}{3}a_0 + a_0, \quad a_3 = -\dfrac{a_0}{36}.$

Substituting $m = 2$ and the above values of the a's in (b), we have

(k) $\qquad\qquad y_1 = a_0 x^2 \left(1 - x + \dfrac{x^2}{3} - \dfrac{x^3}{36} - \cdots \right),$

which are the first four terms of a Frobenius series solution of (a). By (c), f_1 and f_2 are polynomials. Their series representations are therefore, by Comment 37.53, valid for all x. Hence, by Theorem 40.32, y_1 is also valid for all x.

By Comment 40.5, a second solution of (a) will have the form [here N, the difference of the roots of (40.38), is one]

(l) $\qquad\qquad y_2(x) = u(x) - b_1 y_1(x) \log x, \quad x > 0,$

where $u(x)$ is a series of the form

(m) $\qquad\qquad u(x) = x(b_0 + b_1 x + b_2 x^2 + \cdots).$

See Exercise 40, 15.

Case 3. Roots of indicial equation (40.38) equal. If the roots of the indicial equation (40.38) are equal, it is evident that only one set of a's and therefore only one Frobenius series solution of (40.33) can be obtained from (40.37).

Example 40.6. Find a general solution of

(a) $$x^2 y'' + xy' + x^2 y = 0.$$

(Note. This equation is known as Bessel's equation of index zero. A fuller discussion of the general Bessel equation of index n will be found in Lesson 42.]

Solution. Division by x^2 and application of Definitions 40.22 and 40.24, shows that $x = 0$ is a regular singularity of (a). Hence we seek a series solution of the form

(b) $$y = x^m (a_0 + a_1 x + a_2 x^2 + \cdots).$$

Comparing (a) with (40.33), we see that

(c) $$f_1(x) = 1, \quad f_2(x) = x^2,$$

both of which are already in series form, valid, by Comment 37.53, for all x. Comparing (c) with (40.34), we find

(d) $$b_0 = 1, \quad c_0 = 0, \quad c_1 = 0, \quad c_2 = 1.$$

All the remaining b's and c's are zero. The indicial equation (40.38) therefore becomes

(e) $$m(m - 1) + m = 0, \quad m^2 = 0,$$

whose roots are $m = 0$ twice. Hence we can expect only one Frobenius solution from these roots.

Using the root $m = 0$, we obtain from the second coefficient in (40.37),

(f) $$a_1(1 + 0) + a_0(0) = 0, \quad a_1 = 0.$$

Setting the coefficient of x^{m+n} in (40.37) equal to zero, we obtain, with the help of (d), the recursion formula

(g) $$a_n[(n)(n - 1) + (1)(n)] + a_{n-1}(0) + a_{n-2}(1) = 0,$$

which simplifies to

(h) $$n^2 a_n = -a_{n-2}, \quad a_n = \frac{-a_{n-2}}{n^2}, \quad n \geq 2.$$

From (f) and (h), when

(i)
$$n = 2: \quad a_2 = -\frac{1}{2^2} a_0,$$

$$n = 3: \quad a_3 = -\frac{1}{3^2} a_1 = 0,$$

$$n = 4: \quad a_4 = -\frac{1}{4^2} a_2 = \frac{1}{2^2 \cdot 4^2} a_0,$$

$$n = 5: \quad a_5 = -\frac{1}{5^2} a_3 = 0,$$

$$n = 6: \quad a_6 = -\frac{1}{6^2} a_4 = -\frac{1}{2^2 \cdot 4^2 \cdot 6^2} a_0.$$

Substituting in (b) these values of the a's and $m = 0$, we obtain the Frobenius series solution

(j) $\quad y_1 = a_0 \left(1 - \frac{x^2}{2^2} + \frac{x^4}{2^2 \cdot 4^2} - \frac{x^6}{2^2 \cdot 4^2 \cdot 6^2} + \frac{x^8}{2^2 \cdot 4^2 \cdot 6^2 \cdot 8^2} - \cdots \right).$

By Theorem 40.32 this series is valid for all x.

A second solution of (a) can be found by means of the substitution (40.51), with $N = 0$. (Remember N is the difference between the roots.) This solution is the coefficient of c_2 in (k) below. Hence the general solution of (a) is

(k) $\quad y = c_1 y_1 + c_2 \left(\frac{x^2}{2^2} - \frac{1 + \frac{1}{2}}{2^2 \cdot 4^2} x^4 + \frac{1 + \frac{1}{2} + \frac{1}{3}}{2^2 \cdot 4^2 \cdot 6^2} x^6 + \cdots \right.$

$$\left. + (-1)^{n+1} \frac{1 + \frac{1}{2} + \cdots + \frac{1}{n}}{2^2 \cdot 4^2 \cdots (2n)^2} x^{2n} + \cdots + y_1 \log x \right), \quad x > 0,$$

where y_1 is given in (j).

EXERCISE 40

1. Determine the singularities of each of the following differential equations. Also indicate whether they are regular or irregular singularities.

(a) $(x - 1)^3 x^2 y'' - 2(x - 1)xy' - 3y = 0.$
(b) $(x - 1)^2 x^4 y'' + 2(x - 1)xy' - y = 0.$
(c) $(x + 1)^2 y'' + xy' - (x - 1)y = 0.$

Verify that the origin is a regular singularity of each of the equations 2–6 and that the roots of the indicial equation (40.38) do not differ by an integer. Find, by the method of Frobenius, two independent solutions of each equation and intervals of convergence.

2. $x^2 y'' + x(x + \frac{1}{2})y' + xy = 0.$
3. $2x^2 y'' + 3xy' + (2x - 1)y = 0.$
4. $2xy'' + (x + 1)y' + 3y = 0.$
5. $2x^2 y'' - xy' + (1 - x^2)y = 0.$
6. $2(x^2 + x^3)y'' - (x - 3x^2)y' + y = 0.$

Verify that the origin is a regular singularity of each of the equations 7–9 and that the roots of the indicial equation (40.38) differ by an integer. Each equation, however, has two independent Frobenius series solutions. Find these solutions and intervals of convergence.

7. $x^2 y'' - x^2 y' + (x^2 - 2)y = 0.$
8. $x^2 y'' + (1 + x^3)xy' - y = 0.$
9. $x^2 y'' + xy' + (x^2 - \frac{1}{9})y = 0.$

Verify that the origin is a regular singularity of each of the equations 10 and 11, that the roots of the indicial equation (40.38) differ by an integer, and that each equation has only one Frobenius series solution. Find this solution and an interval of convergence.

10. $x^2 y'' + xy' + (x^2 - 1)y = 0.$
11. $x^2 y'' + xy' + (x^2 - 4)y = 0.$

Verify that the origin is a regular singularity of each of the equations 12 and 13 and that there is only one root of the indicial equation (40.38). Hence, there is only one Frobenius series solution. Find this solution and an interval of convergence.

12. $x^2 y'' - 3xy' + 4(x + 1)y = 0.$
13. $xy'' + (1 - x)y' + \frac{1}{2}y = 0.$

14. Prove that the substitution in equation (40.33) of $y_2(x)$, $y_2'(x)$, $y_2''(x)$ as determined by (40.51) will yield (40.53). *Hint.* Use the fact that $y_1(x)$ is a solution of (40.33).

15. In Example 40.531, we found one solution of (a), namely

$$y_1(x) = x^2 \left(1 - x + \frac{x^2}{3} - \frac{x^3}{36} - \cdots\right).$$

By use of (40.51), (40.52), and (40.53), find a second solution of (a). *Hint.* See (l) and (m) of the example for the form of the second solution $y_2(x)$ and of $u(x)$. Substitute u, u', u'', y_1, y_1' in (40.53). The functions f_1 and f_2 are given in (c). Equating coefficients of like powers of x, will determine, in terms of b_0, those values of b_1, b_2, $\cdots$ needed in (l) and (m).

16. Follow the instructions given in problem 15 to find the second solution of (a) of Example 40.6. It is given in (k). *Hint.* Each b with odd subscript is zero. Each b with an even subscript has two in its numerator. Break up the two into $1 + 1$. Two fractions thus result. One gives $b_0 y_1$, where y_1 is the solution of (j); the other gives the coefficient of c_2 in (k).

17. With the aid of the hint given in problem 15, find the general solution of each of the following equations.
  (a) $x^2 y'' - 3xy' + 4(x + 1)y = 0.$ See problem 12 for solution $y_1(x)$.
  (b) $x^2 y'' - x(2 - 5x)y' + (x - 6x^2)y = 0.$

18. The method of Frobenius may also be applied to find a particular solution of the nonhomogeneous linear equation

$$(40.61) \qquad x^2 y'' + x f_1(x)y' + f_2(x)y = Q(x),$$

where $f_1(x)$, $f_2(x)$ are analytic at $x = 0$ and $Q(x)$ is a function that can be expressed as a Frobenius series.

$$(40.62) \qquad Q(x) = x^n(b_0 + b_1 x + b_2 x^2 + \cdots).$$

If neither root of the indicial equation (40.38) exceeds $n - 1$ by a positive integer, then a trial solution $y_p(x)$ will be of the same form as $Q(x)$, namely $y_p(x) = x^n(a_0 + a_1 + a_2x^2 + \cdots)$. Substituting $y_p(x)$, $y_p'(x)$, $y_p''(x)$ in the original equation and equating coefficients of like powers of x, will determine $a_0, a_1, a_2, \cdots$. For each of the following equations find the roots of the indicial equation, then verify that these roots do not exceed $n - 1$ by a positive integer, and finally find a particular solution.

(a) $xy'' + y' - 2xy = 4x + x^2$. *Hint.* Multiply the equation by x to determine n, $f_1(x)$, and $f_2(x)$.

(b) $2x^2y'' + xy' + xy = \sin x + 2x^2 \cos x$. *Hint.* Replace $\sin x$ and $\cos x$ by their Maclaurin series expansions.

19. A series solution, valid in a neighborhood of $x = 0$, may not be of much practical use for large values of x since too many terms in the series may be needed to obtain a desired degree of accuracy. In such cases, it is best to make the substitution

(40.63)
$$u = \frac{1}{x}, \quad x = \frac{1}{u}, \quad \frac{du}{dx} = -\frac{1}{x^2} = -u^2,$$

$$\frac{dy}{dx} = \frac{dy}{du}\frac{du}{dx} = -u^2\frac{dy}{du},$$

$$\frac{d^2y}{dx^2} = -2u\frac{du}{dx}\frac{dy}{du} - u^2\frac{d^2y}{du^2}\frac{du}{dx}$$

$$= 2u^3\frac{dy}{du} + u^4\frac{d^2y}{du^2},$$

and solve the resulting equation for y in powers of u. A solution $y(u)$ for a small value of u in the new equation will then correspond to a solution $y(x)$ for a large value of x in the original equation. Using the above substitutions, find a Frobenius series solution, first in powers of u, then in powers of $1/x$, of each of the following.

(a) $2x^2(x - 1)y'' + x(3x + 1)y' - 2y = 0$.

(b) $2x^3y'' + x^2y' + y = 0$.

(c) $x^2y'' + x\left(5 - \frac{1}{x}\right)y' + \left(7 - \frac{2}{x}\right)y = \frac{8}{x} + \frac{4}{x^2} + \frac{2}{x^3}$, y_p only. See problem 18.

Also determine an interval of convergence in each case.

20. A linear differential equation $F_2(x)y'' + F_1(x)y' + F_0(x)y = 0$ is said to have a **regular singularity at** ∞ if the substitution in it of (40.61) results in an equation which has a regular singularity at $u = 0$. Show that each of the following equations

(a) $x^4y'' - 3x^2y' + (1 - x)y = 0$,

(b) the **Legendre equation** $(1 - x^2)y'' - 2xy' + k(k + 1)y = 0$,

has a regular singularity at ∞. In the case of the Legendre equation, find a Frobenius series solution first in powers of u, then in powers of $1/x$. Show that the series is convergent for $|x| > 1$.

21. The following equation

(40.7) $$x(1-x)y'' + [\gamma - (\alpha + \beta + 1)x]y' - \alpha\beta\gamma = 0,$$

where α, β, γ are constants, is known as **Gauss's equation** or as the **hypergeometric equation.** Verify that:

(a) $x = 0$ is a regular singularity of (40.7).

(b) The roots of the indicial equation (40.38) are $m = 0$ and $m = 1 - \gamma$.

(c) $a_{n+1} = \dfrac{(n+m+\alpha)(n+m+\beta)}{(n+m+1)(n+m+\gamma)} a_n, \quad n = 0, 1, 2, \cdots,$

where $a_0, a_1, a_2, \cdots, a_n, \cdots$ are the coefficients in the Frobenius series $y = x^m(a_0 + a_1x + a_2x^2 + a_3x^3 + \cdots), a_0 \neq 0$. *Hint.* Substitute y and its derivatives directly in (40.7) and equate to zero the coefficient of x^{m+n}.

(d) Solutions, using first the root $m = 0$ and then the root $m = 1 - \gamma$, with $a_0 = 1$, are respectively

(40.71) $$y_1 = F(\alpha,\beta,\gamma; x) = 1 + \frac{\alpha\beta}{\gamma} x + \frac{\alpha(\alpha+1)\beta(\beta+1)}{\gamma(\gamma+1)} \frac{x^2}{2!}$$

$$+ \cdots + \frac{\alpha(\alpha+1)\cdots(\alpha+n-1)\beta(\beta+1)\cdots(\beta+n-1)}{\gamma(\gamma+1)\cdots(\gamma+n-1)} \frac{x^n}{n!} + \cdots,$$

$$\gamma \neq 0, -1, -2, -3, \cdots$$

$$y_2 = x^{1-\gamma}\left[1 + \frac{(\alpha-\gamma+1)(\beta-\gamma+1)}{2-\gamma} x + \cdots\right]$$

$$= x^{1-\gamma}F(\alpha-\gamma+1, \beta-\gamma+1, 2-\gamma; x), \quad \gamma \neq 2, 3, 4, \cdots.$$

The notation $F(\alpha - \gamma + 1, \beta - \gamma + 1, 2 - \gamma; x)$ means, replace in the y_1 equation of (40.71), α by $\alpha - \gamma + 1$, β by $\beta - \gamma + 1$, γ by $2 - \gamma$.

(e) Both solutions are the same when $\gamma = 1$. *Hint.* Note that when $\gamma = 1$, the root $m = 1 - \gamma = 0$.

(f) If γ is not an integer, the general solution of (40.7) is

(40.72) $$y = c_1y_1 + c_2y_2,$$

convergent for $|x| < 1$ except perhaps at $x = 0$. If γ is an integer, then only one of y_1 or y_2 is a solution of (40.7).

The series on the right of y_1 in (40.71) is known as the **hypergeometric series.** By Theorem 37.16, it defines, on the interval $|x| < 1$, a function, called the **hypergeometric function,** which we have designated by $F(\alpha,\beta,\gamma; x)$. If α or β is zero or a negative integer, the hypergeometric series terminates and the hypergeometric function is a polynomial.

22. Using the results obtained in problem 21, find a series solution of each of the following differential equations.

(a) $x(1-x)y'' + (\frac{3}{2} - 2x)y' - \frac{1}{4}y = 0$. *Hint.* In (40.7), $\gamma = \frac{3}{2}$, $\alpha + \beta + 1 = 2$, $\alpha\beta = \frac{1}{4}$. Therefore $\alpha = \frac{1}{2}, \beta = \frac{1}{2}, \gamma = \frac{3}{2}$.

(b) $x(1-x)y'' + (2 - 4x)y' - 2y = 0$. *Hint.* In (40.7), $\gamma = 2$, $\alpha + \beta + 1 = 4$, $\alpha\beta = 2$. Therefore $\alpha = 1, \beta = 2, \gamma = 2$.

23. Show that the substitution

$$(40.73) \quad x = (r_2 - r_1)u + r_1, \qquad u = \frac{x - r_1}{r_2 - r_1}, \qquad \frac{du}{dx} = \frac{1}{r_2 - r_1},$$

$$\frac{dy}{dx} = \frac{dy}{du}\frac{du}{dx} = \frac{1}{r_2 - r_1} \cdot \frac{dy}{du},$$

$$\frac{d^2y}{dx^2} = \frac{d}{dx}\left(\frac{1}{r_2 - r_1}\frac{dy}{du}\right) = \frac{1}{r_2 - r_1}\frac{d^2y}{du^2}\frac{du}{dx}$$

$$= \frac{1}{(r_2 - r_1)^2}\frac{d^2y}{du^2}, \quad r_1 \neq r_2,$$

in the equation

$$(40.74) \qquad (x - r_1)(x - r_2)y'' + \frac{b}{a}(x - r_3)y' + \frac{c}{a}y = 0$$

will transform it into the Gauss equation (40.7), namely

$$(40.75) \qquad u(1 - u)\frac{d^2y}{du^2} + \left[\frac{b}{a}\frac{(r_1 - r_3)}{(r_1 - r_2)} - \frac{b}{a}u\right]\frac{dy}{du} - \frac{c}{a}y = 0.$$

A series solution of (40.75) is, therefore, see problem 21,

$$(40.76) \quad y = c_1 F(\alpha,\beta,\gamma; u) + c_2 u^{1-\gamma}F(\alpha - \gamma + 1, \beta - \gamma + 1, 2 - \gamma; u),$$

where

$$(40.77) \qquad \gamma = \frac{b(r_1 - r_3)}{a(r_1 - r_2)}, \qquad \alpha + \beta + 1 = \frac{b}{a}, \qquad \alpha\beta = c/a.$$

Hence, by (40.73) and (40.76), a solution of (40.74) is

$$(40.78) \qquad y = c_1 F\left(\alpha,\beta,\gamma; \frac{x - r_1}{r_2 - r_1}\right)$$

$$+ c_2 \left(\frac{x - r_1}{r_2 - r_1}\right)^{1-\gamma} F\left(\alpha - \gamma + 1, \beta - \gamma + 1, 2 - \gamma; \frac{x - r_1}{r_2 - r_1}\right).$$

The notation $F\left(\alpha,\beta,\gamma; \frac{x - r_1}{r_2 - r_1}\right)$ means replace x in (40.71) by $\frac{x - r_1}{r_2 - r_1}$.

24. With the aid of problem 23, find a series solution of each of the following.

(a) $(x - 2)(x + 1)y'' + \frac{7}{2}(x + 1)y' - \frac{3}{2}y = 0$.

    Hint. $r_1 = 2, r_2 = -1, r_3 = -1, \dfrac{b}{a} = \dfrac{7}{2}, \dfrac{c}{a} = -\dfrac{3}{2}, x = -3u + 2$.

(b) $(x - 2)(x - 1)y'' + 4xy' + 2y = 0$.

    Hint. $r_1 = 2, r_2 = 1, r_3 = 0, a = 1, b = 4, c = 2, x = -u + 2$.

25. Prove each of the following identities.

(a) $F(1,\beta,\beta; x) = \dfrac{1}{1 - x}$.   *Hint.* See answer to problem 22(b).

(b) $F(\alpha,\beta,\beta; x) = (1 - x)^{-\alpha}$.

(c) $xF(1,1,2; -x) = \log(1 + x)$.

(d) $xF(\frac{1}{2},\frac{1}{2},\frac{3}{2}; x^2) = \sin^{-1} x$.   *Hint.* See answer to problem 22(a).

(e) $F(\frac{1}{2},-\frac{1}{2},\frac{1}{2}; x^2) = (1 - x^2)^{1/2}$.

26. With the aid of problem 23, find a series solution of the following equation, known as **Tschebyscheff's equation,**

$$(40.79) \qquad (x-1)(x+1)y'' + xy' - n^2 y = 0.$$

Hint. $r_1 = 1, r_2 = -1, r_3 = 0, a = 1, b = 1, c - -n^2, x - -2u + 1,$ $\alpha = n, \beta = -n, \gamma = \frac{1}{2}.$

27. With the aid of (40.71) and the answer given in problem 26, find the **Tschebyscheff polynomials** $T_0(x), T_1(x), T_2(x), T_3(x), T_4(x).$

ANSWERS 40

1. (a) $x = 0$ regular singularity, $x - 1$ irregular singularity. (b) $x = 0$ irregular singularity, $x = 1$ regular singularity. (c) $x = -1$ irregular singularity.

2. $y_1 = c_1 x^{1/2} \left(1 - x + \dfrac{x^2}{2!} - \dfrac{x^3}{3!} + \cdots \right)$, all x;

$y_2 = c_2(1 - 2x + \frac{4}{3}x^2 - \frac{8}{15}x^3 + \cdots)$, all x.

3. $y_1 = c_1 x^{1/2}(1 - \frac{2}{5}x + \frac{2}{35}x^2 - \frac{4}{945}x^3 + \cdots)$, all x;

$y_2 = c_2 x^{-1}(1 + 2x - 2x^2 + \frac{4}{9}x^3 + \cdots)$, $x \neq 0$.

4. $y_1 = c_1 \left(1 - 3x + 2x^2 - \dfrac{2x^3}{3} + \cdots \right)$, all x;

$y_2 = c_2 x^{1/2} \left(1 - \dfrac{7x}{6} + \dfrac{21x^2}{40} - \dfrac{11x^3}{80} + \cdots \right)$, all x. .

5. $y_1 = c_1 x \left(1 + \dfrac{x^2}{10} + \dfrac{x^4}{360} + \cdots \right)$, all x;

$y_2 = c_2 x^{1/2} \left(1 + \dfrac{x^2}{6} + \dfrac{x^4}{168} + \cdots \right)$, all x.

6. $y = (c_1 x^{1/2} + c_2 x)(1 - x + x^2 - x^3 + \cdots)$, $|x| < 1$.

7. $y_1 = c_1 x^{-1}(1 + \frac{1}{2}x + \frac{1}{2}x^2 - \frac{1}{8}x^4 - \cdots)$, $x \neq 0$;

$y_2 = c_2 x^2(1 + \frac{1}{2}x + \frac{1}{20}x^2 - \frac{1}{60}x^3 - \cdots)$, all x.

8. $y_1 = c_1 x \left(1 - \dfrac{x^3}{15} + \dfrac{x^6}{180} - \cdots \right)$, all x;

$y_2 = c_2 x^{-1} \left(1 + \dfrac{x^3}{3} - \dfrac{x^6}{36} + \cdots \right)$, $x \neq 0$.

9. $y_1 = c_1 x^{1/3} \left[1 - \dfrac{1}{1 + \frac{1}{3}} \left(\dfrac{x}{2} \right)^2 + \dfrac{1}{2(1 + \frac{1}{3})(2 + \frac{1}{3})} \left(\dfrac{x}{2} \right)^4 - \cdots \right]$, all x;

$y_2 = c_2 x^{-1/3} \left[1 - \dfrac{1}{1 - \frac{1}{3}} \left(\dfrac{x}{2} \right)^2 + \dfrac{1}{2(1 - \frac{1}{3})(2 - \frac{1}{3})} \left(\dfrac{x}{2} \right)^4 - \cdots \right]$, $x \neq 0$.

10. $y = cx \left[1 - \dfrac{1}{2} \left(\dfrac{x}{2} \right)^2 + \dfrac{1}{2(2)(3)} \left(\dfrac{x}{2} \right)^4 + \cdots \right]$, all x.

11. $y = cx^2 \left[1 - \dfrac{1}{3} \left(\dfrac{x}{2} \right)^2 + \dfrac{1}{2(3)(4)} \left(\dfrac{x}{2} \right)^4 + \cdots \right]$, all x.

12. $y(x) = cx^2(1 - 4x + 4x^2 - \frac{16}{9}x^3 + \frac{4}{9}x^4 - \cdots)$, all x.

13. $y = c\left(1 - \dfrac{x}{2} - \dfrac{(\frac{1}{2})(\frac{1}{2})}{2^2} x^2 - \dfrac{(\frac{1}{2})(\frac{1}{2})(\frac{3}{2})}{2^2 \cdot 3^2} x^3 + \cdots\right)$, all x.

15. $y = b_0[x(1 + x - \frac{5}{2}x^2 + \frac{43}{36}x^3 - \cdots) - y_1 \log x]$, $x > 0$.

17. (a) $y = y_1 + c_2[x^2(8x - 12x^2 + \frac{176}{27}x^3 + \cdots) + y_1 \log x]$,   where y_1 is given in problem 12.

(b) $y_1 = c_1 x^3(1 - 4x + 9x^2 - \cdots)$,

$y_2 = c_2[(1 + \frac{1}{2}x - \frac{3}{2}x^2 - \frac{13}{2}x^3 + \cdots) + \frac{13}{2}y_1 \log x]$.

18. (a) $y_p = x^2\left(1 + \dfrac{x}{9} + \dfrac{x^2}{8} + \dfrac{2x^3}{225} + \cdots\right)$.

(b) $y_p = x\left(1 + \dfrac{x}{6} - \dfrac{x^2}{45} - \cdots\right)$.

19. (a) $y = c_1\left(1 + 2u + \dfrac{7u^2}{3} + \dfrac{112u^3}{45} + \cdots\right)$

$\qquad\qquad + c_2 u^{1/2}\left(1 + \dfrac{4u}{3} + \dfrac{22u^2}{15} + \dfrac{484u^3}{315} + \cdots\right)$,

$y = c_1\left(1 + \dfrac{2}{x} + \dfrac{7}{3x^2} + \dfrac{112}{45x^3} + \cdots\right)$

$\qquad\qquad + c_2 x^{-1/2}\left(1 + \dfrac{4}{3x} + \dfrac{22}{15x^2} + \dfrac{484}{315x^3} + \cdots\right)$.

The u series converges for $|u| < 1$. Hence the x series converges for $|x| > 1$.

(b) $y = c_1\left(1 - \dfrac{1}{3x} + \dfrac{1}{30x^2} - \dfrac{1}{630x^3} + \cdots\right)$

$\qquad\qquad + c_2 x^{1/2}\left(1 - \dfrac{1}{x} + \dfrac{1}{6x^2} - \dfrac{1}{90x^3} + \cdots\right)$, $x \neq 0$.

(c) $y_p = \left(\dfrac{2}{x} + \dfrac{2}{x^2} + \dfrac{1}{2x^3} + \cdots\right)$, $x \neq 0$.

20. $y_k(x) = a_0 x^k\left[1 - \dfrac{k(k-1)}{2(2k-1)} x^{-2} + \dfrac{k(k-1)(k-2)(k-3)}{2^2 \cdot 2!(2k-1)(2k-3)} x^{-4} + \cdots\right.$

$\qquad + \dfrac{(-1)^n k(k-1)(k-2)(k-3)\cdots(k-2n+1)}{2^n n!(2k-1)(2k-3)\cdots(2k-2n+1)} x^{-2n} + \cdots\Big]$

$\qquad + a_1 x^{-(k+1)}\left[1 + \dfrac{(k+1)(k+2)}{2(2k+3)} x^{-2}\right.$

$\qquad + \dfrac{(k+1)(k+2)(k+3)(k+4)}{2^2 \cdot 2!(2k+3)(2k+5)} x^{-4} + \cdots$

$\qquad + \dfrac{(k+1)(k+2)(k+3)\cdots(k+2n)}{2^n n!(2k+3)(2k+5)\cdots(2k+2n+1)} x^{-2n} \cdots\Big]$.

22. (a) $y = c_1 F(\frac{1}{2}, \frac{1}{2}, \frac{3}{2}; x) + c_2 x^{-1/2} F(0, 0, \frac{1}{2}; x)$

$\qquad = c_1\left(1 + \dfrac{x}{6} + \dfrac{3x^2}{40} + \dfrac{5x^3}{112} + \cdots\right) + c_2 x^{-1/2}$.

(b) $y_1 = c_1 F(1, 2, 2; x) = c_1(1 + x + x^2 + \cdots) = \dfrac{c_1}{1 - x}$.

You can verify that $y_2 = c_2 x^{-1} F(1,2,2; x) = \dfrac{c^2}{x(1-x)}$ is also a solution.
Here $\alpha = 1, \beta = 2, \gamma = 2$. Note that in this special case, the second and subsequent numerators and denominators of the solution y_2 of (40.71) are both zero.

24. (a) The transformed equation is

$$u(1-u)\frac{d^2 y}{du^2} + \left(\frac{7}{2} - \frac{7}{2}u\right)\frac{dy}{du} + \frac{3}{2}y = 0.$$

Here $\alpha = 3, \beta = -\tfrac{1}{2}, \gamma = \tfrac{7}{2}$.

$$y = c_1 F(3,-\tfrac{1}{2},\tfrac{7}{2}; u) + c_2 u^{1-7/2} F(\tfrac{1}{2},-3,-\tfrac{3}{2}; u)$$

$$y = c_1 F\left(3,-\tfrac{1}{2},\tfrac{7}{2}; \frac{2-x}{3}\right) + c_2 \left(\frac{2-x}{3}\right)^{-5/2} F\left(\tfrac{1}{2},-3,-\tfrac{3}{2}; \frac{2-x}{3}\right).$$

(b) $y = c_1 F(1,2,8; 2-x) + c_2 (2-x)^{-7} F(-6,-5,-6; 2-x)$.

26. $y_n(x) = c_1 F\left(n,-n,\tfrac{1}{2}; \dfrac{1-x}{2}\right)$

$$+ c_2 \left(\frac{1-x}{2}\right)^{1/2} F\left(n+\tfrac{1}{2}, -n+\tfrac{1}{2},\tfrac{3}{2}; \frac{1-x}{2}\right).$$

27. $T_0(x) = 1, T_1(x) = x, T_2(x) = 2x^2 - 1, T_3(x) = 4x^3 - 3x,$
$T_4(x) = 8x^4 - 8x^2 + 1.$

LESSON 41.   The Legendre Differential Equation. Legendre Functions. Legendre Polynomials $P_k(x)$. Properties of Legendre Polynomials $P_k(x)$.

LESSON 41A.   The Legendre Differential Equation.   The linear differential equation

$$(41.1) \qquad (1 - x^2)y'' - 2xy' + k(k+1)y = 0,$$

where k is a real constant, occurs in many physical problems. It is known as the **Legendre equation,** named after the French mathematician A. M. Legendre (1752–1833). Because of its great practical importance, this equation has been studied by many mathematicians and a considerable amount of literature is available on the subject.

You can verify by Definition 40.2, that $x = 0$ is an ordinary point of (41.1). This equation can therefore be solved by the power series method outlined in Lesson 37B. Dividing (41.1) by $1 - x^2$ and then comparing the resulting equation with (37.5), we see that

$$(41.11) \quad f_0(x) = \frac{k(k+1)}{1 - x^2} = k(k+1)(1 + x^2 + x^4 + \cdots),$$

$$f_1(x) = -\frac{2x}{1 - x^2} = -2x(1 + x^2 + x^4 + \cdots).$$

Both series are valid for $|x| < 1$. Hence by Theorem 37.51, (41.1) has a series solution in powers of x which is also valid for $|x| < 1$. We therefore seek a Maclaurin series solution of the form

$$(41.12) \qquad y(x) = a_0 + a_1 x + a_2 x^2 + a_3 x^3 + \cdots$$
$$+ a_n x^n + a_{n+1} x^{n+1} + a_{n+2} x^{n+2} + \cdots.$$

Successive differentiations of (41.12) give

$$(41.13) \quad y'(x) = a_1 + 2a_2 x + 3a_3 x^2 + \cdots$$
$$+ n a_n x^{n-1} + (n+1) a_{n+1} x^n + (n+2) a_{n+2} x^{n+1} + \cdots,$$
$$y''(x) = 2a_2 + 2 \cdot 3 a_3 x + \cdots + n(n-1) a_n x^{n-2}$$
$$+ n(n+1) a_{n+1} x^{n-1} + (n+1)(n+2) a_{n+2} x^n + \cdots.$$

As usual y will be a solution of (41.1), if the substitution in it of (41.12) and (41.13) is an identity in x. Making these substitutions, we obtain

$$(41.14) \qquad (1 - x^2)[2a_2 + 2 \cdot 3 a_3 x + \cdots + n(n-1) a_n x^{n-2}$$
$$+ n(n+1) a_{n+1} x^{n-1} + (n+1)(n+2) a_{n+2} x^n \cdots]$$
$$- 2x[a_1 + 2a_2 x + 3a_3 x^2 + \cdots + n a_n x^{n-1} \cdots]$$
$$+ k(k+1)[a_0 + a_1 x + a_2 x^2 + \cdots + a_n x^n + \cdots] = 0.$$

Expanding and collecting coefficients of like powers of x, we have

$$(41.15) \qquad [2a_2 + k(k+1)a_0] + [3!a_3 - 2a_1 + k(k+1)a_1]x$$
$$+ [12a_4 - 6a_2 + k(k+1)a_2]x^2 + \cdots$$
$$+ [(n+1)(n+2)a_{n+2} - n(n+1)a_n + k(k+1)a_n]x^n = 0.$$

Hence we must choose the set of a's so that

$$(41.16) \qquad 2a_2 + \qquad\qquad k(k+1)a_0 = 0,$$
$$2 \cdot 3 a_3 - \qquad 2a_1 + k(k+1)a_1 = 0,$$
$$3 \cdot 4 a_4 - 2 \cdot 3 a_3 + k(k+1)a_2 = 0,$$
$$4 \cdot 5 a_5 - 3 \cdot 4 a_3 + k(k+1)a_3 = 0,$$
$$5 \cdot 6 a_6 - 4 \cdot 5 a_4 + k(k+1)a_4 = 0,$$
$$\cdots \cdots \cdots \cdots \cdots \cdots \cdots \cdots \cdots$$
$$(n+1)(n+2)a_{n+2} - n(n+1)a_n + k(k+1)a_n = 0.$$

Therefore, by (41.16),

$$(41.17) \qquad a_2 = \frac{-k(k+1)}{2} a_0,$$

$$a_3 = \frac{2 - k(k+1)}{3!} a_1 = - \frac{(k-1)(k+2)}{3!} a_1,$$

$$a_4 = \frac{6 - k(k+1)}{12}\, a_2 = -\frac{(k-2)(k+3)}{12}\, a_2$$

$$= \frac{k(k+1)(k-2)(k+3)}{4!}\, a_0,$$

$$a_5 = \frac{12 - k(k+1)}{20}\, a_3 = -\frac{(k-3)(k+4)}{20}\, a_3$$

$$= \frac{(k-1)(k+2)(k-3)(k+4)}{5!}\, a_1,$$

$$a_6 = \frac{20 - k(k+1)}{30}\, a_4 = -\frac{(k-4)(k+5)}{30}\, a_4$$

$$= \frac{-k(k+1)(k-2)(k+3)(k-4)(k+5)}{6!}\, a_0,$$

. . . .

$$a_{2n} = \frac{(-1)^n k(k+1)(k-2)(k+3)(k-4)(k+5)\cdots(k+2n-1)}{(2n)!}\, a_0,$$

$$a_{2n+1} = \frac{(-1)^n(k-1)(k+2)(k-3)(k+4)(k-5)(k+6)\cdots(k+2n)}{(2n+1)!}\, a_1.$$

Substituting (41.17) in (41.12), we obtain

(41.18) $$y(x) = a_0 S_k(x) + a_1 T_k(x),$$

where a_0 and a_1 are arbitrary, and

(41.2)
$$S_k(x) = \left[1 - \frac{k(k+1)}{2!}\, x^2 + \frac{k(k+1)(k-2)(k+3)}{4!}\, x^4 + \cdots \right.$$
$$\left. + \frac{(-1)^n k(k+1)(k-2)(k+3)(k-4)\cdots(k+2n-1)}{(2n)!}\, x^{2n} + \cdots \right],$$

(41.21)
$$T_k(x) = \left[x - \frac{(k-1)(k+2)}{3!}\, x^3 + \frac{(k-1)(k+2)(k-3)(k+4)}{5!}\, x^5 + \cdots \right.$$
$$\left. + \frac{(-1)^n(k-1)(k+2)(k-3)(k+4)\cdots(k+2n)}{(2n+1)!}\, x^{2n+1} + \cdots \right].$$

LESSON 41B. Comments on the Solution (41.18) of the Legendre Equation (41.1). Legendre Functions. Legendre Polynomials $P_k(x)$. The following comments apply to the solution (41.18).

1. The solution (41.18) is valid for $|x| < 1$. See remarks immediately following (41.11).
2. When k is zero or a positive even integer, 2, 4, 6, $\cdots$, or a negative odd integer, $-1, -3, \cdots$, the series in (41.2) terminates, i.e., it becomes a

polynomial in x. For example, when $k = 2$, the third and all succeeding terms in (41.2) are zero; when $k = -1$, the second and all succeeding terms are zero.

3. When k is a positive odd integer, $1, 3, 5, \cdots$, or a negative even integer, $-2, -4, \cdots$, the series in (41.21) terminates, i.e., it becomes a polynomial in x. For example, when $k = 3$, the third and all succeeding terms in (41.21) are zero. When $k = -2$, the second and all succeeding terms are zero.

4. If k is not an integer or zero, both series (41.2) and (41.21) are nonterminating. By Theorem 37.16, therefore, each defines, for each k, a continuous function on $I: |x| < 1$. These functions are called **Legendre functions.**

5. Since a Taylor series is a special case of a Frobenius series, the Legendre functions (or a Legendre function and a polynomial) defined by the series (41.2) and (41.21) are, by Theorem 40.39, linearly independent. And since each is a solution of (41.1), it follows by Theorem 19.3 and Comment 19.41, that $y(x)$ of (41.18) is a *general solution* of (41.1) on $I: |x| < 1$.

Because of Comments 2 and 3 above, we can write polynomial solutions of the Legendre differential equation (41.1) when k is zero or an integer. All we need do is select the proper series (41.2) or (41.21) for the given k. In Table 41.22, we have listed solutions of (41.1) for a few integer values of k and for $k = 0$.

Table 41.22

If $k =$	The Legendre Equation (41.1) Becomes	A Solution by (41.2) or (41.21) Is
0	$(1 - x^2)y'' - 2xy' = 0$	$y_0(x) = 1$
1	$(1 - x^2)y'' - 2xy' + 2y = 0$	$y_1(x) = x$
2	$(1 - x^2)y'' - 2xy' + 6y = 0$	$y_2(x) = 1 - 3x^2$
3	$(1 - x^2)y'' - 2xy' + 12y = 0$	$y_3(x) = x - \frac{5}{3}x^3$
4	$(1 - x^2)y'' - 2xy' + 20y = 0$	$y_4(x) = 1 - 10x^2 + \frac{35}{3}x^4$

If we solve the last equation in (41.16) for a_{n+2}, we obtain

$$(41.23) \quad a_{n+2} = \frac{[n(n + 1) - k(k + 1)]}{(n + 1)(n + 2)} a_n, \quad n = 0, 1, 2, 3, \cdots.$$

Since the numerator of (41.23) is equal to $-(k - n)(k + n + 1)$, we can write (41.23) as

$$(41.24) \quad a_{n+2} = -\frac{(k - n)(k + n + 1)}{(n + 1)(n + 2)} a_n.$$

Therefore by (41.24), if k is an integer, and

(41.25) $n = k - 2$: $a_k = -\dfrac{2(2k - 1)}{k(k - 1)}\, a_{k-2},$

$\qquad\qquad n = k - 4$: $a_{k-2} = -\dfrac{4(2k - 3)}{(k - 2)(k - 3)}\, a_{k-4},$

$\cdots\cdots\cdots\cdots\cdots\cdots\cdots\cdots\cdots\cdots$

Therefore by (41.25),

(41.26) $a_{k-2} = -\dfrac{k(k - 1)}{2(2k - 1)}\, a_k.$

$\qquad a_{k-4} = -\dfrac{(k - 2)(k - 3)}{4(2k - 3)}\, a_{k-2} = \dfrac{k(k - 1)(k - 2)(k - 3)}{2 \cdot 4(2k - 1)(2k - 3)}\, a_k,$

$\cdots\cdots\cdots\cdots\cdots\cdots\cdots\cdots\cdots\cdots\cdots\cdots$

Substituting (41.26) in (41.12), we obtain polynomial solutions $p_k(x)$ of Legendre's equation (41.1) of the form [write $y(x)$ in descending powers of x, namely $y(x) = \cdots a_k x^k + a_{k-2} x^{k-2} + a_{k-4} x^{k-4} + \cdots$]

(41.27)

$$p_k(x) = a_k\left[x^k - \frac{k(k - 1)}{2(2k - 1)}\, x^{k-2} + \frac{k(k - 1)(k - 2)(k - 3)}{2 \cdot 4(2k - 1)(2k - 3)}\, x^{k-4}\right.$$
$$\left. - \frac{k(k - 1)(k - 2)(k - 3)(k - 4)(k - 5)}{2 \cdot 4 \cdot 6(2k - 1)(2k - 3)(2k - 5)}\, x^{k-6} + \cdots\right],$$
$$k = 0, 1, 2, \cdots,$$

where a_k is an arbitrary constant. Let us take for it the particular value

(41.3)
$$a_k = \frac{(2k)!}{2^k (k!)^2} = \frac{1 \cdot 2 \cdot 3 \cdot 4 \cdots (2k - 1)(2k)}{k!\,2 \cdot 4 \cdot 6 \cdots (2k)} = \frac{1 \cdot 3 \cdot 5 \cdots (2k - 1)}{k!}.$$

(The reason for selecting this value of a_k will be made clear in Lesson 41C-A.) To distinguish between the polynomial solutions $p_k(x)$ of (41.27) and those we shall obtain by giving a_k the value (41.3), we replace p_k by P_k. Hence by (41.3) and (41.27)

(41.31) $P_k(x) = \dfrac{1 \cdot 3 \cdot 5 \cdots (2k - 1)}{k!}\left[x^k - \dfrac{k(k - 1)}{2(2k - 1)}\, x^{k-2}\right.$

$$+ \frac{k(k - 1)(k - 2)(k - 3)}{2 \cdot 4(2k - 1)(2k - 3)}\, x^{k-4}$$

$$- \frac{k(k - 1)(k - 2)(k - 3)(k - 4)(k - 5)}{2 \cdot 4 \cdot 6(2k - 1)(2k - 3)(2k - 5)}\, x^{k-6} + \cdots$$

$$\left. + \frac{(-1)^n k(k - 1)(k - 2) \cdots (k - 2n + 1)}{2^n n!(2k - 1)(2k - 3) \cdots (2k - 2n + 1)}\, x^{k-2n} + \cdots\right],$$

$$k = 0, 1, 2, \cdots.$$

By definition, $s! = 1 \cdot 2 \cdot 3 \cdots s$. Hence,

(41.311) $(s+1)(s+2) \cdots (s+n)$

$$= \frac{1 \cdot 2 \cdot 3 \cdots s(s+1) \cdots (s+n)}{1 \cdot 2 \cdot 3 \cdots s} = \frac{(s+n)!}{s!}.$$

By (41.311)

(41.312) $(2s+2)(2s+4) \cdots (2s+2n)$

$$= 2^n(s+1) \cdots (s+n) = 2^n \frac{(s+n)!}{s!}.$$

By (4.311) and (41.312),

(41.313) $(2s+1)(2s+3) \cdots (2s+2n-1)$

$$= \frac{(2s+1)(2s+2)(2s+3)(2s+4) \cdots (2s+2n-1)(2s+2n)}{(2s+2) \, . \qquad (2s+4) \cdots \qquad\qquad (2s+2n)}$$

$$= \frac{\dfrac{(2s+2n)!}{(2s)!}}{\dfrac{2^n(s+n)!}{s!}} = \frac{s!(2s+2n)!}{2^n(2s)!(s+n)!}.$$

In (41.313), let $s = k - n$. There results

(41.314) $(2k-2n+1)(2k-2n+3) \cdots (2k-1) = \dfrac{(k-n)!(2k)!}{2^n(2k-2n)!k!}.$

By the definition of $k!$,

(41.315) $k(k-1)(k-2) \cdots (k-2n+1) = \dfrac{k!}{(k-2n)!}.$

Substituting (41.315) and (41.314) in the coefficient of x^{k-2n} of (41.31), we obtain for the coefficient

(41.316) $\dfrac{(-1)^n k!}{(k-2n)!} \dfrac{2^n(2k-2n)!k!}{2^n n!(k-n)!(2k)!} = \dfrac{(-1)^n(k!)^2(2k-2n)!}{n!(2k)!(k-2n)!(k-n)!}.$

Since (41.316) is the coefficient of the general term of (41.31), we can write (41.31), with $1 \cdot 3 \cdot 5 \cdots (2k-1)/k!$ replaced by its equivalent factorial form as given in (41.3) after a_k, as

(41.317) $P_k(x) = \displaystyle\sum_{n=0}^{[k/2]} \frac{1}{2^k} \frac{(-1)^n(2k-2n)!}{n!(k-2n)!(k-n)!} x^{k-2n}.$

The symbol $[k/2]$ means the largest integer in $k/2$. For example, if $k = 3$, the largest integer in $k/2$ is one. Hence to evaluate $P_3(x)$, replace k by 3 in (41.317) and sum the two terms obtained with $n = 0$ and $n = 1$. If $k = 4$, the largest integer in $k/2$ is 2.

By (41.31) or (41.317), when

(41.32) $k = 0$: $P_0(x) = 1.$

$k = 1$: $P_1(x) = x.$

$k = 2$: $P_2(x) = \dfrac{1 \cdot 3}{2}\left(x^2 - \dfrac{1}{3}\right) = \dfrac{1}{2}(3x^2 - 1).$

$k = 3$: $P_3(x) = \dfrac{1 \cdot 3 \cdot 5}{3!}\left(x^3 - \dfrac{3x}{5}\right) = \dfrac{1}{2}(5x^3 - 3x).$

$k = 4$: $P_4(x) = \dfrac{1 \cdot 3 \cdot 5 \cdot 7}{4!}\left(x^4 - \dfrac{6x^2}{7} + \dfrac{3}{35}\right)$

$= \dfrac{1}{8}(35x^4 - 30x^2 + 3).$

The polynomials in (41.32) are known as **Legendre polynomials.**

If in the recursion formula (41.24), we let $n = -k - 3, -k - 5$, etc., we obtain, in place of (41.27),

(41.33) $q_k(x) = a_{-k-1}\left[x^{-k-1} + \dfrac{(k + 1)(k + 2)}{2(2k + 3)}x^{-k-3}\right.$

$\left. + \dfrac{(k + 1)(k + 2)(k + 3)(k + 4)}{2 \cdot 4(2k + 3)(2k + 5)}x^{-k-5} + \cdots\right].$

Since a_{-k-1} is an arbitrary constant we may take for it the particular value

(41.34) $a_{-k-1} = \dfrac{2^k(k!)^2}{(2k + 1)!}$

$= \dfrac{k!2 \cdot 4 \cdot 6 \cdots (2k)}{1 \cdot 2 \cdot 3 \cdots (2k)(2k + 1)} = \dfrac{k!}{1 \cdot 3 \cdot 5 \cdots (2k + 1)},$

$k \neq -1, -2, \cdots.$

The functions

(41.35) $Q_k(x) = \dfrac{k!}{1 \cdot 3 \cdot 5 \cdots (2k + 1)}$

$\times\left[x^{-k-1} + \dfrac{(k + 1)(k + 2)}{2(2k + 3)}x^{-k-3}\right.$

$\left. + \dfrac{(k + 1)(k + 2)(k + 3)(k + 4)}{2 \cdot 4(2k + 3)(2k + 5)}x^{-k-5} + \cdots\right],$

$k \neq -1, -2, \cdots,$

obtained from (41.33) by replacing a_{-k-1} by its value in (41.34) are known as **Legendre functions of the second kind.**

LESSON 41C.   Properties of Legendre Polynomials $P_k(x)$. The Legendre polynomials are important primarily in the interval $I: |x| \leq 1$. The graphs of the first five Legendre polynomials in the interval $0 \leq x \leq 1$ are shown in Fig. 41.4.

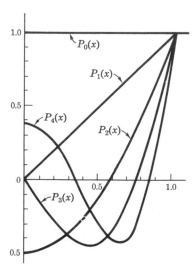

Figure 41.4

It has been proved that a Legendre polynomial of degree n has exactly n distinct real zeros in the interval $-1 \leq x \leq 1$. If we had extended the graphs shown in Fig. 41.4 to include the interval $-1 \leq x \leq 0$, we would have found that $P_0(x)$ has no zeros, $P_1(x)$ has one zero, $P_2(x)$ has two zeros, $P_3(x)$ has three zeros, and $P_4(x)$ has four zeros in the interval $-1 \leq x \leq 1$.

In the remainder of this lesson we shall develop a number of properties of Legendre polynomials which are valid in the interval $I: |x| \leq 1$.

A.   The Legendre polynomials $P_n(x)$ are the coefficients of t^n in the Maclaurin series expansion of $(1 - 2xt + t^2)^{-1/2}$, i.e.,

$$(41.41) \quad (1 - 2xt + t^2)^{-1/2} = P_0(x) + P_1(x)t + P_2(x)t^2 + \cdots$$
$$+ P_n(x)t^n + \cdots,$$

if $|x| \leq 1, |t| < 1.$

Proof. We shall merely give the formal steps of the proof, without attempting to justify each of the steps used. A rigorous proof is beyond the scope of this text. The Maclaurin series expansion of $(1 + u)^k$, called a **binomial series,** is

(a) $\quad (1 + u)^k = 1 + ku + \dfrac{k(k-1)}{2!}\, u^2 + \cdots$

$$+ \dfrac{k(k-1)\cdots(k-n+1)}{n!}\, u^n + \cdots,$$

valid for those values of u and k for which the series converges. If $k = -\frac{1}{2}$, the series converges for $|u| < 1$. Let us now write the left-hand member of (41.41) as $[1 + t(t - 2x)]^{-1/2}$. Hence by (a), if $|t^2 - 2xt| < 1$,

(b) $\quad [1 + t(t - 2x)]^{-1/2} = 1 + \dfrac{1}{2}\, t(2x - t) + \dfrac{1}{2^2}\dfrac{3}{2}\, t^2(2x - t)^2 + \cdots$

$$+ \dfrac{1 \cdot 3 \cdot 5 \cdots (2n-1)}{2^n n!}\, t^n(2x - t)^n + \cdots$$

$$= 1 + \dfrac{1}{2}\, (2xt - t^2) + \dfrac{3}{2^2 2}\, t^2(4x^2 - 4xt + t^2) + \cdots$$

$$+ \dfrac{1 \cdot 3 \cdot 5 \cdots (2n-1)}{2^n n!}\, t^n(2x - t)^n + \cdots.$$

From the last expression on the right of (b), we see that the coefficients of t^0, t, and t^2 are respectively

(c) $\qquad 1, \quad \dfrac{1}{2}\, (2x) = x, \quad -\dfrac{1}{2} + \dfrac{3}{2^2 2}\, (4x^2) = \dfrac{1}{2}\, (3x^2 - 1).$

A comparison of (c) with (41.32) shows that these coefficients are respectively P_0, P_1, P_2. The coefficient of the general term t^n is the sum of the coefficients of t^n in the last term on the right of (b) and of coefficients of t^n in preceding terms. Hence the total coefficient of t^n is

(d) $\quad \dfrac{1 \cdot 3 \cdot 5 \cdots (2n-1)}{2^n n!}\, (2x)^n - \dfrac{1 \cdot 3 \cdot 5 \cdots (2n-3)}{2^{n-1}(n-1)!}\, \dfrac{(n-1)}{1!}\, (2x)^{n-2}$

$$+ \dfrac{1 \cdot 3 \cdot 5 \cdots (2n-5)}{2^{n-2}(n-2)!}\, \dfrac{(n-2)(n-3)}{2!}\, (2x)^{n-4} - \cdots.$$

The last term in (d) can be written as

(e) $\quad \dfrac{1 \cdot 3 \cdot 5 \cdots (2n-5)(2n-3)(2n-1)}{2^{n-2}} \cdot \dfrac{(n)(n-1)}{(2n-3)(2n-1)} \cdot \dfrac{(n)(n-1)}{(n)(n-1)(n-2)!}$

$$\times \dfrac{(n-2)(n-3)}{2!}\, 2^{n-4} x^{n-4}$$

$$= \dfrac{1 \cdot 3 \cdot 5 \cdots (2n-1)}{n!} \left[\dfrac{n(n-1)(n-2)(n-3)}{2 \cdot 4(2n-1)(2n-3)}\, x^{n-4} \right].$$

Hence (d) becomes, with its last term replaced by the expression on the right of (e),

(f)
$$\frac{1 \cdot 3 \cdot 5 \cdots (2n-1)}{n!}$$
$$\times \left[x^n - \frac{n(n-1)}{2(2n-1)} x^{n-2} + \frac{n(n-1)(n-2)(n-3)}{2 \cdot 4(2n-1)(2n-3)} x^{n-4} + \cdots \right].$$

A comparison of (f), which is the coefficient $P_n(x)$ of t^n of (41.41), with the polynomial $P_k(x)$ of (41.31) shows that with k replaced by n, the first three terms of each are alike. If we had used more terms of the binomial series (d), we would have obtained additional terms in (f) in agreement with (41.31). *Note also the similarity of the coefficient in* (f) *with the value of a_k as given in* (41.3). This value was in fact given to a_k in order to obtain the identity (41.41).

B. Values of $P_n(0)$, $P_n(1)$, and $P_n(-1)$. If $x = 1$, the left side of (41.41) simplifies to

(a) $$(1 - 2t + t^2)^{-1/2} = (1 - t)^{-1}.$$

Its series expansion is

(b) $$1 + t + t^2 + t^3 + \cdots + t^n + \cdots, \quad |t| < 1.$$

Comparing the coefficients of this series with the coefficients of the series on the right of (41.41), we see that, with $x = 1$,

(41.42) $P_0(1) = 1,$ $P_1(1) = 1,$ $P_2(1) = 1, \cdots,$ $\boldsymbol{P_n(1) = 1.}$

If $x = 0$, the left side of (41.41) becomes

(c) $$(1 + t^2)^{-1/2},$$

whose series expansion is

(d)
$$1 - \tfrac{1}{2}t^2 + \tfrac{1}{2} \cdot \tfrac{3}{4}t^4 - \cdots + (-1)^n \frac{1 \cdot 3 \cdots (2n-1)}{2 \cdot 4 \cdots (2n)} t^{2n} + \cdots, \quad |t| < 1.$$

A comparison of this series with the right side of (41.41) shows that, with $x = 0$,

(41.43) $P_1(0) = 0,$ $P_3(0) = 0, \cdots,$ $\boldsymbol{P_{2n-1}(0) = 0.}$

(41.44) $P_0(0) = 1,$ $P_2(0) = -\tfrac{1}{2},$ $P_4(0) = \tfrac{1}{2} \cdot \tfrac{3}{4}, \cdots,$

$$\boldsymbol{P_{2n}(0) = (-1)^n \frac{1 \cdot 3 \cdots (2n-1)}{2 \cdot 4 \cdots (2n)}.}$$

If $x = -1$, the left side of (41.41) becomes

(e) $(1 + 2t + t^2)^{-1/2} = (1 + t)^{-1} = 1 - t + t^2 - t^3 + t^4 - \cdots,$
$$|t| < 1.$$

Comparing (e) with the right side of (41.41), we find, with $x = -1$,

(41.45) $P_0(-1) = 1, \quad P_1(-1) = -1, \quad P_2(-1) = 1,$
$$P_3(-1) = -1, \cdots, \quad P_n(-1) = (-1)^n.$$

C.   Recursion formula for $P_n(x)$. Differentiating (41.41) with respect to t, we obtain

(a)
$$-\tfrac{1}{2}(1 - 2xt + t^2)^{-3/2}(-2x + 2t) = P_1 + 2P_2 t + \cdots + nP_n t^{n-1} + \cdots.$$

Multiplying (a) by $1 - 2xt + t^2$, there results

(b) $(x - t)(1 - 2xt + t^2)^{-1/2} = (1 - 2xt + t^2)$
$$\times [P_1 + + 2P_2 t + \cdots + (n - 1)P_{n-1}t^{n-2} + nP_n t^{n-1}$$
$$+ (n + 1)P_{n+1}t^n + \cdots].$$

By (41.41), we can write (b) as

(c) $x[P_0 + P_1 t + \cdots + P_n t^n + \cdots] - [P_0 t + P_1 t^2 + \cdots + P_{n-1}t^n + \cdots]$
$$= [P_1 + 2P_2 t + \cdots + (n + 1)P_{n+1}t^n + \cdots]$$
$$- 2x[P_1 t + \cdots + nP_n t^n + \cdots]$$
$$+ [P_1 t^2 + \cdots + (n - 1)P_{n-1}t^n + \cdots].$$

Equating the coefficient of t^n on both sides of the equal sign, we obtain

(d) $xP_n - P_{n-1} = (n + 1)P_{n+1} - 2xnP_n + (n - 1)P_{n-1},$

which simplifies to the **recursion formula**

(41.46) $(n + 1)P_{n+1}(x) - (2n + 1)xP_n(x) + nP_{n-1}(x) = 0,$
$$n = 1, 2, \cdots.$$

It is easily verified, by (41.32), that

(41.461) $P_1(x) - xP_0(x) \qquad\qquad = 0,$

and, by (41.46), that when

(41.47) $n = 1$: $2P_2(x) - 3xP_1(x) + P_0(x) \;\; = 0,$
$$n = 2: \quad 3P_3(x) - 5xP_2(x) + 2P_1(x) = 0,$$

etc.   We leave it to you as an exercise to verify, by means of (41.32), that the formulas in (41.47) are indeed true equations.

D. Rodrigue's Formula. Another compact formula for expressing the Legendre polynomials $P_n(x)$ is

(41.48) $$P_n(x) = \frac{1}{n!2^n} \frac{d^n}{dx^n} (x^2 - 1)^n.$$

It is known as Rodrigue's formula.

Proof of the Formula. The binomial series expansion of $(x - 1)^n$ can be written as

(a) $$(x - 1)^n = \sum_{k=0}^{n} (-1)^n \frac{n!}{k!(n-k)!} x^{n-k}.$$

Hence

(b) $$(x^2 - 1)^n = \sum_{k=0}^{n} (-1)^n \frac{n!}{k!(n-k)!} (x^2)^{n-k}.$$

Taking n successive derivatives of x^{2n-2k}, we obtain for the nth derivative,

(c) $$\frac{d^n}{dx^n} (x^{2n-2k}) = (2n - 2k)(2n - 2k - 1) \cdots (n - 2k + 1)x^{n-2k},$$

$$= 0, \qquad\qquad \begin{matrix} 2k \leqq n, \\ n < 2k. \end{matrix}$$

By definition of the factorial function, (c) can be written as

(d) $$\frac{d^n}{dx^n} (x^{2n-2k}) = \frac{(2n - 2k)!}{(n - 2k)!} x^{n-2k}, \quad 2k \leqq n,$$

$$= 0, \qquad\qquad n < 2k.$$

Hence by (b) and (d),

(e) $$\frac{d^n}{dx^n} (x^2 - 1)^n = \sum_{k=0}^{[n/2]} (-1)^n \frac{n!}{k!(n-k)!} \frac{(2n - 2k)!}{(n - 2k)!} x^{n-2k}.$$

Therefore by (41.48) and (e),

(f) $$P_n(x) = \sum_{k=0}^{[n/2]} (-1)^n \frac{1}{n!2^n} \frac{n!}{k!(n-k)!} \frac{(2n - 2k)!}{(n - 2k)!} x^{n-2k}$$

$$= \sum_{k=0}^{[n/2]} (-1)^n \frac{(2n - 2k)!}{2^n k!(n - k)!(n - 2k)!} x^{n-2k}.$$

A comparison of (f) with (41.317) shows they are alike. (Interchange n and k in either equation.)

E. Orthogonal Property of Legendre Polynomials.

Definition 41.5. A set of functions $f_1, f_2, \cdots$, is said to be **orthogonal** on an interval $I: a \leqq x \leqq b$, if, for every two distinct functions of the set,

(41.51) $$\int_a^b f_m(x)f_n(x)\,dx = 0, \quad m \neq n.$$

We shall now prove that the set of Legendre polynomials is orthogonal on the interval $I: -1 \leq x \leq 1$, i.e., we shall prove

(41.511) $$\int_{-1}^1 P_m(x)P_n(x)\,dx = 0, \quad m \neq n.$$

Proof. Each Legendre polynomial $P_m(x)$ satisfies the Legendre equation (41.1). Therefore

(a) $$(1 - x^2)P_m''(x) - 2xP_m'(x) + m(m + 1)P_m(x) = 0,$$

which can be written as

(b) $$\frac{d}{dx}\,[(1 - x^2)P_m'(x)] + m(m + 1)P_m(x) = 0.$$

Multiplying (b) by $P_n(x)$, $n \neq m$, and integrating between the limits -1, 1, we obtain

(c)
$$\int_{-1}^1 P_n(x)\,\frac{d}{dx}\,[(1 - x^2)P_m'(x)]\,dx + m(m + 1)\int_{-1}^1 P_n(x)P_m(x)\,dx = 0.$$

The first integral in (c) can be integrated by parts in accordance with the formula

(d) $$\int u\,dv = uv - \int v\,du.$$

In the first integral of (c), let $u = P_n(x)$ and dv equal the rest of the integrand. Then by (d), this integral becomes

(e) $$P_n[(1 - x^2)P_m']\Big|_{-1}^1 - \int_{-1}^1 (1 - x^2)P_m'P_n'\,dx.$$

You can easily verify that the first term in (e) is zero. Therefore, by (e), (c) simplifies to

(f) $$-\int_{-1}^1 (1 - x^2)P_m'P_n'\,dx + m(m + 1)\int_{-1}^1 P_nP_m\,dx = 0.$$

If we had started with $P_n(x)$ in (a) instead of with $P_m(x)$ and followed the steps outlined above, we would have obtained the same result as in (f) with m and n interchanged. Hence also

(g) $\qquad -\int_{-1}^{1} (1 - x^2) P_n' P_m' \, dx + n(n + 1) \int_{-1}^{1} P_m P_n \, dx = 0.$

Subtracting (g) from (f), there results, with $m \neq n$,

(h) $\qquad (m^2 + m - n^2 - n) \int_{-1}^{1} P_m(x) P_n(x) \, dx = 0.$

Since $m^2 + m - n^2 - n = (m - n)(m + n + 1)$ and $m \neq n$, we may divide (h) by this term to obtain (41.511).

F. Other Integral Properties of Legendre Polynomials. The first integral property we shall prove is,

$$(41.52) \qquad\qquad \int_{-1}^{1} R_m(x) P_n(x) \, dx = 0,$$

where $R_m(x)$ is a polynomial of degree m less than n.

Proof of (41.52). Let

(a) $\qquad R_m(x) = a_0 + a_1 x + a_2 x^2 + \cdots + a_{m-1} x^{m-1} + a_m x^m.$

The polynomial $R_m(x)$ can always be written as a linear combination of the Legendre polynomials $P_0, P_1, \cdots, P_m$, i.e., we can always write

(b) $\qquad R_m(x) = a_0 + a_1 x + a_2 x^2 + \cdots + a_{m-1} x^{m-1} + a_m x^m$

$\qquad\qquad\quad = C_0 P_0 + C_1 P_1 + \cdots + C_{m-1} P_{m-1} + C_m P_m.$

The truth of this statement can be demonstrated as follows. P_m is the only polynomial on the right of (b) which has a term in x^m. Hence we can choose C_m so that the coefficient of the x^m term of P_m is equal to a_m. Next we sum the two coefficients of x^{m-1} in P_{m-1} and P_m, and give C_{m-1} a value so that C_{m-1} times this sum equals a_{m-1}, etc. In (41.52) replace $R_m(x)$ by its equal last expression in (b). We thus obtain

(c) $\quad C_0 \int_{-1}^{1} P_0 P_n \, dx + C_1 \int_{-1}^{1} P_1 P_n \, dx + \cdots + C_m \int_{-1}^{1} P_m P_n \, dx, \quad m < n.$

By (41.511), each term in (c) is zero. Hence (41.52) follows.

The second integral property we shall prove is

$$(41.53) \qquad\qquad \int_{-1}^{1} [P_n(x)]^2 \, dx = \frac{2}{2n + 1}.$$

Proof of (41.53). Squaring both sides of (41.41), we obtain

(a) $$(1 - 2xt + t^2)^{-1} = \left[\sum_{n=0}^{\infty} P_n(x)t^n \right]^2 .$$

Integration of (a) with respect to x between the limits -1 and 1 gives

(b) $$\int_{-1}^{1} (1 - 2xt + t^2)^{-1}\, dx = \int_{-1}^{1} \left[\sum_{n=0}^{\infty} P_n(x)t^n \right]^2 dx.$$

When we square the integrand on the right of (b) and perform the integration, the only terms, by (41.511), which are not zero are those in which the subscripts of $P_n(x)$ are the same. Hence (b) simplifies to

(c) $$\int_{-1}^{1} (1 - 2xt + t^2)^{-1}\, dx = \sum_{n=0}^{\infty} t^{2n} \int_{-1}^{1} [P_n(x)]^2\, dx.$$

Integration of the left side of (c) results in (remember t is a constant in this integration and $\log (1 + t) = t - \frac{1}{2}t^2 + \frac{1}{3}t^3 - \cdots$)

(d)
$$-\frac{1}{2t} \log (1 - 2xt + t^2) \Big|_{-1}^{1} = -\frac{1}{2t} \log \frac{(1-t)^2}{(1+t)^2} = \frac{1}{t} \log \frac{1+t}{1-t}$$
$$= \frac{2}{t} \left[t + \frac{t^3}{3} + \frac{t^5}{5} + \frac{t^7}{7} + \cdots \right]$$
$$= 2 \left[1 + \frac{t^2}{3} + \frac{t^4}{5} + \cdots + \frac{t^{2n}}{2n+1} + \cdots \right], \ |t| < 1.$$

Substituting the last expression of (d) for the left side of (c), we have

(e) $$\sum_{n=0}^{\infty} \frac{2t^{2n}}{2n+1} = \sum_{n=0}^{\infty} t^{2n} \int_{-1}^{1} [P_n(x)]^2\, dx.$$

Equating coefficients of t^{2n} in (e), we obtain for each n,

(f) $$\int_{-1}^{1} [P_n(x)]^2\, dx = \frac{2}{2n+1}.$$

EXERCISE 41

1. Find the general solution of each of the following Legendre equations.

   (a) $(1 - x^2)y'' - 2xy' + 2y = 0.$ (b) $(1 - x^2)y'' - 2xy' + 6y = 0.$

2. Find a polynomial solution of the Legendre equation $(1 - x^2)y'' - 2xy' + 30y = 0.$

3. Find the Legendre polynomials $P_5(x)$, $P_6(x)$, $P_7(x)$.

4. Find the general solution of $(1 - x^2)y'' - 2xy' + \frac{3}{4}y = 0$.
5. Verify by direct substitution in (41.32) that
   $P_0(1) = 1, P_1(1) = 1, P_2(1) = 1, P_3(1) = 1, P_4(1) = 1;$
   $P_0(-1) = 1, P_1(-1) = -1, P_2(-1) = 1, P_3(-1) = -1, P_4(-1) = 1;$
   $P_0(0) = 1, P_1(0) = 0, P_2(0) = -\frac{1}{2}, P_3(0) = 0, P_4(0) = \frac{3}{8}.$
6. Verify, by use of (41.32), the accuracy of the equations in (41.47).
7. Express x^3 as a linear combination of Legendre polynomials.
8. Express $x^4 + 2x^3 + 2x^2 - x - 3$ as a linear combination of Legendre polynomials.
9. Assume that a function $f(x)$, defined in the interval $(-1,1)$, can be represented by a series of Legendre polynomials, i.e., assume

(a) $$f(x) = c_0 P_0(x) + c_1 P_1(x) + c_2 P_2(x) + \cdots.$$

   Show that if this series, after multiplication by $P_k(x)$, can be integrated term by term, then the coefficients $c_0, c_1, \cdots$, are given by

   $$c_k = \frac{2k + 1}{2} \int_{-1}^{1} f(x) P_k(x)\, dx.$$

   Hint. Multiply (a) by $P_k(x)$, integrate from -1 to 1, then use (41.511) and (41.53).
10. Prove that $P_{2n}(-x) = P_{2n}(x); P_{2n+1}(-x) = -P_{2n+1}(x)$.
11. Prove each of the following identities:

   (a) $xP_n'(x) - P_{n-1}'(x) = nP_n(x)$. *Hint.* Let

   $$u = (1 - 2xt + t^2)^{-1/2} = P_0(x) + P_1(x)t + \cdots + P_n(x)t^n + \cdots.$$

   First show $(x - t)(\partial u/\partial x)$ is equal to $t(\partial u/\partial t)$. Then perform the indicated operations and equate coefficients of t^n.

   (b) $P_{n+1}'(x) - (n + 1)P_n(x) = xP_n'(x)$. *Hint.* Differentiate (41.46), then substitute for $P_{n-1}'(x)$ its value as given in (a) of this problem.

   (c) $(x^2 - 1)P_n'(x) = nxP_n(x) - nP_{n-1}(x)$. *Hint.* Multiply (a) of this problem by x; replace $n + 1$ in (b) by n, then take the difference between the two resulting equations.

12. Verify that the formula in problem 11(a) is valid for P_2, P_3, P_4.

13. Show that the hypergeometric series $F\left(k + 1, -k, 1; \dfrac{1 - x}{2}\right)$ is a solution of the Legendre equation (41.1). *Hint.* Write (41.1) as $(x - 1)(x + 1)y'' + 2xy' - k(k + 1)y = 0$. Then follow the procedure given in Exercise 40.23. Here $r_1 = 1, r_2 = -1, r_3 = 0, a = 1, b = 2, c = -k(k + 1), x = -2u + 1$. The transformed Gauss equation (40.7) becomes

   $$u(1 - u)\frac{d^2 y}{du^2} + (1 - 2u)\frac{dy}{du} + k(k + 1)y = 0,$$

   whose solution is $F(k + 1, -k, 1; u)$.

14. Verify each of the following.

   (a) $\displaystyle\int_{-1}^{1} [P_3(x)]^2\, dx = \frac{2}{7}$.   (b) $\displaystyle\int_{-1}^{1} [P_4(x)]^2\, dx = \frac{2}{9}$.

15. The following differential equation

$$(41.6) \qquad\qquad y'' - 2xy' + 2ky = 0, \quad k \text{ real,}$$

is known as the **Hermite equation.** Show that its solution, valid for all x, is

$$(41.61)$$

$$y_k(x) = a_0 \left[1 - \frac{2}{2!} kx^2 + \frac{2^2}{4!} k(k-2)x^4 - \frac{2^3}{6!} k(k-2)(k-4)x^6 + \cdots \right]$$

$$+ a_1 \left[x - \frac{2}{3!}(k-1)x^3 + \frac{2^2}{5!}(k-1)(k-3)x^5 \right.$$

$$\left. - \frac{2^3}{7!}(k-1)(k-3)(k-5)x^7 + \cdots \right]$$

$$= a_0 \left[1 + \sum_{n=1}^{\infty} \frac{(-1)^n 2^n}{(2n)!} k(k-2)(k-4) + \cdots + (k-2n+2)x^{2n} \right]$$

$$+ a_1 \left[x + \sum_{n=1}^{\infty} \frac{(-1)^n 2^n}{(2n+1)!}(k-1)(k-3) \cdots (k-2n+1)x^{2n+1} \right].$$

16. Find polynomial solutions of (41.6) for $k = 0, 1, 2, \cdots, 7$. *Hint.* Note that for $k = 0$ or a positive integer, one of the series in (41.61) terminates.

Ans. $y_0(x) = a_0$, $\quad y_1(x) = a_1 x$, $\quad y_2(x) = a_0(1 - 2x^2)$,
$\quad y_3(x) = a_1(x - \frac{2}{3}x^3)$, $\qquad y_4(x) = a_0(1 - 4x^2 + \frac{4}{3}x^4)$,
$\quad y_5(x) = a_1(x - \frac{4}{3}x^3 + \frac{4}{15}x^5)$, $y_6(x) = a_0(1 - 6x^2 + 4x^4 - \frac{8}{15}x^6)$,
$\quad y_7(x) = a_1(x - 2x^3 + \frac{4}{5}x^5 - \frac{8}{105}x^7)$.

17. The following polynomials, known as **Hermite polynomials,**

$$\begin{aligned}
H_0(x) &= 1, \\
H_1(x) &= 2x, \\
H_2(x) &= 4x^2 - 2, \\
H_3(x) &= 8x^3 - 12x, \\
H_4(x) &= 16x^4 - 48x^2 + 12, \\
H_5(x) &= 32x^5 - 160x^3 + 120x, \\
H_6(x) &= 64x^6 - 480x^4 + 720x^2 - 120, \\
H_7(x) &= 128x^7 - 1344^5 + 3360x^3 - 1680x,
\end{aligned}$$

can be obtained from the solutions given in problem 16, by choosing appropriate values for a_0 and a_1. Show that these Hermite polynomials can also be obtained from the formula

$$(41.62) \qquad H_n(x) = (-1)^n e^{x^2} \frac{d^n}{dx^n} e^{-x^2}, \quad n = 0, 1, 2, \cdots.$$

18. In solving (41.6), the following recursion formula results,

$$(a) \qquad\qquad a_{n+2} = \frac{2(n-k)}{(n+1)(n+2)} a_n.$$

Following exactly the steps from (41.24) on, obtain polynomial solutions of the Hermite equation (41.6), of the form (i.e., in descending powers of x)

(b) $h_k(x) = a_k \left[x^k - \dfrac{k(k-1)}{2^2} x^{k-2} + \dfrac{k(k-1)(k-2)(k-3)}{2!} \dfrac{x^{k-4}}{2^4} \right.$

$$- \dfrac{k(k-1)(k-2)\cdots(k-5)}{3!} \dfrac{x^{k-6}}{2^6} + \cdots$$

$$\left. + \dfrac{(-1)^n k(k-1)\cdots(k-2n+1)}{n!} \dfrac{x^{k-2n}}{2^{2n}} \right]$$

$$= a_k \sum_{n=0}^{[k/2]} \dfrac{(-1)^n k!}{n!(k-2n)!} \dfrac{x^{k-2n}}{2^{2n}}, \quad k = 0, 1, 2, \cdots.$$

In (b), a_k is an arbitrary constant. If we take for it the value 2^k, we obtain the Hermite polynomials

(c) $$H_k(x) = \sum_{n=0}^{[k/2]} \dfrac{(-1)^n k!}{n!(k-2n)!} (2x)^{k-2n}, \quad k = 0, 1, 2, \cdots.$$

Verify that the above formula (c) also gives the Hermite polynomials shown in problem 17 above.

19. In the summation in (c) of problem 18 above, and in (41.317), n cannot be larger than $[k/2]$. Why?

20. Show that the coefficient of t^n in the series expansion of e^{2xt-t^2} in powers of t is $H_n(x)/n!$ *Hint.* $e^{2xt-t^2} = e^{2xt}e^{-t^2}$. Write the series expansion of each term of this product, multiply both series, then show the coefficient of t^n is $H_n(x)/n!$.

ANSWERS 41

1. (a) $y = a_0 \left(1 - \dfrac{(1+1)}{2!} x^2 + \dfrac{(1+1)(1-2)(1+3)}{4!} x^4 + \cdots \right) + a_1 x.$

(b) $y = a_0(1 - 3x^2) + a_1 \left(x - \dfrac{(2-1)(2+2)}{3!} x^3 \right.$

$$\left. + \dfrac{(2-1)(2+2)(2-3)(2+4)}{5!} x^5 + \cdots \right).$$

2. $y = x - \frac{14}{3}x^3 + \frac{21}{5}x^5.$

3. $P_5(x) = \frac{1}{8}(63x^5 - 70x^3 + 15x);$

$P_6(x) = \frac{1}{16}(231x^6 - 315x^4 + 105x^2 - 5);$

$P_7(x) = \frac{1}{16}(429x^7 - 693x^5 + 315x^3 - 35x).$

4. $y = a_0 \left(1 - \dfrac{\frac{1}{2}(\frac{1}{2}+1)}{2!} x^2 + \dfrac{\frac{1}{2}(\frac{1}{2}+1)(\frac{1}{2}-2)(\frac{1}{2}+3)}{4!} x^4 + \cdots \right)$

$$+ a_1 \left(x - \dfrac{(\frac{1}{2}-1)(\frac{1}{2}+2)}{3!} x^3 \right.$$

$$\left. - \dfrac{(\frac{1}{2}-1)(\frac{1}{2}+2)(\frac{1}{2}-3)(\frac{1}{2}+4)}{5!} x^5 + \cdots \right).$$

7. $x^3 = \frac{2}{5}P_3(x) + \frac{3}{5}P_1(x).$

8. $\frac{8}{35}P_4(x) + \frac{4}{5}P_3(x) + \frac{40}{21}P_2(x) + \frac{1}{5}P_1(x) - \frac{32}{15}P_0(x)$.

19. For $n > [k/2]$, $k - 2n$ is a negative integer and $(k - 2n)!$ is undefined for negative integers.

LESSON 42. The Bessel Differential Equation. Bessel Function of the First Kind $J_k(x)$. Differential Equations Leading to a Bessel Equation. Properties of $J_k(x)$.

LESSON 42A. The Bessel Differential Equation. Another differential equation which arises frequently in physical problems and about which much has been written is known as the **Bessel equation**, named in honor of the German mathematician F. W. Bessel (1784–1846). It is

$$(42.1) \qquad x^2y'' + xy' + (x^2 - k^2)y = 0,$$

where k is a positive constant or zero. You can verify by definitions 40.22 and 40.24 that $x = 0$ is a regular singularity of (42.1). Hence we seek a Frobenius series solution of the form

$$(42.11) \qquad y = x^m(a_0 + a_1x + a_2x^2 + \cdots + a_nx^n + \cdots).$$

Comparing (42.1) with (40.33), we see that

$$(42.12) \qquad f_1(x) = 1 \quad \text{and} \quad f_2(x) = -k^2 + x^2,$$

both of which are already in series form. Comparing (42.12) with (40.34), we have

$$(42.13) \qquad b_0 = 1, \quad c_0 = -k^2, \quad c_1 = 0, \quad c_2 = 1.$$

All remaining b's and c's are zero. The indicial equation (40.38) therefore becomes

$$(42.14) \qquad m(m - 1) + m - k^2 = 0, \quad m^2 = k^2,$$

whose roots are $\pm k$. As we shall discover, when k is not 0 or an integer, there will be two Frobenius series solutions of (42.1). If k is zero or an integer, there will be only one Frobenius series solution of (42.1). The second independent solution will, by Comment 40.5, be a logarithmic solution and have the form (40.51).

Using the root $m = k$ and the values in (42.13) we obtain, by setting the second coefficient in (40.37) equal to zero,

$$(42.15) \quad a_1[(k + 1)k + (k + 1) - k^2] + 0a_0 = 0, \quad (2k + 1)a_1 = 0.$$

Since k is a positive constant or zero, the second equation in (42.15) will be an identity only if

$$(42.151) \qquad\qquad a_1 = 0.$$

The recursion formula, obtained by setting the coefficient of x^{m+n} in (40.37) equal to zero, becomes with $m = k$ and with the help of (42.13),

$$(42.16) \quad a_n[(k + n)(k + n - 1) + (k + n) - k^2] + 0 + a_{n-2}(1) = 0,$$

which simplifies to

$$(42.17) \qquad\qquad a_n = -\frac{a_{n-2}}{n^2 + 2nk}, \quad n \geqq 2.$$

Since, by (42.151), $a_1 = 0$, we see from (42.17) that

$$(42.18) \quad a_3 = a_5 = a_7 = \cdots = a_{2n+1} = 0, \quad n = 0, 1, 2, \cdots.$$

For even values of n, we have, by (42.17),

(42.19)

$$a_2 = -\frac{a_0}{4 + 4k} = -\frac{(\frac{1}{2})^2}{1 + k} a_0,$$

$$a_4 = -\frac{a_2}{16 + 8k} = \frac{-1}{8(2 + k)} \frac{-(\frac{1}{2})^2}{(1 + k)} a_0 = \frac{(\frac{1}{2})^4}{2!(2 + k)(1 + k)} a_0,$$

$$a_6 = -\frac{a_4}{36 + 12k} = -\frac{(\frac{1}{2})^4}{2 \cdot 12(3 + k)(2 + k)(1 + k)} a_0$$

$$= \frac{-(\frac{1}{2})^6}{3!(3 + k)(2 + k)(1 + k)} a_0,$$

. . . .

$$a_{2n} = \frac{(-1)^n (\frac{1}{2})^{2n}}{n!(1 + k)(2 + k) \cdots (n + k)} a_0.$$

Substituting in (42.11) the root $m = k$ and the above values of the a's, we obtain

$$(42.2) \quad y_k = a_0 x^k \left[1 - \frac{1}{1 + k}\left(\frac{x}{2}\right)^2 + \frac{1}{2!(1 + k)(2 + k)}\left(\frac{x}{2}\right)^4 \right.$$

$$- \frac{1}{3!(1 + k)(2 + k)(3 + k)}\left(\frac{x}{2}\right)^6 + \cdots$$

$$\left. + \frac{(-1)^n}{n!(1 + k)(2 + k) \cdots (n + k)}\left(\frac{x}{2}\right)^{2n} + \cdots\right],$$

which is one solution of (42.1), valid for $k \geqq 0$.

Similarly, by following the above procedure for the root $m = -k$, $k \geqq 0$, we obtain [equivalent to replacing k by $-k$ in (42.2)]

$$(42.21) \quad y_{-k} = a_0 x^{-k}\left[1 - \frac{1}{1 - k}\left(\frac{x}{2}\right)^2 + \frac{1}{2!(1 - k)(2 - k)}\left(\frac{x}{2}\right)^4 \right.$$

$$- \frac{1}{3!(1 - k)(2 - k)(3 - k)}\left(\frac{x}{2}\right)^6 + \cdots$$

$$\left. + \frac{(-1)^n}{n!(1 - k)(2 - k) \cdots (n - k)}\left(\frac{x}{2}\right)^{2n} + \cdots\right],$$

which is a second solution of (42.1), valid for $k \neq 1, 2, 3, \cdots$. (For these values of k, one of the denominators in (24.21) is zero.) Since the functions f_1 and f_2 of (42.12) are polynomials, by Comment 37.53, their series representations are valid for all x. Hence by Theorem 40.39 the series solutions, (42.2) and (42.21), converge for all x except perhaps at $x = 0$.

Note that the above results support the statement we made after (42.14). If $k = 0$, the two solutions (42.2) and (42.21) are, as you can verify, identical, and there is therefore only one Frobenius series solution of (42.1). And if $k = 1, 2, 3, \cdots$, there is again only one Frobenius series solution of (42.1), namely (42.2). For all other positive values of k, y_k, and y_{-k} are, by Theorem 40.39, two linearly independent solutions of (42.1). Hence if $k \neq 0, 1, 2, 3, \cdots$, the general solution of (42.1), by Theorem 19.3 and Comment 19.41, is

$$(42.22) \qquad y(x) = c_1 y_k + c_2 y_{-k}, \quad k \neq 0, 1, 2, 3, \cdots,$$

convergent for all $x \neq 0$, where y_k and y_{-k} are defined by (42.2) and (42.21) respectively.

LESSON 42B. Bessel Functions of the First Kind $J_k(x)$.

In (42.2) a_0 is an arbitrary constant. Let us choose for it the value

$$(42.3) \qquad\qquad a_0 = \frac{1}{2^k k!}, \quad k \geq 0.$$

(We showed in Lesson 27E, that $k!$ is defined for all values of k except negative integers.) The functions resulting from (42.2) when a_0 is given the value (42.3) are designated by $J_k(x)$ and are called **Bessel functions of the first kind of index k.** Hence, with $k!(k + 1)$ replaced by its equal $(k + 1)!$, we obtain, by (42.3) and (42.2),

$$(42.31) \quad J_k(x) = \left(\frac{x}{2}\right)^k \left[\frac{1}{k!} - \frac{1}{(k + 1)!}\left(\frac{x}{2}\right)^2 + \frac{1}{2!(k + 2)!}\left(\frac{x}{2}\right)^4 \right.$$
$$- \frac{1}{3!(k + 3)!}\left(\frac{x}{2}\right)^6 + \cdots$$
$$\left. + \frac{(-1)^n}{n!(k + n)!}\left(\frac{x}{2}\right)^{2n} + \cdots \right],$$

valid for all $k \geq 0$.

If, in (42.31), we replace k by $-k$ [equivalent to choosing $a_0 = \dfrac{1}{2^{-k}(-k)!}$

in (42.21)], we obtain

(42.32)
$$J_{-k}(x) = \left(\frac{x}{2}\right)^{-k} \left[\frac{1}{(-k)!} - \frac{1}{(1-k)!}\left(\frac{x}{2}\right)^2 + \frac{1}{2!(2-k)!}\left(\frac{x}{2}\right)^4 + \cdots \right.$$
$$\left. + \frac{(-1)^n}{n!(n-k)!}\left(\frac{x}{2}\right)^{2n} + \cdots \right],$$

valid for all $k \neq 1, 2, 3, \cdots$.

We showed in Lesson 42A, that if k is zero or a positive integer, then y_k of (42.2) is the only Frobenius series solution of the Bessel equation (42.1). However, if k is not zero or an integer, then the functions y_k and y_{-k} of (42.2) and (42.21) are two linearly independent solutions of (42.1). Since $J_k(x)$ and $J_{-k}(x)$ defined in (42.31) and (42.32) are these same functions y_k and y_{-k} multiplied by an appropriate constant, it follows that if k is not zero or an integer, J_k and J_{-k} are linearly independent solutions of (42.1), each valid for all x except, perhaps for $x = 0$. Hence we can assert, by Theorem 19.3 and Comment 19.41, that

(42.33) $y(x) = c_1 J_k(x) + c_2 J_{-k}(x), \quad k \neq 0, 1, 2, 3, \cdots$

is also a general solution of (42.1), valid for all $x \neq 0$, where $J_k(x)$ and $J_{-k}(x)$ are defined by (42.31) and (42.32).

If however $k = 0, 1, 2, 3, \cdots$, then (42.31) is the only Frobenius solution of (42.1). In this case, a second solution of (42.1) will have the logarithmic form shown in (40.51). The one we shall give below* is designated by $-N_k(x)$ and is called a **Bessel function of the second kind of index k.** It can be shown that if, in (42.1), we make the substitution

$$y_2(x) = u(x) - J_k(x) \log x, \quad x > 0,$$

it becomes (see Exercise 42,2)

$$x^2 u'' + x u' + (x^2 - k^2)u = 2x J_k'(x),$$

where

$$u(x) = x^{-k}(b_0 + b_1 x + b_2 x^2 + \cdots), \quad k > 0.$$

A second solution of (42.1), if k is a positive integer, is then

(42.34)
$$-N_k(x) = \frac{1}{2}\sum_{n=0}^{k-1} \frac{(k-n-1)!}{n!}\left(\frac{x}{2}\right)^{2n-k} + \frac{1}{2}\sum_{n=0}^{\infty} \frac{(-1)^n}{n!(n+k)!}$$
$$\times \left[H_n + \left(1 + \frac{1}{2} + \cdots + \frac{1}{n+k}\right)\right]\left(\frac{x}{2}\right)^{2n+k}$$
$$- J_k(x) \log x, \quad x > 0,$$

---

*There are other forms of the Bessel function of the second kind.

where

$$H_n = 1 + \frac{1}{2} + \frac{1}{3} + \cdots + \frac{1}{n}, \quad \text{if } n \neq 0,$$
$$= 0, \qquad\qquad\qquad\quad \text{if } n = 0.$$

Hence for *positive integer* k, the general solution of (42.1) is

(42.35) $$y(x) = c_1 J_k(x) + c_2 N_k(x).$$

Remark. A second solution, when $k = 0$, can be found at the end of Example 40.6. See (k), page 584.

Note that thus far we have no solution $J_{-k}(x)$ when k is a positive integer. The solution $J_{-k}(x)$ of (42.32) is undefined for $k = 1, 2, 3, \cdots$. To fill in this gap, we define $J_{-k}(x)$, $k = 0, 1, 2, 3, \cdots$, by the relation

(42.36) $$J_{-k}(x) = (-1)^k J_k(x), \quad k = 0, 1, 2, 3, \cdots.$$

The justification for this definition is outlined below.

Justification for (42.36). By (42.31),

(42.37) $$J_k(x) = \sum_{n=0}^{\infty} \frac{(-1)^n}{n!(k+n)!} \left(\frac{x}{2}\right)^{2n+k}, \quad k = 0, 1, 2, 3, \cdots.$$

By (42.32), we write formally

(42.371) $$J_{-k}(x) = \sum_{n=0}^{\infty} \frac{(-1)^n}{n!(n-k)!} \left(\frac{x}{2}\right)^{2n-k}, \quad k = 0, 1, 2, 3, \cdots,$$

even though the series is not defined for $n < k$. [For these values of n, $(n - k)$ is a negative integer and as we showed in Lesson 27E, $(n - k)!$ is undefined for negative integers.] We also showed in Lesson 27E that $(k - 1)! = \Gamma(k) \to \infty$ as $k \to 0, -1, -2, \cdots$. Since a negative integer factorial appears in the denominator of each term of the series (42.371) for which $n < k$, we are led to define each term, for which $n = 0, 1, 2, \cdots, k - 1$, to be zero. Hence we can write (42.371) as

(42.38) $$J_{-k}(x) = \sum_{n=k}^{\infty} \frac{(-1)^n}{n!(n-k)!} \left(\frac{x}{2}\right)^{2n-k}, \quad k = 0, 1, 2, 3, \cdots.$$

Let $p = n - k$. Hence when $n = k$, $p = 0$. This means we can start our summation in (42.38) with $p = 0$ instead of with $n = k$, if at the same time we substitute $p + k$ for its equal n. Therefore (42.38) can be written as

(42.39) $$J_{-k}(x) = \sum_{p=0}^{\infty} \frac{(-1)^{p+k}}{(p+k)!(p+k-k)!} \left(\frac{x}{2}\right)^{2p+k}, \quad k = 0, 1, 2, \cdots.$$

Since the summation in (42.39) is over p, we may remove $(-1)^k$ from inside the summation. It then becomes, after simplification,

$$(42.4) \quad J_{-k}(x) = (-1)^k \sum_{p=0}^{\infty} \frac{(-1)^p}{p!(p+k)!} \left(\frac{x}{2}\right)^{2p+k}, \quad k = 0, 1, 2, \cdots.$$

Comparing the summation in (42.37) with that in (42.4), we see that both are the same. Hence (42.36) follows.

The most frequently encountered Bessel functions of the first kind are $J_0(x)$ and $J_1(x)$. By (42.31), these are

$$(42.41) \quad J_0(x) = 1 - \frac{x^2}{2^2} + \frac{1}{(2!)^2}\frac{x^4}{2^4} - \frac{1}{(3!)^2}\frac{x^6}{2^6} + \cdots$$
$$+ \frac{(-1)^n}{(n!)^2}\frac{x^{2n}}{2^{2n}} + \cdots,$$

$$(42.42) \quad J_1(x) =$$
$$\frac{x}{2}\left(1 - \frac{1}{2!}\frac{x^2}{2^2} + \frac{1}{2!3!}\frac{x^4}{2^4} - \frac{1}{3!4!}\frac{x^6}{2^6} + \cdots + \frac{(-1)^n}{n!(n+1)!}\frac{x^{2n}}{2^{2n}} + \cdots\right).$$

The graphs of these two functions are shown in Fig. 42.43.

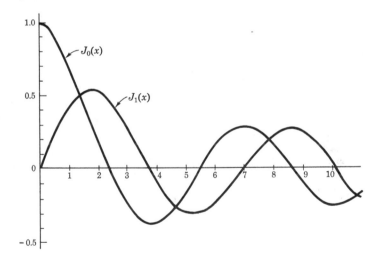

Figure 42.43

Tables of values exist for Bessel functions of the first kind $J_k(x)$, just as they do for $\sin x$, $\log x$ or e^x. Thus if you had to evaluate $J_1(2)$, you could use (42.42) with $x = 2$, or look up its value in a table. Its value is 0.5767. Therefore by (42.36), $J_{-1}(2) = -0.5767$. If you had to evaluate

$J_{-1/2}(3)$, you could use (42.32) with $k = \frac{1}{2}$, $x = 3$ or look up its value in a table. Its value is -0.4560.

LESSON 42C. Differential Equations Which Lead to a Bessel Equation. We give below examples of differential equations whose solutions can be obtained by transforming each into a Bessel equation by means of a suitable substitution.

Example 42.5. Find a solution of

(a) $$u'' + \left(1 + \frac{1 - 4k^2}{4x^2}\right) u = 0, \quad k \text{ real.}$$

Solution. The substitution,

(b) $$u = x^{1/2}y,$$
$$u' = \tfrac{1}{2}x^{-1/2}y + x^{1/2}y',$$
$$u'' = -\tfrac{1}{4}x^{-3/2}y + x^{-1/2}y' + x^{1/2}y'',$$

will transform (a) into the Bessel equation

(c) $$x^2 y'' + xy' + (x^2 - k^2)y = 0.$$

Since $y = J_k(x)$ is a solution of (c), it follows by the first equation in (b) that

(d) $$u = x^{1/2}J_k(x)$$

is a solution of (a).

Example 42.51. Use the result obtained in Example 42.5 to find a solution of

(a) $$u'' + \left(1 - \frac{3}{4x^2}\right)u = 0.$$

Solution. If in (a) of Example 42.5 above, we choose $k = 1$, it becomes the (a) of this example. Hence by (d) above, a solution of (a) is

(b) $$u = x^{1/2}J_1(x).$$

Example 42.52. Find a solution of

(a) $$u'' + x^2 u = 0.$$

Solution. Here two substitutions will be needed. The first substitution is

(b) $$u = x^{1/2}y.$$

Substituting in (a) the value of u as given in (b) and the value of u'' as given in the last equation of (b) of Example 42.5, and then multiplying the result by $x^{3/2}$, we obtain

(c) $$x^2 y'' + x y' + (x^4 - \tfrac{1}{4})y = 0.$$

The second substitution is

(d) $$w = \frac{x^2}{2}, \qquad x = (2w)^{1/2},$$

$$\frac{dw}{dx} = x = (2w)^{1/2}.$$

With the help of (d), we obtain

(e) $$\frac{dy}{dx} = \frac{dy}{dw}\frac{dw}{dx} = (2w)^{1/2}\frac{dy}{dw},$$

$$\frac{d^2 y}{dx^2} = (2w)^{1/2}\frac{d^2 y}{dw^2}\frac{dw}{dx} + (2w)^{-1/2}\frac{dw}{dx}\frac{dy}{dw}$$

$$= (2w)\frac{d^2 y}{dw^2} + \frac{dy}{dw}.$$

Substituting (d) and (e) in (c), it becomes

(f) $$(2w)^2 \frac{d^2 y}{dw^2} + 2w\frac{dy}{dw} + 2w\frac{dy}{dw} + [(2w)^2 - \tfrac{1}{4}]y = 0,$$

which simplifies to

(g) $$w^2 \frac{d^2 y}{dw^2} + w\frac{dy}{dw} + (w^2 - \tfrac{1}{16})y = 0.$$

Since (g) is the Bessel equation (42.1) with $k = \tfrac{1}{4}$, its solution is

(h) $$y = J_{1/4}(w).$$

Replacing w by its value in (d) and then substituting the resulting value of y in (b), we obtain

(i) $$u = x^{1/2} J_{1/4}\left(\frac{x^2}{2}\right),$$

which is a solution of (a).

Comment 42.53. The method used to solve equation (a) of Example 42.52 can be applied to the more general equation

(42.54) $$u'' + b x^m u = 0.$$

The first substitution is the same as (b) of Example 42.52, namely

(a) $$u = x^{1/2} y.$$

The second substitution, however, is more complicated than the one in (d) of Example 42.52. It is

(b)
$$w = \frac{2\sqrt{b}}{m + 2} \sqrt{x^{m+2}}.$$

The substitution of (a) in (42.54) and of (b) in the resulting equation, will yield the Bessel equation

(c)
$$w^2 \frac{d^2y}{dw^2} + w \frac{dy}{dw} + \left(w^2 - \frac{1}{(m+2)^2}\right) y = 0.$$

Its solution is

(d)
$$y = J_{1/(m+2)}(w).$$

Replacing w by its value in (b) and then substituting the resulting value of y in (a), we obtain

(42.55)
$$u = x^{1/2} J_{1/(m+2)}\left(\frac{2\sqrt{b}}{m + 2} \sqrt{x^{m+2}}\right),$$

which is a solution of (42.54). Note that when $m = 2$, $b = 1$, (42.54) reduces to (a) of Example 42.52, and (42.55) reduces to its solution (i).

Example 42.56.   Find a solution of

(a)
$$u'' + 9xu = 0.$$

Solution.   If in (42.54) we let $b = 9$, $m = 1$, it reduces to (a) above. Hence a solution of (a), by (42.55) is

(b)
$$u = x^{1/2} J_{1/3}(2x^{3/2}).$$

Comment 42.57.   The method used to solve equation (a) of Example 42.52 can be applied to a still more general equation than (42.54), namely

(42.58) $$x^2 u'' + (1 - 2a)xu' + (b^2 c^2 x^{2c} + a^2 - k^2 c^2)u = 0.$$

The first substitution

(a)
$$u = x^a y,$$

will change (42.58) into the equation

(b)
$$x^2 \frac{d^2y}{dx^2} + x \frac{dy}{dx} + (b^2 x^{2c} - k^2)c^2 y = 0.$$

Verify it. The second substitution

(c)
$$w = bx^c,$$

will change (b) into the Bessel equation

(d) $$w^2 \frac{d^2 y}{dw^2} + w \frac{dy}{dw} + (w^2 - k^2)y = 0.$$

Its solution is

(e) $$y = J_k(w).$$

Hence by (a), (e), and (c), a solution of (42.58) is

(42.59) $$u = x^a J_k(bx^c).$$

Comment 42.6. If, in (42.58), we make the substitutions

(a) $$a = \tfrac{1}{2}, \qquad c = \frac{m+2}{2}, \qquad k = \frac{1}{m+2},$$

and replace b^2 by $4b/(m+2)^2$, it reduces to

(b) $$x^2 u'' + bx^{m+2} u = 0, \qquad u'' + bx^m u = 0.$$

A solution of (b), by (42.59), is then

(c) $$u = x^{1/2} J_{1/(m+2)} \left(\frac{2\sqrt{b}}{m+2} \sqrt{x^{m+2}} \right).$$

Note that the second equation in (b) is now the same as (42.54). Note too that their respective solutions (c) and (42.55) are, as they should be, also the same. Hence (42.54) is only a special case of (42.58). By assigning different values to a, b, c, and k in (42.58), we can thus obtain many different equations whose solutions will be given by (42.59).

Example 42.61. Find a solution of

(a) $$x^2 u'' + xu' + \left(x - \frac{k^2}{4} \right) u = 0.$$

Solution. If in (42.58), we let

(b) $$a = 0, \qquad b = 2, \qquad c = \tfrac{1}{2},$$

it reduces to (a). Hence a solution of (a) by (42.59) is

(c) $$u = J_k(2\sqrt{x}).$$

Comment 42.62. If k is not zero or a positive integer, then, by (c) above and (42.33),

(d) $$u(x) = c_1 J_k(2\sqrt{x}) + c_2 J_{-k}(2\sqrt{x})$$

is a general solution of (a) of Example 42.61. If k is a positive integer, then by (42.35), a general solution of (a) is

(e) $$u(x) = c_1 J_k(2\sqrt{x}) + c_2 N_k(2\sqrt{x}),$$

where $N_k(x)$ is given by (42.34).

Example 42.63.   Find a solution of

(a) $$x^2 u'' + xu' - (x^2 + k^2)u = 0.$$

Solution.   If in (42.58), we let

(b) $$a - 0, \qquad b^2 - i^2 - -1, \qquad c - 1,$$

it reduces to (a). Hence a solution of (a) by (42.59) is

(c) $$u(x) = J_k(ix).$$

Comment 42.64.   The function obtained by multiplying (c) of Example 42.63 above, by the constant i^{-k} is also a solution of (a) of this example. It is usually written as $I_k(x)$. Hence

(42.65) $$I_k(x) = i^{-k} J_k(ix).$$

The function $I_k(x)$ of (42.65) is called a **modified Bessel function of the first kind.**

Comment 42.66.   We list one more equation which can, by a suitable substitution, be changed into a Bessel equation. It is

(42.67) $$x^2 u'' + x(1 - 2x \tan x)u' - (x \tan x + k^2)u = 0.$$

The substitution $u = y/\cos x$ will transform (42.67) into the Bessel equation

(a) $$x^2 \frac{d^2 y}{dx^2} + x \frac{dy}{dx} + (x^2 - k^2)y = 0.$$

Since a solution of (a) is $y = J_k(x)$, a solution of (42.67) is

(42.68) $$u = \frac{1}{\cos x} J_k(x).$$

LESSON 42D. Properties of Bessel Functions of the First Kind $J_k(x)$.

A. Properties of the Zeros of Bessel Functions $J_k(x)$. We list below, without proof, a number of properties of the zeros of Bessel functions $J_k(x)$ of the first kind.

1. If x_1 and x_2 are two zeros of $J_k(x)$, then in the interval I: $x_1 < x < x_2$, there is a zero of $J_{k-1}(x)$ and $J_{k+1}(x)$.
2. The Bessel function $J_0(x)$ has a zero in each interval of length π.
3. Each Bessel function $J_k(x)$, $k = 1, 2, \cdots$, has an infinite number of real positive zeros in the interval I: $0 < x < \infty$.
4. If $k > \frac{1}{2}$, then the difference between two consecutive zeros of $J_k(x)$ is greater than π.
5. If $k > \frac{1}{2}$, then the difference between two consecutive zeros of $J_k(x)$ approaches π, as x approaches infinity.
6. The first positive zero of $J_k(x)$ is greater than k.
7. A Bessel function $J_k(x)$ has only real zeros.

B. Integral Property of Bessel Functions $J_k(x)$.

Theorem 42.7. *Let r_1, r_2, $\cdots$, be distinct positive zeros of a Bessel function $J_k(x)$, where k is a fixed real number. Then*

$$(42.71) \qquad \int_0^1 xJ_k(r_ix)J_k(r_jx)\, dx = 0, \quad \text{if } r_i \neq r_j,$$
$$= \tfrac{1}{2}[J_k'(r_i)]^2, \quad \text{if } r_i = r_j.$$

Proof. In Example 42.5, we proved that $u(x) = x^{1/2}J_k(x)$ is a solution of $u'' + \left(1 + \dfrac{1 - 4k^2}{4x^2}\right)u = 0$. In a similar manner, it can be shown that

(a) $\quad u_1(x) = x^{1/2}J_k(r_ix)$ is a solution of $u'' + \left(r_i^2 + \dfrac{1 - 4k^2}{4x^2}\right)u = 0$.

$\qquad u_2(x) = x^{1/2}J_k(r_jx)$ is a solution of $u'' + \left(r_j^2 + \dfrac{1 - 4k^2}{4x^2}\right)u = 0$.

[Or you can obtain these solutions by letting $a = \frac{1}{2}$, $b = r$, $c = 1$ in (42.58). It will then reduce to $x^2u'' + (r^2x^2 + \frac{1}{4} - k^2)u = 0$. Division by x^2 will give it the form of the differential equations in (a). A solution by (42.59) will then be

(b) $\qquad\qquad\qquad u = x^{1/2}J_k(rx).]$.

Hence, by (a),

(c) $\qquad\qquad u_1'' + \left(r_i^2 + \dfrac{1 - 4k^2}{4x^2}\right)u_1 = 0,$

$\qquad\qquad\qquad u_2'' + \left(r_j^2 + \dfrac{1 - 4k^2}{4x^2}\right)u_2 = 0.$

Multiplying the first equation in (c) by u_2, the second by $-u_1$ and adding the two resulting equation, we obtain

(d) $\qquad\qquad u_2u_1'' - u_1u_2'' = (r_j^2 - r_i^2)u_1u_2.$

Integrating (d) between the limits 0 and x, and recognizing that the left side of (d) is $(d/dx)(u_2u_1' - u_1u_2')$, there results

(e) $$[u_2u_1' - u_1u_2']_0^x = (r_j{}^2 - r_i{}^2)\int_0^x u_1u_2\,dx.$$

By (a) and its derivatives,

(f) $$u_1(0) = 0, \quad u_2(0) = 0,$$
$$u_1' = r_ix^{1/2}J_k'(r_ix) + \tfrac{1}{2}x^{-1/2}J_k(r_ix),$$
$$u_2' = r_jx^{1/2}J_k'(r_jx) + \tfrac{1}{2}x^{-1/2}J_k(r_jx).$$

Substituting (a) and (f) in (e), and then simplifying the result, we obtain

(g)
$$x[r_iJ_k(r_jx)J_k'(r_ix) - r_jJ_k(r_ix)J_k'(r_jx)] = (r_j{}^2 - r_i{}^2)\int_0^x xJ_k(r_ix)J_k(r_jx)\,dx.$$

If $x = 1$, i.e., if the interval $(0,x)$ is the interval $(0,1)$, then (g) becomes

(h) $$r_iJ_k(r_j)J_k'(r_i) - r_jJ_k(r_i)J_k'(r_j) = (r_j{}^2 - r_i{}^2)\int_0^1 xJ_k(r_ix)J_k(r_jx)\,dx.$$

By hypotheses r_i and r_j are zeros of $J_k(x)$. Therefore, $J_k(r_i) = 0$, $J_k(r_j) = 0$ and the left side of (h) vanishes. Hence if r_i, r_j, are two distinct zeros of $J_k(x)$, we can divide (h) by $r_j{}^2 - r_i{}^2$ to obtain the first equation in (42.71).

We now prove the second equation in (42.71). Differentiating the equation in (g) with respect to r_j, we obtain

(i) $$x[r_ixJ_k'(r_jx)J_k'(r_ix) - J_k(r_ix)J_k'(r_jx) - r_jxJ_k(r_ix)J_k''(r_jx)]$$
$$= 2r_j\int_0^x xJ_k(r_ix)J_k(r_jx)\,dx + (r_j{}^2 - r_i{}^2)\frac{d}{dr_j}\int_0^x xJ_k(r_ix)J_k(r_jx)\,dx.$$

If $r_i = r_j$, then (i) simplifies to

(j) $$x[xr_iJ_k'{}^2(r_ix) - J_k(r_ix)J_k'(r_ix) - r_ixJ_k(r_ix)J_k''(r_ix)]$$
$$= 2r_i\int_0^x xJ_k{}^2(r_ix)\,dx.$$

And if the interval is $(0,1)$ so that $x = 1$, then (j) becomes [remember r_i is a zero of $J_k(x)$; therefore $J_k(r_i) = 0$]

(k) $$r_iJ_k'{}^2(r_i) = 2r_i\int_0^1 xJ_k{}^2(r_ix)\,dx.$$

Hence

(1)
$$\int_0^1 xJ_k^{\,2}(r_ix)\,dx = \tfrac{1}{2}J_k'^{\,2}(r_i),$$

which is the second equation in (42.71).

Comment 42.72. Because of the first equality in (42.71), the set of functions $J_k(r_1x)$, $J_k(r_2x)$, $\cdots$, is said to be orthogonal on $I\colon 0 \leqq x \leqq 1$ with respect to the weight function x. Compare with Definition 41.5 for an orthogonal set of functions.

EXERCISE 42

1. Find general solutions of each of the following Bessel equations.
   (a) $x^2y'' + xy' + (x^2 - 1)y = 0$.
   (b) $x^2y'' + xy' + (x^2 - 4)y = 0$.
   (c) $x^2y'' + xy' + (x^2 - \tfrac{1}{4})y = 0$.
   (d) $x^2y'' + xy' + (x^2 - \tfrac{1}{9})y = 0$.

2. Prove that the substitution in (42.1) of $y_2(x) = u_k(x) - J_k(x)\log x$, $x > 0$, transforms the equation into $x^2u_k'' + xu_k' + (x^2 - k^2)u_k = 2xJ_k'(x)$. *Hint.* Make use of the fact that $J_k(x)$ is a solution of (42.1).

3. Verify, by direct substitution in (42.31), that $J_1(2) = 0.5767$; $J_{-1/2}(3) = -0.4560$. *Hint.* Factor out $(k)!$ and use fact that $(-\tfrac{1}{2})! = \sqrt{\pi}$. Also verify that $J_0(0.3) = 0.9776$; $J_1(0.2) = 0.0995$.

4. Prove that (a) of Example 42.5 is transformed into (c) by the substitution (b).

5. Verify the accuracy of (c) of Example 42.52.

6. Verify the accuracy of (a) of Comment 42.66.

Find a solution of each of the following equations 7–14, by using the appropriate formula given in Lesson 42C.

7. $y'' + 9x^2y = 0$. *Hint.* In (42.54), $b = 9$, $m = 2$.

8. $y'' + \left(1 + \dfrac{3}{16x^2}\right)y = 0$.

9. $y'' + \left(1 - \dfrac{4}{9x^2}\right)y = 0$.

10. $y'' + 4x^3y = 0$.

11. $x^2y'' + xy' + (x - 1)y = 0$.

12. $x^2y'' + xy' - (x^2 + 4)y = 0$.

13. $x^2y'' - xy' + x^2y = 0$.

14. $x^2y'' + \dfrac{x}{2}\,y' + (x^4 - \tfrac{15}{16})y = 0$.

15. By assigning various values to a, b, c, and k in (42.58), obtain at least three different differential equations and their solutions.

16. Show that the substitution $u = 2e^{x/2}$, $e^x = u^2/4$, will transform the equation $y'' + (e^x - m^2)y = 0$ into the Bessel equation $u^2 \dfrac{d^2y}{du^2} + u\dfrac{dy}{du} +$

$(u^2 - 4m^2)y = 0$. Since a solution of the second equation is $y = J_{2m}(u)$, a solution of the original equation is $y(x) = J_{2m}(2e^{x/2})$. *Hint.* Follow the procedure used in making the second substitution after (c) of Example 42.52.

17. With the help of problem 16, find a solution of each of the following differential equations.

(a) $y'' + (e^x - 9)y = 0$.   (b) $y'' + (e^x - \frac{1}{4})y = 0$.
(c) $y'' + (e^x - \frac{1}{9})y = 0$.

18. Assume that a function $f(x)$, defined on the interval $(0,1)$, can be represented by a series of Bessel functions, i.e., assume

(a) $$f(x) = c_0 J_k(r_0 x) + c_1 J_k(r_1 x) + c_2 J_k(r_2 x) + \cdots,$$

where $r_0, r_1, r_2, \cdots$, are the distinct, positive zeros of $J_k(x)$ and k is a fixed real number. Show that the coefficients $c_0, c_1, c_2, \cdots$, are given by

$$[J_k'(r_i)]^2 c_i = 2 \int_0^1 x f(x) J_k(r_i x) \, dx.$$

Hint. Multiply (a) by $x J_k(r_i x)$, integrate from 0 to 1, then use (42.71).

19. Prove each of the following identities.

(a) $\dfrac{d}{dx}[x^k J_k(x)] = x^k J_{k-1}(x)$. *Hint.* Multiply (42.37) by x^k and then take its derivative.

(b) $\dfrac{d}{dx}[x^{-k} J_k(x)] = -x^{-k} J_{k+1}(x)$. See hint in (a).

(c) $J_k'(x) + kx^{-1} J_k(x) = J_{k-1}(x)$. *Hint.* Carry out the differentiation in (a) and then divide by x^k.

(d) $J_k'(x) - kx^{-1} J_k(x) = -J_{k+1}(x)$. Apply hint in (c) to (b). Divide by x^{-k}.

(e) $J_{k-1}(x) - J_{k+1}(x) = 2J_k'(x)$. *Hint.* Add (c) and (d).

(f) $J_{k-1}(x) + J_{k+1}(x) = \dfrac{2k}{x} J_k(x)$.

(g) $\dfrac{d}{dx} J_0(x) = -J_1(x)$. *Hint.* Set $k = 0$ in (a) above, and then make use of (42.36).

(h) $\dfrac{d^2}{dx^2}[J_k(x)] \equiv \frac{1}{4}[J_{k-2}(x) - 2J_k(x) + J_{k+2}(x)]$. *Hint.* Differentiate (e).

Then use (e) again to find $J_{k+1}'(x)$ and $J_{k-1}'(x)$. Substitute these values in the $J_k''(x)$ equation.

(i) $J_{1/2}(x) \equiv \sqrt{\dfrac{2}{\pi x}} \sin x$. *Hint.* $(\frac{1}{2})! = \sqrt{\pi}/2$.

(j) $J_{-1/2}(x) \equiv \sqrt{\dfrac{2}{\pi x}} \cos x$. *Hint.* $(-\frac{1}{2})! = \sqrt{\pi}$.

20. Show that the coefficient of t^n in the series expansion of $e^{(x/2)[t-(1/t)]}$ in powers of t is $J_n(x)$. *Hint.* $e^{(x/2)[t-(1/t)]} = e^{xt/2} e^{-x/2t}$. Write the series expansion of each term of this product, multiply both series, then show that the coefficient of $t^n = J_n(x)$.

ANSWERS 42

1. Substitute: (a) $k = 1$ in (42.2) and (42.34); (b) $k = 2$ in (42.2) and (42.34);
   (c) $k = \frac{1}{2}$ in (42.2) and (42.21); (d) $k = \frac{1}{3}$ in (42.2) and (42.21).

7. $y = x^{1/2}J_{1/4}(\frac{3}{2}x^2)$.   11. $y = J_2(2\sqrt{x})$.

8. $y = x^{1/2}J_{1/4}(x)$.   12. $y = J_2(ix)$.

9. $y = x^{1/2}J_{5/6}(x)$.   13. $y = xJ_1(x)$.

10. $y = x^{1/2}J_{1/5}(\frac{4}{5}x^{5/2})$.   14. $y = x^{1/4}J_{1/2}\left(\dfrac{x^2}{2}\right)$.

17. (a) $y = J_6(2e^{x/2})$.   (b) $y = J_1(2e^{x/2})$.   (c) $y = J_{2/3}(2e^{x/2})$.

LESSON 43.   The Laguerre Differential Equation. Laguerre Polynomials $L_k(x)$. Properties of $L_k(x)$.

LESSON 43A.   The Laguerre Differential Equation and Its Solution. The differential equation

$$(43.1) \qquad xy'' + (1 - x)y' + ky = 0, \quad k \text{ real,}$$

is called the **Laguerre** equation, after E. Laguerre (1834–1866). It is of interest only when k is an integer and the interval is $x \geqq 0$. You can verify by Definitions 40.22 and 40.24 that $x = 0$ is a regular singularity of (43.1). Hence we seek a Frobenius series solution of the form

$$(43.11) \qquad y = x^m(a_0 + a_1x + a_2x^2 + \cdots).$$

Multiplying (43.1) by x and comparing the resulting equation with (40.33), we find that

$$(43.12) \qquad f_1(x) = 1 - x, \qquad f_2(x) = kx,$$

both of which are already in series form and valid, by Comment 37.53, for all x. Hence, by Theorem 40.32, a Frobenius series solution of (43.1) will be valid for all x, except perhaps at $x = 0$. Comparing (43.12) with (40.34), we find

$$(43.13) \qquad b_0 = 1, \qquad b_1 = -1, \qquad c_0 = 0, \qquad c_1 = k.$$

All remaining b's and c's are zero. The indicial equation (40.38) therefore becomes

$$(43.14) \qquad m^2 - m + m = 0, \quad m^2 = 0,$$

whose roots are $m = 0$ twice. We can therefore expect only one Frobenius series solution of (43.1).

Using the root $m = 0$ and the values in (43.13), we obtain, by setting the second coefficient in (40.37) equal to zero,

$$(43.15) \qquad a_1 + a_0 k = 0, \qquad a_1 = -ka_0.$$

The recursion formula, obtained by setting the coefficient of x^{m+n} in (40.37) equal to zero, becomes with $m = 0$ and the help of (43.13),

$$(43.16) \qquad a_n[n(n-1) + n] + a_{n-1}[-(n-1) + k] = 0,$$

which simplifies to

$$(43.17) \qquad a_n = \frac{(n-1) - k}{n^2}\, a_{n-1}.$$

By (43.17) and (43.15),

$$(43.18) \qquad a_2 = \frac{1-k}{2^2}\, a_1 = \frac{-k(1-k)}{2^2}\, a_0 = \frac{k(k-1)}{2^2}\, a_0,$$

$$a_3 = \frac{2-k}{3^2}\, a_2 = \frac{-k(k-1)(k-2)}{2^2 \cdot 3^2}\, a_0,$$

$$a_4 = \frac{3-k}{4^2}\, a_3 = \frac{k(k-1)(k-2)(k-3)}{2^2 \cdot 3^2 \cdot 4^2}\, a_0,$$

$$\cdots \cdots \cdots \cdots \cdots \cdots \cdots \cdots \cdots \cdots \cdots \cdots$$

$$a_n = \frac{(-1)^n k(k-1)(k-2) \cdots (k-n+1)}{2^2 \cdot 3^2 \cdot 4^2 \cdots n^2}\, a_0,$$

$$= \frac{(-1)^n k!}{(n!)^2(k-n)!}\, a_0, \quad n = 0, 1, 2, \cdots.$$

Substituting $m = 0$ and the above values of the a's in (43.11), we obtain

$$(43.19) \qquad y_k(x) = a_0 \left(1 - kx + \frac{k(k-1)}{2^2}\, x^2 - \frac{k(k-1)(k-2)}{2^2 \cdot 3^2}\, x^3 \right.$$

$$+ \frac{k(k-1)(k-2)(k-3)}{2^2 \cdot 3^2 \cdot 4^2}\, x^4 + \cdots$$

$$\left. + \frac{(-1)^n k!}{(n!)^2(k-n)!}\, x^n + \cdots \right),$$

which is a series solution of (43.1), valid for all x. A second solution will have the logarithmic form shown in (40.51).

LESSON 43B. The Laguerre Polynomial $L_k(x)$. If $k = 0, 1, 2, 3, \cdots$, the series (43.19) terminates. The resulting polynomials, with $a_0 = k!$, are known as **Laguerre polynomials** and are designated by

$L_k(x)$. Hence, by (43.19), with $a_0 = k!$,

(43.2)
$$L_0(x) = 1,$$
$$L_1(x) = 1 - x,$$
$$L_2(x) = 2 - 4x + x^2,$$
$$L_3(x) = 6 - 18x + 9x^2 - x^3,$$
$$L_4(x) = 24 - 96x + 72x^2 - 16x^3 + x^4.$$
. . . .

$$L_k(x) = (k!)^2 \sum_{n=0}^{k} \frac{(-1)^n}{(n!)^2(k-n)!} x^n.$$

We may place $(k!)^2$ outside the summation sign, since the summation is over n. Graphs of the first four Laguerre polynomials are shown in Fig. 43.21.

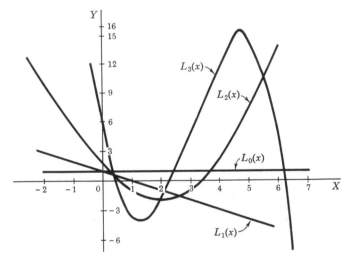

Figure 43.21

It has been proved that a Laguerre polynomial of degree n has exactly n real zeros in the interval $0 < x < \infty$. Note in Fig. 43.21 that $L_0(x)$ has no zeros, $L_1(x)$ has one zero, $L_2(x)$ has two zeros, and $L_3(x)$ has three zeros in the interval $x > 0$.

We give below two equations which can be transformed into Laguerre equations by a proper substitution.

1. The substitution

(43.22) $u = e^{-x}y$

in the equation

(43.23) $$xu'' + (1 + x)u' + (k + 1)u = 0$$

will transform it into the Laguerre equation (43.1). Since $y = L_k(x)$ is a solution of (43.1), it follows by (43.22) that

(43.24) $$u = e^{-x}L_k(x)$$

is a solution of (43.23).

2. Similarly, the substitution

(43.25) $$u = e^{-x/2}x^{1/2}y$$

in the equation

(43.26) $$u'' + \left(\frac{1}{4x^2} + \frac{2k + 1}{2x} - \frac{1}{4}\right)u = 0,$$

will transform it into the Laguerre equation (43.1). Since $y = L_k(x)$ is a solution of (43.1), it follows by (43.25) that

(43.27) $$u = e^{-x/2}x^{1/2}L_k(x)$$

is a solution of (43.26).

LESSON 43C.   Some Properties of Laguerre Polynomials $L_k(x)$.

A.   Analog of Rodrigue's Formula for the Legendre Polynomial.

(43.3) $$L_n(x) = e^x \frac{d^n}{dx^n}(x^n e^{-x}).$$

Proof. We shall show first that the formula is valid for the first three Laguerre polynomials. By (43.3),

(a) $$L_0(x) = e^x x^0 e^{-x} = 1,$$

$$L_1(x) = e^x \frac{d}{dx}(xe^{-x}) = e^x(e^{-x} - xe^{-x}) = 1 - x,$$

$$L_2(x) = e^x \frac{d^2}{dx^2}(x^2 e^{-x}) = e^x \frac{d}{dx}(2xe^{-x} - x^2 e^{-x})$$

$$= e^x(2e^{-x} - 4xe^{-x} + x^2 e^{-x})$$

$$= 2 - 4x + x^2.$$

$$L_3(x) = e^x \frac{d^3}{dx^3}(x^3 e^{-x}) = e^x \frac{d^2}{dx^2}(3x^2 e^{-x} - x^3 e^{-x})$$

$$= e^x \frac{d}{dx}(6xe^{-x} - 6x^2 e^{-x} + x^3 e^{-x})$$

$$= e^x(6e^{-x} - 18xe^{-x} + 9x^2 e^{-x} - x^3 e^{-x})$$

$$= 6 - 18x + 9x^2 - x^3.$$

You can verify that each of the formulas in (a) agrees with those in (43.2).

The proof of (43.3) for the general case follows. First verify that

$$(43.31) \qquad [f(x)g(x)]^{(n)} = \sum_{k=0}^{n} \frac{n!}{k!(n-k)!} f^{(k)}(x)g^{(n-k)}(x),$$

i.e., the coefficients of the nth derivative of $f(x)g(x)$ are the same as the coefficients in the binomial expansion of $(x+1)^n$. In (43.31), let $f(x) = x^n$ and $g(x) = e^{-x}$. It then becomes

$$(43.32) \qquad \frac{d^n}{dx^n} (x^n e^{-x}) = \sum_{k=0}^{n} \frac{n!}{k!(n-k)!} (x^n)^{(k)} (e^{-x})^{(n-k)}.$$

When

$$(43.33) \quad n = 1: \; \frac{d}{dx} (x^n) = nx^{n-1},$$

$$n = 2: \frac{d^2}{dx^2} (x^n) = n(n-1)x^{n-2},$$

$$n = 3: \frac{d^3}{dx^3} (x^n) = n(n-1)(n-2)x^{n-3},$$

$$n = k: \frac{d^k}{dx^k} (x^n) = n(n-1)(n-2) \cdots (n-k+1)x^{n-k}$$

$$= \frac{n!}{(n-k)!} x^{n-k}.$$

Also

$$(43.34) \qquad \frac{d^{n-k}}{dx^{n-k}} (e^{-x}) = (-1)^{n-k} e^{-x}.$$

Substituting in (43.32) the last equality of (43.33), the equality (43.34) and then multiplying the result by e^x, we have

$$(43.35) \qquad e^x \frac{d^n}{dx^n} (x^n e^{-x}) = (n!)^2 \sum_{k=0}^{n} \frac{(-1)^{n-k}}{k![(n-k)!]^2} x^{n-k}.$$

You can verify that the right side of (43.35) also will give the first four Laguerre polynomials in (43.2). It is, in fact, another form of writing the polynomial solutions of the Laguerre equation. We shall now show that the right side of (43.35) and of the last equation of (43.2) are equivalent. In the summation of (43.35), let $p = n - k$. Therefore $k = n - p$ and when $k = 0$, $p = n$ and when $k = n$, $p = 0$. Hence the right side of (43.35) can be written as

$$(43.36) \qquad (n!)^2 \sum_{p=n}^{0} \frac{(-1)^p}{(n-p)!(p!)^2} x^p.$$

Therefore, by (43.36), with $n = k$,

$$(43.37) \qquad (k!)^2 \sum_{p=k}^{0} \frac{(-1)^p}{(k-p)!(p!)^2} x^k.$$

The summation in (43.37) is now the same as the right side of $L_k(x)$ of (43.2).

B. Integral Property of Laguerre Polynomials.

Theorem 43.4. *Let* $L_0(x)$, $L_1(x)$, $L_2(x)$, $\cdots$, *be Laguerre polynomial solutions of* (43.1). *Then*

$$(43.41) \qquad \int_0^\infty e^{-x} L_m(x) L_n(x)\, dx = 0, \quad \text{if } m \neq n.$$

Proof. Let u_m and u_n be two solutions of (43.26). Therefore

(a)
$$u_m'' + \left(\frac{1}{4x^2} + \frac{2m+1}{2x} - \frac{1}{4} \right) u_m = 0,$$

$$u_n'' + \left(\frac{1}{4x^2} + \frac{2n+1}{2x} - \frac{1}{4} \right) u_n = 0.$$

Multiplying the first by u_n, the second by $-u_m$ and adding the resulting equations, we obtain

(b)
$$u_n u_m'' - u_m u_n'' = \left(\frac{n-m}{x} \right) u_m u_n.$$

Integrating between the limits 0, ∞, and recognizing that the left side of (b) is $(d/dx)(u_n u_m' - u_m u_n')$, there results

(c)
$$\lim_{h \to \infty} [u_n u_m' - u_m u_n']_0^h = (n - m) \int_0^\infty \frac{1}{x} u_m u_n\, dx, \quad m \neq n.$$

By (43.27), solutions of (a) and their respective derivatives are,

(d)
$$u_m = e^{-x/2} x^{1/2} L_m,$$
$$u_m' = -\tfrac{1}{2} e^{-x/2} x^{1/2} L_m + \tfrac{1}{2} e^{-x/2} x^{-1/2} L_m + e^{-x/2} x^{1/2} L_m',$$
$$u_n = e^{-x/2} x^{1/2} L_n,$$
$$u_n' = -\tfrac{1}{2} e^{-x/2} x^{1/2} L_n + \tfrac{1}{2} e^{-x/2} x^{-1/2} L_n + e^{-x/2} x^{1/2} L_n'.$$

Inserting the above values in (c), we obtain

(e)
$$\lim_{h \to \infty} [e^{-x} x L_n(x) L_m'(x) - e^{-x} x L_m(x) L_n'(x)]_0^h$$
$$= (n - m) \int_0^\infty e^{-x} L_m(x) L_n(x)\, dx.$$

The left side of (e) is zero when $x = 0$, and, by (27.113)(d), it approaches zero for positive h as $h \to \infty$. By hypothesis $m \neq n$. Therefore (e) simplifies to

(f)
$$\int_0^\infty e^{-x} L_m(x) L_n(x)\, dx = 0.$$

Because of (f), the set of Laguerre polynomials is said to be orthogonal on I: $0 \leq x < \infty$ with respect to the weight function e^{-x}. Compare with Definition 41.5 for an orthogonal set of functions.

EXERCISE 43

1. Find a series solution of each of the following differential equations.

    (a) $xy'' + (1 - x)y' + \frac{1}{2}y = 0$.   (b) $xy'' + (1 - x)y' + 1.2y = 0$.

2. Verify that the substitution (43.22) in (43.23) gives the Laguerre equation (43.1).
3. Verify that the substitution (43.25) in (43.26) gives the Laguerre equation (43.1).
4. Use formula (43.3) to find $L_4(x)$.
5. Find a series solution of each of the following.

    (a) $xy'' + (1 + x)y' + y = 0$. *Hint.* See (43.23).
    (b) $xy'' + (1 + x)y' + 2y = 0$.   (c) $xy'' + (1 + x)y' + \frac{3}{2}y = 0$.

6. Find a series solution of each of the following.

    (a) $y'' + \left(\dfrac{1}{4x^2} + \dfrac{1}{2x} - \dfrac{1}{4} \right) y = 0$. *Hint.* See (43.26).

    (b) $y'' + \left(\dfrac{1}{4x^2} + \dfrac{3}{2x} - \dfrac{1}{4} \right) y = 0$.   (c) $y'' + \left(\dfrac{1}{4x^2} + \dfrac{1}{x} - \dfrac{1}{4} \right) y = 0$.

7. Assume that a function $f(x)$, defined on the interval $(0, \infty)$, can be represented by a series of Laguerre polynomials, i.e., assume

    (a) $f(x) = c_0 L_0(x) + c_1 L_1(x) + c_2 L_2(x) + \cdots$.

Show that the coefficients $c_0, c_1, c_2, \cdots$ are given by

$$c_k = \frac{\displaystyle\int_0^\infty e^{-x} L_k(x) f(x)\, dx}{\displaystyle\int_0^\infty e^{-x} [L_k(x)]^2\, dx}.$$

Hint. Multiply (a) by $e^{-x}L_k(x)$, integrate from 0 to ∞, then use (43.41).

ANSWERS 43

1. (a) replace k by $\frac{1}{2}$ in (43.19).   (b) replace k by 1.2 in (43.19).
5. (a) $y = e^{-x}L_0(x) = e^{-x}$.   (b) $y = e^{-x}L_1(x) = e^{-x}(1 - x)$.
    (c) $y = e^{-x}L_{1/2}(x)$.
6. (a) $y = e^{-x/2}x^{1/2}L_0(x) = e^{-x/2}x^{1/2}$.
    (b) $y = e^{-x/2}x^{1/2}L_1(x) = e^{-x/2}x^{1/2}(1 - x)$.
    (c) $y = e^{-x/2}x^{1/2}L_{1/2}(x)$.

Chapter 10

Numerical Methods

Introduction. In many practical problems involving differential equations, what is often wanted is a table of values of a solution $y = y(x)$, satisfying given initial conditions, for a limited range of values of x near the initial point x_0. For example, we may want values of $y(x)$ when $x = x_0 + h$, $x_0 + 2h$, $x_0 + 3h$, etc., where h is 0.05 or 0.1 or 0.2, etc. Even when a solution $y = y(x)$ of a differential equation can be written in terms of elementary functions, it may at times be easier to obtain this limited table of values by the numerical methods we shall describe in this chapter, rather than from the analytic solution itself. This statement is especially true when the solution is an implicit one. As we have remarked on numerous occasions, implicit solutions are usually such complicated expressions that it is almost impossible to find the needed function $g(x)$ which it implicitly defines, or to calculate values of y for given values of x. Moreover, in a great many problems, a solution of a differential equation cannot be expressed in terms of elementary functions for the very good reason that the differential equation does not have any such solution. For example, the equation $y' = 1/\sqrt{x^3 + 1}$ does not have a solution in terms of elementary functions.

By a numerical solution of a differential equation, we shall mean a table of values such that for each x there is a corresponding value of $y(x)$. In this sense, even an explicit solution in terms of an elementary function, such as $y = \sin x$ or in terms of a nonelementary function such as $y = J_0(x)$, is a numerical solution. For each x, we can look in a table and find a value of $\sin x$ or of the Bessel function $J_0(x)$.

You will soon discover that the work involved in computing a table of values even when a moderate degree of accuracy is needed is laborious and tedious. However, with the current increased availability of high-speed computing machines, it is possible to have them make many burdensome calculations for you. But it will still be necessary for you to know how and what to feed these machines.

In the following lessons of this chapter, we shall explain various methods by which a numerical solution of a differential equation can be obtained.

To keep the arithmetical calculations within reasonable bounds, and also to be able to check the accuracy of our results, we have selected simple differential equations for our examples, ones which can be solved explicitly in terms of elementary functions. You must keep in mind, however, that we are using them only to illustrate a method. These same methods can be employed to find numerical solutions of more complicated equations.

We illustrate the methods we shall develop for finding a numerical solution of a first order differential equation by applying them to an equation of the form $y' = f(x,y)$ for which $y(x_0) = y_0$. We assume in our discussion that a unique particular solution of this equation, satisfying the given initial condition, exists. (For criteria which will give a sufficient condition for the existence of this unique particular solution, see Theorem 58.5. If the criteria of the existence theorem are too difficult to apply, a practical man will usually know from his experience and from the nature of the physical problem which gave rise to the differential equation whether a solution exists.)

The methods we shall develop for finding a numerical solution of a differential equation have been divided into three categories. In one category, we include those methods which need only the given equation $y' = f(x,y)$ and the initial condition $y(x_0) = y_0$ in order to start the construction of a table of values of y for given values of x. They are therefore called appropriately **starting methods.** In a second category, we include those methods which need more values of y than only the initial condition $y(x_0) = y_0$ before they can be used. These methods are therefore called appropriately **continuing methods** since they can be used to continue the construction of the table only after the needed preliminary values have been obtained by starting methods. In a third category we include those methods whose only purpose is to correct values of y obtained by starting and continuing methods. These methods are therefore called appropriately **corrector methods.**

LESSON 44. Starting Method. Polygonal Approximation.

In this lesson we shall show, by a method called the **polygonal method,** how to start the construction of a table of approximate values of $y(x_0 + h)$, $y(x_0 + 2h)$, $\cdots$, where h is a constant and $y(x)$ is the unique particular solution of

$$(44.1) \qquad\qquad y' = f(x,y)$$

satisfying the initial condition

$$(44.11) \qquad\qquad y(x_0) = y_0.$$

We proceed as follows (see Fig. 44.12). By (44.1) and (44.11), we determine

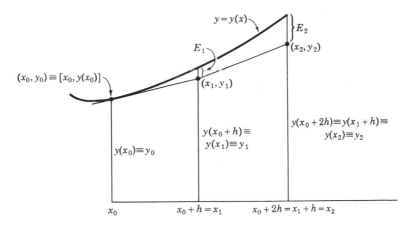

Figure 44.12

the value of y' at (x_0,y_0). The equation of the tangent to the integral curve $y(x)$ at the point (x_0,y_0) is therefore

$$(44.13) \qquad y - y(x_0) = (x - x_0)y'(x_0).$$

This tangent line will intersect the line $x = x_0 + h$ in a point whose ordinate is [in (44.13), replace x by $x_0 + h$],

$$(44.14) \qquad y(x_0 + h) = y(x_0) + y'(x_0)h.$$

In our table, we can now record the value of $y(x_0 + h)$ obtained by (44.14). It is an approximation to the actual value of $y(x_0 + h)$. The error in this computation is shown as E_1 in Fig. 44.12.

For convenience we write (x_1,y_1) for the point $[x_0 + h, y(x_0 + h)]$. At $[x_1,y_1]$ we repeat the above procedure as if (x_1,y_1) were actually on the integral curve. We find the equation of a line through (x_1,y_1) having a slope obtained by (44.1), with $x = x_1$, $y = y_1$. If we call this slope $y'(x_1)$, then the equation of this line is

$$(44.15) \qquad y - y(x_1) = y'(x_1)(x - x_1).$$

Its intersection with the line $x = x_0 + 2h \equiv x_1 + h$ is

$$(44.16) \qquad y(x_0 + 2h) \equiv y(x_1 + h) = y(x_1) + y'(x_1)h.$$

In our table, we can therefore now record the approximate value of $y(x_0 + 2h)$. The error in the computation is shown as E_2 in Fig. 44.12.

Continuing in this manner, we find

$$(44.17) \qquad y(x_0 + 3h) \equiv y(x_2 + h) = y(x_2) + y'(x_2)h,$$

which is an approximate value of $y(x_0 + 3h)$. And in general, we find

(44.18) $y(x_n + h) = y(x_n) + y'(x_n)h,$

which is an approximate value of $y[x_0 + (n + 1)h]$.

Example 44.2. By means of the polygonal method, find approximate values when $x = 0.1, 0.2, 0.3$ of the particular solution of the differential equation

(a) $y' = x^2 + y,$

for·which $y(0) = 1$. Take $h = 0.1$ and $h = 0.05$.

Solution. Comparing the initial condition with (44.11), we see that $x_0 = 0, y_0 = 1$. By (44.18), with $h = 0.1$ and x_n taking on the values $0, 0.1, 0.2$, we obtain

(b) $y(0 + 0.1) = y(0.1) = y(0) + y'(0)(0.1),$

(c) $y(0.1 + 0.1) = y(0.2) = y(0.1) + y'(0.1)(0.1),$

(d) $y(0.2 + 0.1) = y(0.3) = y(0.2) + y'(0.2)(0.1).$

By (a) and the initial conditions, $y'(0) = 0 + 1 = 1$. Therefore (b) becomes

(e) $y(0.1) = 1 + 1(0.1) = 1.1.$

By (a), when $x = 0.1, y = 1.1$, we find $y'(0.1) = (0.1)^2 + 1.1 = 1.11$. Therefore (c) becomes

(f) $y(0.2) = 1.1 + 1.11(0.1) = 1.211.$

With $x = 0.2, y = 1.211$, we find by (a), $y'(0.2) = (0.2)^2 + 1.211 = 1.251$, and by (d)

(g) $y(0.3) = 1.211 + 1.251(0.1) = 1.336.$

In numerical solutions, it is usually desirable to construct a table in which all relevant computations are systematically recorded. For the above example, the table has the appearance of Table 44.21. The actual

Table 44.21

$x_n =$	$y(x_n) =$	$y'(x_n) =$	$hy'(x_n) =$	$y(x_n + h) =$	Actual Values of $y(x_n) =$
0.0	1.0	1.000	0.1	1.1	1.000
0.1	1.1	1.110	0.111	1.211	1.106
0.2	1.211	1.251	0.125	1.336	1.224
0.3	1.336				1.360

solution of (a) satisfying $y(0) = 1$ is $y = 3e^x - x^2 - 2x - 2$. By it we obtained the figures in the last column of Table 44.21.

With $h = 0.05$, our table of values becomes, by (a) and (44.18), Table 44.22.

Table 44.22

x_n	$y(x_n)$	$y'(x_n)$	$hy'(x_n)$	$y(x_n + h)$	Actual Values of $y(x_n)$
0.0	1.0000	1.0000	0.0500	1.0500	1.0000
0.05	1.0500	1.0525	0.0526	1.1026	1.0513
0.1	1.1026	1.1126	0.0556	1.1582	1.1055
0.15	1.1582	1.1807	0.0590	1.2172	1.1630
0.2	1.2172	1.2572	0.0629	1.2801	1.2242
0.25	1.2801	1.3426	0.0671	1.3472	1.2896
0.3	1.3472				1.3596

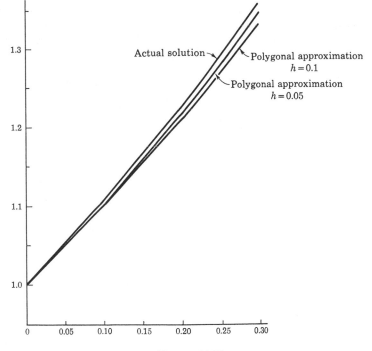

Figure 44.23

A graph of the actual solution and of the polygonal approximations are shown in Fig. 44.23.

General Comment on Errors in a Numerical Computation. The question of determining the errors in a table of numerical values is an extremely complex one. There are in general four types of errors.

1. *Arithmetical errors* made by the individual or due to the misbehavior of a calculator.
2. *Rounding off errors* due to stopping with a certain decimal place.
3. *Formula errors* due to the use of an approximating formula to obtain a numerical answer.
4. *Cumulative errors.* At each step in a lengthy process an error occurs that is carried along to the next stage.

We shall assume that the error due to the rounding off of a decimal can be compensated for by retaining a sufficient number of decimal places at each step, so that the accuracy desired in the last tabulated value of $y(x_0 + nh)$ will not be affected even in the most unfavorable circumstances, as for example when we have to drop the 49 in 0.3265349 in order to round off the decimal to five places. If, however, too many steps are required to reach $y(x_0 + nh)$, it may not always be possible to attain this desired objective, but then the loss in rounding off at one step may be offset by a gain at another. As regards the other types of errors, we shall comment on each of them at the appropriate time.

Comment 44.3.   Comment on Error in Polygonal Method.

1. In this method, we start with a point and a slope that agree with the solution of (44.1) satisfying (44.11). But since the straight line drawn at this point (x_0,y_0) may not be the actual integral curve $y = y(x)$, a *formula* error is introduced in the first step. It is represented by E_1 in Fig. 44.12. At each successive point (x_2,y_2), (x_3,y_3), $\cdots$, at least two errors are introduced, a starting error and a formula error, so that the cumulative error may soon become large. Hence this method is useful only if too great accuracy is not required or if h is very small.

2. Comparing (44.14) with the Taylor series formula, see (37.35),

$$y(x_0 + h) = y(x_0) + y'(x_0)h + \frac{y''(x_0)}{2!} h^2 + \cdots$$
$$+ \frac{y^n(x_0)}{n!} h^n + \frac{y^{n+1}(X)}{(n+1)!} h^{n+1},$$

we see that (44.14) contains the first two terms of a Taylor series. Its remainder or error term is therefore, see (37.36),

(44.31) $$E = \frac{y''(X)}{2!} h^2,$$

where X is a value of x in the interval under consideration of width h. Let

(44.32) $E(x_0 + h)$ = error in computing $y(x_0 + h)$ by (44.14).

Then, by (44.31),

(44.33) $E(x_0 + h) = ch^2,$

where $c = y''(X)/2!$. Let us now divide the h interval in half and compute $y(x_0 + h)$ in two steps. If we have some basis for believing that for a small h, $y''(x)$ changes rather slowly in this h interval so that the variation in the value of $y''(X)$ in each $h/2$ interval is negligible, then we commit a small error in using the same c of (44.33) for each half interval. Therefore, by (44.33), the error in computing $y(x_0 + h/2)$ for a half interval $h/2$ is approximately

(44.34) $E\left(x_0 + \dfrac{h}{2}\right) = c\left(\dfrac{h}{2}\right)^2 = \dfrac{ch^2}{4} = \tfrac{1}{4}E(x_0 + h).$

This means that the error in $y\left(x_0 + \dfrac{h}{2}\right)$ is approximately equal to one-fourth the error of $y(x_0 + h)$. Hence the error in computing in two steps the value of $y(x_0 + h)$, which we write as $y\left[\left(x_0 + \dfrac{h}{2}\right) + \dfrac{h}{2}\right]$, will have an inherited error in the starting value of $y\left(x_0 + \dfrac{h}{2}\right)$ plus its own formula error. Since each error equals one-fourth the error in $y(x_0 + h)$, the total error in $y\left[\left(x_0 + \dfrac{h}{2}\right) + \dfrac{h}{2}\right]$, i.e., the error in computing $y(x_0 + h)$, in two steps, is approximately equal to one-half the error in $y(x_0 + h)$ computed in one step. Hence

(44.35) $E\left[\left(x_0 + \dfrac{h}{2}\right) + \dfrac{h}{2}\right] = \tfrac{1}{2}E(x_0 + h),$

approximately. Let $Y(x_0 + h)$ be the actual value of the solution. Then

(44.36) $Y(x_0 + h) - y(x_0 + h) = E(x_0 + h),$

$Y(x_0 + h) - y\left[\left(x_0 + \dfrac{h}{2}\right) + \dfrac{h}{2}\right] = E\left[\left(x_0 + \dfrac{h}{2}\right) + \dfrac{h}{2}\right].$

Subtracting the second equation in (44.36) from the first, we obtain

(44.37)
$y\left[\left(x_0 + \dfrac{h}{2}\right) + \dfrac{h}{2}\right] - y(x_0 + h) = E(x_0 + h) - E\left[\left(x_0 + \dfrac{h}{2}\right) + \dfrac{h}{2}\right].$

Substituting (44.35) in the right side of (44.37), the following equations result.

(44.38) (a) $E\left[\left(x_0 + \dfrac{h}{2}\right) + \dfrac{h}{2}\right] = y\left[\left(x_0 + \dfrac{h}{2}\right) + \dfrac{h}{2}\right] - y(x_0 + h),$

(b) $\frac{1}{2}E(x_0 + h) = y\left[\left(x_0 + \dfrac{h}{2}\right) + \dfrac{h}{2}\right] - y(x_0 + h).$

The first formula says that the error in the value of $y(x_0 + h)$ computed in two steps is equal to the difference in values of $y(x_0 + h)$ computed in two steps and in one step.

Formulas (44.38) give us a means of estimating errors at each step. Their accuracy does not depend on a knowledge of the size of $y''(X)$, which we do not know, but only on the *variation* of $y''(X)$ over a small interval. We have assumed this variation to be negligible, an assumption which is not unreasonable if $y''(x)$ changes slowly over h. For example, from Tables 44.22 and 44.21, with $x_0 = 0$, $h = 0.1$,

(a) $y\left[\left(x_0 + \dfrac{h}{2}\right) + \dfrac{h}{2}\right] = y(0.05 + 0.05) = y(0.1) = 1.1026,$

$y(x_0 + h) = y(0.1) = 1.1.$

Therefore by (44.38)(a),

(b) $E(0.05 + 0.05) = 1.1026 - 1.1 = 0.0026,$

which is the approximate error of $y(0.1)$ computed in two steps. The actual error by Table 44.22 is $1.1055 - 1.1026 = 0.0029$.

3. A check on errors, which practical people frequently use, and which seems to work, is to make all calculations over again with an h half the size of the original one. If the results obtained with the smaller h agree with those obtained with the larger h to k decimal places, after being properly rounded off, then it is assumed that their common numerical value has k decimal place accuracy. For example, by Tables 44.21 and 44.22, $y(0.3)$ computed with $h = 0.1$ and with $h = 0.05$ agree to one decimal place. Hence we assume that $y(0.3) = 1.3$ has one-decimal accuracy.

4. This method does not give a check on arithmetical errors. However, if there is little agreement between the values obtained by using h and $h/2$, all arithmetical computations should be checked. If the arithmetic is correct, h should be reduced.

Comment 44.39. We have headed this lesson "Starting Method." It can also be used as a continuing one. For instance, we can, in Example 44.2, use this method to find $y(0.4)$, $y(0.5)$, etc. However its low order of accuracy makes it a poor continuing method. In the next and succeeding lessons, we shall present better ones.

EXERCISE 44

1. Add to Table 44.21, values of y when $x = 0.4$ and 0.5. Also add actual values obtained from the solution $y = 3e^x - x^2 - 2x - 2$.

2. Add to Table 44.22, values of y when $x = 0.35, 0.4, 0.45, 0.5$. Also add actual values obtained from the solution $y = 3e^x - x^2 - 2x - 2$.

3. Reconstruct Table 44.21 by applying error formulas (44.38) to correct the entries at each step before proceeding to the next one. For example, by (44.38)(b), with $x_0 = 0$, $h = 0.1$,

$$E(0.1) = 2[y(0.05 + 0.05) - y(0.1)] = 2(1.1026 - 1.1) = 0.0052.$$

Hence the corrected value of $y(0.1) = 1.1 + 0.0052 = 1.1052$. Record this new value of $y(0.1)$ in your reconstructed Table 44.21. Starting with this corrected value of $y(0.1) = 1.1052$, compute $y(0.2)$ in one step and two steps. Then apply error formulas (44.38) to correct $y(0.2)$, etc. Compare with previous results and with actual values of the solution.

4. Find approximate values when $x = 0.05, 0.1, 0.15, 0.2$ of the particular solution of the equation $y' = x + y^2$ for which $y(0) = 1$. Take $h = 0.05$.

5. In problem 4, compute $y(0.1)$ and $y(0.2)$ with $h = 0.1$. Compare with the value of $y(0.2)$ obtained in 4. In the absence of a solution, how many decimal place accuracy could you assume in the value of $y(0.2)$? *Hint.* See Comment 44.3–3.

6. By use of error formulas (44.38), correct the value of $y(0.1)$ obtained in problems 4 and 5. Using this corrected figure, proceed to find $y(0.2)$ in two steps and in one step. Then correct $y(0.2)$.

7. Find approximate values when $x = 1.1, 1.2, 1.3$ of the particular solution of the differential equation $y' = x^2 + y^2$ for which $y(1) = 1$. Take $h = 0.1$. What is the approximate formula error in $y(1.1)$? [*Hint.* Calculate $y(1.1)$ in two steps and then use error formula (44.38).] In the absence of a solution or error formula, how could you estimate the error in $y(1.3)$?

8. Find an approximate value when $x = 1$, of the particular solution of the equation $y' = 1/(1 + x^2)$ for which $y(0) = 0$. Use $h = 0.2$ and proceed as follows. Find $y(0.2)$ in one step and in two steps. Correct $y(0.2)$ by means of (44.38). Then find $y(0.4)$ in one step and in two steps. Correct $y(0.4)$ by means of (44.38). Continue in this way until you reach $y(1)$. Show how this value of $y(1)$ can be used to approximate π. Compare with actual value of π. *Hint.* The solution of $y' = 1/(1 + x^2)$ for which $y(0) = 0$ is $y = $ Arc tan x. Therefore $y(1) = \pi/4$ so that $\pi = 4y(1)$.

9. Follow the procedure outlined in problem 8 to find an approximate value when $x = 1$, of the particular solution of the equation $y' = y$ for which $y(0) = 1$. Show how this value of $y(1)$ can be used to approximate e. Compare with actual value of e. *Hint.* The solution of $y' = y$ for which $y(0) = 1$ is $y = e^x$. Therefore $y(1) = e$.

10. Find an approximate value when $x = 2$ of the particular solution of the equation $y' = 1/x$ for which $y(1) = 0$. Take $h = 0.2$ and follow the procedure outlined in problem 8. Show how this value of $y(2)$ can be used to approximate log 2. *Hint.* The solution of $y' = 1/x$ for which $y(1) = 0$ is $y = \log x$. Therefore $y(2) = \log 2$.

11. By formulas (44.14) and (44.31), we have

$$y(x_0 + h) = y(x_0) + hy'(x_0) + E,$$

where $E = y''(X)h^2/2$ and X is a value of x in the interval $(x_0, x_0 + h)$. E is the formula error due to stopping with the $y'(x_0)$ term in a Taylor series.

If we knew which value of x to choose for X in this interval, we would know the exact value of E. But we don't. However, if M is the maximum value of $|y''(x)|$ in the interval $(x_0, x_0 + h)$, then $|E| \leqq Mh^2/2$. We can thus establish an upper bound of the error, in an interval of width h, in using formula (44.14) to compute $y(x_0 + h)$. Call this error E_1 and call M_1 the maximum value of $y''(x)$ in this interval. Designate by E_2, E_3, $\cdots$, E_n, the errors in each additional interval of width h, and M_2, $\cdots$, M_n the maximum value of $|y''(x)|$ in each such additional interval. Then the upper bound of the total error E for all intervals is

$$(44.4) \qquad |E| \leqq |E_1| + |E_2| + \cdots + |E_n| \leqq \frac{h^2}{2}(M_1 + M_2 + \cdots + M_n).$$

Let M be the largest of the numbers M_1, M_2, $\cdots$, M_n. Then, by (44.4),

$$(44.41) \qquad\qquad |E| \leqq \frac{h^2}{2} nM,$$

where M is the maximum value of $|y''(x)|$ in the interval $(x_0, x_0 + nh)$.

(a) Use formula (44.41) to compute the upper bound of the error in the value of log 2 if in problem 10 we had used $h = 0.2$ without corrections. *Hint.* $h = 0.2$, $n = 5$, $M = \max. |y''| = \max. |-1/x^2| = 1$ in the interval $(1,2)$.

(b) What is the largest value of h that can be used to insure that the upper bound of the error in the computation of log 2 in problem 10 is 0.005, if no corrections were made? *Hint.* By (44.41), we want an h such that $h^2 nM/2 \leqq 0.005$. Remember $hn = 1$.

ANSWERS 44

1. $y(0.4) = 1.471$, $y(0.5) = 1.634$. Actual values: 1.51547, 1.69616.

2. $y(0.35) = 1.4191$, $y(0.4) = 1.4962$, $y(0.45) = 1.5790$, $y(0.5) = 1.6681$. Actual values 1.43470, 1.51547, 1.60244, 1.69616.

3. $y(0.2) = 1.2237$, $y(0.3) = 1.3557$, $y(0.4) = 1.5107$, $y(0.5) = 1.6902$.

4. $y(0.05) = 1.0500$, $y(0.1) = 1.1076$, $y(0.15) = 1.1739$, $y(0.2) = 1.2503$.

5. $y(0.1) = 1.100$, $y(0.2) = 1.231$.
Can assume only zero decimal place accuracy if rounded off to one decimal.

6. $y(0.1) = 1.1152$, $y[(0.1 + 0.05) + 0.05] = 1.2598$, $y(0.1 + 0.1) = 1.2496$, $y(0.2) = 1.2700$.

7. $y(1.1) = 1.2000$, $y(1.2) = 1.4650$, $y(1.3) = 1.8236$, $E(1.1) = 0.0312$. Would need to make all calculations over again with $h = 0.05$, see Comment 44.3–3.

8. $y(0.2) = 0.1980$, $y(0.4) = 0.3815$, $y(0.6) = 0.5415$, $y(0.8) = 0.6756$, $y(1) = 0.7860$. Actual value: $\pi = 3.14159$.

9. $y(0.2) = 1.2200$, $y(0.4) = 1.4884$, $y(0.6) = 1.8158$, $y(0.8) = 2.2152$, $y(1) = 2.7026$. Actual value: $e = 2.71828$.

10 $y(1.2) = 0.1818$, $y(1.4) = 0.3355$, $y(1.6) = 0.4688$, $y(1.8) = 0.5864$, $y(2) = 0.6917$. Actual value: log 2 = 0.6931.

11. (a) $|E| \leqq 0.1$.   (b) $h = 0.01$.

LESSON 45.   An Improvement of the Polygonal Starting Method.

Let $y(x)$ be a particular solution of

(45.1) $$y' = f(x,y)$$

satisfying the initial condition

(45.11) $$y(x_0) = y_0.$$

In the polygonal method, we found the approximation, see (44.14),

(45.12) $$y(x_0 + h) = y(x_0) + y'(x_0)h.$$

It is possible to improve this estimate using a method similar to the one of the previous lesson. In this previous lesson, we found $y(x_0 + h)$ of (45.12) by drawing, at (x_0, y_0), a tangent line to the integral curve $y(x)$, and determining its intersection with the line $x = x_0 + h$. It is marked R in Fig. 45.13. The error in $y(x_0 + h)$ is shown as E_1. The point P in

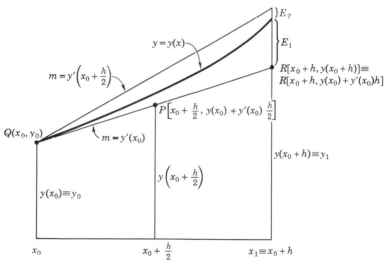

Figure 45.13

the figure is the mid-point of the segment of this tangent line between Q and R. The coordinates of P were obtained by using the mid-point formula of analytic geometry. Substituting the coordinates of P in (45.1), we find

(45.14) $$y'\left(x_0 + \frac{h}{2}\right) = f\left[x_0 + \frac{h}{2}, \quad y(x_0) + y'(x_0)\frac{h}{2}\right].$$

The equation of the line through (x_0, y_0) with slope (45.14) is

(45.15) $y = y(x_0) + (x - x_0)y'\left(x_0 + \dfrac{h}{2}\right)$

$= y(x_0) + (x - x_0)f\left[x_0 + \dfrac{h}{2}, \quad y(x_0) + y'(x_0)\,\dfrac{h}{2}\right].$

Its intersection with the line $x = x_0 + h$ is

(45.16) $y(x_0 + h) = y(x_0) + hf\left[x_0 + \dfrac{h}{2}, \quad y_0 + \dfrac{h}{2}\,y'(x_0)\right].$

We shall now prove that the right side of (45.16) is equivalent to the first three terms of a Taylor series instead of only the first two as in (45.12). Hence the error in the approximate value of $y(x_0 + h)$ of (45.16), shown as E_2 in Fig. 45.13, usually will be smaller than the error E_1.*

Proof. By (38.12) and (38.13), a Taylor series expansion of a function of two variables is, with $x = x_0 + a$, $y = y_0 + b$,

(a) $f(x_0 + a, y_0 + b) = f(x_0, y_0) + a\,\dfrac{\partial f(x_0, y_0)}{\partial x} + b\,\dfrac{\partial f(x_0, y_0)}{\partial y} + \cdots.$

Therefore with $a = \dfrac{h}{2}$, $b = \dfrac{h}{2}\,y'(x_0)$,

(b) $f\left[x_0 + \dfrac{h}{2}, y_0 + \dfrac{h}{2}\,y'(x_0)\right] = f(x_0, y_0) + \dfrac{h}{2}\,\dfrac{\partial f(x_0, y_0)}{\partial x}$

$+ \dfrac{h}{2}\,y'(x_0)\,\dfrac{\partial f(x_0, y_0)}{\partial y} + \cdots.$

Substituting (b) in (45.16), and replacing $f(x_0, y_0)$ by its equal $y'(x_0)$ of (45.1), we obtain

(c) $y(x_0 + h) = y(x_0) + hy'(x_0)$

$+ \dfrac{h^2}{2}\left[\dfrac{\partial f(x_0, y_0)}{\partial x} + y'(x_0)\,\dfrac{\partial f(x_0, y_0)}{\partial y}\right] + \cdots.$

By differentiation of (45.1)—we assume the derivatives exist—we obtain

(d) $y'' = \dfrac{\partial f(x, y)}{\partial x} + \dfrac{\partial f(x, y)}{\partial y}\,\dfrac{dy}{dx}.$

---

*By (37.36), the error or remainder term $E_1 = \dfrac{y''(X_1)}{2}\,h^2$ if the series includes the first two terms of a Taylor series; the error term $E_2 = \dfrac{y'''(X_2)}{6}\,h^3$ if the series includes the first three terms. Hence $E_2 < E_1$ if $\dfrac{y'''(X_2)}{6}\,h^3 < \dfrac{y''(X_1)}{2}\,h^2$, i.e., if $y'''(X_2) < \dfrac{3}{h}\,y''(X_1)$. For a small h, $3/h$ is large.

Hence

(e) $$y''(x_0) = \frac{\partial f(x_0,y_0)}{\partial x} + \frac{\partial f(x_0,y_0)}{\partial y}\, y'(x_0).$$

Therefore by (e), we can write (c) as

(f) $$y(x_0 + h) = y(x_0) + hy'(x_0) + \frac{h^2}{2}\, y''(x_0) + \cdots.$$

A comparison of (f) with the Taylor series (37.27) shows, with $x = x_0 + h$, that the first three terms of each are the same.

NOTE. *Use of this method is therefore permissible only if* $\dfrac{\partial f(x,y)}{\partial x}$ *and* $\dfrac{\partial f(x,y)}{\partial y}$ *exist at* (x_0,y_0).

Example 45.2. Find, by the method of this lesson, an approximate value, when $x = 0.1$, of the particular solution of

(a) $$y' = x^2 + y$$

for which $y(0) = 1$.

Solution. Comparing the initial condition with (45.11), we see that $x_0 = 0$, $y_0 = 1$. Therefore by (a), $y'(0) = 0 + 1 = 1$. Hence with $x_0 = 0$, $h = 0.1$, (45.16) becomes

(b) $$y(0.1) = 1 + 0.1 f(0.05, 1 + 0.05).$$

Here $f(x,y) = x^2 + y$. Therefore

(c) $$f(0.05,\, 1.05) = (0.05)^2 + 1.05 = 1.0525.$$

Substituting (c) in (b), we obtain

(d) $$y(0.1) = 1 + 0.1053 = 1.1053.$$

The actual value of $y(0.1)$ is 1.1055. In the polygonal method using two steps, we obtained a value of 1.1026. Note the greater accuracy of the above method.

EXERCISE 45

Apply the method of this lesson to solve the problems which follow.

1. Starting with $y(0.1) = 1.1053$, compute $y(0.2)$ and $y(0.3)$ of Example 45.2. Take $h = 0.1$. Compare with the values found in Lesson 44 and in Exercise 44,3.
2. Find $y(0.1)$ of Example 45.2 in two steps, i.e., first find $y(0.05)$ and then $y(0.05 + 0.05)$. Compare with actual value of $y(0.1) = 1.1055$. Do the same for $y(0.2)$.

3. Following the steps in Comment 44.3–2, develop error formulas, comparable to those in (44.38), for the numerical method of this lesson. *Hint.* Since formula (45.16) is equivalent to the first three terms of a Taylor series, the remainder or error term, by (37.36), is $y'''(X)h^3/3!$, where X is a value of x in the interval under consideration of width h. Ans.

$$(45.3) \quad E\left[\left(x_0 + \frac{h}{2}\right) + \frac{h}{2}\right] = \frac{1}{3}\left\{y\left[\left(x_0 + \frac{h}{2}\right) + \frac{h}{2}\right] - y(x_0 + h)\right\},$$

$$E(x_0 + h) = \frac{4}{3}\left\{y\left[\left(x_0 + \frac{h}{2}\right) + \frac{h}{2}\right] - y(x_0 + h)\right\}.$$

4. Find approximate values when $x = 0.05$, 0.1, 0.15, 0.2 of the particular solution of the differential equation $y' = x + y$, for which $y(0) = 1$. Take $h = 0.05$.

5. In problem 4, compute $y(0.1)$ and $y(0.2)$ with $h = 0.1$. Compare with the value of $y(0.2)$ obtained in 4. In the absence of a solution in terms of elementary function or an error formula, how many decimal place accuracy could you assume in your value of $y(0.2)$. *Hint.* See Comment 44.3–3 Solve the equation and compare with actual value of $y(0.2)$.

6. Find approximate values when $x = 1.1$, 1.2, 1.3 of the particular solution of the equation $y' = x^2 + y^2$ for which $y(1) = 1$. Take $h = 0.1$. What is the approximate formula error in $y(1.1)$? *Hint.* Calculate $y(1.1)$ in two steps and then apply error formula (45.3).

7. Find approximate values when $x = 0.2$, 0.4, 0.6, 0.8, 1 of the particular solution of the equation $y' = 1/(1 + x^2)$ for which $y(0) = 0$. Take $h = 0.2$. Use this value of $y(1)$ to approximate π. *Hint.* See Exercise 44,8. Compare results.

8. Find an approximate value when $x = 1$ of the particular solution of the differential equation $y' = y$ for which $y(0) = 1$. Take $h = 0.1$. Use the value of $y(1)$ to approximate e. *Hint.* See Exercise 44,9. Compare results.

9. Find an approximate value when $x = 2$ of the particular solution of the equation $y' = 1/x$ for which $y(1) = 0$. Take $h = 0.25$. Use this value of $y(2)$ to approximate log 2. *Hint.* See Exercise 44,10. Compare results.

10. By formulas (45.16) and the error term as given in problem 3, we have

$$y(x_0 + h) = y(x_0) + hf\left[x_0 + \frac{h}{2}, \quad y_0 + \frac{h}{2}y'(x_0)\right] + E,$$

where $E = y'''(X)h^3/3!$ Following the procedure outlined in Exercise 44,11, show that the upper bound of the error E in the interval $(x_0, x_0 + nh)$ is

$$(45.31) \quad |E| \leq \frac{h^3}{3!}nM,$$

where M is the maximum value of $y'''(x)$ in the interval $(x_0, x_0 + nh)$.

(a) Use formula (45.31) to compute the upper bound of the error in the value of log 2 as found in problem 9 above. *Hint.* $h = 0.25$, $n = 4$,

$$M = \max|y'''| = \max\left|\frac{2}{x^3}\right| = 2 \text{ in interval } (1,2).$$

(b) What is the largest value of h which can be used to insure that the upper bound of the error in the computation of log 2 is less than 0.005? *Hint.* By (45.31) we want an h such that $h^3nM/3! < 0.005$ and remember $nh = 1$.

ANSWERS 45

1. $y(0.2) = 1.2237$, $y(0.3) = 1.3586$.

2. $y(0.05) = 1.0513$, $y(0.05 + 0.05) = 1.1055$, $y(0.15) = 1.1630$,
$y(0.15 + 0.05) = 1.2242$.

4. $y(0.05) = 1.0525$, $y(0.1) = 1.1103$, $y(0.15) = 1.1736$, $y(0.2) = 1.2427$.

5. $y(0.1) = 1.11$, $y(0.2) = 1.2421$. Can assume two decimal place accuracy.
Actual value: $y(0.2) = 1.2428$.

6. $y(1.1) = 1.2313$, $y(1.2) = 1.5506$, $y(1.3) = 2.0106$, $y(1.05) = 1.1077$,
$y(1.05 + 0.05) = 1.2334$; $E(1.1) = 0.0028$.

7. $y(0.2) = 0.1980$, $y(0.4) = 0.3815$, $y(0.6) = 0.5415$, $y(0.8) = 0.6757$,
$y(1) = 0.7862$.

8. $y(0.1) = 1.105$, $y(0.2) = 1.2210$, $y(0.3) = 1.3492$, $y(0.4) = 1.4909$,
$y(0.5) = 1.6474$, $y(0.6) = 1.8204$, $y(0.7) = 2.0115$, $y(0.8) = 2.2227$,
$y(0.9) = 2.4561$, $y(1) = 2.7140$.

9. $y(1.25) = 0.2222$, $y(1.5) = 0.4040$, $y(1.75) = 0.5578$, $y(2) = 0.6911$.

10. (a) $|E| \leqq 0.0209$.   (b) $h = 0.12$ to two decimal places.

LESSON 46. Starting Method—Taylor Series.

In Lesson 44, we found a numerical solution of the differential equation

$$(46.1) \qquad\qquad y' = f(x,y),$$

for which

$$(46.11) \qquad\qquad y(x_0) = y_0,$$

by a method which is equivalent to using a Taylor series to terms of the first order; in Lesson 45 by a method which is equivalent to using a Taylor series to terms of the second order. These methods suggest that greater accuracy may be achieved if, for a starting method, we use a Taylor series to terms of order greater than two. There are, however, two practical difficulties to the use of a Taylor series.

1. The function $f(x,y)$ may not have a Taylor series expansion over the interval in which a solution is desired. For example, if $y' = f(x,y) = \sqrt{x} + y^2$, then y'' and higher derivatives do not exist at $x = 0$.

2. If $f(x,y)$ has a Taylor series expansion, it may be extremely difficult to obtain the derivatives needed in formula (37.27). For example, try taking a few derivatives of $f(x,y) = \sqrt{x^3 y + xy^3}$.

If these two difficulties are not present, then a Taylor series is indeed a good starting method. By (37.27), a Taylor series has the form

$$(46.12) \quad y(x_0 + h) = y(x_0) + y'(x_0)h + \frac{y''(x_0)}{2!}h^2 + \frac{y'''(x_0)}{3!}h^3$$
$$+ \frac{y^{(4)}(x_0)}{4!}h^4 + \cdots.$$

It says in effect that if *one knows the values of the function $y(x)$ and its derivatives at a point $x = x_0$, then one can find the value of the function for a neighboring point h units away*. By direct substitution in (46.12), we can therefore find $y(x_0 + 0.1)$, $y(x_0 + 0.2)$, $y(x_0 + 0.3)$, etc., provided the series converges for these values of x.

LESSON 46A. Numerical Solution of $y' = f(x,y)$ by Direct Substitution in a Taylor Series. We illustrate this method by means of an example.

Example 46.2. Find approximate values when $x = 0.1, 0.2, 0.3, 0.4$ of a particular solution of the differential equation

(a) $$y' = x^2 + y,$$

for which $y(0) = 1$.

Solution. From (a) we obtain

(b) $y' = x^2 + y, \qquad y'' = 2x + y', \qquad y''' = 2 + y'', \qquad y^{(4)} = y'''.$

Hence when $x = 0$ and $y = 1$, we find from (b)

(c) $y'(0) = 1, \qquad y''(0) = 0 + 1 = 1, \qquad y'''(0) = 2 + 1 = 3, \qquad y^{(4)}(0) = 3.$

By (46.12), with $x_0 = 0$, we have

(d) $$y(h) = y(0) + y'(0)h + \frac{y''(0)}{2!} h^2 + \frac{y'''(0)}{3!} h^3 + \frac{y^{(4)}(0)}{4!} h^4 + \cdots.$$

Substituting in (d), the initial condition and the values found in (c), we obtain

(e) $$y(h) = 1 + h + \frac{h^2}{2} + \frac{h^3}{2} + \frac{h^4}{8} + \cdots.$$

By direct substitution in (e), we have, using only terms to $h^4/8$,

(f) $y(0.1) = 1 + 0.1 + \frac{1}{2}(0.1)^2 + \frac{1}{2}(0.1)^3 + \frac{1}{8}(0.1)^4 = 1.1055125.$

   $y(0.2) = 1 + 0.2 + \frac{1}{2}(0.2)^2 + \frac{1}{2}(0.2)^3 + \frac{1}{8}(0.2)^4 = 1.2242000.$

   $y(0.3) = 1 + 0.3 + \frac{1}{2}(0.3)^2 + \frac{1}{2}(0.3)^3 + \frac{1}{8}(0.3)^4 = 1.3595125.$

   $y(0.4) = 1 + 0.4 + \frac{1}{2}(0.4)^2 + \frac{1}{2}(0.4)^3 + \frac{1}{8}(0.4)^4 = 1.5152000.$

LESSON 46B. Numerical Solution of $y' = f(x,y)$ by the "Creeping up" Process. By the direct substitution method, more and more terms of the series must be included, as h increases, in order to maintain a desired degree of accuracy. These terms, if the derivatives of $f(x,y)$ are complicated functions, may be difficult to obtain. A more accurate method of using a Taylor series with the *same* number of terms is to keep *h fixed* and "creep up" to the value of $y(x_0 + nh)$ in successive steps. We shall demonstrate by the example below how this method works. You will soon discover that the greater accuracy is purchased at a price—more labor.

Example 46.21. Solve the problem of Example 46.2 by the "creeping up" process.

Solution. Using our basic equation (46.12), with $h = 0.1$ and x_0 equal successively 0, 0.1, 0.2, 0.3, we obtain

(a) $y(0 + 0.1) = y(0.1)$

$$= y(0) + y'(0)(0.1) + \frac{y''(0)}{2!}(0.1)^2 + \frac{y'''(0)}{3!}(0.1)^3$$
$$+ \frac{y^{(4)}(0)}{4!}(0.1)^4 + \cdots,$$

$y(0.1 + 0.1) = y(0.2)$

$$= y(0.1) + y'(0.1)(0.1) + \frac{y''(0.1)}{2!}(0.1)^2$$
$$+ \frac{y'''(0.1)}{3!}(0.1)^3 + \frac{y^{(4)}(0.1)}{4!}(0.1)^4 + \cdots,$$

$y(0.2 + 0.1) = y(0.3)$

$$= y(0.2) + y'(0.2)(0.1) + \frac{y''(0.2)}{2!}(0.1)^2$$
$$+ \frac{y'''(0.2)}{3!}(0.1)^3 + \frac{y^{(4)}(0.2)}{4!}(0.1)^4 + \cdots,$$

$y(0.3 + 0.1) = y(0.4)$

$$= y(0.3) + y'(0.3)(0.1) + \frac{y''(0.3)}{2!}(0.1)^2$$
$$+ \frac{y'''(0.3)}{3!}(0.1)^3 + \frac{y^{(4)}(0.3)}{4!}(0.1)^4 + \cdots.$$

The first equation in (a) is the same as (d) of Example 46.2 with $h = 0.1$. Hence by (f) of that example

(b) $$y(0.1) = 1.1055125.$$

By (b) of Example 46.2, the values of the derivatives of $y(x)$ when $x = 0.1$ and $y = 1.1055125$ are

(c) $$y'(0.1) = (0.1)^2 + 1.1055125 = 1.1155125,$$
$$y''(0.1) = 0.2 + 1.1155125 = 1.3155125,$$
$$y'''(0.1) = 2 + 1.3155125 = 3.3155125,$$
$$y^{(4)}(0.1) = 3.3155125.$$

Substituting (b) and (c) in the second line of (a) gives

(d) $y(0.2) = 1.1055125 + (0.1)(1.1155125) + \dfrac{0.01}{2} (1.3155125)$

$$+ \dfrac{0.001}{6} (3.3155125)$$

$$+ \dfrac{0.0001}{24} (3.3155125) = 1.2242077.$$

When $x = 0.2$ and $y(0.2) = 1.2242077$, we find from (b) of Example 46.2,

(e) $\qquad\qquad y'(0.2) = 1.2642077,$

$\qquad\qquad\qquad y''(0.2) = 1.6642077,$

$\qquad\qquad\qquad y'''(0.2) = 3.6642077,$

$\qquad\qquad\qquad y^{(4)}(0.2) = 3.6642077.$

Substituting (d) and (e) in the third line of (a), it becomes

(f) $y(0.3) = 1.2242077 + (0.1)(1.2642077) + (0.005)(1.6642077)$

$$+ \dfrac{0.001}{6} (3.6642077)$$

$$+ \dfrac{0.0001}{24} (3.6642077) = 1.3595755.$$

And when $x = 0.3$, $y(0.3) = 1.3595755$, we find from (b), Example 46.2,

(g) $\qquad\qquad y'(0.3) = 1.4495755,$

$\qquad\qquad\qquad y''(0.3) = 2.0495755,$

$\qquad\qquad\qquad y'''(0.3) = 4.0495755,$

$\qquad\qquad\qquad y^{(4)}(0.3) = 4.0495755.$

Substituting (f) and (g) in the fourth line of (a), we obtain

(h) $y(0.4) = 1.3595755 + (0.1)(1.4495755) + (0.005)(2.0495755)$

$$+ \dfrac{0.001}{6} (4.0495755)$$

$$+ \dfrac{0.0001}{24} (4.0495755) = 1.5154727.$$

Comment 46.3. The particular solution of (a) of Example 46.2 for which $y(0) = 1$ is

(i) $\qquad\qquad y(x) = 3e^x - x^2 - 2x - 2.$

We can therefore compare the actual values of y with those obtained by the direct substitution method and by the creeping up process, see Table 46.31. An examination of the table discloses that the direct substitution

Table 46.31

	Direct Substitution See (f) of Example 46.2	Creeping Up Process See (d), (f), (h) of Example 46.21	Actual Value Obtained from (i)
$y(0.1) =$	1.1055125	1.1055125	1.1055128
$y(0.2) =$	1.2242000	1.2242077	1.2242083
$y(0.3) =$	1.3505125	1.3595755	1.3595764
$y(0.4) =$	1.5152000	1.5154727	1.5154741

method has given six decimal accuracy for $y(0.1)$, four for $y(0.2)$, and three for $y(0.3)$ and $y(0.4)$, after being rounded off to these respective number of places. On the other hand, the creeping up method has given six decimal accuracy for $y(0.1)$, $y(0.2)$, $y(0.3)$, and five decimal accuracy for $y(0.4)$.

Comment 46.4.   Comment on Error in Taylor Series Method.

1. By Theorem 37.34, the remainder or error term of a Taylor series is

$$(46.41) \qquad E = \frac{y^{(n+1)}(X)}{(n+1)!} h^{n+1},$$

where X is a value of x in the interval under consideration of width h. In Example 46.2, we stopped with $y^{(4)}(x)$. Therefore by (46.41), the limit of error in using $h = 0.4$ is, with $n = 4$,

$$(46.42) \qquad |E| \leqq \frac{(0.4)^5}{5!} M = 0.000085M,$$

where M is the maximum value of $|y^{(5)}(x)|$ in the interval $(0,0.4)$ of width 0.4. In the creeping up process, the limit of error in each calculation is, by (46.41), with $h = 0.1$,

$$(46.43) \qquad |E| \leqq \frac{(0.1)^5}{5!} M.$$

The total upper limit of error for four steps due to formula only, i.e., the limit of error in an interval of width 0.4 by the creeping up process, excluding cumulative errors, is therefore,

$$(46.44) \qquad E \leqq 4 \frac{(0.1)^5}{5!} M = 0.000000333M.$$

A comparison of (46.44) with (46.42) shows approximately how much more accurate the creeping up method is.

2. Let

(46.5) $E(x_0 + h) =$ the error in computing $y(x_0 + h)$ by use of formula (46.12) including terms to h^4.

Then by (46.41), with $n = 4$,

(46.51) $E(x_0 + h) = ch^5,$

where $c = y^{(5)}(X)/5!$ Let us divide the h interval in half and compute $y(x_0 + h)$ in two steps. If we have some basis for believing that $y^{(5)}(x)$ changes rather slowly in this small h interval, so that the variation in the value of $y^{(5)}(X)$ in each $h/2$ interval is negligible, then we commit a small error by using the same c of (46.51) for each half interval. Therefore, by (46.51), the error in computing $y(x_0 + h/2)$ for a half interval $h/2$ is approximately

(46.52) $E\left(x_0 + \dfrac{h}{2}\right) = c\left(\dfrac{h}{2}\right)^5 = \dfrac{1}{32}ch^5 = \dfrac{1}{32}E(x_0 + h).$

This means that the error in $y\left(x_0 + \dfrac{h}{2}\right)$ is approximately equal to one thirty-second the error in $y(x_0 + h)$. Hence the error in computing in two steps the value of $y(x_0 + h)$, which we write as $y\left[\left(x_0 + \dfrac{h}{2}\right) + \dfrac{h}{2}\right]$, will have an inherited error in the starting value of $y\left(x_0 + \dfrac{h}{2}\right)$ plus its own formula error. Since each error equals one thirty-second the error in $y(x_0 + h)$, the total error in $y(x_0 + h)$ computed in two steps, i.e., the error in $y\left[\left(x_0 + \dfrac{h}{2}\right) + \dfrac{h}{2}\right]$, will be approximately equal to one-sixteenth the error in $y(x_0 + h)$ computed in one step. Hence

(46.53) $E\left[\left(x_0 + \dfrac{h}{2}\right) + \dfrac{h}{2}\right] = \dfrac{1}{16}E(x_0 + h),$

approximately. Let $Y(x_0 + h)$ be the actual value of the solution. Then

(46.54) $Y(x_0 + h) - y(x_0 + h) = E(x_0 + h),$

$$Y(x_0 + h) - y\left[\left(x_0 + \dfrac{h}{2}\right) + \dfrac{h}{2}\right] = E\left[\left(x_0 + \dfrac{h}{2}\right) + \dfrac{h}{2}\right].$$

Subtracting the second equation in (46.54) from the first, we obtain

(46.55)
$$y\left[\left(x_0 + \dfrac{h}{2}\right) + \dfrac{h}{2}\right] - y(x_0 + h) = E(x_0 + h) - E\left[\left(x_0 + \dfrac{h}{2}\right) + \dfrac{h}{2}\right].$$

Substituting (46.53) in the right side of (46.55), the following equations result.

$$(46.56) \quad y\left[\left(x_0 + \frac{h}{2}\right) + \frac{h}{2}\right] - y(x_0 + h)$$
$$= 16E\left[\left(x_0 + \frac{h}{2}\right) + \frac{h}{2}\right] - E\left[\left(x_0 + \frac{h}{2}\right) + \frac{h}{2}\right],$$

and

$$y\left[\left(x_0 + \frac{h}{2}\right) + \frac{h}{2}\right] - y(x_0 + h) = E(x_0 + h) - \tfrac{1}{16}F(x_0 + h).$$

Simplification of (46.56) gives

$$(46.57) \quad E\left[\left(x_0 + \frac{h}{2}\right) + \frac{h}{2}\right] = \frac{1}{15}\left\{y\left[\left(x_0 + \frac{h}{2}\right) + \frac{h}{2}\right] - y(x_0 + h)\right\},$$

and

$$E(x_0 + h) = \frac{16}{15}\left\{y\left[\left(x_0 + \frac{h}{2}\right) + \frac{h}{2}\right] - y(x_0 + h)\right\}.$$

The first formula says that the error in the value of $y(x_0 + h)$ computed in two steps is equal to one-fifteenth the difference in values of $y(x_0 + h)$ computed in two steps and in one step.

Formulas (46.57) give us a means of estimating errors at each step. Their accuracy does not depend on a knowledge of the size of $y^{(5)}(X)$ which we do not know, but only on the variation of $y^{(5)}(X)/5!$ over a small interval. We have assumed this variation to be negligible, an assumption which is not unreasonable if $y^{(5)}(x)$ changes slowly over h. For example, from the column headed "creeping up" process in Table 46.31, we have with $x_0 = 0$, $h = 0.2$,

(a) $y\left[\left(x_0 + \frac{h}{2}\right) + \frac{h}{2}\right] = y(0.1 + 0.1) = y(0.2) = 1.2242077,$

and from the direct substitution column,

(b) $y(x_0 + h) = y(0.2) = 1.2242000.$

Hence by the first equation in (46.57),

(c) $E\left[\left(x_0 + \frac{h}{2}\right) + \frac{h}{2}\right] = E(0.1 + 0.1) = \tfrac{1}{15}(1.2242077 - 1.224200)$
$$= 0.0000005,$$

which is the approximate error of $y(0.2)$ computed in two steps. The actual error is $1.2242083 - 1.2242077 = 0.0000006$.

3. A check on errors which practical people frequently use, and which seems to work, is to make all calculations over again with an h half the size of the original one. If the results obtained with the smaller h agree with those obtained with the larger h to k decimal places, after being properly rounded off, then it is assumed that the common numerical value has k decimal place accuracy. This method is, of course, meaningful only if the creeping up process is used.

4. This method does not contain a check on arithmetical errors. However, if there is little agreement between the values obtained by using h and $h/2$, all arithmetical computations should be checked. If arithmetical computations are correct, h should be reduced.

Comment 46.58. We have headed this lesson "Starting Formula—Taylor Series." It is evident that it may also be used as a continuing method to find $y(0.5)$, $y(0.6)$, etc. There is, however, a practical objection to its continued use. To insure a desired degree of accuracy, more and more terms of the series will be needed as the distance from the initial point x_0 increases. This means we must evaluate higher and higher order derivatives. In many practical cases, these derivatives, as remarked earlier, may become extremely complicated or may be even so right from the start. Hence, we must seek other starting and continuing methods which do not depend on our ability to obtain derivatives. If, however, it is not difficult to obtain many derivatives, then a Taylor series is not only a good starting method but is also a good continuing method. It can be used as long as the x values remain within the interval of convergence and as long as you are reasonably sure that you have enough terms in the series to give you values that are accurate to the desired number of decimal places. A high-speed calculating machine can, for example, add with ease 100 terms of a series.

EXERCISE 46

Apply the methods of this lesson to solve the problems which follow. Use Taylor series to terms of the fourth order.

1. Use the direct substitution method to find approximate values when $x = 0.1$, 0.2, 0.3, 0.4 of a particular solution of the differential equation $y' = x + y$ for which $y(0) = 1$. Take $h = 0.1$. Solve the equation and compare your results with actual values.

2. (a) Solve problem 1 by using the creeping up method. In the absence of a solution in terms of elementary functions or of an error formula, how could you determine the accuracy of $y(0.4)$?

   (b) Why could not one use error formula (46.41) to determine the upper bound of the error in $y(0.4)$, just as we did in Exercises 44,11 and 45,10 to find an upper bound of the error in the numerical value of $y(2)$? *Ans.* Here y' is a function of x and y; so also is $y^{(n+1)}(x)$. For example, $y^{(4)} = y''' = y'' = 1 + y' = 1 + x + y$. Hence one cannot determine the maximum value of $y^{(n+1)}(x)$ in an interval of width h without knowing $y(x)$. But $y(x)$ is the solution we seek and do not know. In the previous problems y' was a function only of x.

3. Start again with the equation $y' = x + y$ for which $y(0) = 1$. Using the values of $y(0.2)$ found in one step in 1 and in two steps in 2, apply formula (46.57) to correct $y(0.2)$. (NOTE. Formula (46.57) is based on using terms in a Taylor series to order four.) Starting with the corrected figure of $y(0.2)$, compute $y(0.4)$ in one step and in two steps. Correct $y(0.4)$ by means of (46.57). Compare results with those obtained in 1 and 2 and with actual values of the solution.

4. Error formulas (46.57) are based on stopping with the fourth order term in a Taylor series. Find comparable error formulas if one stopped with a third order term, with a fifth order term.

5. Using the direct substitution method, find approximate values when $x = 0.1$, $0.2, 0.3$ of a particular solution of the equation $y' = -xy$ for which $y(0) = 1$.

6. Solve problem 5, using the creeping up method. How could you determine the accuracy of $y(0.3)$? *Hint.* See answer to problem 2(a) above.

7. Can the direct substitution method be used in 5 to find approximate values of $y(0.4)$, $y(0.5)$, $y(0.6)$, $y(0.7)$, $y(0.8)$, $\cdots$?

ANSWERS 46

1. 1.1103417, 1.2428000, 1.3996750, 1.5834667.   Actual values: 1.1103418, 1.2428055, 1.3997176, 1.5836494.

2. (a) 1.1103417, 1.2428052, 1.3997171, 1.5836486.   Compute $y(0.4)$ by the creeping up method with $h = 0.05$. Then make use of Comment 46.4–3.

3. Corrected values: $y(0.2) = 1.2428055$, $y(0.4) = 1.5836493$.

4. Third order term:

$$E\left[\left(x_0 + \frac{h}{2}\right) + -\frac{h}{2}\right] = \frac{1}{7}\left\{y\left[\left(x_0 + \frac{h}{2}\right) + \frac{h}{2}\right] - y(x_0 + h)\right\},$$

$$E(x_0 + h) = \frac{8}{7}\left\{y\left[\left(x_0 + \frac{h}{2}\right) + \frac{h}{2}\right] - y(x_0 + h)\right\}.$$

Fifth order term:

$$E\left[\left(x_0 + \frac{h}{2}\right) + \frac{h}{2}\right] = \frac{1}{31}\left\{y\left[\left(x_0 + \frac{h}{2}\right) + \frac{h}{2}\right] - y(x_0 + h)\right\},$$

$$E(x_0 + h) = \frac{32}{31}\left\{y\left[\left(x_0 + \frac{h}{2}\right) - \frac{h}{2}\right] - y(x_0 + h)\right\}.$$

5. 0.99501, 0.98020, 0.95601.

6. 0.99501, 0.98019, 0.95598.

7. Read Theorem 38.14. The maximum interval of convergence given by the theorem is $|x| < \frac{1}{3}$. Hence, in the absence of additional information, the series found in 5 cannot be used for $|x| > \frac{1}{3}$.

LESSON 47. Starting Method—Runge-Kutta Formulas.

The starting method we shall develop in this lesson for finding a numerical solution of

(47.1) $$y' = f(x,y)$$

satisfying the initial condition

(47.11) $y(x_0) = y_0,$

was developed by the two people after whom it is named, Runge (1856–1927) and Kutta (1867–1944). It has an advantage over the Taylor series method in that it does not use derivatives. As mentioned in the previous lesson, if the function $f(x,y)$ is very complicated, it may be extremely difficult to obtain its second derivative, let alone its third and fourth. Because of this fact and its high degree of accuracy, it is in practice probably the most widely used starting method.

By (46.12),

$$(47.12)\quad y(x_0 + h) = y(x_0) + y'(x_0)h + \frac{y''(x_0)}{2!} h^2 + \frac{y'''(x_0)}{3!} h^3$$
$$+ \frac{y^{(4)}(x_0)}{4!} h^4 + \cdots.$$

Differentiating (47.1), we obtain

$$(47.13)\qquad y''(x) = \frac{\partial}{\partial x} f(x,y) + \frac{\partial}{\partial y} f(x,y)y'(x).$$

Substituting (47.1) and (47.13) in (47.12), and stopping with the h^2 term, we have

$$(47.14)\qquad y(x_0 + h) = y(x_0) + f(x_0,y_0)h$$
$$+ \left[\frac{\partial f(x_0,y_0)}{\partial x} + \frac{\partial f(x_0,y_0)}{\partial y} f(x_0,y_0) \right] \frac{h^2}{2}.$$

Our aim will be to rewrite the right side of (47.14) so that it will have the form

$$(47.15)\quad y(x_0 + h)$$
$$= y(x_0) + Ahf(x_0,y_0) + Bhf[x_0 + Ch,\ y_0 + Dhf(x_0,y_0)],$$

and then find values of A, B, C, D so that the right side of (47.15) will actually equal the right side of (47.14).

By (38.12) and (38.13), the Taylor series expansion of a function of two variables about a point (x_0,y_0), is, with $x = x_0 + a$, $y = y_0 + b$,

$$(47.16)$$
$$f(x_0 + a,\ y_0 + b) = f(x_0,y_0) + a \frac{\partial f(x_0,y_0)}{\partial x} + b \frac{\partial f(x_0,y_0)}{\partial y} + \cdots.$$

In (47.16) let $a = Ch$ and $b = Dhf(x_0,y_0)$. There results

$$(47.17)\qquad f[x_0 + Ch,\ y_0 + Dhf(x_0,y_0)] = f(x_0,y_0)$$
$$+ Ch \frac{\partial f(x_0,y_0)}{\partial x} + Dhf(x_0,y_0) \frac{\partial f(x_0,y_0)}{\partial y} + \cdots.$$

Substituting (47.17) in (47.15) and simplifying the result, we obtain

$$(47.18) \qquad y(x_0 + h) = y(x_0) + (A + B)hf(x_0,y_0)$$
$$+ Bh^2\left[C\,\frac{\partial f(x_0,y_0)}{\partial x} + Df(x_0,y_0)\,\frac{\partial f(x_0,y_0)}{\partial y}\right].$$

Comparing (47.18) with (47.14), we see that both right sides will be alike if, for example,*

$$(47.19) \qquad A + B = 1, \qquad B = \tfrac{1}{2}, \qquad C = 1, \qquad D = 1,$$

from which we obtain

$$(47.2) \qquad A = \tfrac{1}{2}, \qquad B = \tfrac{1}{2}, \qquad C = 1, \qquad D = 1.$$

Substituting these values in (47.15), we have

$$(47.21) \quad y(x_0 + h) = y(x_0) + \tfrac{1}{2}hf(x_0,y_0) + \tfrac{1}{2}hf[x_0 + h,\, y_0 + hf(x_0,y_0)].$$

Formula (47.21) can be written in the simpler form

$$(47.3) \qquad y(x_0 + h) = y(x_0) + \tfrac{1}{2}(u_1 + u_2),$$

where

$$(47.31) \qquad u_1 = hf(x_0,y_0),$$
$$u_2 = hf(x_0 + h,\, y_0 + u_1).$$

If the first four terms of the series (47.12) are used, then the Runge-Kutta formula becomes

$$(47.32) \qquad y(x_0 + h) = y(x_0) + \tfrac{1}{6}(v_1 + 4v_2 + v_3),$$

where

$$(47.33) \qquad v_1 = hf(x_0,y_0) = u_1,$$
$$v_2 = hf(x_0 + \tfrac{1}{2}h,\, y_0 + \tfrac{1}{2}v_1),$$
$$v_3 = hf(x_0 + h,\, y_0 + 2v_2 - v_1).$$

And if the first five terms of the series (47.12) are used, then the Runge-Kutta formula becomes

$$(47.34) \qquad y(x_0 + h) = y(x_0) + \tfrac{1}{6}(w_1 + 2w_2 + 2w_3 + w_4),$$

where

$$(47.35) \qquad w_1 = hf(x_0,y_0) = v_1,$$
$$w_2 = hf(x_0 + \tfrac{1}{2}h,\, y_0 + \tfrac{1}{2}w_1) = v_2,$$
$$w_3 = hf(x_0 + \tfrac{1}{2}h,\, y_0 + \tfrac{1}{2}w_2),$$
$$w_4 = hf(x_0 + h,\, y_0 + w_3).$$

---

*There are other possible choices of A, B, C, D, but these are the simplest.

Since (47.3), (47.32), and (47.34) were obtained by including the second, third, and fourth order terms respectively of a Taylor series, these formulas have been called respectively the **second, third, and fourth order forms.**

As in Taylor series, two methods are available for finding $y(x_0 + 0.1)$, $y(x_0 + 0.2)$, $y(x_0 + 0.3)$ by the Runge-Kutta method, one by direct substitution in formula (47.3) or (47.32) or (47.34), the other by creeping up to $y(x_0 + 0.2)$ and then to $y(x_0 + 0.3)$ by keeping $h = 0.1$ *fixed* and using any one of the formulas in succession.

Example 47.36. Use the fourth order form and the creeping up process to find an approximate value when $x = 0.2$ of the particular solution of

(a) $$y' = x^2 + y,$$

for which $y(0) = 1$.

Solution. Comparing the initial condition with (47.11), we see that $x_0 = 0$, $y_0 = y(x_0) = 1$. Hence, by (47.34), with $x_0 = 0$, $h = 0.1$,

(b) $$y(0.1) = y(0) + \tfrac{1}{6}(w_1 + 2w_2 + 2w_3 + w_4).$$

By (47.35), with $x_0 = 0$, $y_0 = y(x_0) = 1$, $h = 0.1$ and $f(x,y) = x^2 + y$, we have

$$w_1 = 0.1f(0,1) = 0.1(1) = 0.1,$$
$$w_2 = 0.1f(0.05, 1 + 0.05) = 0.1[(0.05)^2 + 1.05] = 0.10525,$$
$$w_3 = 0.1f(0.05, 1 + 0.052625) = 0.1(0.05^2 + 1.052625)$$
$$= 0.1055125,$$
$$w_4 = 0.1f(0.1, 1 + 0.1055125) = 0.1(0.1^2 + 1.1055125)$$
$$= 0.11155125.$$

Hence, by the above values, the initial condition and (b), an approximate value of $y(0.1)$ is

(c) $y(0.1) = 1 + \tfrac{1}{6}(0.1 + 0.2105 + 0.211025 + 0.11155125) = 1.1055127.$

With $x_0 = 0.1$ and $h = 0.1$, (47.34) now becomes

(d) $$y(0.2) = y(0.1) + \tfrac{1}{6}(w_1 + 2w_2 + 2w_3 + w_4).$$

By (47.35), with $x_0 = 0.1$, $y_0 \equiv y(x_0) = y(0.1) = 1.1055127$ and $f(x,y) = x^2 + y$,

$$w_1 = 0.1f(0.1, 1.1055127) = 0.1(0.1^2 + 1.1055127) = 0.11155127,$$
$$w_2 = 0.1f(0.15, 1.1055127 + 0.0557756)$$
$$= 0.1(0.15^2 + 1.1612883) = 0.1183788,$$

$$w_3 = 0.1f(0.15, 1.1055127 + 0.0591894)$$
$$= 0.1(0.15^2 + 1.1647021) = 0.1187202,$$
$$w_4 = 0.1f(0.2, 1.1055127 + 0.1187202) = 0.1264233.$$

Hence by (c) and (d), an approximation of $y(0.2)$ is

(e) $y(0.2)$
$$= 1.1055127 + \tfrac{1}{6}(0.1115513 + 0.2367576 + 0.2374404 + 0.1264233)$$
$$= 1.1055127 + 0.1186954 = 1.2242081.$$

Comment 47.37. If we had found $y(0.2)$ in one step, formula (47.34) would have become, with $h = 0.2$, $x_0 = 0$, $y(x_0) = y_0 = 1$,

(f) $$y(0.2) = 1 + \tfrac{1}{6}(w_1 + 2w_2 + 2w_3 + w_4),$$

where

(g) $w_1 = 0.2f(0,1) = 0.2(1) = 0.2,$
$w_2 = 0.2f(0.1, 1 + 0.1) = 0.2[(0.1)^2 + 1.1] = 0.222,$
$w_3 = 0.2f(0.1, 1 + 0.111) = 0.2[(0.1)^2 + 1.111] = 0.2242,$
$w_4 = 0.2f(0.2, 1 + 0.2242) = 0.2[(0.2)^2 + 1.2242] = 0.25284.$

Hence by (f) and (g)

(h) $$y(0.2) = 1 + \tfrac{1}{6}(0.2 + 0.444 + 0.4484 + 0.25284)$$
$$= 1 + \tfrac{1}{6}(1.34524) = 1.2242067.$$

Comment 47.4.   Comment on Error in Runge-Kutta Method.

1. It has been proved that the error for an interval of width h, in using the Runge-Kutta fourth order form, is

(47.41) $$E(x_0 + h) = Ch^5,$$

where C is a constant. Although the constant C in (47.41) is not the same as the c in (46.51), if we start with (47.41) and follow the steps after (46.51), we arrive at the same conclusion (46.57). Hence by (46.57),

(47.42) $$E\left[\left(x_0 + \frac{h}{2}\right) + \frac{h}{2}\right] = \frac{1}{15}\left\{y\left[\left(x_0 + \frac{h}{2}\right) + \frac{h}{2}\right] - y(x_0 + h)\right\},$$
$$E(x_0 + h) = \frac{16}{15}\left\{y\left[\left(x_0 + \frac{h}{2}\right) + \frac{h}{2}\right] - y(x_0 + h)\right\}.$$

In Example 47.36, we found, with $x_0 = 0$, $h = 0.2$, [see (e)]

(a) $$y\left[\left(x_0 + \frac{h}{2}\right) + \frac{h}{2}\right] = y[0.1 + 0.1] = y(0.2) = 1.2242081,$$

and, in Comment 47.37,

(b) $$y(x_0 + h) = y(0.2) = 1.2242067.$$

Therefore by (47.42), with $x_0 = 0$, $h = 0.2$,

(c) $$E\left[\left(x_0 + \frac{h}{2}\right) + \frac{h}{2}\right] = E(0.1 + 0.1) = \tfrac{1}{15}(1.2242081 - 1.2242067)$$
$$= 0.0000001,$$

which is the approximate error in $y(0.2)$ obtained in two steps. The actual error in $y(0.2)$ obtained in two steps is 0.0000002, see Table 46.31.

2. All other remarks on errors in Comment 46.4 carry over without change to the Runge-Kutta method.

Comment 47.5. This method could also be used as a continuing one. The objection to its use for this purpose lies in the large number of computations which must be made at each step. We shall later give a continuing method which is less time consuming and just as accurate.

EXERCISE 47

In solving the following problems, use the *fourth order* Runge-Kutta method.

1. Prove error formulas (47.42).
2. Using direct substitution, find approximate values when $x = 0.1$ and 0.2 of a particular solution of the equation $y' = x + y$ for which $y(0) = 1$. Take $h = 0.1$. Compare results with actual values.
3. Solve problem 2 by using the creeping up method. In the absence of a solution in terms of elementary functions or of an error formula, how many decimal place accuracy could you assume in the value of $y(0.2)$? Compare with actual value.
4. Using the values of $y(0.2)$ found in one step in 2 and in two steps in 3, apply formula (47.42) to approximate the error in $y(0.1 + 0.1)$. Compare with actual error.
5. Using direct substitution, find approximate values when $x = 0.1$ and 0.2 of a particular solution of the equation $y' = x^2 + y^2$ for which $y(0) = 1$.
6. Solve problem 5, using the creeping up method.
7. Using the values of $y(0.2)$ found in 5 and 6, determine, by means of formula (47.42), the approximate error in $y(0.1 + 0.1)$.

ANSWERS 47

2. $y(0.1) = 1.1103417$, $y(0.2) = 1.2428000$. Actual values: 1.1103418, 1.2428055.
3. $y(0.2) = 1.2428052$. Can assume five decimal accuracy, see Comment 46.4–3.
4. $E(0.1 + 0.1) = 0.0000003$. Actual error $= 0.0000003$.
5. $y(0.1) = 1.1114629$, $y(0.2) = 1.2529908$.
6. $y(0.15) = 1.1776978$, $y(0.2) = 1.2530162$.
7. $E(0.1 + 0.1) = 0.0000017$.

LESSON 48.   Finite Differences.   Interpolation.

Before we can develop a continuing method which is just as accurate as Runge-Kutta but less time consuming, we shall need additional information. We momentarily digress, therefore, to develop this needed information.

LESSON 48A.   Finite Differences.

Definition 48.1. The **first difference** of a function $f(x)$, written as $\Delta f(x)$ (read "delta f of x") is defined as

(48.11) $\Delta f(x) = f(x + h) - f(x)$,

where h is a fixed constant. It is the difference in values of the function for two neighboring values of x, h units apart.

The **second difference** $\Delta^2 f(x)$ is defined as the difference of the first difference of $f(x)$ for two neighboring values of x, h units apart. By definition, therefore,

(48.12) $\Delta^2 f(x) = \Delta[\Delta f(x)] = \Delta[f(x + h) - f(x)] = \Delta f(x + h) - \Delta f(x)$.

Similarly the **third difference** $\Delta^3 f(x)$ is defined as the difference of the second difference of $f(x)$ for two neighboring values of x, h units apart. Therefore

(48.13) $\Delta^3 f(x) = \Delta[\Delta^2 f(x)]$.

And in general, we define

(48.14) $\Delta^n f(x) = \Delta[\Delta^{n-1} f(x)], \quad n = 1, 2, \cdots$.

With the aid of the above definitions, we shall now explain how to construct a table of differences $\Delta f(x), \Delta^2 f(x), \cdots, \Delta^n f(x)$.

Example 48.15. If

(a) $f(x) = e^x$

and $h = 0.1$, construct a table of differences of e^x for values of x from 0 to 0.5.

Solution. (See Table 48.16). Since $h = 0.1$, we shall need values of $f(x) = e^x$ that, beginning with $x = 0$, are 0.1 unit apart. In the first column of our table, we therefore write $x = 0, 0.1, 0.2, 0.3, 0.4, 0.5$. In the second column we write the values of $f(x) = e^x$ when $x = 0, 0.1, \cdots, 0.5$.* By (48.11) $\Delta f(x)$ is the difference in the values of $f(x)$ for two

---

*Various texts are available that contain tables of values of e^x, e^{-x}, $\log x$, $\sin x$, $\cos x$, etc. Excellent tables of values of these and other functions, correct to twelve and more decimal places, have been published under the sponsorship of the National Bureau of Standards, Mathematical tables project.

Table 48.16

$x =$	$f(x) =$ $e^x =$	$\Delta f(x) =$ $\Delta e^x =$	$\Delta^2 f(x) =$ $\Delta^2 e^x =$	$\Delta^3 f(x) =$ $\Delta^3 e^x =$	$\Delta^4 f(x) =$ $\Delta^4 e^x =$	$\Delta^5 f(x) =$ $\Delta^5 e^x =$
0	1.00000	0.10517	0.01106	0.00117	0.00010	0.00007
0.1	1.10517	0.11623	0.01223	0.00127	0.00017	
0.2	1.22140	0.12846	0.01350	0.00144		
0.3	1.34986	0.14196	0.01494			
0.4	1.49182	0.15690				
0.5	1.64872					

neighboring values of x, 0.1 unit apart. Hence we obtain the third column of the table by taking the differences of the second column. For example, by (48.11), $\Delta e^{0.2} = e^{0.3} - e^{0.2} = 1.34986 - 1.22140 = 0.12846$. Analogously $\Delta^2 f(x)$, by (48.12), is the difference of $\Delta f(x)$ for two neighboring values of x, 0.1 unit apart. Therefore we obtain the fourth column by taking differences of the third column. For example, by (48.12), $\Delta^2 e^{0.2} = \Delta(\Delta e^{0.2}) = \Delta e^{0.3} - \Delta e^{0.2} = 0.14196 - 0.12846 = 0.01350$. The elements of the fifth column $\Delta^3 e^x$ are the differences of the fourth column, etc.

Comment 48.17. If you want the fifth difference of $f(x)$, i.e., $\Delta^5 f(x)$, when $x = a$, you must know $f(a), f(a + h), \cdots, f(a + 5h)$. And in general if you want the mth difference of $f(x)$ when $x = a$, you must know $f(a)$, $f(a + h), \cdots, f(a + mh)$. In the above example, $a = 0$, and we therefore had to know $f(0), f(0.1), \cdots, f(0.5)$ in order to evaluate the fifth difference of e^x, i.e., $\Delta^5 e^x$, when $x = 0$.

Comment 48.18. Call $y_0, y_1, \cdots, y_n$ the values of $f(x)$ corresponding to $f(x_0), f(x_0 + h), \cdots, f(x_0 + nh)$. By Definition 48.11,

$$(48.181) \qquad\qquad \Delta y_0 = y_1 - y_0.$$

By (48.12), (48.181) and Definition 48.11,

$$
\begin{aligned}
(48.182) \quad \Delta^2 y_0 &= \Delta[\Delta y_0] = \Delta(y_1 - y_0) = \Delta y_1 - \Delta y_0 \\
&= y_2 - y_1 - y_1 + y_0 = y_2 - 2y_1 + y_0.
\end{aligned}
$$

By (48.13), (48.181) and (48.182),

$$
\begin{aligned}
(48.183) \quad \Delta^3 y_0 &= \Delta[\Delta^2 y_0] = \Delta(y_2 - 2y_1 + y_0) = \Delta y_2 - 2\Delta y_1 + \Delta y_0 \\
&= y_3 - y_2 - 2y_2 + 2y_1 + y_1 - y_0 = y_3 - 3y_2 + 3y_1 - y_0.
\end{aligned}
$$

A comparison of the coefficients on the right sides of (48.182) and (48.183) with the respective coefficients in the expansion of $(x - 1)^2$ and $(x - 1)^3$

shows that they are alike. This leads one to suspect that the fourth difference of y_0 is

(48.184) $\Delta^4 y_0 = y_4 - 4y_3 + 6y_2 - 4y_1 + y_0.$

You can verify that this suspicion is indeed correct. In fact it can be proved by induction, using an argument similar to the one given in Lesson 49A, that the coefficients in the expansion of $\Delta^n y_0$ are the same as the coefficients in the expansion of $(x - 1)^n$. Therefore

(48.19) $$\Delta^n y_0 = \sum_{k=0}^{n} \frac{(-1)^k n!}{k!(n-k)!}\, y_{n-k},^*$$

where it is understood that $\Delta^0 y_0 = y_0$. Hence, in terms of $f(x)$, (48.19) becomes

(48.191) $$\Delta^n f(x_0) = \sum_{k=0}^{n} \frac{(-1)n!}{k!(n-k)!}\, f[x_0 + (n-k)h].$$

LESSON 48B. Polynomial Interpolation. Let us assume that the only information we have about a function is its values when $x = x_0$, $x_0 + h$, $x_0 + 2h$, $\cdots$, $x_0 + mh$, and that we wish to find a value of the function when $x_0 = x_0 + dh$, $d \neq 0$, 1, 2, $\cdots$. This situation occurs, for example, when we need the value of sin 0.5245 and the table we consult gives values of sin 0.5236, sin 0.5265, sin 0.5294, etc., i.e., it gives values over intervals of width $h = 0.0029$. The usual procedure in this case is to add to the value of sin 0.5236, $\frac{9}{29}$ of the difference in values between sin 0.5236 and sin 0.5265. When we do this, we are implicitly assuming that the graph of the sine function between these two points is approximated by a straight line. We are thus in effect using a linear interpolation to approximate the sine function between two points of its graph.

A better method of interpolating the value of sin 0.5245 from the tabular values is to approximate the sine function by a polynomial $f(x)$ whose graph coincides with the sine function at more than the two points sin 0.5236 and sin 0.5265. If it coincides with three values of the sine function, we get a second degree polynomial to approximate sin x; if it coincides with $m + 1$ values, we get a polynomial of degree m to approximate sin x. This approximating polynomial $f(x)$ will then give, when $x = 0.5245$, a more accurate value of sin 0.5245 than will the linear interpolation method described above. Two questions now arise. Given a set of values of a function $f(x)$ at x_0, x_1, $\cdots$, x_m. Does a polynomial exist that can take on these values at the given points, and if there is such a

---

*The fraction $n!/[k!(n-k)!]$ is also frequently written as $\binom{n}{k}$.

polynomial, is it unique? The answer to both questions is given by the following existence and uniqueness theorem which we state without proof.

Theorem 48.2. Let $f(x_0), f(x_1), \cdots, f(x_m)$, be $m + 1$ *distinct values of a function* $f(x)$. *Then there is one and only one polynomial* $F(x)$ *of degree less than or equal to* m *which coincides with these* $m + 1$ *values of* $f(x)$.

Definition 48.21. The unique polynomial $F(x)$ of degree less than or equal to m, whose values coincide with $m + 1$ distinct points of a function $f(x)$, is called a **polynomial interpolating function of** $f(x)$.

Comment 48.22. We shall be interested only in a polynomial interpolating function, although functions other than polynomials may be used for interpolation purposes.

Comment 48.23. Very few of you have used nonlinear interpolation to approximate the value of a function $f(x)$ that is not given in tables, from those which are. To use this method, it is necessary to find an interpolating polynomial $F(x)$ which agrees with $f(x)$ at three or more points and is of degree $\geqq 2$. It would be troublesome indeed, if each time we wished to use polynomial interpolation, we had to find the approximating function $F(x)$. We are indebted to Newton (1642–1727) for formulas by which we can readily make polynomial interpolations whenever the abscissa points, for which the functional values are known, are evenly spaced. These formulas are derived in the next lesson.

EXERCISE 48

1. If $f(x) = \log x$ and $h = 0.1$, construct a table of differences of $\log x$ for values of x from 1 to 1.5.
2. If $f(\theta) = \sin(\theta)$ and $h = 5°$, construct a table of differences of $\sin \theta$ for values of θ from 10° to 35°.

ANSWERS 48

1.

$x =$	$f(x) =$ $\log x =$	$\Delta f(x) =$ $\Delta \log x =$	$\Delta^2 f(x) =$ $\Delta^2 \log x =$	$\Delta^3 f(x) =$ $\Delta^3 \log x =$	$\Delta^4 f(x) =$ $\Delta^4 \log x =$	$\Delta^5 f(x) =$ $\Delta^5 \log x =$
1.1	0.09531	0.08701	−0.00697	0.00104	−0.00022	0.00004
1.2	0.18232	0.08004	−0.00593	0.00082	−0.00018	
1.3	0.26236	0.07411	−0.00511	0.00064		
1.4	0.33647	0.06900	−0.00447			
1.5	0.40547	0.06453				
1.6	0.47000					

2.

$\theta =$	$f(\theta) =$ $\sin\theta =$	$\Delta f(\theta) =$ $\Delta\sin\theta =$	$\Delta^2 f(\theta) =$ $\Delta^2\sin\theta =$	$\Delta^3 f(\theta) =$ $\Delta^3\sin\theta =$	$\Delta^4 f(\theta) =$ $\Delta^4\sin\theta =$	$\Delta^5 f(\theta) =$ $\Delta^5\sin\theta =$
10°	0.17365	0.08517	−0.00197	−0.00063	0.00001	0.00003
15°	0.25882	0.08320	−0.00260	−0.00062	0.00004	
20°	0.34202	0.08060	−0.00322	−0.00058		
25°	0.42262	0.07738	−0.00380			
30°	0.50000	0.07358				
35°	0.57358					

LESSON 49. Newton's Interpolation Formulas.

LESSON 49A. Newton's (Forward) Interpolation Formula. We assume that $f(x)$ is a function which is defined on an interval $(x_0, x_0 + mh)$, and that its values have been given only for the $m + 1$ distinct abscissa points: $x_0, x_0 + h, x_0 + 2h, \cdots, x_0 + mh$, as happens, for example, when we consult a table of values of the function or when these isolated, distinct values have been found experimentally, Fig. 49.1. Our object will be to

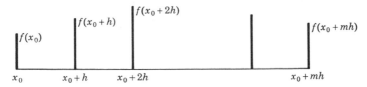

Figure 49.1

find a polynomial $F(x)$, of degree $\leq m$ that agrees with $f(x)$ at these $m + 1$ points. By Theorem 48.2, we know that there is such a polynomial $F(x)$ and that it is unique. By Definition 48.21, this unique polynomial $F(x)$ is an interpolating function of $f(x)$, by which we mean that it will give an approximate value of the function $f(x)$ for an x which does not coincide with the points $x_0, x_0 + h, \cdots, x_0 + mh$. To obtain a formula for $F(x)$, we proceed as follows.

By Definition 48.1

$$\Delta f(x_0) = f(x_0 + h) - f(x_0).$$

Solving for $f(x_0 + h)$, we have

(49.11) $$f(x_0 + h) = f(x_0) + \Delta f(x_0).$$

Applying Δ to both sides of (49.11) gives

(a) $$\Delta f(x_0 + h) = \Delta f(x_0) + \Delta^2 f(x_0).$$

Again by Definition 48.1, we obtain

(b) $\qquad\qquad \Delta f(x_0 + h) = f(x_0 + 2h) - f(x_0 + h).$

Since the left sides of (a) and (b) are the same, we can equate their right sides. Hence

(c) $\qquad\qquad f(x_0 + 2h) = f(x_0 + h) + \Delta f(x_0) + \Delta^2 f(x_0).$

By replacing in (c) the value of $f(x_0 + h)$ as given in (49.11), we have finally

(49.12) $\qquad f(x_0 + 2h) = f(x_0) + 2\Delta f(x_0) + \Delta^2 f(x_0).$

Applying Δ to both sides of (49.12) gives

(d) $\qquad\qquad \Delta f(x_0 + 2h) = \Delta f(x_0) + 2\Delta^2 f(x_0) + \Delta^3 f(x_0).$

By Definition 48.1,

(e) $\qquad\qquad \Delta f(x_0 + 2h) = f(x_0 + 3h) - f(x_0 + 2h).$

Hence equating the right sides of (d) and (e), we obtain

(f) $\qquad f(x_0 + 3h) = f(x_0 + 2h) + \Delta f(x_0) + 2\Delta^2 f(x_0) + \Delta^3 f(x_0).$

Replacing in (f) the value of $f(x_0 + 2h)$ as given in (49.12), we have finally

(49.13) $\quad f(x_0 + 3h) = f(x_0) + 3\Delta f(x_0) + 3\Delta^2 f(x_0) + \Delta^3 f(x_0).$

If you will compare the constant coefficients in the final expressions for $f(x_0 + h), f(x_0 + 2h), f(x_0 + 3h)$ as given in (49.11), (49.12), and (49.13), with the coefficients in the respective expansions of $(x + 1)$, $(x + 1)^2$, $(x + 1)^3$, you will discover that they are the same. This observation leads one to suspect that the coefficients in the expansion of $f(x_0 + nh)$ will be the same as those in the expansion of $(x + 1)^n$. We shall now prove by induction that this suspicion is indeed correct.

Proof. We assume that the coefficients in the expansion of $f[x_0 + (k - 1)h]$ are the same as the coefficients in the expansion of $(x + 1)^{k-1}$. That is we assume

(a) $f[x_0 + (k - 1)h] = f(x_0) + (k - 1)\,\Delta f(x_0)$

$\qquad\qquad + \dfrac{(k - 1)(k - 2)}{2!}\,\Delta^2 f(x_0)$

$\qquad\qquad + \dfrac{(k - 1)(k - 2)(k - 3)}{3!}\,\Delta^3 f(x_0) + \cdots + \Delta^{k-1} f(x_0)$

is a valid equation. We must then prove, by Comment 24.1, that the coefficients in the expansion of $f(x_0 + kh)$ are the same as the coefficients in the expansion of $(x + 1)^k$. Applying Δ to both sides of (a), we obtain

(b) $\Delta f[x_0 + (k - 1)h] = \Delta f(x_0) + (k - 1) \Delta^2 f(x_0)$

$$+ \frac{(k - 1)(k - 2)}{2!} \Delta^3 f(x_0) + \cdots + \Delta^k f(x_0).$$

Adding the *left* sides of (a) and (b) there results, by (48.11),

(c) $f[x_0 + (k - 1)h] + \Delta f[x_0 + (k - 1)h]$

$$= f[x_0 + (k - 1)h] + f(x_0 + kh) - f[x_0 + (k - 1)h]$$

$$= f(x_0 + kh).$$

The addition of the *right* sides of (a) and (b) gives

(d) $f(x_0) + k \Delta f(x_0) + \dfrac{(k - 1)(k - 2) + 2(k - 1)}{2!} \Delta^2 f(x_0)$

$$+ \frac{(k - 1)(k - 2)(k - 3) + 3(k - 1)(k - 2)}{3!} \Delta^3 f(x_0) + \cdots$$

$$+ \frac{(k - 1)(k - 2) \cdots (k - m) + m(k - 1)(k - 2) \cdots (k - m + 1)}{m!}$$

$$\times \Delta^m f(x_0) + \cdots + \Delta^k f(x_0),$$

which simplifies to

(e) $\qquad f(x_0) + k \Delta f(x_0) + \dfrac{k(k - 1)}{2!} \Delta^2 f(x_0) + \dfrac{k(k - 1)(k - 2)}{3!} \Delta^3 f(x_0)$

$$+ \cdots + \frac{k(k - 1)(k - 2) \cdots (k - m + 1)}{m!} \Delta^m f(x_0) + \cdots + \Delta^k f(x_0).$$

Since (c) is the sum of the left sides of (a) and (b), and (e) is the sum of their right sides, we can equate these two to obtain

(f) $f(x_0 + kh) = f(x_0) + k \Delta f(x_0) + \dfrac{k(k - 1)}{2!} \Delta^2 f(x_0)$

$$+ \frac{k(k - 1)(k - 2)}{3!} \Delta^3 f(x_0) + \cdots$$

$$+ \frac{k(k - 1)(k - 2) \cdots (k - m + 1)}{m!} \Delta^m f(x_0) + \cdots + \Delta^k f(x_0).$$

The coefficients in the expression on the right side are the same as those in the expansion of $(x + 1)^k$. Hence our suspicion is proved.

We therefore can now write, by (f),

(49.14) $\qquad f(x_0 + nh) = f(x_0) + n \Delta f(x_0) + \dfrac{n(n - 1)}{2!} \Delta^2 f(x_0)$

$$+ \frac{n(n - 1)(n - 2)}{3!} \Delta^3 f(x_0) + \cdots$$

$$+ \frac{n(n - 1)(n - 2) \cdots (n - m + 1)}{m!} \Delta^m f(x_0),$$

$$n = 0, 1, 2, \cdots, m.$$

[Note that when $n = m$, the last term simplifies to $\Delta^m f(x_0)$.] For each $n = 1, 2, \cdots, m$, the right side of (49.14) gives a formula for computing the respective ordinates $f(x_0)$, $f(x_0 + h)$, $\cdots$, $f(x_0 + nh)$ in terms of $f(x_0)$ and the differences of $f(x_0)$.*

Since the right side of (49.14) is meaningful even when n is not zero or an integer, we can let

$$(49.15) \quad F(x) = f(x_0) + n\,\Delta f(x_0) + \frac{n(n-1)}{2!}\,\Delta^2 f(x_0)$$

$$+ \frac{n(n-1)(n-2)}{3!}\,\Delta^3 f(x_0) + \cdots$$

$$+ \frac{n(n-1)(n-2)\cdots(n-m+1)}{m!}\,\Delta^m f(x_0),$$

where n is *any* number between 0 and m, and where

$$(49.16) \quad\quad\quad x = x_0 + nh, \quad n = \frac{x - x_0}{h}.$$

Since the right side of (49.15) is the same as the right side of (49.14), we know that

$$(49.161) \quad F(x_0 + nh) = f(x_0 + nh), \quad n = 0, 1, 2, \cdots, m.$$

Substituting the second equality of (49.16) in (49.15), we obtain

$$(49.17) \quad F(x) = f(x_0) + \frac{\Delta f(x_0)}{h}\,(x - x_0) + \frac{\Delta^2 f(x_0)}{2!h^2}\,(x - x_0)(x - x_0 - h)$$

$$+ \frac{\Delta^3 f(x_0)}{3!h^3}\,(x - x_0)(x - x_0 - h)(x - x_0 - 2h) + \cdots$$

$$+ \frac{\Delta^m f(x_0)}{m!h^m}\,(x - x_0)(x - x_0 - h)$$

$$\times\,(x - x_0 - 2h) \cdots (x - x_0 - [m-1]h).$$

By (49.17), we see that $F(x)$ is a polynomial of degree $\leqq m$.

We have thus proved, by (49.161) and (49.17), that:

1. The function $F(x)$ defined by (49.15) or (49.17) agrees with the function $f(x)$ at the $m + 1$ points for which $x = x_0$, $x_0 + h$, $x_0 + 2h$, $\cdots, x_0 + mh$,
2. $F(x)$ is a polynomial of degree $\leqq m$.

Therefore, by Theorem 48.2 and Definition 48.21, $F(x)$ is the unique

---

*Note the resemblance between this formula and the Taylor series formula (37.27) by which one can compute a value of $f(x_0 + nh)$ in terms of $f(x_0)$ and the derivatives of $f(x)$ at $x = x_0$.

polynomial interpolating function of the given function $f(x)$. This means that if we wish to find an approximate value of $f(x)$, when

$$(49.18) \qquad x = x_0 + nh, \qquad n = \frac{x - x_0}{h}, \qquad n \text{ real,}$$

we can use the function $F(x)$ defined by either of the two formulas (49.15) or (49.17). These formulas are known as **Newton's (forward) interpolation formulas.**

Example 49.19. Use Newton's (forward) interpolation formulas (49.15) and (49.17) and the following table of values

(a)
$$e^{0.1} = 1.105\ 17,$$
$$e^{0.2} = 1.221\ 40,$$
$$e^{0.3} = 1.349\ 86,$$
$$e^{0.4} = 1.491\ 82,$$
$$e^{0.5} = 1.648\ 72,$$

to find an approximate value of $e^{0.14}$.

Solution. *By use of (49.15).* Here $f(x) = e^x$, $x_0 = 0.1$ and $h = 0.1$. Since we have been given five equally spaced values of the function e^x, by Comment 48.17, we shall be able to find only the fourth difference, namely $\Delta^4 f(x_0)$, in formula (49.15) or (49.17). We shall therefore have to stop with $m = 4$. And since we seek the value of $e^{0.14}$, the x in these formulas is 0.14. Therefore by (49.18), $n = \dfrac{0.14 - 0.1}{0.1} = 0.4$. Substituting the above values in (49.15), it becomes

(b) $$F(0.14) = f(0.1) + (0.4)\,\Delta f(0.1) + \frac{(0.4)(0.4 - 1)}{2!}\,\Delta^2 f(0.1)$$

$$+ \frac{(0.4)(0.4 - 1)(0.4 - 2)}{3!}\,\Delta^3 f(0.1)$$

$$+ \frac{(0.4)(0.4 - 1)(0.4 - 2)(0.4 - 3)}{4!}\,\Delta^4 f(0.1).$$

The needed differences have already been calculated in Table 48.16 in the row beginning with 0.1. Therefore (b) becomes

(c) $$F(0.14) = 1.105\ 17 + (0.116\ 23)(0.4) - (0.012\ 23)(0.12)$$

$$+ (0.001\ 27)(0.064) - (0.000\ 17)(0.0416)$$

$$= 1.150\ 27,$$

which is the approximate value of $e^{0.14}$.

Solution. *By use of (49.17).* In this formula $x - x_0 = 0.14 - 0.1 = 0.04$. As before $h = 0.1$, $m = 4$. Substituting these values in (49.17), it becomes

(d) $F(0.14) = f(0.1) + \dfrac{0.04}{0.1}\, \Delta f(0.1) + \dfrac{\Delta^2 f(0.1)}{2!(0.1)^2}\,(0.04)(0.04 - 0.1)$

$\qquad + \dfrac{\Delta^3 f(0.1)}{3!(0.1)^3}\,(0.04)(0.04 - 0.1)(0.04 - 0.2)$

$\qquad + \dfrac{\Delta^4 f(0.1)}{4!(0.1)^4}\,(0.04)(0.04 - 0.1)(0.04 - 0.2)(0.04 - 0.3),$

which is equivalent to (b) above.

Comment 49.191. The actual value of $e^{0.14}$ to five decimals is 1.150 27, the same as that obtained in (c) above by using a fourth degree polynomial interpolating function. If you had used linear interpolation, i.e., obtained the value of $e^{0.14}$ by adding to $e^{0.1}$ two-fifths of the difference in values between $e^{0.1}$ and $e^{0.2}$, you would have obtained 1.151 66 which is correct to only two decimal places.

LESSON 49B. Newton's (Backward) Interpolation Formula. Newton also gave a formula by which it is possible to interpolate backward instead of forward. Let ∇, which is an upside down delta, called "del," be defined by

(49.2) $\qquad\qquad \nabla f(x) = f(x) - f(x - h).$

As before we assume we have been given values of $f(x)$ only for those points whose abscissas are x_0, $x_0 - h$, $x_0 - 2h$, $\cdots$, $x_0 - mh$, Fig. 49.21.

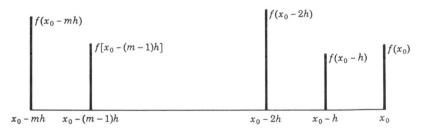

Figure 49.21

Let $F(x)$, of degree $\leqq m$, be a polynomial interpolating function of $f(x)$. By following the procedure outlined in Lesson 49A, you will obtain, in place of (49.15),

$$(49.22) \quad F(x) = f(x_0) - n\nabla f(x_0) + \frac{n(n-1)}{2!} \nabla^2 f(x_0)$$
$$- \frac{n(n-1)(n-2)}{3!} \nabla^3 f(x_0) + \cdots$$
$$+ \frac{n(n-1)(n-2)\cdots(n-m+1)}{m!} (-1)^m \nabla^m f(x_0),$$

where

$$(49.23) \quad x = x_0 - nh, \qquad n = \frac{x_0 - x}{h} = -\frac{x - x_0}{h}, \quad n \text{ real.}$$

We have left the proof to you as an exercise.

If, in (49.22), we replace n by its value as given in (49.23), we obtain the alternate form

$$(49.24) \quad F(x) = f(x_0) + \frac{\nabla f(x_0)}{h}(x - x_0) + \frac{\nabla^2 f(x_0)}{2!h^2}(x - x_0)(x - x_0 + h)$$
$$+ \frac{\nabla^3 f(x_0)}{3!h^3}(x - x_0)(x - x_0 + h)(x - x_0 + 2h) + \cdots$$
$$+ \frac{\nabla^m f(x_0)}{m!h^m}(x - x_0)(x - x_0 + h)$$
$$\times (x - x_0 + 2h)\cdots(x - x_0 + [m-1]h).$$

Formulas (49.22) and (49.24) are known as **Newton's (backward) interpolation formulas.**

Example 49.25. Use the values of e^x given in (a) of Example 49.19, and Newton's backward interpolation formula to find an approximate value of $e^{0.46}$.

Solution. As in the previous Example 49.19, $f(x) = e^x$, $h = 0.1$ and since we have been given five evenly spaced values of e^x, the m in (49.22) is four. Here, however, $x_0 = 0.5$, $x = 0.46$ and by (49.23), $n = \dfrac{0.5 - 0.46}{0.1} = 0.4$. Substituting these values in (49.22), we obtain

(a) $\quad F(0.46) = f(0.5) - 0.4\nabla f(0.5) + \dfrac{0.4(-0.6)}{2!} \nabla^2 f(0.5)$

$\quad - \dfrac{0.4(-0.6)(-1.6)}{3!} \nabla^3 f(0.5) + \dfrac{(0.4)(-0.6)(-1.6)(-2.6)}{4!} \nabla^4 f(0.5).$

The needed differences are given in Table 49.26. They are slightly different in this case from those in Table 48.16. Here $x_0 = 0.5$. We must therefore find differences of $f(0.5)$ instead of $f(0.1)$, and remember $\nabla e^{0.5} = e^{0.5} - e^{0.4}$, $\nabla e^{0.4} = e^{0.4} - e^{0.3}$, etc. Substituting in (a) the

Table 49.26

$x =$	$f(x) =$ $e^x =$	$\nabla f(x) =$ $\nabla e^x =$	$\nabla^2 f(x) =$ $\nabla^2 e^x =$	$\nabla^3 f(x) =$ $\nabla^3 e^x =$	$\nabla^4 f(x) =$ $\nabla^4 e^x =$
0.1	1.10517				
0.2	1.22140	0.11623			
0.3	1.34986	0.12846	0.01223		
0.4	1.49182	0.14196	0.01350	0.00127	
0.5	1.64872	0.15690	0.01494	0.00144	0.00017

figures shown in the row beginning with 0.5 of Table 49.26, we obtain

(b) $\quad F(0.46) = 1.64872 - 0.4(0.15690) - (0.12)(0.01494)$
$$- (0.064)(0.00144) - (0.0416)(0.00017)$$
$$= 1.64872 - 0.06465$$
$$= 1.58407,$$

which is the approximate value of $e^{0.46}$. Its actual value to five places is 1.58407 which agrees with (b) to five places. By a linear interpolation, you would have obtained 1.58596.

LESSON 49C. The Error in Polynomial Interpolation. When we use a polynomial $F(x)$ of degree $\leq m$ as an interpolating function for $f(x)$, we know only that both agree at those points whose abscissas are x_0, $x_0 + h$, $\cdots$, $x_0 + mh$. For other values of x, the difference between $F(x)$ and $f(x)$ may be very small or very large depending on what the graph of $f(x)$ looks like for these intermediate values.

The following error formula (49.31), which we state without proof, is based on the assumption that $f(x)$ has a continuous derivative of order $m + 1$ in an interval $I: a \leq x \leq b$, which contains the points $x_0, x_0 + h$, $\cdots, x_0 + mh$. Let

(49.3) $\qquad E(x) = f(x) - F(x),$

where $E(x)$ is the **error function** or the **remainder function**. It is the difference between the value of the function $f(x)$ and the interpolating function $F(x)$. Then

(49.31) $\qquad E(x) = \dfrac{(x - x_0)(x - x_1) \cdots (x - x_m)}{(m + 1)!} f^{(m+1)}(X),$

where x_0, $x_1 = x_0 + h$, $x_2 = x_0 + 2h$, $\cdots$, $x_m = x_0 + mh$ are the values of x for which $f(x)$ and $F(x)$ agree, and X is a number between the

smallest and largest values of x_0, x_1, $\cdots$, x_m and x. Unfortunately the
formula does not tell us how to choose X. However, if we can determine
the minimum and maximum values of $|f^{(m+1)}(x)|$ in the given interval I,
then $|f^{(m+1)}(X)|$ must be between them. Therefore by (49.31),

$$(49.32) \qquad \left| \frac{(x - x_0)(x - x_1) \cdots (x - x_m)}{(m + 1)!} \right| [\min |f^{(m+1)}(x)|] \leqq |E(x)|$$

$$\leqq \left| \frac{(x - x_0)(x - x_1) \cdots (x - x_m)}{(m + 1)!} \right| [\max |f^{(m+1)}(x)|], \quad x \text{ in } I.$$

Formula (49.32) is also valid if $x_1 = x_0 - h$, $x_2 = x_0 - 2h$, $\cdots$,
$x_m = x_0 - mh$. Hence (40.32) is valid for both Newton's forward and
backward interpolation formulas.

Example 49.33. Find the maximum error incurred in using Newton's
forward interpolation formula in Example 49.19.

Solution. In Example 49.19, $h = 0.1$, $x_0 = 0.1$, $x = 0.14$, $f(x) = e^x$,
$m = 4$. Hence $x_1 = x_0 + h = 0.2$, $x_2 = x_0 + 2h = 0.3$, $x_3 = x_0 + 3h = 0.3$,
$x_4 = x_0 + 4h = 0.4$. Since $m = 4$, we must, in (49.31), stop with $x - x_4$.
Hence $f^{(m+1)}e^x = f^{(5)}e^x = e^x$. In the interval $(0.1, 0.5)$, e^x has the largest
value when $x = 0.5$. Therefore by (49.32),

$$(a) \quad |E(x)| \leqq \left| \frac{(0.14 - 0.1)(0.14 - 0.2)(0.14 - 0.3)}{5!} \times (0.14 - 0.4)(0.14 - 0.5) \, e^{0.5} \right|$$

$$< 0.0000005.$$

EXERCISE 49

1. Use Newton's (forward) interpolation formulas (49.15) and (49.17) and the
following table of values, $e^{0.3} = 1.34986$, $e^{0.4} = 1.49182$, $e^{0.5} = 1.64872$,
$e^{0.6} = 1.82212$, $e^{0.7} = 2.01375$, to find an approximate value of $e^{0.35}$.
Compare with actual value and value obtained by a linear interpolation.

2. Use (49.15) and (49.17) and the following table of values, $\log 1.1 = 0.09531$,
$\log 1.2 = 0.18232$, $\log 1.3 = 0.26236$, $\log 1.4 = 0.33647$, $\log 1.5 = 0.40547$,
$\log 1.6 = 0.47000$, to approximate $\log 1.12$. Compare with actual value and
value obtained by a linear interpolation. *Hint.* The needed table of differences is given in answer to Exercise 48,1.

3. The following temperatures were recorded at hourly intervals: 6 A.M., $2°$;
7 A.M., $12°$; 8 A.M., $17°$; 9 A.M., $20°$; 10 A.M., $22°$. What was the approximate
temperature at 6:30 A.M.?

4. Prove formula (49.22). *Hint.* Follow the procedure used to obtain (49.15).

5. Use Newton's (backward) interpolation formulas (49.22) and (49.24) and
the table of values given in 1 and 2 to approximate respectively (a) $e^{0.65}$,
(b) $\log 1.53$. Compare with actual values and with those obtained by a
linear interpolation.

ANSWERS 49

1. $e^{0.35} = 1.41906$. Actual value: 1.41907. Linear interpolation: 1.42084.
2. $\log 1.12 = 0.11333$. Actual value: 0.11333. Linear interpolation 0.11271.
3. Approximately 7.9°.
5. (a) $e^{0.65} = 1.91554$. Actual value 1.91554.   Linear interpolation 1.91794.
   (b) $\log 1.53 = 0.42527$. Actual value 0.42527. Linear interpolation: 0.42483.

LESSON 50.   Approximation Formulas Including Simpson's Rule and Weddle's Rule.

We are now ready to continue our study of the problem of finding a numerical solution of

(50.1) $$y' = f(x,y)$$

satisfying the initial condition

(50.11) $$y(x_0) = y_0.$$

Let $y(x)$ be a solution of (50.1) fulfilling (50.11). Then, by (50.1),

(50.12) $$y' = f[x,y(x)],$$

where f is now a function of x alone. Integration of (50.12) and insertion of the initial condition gives

(50.13) $$\int_{y=y_0}^{y} dy = \int_{x=x_0}^{x} f[x,y(x)]\,dx, \qquad y = y_0 + \int_{x_0}^{x} f[x,y(x)]\,dx.$$

If $x = x_0 + nh$, (50.13) becomes, with y_0 replaced by its equal $y(x_0)$ of (50.11),

(50.14) $$y(x_0 + nh) = y(x_0) + \int_{x_0}^{x_0+nh} f[x,y(x)]\,dx.$$

Since $y(x)$ is the unknown solution of (50.1), the function $f[x,y(x)]$ is also unknown. Hence we cannot possibly hope to perform the integration in (50.14). However, we can use a polynomial interpolating function $F(x)$ for $f[x,y(x)]$ just as we did in the previous lesson for $f(x)$.

Let $F(x)$ be such a polynomial interpolating function, agreeing with $f[x,y(x)]$ at the $m + 1$ points whose abscissas are x_0, $x_0 + h$, $x_0 + 2h$, $\cdots$, $x_0 + mh$. Therefore, by (49.17), with $f(x_0)$ replaced by $f[x_0,y(x_0)]$ which, by (50.12), is equal to $y'(x_0)$,

(50.15) $$F(x) = \left[1 + \frac{x - x_0}{h}\,\Delta + \frac{(x - x_0)(x - x_0 - h)}{2!h^2}\,\Delta^2 \right.$$
$$+ \frac{(x - x_0)(x - x_0 - h)(x - x_0 - 2h)}{3!h^3}\,\Delta^3 + \cdots$$
$$\left. + \frac{(x - x_0)(x - x_0 - h)(x - x_0 - 2h)\cdots(x - x_0 - [m-1]h)}{m!h^m}\,\Delta^m \right] y'(x_0),$$

where for convenience in writing we have placed $y'(x_0)$ outside the brackets.

Comment 50.16. If we stop with the Δ^2 term in (50.15), then $m = 2$ and $F(x)$ becomes a polynomial of degree less than or equal to two agreeing with $f[x,y(x)] = y'(x)$ at the three points whose abscissas are x_0, $x_0 + h$, $x_0 + 2h$, see 1 after (49.17). If we stop with the Δ^3 term, then $m = 3$ and $F(x)$ becomes a polynomial of degree less than or equal to three that agrees with $f[x,y(x)] = y'(x)$ at the four points whose abscissas are $x_0, \cdots, x_0 + 3h$, etc.

By (50.15), an approximation to the integral in (50.14) is, therefore,

$$
(50.2) \quad \int_{x_0}^{x_0+nh} F(x)\, dx
$$

$$
= \int_{x_0}^{x_0+nh} \left[1 + \frac{x - x_0}{h}\,\Delta + \frac{(x - x_0)(x - x_0 - h)}{2!h^2}\,\Delta^2 + \cdots \right.
$$

$$
\left. + \frac{\begin{Bmatrix} (x - x_0)(x - x_0 - h)(x - x_0 - 2h) \cdots \\ (x - x_0 - [m-1]h) \end{Bmatrix}}{m!h^m}\,\Delta^m \right] y'(x_0)\, dx.
$$

To simplify integrating the right side of (50.2), we make the substitution $u = x - x_0$. Therefore $du = dx$, $u = 0$ when $x = x_0$ and $u = nh$ when $x = x_0 + nh$. Hence we can replace (50.2) by

$$
(50.21) \quad \int_{x_0}^{x_0+nh} F(x)\, dx
$$

$$
= \int_0^{nh} \left[1 + \frac{u}{h}\,\Delta + \frac{u(u - h)}{2!h^2}\,\Delta^2 + \frac{u(u - h)(u - 2h)}{3!h^3}\,\Delta^3 + \cdots \right.
$$

$$
\left. + \frac{u(u - h)(u - 2h) \cdots (u - [m-1]h)}{m!h^m}\,\Delta^m \right] y'(x_0)\, du
$$

$$
= \int_0^{nh} \left(1 + \frac{u}{h}\,\Delta + \frac{u^2 - hu}{2!h^2}\,\Delta^2 + \frac{u^3 - 3hu^2 + 2h^2u}{3!h^3}\,\Delta^3 \right.
$$

$$
+ \frac{u^4 - 6hu^3 + 11h^2u^2 - 6h^3u}{4!h^4}\,\Delta^4
$$

$$
+ \frac{u^5 - 10hu^4 + 35h^2u^3 - 50h^3u^2 + 24h^4u}{5!h^5}\,\Delta^5
$$

$$
+ \frac{\begin{Bmatrix} u^6 - 15hu^5 + 85h^2u^4 - 225h^3u^3 \\ + 274h^4u^2 - 120h^5u \end{Bmatrix}}{6!h^6}\,\Delta^6
$$

$$
\left. + \cdots \right) y'(x_0)\, du.
$$

Integration of (50.21) gives

(50.22) $\displaystyle\int_{x_0}^{x_0+nh} F(x)\,dx =$

$$\left[nh + \frac{1}{h}\frac{n^2h^2}{2}\Delta + \frac{1}{2h^2}\left(\frac{n^3h^3}{3} - \frac{n^2h^3}{2}\right)\Delta^2 \right.$$

$$+ \frac{1}{6h^3}\left(\frac{n^4h^4}{4} - n^3h^4 + n^2h^4\right)\Delta^3$$

$$+ \frac{1}{24h^4}\left(\frac{n^5h^5}{5} - \frac{3n^4h^5}{2} + \frac{11n^3h^5}{3} - 3n^2h^5\right)\Delta^4$$

$$+ \frac{1}{120h^5}\left(\frac{n^6h^6}{6} - 2n^5h^6 + \frac{35n^4h^6}{4} - \frac{50n^3h^6}{3} + 12n^2h^6\right)\Delta^5$$

$$+ \frac{1}{720h^6}\left(\frac{n^7h^7}{7} - \frac{5n^6h^7}{2} + 17n^5h^7 - \frac{225n^4h^7}{4}\right.$$

$$\left.\left. + \frac{274n^3h^7}{3} - 60n^2h^7\right)\Delta^6 + \cdots \right] y'(x_0).$$

Simplification of (50.22) results in

(50.23) $\displaystyle\int_{x_0}^{x_0+nh} F(x)\,dx$

$$= nh\left[1 + \frac{n}{2}\Delta + \frac{1}{2}\left(\frac{n^2}{3} - \frac{n}{2}\right)\Delta^2 + \frac{1}{6}\left(\frac{n^3}{4} - n^2 + n\right)\Delta^3 \right.$$

$$+ \frac{1}{24}\left(\frac{n^4}{5} - \frac{3n^3}{2} + \frac{11n^2}{3} - 3n\right)\Delta^4$$

$$+ \frac{1}{120}\left(\frac{n^5}{6} - 2n^4 + \frac{35n^3}{4} - \frac{50n^2}{3} + 12n\right)\Delta^5$$

$$+ \frac{1}{720}\left(\frac{n^6}{7} - \frac{5n^5}{2} + 17n^4 - \frac{225n^3}{4} + \frac{274n^2}{3} - 60n\right)\Delta^6$$

$$\left. + \cdots \right] y'(x_0).$$

If $n = 1$ in (50.23), we obtain

(50.24) $\displaystyle\int_{x_0}^{x_0+h} F(x)\,dx = h\left[1 + \frac{1}{2}\Delta - \frac{1}{12}\Delta^2 + \frac{1}{24}\Delta^3 - \frac{19}{720}\Delta^4 \right.$

$$\left. + \frac{3}{160}\Delta^5 - \frac{863}{60,480}\Delta^6 + \cdots \right] y'(x_0).$$

Therefore, by (50.14) and (50.24), an approximation to $y(x_0 + h)$ in terms of forward differences of $y'(x_0)$ is

(50.25) $$y(x_0 + h) = y(x_0) + H,$$

where H is the expression on the right of (50.24).

If in (50.23), we stop with the Δ term, then by Comment 50.16, the polynomial interpolating function $F(x)$ is of degree less than or equal to one, agreeing with $f[x,y(x)]$ at the two points for which $x = x_0$ and $x_0 + h$. With $n = 1$ and using only terms including Δ of (50.23), we obtain, with the help of (48.11),

(50.3)
$$\int_{x_0}^{x_0+h} F(x)\, dx = h[1 + \tfrac{1}{2}\Delta]\, y'(x_0)$$
$$= h\{y'(x_0) + \tfrac{1}{2}[y'(x_0 + h) - y'(x_0)]\}.$$
$$= \frac{h}{2}\, [y'(x_0) + y'(x_0 + h)].$$

Therefore, by (50.14) and (50.3), an approximation of $y(x_0 + h)$ is

(50.31) $$y(x_0 + h) = y(x_0) + \frac{h}{2}\, [y'(x_0) + y'(x_0 + h)].$$

Formula (50.31) is known as the **trapezoidal rule** for approximating a numerical solution of (50.1) satisfying (50.11).

If in (50.23), we stop with the Δ^2 term, then by Comment 50.16, $F(x)$ is the polynomial interpolation function of $f[x,y(x)]$ of degree less than or equal to two, agreeing with f at the three points for which $x = x_0, x_0 + h$, $x_0 + 2h$. With $n = 2$, and using only terms including Δ^2 of (50.23), we obtain, with the help of Comment 48.18,

(50.32)
$$\int_{x_0}^{x_0+2h} F(x)\, dx = 2h[1 + \Delta + \tfrac{1}{2}(\tfrac{4}{3} - 1)\Delta^2]\, y'(x_0)$$
$$= 2h[y'(x_0) + \Delta y'(x_0) + \tfrac{1}{6}\Delta^2 y'(x_0)]$$
$$= 2h\{y'(x_0) + y'(x_0 + h) - y'(x_0)$$
$$+ \tfrac{1}{6}[y'(x_0 + 2h) - 2y'(x_0 + h) + y'(x_0)]\}$$
$$= \frac{h}{3}\, [y'(x_0) + 4y'(x_0 + h) + y'(x_0 + 2h)].$$

Therefore by (50.14) and (50.32), an approximation to $y(x_0 + 2h)$ is

(50.33) $$y(x_0 + 2h) = y(x_0) + \frac{h}{3}\, [y'(x_0) + 4y'(x_0 + h) + y'(x_0 + 2h)].$$

Formula (50.33) is known as **Simpson's rule** for approximating a numerical solution of (50.1) satisfying (50.11).

If in (50.23), we stop with the Δ^3 term, then by Comment 50.16, the polynomial interpolation function $F(x)$ is of degree less than or equal to three, agreeing with $f[x,y(x)]$ at the four points for which $x = x_0$, $x_0 + h$, $x_0 + 2h$, $x_0 + 3h$. With $n = 4$,* and using only terms including Δ^3 of (50.23), we obtain, with the help of Comment 48.18,

$$(50.34) \quad \int_{x_0}^{x_0+4h} F(x)\, dx$$

$$= 4h[1 + 2\Delta + \tfrac{1}{2}(\tfrac{16}{3} - 2)\Delta^2 + \tfrac{1}{6}(16 - 16 + 4)\Delta^3]y'(x_0)$$

$$= 4h[y'(x_0) + 2\Delta y'(x_0) + \tfrac{5}{3}\Delta^2 y'(x_0) + \tfrac{2}{3}\Delta^3 y'(x_0)]$$

$$= 4h\big\{ y'(x_0) + 2[y'(x_0 + h) - y'(x_0)]$$

$$+ \tfrac{5}{3}[y'(x_0 + 2h) - 2y'(x_0 + h) + y'(x_0)]$$

$$+ \tfrac{2}{3}[y'(x_0 + 3h) - 3y'(x_0 + 2h)$$

$$+ 3y'(x_0 + h) - y'(x_0)]\big\}$$

$$= \frac{4h}{3}[2y'(x_0 + h) - y'(x_0 + 2h) + 2y'(x_0 + 3h)].$$

Therefore, by (50.14) and (50.34), an approximation of $y(x_0 + 4h)$ is

$$(50.35) \quad y(x_0 + 4h) = y(x_0)$$

$$+ \frac{4h}{3}[2y'(x_0 + h) - y'(x_0 + 2h) + 2y'(x_0 + 3h)].$$

Proceeding as we did above, we can obtain approximating polynomials $F(x)$ of still higher degrees, agreeing with $f[x,y(x)]$ at more and more points. The only other two approximating formulas, however, that will interest us are those in which $F(x)$ is of degree five and six; see Exercise 50,2 for the case when $F(x)$ is of degree four. For a fifth degree polynomial interpolating function, (50.23) will give

$$(50.36) \quad \int_{x_0}^{x_0+5h} F(x)\, dx$$

$$= \frac{5h}{288}[19y'(x_0) + 75y'(x_0 + h) + 50y'(x_0 + 2h) + 50y'(x_0 + 3h)$$

$$+ 75y'(x_0 + 4h) + 19y'(x_0 + 5h)].$$

Therefore, by (50.14), an approximation of $y(x_0 + 5h)$ is

$$(50.37) \quad y(x_0 + 5h) = y(x_0) + K,$$

where K is the expression on the right of (50.36).

---

*The reason for taking $n = 4$ instead of $n = 3$ will become apparent when we use the resulting Formula (50.35). For the case when $n = 3$, see Exercise 50,1.

For a sixth degree interpolating function, (50.23) will give

$$(50.38) \qquad \int_{x_0}^{x_0+6h} F(x)\,dx$$

$$= \frac{3h}{420}\big[41y'(x_0) + 216y'(x_0 + h) + 27y'(x_0 + 2h)$$
$$+ 272y'(x_0 + 3h) + 27y'(x_0 + 4h)$$
$$+ 216y'(x_0 + 5h) + 41y'(x_0 + 6h)\big].$$

Hence, by (50.14), an approximation of $y(x_0 + 6h)$ is

$$(50.39) \qquad y(x_0 + 6h) = y(x_0) + G,$$

where G is the expression on the right of (50.38).

Formula (50.39) is usually replaced by the following formula,

$$(50.4) \qquad y(x_0 + 6h) = y(x_0)$$

$$+ \frac{3h}{420}\big[42y'(x_0) + 210y'(x_0 + h) + 42y'(x_0 + 2h)$$
$$+ 252y'(x_0 + 3h) + 42y'(x_0 + 4h) + 210y'(x_0 + 5h) + 42y'(x_0 + 6h)\big].$$

Note that it adds a little to some terms in (50.38) and subtracts a little from others, so that its overall accuracy is very close to that given by (50.38). Its advantage over (50.38) lies in its being reducible to the simpler form

$$(50.41) \qquad y(x_0 + 6h) = y(x_0)$$

$$+ \frac{3h}{10}\big[y'(x_0) + 5y'(x_0 + h) + y'(x_0 + 2h) + 6y'(x_0 + 3h)$$
$$+ y'(x_0 + 4h) + 5y'(x_0 + 5h) + y'(x_0 + 2h)\big].$$

This last formula is known as **Weddle's rule.**

We obtained (50.23) by using Newton's forward interpolation formula (49.17). As an exercise, start with Newton's backward interpolation formula (49.24) and show that

$$(50.5) \qquad \int_{x_0}^{x_0+nh} F(x)\,dx = nh\bigg[1 + \frac{n}{2}\nabla + \frac{1}{2}\bigg(\frac{n^2}{3} + \frac{n}{2}\bigg)\nabla^2$$

$$+ \frac{1}{6}\bigg(\frac{n^3}{4} + n^2 + n\bigg)\nabla^3 + \frac{1}{24}\bigg(\frac{n^4}{5} + \frac{3n^3}{2} + \frac{11n^2}{3} + 3n\bigg)\nabla^4$$

$$+ \frac{1}{120}\bigg(\frac{n^5}{6} + 2n^4 + \frac{35n^3}{4} + \frac{50n^2}{3} + 12n\bigg)\nabla^5 + \cdots\bigg]y'(x_0).$$

For convenient reference we collect the approximation formulas developed in this lesson.

	Interpolating Polynomial	Approximate Numerical Solution of (50.1) Satisfying (50.11)	Reference
(50.6)	$a + bx$ (trapezoidal rule)	$y(x_0 + h) = y(x_0) + \dfrac{h}{2}\left[y'(x_0) + y'(x_0 + h)\right]$	(50.31)
(50.61)	$a + bx + cx^2$ (Simpson's rule)	$y(x_0 + 2h) = y(x_0) + \dfrac{h}{3}\left[y'(x_0) + 4y'(x_0 + h) + y'(x_0 + 2h)\right]$	(50.33)
(50.62)	Third degree polynomial	$y(x_0 + 4h) = y(x_0) + \dfrac{4h}{3}\left[2y'(x_0 + h) - y'(x_0 + 2h) + 2y'(x_0 + 3h)\right]$	(50.35)
(50.63)	Fifth degree polynomial	$y(x_0 + 5h) = y(x_0) + \dfrac{5h}{288}\left[19y'(x_0) + 75y'(x_0 + h) + 50y'(x_0 + 2h) \right.$ $\left. + 50y'(x_0 + 3h) + 75y'(x_0 + 4h) + 19y'(x_0 + 5h)\right]$	(50.37)
(50.64)	Weddle's rule	$y(x_0 + 6h) = y(x_0) + \dfrac{3h}{10}\left[y'(x_0) + 5y'(x_0 + h) + y'(x_0 + 2h) \right.$ $\left. + 6y'(x_0 + 3h) + y'(x_0 + 4h) + 5y'(x_0 + 5h) + y'(x_0 + 6h)\right]$	(50.41)

We also collect below for convenience, the error term associated with each of the above formulas.*

	Interpolating Polynomial	*Error*
(50.7)	$a + bx$ (trapezoidal rule)	$-\dfrac{h^3}{12}\, y'''(X)$
(50.71)	$a + bx + cx^2$ (Simpson's rule)	$-\dfrac{h^5}{90}\, y^{(5)}(X)$
(50.72)	Third degree polynomial	$\dfrac{28}{90}\, h^5 y^{(5)}(X)$
(50.73)	Fifth degree polynomial	$-\dfrac{275}{12,096}\, h^7 y^{(7)}(X)$
(50.74)	Weddle's rule	$-\dfrac{h^7}{140}\, y^{(7)}(X_1) - \dfrac{9h^9}{1400}\, y^{(9)}(X_2)$

In all the above error formulas, X is a value of x in the interval $(x_0, x_0 + nh)$, corresponding to the $x_0 + nh$ in the respective formulas (50.6) to (50.64).

Comment on Approximation Formulas (50.6) to (50.64). We assume in the discussion which follows that $y(x)$ is a solution of

$$(50.8) \qquad\qquad y' = f(x,y)$$

satisfying the initial condition

$$(50.81) \qquad\qquad y(x_0) = y_0.$$

Hence when $x = x_0$, y' by (50.8), has the value

$$(a) \qquad\qquad y'(x_0) = f[x_0,y(x_0)].$$

Since $y(x_0)$ is the initial condition given in (50.81), we can calculate $y'(x_0)$ of (a). Again by (50.8), when $x = x_0 + h$,

$$(b) \qquad\qquad y'(x_0 + h) = f[x_0 + h, y(x_0 + h)].$$

To determine $y'(x_0 + h)$ of (b), we must therefore know the value of $y(x_0 + h)$. But we cannot know $y(x_0 + h)$ unless we know $y(x)$. And since $y(x)$ is the very solution we seek, we cannot hope to evaluate $y'(x_0 + h)$ by use of (b). It therefore follows that we cannot, for example, use formula (50.6) unless we have other means available by which we can estimate a value of the $y'(x_0 + h)$ term which appears in it.

---

*William Edmund Milne, *Numerical Calculus*, Princeton University Press, Princeton, N.J., 1949; *Numerical Solutions of Differential Equations*, John Wiley & Sons, New York, 1953.

All the remaining formulas (50.61) to (50.64) have a similar drawback. Formula (50.61), for example, can be used only after we have been able to obtain, in addition to $y(x_0)$, estimated values of $y(x_0 + h)$ and $y(x_0 + 2h)$. With them, we can then, by (50.8), estimate values of the terms

$$y'(x_0 + h) = f[x_0 + h, y(x_0 + h)]$$

and $$y'(x_0 + 2h) = f[x_0 + 2h, y(x_0 + 2h)]$$

that appear in the formula. The last formula (50.64) requires six preliminary estimates in addition to $y(x_0)$, before it can be used, namely estimates of $y(x_0 + h), \cdots, y(x_0 + 6h)$. By (50.8), we can then estimate values of $y'(x_0 + h), \cdots, y'(x_0 + 6h)$.

With the exception of (50.62), there is another unusual feature about all these formulas. Consider for example formula (50.61). After the needed two preliminary values of $y'(x_0 + h)$ and $y'(x_0 + 2h)$ have been estimated by finding approximate values of $y(x_0 + h)$ and $y(x_0 + 2h)$ by other means, all (50.61) will do is give a value of $y(x_0 + 2h)$ all over again. Why then is the formula necessary at all? If we have approximated $y(x_0 + 2h)$ by some other method, why calculate $y(x_0 + 2h)$ once more? It turns out that formula (50.61) is actually a corrector formula, i.e., repeated application of the formula will improve an estimated approximation of $y(x_0 + 2h)$ computed by a less accurate formula than itself (as determined by their respective error terms). We shall clarify this point at the time *we use the formula*. It is therefore called appropriately a **corrector formula**. If you will carefully examine all the other formulas in this list, you will find that, with the exception of (50.62), all are corrector formulas. All they will do is correct the last estimated approximation of $y(x_0 + nh)$ [needed to approximate $y'(x_0 + nh)$ in the formula] obtained by other less accurate methods.

Formula (50.62) on the other hand is a **continuing formula**. With it, we can evaluate $y(x_0 + 4h)$ provided we have obtained estimated values of $y(x_0)$ and $y'(x_0 + h)$, $y'(x_0 + 2h)$, $y'(x_0 + 3h)$.

It is evident, therefore, that we cannot start to use any of the formulas (50.6) to (50.64), whether to correct or to continue, unless we have a certain number of preliminary estimates. It is for this reason that the formulas developed in previous lessons, since they do not require preliminary estimates for their use, have been called appropriately **starting formulas.**

There are, therefore, as mentioned in the introduction to this chapter, three types of formulas in numerical methods.

1. Starting formulas.
2. Continuing formulas.
3. Corrector formulas.

We have already developed various starting formulas. In the next lesson, we shall describe a simple continuing and corrector combination formula.

EXERCISE 50

1. In (50.23), take $n = 3$ and stop with the Δ^3 term. The polynomial interpolating function $F(x)$ is therefore of degree less than or equal to three, agreeing with $f[x,y(x)]$ of (50.1) at four points. Prove that

$$\int_{x_0}^{x_0+3h} F(x)\,dx = \frac{3h}{8}\,[y'(x_0) + 3y'(x_0 + h) + 3y'(x_0 + 2h) + y'(x_0 + 3h)].$$

This formula is known as the **three-eighths rule**.

2. In (50.23), take $n = 4$ and stop with the Δ^4 term. The polynomial interpolating function $F(x)$ is therefore of degree less than or equal to four, agreeing with $f[x,y(x)]$ of (50.1) at five points. Prove that

$$(50.82) \qquad \int_{x_0}^{x_0+4h} F(x)\,dx = \frac{2h}{45}\,[7y'(x_0) + 32y'(x_0 + h) + 12y'(x_0 + 2h)$$
$$+\, 32y'(x_0 + 3h) + 7y'(x_0 + 4h)].$$

3. Prove (50.36) by taking $n = 5$ and stopping with the Δ^5 term in (50.23).

4. Prove (50.38) by taking $n = 6$ and stopping with the Δ^6 term in (50.23).

5. Prove (50.5). *Hint.* Start with Newton's backward interpolation formula (49.24) and follow the procedure used in the text to arrive at (50.23).

6. Prove that, if $n = 1$, (50.5) reduces to

$$(50.83) \qquad \int_{x_0}^{x_0+h} F(x)\,dx$$
$$= h[1 + \tfrac{1}{2}\nabla + \tfrac{5}{12}\nabla^2 + \tfrac{3}{8}\nabla^3 + \tfrac{251}{720}\nabla^4 + \tfrac{95}{288}\nabla^5 + \cdots]y'(x_0).$$

Hence an approximation to $y(x_0 + nh)$ of (50.14), with $n = 1$, in terms of backward differences, is

$$(50.84) \qquad\qquad y(x_0 + h) = y(x_0) + k,$$

where k is the expression on the right (50.83). The method which makes use of formula (50.84) is known as **Adams' method** of approximating a numerical solution of $y' = f(x,y)$ satisfying $y(x_0) = y_0$. To calculate $\nabla^m y'(x_0)$, you must know $m + 1$ evenly spaced values of $y'(x)$, namely $y'(x_0)$, $y'(x_0 - h), \cdots, y'(x_0 - mh)$, see Lesson 49B. These values can be obtained from the given differential equation $y' = f(x,y)$ only after corresponding values of $y(x_0), \cdots, y(x_0 - mh)$ have been found by starting formulas. Hence (50.84) is a continuing formula.

By means of Adams' method, find an approximate value, when $x = 0.5$, of the particular solution of the differential equation $y' = x^2 + y$ for which $y(0) = 1$. Take $h = 0.1$ and stop with the ∇^4 term in (50.84). *Hint.* By (50.84), with $x_0 = 0.4$ and $h = 0.1$,

$$(50.85) \qquad y(0.5) = y(0.4) + 0.1[y'(0.4) + \tfrac{1}{2}\nabla y'(0.4) + \tfrac{5}{12}\nabla^2 y'(0.4)$$
$$+\, \tfrac{3}{8}\nabla^3 y'(0.4) + \tfrac{251}{720}\nabla^4 y'(0.4)].$$

To calculate $\nabla^4 y'(0.4)$, you must know $y'(0.4)$, $y'(0.3)$, $y'(0.2)$, $y'(0.1)$, $y'(0)$. From the given equation, when $x = 0$, $y = 1$, $y'(0) = 0 + 1 = 1$. Use Table 46.31, creeping up process column, to find the other needed values of y'.* Then construct a table of differences of $y'(0.4)$. Head the first column $y' = x^2 + y$ and enter the values of $y'(0)$, $y'(0.1)$, $y'(0.2)$, $y'(0.3)$,

---

*The actual values will, of course, give a more accurate result for $y(0.5)$. However, we assume these actual values are unknown.

$y'(0.4)$. Head the second column $\nabla y'$ and enter in it the differences of the first column—remember $\nabla y'(0.1) = y'(0.1) - y'(0.0), \nabla y'(0.2) = y'(0.2) - y'(0.1)$. Head the third column $\nabla^2 y'$ and enter in it differences of the second column. Head the fourth column $\nabla^3 y'$ and enter in it differences of the third column. Head the last column $\nabla^4 y'$ and enter in it differences of the fourth column. The last row should then contain all the differences needed in the above formula (50.85).

7. In (50.1), let $f(x,y)$ be a function only of x so that $y' = f(x)$ for which $y(x_0) = y_0$. Hence, by (50.14),

$$(50.9) \qquad y(x_0 + nh) = y(x_0) + \int_{x_0}^{x_0+nh} f(x)\, dx.$$

Let $F(x)$ be the interpolating function for $f(x)$, agreeing with $f(x)$ at the $n + 1$ points whose abscissas are $x_0, \ x_0 + h, \ x_0 + 2h, \cdots, \ x_0 + nh$,

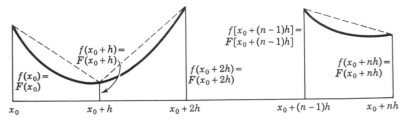

Figure 50.91

Fig. 50.91. Let the graph of $F(x)$ be the straight line joining the end points of these ordinates. Then an approximation of $y(x_0 + nh)$ of (50.9) is

$$(50.92) \qquad y(x_0 + nh) = y(x_0) + \int_{x_0}^{x_0+nh} F(x)\, dx,$$

where $F(x)$ is the function whose graph consists of these straight lines.

By (50.6) and (50.7), with y' replaced by its equal $f(x)$, we have

$$(50.93) \qquad y(x_0 + h) = y(x_0) + \frac{h}{2}\, [f(x_0) + f(x_0 + h)] + E,$$

where $E = -h^3 f''(X)/12$, and X is a value of x in the interval $(x_0, x_0 + h)$. The term $\frac{h}{2} [f(x_0) + f(x_0 + h)]$ may also be looked at as the area of a trapezoid of width h and heights $f(x_0)$, $f(x_0 + h)$, Fig. 50.91; hence the name **trapezoidal rule.**

(a) By adding the areas of each trapezoid in Fig. 50.91, show that

$$(50.94) \qquad \int_{x_0}^{x_0+nh} F(x)\, dx = \frac{h}{2}\, \{f(x_0) + 2f(x_0 + h) + 2f(x_0 + 2h) + \cdots$$
$$+ 2f[x_0 + (n-1)h] + f(x_0 + nh)\}.$$

Hence an approximation to (50.9) is

$$(50.95) \qquad y(x_0 + nh) = y(x_0) + H,$$

where H is the right side of (50.94).

(b) Following the procedure outlined in Exercise 44,11, show that the upper bound of the error in formula (50.95) due to using $F(x)$ to approximate $f(x)$ in (50.9), is

(50.96)
$$|E| \leqq \frac{nh^3 M}{12},$$

where M is the maximum value of $|f''(x)|$ in the interval $(x_0, x_0 + nh)$.

8. (a) Use formula (50.95), with $n = 4$, $h = \frac{1}{4}$, to obtain an approximate value, when $x = 2$, of the particular solution of the equation $y' = 1/x$ for which $y(1) = 0$. Show how this value of $y(2)$ can be used to approximate $\log 2$. *Hint.* Here $x_0 = 1$, $y(x_0) = 0$, $nh = 1$. The solution of $y' = 1/x$ for which $y(1) = 0$ is $y = \log x$. Therefore $y(2) = \log 2$.

(b) Use (50.96) to find an upper bound to the error in the result obtained in (a). Compare with actual error. *Hint.* $n - 4$, $h - \frac{1}{4}$, $f(x) = 1/x$, $M = \max. |f''(x)| = \max. |2/x^3| = 2$ in interval $(1,2)$.

(c) What is the largest value of h which can be used to insure that the error E in the computation of $\log 2$ is less than 0.0005? Into how many parts would it be necessary to divide the interval $(1,2)$?

9. In (50.1), let $f(x,y)$ be a function only of x, so that $y' = f(x)$ for which $y(x_0) = y_0$. Hence, by (50.14),

(50.97)
$$y(x_0 + nh) = y(x_0) + \int_{x_0}^{x_0+nh} f(x)\, dx.$$

Let $F(x)$ be the interpolating function for $f(x)$, agreeing with $f(x)$ at the $n + 1$ points whose abscissas are x, $x_0 + h$, $x_0 + 2h$, $\cdots$, $x_0 + nh$, where n is *even*. Let the graph of $F(x)$ consist of the parabolic arc connecting the three ordinates $f(x_0)$, $f(x_0 + h)$, $f(x_0 + 2h)$, solid curve in Fig. 50.91, plus the parabolic arc connecting the three ordinates $f(x_0 + 2h)$, $f(x_0 + 3h)$, $f(x_0 + 4h)$, etc. Then an approximation to $y(x_0 + nh)$ of (50.97) is

(50.98)
$$y(x_0 + nh) = y(x_0) + \int_{x_0}^{x_0+nh} F(x)\, dx,$$

where $F(x)$ is the function whose graph consists of these parabolic arcs.

By (50.61) and (50.71), with y' replaced by its equal $f(x)$, we have

(50.99)
$$y(x_0 + 2h) = y(x_0) + \frac{h}{3}\,[f(x_0) + 4f(x_0 + h) + f(x_0 + 2h)] + E,$$

where $E = -h^5 f^{(4)}(X)/90$, and X is a value of x in the interval $(x_0, x_0 + 2h)$. The term $\frac{h}{3}\,[f(x_0) + 4f(x_0 + h) + f(x_0 + 2h)]$, it has been proved, see a calculus text, is the area under a parabolic arc joining the ends of three ordinates $f(x_0)$, $f(x_0 + h)$, $f(x_0 + 2h)$.

(a) By adding the areas under each parabolic arc in Fig. 50.91, show that

(50.991)
$$\int_{x_0}^{x_0+nh} F(x)\, dx = \frac{h}{3}\,\Big\{ f(x_0) + 4f(x_0 + h) + 2f(x_0 + 2h)$$
$$+ 4f(x_0 + 3h) + 2f(x_0 + 4h) + \cdots$$
$$+ 2f[x_0 + (n - 2)h]$$
$$+ 4f[x_0 + (n - 1)h] + f(x_0 + nh) \Big\}.$$

Hence an approximation to (50.97) is

(50.992) $$y(x_0 + nh) = y(x_0) + k,$$

where k is the right side of (50.991). Formula (50.992) is known as **Simpson's rule.**

(b) Following the procedure outlined in Exercise 44,11 [note that here the interval is $(x_0, x_0 + 2h)$ instead of $(x_0, x_0 + h)$], show that the upper bound of the error in formula (50.992) due to using $F(x)$ to approximate $f(x)$ in (50.97), is

(50.993) $$|E| \leq \frac{h^5}{180} nM,$$

where M is the maximum value of $|f^{(4)}(x)|$ in the interval $(x_0, x_0 + nh)$.

10. Use formulas (50.992) and (50.993), with $n = 4$, $h = \frac{1}{4}$ to answer questions (a), (b), (c) of problem 8. *Hint.* Here $M = $ max. $|f^{(4)}(x)| = $ max. $|24/x^5| = 24$ in interval $(1,2)$.

ANSWERS 50

6. $y(0.5) = 1.6961610$.

8. (a) $y(2) = 0.69702$. Actual value: $\log 2 = 0.69315$.   (b) $|E| < 0.01042$. Actual error $= 0.00387$.   (c) $h = 0.05$ to two decimal places. Therefore we need to divide the interval $(1,2)$ into twenty parts.

10. (a) $y(0.2) = 0.69325$.   (b) $|E| < 0.00052$.   Actual error $= 0.00010$. (c) $h < 0.24$. Since n must be even, we need to divide the interval $(1,2)$ into six parts. Compare with 8(c).

LESSON 51. Milne's Method of Finding an Approximate Numerical Solution of $y' = f(x,y)$.

The method we are about to describe for finding a numerical solution of the differential equation $y' = f(x,y)$ satisfying the initial condition $y(x_0) = y_0$ is probably the one most widely used today. It is simple in form and has a relatively high degree of accuracy. The method uses continuing formula (50.62) to estimate or predict a value of $y(x_0 + 4h)$ and then employs Simpson's formula (50.61) to correct it. For convenience we rewrite these formulas below, using a subscript p for the formula we shall use as a predictor or estimator and a subscript c for the one we shall use as a corrector.

(51.1) $$y_p(x_0 + 4h) = y(x_0)$$
$$+ \frac{4h}{3}[2y'(x_0 + h) - y'(x_0 + 2h) + 2y'(x_0 + 3h)],$$

(51.11) $$y_c(x_0 + 4h) = y(x_0 + 2h)$$
$$+ \frac{h}{3}[y'(x_0 + 2h) + 4y'(x_0 + 3h) + y_p'(x_0 + 4h)].$$

When these two formulas are used in combination, they are known collectively as **Milne's method.** In order to use (51.1) we must know

$y'(x_0 + h)$, $y'(x_0 + 2h)$, $y'(x_0 + 3h)$. We can determine these values from the given differential equation $y' = f[x,y(x)]$ only after we know the corresponding values of $y(x_0 + h)$, $y(x_0 + 2h)$, $y(x_0 + 3h)$. Hence we must use starting formulas in order to obtain these needed preliminary estimates. Formula (51.1) will then predict or estimate a value of $y(x_0 + 4h)$; formula (51.11) will correct this estimate. The new value of $y(x_0 + 4h)$ thus found can in its turn be used as a new estimated value and (51.11) used over again to correct it. It has been proved that, if the original estimate is not too far away from the true value and if h is sufficiently small, the repeated use of (51.11) will give a sequence of values of $y(x_0 + 4h)$ which will eventually converge.

Example 51.12. Use Milne's method to find an approximate value when $x = 0.5$ of the particular solution of the differential equation

(a)
$$y' = x^2 + y,$$

for which $y(0) = 1$.

Solution. In Example 46.21, we found by Taylor series methods,

(b)
$$y(0.1) = 1.1055125, \quad y'(0.1) = 1.1155125;$$
$$y(0.2) = 1.2242077, \quad y'(0.2) = 1.2642077;$$
$$y(0.3) = 1.3595755, \quad y'(0.3) = 1.4495755.$$

By (51.1) and (51.11) with $x_0 = 0$, $h = 0.1$,

(c)
$$y_p(0.4) = y(0) + \frac{0.4}{3}[2y'(0.1) - y'(0.2) + 2y'(0.3)],$$

$$y_c(0.4) = y(0.2) + \frac{0.1}{3}[y'(0.2) + 4y'(0.3) + y_p'(0.4)].$$

Substituting the values of (b) and the initial condition $y(0) = 1$, in the first equation of (c), we obtain

(d)
$$y_p(0.4) = 1 + \frac{0.4}{3}[2(1.1155125) - 1.2642077 + 2(1.4495755)]$$
$$= 1.5154624.$$

By (a) and (d),

$$y_p'(0.4) = 0.16 + 1.5154624 = 1.6754624.$$

Hence the second equation of (c) becomes

(e)
$$y_c(0.4) = 1.2242077 + \frac{0.1}{3}[1.2642077 + 4(1.4495755)$$
$$+ 1.6754624] = 1.2242077 + 0.2912657 = 1.5154734.$$

We now make use of (51.11) once more to see whether it will correct the value in (e). Using the value in (e) as a new estimate, we have

(f) $$y_p'(0.4) = 0.4^2 + 1.5154734 = 1.6754734.$$

Our corrector formula (c) becomes, with the help of (b) and (f),

(g) $$y_c(0.4) = 1.2242077 + \frac{0.1}{3} [1.2642077 + 4(1.4495755)$$
$$+ 1.6754734] = 1.5154738.$$

You can verify that a third application of (51.11), using the value in (g) as a new estimate, will not change this value. Since our corrector will make no more corrections, we accept this value of $y(0.4)$.

From here on we continue to make use of our predictor and corrector formulas, using the first to estimate an approximation, the second to correct this estimate; the corrector formula being used repeatedly until no further correction results. By (51.1), with $h = 0.1$ and $x_0 = 0.1$,

(h) $$y_p(0.5) = y(0.1) + \frac{0.4}{3} [2y'(0.2) - y'(0.3) + 2y'(0.4)].$$

By (a) and (g)

(i) $$y'(0.4) = 0.16 + 1.5154738 = 1.6754738.$$

Therefore by (b) and (i), (h) becomes

(j) $$y_p(0.5) = 1.1055125 + \frac{0.4}{3} [2(1.2642077) - (1.4495755)$$
$$+ 2(1.6754738)] = 1.6961508,$$

which is an estimated value of $y(0.5)$. Using this value in (a), we obtain

(k) $$y_p'(0.5) = 0.25 + 1.6961508 = 1.9461508.$$

By (51.11), with $h = 0.1$, $x_0 = 0.1$,

(l) $$y_c(0.5) = y(0.3) + \frac{h}{3} [y'(0.3) + 4y'(0.4) + y_p'(0.5)].$$

By (b), (i), and (k), (l) becomes

(m) $$y_c(0.5) = 1.3595755 + \frac{0.1}{3} [1.4495755 + 4(1.6754738)$$
$$+ 1.9461508] = 1.6961629.$$

Using the value $y(0.5) = 1.6961629$ as a new estimate, we find by (a),

(n) $$y_p'(0.5) = 0.25 + 1.6961629 = 1.9461629,$$

and by (51.11)

(o) $y_c(0.5) = 1.3595755 + \dfrac{0.1}{3} [1.4495755 + 4(1.6754738)$

$+ 1.9461629] = 1.6961633.$

You can verify that further use .of (51.11) will not change the value in (o). We therefore accept this estimated value of $y(0.5)$.

Comment 51.2. In this manner, alternately using predictor formula (51.1) and corrector formula (51.11) as many times as needed, you can obtain $y(0.6)$, $y(0.7)$, $y(0.8)$, etc. It will be found desirable to construct a table in which to record all relevant calculations as they are found. Its form is given in Table 51.22 below. The letter D which appears in it is defined as

(51.21) $D(x_0 + 4h) = y_c(x_0 + 4h) - y_p(x_0 + 4h),$

where y_p is the estimate computed by (51.1) and y_c is the *first* correction of this y_p computed by (51.11). Its important purpose will be explained later. We have left it to you as an exercise to complete the line beginning with 0.6.

Table 51.22

x	y_p of (51.1)	y_c of (51.11)	y_p'	y_c'	D of (51.21)	Actual Value
0.4	1.5154624	1.5154734	1.6754624	1.6754734	0.0000110	
		1.5154738		1.6754738		1.5154741
0.5	1.6961508	1.6961629	1.9461508	1.9461629	0.0000121	
		1.6961633		1.9461633		1.6961638
0.6						1.9063564

Comment 51.23. Formula (51.1) is a continuing formula. However, it is rarely, if ever, used for this purpose since, as we shall show below, it is considerably less accurate than corrector formula (51.11). By (50.72), the error term E_p of (51.1) is

(51.24) $E_p = \frac{28}{90} h^5 y^{(5)}(X_1).$

By (50.71) the error term E_c of (51.11) is

(51.25) $E_c = -\frac{1}{90} h^5 y^{(5)}(X_2).$

If we may assume that h is sufficiently small so that the variation in value of $y^{(5)}(X_1)$ and $y^{(5)}(X_2)$ in an interval of width $4h$ [X_1 is a value of x in the interval $(x_0, x_0 + 4h)$; X_2 is a value of x in the interval $(x_0 + 2h, x_0 + 4h)$] is negligible, then we see from error formulas (51.24) and (51.25) that the corrector formula at each single step is approximately 28 times as accurate as the predictor formula.

Comment 51.3. Comment on Error in Milne's Method.
1. Call $E_p(x_0 + 4h)$ the error in computing $y_p(x_0 + 4h)$ by using formula (51.1); call $E_c(x_0 + 4h)$ the error in computing $y_c(x_0 + 4h)$ by using formula (51.11) to correct this predicted value of $y_p(x_0 + 4h)$. Let $Y(x_0 + 4h)$ be the actual value of $y(x_0 + 4h)$. Then

$$(51.31) \qquad E_c(x_0 + 4h) = Y(x_0 + 4h) - y_c(x_0 + 4h),$$
$$E_p(x_0 + 4h) = Y(x_0 + 4h) - y_p(x_0 + 4h).$$

Subtracting the first equation from the second, we have

$$(51.32) \quad y_c(x_0 + 4h) - y_p(x_0 + 4h) = E_p(x_0 + 4h) - E_c(x_0 + 4h).$$

If we may assume that h is sufficiently small so that the variation between $y^{(5)}(X_1)$ of error formula (51.24) and $y^{(5)}(X_2)$ of error formula (51.25) is negligible, then we commit a small error by using $y^{(5)}(X)$ as their approximate common value. Hence subtracting (51.25) from (51.24) and replacing $y^{(5)}(X_1)$ and $y^{(5)}(X_2)$ by $y^{(5)}(X)$, we obtain

$$(51.33) \qquad E_p(x_0 + 4h) - E_c(x_0 + 4h) = \tfrac{29}{90}h^5 y^{(5)}(X).$$

By (51.21), the left side of (51.32) is $D(x_0 + 4h)$. Hence by (51.32), (51.33), and (51.25), we have

$$(51.34) \qquad D(x_0 + 4h) = \tfrac{29}{90}h^5 y^5(X) = -29 E_c(x_0 + 4h).$$

Therefore by (51.34),

$$(51.35) \qquad E_c(x_0 + 4h) = -\frac{D(x_0 + 4h)}{29}.$$

Formula (51.35) tells us that the approximate formula error in the *first* corrected value of $y_c(x_0 + 4h)$ is $-\frac{1}{29}$ the difference between the first corrector and predictor values of $y(x_0 + 4h)$.

2. The column headed D in Table 51.22 thus serves a very useful purpose. Dividing it by 29 will give an estimate of the formula error in the first corrected value of y_c over one h step. If therefore we want $|E| <$ 0.000005, then $|D/29|$ must be <0.000005 or $|D|$ must be <0.000145. As long as $|D|$ remains less than this figure, we assume that the error in our first corrected estimate y_c, over each h step, is less than 0.000005.

3. As long as $D/29$ is less than the desired accuracy, we may continue to use the same h interval. As soon as $D/29$ becomes greater than desired accuracy, we must reduce h. Conversely, if $D/29$ is considerably smaller than desired accuracy, we can safely increase h. The method of reducing and increasing h in the course of an extended computation is discussed in the next Lesson 52.

4. The corrector formula may be used repeatedly at each step, until there is no difference between two successive values of $y(x_0 + 4h)$. However, as the difference between predictor and first corrector values increases, you will find that you must use the corrector more and more times at each step. When this happens, even though $D/29$ may still be less than desired accuracy, it will usually be found best to reduce h.

5. The column headed D serves another useful purpose. It controls in some measure arithmetical accuracy at each step. Whenever an entry in D shows a sudden change from a definite behavior pattern, the preceding and current calculations should be checked.

6. If predictor and corrector values agree to k decimal places after being properly rounded off, then k decimal accuracy is assumed.

EXERCISE 51

1. Complete line 0.6 in Table 51.22.

2. Replace (b) of Example 51.12, by the actual values of $y(0.1)$, $y(0.2)$, $y(0.3)$ as given in Table 46.31. Following the method of this lesson, compute $y_p(0.4)$, $y_c(0.4)$, and $D(0.4)$. Then use (51.35) to correct $y_c(0.4)$. Add this corrected value of $y(0.4)$ to your table. Now proceed to compute $y_p(0.5)$, $y_c(0.5)$, and $D(0.5)$. Use (51.35) to correct $y_c(0.5)$. Add this corrected value of $y(0.5)$ to your table. Calculate $y_p(0.6)$, $y_c(0.6)$, and $D(0.6)$. Correct $y_c(0.6)$. Compare these values with those obtained previously and with actual values given in Table 51.22.

3. Using the six preliminary results obtained in 2, compute $y_w(0.6)$ by means of Weddle's rule (50.64). In the absence of a solution, how many decimal place accuracy could you assume in $y(0.6)$? *Hint.* See 3 after "**3.** Cumulative Errors," Lesson 52A.

4. Find numerical approximations when $x = 0.4$, 0.5, 0.6 of the particular solution of the differential equation $y' = x + y$ for which $y(0) = 1$. Take $h = 0.1$. Follow the method employed in Example 51.12, using first the predictor formula, then the corrector formula as often as is necessary until no further correction results. For preliminary values, take $y(0.1) = 1.1103418$, $y(0.2) = 1.2428055$, $y(0.3) = 1.3997176$. Solve the equation and compare your results with actual values.

5. Using the six preliminary results obtained in 4, compute $y_w(0.6)$ by means of Weddle's rule (50.64). In the absence of a solution, how many decimal place accuracy could you assume in $y(0.6)$? See *Hint* in 3.

6. Find numerical approximations when $x = 1.4$, 1.5 of the particular solution of the differential equation $y' = xy$ for which $y(1) = 1$. Take $h = 0.1$. Follow the method used in Example 51.12, employing first the predictor and then the corrector formula as often as necessary. For preliminary values take $y(1.1) = 1.11071$, $y(1.2) = 1.24608$, $y(1.3) = 1.41199$. Solve the equation and compare results with actual values.

7. Using the results obtained in problem 6, compute $y(1.5)$ by means of corrector formula (50.63). Here $x_0 = 1$, $y(x_0) = 1$, $h = 0.1$. In the absence of a solution, how many decimal place accuracy could you assume? See *Hint* in 3.

ANSWERS 51

1. $y_p(0.6) = 1.9063415$, $y_c(0.6) = 1.9063561$, $D(0.6) = 0.0000146$.
2. $y_p(0.4) = 1.5154627$, $y_c(0.4) = 1.5154742$, $D(0.4) = 0.0000115$.
   $y(0.4) = 1.5154738$, $y_p(0.5) = 1.6961512$, $y_c(0.5) = 1.6961638$,
   $D(0.5) = 0.0000126$, $y(0.5) = 1.6961634$, $y_p(0.6) = 1.9063424$,
   $y_c(0.6) = 1.9063561$, $D(0.6) = 0.0000137$, $y(0.6) = 1.9063556$.
3. $y_w(0.6) = 1.9063563$. Can assume six decimal place accuracy if rounded
   off to six decimals.
4. $y_c(0.4) = 1.5836497$, $y_c(0.5) = 1.7974429$, $y_c(0.6) = 2.0442384$. Actual
   values: 1.5836494, 1.7974425, 2.0442376.
5. $y_w(0.6) = 2.0442377$. Can assume six decimal place accuracy.
6. $y_c(1.4) = 1.61609$, $y_c(1.5) = 1.86826$. Actual values: 1.61607, 1.86825.
7. $y(1.5) = 1.86825$. Can assume four decimal place accuracy if rounded off to
   four decimals.

LESSON 52. General Comments. Selecting h. Reducing h. Summary and an Example.

LESSON 52A. Comment on Errors.

1. *Formula Errors.* With each approximating formula we have also given a companion error term which measures the magnitude of the error. Since these error formulas cannot usually be used, we have also suggested practical means by which the magnitude of error can be estimated. Unless a numerical method has associated with it a useful error formula, the method is of little value.

2. *Rounding off Errors.* In a practical problem, the number of steps needed will usually be known, also the accuracy desired. It will thus be possible to estimate the number of decimal places that should be used at the start in order to offset rounding off errors under the most unfavorable circumstances. For example, suppose you want four decimal accuracy and expect to use eight steps. If you round off to six decimals, then the maximum possible absolute value of the error due only to rounding off, i.e., due to dropping the seventh and later decimals, is, after eight steps, $0.0000005 \times 8 = 0.000004$. It must be remembered, however, that the use of a formula may considerably magnify, at each step, the effect of the rounding off error.

3. *Cumulative Errors.* At each step of a building up process, two errors occur. One, because we start off with an inherited error, and two, because we are using an approximation formula. Since each succeeding step depends on the previous estimate, it will be unusual indeed if we obtain increasing accuracy as we proceed. It may happen in a rare case that one error may be offset by a succeeding one. In the usual case, it will not happen, and accuracy will decrease at each step.

Although formulas exist which will give the upper limit of error at the end of each step due both to formula and cumulative errors, they are not easy to use. We give below three practical suggestions for estimating accuracy, ones which have been found adequate in most cases.

1. Start with many more decimals than you need.
2. Make all calculations over again with an h equal to one half its previous value. If the new final result agrees with the previous one to k decimal places, after being properly rounded off, then k decimal place accuracy is assumed.
3. Apply corrector formula (50.63) after five preliminary estimates have been obtained or Weddle's corrector (50.64) after six steps. If the result obtained by these corrector formulas agrees with the last estimate used *in* the formula to k decimal places, after being properly rounded off, then k decimal place accuracy is assumed. Weddle's rule, in particular, is simple in structure and is extremely good at discovering errors. If there is little agreement between the last estimated value and its corrected value, then either an arithmetical error has been made or h is too large.

A final word of caution. In most problems, the practical approach to errors as outlined above, will within reasonable certainty, assure you of a result which is correct to k decimal places. However, only a formula which gives the upper bound of the error due to rounding off, formula and cumulative errors, can give, with certainty, the magnitude of the error in a numerical computation.

LESSON 52B.   Choosing the Size of h.   You may have been saying to yourself, "How does one know what size h to select at the start?" We used $h = 0.1$ in our examples, but what made us pick 0.1 instead of 0.2 or 0.05 or 0.3? If it is possible to use an error formula for a given method, then it is also possible to determine h so that the error due to the approximating formula will remain within the desired limits. Frequently a knowledge of the problem plus practical experience will determine a starting value of h.

In the absence of a useful error formula or practical information, then all you can do is to start with an h which seems reasonable. Say you decide to start with $h = 0.3$ and to use the Runge-Kutta method to get your first approximations. Calculate $y(0.3)$ in one step and then in two steps, i.e., calculate $y(0.3)$ and $y(0.15 + 0.15)$. Formula (47.42) will then give you an approximate value of the magnitude of the error. It states that the approximate error in $y(0.15 + 0.15)$ is one-fifteenth the difference between $y(0.15 + 0.15)$ and $y(0.3)$. If one-fifteenth this difference is greater than the desired error, you must reduce h; if it is reasonably less than the desired error, you may retain h; if it is very much less than the desired error, you may increase h.

Let us assume $h = 0.3$ is satisfactory. Starting with the value of $y(0.3)$ found in two steps, we then proceed to find $y(0.6)$ and $y(0.9)$. The initial condition plus these three preliminary estimates, $y(0.3)$, $y(0.6)$ and $y(0.9)$, will enable us to switch to Milne's method. From here on, we begin to watch the size of the column D in our table. By (51.21), it is the difference between the predictor value computed by using (51.1) and the first corrected value computed by using (51.11). Dividing D by 29 will keep us posted as to the approximate magnitude of the error in the first y_c figure in our table. When it becomes larger than the desired error, we must reduce h. This brings up the question of how to reduce h in the course of an extended computation; also how to increase h.

LESSON 52C.   Reducing and Increasing h.   As we proceed step by step, an h which is satisfactory in early stages may become too large in later ones. Suppose, for example, we are satisfied that the values of $y(0.1)$, $y(0.2)$, $y(0.3)$ found by a starting method have the desired accuracy. We switch to the Milne method and determine that the values of $y(0.4)$, $y(0.5)$, $y(0.6)$, $y(0.7)$ still have the desired accuracy. However to obtain $y(0.8)$, we find it necessary to reduce h to 0.05. The next value we must find is, therefore, $y(0.75)$. To use Milne's predictor formula (51.1) with $h = 0.05$, $x_0 = 0.55$, we must know $y(0.55)$, $y(0.6)$, $y(0.65)$, $y(0.7)$. We already know $y(0.6)$ and $y(0.7)$. How do we find $y(0.55)$ and $y(0.65)$ without the necessity of starting from the beginning all over again with $h = 0.05$?

One way of obtaining $y(0.55)$ and $y(0.65)$ is by use of Newton's backward interpolation formula (49.22). Since we know eight evenly spaced values of $y(x)$, 0.1 unit apart, namely $y(0)$, $y(0.1)$, $\cdots$, $y(0.6)$, $y(0.7)$, we can use terms in this formula to $\nabla^7 y(0.7)$. By (49.23), with $x_0 = 0.7$ $x = 0.65$ and $h = 0.1$, we find $n = (7 - 0.65)/0.1 = \frac{1}{2}$. With $n = \frac{1}{2}$

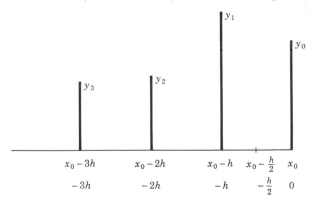

Figure 52.1

and the needed values in the formula obtained by constructing a table of differences of $y(0.7)$, we can determine $y(0.65)$. Similarly, by taking $x = 0.55$ so that $n = \frac{3}{2}$, we can approximate $y(0.55)$.

A second and perhaps easier way to obtain these values is by use of a formula which has approximately the same order of accuracy as Simpson's rule. Let the third degree polynomial $F(x) = a + bx + cx^2 + dx^3$ be an interpolating function for $y(x)$. Since $F(x)$ is a third degree polynomial, we may assume by Theorem 48.2, that it agrees with $y(x)$ at four points whose abscissa values are h units apart. Call these four abscissas x_0, $x_0 - h$, $x_0 - 2h$, $x_0 - 3h$, and their respective ordinates y_0, y_1, y_2, y_3, Fig. 52.1. For convenience in calculation, we take the origin at x_0, so that the coordinates of the points at which $F(x)$ and $y(x)$ agree are $(0, y_0)$, $(-h, y_1)$, $(-2h, y_2)$, $(-3h, y_3)$. Since each of these points satisfies the equation

$$(52.11) \qquad F(x) = a + bx + cx^2 + dx^3,$$

we have

$$(52.12) \qquad \begin{aligned} y_0 &= a, \\ y_1 &= a - hb + h^2 c - h^3 d, \\ y_2 &= a - 2hb + 4h^2 c - 8h^3 d, \\ y_3 &= a - 3hb + 9h^2 c - 27h^3 d. \end{aligned}$$

Solving (52.12), for a, b, c, d, we obtain

$$(52.13) \qquad \begin{aligned} a &= y_0, \\ b &= \frac{11y_0 - 18y_1 + 9y_2 - 2y_3}{6h}, \\ c &= \frac{2y_0 - 5y_1 + 4y_2 - y_3}{2h^2}, \\ d &= \frac{y_0 - 3y_1 + 3y_2 - y_3}{6h^3}. \end{aligned}$$

Substituting (52.13) in (52.11) will give the equation of the interpolating function $F(x)$ for $y(x)$. Therefore when $x = -h/2$, $F(-h/2)$ will give the approximate value of $y(-h/2)$. Making the substitution (52.13) in (52.11) and replacing x by $-h/2$, we have

$$(52.14) \quad F\left(-\frac{h}{2}\right) = y_0 - \frac{11y_0 - 18y_1 + 9y_2 - 2y_3}{12}$$
$$+ \frac{2y_0 - 5y_1 + 4y_2 - y_3}{8} - \frac{y_0 - 3y_1 + 3y_2 - y_3}{48}$$
$$= \frac{5y_0 + 15y_1 - 5y_2 + y_3}{16},$$

which is an approximation of $y(-h/2)$. Hence in terms of our original abscissas, see Fig. 52.1, we obtain the approximation formula,

$$(52.15) \quad y\left(x_0 - \frac{h}{2}\right) =$$
$$\tfrac{1}{16}[5y(x_0) + 15y(x_0 - h) - 5y(x_0 - 2h) + y(x_0 - 3h)].$$

Comment 52.16. To double h is a simple matter. If you decide that the h you have been using can be safely doubled, all you need do is to take every other preceding estimate. For example, if you have been using $h = 0.1$ and find $h = 0.2$ will be satisfactory, you need use only the previous estimates, $y(x_0 + 0.2)$, $y(x_0 + 0.4)$, etc.

LESSON 52D. Summary and an Illustrative Example. We have given various starting, continuing, predictor, and corrector formulas. Which ones you should choose in a particular problem will be determined by the degree of accuracy desired, the relative difficulty of the method, and the amount of labor involved. If you have many occasions to use numerical methods, experience will be your best guide.

It is the usual practice to start with the Runge-Kutta method. After the needed preliminary estimates have been obtained, it is then customary to switch to the Milne method. In solving problem 52.2 below, we shall assume that we know nothing about its solution. Hence we shall have to depend for a determination of the approximate accuracy of our result on the suggestions made in Lesson 52A.

Example 52.2. Find an approximate value when $x = 0.6$ of the particular solution of the differential equation

(a) $$y' = x^2 + y,$$

for which $y(0) = 1$. Assume we wish the error to be less than 0.00005.

Solution. To bring more methods into the discussion, we shall start with Taylor series instead of with the usual Runge-Kutta formulas, then switch to Runge-Kutta and end finally with Milne.

First we must choose an h. We decide to try $h = 0.1$, and proceed to calculate $y(0.1)$ in two steps and in one step in order to see how much agreement there is in the two results. Using our basic equation (46.12) with $h = 0.05$ and x_0 equal successively to 0 and 0.05, we obtain

(b) $$y(0 + 0.05) = y(0.05) = y(0) + y'(0)(0.05)$$
$$+ \frac{y''(0)}{2}(0.05)^2 + \frac{y'''(0)}{3!}(0.05)^3 + \frac{y^{(4)}(0)}{4!}(0.05)^4,$$

$$y(0.05 + 0.05) = y(0.1) = y(0.05) + y'(0.05)(0.05)$$
$$+ \frac{y''(0.05)}{2}(0.05)^2 + \frac{y'''(0.05)}{3!}(0.05)^3 + \frac{y^{(4)}(0.05)}{4!}(0.05)^4.$$

The derivatives of (a) to order four are

(c) $y' = x^2 + y,$ $y'' = 2x + y',$ $y''' = 2 + y'',$ $y^{(4)} = y'''.$

Hence when $x = 0$ and $y(0) = 1$, we find from (c)

(d) $y'(0) = 1,$ $y''(0) = 1,$ $y'''(0) = 3,$ $y^{(4)}(0) = 3.$

Substituting these values in the first equation of (b), we obtain

(e) $y(0.05) = 1 + 0.05 + \frac{1}{2}(0.05)^2 + \frac{1}{2}(0.05)^3 + \frac{1}{8}(0.05)^4$
$= 1.05 + 0.00125 + 0.0000625 + 0.0000008 = 1.0513133.$

With $x = 0.05$, $y(0.05) = 1.0513133$, we find from (c)

(f) $y'(0.05) = 1.0538133,$ $y''(0.05) = 1.1538133,$
$y'''(0.05) = 3.1538133,$ $y^{(4)}(0.05) = 3.1538133.$

Substituting (e) and (f) in the second equation of (b), we obtain

(g) $y(0.05 + 0.05)$
$= 1.0513133 + 1.0538133(0.05) + \frac{1}{2}(1.1538133)(0.05)^2$
$+ \frac{1}{6}(3.1538133)(0.05)^3 + \frac{1}{24}(3.1538133)(0.05)^4$
$= 1.0513133 + 0.05269067 + 0.00144227 + 0.00006570$
$+ 0.00000082 = 1.1055128.$

In Example 46.2, we found the value of $y(0.1)$ in one step. By (f) of that example, $y(0.1) = 1.1055125$. We now note that the value of $y(0.1)$ obtained in one step agrees, when rounded off, to six decimal places with the value of $y(0.1)$ obtained in two steps. We note also by (46.57), that

(h) $E(0.1) = \frac{16}{15}[y(0.5 + 0.5) - y(0.1)] = \frac{16}{15}(0.0000003) = 0.00000032.$

Since this error is sufficiently smaller than the desired one, and because $y(0.1)$ and $y(0.05 + 0.05)$ agree to six decimals, we accept $h = 0.1$ as a proper starting value.

Our next task is to determine how many decimal places to carry. Since $h = 0.1$ and we want $y(0.6)$, there will be, if h does not need to be reduced, a total of six steps. If therefore we carry seven decimal places, dropping the eighth, the absolute value of the error due only to rounding off, in the most unfavorable circumstances, is less than $6(0.00000005) = 0.0000003$, which does not affect the sixth place. Since we want our error to be <0.00005, we decide that seven decimals will give us a sufficient margin of safety. We are now ready to calculate $y(0.2)$. With $x_0 = 0.1$

and $h = 0.1$, our basic equation (46.12) becomes

(i) $y(0.2) = y(0.1) + y'(0.1) (0.1) + \dfrac{y''(0.1)}{2!} (0.1)^2$

$$+ \dfrac{y'''(0.1)}{3!} (0.1)^3 + \dfrac{y^{(4)}(0.1)}{4!} (0.1)^4 + \cdots.$$

Although the value of $y(0.1) = 1.1055128$ found in two steps with $h = 0.05$ is more accurate than $y(0.1) = 1.1055125$ found in one step with $h = 0.1$, we shall use the last figure because of the adoption of a 0.1 interval. We shall then be able to use those error formulas that are based on $h = 0.1$.

By (c), with $x = 0.1$, $y = 1.1055125$.

(j) $y'(0.1) = 1.1155125, \qquad y''(0.1) = 1.3155125,$

$y'''(0.1) = 3.3155125, \qquad y^{(4)}(0.1) = 3.3155125.$

Substituting the above values in (i), we obtain

(k) $y(0.2) = 1.1055125 + (0.1)(1.1155125) + \dfrac{0.01}{2} (1.3155125)$

$$+ \dfrac{0.001}{6} (3.3155125) + \dfrac{0.0001}{24} (3.3155125) = 1.2242077.$$

Before proceeding to find $y(0.3)$ we shall check the accuracy of $y(0.2) = 1.2242077$ as given in (k). By (46.57),

(l) $E(0.1 + 0.1) = \tfrac{1}{15}[y(0.1 + 0.1) - y(0.2)].$

By (k)

(m) $y(0.1 + 0.1) = 1.2242077.$

By (f) of Example 46.2

(n) $y(0.2) = 1.2242000.$

Hence

$$E(0.1 + 0.1) = \tfrac{1}{15}(1.2242077 - 1.224200) = 0.0000005,$$

which is the approximate error in $y(0.2)$ of (k) found in two steps. Since this error is still much less than the desired one of 0.00005, we continue to use $h = 0.1$.

To find $y(0.3)$, we switch to the Runge-Kutta method. By (47.34), with $x_0 = 0.2$, $h = 0.1$, $y(0.2) = 1.2242077$,

(o) $y(0.3) = 1.2242077 + \tfrac{1}{6}(w_1 + 2w_2 + 2w_3 + w_4).$

By (47.35), with $h = 0.1$, $x_0 = 0.2$, $y_0 = y(x_0) = y(0.2) = 1.2242077$,

$$f(x,y) = x^2 + y,$$

(p) $\qquad w_1 = 0.1f(0.2, 1.2242077) = 0.1(0.2^2 + 1.2242077)$
$\qquad\qquad = 0.12642077,$

$\qquad w_2 = 0.1f(0.25, 1.2242077 + 0.0632104)$
$\qquad\qquad = 0.1(0.25^2 + 1.2874181) = 0.1349918,$

$\qquad w_3 = 0.1f(0.25, 1.2242077 + 0.0674959)$
$\qquad\qquad = 0.1(0.25^2 + 1.2917036) = 0.1354204,$

$\qquad w_4 = 0.1f(0.3, 1.2242077 + 0.1354204)$
$\qquad\qquad = 0.1(0.3^2 + 1.3596281) = 0.1449628.$

Substituting (p) in (o), we have

(q) $\ y(0.3)$

$\qquad = 1.2242077 + \tfrac{1}{6}(0.1264208 + 0.2699836 + 0.2708408 + 0.1449628)$
$\qquad = 1.2242077 + 0.1353680$
$\qquad = 1.3595757.$

We are now ready to proceed with Milne's method. For convenience, we collect the results thus far obtained.

(52.21) $\quad y(0) \ \ = 1.0000000, \qquad y'(0) \ \ = 1.0000000;$
$\qquad\qquad y(0.1) = 1.1055125, \qquad y'(0.1) = 1.1155125;$
$\qquad\qquad y(0.2) = 1.2242077, \qquad y'(0.2) = 1.2642077 \text{ [by (a) and (k)]};$
$\qquad\qquad y(0.3) = 1.3595757, \qquad y'(0.3) = 1.4495757 \text{ [by (a) and (q)]}.$

The Milne formulas are, by (51.1) and (51.11),

(52.22) $\quad y_p(x_0 + 4h) = y(x_0)$
$$+ \frac{4h}{3}[2y'(x_0 + h) - y'(x_0 + 2h) + 2y'(x_0 + 3h)],$$

(52.23) $\quad y_c(x_0 + 4h) = y(x_0 + 2h)$
$$+ \frac{h}{3}[y'(x_0 + 2h) + 4y'(x_0 + 3h) + y_p{}^{\boldsymbol{\cdot}}(x_0 + 4h)].$$

Using (52.22) and the values in (52.21), we have with $x_0 = 0$, $h = 0.1$,

(r) $\quad y_p(0.4) = y(0) + \dfrac{0.4}{3}[2y'(0.1) - y'(0.2) + 2y'(0.3)]$

$$= 1.0 + \frac{0.4}{3}[2(1.1155125) - 1.2642077 + 2(1.4495757)]$$

$$= 1.5154625.$$

By (a) and (r),

(s) $$y_p'(0.4) = 1.6754625.$$

By (52.23) with $x_0 = 0$, $h = 0.1$,

(t) $y_c(0.4) = y(0.2) + \dfrac{0.1}{3}\,[y'(0.2) + 4y'(0.3) + y_p'(0.4)]$

$$= 1.2242077 + \frac{0.1}{3}\,[1.2642077 + 4(1.4495757) + 1.6754625]$$

$$= 1.5154735.$$

And by (a) and (t),

(u) $$y_c'(0.4) = 1.6754735.$$

Using the value in (u) as a new estimated value in Simpson's formula (52.23), we have, with the help of (52.21),

(v) $y_c(0.4) = 1.2242077 + \dfrac{0.1}{3}\,[1.2642077 + 4(1.4495757) + 1.6754735]$

$$= 1.5154738.$$

A third application of Simpson's formula will not change the value in (v). By (a) and (v),

(w) $$y_c'(0.4) = 1.6754738.$$

In Table 52.24, we start to record the Milne values, keeping a careful watch on the column headed D, which by (51.21), is the difference between

Table 52.24

x	y_p of (52.22)	y_c of (52.23)	y_p'	y_c'	$D =$ first $y_c - y_p$	$\lvert D/29 \rvert <$
0.4	1.5154625	1.5154735	1.6754625	1.6754735	0.0000110	0.0000004
		1.5154738		1.6754738		
0.5	1.6961508	1.6961631	1.9461508	1.9461631	0.0000123	0.0000005
		1.6961635		1.9461635		
0.55	1.7972722	1.7972764	2.0997722	2.0997764	0.0000042	0.0000002
		1.7972765		2.0997765		
0.6	1.9063602	1.9063573	2.2663602	2.2663573	−0.0000029	0.0000001

the predictor value and the first corrector value. And, by (51.35), the error in the first corrector value for one step is approximately $-D/29$. Since, for $y(0.4)$, $\lvert D/29 \rvert$ is less than the desired error of 0.00005, we continue with $h = 0.1$.

By (52.22) with $h = 0.1$, $x_0 = 0.1$,

(x) $y_p(0.5) = y(0.1) + \dfrac{0.4}{3}[2y'(0.2) - y'(0.3) + 2y'(0.4)]$

$\qquad = 1.1055125 + \dfrac{0.4}{3}[2(1.2642077) - 1.4495757 + 2(1.6754738)]$

$\qquad = 1.6961508.$

By (a) and (x)

(aa) $\qquad\qquad\qquad y_p'(0.5) = 1.9461508.$

Therefore by (52.23) with $h = 0.1$, $x_0 = 0.1$,

(bb) $y_c(0.5) = y(0.3) + \dfrac{0.1}{3}[y'(0.3) + 4y'(0.4) + y_p'(0.5)]$

$\qquad = 1.3595757 + \dfrac{0.1}{3}[1.4495757 + 4(1.6754738) + 1.9461508]$

$\qquad = 1.6961631.$

A second application of Simpson's corrector will change the value in (bb) to

(cc) $\qquad\qquad\qquad y_c(0.5) = 1.6961635.$

A third application of (52.22) will not change the value in (cc). With this value of $y_c(0.5)$, we find from (a), $y_c'(0.5) = 1.9461635$. Although $|D/29|$ is still much less than desired accuracy, and we may safely continue to use $h = 0.1$, we shall calculate $y(0.6)$ by reducing h to 0.05 in order to demonstrate how to use formula (52.15).

With h now equal to 0.05, we cannot use the predictor formula (52.22) to find $y_p(0.55)$ unless we know four preceding values of y, 0.05 unit apart, namely $y(0.5)$, $y(0.45)$, $y(0.4)$, and $y(0.35)$. We already know $y(0.5)$ and $y(0.4)$. To find $y(0.45)$ and $y(0.35)$, we make use of formula (52.15). With $h = 0.1$, the formula becomes

(dd) $y(0.45)$

$\qquad = y(0.5 - 0.05)$

$\qquad = \tfrac{1}{16}[5y(0.5) + 15y(0.4) - 5y(0.3) + y(0.2)]$

$\qquad = \tfrac{1}{16}[5(1.6961635) + 15(1.5154738) - 5(1.3595757) + 1.2242077]$

$\qquad = 1.6024534,$

$y(0.35)$

$\qquad = y(0.4 - 0.05)$

$\qquad = \tfrac{1}{16}[5y(0.4) + 15y(0.3) - 5y(0.2) + y(0.1)]$

$\qquad = \tfrac{1}{16}[5(1.5154738) + 15(1.3595757) - 5(1.2242077) + 1.1055125]$

$\qquad = 1.4347174.$

By (a) and (dd),

(ee) $\qquad y'(0.45) = (0.45)^2 + 1.6024534 = 1.8049534,$

We now have the needed preliminary values to use the predictor formula (52.22) to estimate $y_p(0.55)$. With $h = 0.05$, $x_0 = 0.35$, it becomes

(ff) $\ y_p(0.55)$

$$= y_p(0.35 + 0.20)$$

$$= y(0.35) + \frac{4(0.05)}{3}\,[2y'(0.4) - y'(0.45) + 2y'(0.5)]$$

$$= 1.4347174 + \frac{0.2}{3}\,[2(1.6754738) - 1.8049534 + 2(1.9461635)]$$

$$= 1.7972721.$$

By (a) and (ff)

(gg) $\qquad y_p'(0.55) = (0.55)^2 + 1.7972721 = 2.0997721.$

The corrector formula (52.23), with $h = 0.05$, $x_0 = 0.35$, now becomes

(hh) $\ y_c(0.55)$

$$= y_c(0.35 + 0.20)$$

$$= y(0.45) + \frac{0.05}{3}\,[y'(0.45) + 4y'(0.5) + y_p'(0.55)]$$

$$= 1.6024534 + \frac{0.05}{3}\,[1.8049534 + 4(1.9461635) + 2.0997721]$$

$$= 1.7972764.$$

A second application of the corrector formula changes the value in (hh) to

(ii) $\qquad\qquad\qquad y_c(0.55) = 1.7972765.$

With this value of $y_c(0.55)$, we find from (a), $y_c'(0.55) = 2.0997765$.

Returning to the predictor formula (52.22) with $h = 0.05$, $x_0 = 0.4$,

(jj) $\ y_p(0.6)$

$$= y_p(0.4 + 0.2)$$

$$= y(0.4) + \frac{4(0.05)}{3}\,[2y'(0.45) - y'(0.5) + 2y'(0.55)]$$

$$= 1.5154738 + \frac{0.2}{3}\,[2(1.8049534) - 1.9461635 + 2(2.0997765)]$$

$$= 1.9063602.$$

By (a) and (jj)

(kk) $\qquad y_p'(0.6) = (0.6)^2 + 1.9063602 = 2.2663602.$

By (52.23) with $h = 0.05$, $x_0 = 0.4$,

(ll) $y_c(0.6) = y(0.4 + 0.2)$

$$= y(0.5) + \frac{0.05}{3} [y'(0.5) + 4y'(0.55) + y_p'(0.6)]$$

$$= 1.6961635 + \frac{0.05}{3} [1.9461635 + 4(2.0997765) + 2.2663602]$$

$$= 1.9063573.$$

A second application of formula (52.23) does not change the value in (ll). Hence, by (a),

(mm) $y_c'(0.6) = 2.2663573$

As a final check on the overall accuracy of our result, we make use of Weddle's corrector formula (50.64). With $h = 0.1$, $x_0 = 0$, it becomes (we use the subscript w for Weddle)

(nn) $y_w(0.6) = y(0) + \frac{0.3}{10} [y'(0) + 5y'(0.1) + y'(0.2) + 6y'(0.3)$

$$+ y'(0.4) + 5y'(0.5) + y'(0.6)]$$

$$= 1 + 0.03[1 + 5(1.1155125) + 1.2642077 + 6(1.4495757)$$

$$+ 1.6754738 + 5(1.9461635) + 2.2663573]$$

$$= 1.9063562,$$

which is a closer approximation to the true value of $y(0.6)$ than is the one in (ll). Since the difference between $y_c(0.6)$ and $y_w(0.6)$ is not significant, we assume their common value 1.90636 rounded off to five decimals is accurate to five places and its error is therefore < 0.000005. [The actual value of $y(0.6)$ is 1.9063564 so that the error in the value of $y(0.6)$ in (nn) is 0.0000002. Rounded off to five decimal places, the actual value of $y(0.6)$ is 1.90636, a figure which agrees with our final result rounded off to five decimal places.]

EXERCISE 52

1. In Example 52.2, we found $y(0.1) = 1.1055125$ and $y(0.05 + 0.05) = 1.1055128$. Use formula (46.57) to correct $y(0.1)$. Using this corrected figure of $y(0.1)$, compute by series methods $y(0.2)$ and $y(0.1 + 0.1)$, i.e., compute the value of $y(0.2)$ in one step and in two steps. The value of $y(0.2) = 1.2242000$, computed in one step can be found in (n) of this example. Use formula (46.57), with $x_0 = 0$, $h = 0.2$, to correct $y(0.2)$. Starting with this corrected value of $y(0.2)$, switch to the Runge-Kutta method, fourth order form, and compute $y(0.2 + 0.1)$ and $y[(0.2 + 0.05) + 0.05]$, i.e., compute $y(0.3)$ in one step and in two steps. Apply formula (47.42) to correct $y(0.3)$. Replace (52.21) by these new corrected figures just obtained. Use Milne's method to find $y_p(0.4)$, $y_c(0.4)$, $D(0.4)$. Apply formula (51.35) to correct $y_c(0.4)$ and add this

corrected figure to your table. Compute $y_p(0.5)$, $y_c(0.5)$, $D(0.5)$. Correct $y_c(0.5)$ by (51.35). Do the same for $y_c(0.6)$. Finally apply Weddle's rule (50.64) to compute $y_w(0.6)$. How many decimal place accuracy do you now have? Compare all results with previous figures and with actual values.

2. Use the method of this lesson to find an approximate value when $x = 0.6$ of the particular solution of $y' = x + y$ for which $y(0) = 1$. Assume you know nothing about the solution and that you wish the error to be less than 0.00005. Use Taylor series for the first two approximations, Runge-Kutta for the third approximation. Then switch to Milne's method. In the course of your computations reduce h by one-half even though it may not be essential. Use formula (50.63) to check the accuracy of $y(0.5)$ and Weddle's rule (50.64) to check the accuracy of $y(0.6)$. In the absence of a solution, how many decimal place accuracy could you assume in $y(0.5)$, in $y(0.6)$? See 3 after "3. Cumulative Errors," Lesson 52A. Solve the equation and compare your results with actual value of $y(0.5)$ and $y(0.6)$.

3. Follow the instructions in 2 to find an approximate value when $x = 1.6$, of the particular solution of $y' = xy$ for which $y(1) = 1$.

ANSWERS 52

1. $y(0.1) = 1.1055128$, $y(0.2) = 1.2242085$, $y(0.3) = 1.3595771$,
$y_p(0.4) = 1.5154628$, $y_p(0.5) = 1.6961512$, $y_p(0.6) = 1.9063429$,
$y_c(0.4) = 1.5154745$, $y_c(0.5) = 1.6961646$, $y_c(0.6) = 1.9063565$,
$D(0.4) = 0.0000117$, $D(0.5) = 0.0000134$, $D(0.6) = 0.0000136$,
$y(0.4) = 1.5154741$, $y(0.5) = 1.6961641$, $y(0.6) = 1.9063560$,
$y_w(0.6) = 1.9063566$. Can assume five decimal place accuracy. Actual value: $y(0.6) = 1.9063564$.

2. $y = 2e^x - x - 1$. Actual values: $y(0.5) = 1.7974425$,
$y(0.6) = 2.0442376$.

3. $y = e^{(x^2-1)/2}$. Actual value: $y(1.6) = 2.1814723$.

LESSON 53. Numerical Methods Applied to a System of Two First Order Equations.

We consider a system of two first order equations

(53.1)
$$\frac{dx}{dt} = f(t,x,y),$$

$$\frac{dy}{dt} = g(t,x,y),$$

for which

(53.11) $x(t_0) = x_0, \qquad y(t_0) = y_0.$

In Lesson 39, Example 39.17, we showed how to find a series solution of a system such as (53.1). Hence we can use this method to obtain starting approximations, provided the derivatives can be obtained without excessive difficulty. If they cannot, we can use the following Runge-Kutta fourth order formulas.

(53.12) $x(t_0 + h) = x(t_0) + \frac{1}{6}(v_1 + 2v_2 + 2v_3 + v_4),$

$y(t_0 + h) = y(t_0) + \frac{1}{6}(w_1 + 2w_2 + 2w_3 + w_4),$

where

(53.13)
$$v_1 = hf(t_0, x_0, y_0),$$
$$w_1 = hg(t_0, x_0, y_0),$$
$$v_2 = hf(t_0 + \tfrac{1}{2}h, x_0 + \tfrac{1}{2}v_1, y_0 + \tfrac{1}{2}w_1),$$
$$w_2 = hg(t_0 + \tfrac{1}{2}h, x_0 + \tfrac{1}{2}v_1, y_0 + \tfrac{1}{2}w_1),$$
$$v_3 = hf(t_0 + \tfrac{1}{2}h, x_0 + \tfrac{1}{2}v_2, y_0 + \tfrac{1}{2}w_2),$$
$$w_3 = hg(t_0 + \tfrac{1}{2}h, x_0 + \tfrac{1}{2}v_2, y_0 + \tfrac{1}{2}w_2),$$
$$v_4 = hf(t_0 + h, x_0 + v_3, y_0 + w_3),$$
$$w_4 = hg(t_0 + h, x_0 + v_3, y_0 + w_3).$$

The Milne predictor and corrector formulas for the system (53.1) are

$$(53.14) \quad x_p(t_0 + 4h) = x(t_0) + \frac{4h}{3}[2x'(t_0+h) - x'(t_0+2h) + 2x'(t_0+3h)],$$

$$y_p(t_0 + 4h) = y(t_0) + \frac{4h}{3}[2y'(t_0+h) - y'(t_0+2h) + 2y'(t_0+3h)],$$

$$(53.15) \quad x_c(t_0 + 4h)$$
$$= x(t_0 + 2h) + \frac{h}{3}[x'(t_0+2h) + 4x'(t_0+3h) + x_p'(t_0+4h)],$$

$$y_c(t_0 + 4h)$$
$$= y(t_0 + 2h) + \frac{h}{3}[x'(t_0+2h) + 4y'(t_0+3h) + y_p'(t_0+4h)].$$

All comments and formulas in regard to errors, made in connection with the numerical solution of the equation $y' = f(x,y)$, apply to each function making up the solution of a system. The fifth degree formula (50.63) or Weddle's formula (50.64) can be used to check the overall accuracy of the values of $x(t)$ and of $y(t)$.

Example 53.16.   Find approximate values when $x = 0.4$ and $y = 0.4$ of a particular solution of the system

(a)
$$x'(t) = ty,$$
$$y'(t) = xy,$$

for which $x(0) = 1$ and $y(0) = 1$.

Solution.   To bring into the discussion all the above formulas, we begin by using Taylor series.*   The series solution of (a) has already been

---

*By (f) of Example 39.17, the interval of convergence is $|t| < 0.0625$. We, therefore, cannot use Taylor series to find $x(0.1)$, $y(0.1)$, $x(0.2)$, $y(0.2)$ unless we can establish a larger interval of convergence. The chances are that the series solution does actually have such a larger interval since the theorem gives only a minimum interval. But unless we can prove both series converge for $t = 0.1$, $t = 0.2$, we would need to use Runge-Kutta or some other starting method in place of Taylor series. We have used Taylor series only for illustrative purposes.

found in Example 39.17. It is [see (k)]

(b) $$x(t) = 1 + \frac{t^2}{2} + \frac{t^3}{3} + \frac{t^4}{8} + \cdots,$$

$$y(t) = 1 + t + \frac{t^2}{2} + \frac{t^3}{3} + \frac{7t^4}{24} + \cdots.$$

Hence, by (b), with $t = 0.1$ and 0.2 respectively, we obtain

(c) $x(0.1) = 1 + \dfrac{(0.1)^2}{2} + \dfrac{(0.1)^3}{3} + \dfrac{(0.1)^4}{8} + \cdots = 1.0053,$

$y(0.1) = 1 + 0.1 + \dfrac{(0.1)^2}{2} + \dfrac{(0.1)^3}{3} + \dfrac{7(0.1)^4}{24} + \cdots = 1.1054,$

$x(0.2) = 1 + \dfrac{(0.2)^2}{2} + \dfrac{(0.2)^3}{3} + \dfrac{(0.2)^4}{8} + \cdots = 1.0229,$

$y(0.2) = 1 + (0.2) + \dfrac{(0.2)^2}{2} + \dfrac{(0.2)^3}{3} + \dfrac{7(0.2)^4}{24} + \cdots = 1.2231,$

Remark. We have used direct substitution to find $x(0.2)$ and $y(0.2)$. It would have been more accurate, but have involved much more labor, if we had used the creeping up method.

To find $x(0.3)$ and $y(0.3)$, we switch to the Runge-Kutta method. With $t_0 = 0.2$, $h = 0.1$, $x(0.2) = 1.0229$, $y(0.2) = 1.2231$, (53.12) becomes

(d) $$x(0.3) = 1.0229 + \tfrac{1}{6}(v_1 + 2v_2 + 2v_3 + v_4),$$

$$y(0.3) = 1.2231 + \tfrac{1}{6}(w_1 + 2w_2 + 2w_3 + w_4).$$

With $h = 0.1$, $t_0 = 0.2$, $x_0 = x(t_0) = x(0.2) = 1.0229$, $y_0 = y(t_0) = y(0.2) = 1.2231$, $f(t,x,y) = ty$ and $g(t,x,y) = xy$, (53.13) becomes

(e) $v_1 = 0.1f(0.2, 1.0229, 1.2231) = 0.1(0.2)(1.2231) = 0.02446,$

$w_1 = 0.1g(0.2, 1.0229, 1.2231) = 0.1(1.0229)(1.2231) = 0.1251,$

$v_2 = 0.1f(0.25, 1.0229 + 0.0122, 1.2231 + 0.0626)$
$\quad = 0.1(0.25)(1.2857) = 0.0321,$

$w_2 = 0.1g(0.25, 1.0229 + 0.0122, 1.2231 + 0.0626)$
$\quad = 0.1(1.0351)(1.2857) = 0.1331,$

$v_3 = 0.1f(0.25, 1.0229 + 0.0161, 1.2231 + 0.0666)$
$\quad = 0.1(0.25)(1.2897) = 0.0322,$

$w_3 = 0.1g(0.25, 1.0229 + 0.0161, 1.2231 + 0.0666)$
$\quad = 0.1(1.0390)(1.2897) = 0.1340,$

$$v_4 = 0.1f(0.3, 1.0229 + 0.0322, 1.2231 + 0.1340)$$
$$= .0.1(0.3)(1.3571) = 0.0407,$$

$$w_4 = 0.1g(0.3, 1.0229 + 0.0322, 1.2231 + 0.1340)$$
$$= 0.1(1.0551)(1.3571) = 0.1432.$$

Hence (d) becomes

(f) $x(0.3) = 1.0229 + \frac{1}{6}[0.0245 + 2(0.0321) + 2(0.0322) + 0.0407]$
$$= 1.0552,$$

$\qquad y(0.3) = 1.2231 + \frac{1}{6}[0.1251 + 2(0.1331) + 2(0.1340) + 0.1432]$
$$= 1.3568.$$

Since we now have the needed number of preliminary estimates, we switch to Milne's method to find $x(0.4)$ and $y(0.4)$. Before we can use formulas (53.14) and (53.15), however, we must know $x'(0.1)$, $x'(0.2)$, $x'(0.3)$ and $y'(0.1)$, $y'(0.2)$, $y'(0.3)$. We obtain these values by use of (a) and the values of $x(0.1)$, $x(0.2)$, $x(0.3)$, $y(0.1)$, $y(0.2)$, $y(0.3)$ as found in (c) and (f) above. Therefore

(g) $\qquad x'(0.1) = 0.1y(0.1) = (0.1)(1.1054) = 0.1105,$

$\qquad\qquad x'(0.2) = 0.2y(0.2) = (0.2)(1.2231) = 0.2446,$

$\qquad\qquad x'(0.3) = 0.3y(0.3) = (0.3)(1.3568) = 0.4070,$

$\qquad\qquad y'(0.1) = x(0.1)y(0.1) = (1.0053)(1.1054) = 1.1113,$

$\qquad\qquad y'(0.2) = x(0.2)y(0.2) = (1.0229)(1.2231) = 1.2511,$

$\qquad\qquad y'(0.3) = x(0.3)y(0.3) = (1.0552)(1.3568) = 1.4317.$

Hence with $t_0 = 0$, $h = 0.1$, and with the initial conditions $x(0) = 1$, $y(0) = 1$, we can write (53.14) as

(h) $x_p(0.4) = 1 + \dfrac{0.4}{3}[2x'(0.1) - x'(0.2) + 2x'(0.3)]$

$$= 1 + \frac{0.4}{3}[2(0.1105) - 0.2446 + 2(0.4070)] = 1.1054,$$

$\qquad y_p(0.4) = 1 + \dfrac{0.4}{3}[2y'(0.1) - y'(0.2) + 2y'(0.3)]$

$$= 1 + \frac{0.4}{3}[2(1.1113) - 1.2511 + 2(1.4317)] = 1.5113.$$

By (a) and (h), we obtain

(i) $\qquad x_p'(0.4) = 0.4y(0.4) = 0.4(1.5113) = 0.6045,$

$\qquad\qquad y_p'(0.4) = x(0.4)y(0.4) = (1.1054)(1.5113) = 1.6706.$

The corrector formulas (53.15) are therefore

(j) $x_c(0.4) = x(0.2) + \dfrac{0.1}{3}\,[x'(0.2) + 4x'(0.3) + x_p'(0.4)]$

           $= 1.0229 + \dfrac{0.1}{3}\,[0.2446 + 4(0.4070) + 0.6045]$

           $= 1.1055,$

     $y_c(0.4) = y(0.2) + \dfrac{0.1}{3}\,[y'(0.2) + 4y'(0.3) + y_p'(0.4)]$

           $= 1.2231 + \dfrac{0.1}{3}\,[1.2511 + 4(1.4317) + 1.6706]$

           $= 1.5114.$

EXERCISE 53

1. Using the values of $x(0.4)$ and $y(0.4)$, given in (j) of Example 53.16, as new predictor values, use corrector formulas (53.15) to see if they will correct these results. If they do, repeat the process until two successive values of $x(0.4)$ and $y(0.4)$ agree. Then compute $x(0.5)$ and $y(0.5)$ by means of the predictor formulas and the repeated use of the corrector formulas. With five values of $x(t)$ known in addition to the initial condition, use corrector formula (50.63) to check the accuracy of $x(0.5)$. Do the same for $y(0.5)$. How many decimal place accuracy can you assume in $x(0.5)$, $y(0.5)$? *Hint.* See 3 after "**3.** *Cumulative Errors*," Lesson 52A.

2. Find approximate values when $x = 0.1, 0.2, \cdots, 0.5$, $y = 0.1, 0.2, \cdots,$ 0.5 of the particular solution of the first order linear system

$$x'(t) = x - y, \qquad y'(t) = -4x + y,$$

for which $x(0) = 1$, $y(0) = 1$. Take $h = 0.1$. Use Taylor series to the fourth order, direct substitution method, to calculate $x(0.1)$, $y(0.1)$, $x(0.2)$, $y(0.2)$; Runge-Kutta method to calculate $x(0.3)$, $y(0.3)$; Milne's method to calculate $x(0.4)$, $y(0.4)$, $x(0.5)$, $y(0.5)$. Finally apply corrector formula (50.63) to compute $x(0.5)$, $y(0.5)$. How many place accuracy can you assume in $x(0.5)$, $y(0.5)$? Now solve the system and compare results.

3. In problem **2**, compute $x(0.2)$, $y(0.2)$ by the creeping up method. Then apply (46.57) to correct $x(0.2)$, $y(0.2)$. With these corrected values of $x(0.2)$, $y(0.2)$, use Runge-Kutta method to compute $x[(0.2 + 0.05) + 0.05]$, $x(0.2 + 0.1)$. Do the same for $y(0.3)$. Correct $x(0.3)$, $y(0.3)$ by means of (47.42). Compute $x_p(0.4)$, $x_c(0.4)$, $D(0.4)$. Correct $x_c(0.4)$ by means of (51.35). Do the same for $y_c(0.4)$. Compute $x_p(0.5)$, $x_c(0.5)$, $D(0.5)$. Correct $x_c(0.5)$. Do the same for $y_c(0.5)$. Finally apply corrector formula (50.63) to compute $x(0.5)$, $y(0.5)$. How many decimal place accuracy can you assume in your results? Compare with results obtained in **2** and with actual solution.

ANSWERS 53

1. $x(0.4) = 1.1055$, $y(0.4) = 1.5114$; $x(0.5) = 1.1776$, $y(0.5) = 1.6938$. Corrector (50.63): $x(0.5) = 1.1776$, $y(0.5) = 1.6939$. Can assume four and three decimal place accuracy.

2. $x(t)$ values: 1.01609, 1.06940, 1.17026, 1.33246, 1.57490. $y(t)$ values: 0.68234, 0.31740, -0.11807, -0.65395, -1.33021. Corrector values (50.63): $x(0.5) = 1.57504$, $y(0.5) = -1.33050$. Can assume three and two decimal place accuracy respectively, if we round off to this number of decimals.
Solution: $x = \frac{1}{4}(e^{3t} + 3e^{-t})$, $y = \frac{1}{2}(-e^{3t} + 3e^{-t})$.
Actual values: $x(0.5) = 1.5753203$, $y(0.5) = -1.3310485$.

3. $x(0.2) = 1.06958$, $y(0.2) = 0.31705$;
$x(0.3) = 1.17051$, $y(0.3) = -0.11855$;
$x(0.4) = 1.33274$, $y(0.4) = -0.65450$;
$x(0.5) = 1.57525$, $y(0.5) = -1.33091$.
Corrector values (50.63): $x(0.5) = 1.57530$, $y(0.5) = -1.33101$.

LESSON 54. Numerical Solution of a Second Order Differential Equation.

In the proof of Theorem 62.22, we show how a second order differential equation $y'' = f(x,y,y')$ can be reduced to a system of two first order equations. A numerical solution of this equation can therefore be found by the method of Lesson 53. However, it is also possible to find a numerical solution of such an equation without the necessity of reducing it to a system. We illustrate the method by an example.

Example 54.1. Find an approximate value when $x = 0.4$ of a particular solution of the equation

(a) $$y'' = 2x + 2y - y',$$

for which $y(0) = 1$, $y'(0) = 1$.

Solution. In order to bring into the discussion Taylor series, Runge-Kutta and Milne methods, we shall find $y(0.1)$ and $y(0.2)$ by Taylor series method, $y(0.3)$ by Runge-Kutta's method and $y(0.4)$ by Milne's method.
By Theorem 37.2, if

$$(54.11) \quad y(x_0 + h) = y(x_0) + y'(x_0)h$$
$$+ \frac{y''(x_0)}{2!} h^2 + \frac{y'''(x_0)}{3!} h^3 + \frac{y^{(4)}(x_0)}{4!} h^4 + \cdots,$$

then

$$(54.12) \quad y'(x_0 + h) = y'(x_0) + y''(x_0)h + \frac{y'''(x_0)}{2!} h^2 + \frac{y^{(4)}(x_0)}{3!} h^4 + \cdots.$$

From (a), and by differentiation of (a), we obtain

(b) $\quad y'' = 2x + 2y - y', \quad y''' = 2 + 2y' - y'', \quad y^{(4)} = 2y'' - y'''.$

By the initial conditions, $x = 0$, $y = 1$, $y' = 1$. Substituting these values in (b), there results

(c) $\quad y''(0) = 1, \quad y'''(0) = 2 + 2 - 1 = 3, \quad y^{(4)}(0) = 2 - 3 = -1.$

Substituting the initial conditions and (c) in (54.11) it becomes, with $x_0 = 0$,

(d) $$y(h) = 1 + h + \frac{h^2}{2} + \frac{h^3}{2} - \frac{h^4}{24} + \cdots.$$

As in the first order case, two methods are available to us to find $y(0.1)$ and $y(0.2)$. By direct substitution, we obtain from (d),

(e) $y(0.1) = 1 + 0.1 + \frac{1}{2}(0.1)^2 + \frac{1}{2}(0.1)^3 - \frac{1}{24}(0.1)^4 = 1.105496,$

$y(0.2) = 1 + 0.2 + \frac{1}{2}(0.2)^2 + \frac{1}{2}(0.2)^3 - \frac{1}{24}(0.2)^4 = 1.223933.$

The creeping up method involves much more arithmetic, but insures greater accuracy. By our basic equations (54.11) and (54.12), with $h = 0.1$ and x_0 equal successively 0.0, 0.1, we find

(f) $y(0 + 0.1) = y(0.1)$

$$= y(0) + y'(0)(0.1) + \frac{y''(0)}{2!}(0.1)^2$$

$$+ \frac{y'''(0)}{3!}(0.1)^3 + \frac{y^{(4)}(0)}{4!}(0.1)^4 + \cdots,$$

$y'(0 + 0.1) = y'(0.1)$

$$= y'(0) + y''(0)(0.1) + \frac{y'''(0)}{2!}(0.1)^2$$

$$+ \frac{y^{(4)}(0)}{3!}(0.1)^3 + \cdots,$$

$y(0.1 + 0.1) = y(0.2)$

$$= y(0.1) + y'(0.1)(0.1) + \frac{y''(0.1)}{2!}(0.1)^2$$

$$+ \frac{y'''(0.1)}{3!}(0.1)^3 + \frac{y^{(4)}(0.1)}{4!}(0.1)^4 + \cdots,$$

$y'(0.1 + 0.1) = y'(0.2)$

$$= y'(0.1) + y''(0.1)(0.1) + \frac{y'''(0.1)}{2!}(0.1)^2$$

$$+ \frac{y^{(4)}(0.1)}{3!}(0.1)^3 + \cdots.$$

Substituting the initial conditions and (c) in the first two lines of (f), we obtain

(g) $y(0.1) = 1 + (1)(0.1) + \frac{1}{2}(0.1)^2 + \frac{3}{6}(0.1)^3 - \frac{1}{24}(0.1)^4 = 1.105496,$

$y'(0.1) = 1 + (1)(0.1) + \frac{3}{2}(0.1)^2 - \frac{1}{6}(0.1)^3 = 1.114833.$

Therefore with $x = 0.1$, $y(0.1) = 1.105496$, $y'(0.1) = 1.114833$, we find by (b),

(h) $y''(0.1) = 2(0.1) + 2(1.105496) - 1.114833 = 1.296159,$

   $y'''(0.1) = 2 + 2(1.114833) - 1.296159 = 2.933507,$

   $y^{(4)}(0.1) = 2(1.296159) - 2.933507 = -0.341189.$

Substituting (g) and (h) in the third and fourth lines of (f), there results

(i) $y(0.2) = 1.105496 + (1.114833)(0.1) + \dfrac{1.296159}{2}(0.1)^2$

   $+ \dfrac{2.933507}{6}(0.1)^3 - \dfrac{0.341189}{24}(0.1)^4 = 1.223948,$

   $y'(0.2) = 1.114833 + (1.296159)(0.1) + \dfrac{2.933507}{2}(0.1)^2$

   $- \dfrac{0.341189}{6}(0.1)^3 = 1.259060.$

We now switch to the Runge-Kutta method to find $y(0.3)$. The fourth order form for the second order equation $y'' = f(x,y,y')$ for which $y(x_0) = y_0$, $y'(x_0) = y_1$ is

(54.13) $y(x_0 + h) = y(x_0) + \frac{1}{6}(v_1 + 2v_2 + 2v_3 + v_4),$

   $y'(x_0 + h) = y'(x_0) + \frac{1}{6}(w_1 + 2w_2 + 2w_3 + w_4),$

where, with $y_0' = y'(x_0)$,

(54.14) $v_1 = hy_0',$

   $w_1 = hf(x_0, y_0, y_0'),$

   $v_2 = h(y_0' + \frac{1}{2}w_1),$

   $w_2 = hf(x_0 + \frac{1}{2}h,\ y_0 + \frac{1}{2}v_1,\ y_0' + \frac{1}{2}w_1),$

   $v_3 = h(y_0' + \frac{1}{2}w_2),$

   $w_3 = hf(x_0 + \frac{1}{2}h,\ y_0 + \frac{1}{2}v_2,\ y_0' + \frac{1}{2}w_2),$

   $v_4 = h(y_0' + w_3),$

   $w_4 = hf(x_0 + h,\ y_0 + v_3,\ y_0' + w_3).$

Therefore with $x_0 = 0.2$, $h = 0.1$, $y(x_0) = y(0.2) = 1.223948$, $y'(0.2) = 1.259060$, (54.13) becomes

(j) $y(0.3) = y(0.2 + 0.1) = 1.223948 + \frac{1}{6}(v_1 + 2v_2 + 2v_3 + v_4),$

   $y'(0.3) = y'(0.2 + 0.1) = 1.259060 + \frac{1}{6}(w_1 + 2w_2 + 2w_3 + w_4).$

With $h = 0.1$, $y_0' = y'(x_0) = y'(0.2) = 1.259060$, $y_0 = y(x_0) = y(0.2) =$

1.223948, and $f(x,y,y') = 2x + 2y - y'$, (54.14) becomes

(k) $v_1 = 0.1(1.259060) = 0.125906,$

$w_1 = 0.1f(0.2, 1.223948, 1.259060)$
$= 0.1[2(0.2) + 2(1.223948) - 1.259060] = 0.158836,$

$v_2 = 0.1(1.259060 + 0.0794418) = 0.1338502,$

$w_2 = 0.1f(0.25, 1.223948 + 0.062953, 1.259060 + 0.079418)$
$= 0.1(0.5 + 2.573802 - 1.338502) = 0.173530,$

$v_3 = 0.1(1.259060 + 0.086765) = 0.1345825,$

$w_3 = 0.1f(0.25, 1.223948 + 0.066925, 1.259060 + 0.086765)$
$= 0.1(0.5 + 2.581746 - 1.345825) = 0.173592,$

$v_4 = 0.1(1.259060 + 0.173592) = 0.1432652,$

$w_4 = 0.1f(0.3, 1.223948 + 0.1345825, 1.259060 + 0.173592)$
$= 0.1(0.6 + 2.717061 - 1.432652) = 0.188441.$

Substituting (k) in (j), there results

(l) $y(0.3) = 1.223948 + \frac{1}{6}(0.125906 + 0.267700 + 0.269165 + 0.143265)$
$= 1.223948 + \frac{1}{6}(0.806036) = 1.358287,$

$y'(0.3) = 1.259060 + \frac{1}{6}(0.158836 + 0.347060 + 0.347184 + 0.188441)$
$= 1.259060 + \frac{1}{6}(1.041521) = 1.432647.$

We summarize the results thus far obtained

(m) $y(0) = 1.000000, \quad y'(0) = 1.000000;$

$y(0.1) = 1.105496, \quad y'(0.1) = 1.114833;$

$y(0.2) = 1.223948, \quad y'(0.2) = 1.259060;$

$y(0.3) = 1.358287, \quad y'(0.3) = 1.432647.$

The Milne predictor formulas for the second order equation $y'' = f(x,y,y')$ for which $y(x_0) = y_0$, $y'(x_0) = y_1$ are

(54.15)   (a) $y_p'(x_0 + 4h) = y'(x_0)$
$$+ \frac{4h}{3}[2y''(x_0 + h) - y''(x_0 + 2h) + 2y''(x_0 + 3h)],$$

(b) $y_p(x_0 + 4h) = y(x_0 + 2h)$
$$+ \frac{h}{3}[y'(x_0 + 2h) + 4y'(x_0 + 3h) + y_p'(x_0 + 4h)].$$

The corrector formulas are

(54.16)   (a) $y_c'(x_0 + 4h) = y'(x_0 + 2h)$
$$+ \frac{h}{3}[y''(x_0 + 2h) + 4y''(x_0 + 3h) + y_p''(x_0 + 4h)],$$

(b) $y_c(x_0 + 4h) = y(x_0 + 2h)$
$$+ \frac{h}{3} [y'(x_0 + 2h) + 4y'(x_0 + 3h) + y_c'(x_0 + 4h)].$$

As in the first order case, formulas (54.16) can be used again and again until two successive values of $y_c(x_0 + 4h)$ agree.

By (54.15) and (54.16), with $x_0 = 0$, $h = 0.1$,

(n) $y_p'(0.4) = y'(0) + \dfrac{0.4}{3} [2y''(0.1) - y''(0.2) + 2y''(0.3)],$

  $y_p(0.4) = y(0.2) + \dfrac{0.1}{3} [y'(0.2) + 4y'(0.3) + y_p'(0.4)],$

  $y_c'(0.4) = y'(0.2) + \dfrac{0.1}{3} [y''(0.2) + 4y''(0.3) + y_p''(0.4)],$

  $y_c(0.4) = y(0.2) + \dfrac{0.1}{3} [y'(0.2) + 4y'(0.3) + y_c'(0.4)].$

Before we can use these formulas, we must know the value of y'' when $x = 0.1, 0.2, 0.3$. By (a) and (m), these needed values are

(o) $y''(0.1) = 2(0.1) + 2(1.105496) - 1.114833 = 1.296159,$
  $y''(0.2) = 2(0.2) + 2(1.223948) - 1.259060 = 1.588836,$
  $y''(0.3) = 2(0.3) + 2(1.358287) - 1.432647 = 1.883927.$

Substituting in (n), the initial conditions and the values in (m) and (o), we obtain

(p) $y_p'(0.4) = 1 + \dfrac{0.4}{3} [2(1.296159) - 1.588836 + 2(1.883927)]$

  $= 1.636178,$

  $y_p(0.4) = 1.223948 + \dfrac{0.1}{3} [1.259060 + 4(1.432647) + 1.636178]$

  $= 1.511476.$

To use the third formula in (n), we need to know $y_p''(0.4)$. With $x = 0.4$ and $y_p(0.4)$, $y_p'(0.4)$ having the values in (p), we obtain from (a),

(q) $y_p''(0.4) = 2(0.4) + 2(1.511476) - 1.636178 = 2.186774.$

Hence the last two formulas in (n) become

(r) $y_c'(0.4) = 1.259060 + \dfrac{0.1}{3} [1.588836 + 4(1.883927) + 2.186774]$

  $= 1.636104,$

  $y_c(0.4) = 1.223948 + \dfrac{0.1}{3} [1.259060 + 4(1.432647) + 1.636104]$

  $= 1.511473.$

EXERCISE 54

1. Using the value of $y(0.4)$, given in (r) of Example 54.1, as a new predictor value, apply corrector formulas (54.16) repeatedly until two successive values of $y(0.4)$ agree. Then find $y(0.5)$ by means of the predictor formulas and repeated use of the corrector formulas. Do the same for $y(0.6)$. Finally apply Weddle's rule (50.64) to compute $y_w(0.6)$. How many decimal place accuracy can you assume in $y(0.6)$? See 3 after "3. *Cumulative Errors*," Lesson 52A. Solve (a) of Example 54.1 and compare your results with actual values.

2. Apply the method of this lesson to find an approximate value when $x = 0.5$ of a particular solution of the equation $y'' = x + 2y + y'$ for which $y(0) = 1$, $y'(0) = 1$. Apply corrector formula (50.63) to evaluate $y(0.5)$. How many decimal place accuracy can you assume? Solve the equation and compare your results with actual values.

3. The Adams method, see Exercise 50,6, can also be used as a continuing formula for finding a numerical solution of a second order equation. The needed formulas are

(54.2)   (a) $y(x_0 + h) = y(x_0)$
$$+ h[y'(x_0) + \tfrac{1}{2}\nabla y'(x_0) + \tfrac{5}{12}\nabla^2 y'(x_0) + \tfrac{3}{8}\nabla^3 y'(x_0)].$$

    (b) $y'(x_0 + h) = y'(x_0)$
$$+ h[y''(x_0) + \tfrac{1}{2}\nabla y''(x_0) + \tfrac{5}{12}\nabla^2 y''(x_0) + \tfrac{3}{8}\nabla^3 y''(x_0)].$$

Use these formulas to find approximate values of $y(0.4)$ and $y'(0.4)$ of the particular solution of the equation $y'' = 2x + 2y - y'$ for which $y(0) = 1$, $y'(0) = 1$. *Hint.* With $x_0 = 0.3$, $h = 0.1$, you must know $y'(0)$, $y'(0.1)$, $y'(0.2)$, $y'(0.3)$; $y''(0)$, $y''(0.1)$, $y''(0.2)$, $y''(0.3)$ in order to set up the needed tables of differences. You will find these values in Example 54.1.

4. Use Milne's method to find an approximate value of $y(0.4)$ of the particular solution of the equation $y''' = y'' + xy' + 2y$ for which $y(0) = y'(0) = y''(0) = 1$. The necessary Milne's formulas are

$$y_p''(0.4) = y''(0) + \frac{0.4}{3}[2y'''(0.1) - y'''(0.2) + 2y'''(0.3)],$$

$$y_p'(0.4) = y'(0.2) + \frac{0.1}{3}[y''(0.2) + 4y''(0.3) + y_p''(0.4)],$$

$$y_c(0.4) = y(0.2) + \frac{0.1}{3}[y'(0.2) + 4y'(0.3) + y_p'(0.4)].$$

Obtain the needed preliminary values by means of Taylor series, direct substitution.

ANSWERS 54

1. $y(0.4) = 1.511473$, $y(0.5) = 1.686538$, $y(0.6) = 1.886643$. $y_w(0.6) = 1.886640$. Can assume five decimal place accuracy. Solution: $y = \tfrac{5}{3}e^x - \tfrac{1}{6}e^{-2x} - x - \tfrac{1}{2}$. Actual value: $y(0.6) = 1.886666$.

2. $y = \dfrac{3}{4}e^{2x} - \dfrac{x}{2} + \dfrac{1}{4}$. Actual value: $y(0.5) = 2.038711$.

3. $y(0.4) = 1.51145$, $y'(0.4) = 1.63612$.

4. $y(0.4) = 1.520$.

LESSON 55. Perturbation Method. First Order Equation.

In physical problems, we frequently encounter a differential equation, as for example, the differential equation

(55.1) $$y' + y^2 = 0, \quad y(1) = 1,$$

which has been disturbed by a small effect, so that (55.1) has to be modified to read

(55.2) $$y' + y^2 = \epsilon x, \quad y(1) = 1,$$

where ϵ is small. It then becomes necessary to determine by how much the solution of (55.1) has been altered because of the presence of the disturbing function ϵx. We refer to this change in the solution as a **perturbation.**

A precise perturbation theory is extremely difficult. In this lesson, we shall aim to give only a rough outline of a method by which this problem can be handled. Call $y_0(x)$ a solution of (55.1) satisfying $y(1) = 1$, and denote the solution of (55.2) by

(55.3) $$y(x) = y_0(x) + p(x)$$

where $p(x)$ is the perturbation. We next expand $y(x)$ in a series in powers of ϵ, so that

(55.4) $$y(x) = y_0(x) + \epsilon y_1(x) + \epsilon^2 y_2(x) + \epsilon^3 y_3(x) + \cdots .$$

Comparing (55.3) with (55.4), we see that

(55.5) $$p(x) = \epsilon y_1(x) + \epsilon^2 y_2(x) + \epsilon^3 y_3(x) + \cdots .$$

The first term $\epsilon y_1(x)$ is called the **first order perturbation;** the second term $\epsilon^2 y_2(x)$ is called the **second order perturbation,** etc.

Substituting (55.4) in (55.2), we obtain

(55.6) $$y_0' + \epsilon y_1' + \epsilon^2 y_2' + \epsilon^3 y_3' + \cdots$$
$$+ (y_0 + \epsilon y_1 + \epsilon^2 y_2 + \epsilon^3 y_3 + \cdots)^2 = \epsilon x.$$

Carrying out the indicated multiplication, then collecting coefficients of like powers of ϵ, we have

(55.7) $$(y_0' + y_0{}^2) + (y_1' + 2y_0 y_1)\epsilon$$
$$+ (y_2' + 2y_0 y_2 + y_1{}^2)\epsilon^2 + (\cdots\cdots)\epsilon^3 + \cdots = \epsilon x.$$

Next we equate like powers of ϵ. There results

(55.8) $$y_0' + y_0{}^2 = 0,$$
$$y_1' + 2y_0 y_1 = x,$$
$$y_2' + 2y_0 y_2 + y_1{}^2 = 0,$$
$$\cdot \; \cdot \; \cdot \; \cdot \; \cdot \; \cdot \; \cdot \; \cdot \; \cdot \; \cdot \; \cdot \; \cdot \; \cdot$$

By solving each equation of (55.8) in *succession*, we can thus determine the functions $y_1(x)$, $y_2(x)$, $\cdots$ in (55.4). Each of these functions, however, must satisfy an initial condition. Since the initial condition associated with the original equation (55.1) is $y(1) = 1$, and since y_0 is a solution of (55.1) so that $y_0(1) = 1$, this initial condition will be satisfied if, in (55.4), we assume

(55.81) $y_0(1) = 1,$ $y_1(1) = 0,$ $y_2(1) = 0, \cdots$.

We illustrate the details of the above method by solving Example 55.9 below. In practice the first and second order perturbation terms of (55.5) are usually sufficient.

Example 55.9.   Find the first and second order perturbation terms in the solution of

(a) $y' + y^2 = 0,$ for which $y(1) = 1,$

due to the presence of a disturbing function ϵx, where ϵ is small.

Solution.   Because of the disturbing function ϵx, (a) must be modified to read

(b) $y' + y^2 = \epsilon x,$

for which $y(1) = 1$. Following the procedure outlined above we let, see (55.4) and (55.81),

(c) $y(x) = y_0(x) + \epsilon y_1(x) + \epsilon^2 y_2(x) + \cdots,$

with initial conditions

(d) $y_0(1) = 1,$ $y_1(1) = 0,$ $y_2(1) = 0, \cdots$.

Substituting (c) in (b), we obtain, see (55.7),

(e) $(y_0' + y_0^2) + (y_1' + 2y_0 y_1)\epsilon + (y_2' + 2y_0 y_2 + y_1^2)\epsilon^2 + \cdots = \epsilon x.$

Equating coefficients of like powers of ϵ^0, ϵ, ϵ^2, we obtain from (e) the system of equations

(f) $y_0' + y_0^2 = 0,$ $y_1' + 2y_0 y_1 = x,$ $y_2' + 2y_0 y_2 + y_1^2 = 0.$

A solution of the first equation of (f), satisfying the initial condition $y_0(1) = 1$ of (d), is

(g) $y_0 = \dfrac{1}{x}.$

Substituting (g) in the second equation of (f), we obtain

(h) $$y_1' + \frac{2}{x} y_1 = x.$$

A solution of (h) satisfying $y_1(1) = 0$ of (d) is

(i) $$y_1 = \frac{1}{4}\left(x^2 - \frac{1}{x^2}\right).$$

Substituting (g) and (i) in the third equation of (f), we obtain

(j) $$y_2' + \frac{2}{x} y_2 = -\frac{1}{16}\left(x^4 - 2 + \frac{1}{x^4}\right).$$

A solution of (j) satisfying $y_2(1) = 0$ of (d), is

(k) $$y_2 = -\frac{1}{16}\left(\frac{x^5}{7} - \frac{2x}{3} - \frac{1}{x^3}\right) - \frac{2}{21x^2}.$$

Substituting (g), (i), (k) in (c), we obtain

(l) $$y = \frac{1}{x} + \frac{\epsilon}{4}\left(x^2 - \frac{1}{x^2}\right) - \frac{\epsilon^2}{336}\left(3x^5 - 14x - \frac{21}{x^3} + \frac{32}{x^2}\right).$$

The solution of (a) satisfying $y(1) = 1$, i.e., its solution if there were no disturbing function ϵx present, is $1/x$. Because of the disturbing function ϵx, the first and second order perturbation terms are, respectively, the second and third terms in (l).

EXERCISE 55

1. Find the first and second order perturbation terms in the solution of $y' + y^2 = 0$, for which $y(1) = 1$, due to the presence of a disturbing function ϵx^2, where ϵ is small.

ANSWERS 55

1. First order: $\dfrac{\epsilon}{5}\left(x^3 - \dfrac{1}{x^2}\right)$; second order: $-\dfrac{\epsilon^2}{450}\left(2x^7 - 9x^2 - \dfrac{18}{x^3} + \dfrac{25}{x^2}\right).$

LESSON 56. Perturbation Method. Second Order Equation.

In Lesson 55, we outlined a method of determining the perturbation of a solution of a first order differential equation due to a small disturbance. We shall now apply this method to determine the perturbation of a solution of a second order equation. Consider the differential equation

(56.01) $$y'' + y = 0,$$

for which

(56.02) $$y(0) = 0, \qquad y'(0) = 1.$$

Note that (56.01) is the differential equation of simple harmonic motion. Let the disturbing function be $-2\epsilon(y')^2$, where ϵ is small. Therefore (56.01) becomes

(56.1) $$y'' + y = -2\epsilon(y')^2$$

for which

(56.11) $$y(0) = 0, \qquad y'(0) = 1.$$

We wish to find the first and second order perturbation terms of a solution of (56.01) satisfying (56.02) resulting from the presence of a disturbing function $-2\epsilon(y')^2$.

Call $y_0(x)$ a solution of (56.1) satisfying (56.11). Let

(56.12) $$y = y_0 + \epsilon y_1 + \epsilon^2 y_2 + \cdots;$$

therefore

$$y'' = y_0'' + \epsilon y_1'' + \epsilon^2 y_2'' + \cdots.$$

In order that (56.12) may satisfy the initial conditions (56.02), we assume

(56.13) $$y_0(0) = 0, \qquad y_1(0) = 0, \qquad y_2(0) = 0, \cdots.$$
$$y_0'(0) = 1, \qquad y_1'(0) = 0, \qquad y_2'(0) = 0, \cdots.$$

Substituting (56.12) in (56.1), we obtain, using only terms to ϵ^2,

(56.14) $$y_0'' + \epsilon y_1'' + \epsilon^2 y_2'' + y_0 + \epsilon y_1 + \epsilon^2 y_2$$
$$= -2\epsilon[(y_0')^2 + \epsilon^2(y_1')^2 + \epsilon^4(y_2')^2 + 2\epsilon y_0' y_1'$$
$$+ 2\epsilon^2 y_0' y_2' + 2\epsilon^3 y_1' y_2'].$$

Collecting coefficients of like powers of ϵ, we have

(56.15) $$(y_0'' + y_0) + (y_1'' + y_1)\epsilon + (y_2'' + y_2)\epsilon^2$$
$$= -2(y_0')^2\epsilon - 4y_0' y_1'\epsilon^2.$$

Equating coefficients of like powers of ϵ^0, ϵ, ϵ^2, we obtain from (56.15),

(56.16) $$y_0'' + y_0 = 0, \qquad y_1'' + y_1 = -2(y_0')^2, \qquad y_2'' + y_2 = -4y_0' y_1'.$$

A solution of the first equation of (56.16) satisfying $y_0(0) = 0$, $y_0'(0) = 1$ is

(56.2) $$y_0 = \sin x, \qquad y_0' = \cos x.$$

Substituting (56.2) in the second equation of (56.16), we obtain

$$(56.21) \qquad y_1'' + y_1 = -2\cos^2 x.$$

A general solution of (56.21) is

$$(56.22) \qquad y_1 = c_1 \sin x + c_2 \cos x - \tfrac{4}{3}\sin^2 x - \tfrac{2}{3}\cos^2 x,$$
$$y_1' = c_1 \cos x - c_2 \sin x - \tfrac{4}{3}\sin x \cos x.$$

A particular solution of (56.22) satisfying $y_1(0) = 0$, $y_1'(0) = 0$ is

$$(56.23) \qquad y_1 = \tfrac{2}{3}\cos x - \tfrac{4}{3}\sin^2 x - \tfrac{2}{3}\cos^2 x,$$
$$y_1' = -\tfrac{2}{3}\sin x - \tfrac{8}{3}\sin x \cos x + \tfrac{4}{3}\sin x \cos x$$
$$= -\tfrac{2}{3}\sin x - \tfrac{4}{3}\sin x \cos x.$$

Substituting in the third equation of (56.16), the values of y_0' and y_1' as given in (56.2) and (56.23), we obtain

$$(56.24) \qquad y_2'' + y_2 = -4(-\tfrac{2}{3}\sin x \cos x - \tfrac{4}{3}\sin x \cos^2 x)$$
$$= \tfrac{8}{3}\sin x \cos x + \tfrac{16}{3}\sin x \cos^2 x$$
$$= \tfrac{8}{3}\sin x \cos x + \tfrac{16}{3}\sin x - \tfrac{16}{3}\sin^3 x.$$

The complementary function of (56.24) is

$$(56.25) \qquad y_c = c_1 \sin x + c_2 \cos x.$$

A particular solution

(56.251) of $y_2'' + y_2 = \tfrac{8}{3}\sin x \cos x$, is $y_2 = -\tfrac{8}{9}\sin x \cos x$,

         of $y_2'' + y_2 = \tfrac{16}{3}\sin x$,   is $y_2 = -\tfrac{8}{3}x \cos x$,

         of $y_2'' + y_2 = \tfrac{16}{3}\sin^3 x$, by Example 21.32,

                         is $y_2 = \tfrac{1}{6}\sin 3x - 2x \cos x.$

Hence a general solution of (56.24), by (56.25) and (56.251), is

$$(56.26) \quad y_2 = c_1 \sin x + c_2 \cos x - \tfrac{8}{9}\sin x \cos x - \tfrac{2}{3}x \cos x - \tfrac{1}{6}\sin 3x,$$
$$y_2' = c_1 \cos x - c_2 \sin x + \tfrac{8}{9}(\sin^2 x - \cos^2 x)$$
$$+ \tfrac{2}{3}(x \sin x - \cos x) - \tfrac{1}{2}\cos 3x.$$

By (56.13), $y_2(0) = 0$, $y_2'(0) = 0$. Inserting these values in (56.26), we find

$$(56.27) \qquad 0 = c_2, \qquad c_1 = \tfrac{8}{9} + \tfrac{2}{3} + \tfrac{1}{2} = \tfrac{37}{18}.$$

Therefore by (56.26) and (56.27), a particular solution of (56.24) satisfying $y_2(0) = 0$, $\dot{y}_2'(0) = 0$ is

$$(56.28) \quad y_2 = \tfrac{37}{18}\sin x - \tfrac{8}{9}\sin x \cos x - \tfrac{2}{3}x \cos x - \tfrac{1}{6}\sin 3x.$$

Substituting, in (56.12), the values of y_0, y_1, and y_2 as found in (56.2), (56.23), and (56.28), we obtain

$$(56.29) \quad y = \sin x + \epsilon(\tfrac{2}{3} \cos x - \tfrac{4}{3} \sin^2 x - \tfrac{2}{3} \cos^2 x)$$
$$+ \epsilon^2(\tfrac{37}{18} \sin x - \tfrac{8}{9} \sin x \cos x - \tfrac{2}{3}x \cos x - \tfrac{1}{6} \sin 3x).$$

The solution of (56.01) satisfying (56.11), i.e., its solution if there were no disturbing function present, is $\sin x$. Because of the disturbance function $-2\epsilon(y')^2$, the first and second order perturbation terms are, respectively, the second and third terms in (56.29).

EXERCISE 56

1. Find the first and second order perturbation terms in the solution of $y'' - y = 0$ for which $y(0) = 0$, $y'(0) = 2$ due to the presence of a disturbing function $\epsilon y'$, where ϵ is small.

ANSWERS 56

1. First order: $\epsilon\left(\dfrac{xe^x}{2} - \dfrac{xe^{-x}}{2}\right)$;

second order: $\epsilon^2\left[\dfrac{e^x}{8}(-1 + x + x^2) + \dfrac{e^{-x}}{8}(1 + x - x^2)\right].$

Existence and Uniqueness Theorem
for the First Order Differential
Equation $y' = f(x,y)$. Picard's Method.
Envelopes. Clairaut Equation.

Introductory Remarks. As we have repeatedly emphasized, differential equations whose solutions can be expressed explicitly or implicitly in terms of elementary functions are relatively few in number. Even a first order differential equation

$$(57.1) \qquad\qquad y' = f(x,y)$$

will usually not have an elementary solution. In these cases, it is desirable to have theorems which will answer the following questions for us.

1. Does (57.1) have a 1-parameter family of solutions? See Examples 4.21 and 4.22 for differential equations which have no solutions; Example 4.2 for one which has only one solution; (4.652) or Example 5.3 for one which has two 1-parameter family of solutions.
2. If (57.1) has a 1-parameter family of solutions, is it a general solution as we defined this term in Definition 4.7, i.e., does it contain every particular solution? See the first example in Lesson 4C for a 1-parameter family which does not contain every particular solution.
3. Is there a particular solution of (57.1) valid on some interval and satisfying a given initial condition $y(x_0) = y_0$?
4. Is a particular solution satisfying an initial condition $y(x_0) = y_0$ unique? See (b) of Example 5.3 where the point $(0,1)$ lies on an infinite number of particular solutions.

Fortunately there are theorems which will give us, under appropriate hypotheses, the answers to these questions. A theorem which answers questions 1, 2, and 3, i.e., one which tells us whether a solution exists is called an **existence theorem.** A theorem which answers question 4, i.e., one which tells us whether a solution is unique is called a **uniqueness theorem.**

It should be emphasized that an existence and uniqueness theorem only guarantees or assures the existence and uniqueness of a solution. It will not tell you whether the solution can or cannot be expressed in terms of elementary functions, or help you to find the solution. For example, the particular solution of the differential equation

$$(57.11) \qquad\qquad y' = e^{-x^2},$$

for which $y(x_0) = y_0$ is

$$(57.12) \qquad\qquad y(x) = y_0 + \int_{x_0}^{x} e^{-x^2}\, dx.$$

It has been proved that this integral cannot be expressed in terms of elementary functions. However, the existence and uniqueness theorem, which we shall state later, will not help you to discover this fact. All the theorem will tell you, is that since the function e^{-x^2} and the point (x_0,y_0) satisfy its hypotheses, an unique particular solution satisfying (57.11) and the initial condition exists.

There are several ways of proving the existence and uniqueness theorem for the first order differential equation $y' = f(x,y)$, satisfying the condition $y(x_0) = y_0$. The one we shall use is dependent on a *method* which is known as **Picard's approximation method,** named after the French mathematician, Charles Émile Picard (1856–1941). Hence before we can prove the theorem, we shall first need to explain Picard's method.

LESSON 57. Picard's Method of Successive Approximations.

We assume for the moment that a unique particular solution of the differential equation

$$(57.2) \qquad\qquad y' = f(x,y),$$

satisfying the initial condition

$$(57.21) \qquad\qquad y(x_0) = y_0,$$

exists. Let $y(x)$ be the required particular solution. Then by (57.2), $y' = f[x,y(x)]$, where $f[x,y(x)]$ is now a function only of x. Integrating this equation between the limits x_0 and x and noting, by (57.21), that when $x = x_0$, $y = y_0$, we obtain

$$(57.22) \qquad \int_{y_0}^{y} dy = \int_{x_0}^{x} f[t,y(t)]\, dt, \qquad y(x) = y_0 + \int_{x_0}^{x} f[t,y(t)]\, dt.$$

In employing numerical methods to approximate the particular solution $y(x)$ of (57.22), we used a polynomial interpolating function in place of the integrand, $f[x,y(x)]$. In Picard's method we also use the idea of an approximation but of a totally different kind. In this method, we obtain a *sequence* of functions $y_0(x)$, $y_1(x)$, $\cdots$, $y_n(x)$, each of which satisfies the initial condition (57.21). The existence and uniqueness Theorem 58.5 which follows will then give the conditions that $f(x,y)$ of (57.2) must fulfill in order that an interval about x_0 exist, on which, as $n \to \infty$, this sequence of functions approach the particular solution $y(x)$ of (57.22). Furthermore, each function in the sequence is an approximation of the particular solution $y(x)$: a later one, in general, being a better approximation than a preceding one. Hence the name successive approximations.

We illustrate the method by means of examples. We concentrate for the moment only on the mechanics of the method without considering the size of the interval about x_0 for which the sequence of functions thus obtained converges to the particular solution $y(x)$ of (57.22). (For the meaning of the convergence of a *sequence* of functions, see Definition 58.1 and Example 58.13.)

The first approximation of a solution of (57.2) satisfying (57.21) is called $y_0(x)$. The function $y_0(x)$ may, as we shall show later, be *any* arbitrary continuous function defined in a neighborhood of x_0. In the absence of additional information, it is usually taken to be the constant function

$$(57.23) \qquad y_0(x) = y_0,$$

where y_0 is the initial value given in (57.21). It is evident that this approximation to the solution is not a very satisfactory one. It is the equation of a straight line parallel to the x axis and y_0 units from it.

The subsequent members of the sequence of approximating solutions of (57.2) satisfying (57.21) are called $y_1(x)$, $y_2(x)$, $\cdots$, $y_n(x)$, $\cdots$, and are obtained in the following manner.

$$(57.24) \qquad y_1(x) = y_0 + \int_{x_0}^{x} f[x,y_0(x)]\,dx,$$

$$y_2(x) = y_0 + \int_{x_0}^{x} f[x,y_1(x)]\,dx,$$

$$y_3(x) = y_0 + \int_{x_0}^{x} f[x,y_2(x)]\,dx,$$

$$\cdots\cdots\cdots\cdots\cdots\cdots\cdots\cdots$$

$$y_n(x) = y_0 + \int_{x_n}^{x} f[x,y_{n-1}(x)]\,dx,$$

where x_0 and y_0 are given in (57.21). (Remember $f[x,y_0(x)]$ means replace y in $f(x,y)$ by $y_0(x)$; $f[x,y_1(x)]$ means replace y in $f(x,y)$ by $y_1(x)$, etc.)

Comment 57.241. For different permissible starting approximations $y_0(x)$, different sequences $y_0(x)$, $y_1(x)$, $\cdots$, $y_n(x)$ will result. However, each will have the property that for an x in an interval about x_0,

$$\lim_{n \to \infty} y_n(x) = y(x),$$

where $y(x)$ is the solution of (57.2) satisfying (57.21). The rapidity with which the sequence of approximations will converge to the solution $y(x)$ will depend on how closely the starting solution $y_0(x)$ approximates the actual solution $y(x)$; the closer the approximation, the quicker the convergence.

Example 57.25. Find the first four Picard approximations if

(a) $$y' = xy,$$

and $y(0) = 1$.

Solution. Comparing (a) and the initial condition with (57.2) and (57.21), we see that $f(x,y) = xy$, $x_0 = 0$, $y_0 = 1$. By (57.23), our first approximation is therefore $y_0(x) = 1$. The succeeding approximations, by (57.24), are

(b) $$y_1(x) = 1 + \int_0^x f(x,y_0)\,dx = 1 + \int_0^x x\,dx = 1 + \frac{x^2}{2},$$

$$y_2(x) = 1 + \int_0^x f(x,y_1)\,dx$$

$$= 1 + \int_0^x x\left(1 + \frac{x^2}{2}\right)dx = 1 + \frac{x^2}{2} + \frac{x^4}{8},$$

$$y_3(x) = 1 + \int_0^x f(x,y_2)\,dx = 1 + \int_0^x x\left(1 + \frac{x^2}{2} + \frac{x^4}{8}\right)dx,$$

$$= 1 + \frac{x^2}{2} + \frac{x^4}{8} + \frac{x^6}{48}.$$

Comment 57.251. If we solve (a) by the method of separation of variables, we obtain the particular solution $y = e^{x^2/2}$ whose series expansion is

$$1 + \frac{x^2}{2} + \frac{x^4}{8} + \frac{x^6}{48} + \cdots.$$

Note that each succeeding function in the sequence y_0, y_1, y_2, y_3, $\cdots$ is a closer approximation to the actual solution than is the previous one.

Example 57.26. Find the first three Picard approximations if

(a) $$y' = x^2 - y$$

and $y(1) = 2$.

Solution. Comparing (a) and the initial condition with (57.2) and (57.21), we see that $f(x,y) = x^2 - y$, $x_0 = 1$, $y_0 = 2$. By (57.23) our first approximation is therefore $y_0(x) = 2$. The succeeding approximations are, by (57.24),

(b)
$$y_1(x) = 2 + \int_1^x f(x,y_0)\, dx$$

$$= 2 + \int_1^x (x^2 - 2)\, dx = \frac{x^3}{3} - 2x + \frac{11}{3},$$

$$y_2(x) = 2 + \int_1^x f(x,y_1)\, dx$$

$$= 2 + \int_1^x \left(x^2 - \frac{x^3}{3} + 2x - \frac{11}{3} \right) dx,$$

$$= \frac{x^3}{3} - \frac{x^4}{12} + x^2 - \frac{11}{3} x + \frac{53}{12}.$$

Picard's Method Applied to a System of Two First Order Equations. Picard's method of successive approximations can also be applied to a system of first order equations. We illustrate the method for the pair of first order equations

(57.3) $$\frac{dx}{dt} = f_1(t,x,y), \qquad \frac{dy}{dt} = f_2(t,x,y),$$

satisfying the initial conditions.

(57.31) $$x(t_0) = x_0, \qquad y(t_0) = y_0.$$

The first approximations of the solution of the system (57.3) satisfying (57.31) are called $x_0(t)$ and $y_0(t)$. In the absence of additional information, they are usually taken to be the constant functions

(57.32) $$x_0(t) = x_0, \qquad y_0(t) = y_0,$$

where x_0 and y_0 are given in (57.31). The subsequent approximations are

(57.33) 1. $$x_1(t) = x_0 + \int_{t_0}^t f_1[t,x_0(t),y_0(t)]\, dt,$$

$$y_1(t) = y_0 + \int_{t_0}^t f_2[t,x_0(t),y_0(t)]\, dt.$$

2. $$x_2(t) = x_0 + \int_{t_0}^t f_1[t,x_1(t),y_1(t)]\, dt,$$

$$y_2(t) = y_0 + \int_{t_0}^t f_2[t,x_1(t),y_1(t)] \, dt.$$

. . . .

$$n. \quad x_n(t) = x_0 + \int_{t_0}^t f_1[t,x_{n-1}(t),y_{n-1}(t)] \, dt,$$

$$y_n(t) = y_0 + \int_{t_0}^t f_2[t,x_{n-1}(t),y_{n-1}(t)] \, dt.$$

Picard's method thus yields two sequences of functions, x_0, x_1, $\cdots$, x_n, $\cdots$, and y_0, y_1, $\cdots$, y_n, $\cdots$, each satisfying the appropriate initial condition in (57.31). The existence Theorem 62.12 which follows, will then give the conditions which f_1 and f_2 of (57.3) must fulfill in order that an interval I about t_0 exist, on which, as $n \to \infty$, the first sequence approach a limiting function $x(t)$ and the second sequence approach a limiting function $y(t)$. This pair of functions is, on I, the unique solution of (57.3) satisfying (57.31). For different permissible starting approximations, different sequences will result but each will have the property that on I,

$$\lim_{n\to\infty} x_n(t) = x(t) \quad \text{and} \quad \lim_{n\to\infty} y_n(t) = y(t).$$

The rapidity with which each sequence will converge to its limiting function will depend on how close the starting approximations are to the actual solutions.

Example 57.34. Find the first three Picard approximations if the first order system is

(a) $$\frac{dx}{dt} = t + x, \qquad \frac{dy}{dt} = t - x,$$

and $x(0) = 1$, $y(0) = -1$.

Solution. Comparing (a) and the initial conditions with (57.3) and (57.31), we see that $f_1(t,x,y) = t + x$, $f_2(t,x,y) = t - x$, $t_0 = 0$, $x_0 = 1$, $y_0 = -1$. By (57.32) our first approximations are $x_0(t) = 1$, $y_0(t) = -1$. By (57.33), the succeeding approximations are

(b) 1. $$x_1(t) = 1 + \int_0^t f_1(t,x_0,y_0) \, dt = 1 + \int_0^t (t + 1) \, dt$$

$$= 1 + \frac{t^2}{2} + t,$$

$$y_1(t) = -1 + \int_0^t f_2(t,x_0,y_0) \, dt = -1 + \int_0^t (t - 1) \, dt$$

$$= -1 + \frac{t^2}{2} - t.$$

2. $x_2(t) = 1 + \displaystyle\int_0^t f_1(t,x_1,y_1)\, dt = 1 + \int_0^t \left(t + 1 + \frac{t^2}{2} + t\right) dt$

$\qquad = 1 + t + t^2 + \dfrac{t^3}{3!},$

$y_2(t) = -1 + \displaystyle\int_0^t f_2(t,x_1,y_1)\, dt$

$\qquad = -1 + \displaystyle\int_0^t \left(t - 1 - \frac{t^2}{2} - t\right) dt$

$\qquad = -1 - t - \dfrac{t^3}{3!}.$

Example 57.35. Find the first four Picard approximations if the first order system is

(a) $\qquad\qquad\qquad \dfrac{dx}{dt} = ty, \qquad \dfrac{dy}{dt} = xy,$

and $x(0) = 1$, $y(0) = 1$.

Solution. Comparing (a) and the initial conditions with (57.3) and (57.31), we see that $f_1(t,x,y) = ty$, $f_2(t,x,y) = xy$, $t_0 = 0$, $x_0 = 1$, $y_0 = 1$. By (57.32), our first approximations are $x_0(t) = 1$, $y_0(t) = 1$. By (57.33), the succeeding approximations are

(b) 1. $x_1(t) = 1 + \displaystyle\int_0^t f_1(t,x_0,y_0)\, dt = 1 + \int_0^t t\, dt = 1 + \frac{t^2}{2},$

$\qquad y_1(t) = 1 + \displaystyle\int_0^t f_2(t,x_0,y_0)\, dt = 1 + \int_0^t dt = 1 + t.$

2. $x_2(t) = 1 + \displaystyle\int_0^t f_1(t,x_1,y_1)\, dt = 1 + \int_0^t (t + t^2)\, dt$

$\qquad = 1 + \dfrac{t^2}{2} + \dfrac{t^3}{3},$

$\qquad y_2(t) = 1 + \displaystyle\int_0^t f_2(t,x_1,y_1)\, dt = 1 + \int_0^t \left(1 + t + \frac{t^2}{2} + \frac{t^3}{2}\right) dt$

$\qquad = 1 + t + \dfrac{t^2}{2} + \dfrac{t^3}{6} + \dfrac{t^4}{8},$

3. $x_3(t) = 1 + \displaystyle\int_0^t \left(t + t^2 + \frac{t^3}{2} + \frac{t^4}{6} + \frac{t^5}{8}\right) dt$

$\qquad = 1 + \dfrac{t^2}{2} + \dfrac{t^3}{3} + \dfrac{t^4}{8} + \dfrac{t^5}{30} + \dfrac{t^6}{48},$

$$y_3(t) = 1 + \int_0^t \left(1 + t + t^2 + t^3 + \frac{17t^4}{24} + \frac{t^5}{4} + \frac{17t^6}{144} + \frac{t^7}{24}\right) dt$$

$$= 1 + t + \frac{t^2}{2} + \frac{t^3}{3} + \frac{t^4}{4} + \frac{17t^5}{120} + \frac{t^6}{24} + \frac{17t^7}{1008} + \frac{t^8}{192}.$$

EXERCISE 57

Find the first k Picard approximations after $y_0(x)$, of the particular solution of each of the following equations 1–5, where k is the number shown alongside each equation.

1. $y' = x - y$, $y(0) = 1$, $k = 4$.
2. $y' = x^2 + y$, $y(1) = 3$, $k = 3$.
3. $y' = x + y^2$, $y(0) = 0$, $k = 3$.
4. $y' = 1 + xy$, $y(1) = 2$, $k = 3$.
5. $y' = e^x + y$, $y(0) = 0$, $k = 4$.

Find the first k Picard approximations after $x_0(t)$, $y_0(t)$, of the particular solution of each of the following systems, where k is the number shown alongside each system.

6. $\dfrac{dx}{dt} = t + y$, $\dfrac{dy}{dt} = t - x^2$, $x(0) = 2$, $y(0) = 1$, $k = 3$.

7. $\dfrac{dx}{dt} = t + y^2$, $\dfrac{dy}{dt} = x - t$, $x(0) = 0$, $y(0) = 1$, $k = 3$.

8. $\dfrac{dx}{dt} = xt$, $\dfrac{dy}{dt} = x - e^t$, $x(0) = 1$, $y(0) = -1$, $k = 3$.

9. Picard's method of successive approximations can also be applied to a system of first order equations greater than two. For the system of three first order equations: $dx/dt = f_1(t,x,y,z)$, $dy/dt = f_2(t,x,y,z)$, $dz/dt = f_3(t,x,y,z)$ for which $x(t_0) = x_0$, $y(t_0) = y_0$, $z(t_0) = z_0$, the successive approximations are

(57.4)   0. $x_0(t) = x_0$, $y_0(t) = y_0$, $z_0(t) = z_0$.

    1. $x_1(t) = x_0 + \displaystyle\int_{t_0}^t f_1(t,x_0,y_0,z_0)\, dt$,

      $y_1(t) = y_0 + \displaystyle\int_{t_0}^t f_2(t,x_0,y_0,z_0)\, dt$,

      $z_1(t) = z_0 + \displaystyle\int_{t_0}^t f_3(t,x_0,y_0,z_0)\, dt$.

    2. $x_2(t) = x_0 + \displaystyle\int_{t_0}^t f_1(t,x_1,y_1z_1,)\, dt$,

      $y_2(t) = y_0 + \displaystyle\int_{t_0}^t f_2(t,x_1,y_1,z_1)\, dt$,

$$z_2(t) = z_0 + \int_{t_0}^{t} f_3(t, x_1, y_1, z_1) \, dt.$$

. . . .

$n.$ $\displaystyle x_n(t) = x_0 + \int_{t_0}^{t} f_1(t, x_{n-1}, y_{n-1}, z_{n-1}) \, dt,$

$$y_n(t) = y_0 + \int_{t_0}^{t} f_2(t, x_{n-1}, y_{n-1}, z_{n-1}) \, dt,$$

$$z_n(t) = z_0 + \int_{t_0}^{t} f_3(t, x_{n-1}, y_{n-1}, z_{n-1}) \, dt.$$

Use (57.4) to find the Picard approximations x_3, y_3, z_3 of the particular solution of the system

$$\frac{dx}{dt} = y^2, \qquad \frac{dy}{dt} = x + z, \qquad \frac{dz}{dt} = z - y,$$

for which $x(0) = 1$, $y(0) = 0$, $z(0) = 1$.

ANSWERS 57

1. $y_4 = 1 - x + x^2 - \dfrac{x^3}{3} + \dfrac{x^4}{12} - \dfrac{x^5}{120}.$

2. $y_3 = \dfrac{49}{60} + \dfrac{17}{12}x - \dfrac{x^2}{6} + \dfrac{5x^3}{6} + \dfrac{x^4}{12} + \dfrac{x^5}{60}.$

3. $y_3 = \dfrac{x^2}{2} + \dfrac{x^5}{20} + \dfrac{x^8}{160} + \dfrac{x^{11}}{4400}.$

4. $y_3 = \dfrac{7}{20} + x + \dfrac{5}{24}x^2 + \dfrac{x^3}{3} + \dfrac{x^5}{15} + \dfrac{x^6}{24}.$

5. $y_4 = 4e^x - \dfrac{x^3}{6} - x^2 - 3x - 4.$

6. $x_3 = 2 + t - \tfrac{3}{2}t^2 - \tfrac{1}{2}t^3 - \tfrac{1}{4}t^4 - \tfrac{1}{20}t^5 - \tfrac{1}{120}t^6,$

$\qquad y_3 = 1 - 4t - \tfrac{3}{2}t^2 + \tfrac{5}{3}t^3 + \tfrac{7}{12}t^4 - \tfrac{31}{60}t^5 + \tfrac{1}{12}t^6 - \tfrac{1}{252}t^7.$

7. $x_3 = t + \dfrac{t^2}{2} + \dfrac{t^4}{12} + \dfrac{t^7}{252},\ y_3 = 1 + \dfrac{t^3}{3} - \dfrac{t^4}{12} + \dfrac{t^6}{120}.$

8. $x_3 = 1 + \dfrac{t^2}{2} + \dfrac{t^4}{8} + \dfrac{t^6}{48},\ y_3 = t + \dfrac{t^3}{6} + \dfrac{t^5}{40} + \dfrac{t^7}{252} - e^t.$

9. $x_3 = 1 + \dfrac{4t^3}{3} + \dfrac{t^4}{2} + \dfrac{t^5}{20},\ y_3 = 2t + \dfrac{t^2}{2} - \dfrac{t^3}{6} + \dfrac{t^4}{3},\ z_3 = 1 + t - \dfrac{t^2}{2} - \dfrac{t^3}{3}.$

LESSON 58.   **An Existence and Uniqueness Theorem for the First Order Differential Equation $y' = f(x,y)$ Satisfying $y(x_0) = y_0$.**

For a clearer understanding of the proof of the existence and uniqueness Theorem 58.5 which follows, it will be necessary to know the meaning of the convergence of a sequence of functions, the meaning of the uniform convergence of a sequence of functions, the meaning of a Lipschitz condition, and to be acquainted with certain theorems from analysis. Hence, before beginning the proof, we shall briefly discuss these topics and list the needed theorems.

LESSON 58A.   **Convergence and Uniform Convergence of a Sequence of Functions. Definition of a Continuous Function.**

Definition 58.1.   A **sequence** of functions

(58.11) $$f_1(x), f_2(x), \cdots, f_n(x), \cdots,$$

each defined on a common set S is said to **converge** to a function $f(x)$ on S, if for each x in S and for each fixed $\epsilon > 0$, no matter how small, there is an N such that

(58.12) $$|f_n(x) - f(x)| < \epsilon, \quad \text{when } n > N.$$

In words the definition says the following. Pick any x you wish in this set S and choose any positive number ϵ as small as you like. For this x calculate $f_1(x), f_2(x), f_3(x), \cdots$ and $f(x)$. If the sequence of functions $f_1(x), f_2(x), \cdots, f_n(x), \cdots$ converges to $f(x)$, then according to the definition you must eventually reach a function $f_{N+1}(x)$ in the sequence, such that $|f_{N+1}(x) - f(x)|$ is less than this chosen positive ϵ. Further *every* function in the sequence *after* $f_{N+1}(x)$ must also differ from $f(x)$ in absolute value by an amount less than ϵ.

Example 58.13.   Show that the sequence of functions

(a) $f_1(x) = \dfrac{1}{1 + x}, \quad f_2(x) = \dfrac{1}{1 + 2x}, \quad f_3(x) = \dfrac{1}{1 + 3x}, \cdots,$

$f_n(x) = \dfrac{1}{1 + nx}, \cdots,$

converges to the function $f(x) = 0$ on $I: 0 < x \leqq 1$.

Solution.   Let x be any number in the interval $I: 0 < x \leqq 1$, and let ϵ be any positive number. Here $f_n(x)$ of (58.12) is $1/(1 + nx)$ and $f(x) = 0$. Hence by Definition 58.1, we must show that there exists an N such that

(b) $$\left| \frac{1}{1 + nx} - 0 \right| < \epsilon, \quad \text{when } n > N.$$

The inequality (b) is equivalent to

(c) $\dfrac{1}{\epsilon} < 1 + nx, \quad \dfrac{1}{\epsilon} - 1 < nx, \quad n > \dfrac{1}{x}\left(\dfrac{1}{\epsilon} - 1\right).$

If, therefore,

(d) $$N = \left[\dfrac{1}{x}\left(\dfrac{1}{\epsilon} - 1\right)\right],$$

then (b) will hold when $n > N$. [Note. If $1/\epsilon - 1$ is negative, take $N = 0$.] For example, if we choose $x = \frac{1}{2}$ and $\epsilon = 0.02$, then by (d), $N = 2(50 - 1) = 98$. It should therefore follow that the value of each function after the 98th, namely, $f_{00}(\frac{1}{2}) = \dfrac{1}{1 + 99/2}$, $f_{100}(x) = \dfrac{1}{1 + 50}$, $\cdots$ should be less than $\epsilon = 0.02$. You can easily verify that each such function is indeed less than 0.02; remember, with a fixed numerator, the value of a fraction decreases as the denominator increases. And if we choose $x = 1/100$, then by (d), $N = 100(50 - 1) = 4900$. We would therefore need to go to the 4901th function in the sequence before coming to the first one whose value differs from zero by less than $\epsilon = 0.02$. And if we pick $x = 1/1,000,000$, then by (d), $N = 1,000,000(50 - 1) = 49,000,000$, i.e., we shall need to go to the 49,000,001st function in the sequence before coming to the first one whose value differs from zero by less than 0.02.

It is evident from (d) and the above examples, that for each fixed positive ϵ, and for each different x, a different N will be required in the sequence to insure the validity of (58.12), i.e., N depends on both the values of x and ϵ. This dependency of N on both x and ϵ is common to a great many sequences. On the other hand, there are certain sequences where N does not depend on x but only on the value of ϵ, i.e., for a fixed $\epsilon > 0$, no matter how small, it may be possible to find an N such that (58.12) will be valid for *every* x in I when $n > N$. In other words, we do not, in this case, need to hunt for a different N for each different x. We find an N once and for all which will hold for every x in I. In this event, we say the sequence of functions (58.11) *converges uniformly* on I to the function $f(x)$.

Definition 58.14. A **sequence** of functions (58.11), each defined on a common set S is said to **converge uniformly** on S to a function $f(x)$, if for a fixed positive ϵ, no matter how small, there is an N such that

(58.15) $|f_n(x) - f(x)| < \epsilon$ when $n > N$,

for *every* x in S.

We emphasize once more, that here for a fixed $\epsilon > 0$, we pick an N only once. For *each* x in S, the absolute value of the difference of each of the functions $f_{N+1}(x), f_{N+2}(x), \cdots$ and $f(x)$ will be less than ϵ.

Example 58.16. Show that the sequence of functions

(a) $f_1(x) = \dfrac{1}{1 + x}$, $\quad f_2(x) = \dfrac{1}{1 + 2x}$, $\quad f_3(x) = \dfrac{1}{1 + 3x}$, $\cdots$,

$\qquad f_n(x) = \dfrac{1}{1 + nx}$, $\cdots$,

converges *uniformly* to the function $f(x) = 0$ on $I: 1 \leqq x$.

Solution. Note that this sequence of functions is the same as that of Example 58.13, but that the *interval* is different. Here we must show, by Definition 58.14, that for a given $\epsilon > 0$,

(b) $\qquad\qquad \left| \dfrac{1}{1 + nx} - 0 \right| < \epsilon$, $\quad$ when $n > N$,

for *every* x in $I: 1 \leqq x < \infty$. By (c) of the previous example,

(c) $\qquad\qquad\qquad n > \dfrac{1}{x}\left(\dfrac{1}{\epsilon} - 1\right).$

For a fixed $\epsilon > 0$ and for an x in $I: 1 \leqq x$, the expression on the right of (c) will have its largest value when $x = 1$. And when $x = 1$, (c) becomes $n > \dfrac{1}{\epsilon} - 1$. If therefore

(d) $\qquad\qquad\qquad N = \left[\dfrac{1}{\epsilon} - 1\right],$

then (b) will hold for each x in I when $n > N$. For example, if $\epsilon = 0.02$, then by (d), $N = 49$. You can verify that each functional value in the sequence after the 49th, namely, $f_{50}(1) = \dfrac{1}{1 + 50}$, $f_{51}(1) = \dfrac{1}{1 + 51}$, $\cdots$ is less than 0.02. Hence for $x > 1$, each functional value in the sequence after the 49th is surely less than 0.02. Our given sequence of functions, therefore, converges uniformly on $I: 1 \leqq x < \infty$ to the function $f(x) = 0$.

Definition 58.17. **A function $f(x)$ is continuous at a point $x = a$** if $f(a)$ exists and if

(58.171) $\qquad\qquad\qquad \lim_{x \to a} f(x) = f(a).$

Definition 58.172. **A function $f(x)$ is continuous on, or in, an interval I** if it is continuous at every point of I.

Definition 58.18. **A function $f(x,y)$ is continuous at a point (a,b),** if $f(a,b)$ exists and

(58.181) $\qquad\qquad\qquad \lim_{x,y \to a,b} f(x,y) = f(a,b).$

Definition 58.19. **A function $f(x,y)$ is continuous on, or in, a region S** if it is continuous at every point of S.

LESSON 58B. Lipschitz Condition. Theorems from Analysis.

(The theorems from analysis which we shall need have been stated without proof. Their proofs can be found in advanced calculus text books.)

Definition 58.2. If $f(x,y)$ is a function of x and y in a region S such that, for *every* two points (x,y) and $(x,\bar{y})$ in S,

$$(58.21) \qquad |f(x,y) - f(x,\bar{y})| \leqq N|y - \bar{y}|,$$

where N is a positive constant, then $f(x,y)$ is said to satisfy a **Lipschitz condition** in S (see Fig. 58.57).

Theorem 58.22. **Law of the mean.** See Fig. 58.57. *If $f(x,y)$ is a function of x and y that has a continuous partial derivative with respect to y in a region S, then for each x there exists a number Y such that*

$$(58.23) \qquad \frac{f(x,y) - f(x,\bar{y})}{y - \bar{y}} = \frac{\partial}{\partial y} f(x,Y),$$

where (x,y) and $(x,\bar{y})$ are any two points in S and Y lies between y and $\bar{y}$.

Remark. The notation $\dfrac{\partial}{\partial y} f(x,Y)$ means the value of the partial derivative of the function $f(x,y)$ with respect to y when $y = Y$.

Comment 58.24. If $f(x,y)$ has a continuous partial derivative with respect to y in a region S, and if this partial derivative as a function of the two variables x,y is bounded in S, then $f(x,y)$ satisfies a Lipschitz condition. The proof proceeds as follows. Since $\partial f(x,y)/\partial y$ is bounded in S, there exists a constant N such that

$$(\text{a}) \qquad \left| \frac{\partial}{\partial y} f(x,y) \right| \leqq N,$$

for *every* point (x,y) in S. Let (x,y) and $(x,\bar{y})$ be any two points in S. Then by Theorem 58.22 there exists a number Y between y and $\bar{y}$, such that

$$(\text{b}) \qquad \left| \frac{f(x,y) - f(x,\bar{y})}{y - \bar{y}} \right| = \left| \frac{\partial}{\partial y} f(x,Y) \right|.$$

Since (x,Y) is a point in S, we can substitute the inequality (a) for the right side of (b) and thus obtain (58.21).

Theorem 58.25. *If $f(x)$ is a Riemann-integrable function on $I: a \leqq x \leqq b$, then*

$$(58.26) \qquad \left| \int_a^b f(x)\, dx \right| \leqq \int_a^b |f(x)|\, dx.$$

Theorem 58.3. *If $f(x)$ is a continuous function on the interval I: $a \leq x \leq b$ and if*

$$(58.31) \qquad F(x) = \int_a^x f(t)\, dt, \quad a \leq x \leq b,$$

then $F(x)$ is continuous on I, and

$$(58.32) \qquad F'(x) = f(x), \quad a < x < b.$$

Theorem 58.4. *If a **sequence** of continuous functions $f_1(x), f_2(x), \cdots,$ $f_n(x), \cdots,$ each defined on a common interval I, **converges uniformly** on I, to a function $f(x)$, then $f(x)$ is continuous on I.*

Comment 58.41. The preceding theorem is not true if, on I, the sequence of functions merely converges to $f(x)$, but not uniformly. For instance, the sequence of continuous functions of Example 58.13 converges on I: $0 \leq x \leq 1$, but not uniformly. The sequence converges on $0 < x \leq 1$ to the function $f(x) = 0$. But when $x = 0$, each function in the sequence has the value one. Hence on the interval I: $0 \leq x \leq 1$, the limiting function of the sequence is

$$(\text{a}) \qquad f(x) = \begin{cases} 0, & 0 < x \leq 1, \\ 1, & x = 0. \end{cases}$$

The function $f(x)$ is therefore discontinuous on I.

Theorem 58.42. *If a **sequence** of continuous functions $f_1(x), f_2(x), \cdots,$ $f_n(x), \cdots,$ each defined on a common interval I, **converges uniformly** on I to a function $f(x)$, then*

$$(58.43) \qquad \lim_{n \to \infty} \int_{x_0}^x f_n(x)\, dx = \int_{x_0}^x \lim_{n \to \infty} f_n(x)\, dx = \int_{x_0}^x f(x)\, dx,$$

where the interval (x_0, x) is contained in I.

Definition 58.44. A **series** of functions

$$f_1(x) + f_2(x) + \cdots + f_n(x) + \cdots,$$

each defined on a common interval I, is said to **converge uniformly** on I to a function $f(x)$, if the sequence of partial sums $F_1(x), F_2(x), \cdots, F_n(x)$, where

$$F_n(x) = f_1(x) + \cdots + f_n(x),$$

converges uniformly on I to $f(x)$.

Theorem 58.45. *If each function $f_1(x), f_2(x), \cdots, f_n(x)$ is defined and bounded on a common interval I, i.e., if*

$$(58.46) \qquad |f_i(x)| \leq M_i, \quad i = 1, 2, \cdots, n, \cdots,$$

and if the infinite series of positive terms

(58.47) $M_1 + M_2 + \cdots + M_n + \cdots$

converges, then the **series**

(58.48) $f_1(x) + f_2(x) + \cdots + f_n(x) + \cdots$

converges uniformly *on I to a function f(x).*

   Comment 58.481.   By the above theorem, if each $|f_i(x)| \leqq M_i$, and $\sum M_i$ converges, then for a given $\epsilon > 0$, there is a positive N such that

$$[f_1(x) + f_2(x) + \cdots + f_n(x) - f(x)| < \epsilon, \quad \text{when } n > N,$$

for *every* x in I.

   Example 58.49.   Show that the series of functions,

(a) $\displaystyle\sum_i f_i(x) = \sum_{i=1}^{\infty} \frac{1}{2^i + x}, \quad I: 0 \leqq x \leqq 1,$

converges uniformly on I.

   Solution.   Take $M_i = 1/2^i$. Therefore

(b) $\displaystyle\sum_{i=1}^{\infty} M_i = M_1 + M_2 + M_3 + \cdots = \frac{1}{2} + \frac{1}{2^2} + \frac{1}{2^3} + \cdots.$

By (a), and for each x such that $0 \leqq x \leqq 1$,

(c) $\displaystyle\sum_{i=1}^{\infty} f_i(x) = \frac{1}{2 + x} + \frac{1}{2^2 + x} + \frac{1}{2^3 + x} + \cdots$

$$\leqq \frac{1}{2} + \frac{1}{2^2} + \frac{1}{2^3} + \cdots,$$

The last series on the right of (c) is a geometric series which converges to one. Hence the first series of functions, which is the given series (a), by Theorem 58.45, converges uniformly on I to a function $f(x)$.

LESSON 58C.   Proof of the Existence and Uniqueness Theorem for the First Order Differential Equation $y' = f(x,y)$.   In Theorem 38.14, we gave a sufficient condition for the existence of a *power series* solution of $y' = f(x,y)$ satisfying an initial condition $y(x_0) = y_0$. Compare it with the theorem we now state, which gives a sufficient condition for the existence and uniqueness of a solution of $y' = f(x,y)$ for which $y(x_0) = y_0$.

Theorem 58.5. See Fig. 58.57. *Let $f(x,y)$ be a bounded, continuous function of x and y in a region S of the xy plane and let (x_0,y_0) be a point of S. In S, let the function f satisfy the Lipschitz condition (58.21), namely*

(58.51) $$|f(x,y) - f(x,\bar{y})| \leq N|y - \bar{y}|,$$

for every two points (x,y) and $(x,\bar{y})$ in S.

Then an interval

(58.52) $$I_0 : |x - x_0| < h, \ \cdot h > 0,$$

exists on which there is one and only one continuous function $y(x)$, with a continuous derivative on I_0, satisfying the differential equation

(58.53) $$y' = f(x,y)$$

and the initial condition

(58.54) $$y(x_0) = y_0.$$

Remark. The above theorem is a special case of the more general Theorem 62.12 on systems of first order equations.

Proof. The proof of the theorem will be based on Picard's methods of successive approximations. By (57.23) and (57.24), these approximating functions are

(58.55)
$$y_0(x) = y_0,$$
$$y_1(x) = y_0 + \int_{x_0}^{x} f[t,y_0(t)]\, dt,$$
$$y_2(x) = y_0 + \int_{x_0}^{x} f[t,y_1(t)]\, dt,$$
$$\cdot\ \cdot\ \cdot\ \cdot\ \cdot\ \cdot\ \cdot\ \cdot\ \cdot\ \cdot\ \cdot\ \cdot\ \cdot\ \cdot\ \cdot\ \cdot\ \cdot$$
$$y_n(x) = y_0 + \int_{x_0}^{x} f[t,y_{n-1}(t)]\, dt,$$

where $f(x,y)$ is the continuous function of (58.53), y_0 is the constant of (58.54) and the interval (x_0,x) is contained in S. Because the proof is long, we have divided it into four parts. NOTE. We shall use $I: |x - x_0| \leq h$ to denote the closed interval (58.52).

A. First we shall show how to determine the interval I_0 of (58.52).

B. Second we shall prove that on I each of the functions $y_0(x)$, $y_1(x)$, $\cdots$, $y_n(x)$ of (58.55) is continuous and its graph lies in a rectangle R contained in the given region S.

C. Third we shall prove that this sequence of functions y_0, y_1, $\cdots$, y_n of (58.55) converges uniformly on I to a function $y(x)$ which, on I_0, is a solution of (58.53) satisfying (58.54). We thus establish the existence of a particular solution $y(x)$.

D. Finally we shall prove that this solution $y(x)$ is, on I_0, the unique particular solution of (58.53) satisfying (58.54).

A. *Determining an interval I.* Since the given function $f(x,y)$ is, by hypothesis, bounded in S, there exists a positive constant M such that

$$(58.56) \qquad\qquad |f(x,y)| < M,$$

for every point (x,y) in S. The point (x_0,y_0) of (58.54) is by hypothesis a point of a region S. (See Fig. 58.57 and keep referring to it.) Hence we

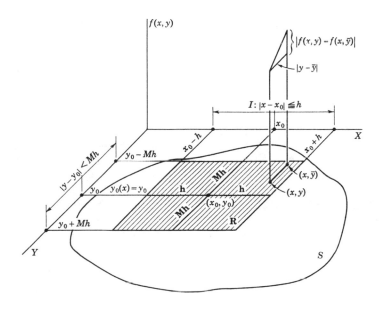

Figure 58.57

can find a positive number h such that the rectangle of dimensions $|x - x_0| \leq h$, $|y - y_0| < Mh$ which has this point (x_0,y_0) at its center, and where M is the constant in (58.56), lies entirely in S. Call this rectangle R. Therefore all points in R (since they are also in S) whose x coordinates are in the interval

$$(58.58) \qquad I: x_0 - h \leq x \leq x_0 + h, \quad |x - x_0| \leq h,$$

satisfy (58.51) and (58.56). This interval I, without its endpoints, is the interval I_0 of (58.52) referred to in the theorem.

Proof of B. *Each function $y_0(x)$, $y_1(x)$, $\cdots$, $y_n(x)$ of (58.55) is, on the interval I of (58.58), continuous and its graph lies in R.* Consider the first function $y_0(x) = y_0$. Its graph is a straight line parallel to the x axis and y_0 units distant from it. Hence for all x in I, the graph of $y_0(x) = y_0$ lies in R (keep referring to Fig. 58.57). To complete the proof, we use the inductive method of reasoning described in Lesson 24A. We assume the graph of the function $y_{n-1}(x)$ lies in R and must then show that the graph of the function $y_n(x)$ is also in R.

Hence we assume that for all x in I of (58.58), the graph of $y_{n-1}(x)$ lies in R. This means we assume that for each x in I, $y_{n-1}(x)$ will give a value of y which is in R. Therefore $[x,y_{n-1}(x)]$ is a point of R. Hence by (58.56)

(a) $$|f[x,y_{n-1}(x)]| < M.$$

By the last equation of (58.55), by (58.26), (a) and (58.58), in the order listed, we obtain

(b) $$|y_n(x) - y_0| = \left| \int_{x_0}^x f[t,y_{n-1}(t)]\, dt \right| \le \left| \int_{x_0}^x |f[t,y_{n-1}(t)]|\, dt \right|$$
$$< M \left| \int_{x_0}^x dt \right| = M|x - x_0| \le Mh.$$

Equation (b) says that for each x in I, the distance from $y_n(x)$ to $y_0 < Mh$ (see Fig. 58.57). Therefore the graph of $y_n(x)$ lies in R.

We have thus proved that for each x in I of (58.58), $[x,y_k(x)]$, $k = 0$, $1, 2, \cdots, n, \cdots$ is a point of R contained in S. Hence, by the hypothesis of the theorem, each integrand $f[x,y_k(x)]$, $k = 0, 1, 2, \cdots, n, \cdots$ of (58.55) is, in R, and therefore on I, a continuous function. It therefore follows by Theorem 58.3, that each function $y_1(x), \cdots, y_n(x), \cdots$ of (58.55) is also a continuous function on I. And since $y_0(x) = y_0$, a constant, it too is a continuous function on I.

Proof of C. *There exists a particular solution of (58.53) satisfying (58.54).* By the second equation in (58.55) and Theorem 58.25,

(c) $$|y_1(x) - y_0(x)| = \left| \int_{x_0}^x f[t,y_0(t)]\, dt \right| \le \left| \int_{x_0}^x |f[t,y_0(t)]|\, dt \right| .$$

In B, we proved that for each x in I of (58.58), the point $[x,y_0(x)]$ is in R. Therefore by (58.56) and (58.58), (c) becomes

(d) $$|y_1(x) - y_0(x)| < M \left| \int_{x_0}^x dt \right| = M|x - x_0| \le Mh.$$

Subtracting the second equation of (58.55) from the third, we obtain

(e) $$|y_2 - y_1| = \left| \int_{x_0}^x \{f[t,y_1(t)] - f[t,y_0(t)]\}\, dt \right|$$
$$\le \left| \int_{x_0}^x |f[t,y_1(t)] - f[t,y_0(t)]|\, dt \right| .$$

For each x in I, both points $[x, y_1(x)]$ and $[x, y_0(x)]$ are, by B, in R. Hence by (58.51), we can write (e) as

(f) $$|y_2 - y_1| \leqq N \left| \int_{x_0}^{x} |y_1(t) - y_0(t)|\, dt \right|.$$

Replacing the integrand in (f) by its value $M|t - x_0|$ as given in (d), we obtain with the help of (58.58),

(g) $$|y_2 - y_1| < MN \left| \int_{x_0}^{x} |t - x_0|\, dt \right| = MN\, \frac{|x - x_0|^2}{2!} \leqq MN\, \frac{h^2}{2!}.$$

Repeating the above procedure, we find, with the help of (58.55), (58.51), (g), and (58.58), in the order listed,

(h) $$|y_3 - y_2| \leqq \left| \int_{x_0}^{x} |f[t, y_2(t)] - f[t, y_1(t)]|\, dt \right|$$

$$\leqq N \left| \int_{x_0}^{x} |y_2(t) - y_1(t)|\, dt \right|$$

$$< MN^2 \left| \int_{x_0}^{x} \frac{|t - x_0|^2}{2!}\, dt \right| = MN^2\, \frac{|x - x_0|^3}{3!}$$

$$< MN^2\, \frac{h^3}{3!}.$$

And in general, it can be shown by the same type of inductive argument we used in B, that for every x in I and every n,

(i) $$|y_n(x) - y_{n-1}(x)| < MN^{n-1}\, \frac{|x - x_0|^n}{n!}$$

$$< MN^{n-1}\, \frac{h^n}{n!} = \frac{M}{N}\, \frac{(Nh)^n}{n!}.$$

By (i)

(j) $$|y_1 - y_0| < \frac{M}{N}\, \frac{Nh}{1!},$$

$$|y_2 - y_1| < \frac{M}{N}\, \frac{(Nh)^2}{2!},$$

$$\cdots\cdots\cdots\cdots\cdots$$

$$|y_n - y_{n-1}| < \frac{M}{N}\, \frac{(Nh)^n}{n!}.$$

The sum of y_0 and the positive terms on the right side of (j) is

$$y_0 + \frac{M}{N} \left[\frac{Nh}{1!} + \frac{(Nh)^2}{2!} + \cdots + \frac{(Nh)^n}{n!} + \cdots \right],$$

a series which, by (37.42), converges to $y_0 + \dfrac{M}{N}(e^{Nh} - 1)$. Hence, by Theorem 58.45, the series

(k) $\qquad y_0 + (y_1 - y_0) + (y_2 - y_1) + \cdots + (y_n - y_{n-1}),$

which is the sum of y_0 and the functions on the left side of (j), converges uniformly on I to a function $y(x)$. But the sum (k) is y_n. We have thus proved that on the interval I of (58.58),

(l) $\qquad\qquad\qquad \lim_{n\to\infty} y_n(x) = y(x)$

uniformly, i.e., the sequence of continuous functions $y_0(x),\, y_1(x),\, \cdots,$ $y_n(x)$, defined in (58.55), converges uniformly on I to a function $y(x)$. By Theorem 58.4, therefore, this function $y(x)$ is continuous on I. Moreover, it follows by Definition 58.14 since the graph of $y_n(x)$ is in R, that the graph of $y(x)$ must also be in R.

Since for every x in I, $[x,y(x)]$ and $[x,y_n(x)]$ are points of R, we have, by (58.51),

(m) $\qquad\qquad |f[x,y_n(x)] - f[x,y(x)]| \leqq N|y_n(x) - y(x)|.$

By (l), the sequence $y_n(x) \to y(x)$ uniformly on I. Therefore, by Definition 58.14, there exists an index P such that $|y_n(x) - y(x)| < \epsilon/N$, when $n > P$ for every x in I. Hence, by (m),

$$|f[x,y_n(x)] - f[x,y(x)]| < \epsilon, \quad \text{when } n > P,$$

for every x in I. Therefore, by Definition 58.14, the sequence of continuous functions $f[x,y_0(x)],\, f[x,y_1(x)],\, \cdots,\, f[x,y_n(x)],\, \cdots$, converges uniformly on I to $f[x,y(x)]$, i.e.,

$$\lim_{n\to\infty} f[x,y_n(x)] = f[x,y(x)]$$

uniformly. Therefore, by Theorem 58.4, $f[x,y(x)]$ is a continuous function on I.

Hence by (l), the last equation of (58.55), Theorem 58.42, and the equation immediately above, in the order listed,

(n) $\qquad y(x) = \lim_{n\to\infty} y_n(x) = y_0 + \lim_{n\to\infty} \int_{x_0}^{x} f[t, y_{n-1}(t)]\, dt$

$$= y_0 + \int_{x_0}^{x} \lim_{n\to\infty} f[t, y_{n-1}(t)]\, dt$$

$$= y_0 + \int_{x_0}^{x} f[t, y(t)]\, dt.$$

We showed above that $f[x,y(x)]$ is a continuous function on $I: |x - x_0| \leqq h$. It therefore follows by (n) and Theorem 58.3, that

$$y'(x) = f[x,y(x)], \qquad I_0: |x - x_0| < h.$$

By (n), we see also that

$$y(x_0) = y_0 + \int_{x_0}^{x_0} f[t,y(t)]\, dt = y_0 + 0 = y_0.$$

We have thus proved the existence on I_0 of a particular solution of (58.53) satisfying (58.54). It is the limiting function $y(x)$ of (1). We must still prove that this particular solution $y(x)$ is unique.

Proof of D. *The function $y(x)$ of (1) is the unique particular solution of (58.53) satisfying (58.54).* Assume $g(x)$ is another particular solution of (58.53) satisfying (58.54). Therefore we assume

(o) $$g'(x) = f[x,g(x)],$$

for which $g(x_0) = y_0$. By integrating (o) and inserting the initial conditions $g(x_0) = y_0$, we obtain

$$g(x) = y_0 + \int_{x_0}^{x} f[t,g(t)]\, dt.$$

Subtracting (n) from the equation immediately above, we have

(p) $$|g(x) - y(x)| = \left| \int_{x_0}^{x} \{f[t,g(t)] - f[t,y(t)]\}\, dt \right|$$

$$\leqq \left| \int_{x_0}^{x} |f[t,g(t)] - f[t,y(t)]|\, dt \right| \cdot$$

We proved in B, that for every x in I of (58.58), the graph of each function $y_0, y_1, \cdots, y_n$ of (58.55) lies in a rectangle R contained in S. We proved in C that this sequence of functions converges uniformly on I, to a continuous function $y(x)$ whose graph also lies in R. Similarly, it can be shown that $g(x)$ is a continuous function on I whose graph lies in a rectangle contained in S. Therefore for each x in I of (58.58), the points $[x,y(x)]$ and $[x,g(x)]$ are in S. We may therefore apply (58.51) to the last expression on the right side of (p). We thus obtain

(q) $$|g(x) - y(x)| \leqq N \left| \int_{x_0}^{x} |(g(t) - y(t))|\, dt \right|.$$

For every x in I, the graphs of $g(x)$ and $y(x)$ are in bounded rectangles contained in S. Call D the maximum value of $|g(x) - y(x)|$ for x in I.

Then by (q)

(r) $$|g(x) - y(x)| \leqq DN \left| \int_{x_0}^{x} dt \right| = DN|x - x_0|.$$

Substituting (r) for the integrand in (q), we obtain

(s) $$|g(x) - y(x)| \leqq DN^2 \left| \int_{x_0}^{x} |t - x_0| \, dt \right| = DN^2 \frac{|x - x_0|^2}{2!}.$$

Again substituting (s) for the integrand in (q), we obtain

(t) $$|g(x) - y(x)| \leqq DN^3 \left| \int_{x_0}^{x} \frac{|t - x_0|^2}{2!} \, dt \right| = DN^3 \frac{|x - x_0|^3}{3!}.$$

Continuing in this manner, we find, with the help of (58.58),

(u) $$|g(x) - y(x)| \leqq DN^n \frac{|x - x_0|^n}{n!} \leqq \frac{DN^n h^n}{n!} = D \frac{(Nh)^n}{n!}.$$

In the last term of (u), D is a constant and $(Nh)^n/n!$ is, by (37.42), the general term of the series expansion of e^{Nh} which converges for all Nh. Its nth term, therefore, approaches zero as $n \rightarrow \infty$. Hence $D(Nh)^n/n!$ approaches zero as n approaches ∞. Since $|g(x) - y(x)|$ can be made less than any number no matter how small, it follows that

(v) $$g(x) - y(x) = 0, \qquad g(x) = y(x).$$

We have thus proved finally that the particular solution $y(x)$ of (1) is unique. The assumed second solution $g(x)$ is the same as $y(x)$.

Comment 58.59. It is now evident why in Picard's method the first starting approximation $y_0(x)$ need not be the line $y_0(x) = y_0$, where y_0 is the initial condition given in (58.54). It can be any continuous function through the point (x_0, y_0) whose graph is in R. Each such starting function will determine a sequence of functions whose limiting function is a particular solution of the given equation satisfying the given initial condition, valid on the interval I of (58.52). By the theorem there can be only one such particular solution.

Comment 58.6. Let x_0 be a fixed value of x. Then each point (x_0, c) of S, where c is an arbitrary constant, determines a unique solution $y(x)$ of (58.53). Hence the solution $y(x)$ is a function not only of x but also of the ordinate c, i.e., $y = Y(x,c)$. There is therefore a 1-parameter family of solutions of (58.53) cutting across each line $x = x_0$.

Comment 58.61. Theorem 58.5 gives only a *sufficient* condition for the existence and uniqueness of a particular solution of a differential equation $y' = f(x,y)$ for which $y(x_0) = y_0$. It is not a *necessary* condition.

The sufficient condition means that if a region S contains only points which fulfill (58.51) and (58.56), then the conclusion of the theorem *must* follow, i.e., each point of S lies on one and only one particular solution of (58.53). That the condition is not necessary implies its conclusion *may* still be true for points in S which do not fulfill (58.51) and (58.56), i.e., these points may still lie on one and only one particular solution.

Comment 58.62. Theorem 58.5 serves another useful purpose. By Definition 5.4, an ordinary point of a first order equation $y = f(x,y)$ lies on one and only one integral curve. Hence every point (x_0,y_0) of a region S that fulfills (58.51) and (58.56) must be an ordinary point since, by the theorem, each such point lies on one and only one integral curve. Assume that S contains only ordinary points and that we are able to write a 1-parameter family of solutions of a differential equation explicitly or implicitly in terms of elementary functions. If for each point (x_0,y_0) of S, the 1-parameter family yields a particular solution through it—and by the theorem there is only one such solution—then this family, by Definition 4.7, must be a general solution since it contains every particular solution of the differential equation. In this special case, therefore, the theorem makes us aware whether our n-parameter family of solutions is a general one or not.

Example 58.63. Obtain four Picard approximations if

(a) $$y' = 1 + y^2$$

and $y(0) = 0$. Find an interval for which the sequence of Picard approximations will converge to the actual solution.

Solution. Following the method outlined in Lesson 57B, we find

(b) $y_0(x) = 0,$

$$y_1(x) = \int_0^x dx = x,$$

$$y_2(x) = \int_0^x (1 + x^2)\, dx = x + \frac{x^3}{3},$$

$$y_3(x) = \int_0^x 1 + \left(x + \frac{x^3}{3}\right)^2 dx = x + \frac{x^3}{3} + \frac{2}{15}x^5 + \frac{1}{63}x^7.$$

Here $f(x,y)$ of Theorem 58.5 is $1 + y^2$ and $\dfrac{\partial f(x,y)}{\partial y} = 2y$. Therefore in any bounded region S, no matter how large, $f(x,y)$ is continuous and satisfies (58.56). Its partial derivative is also continuous and bounded in S. Hence, by Comment 58.24, $f(x,y)$ satisfies (58.51). Let us take for S the region $-5 < y < 5$ (see Fig. 58.64). Then the M of (58.56) is found as follows.

(c) $$|f(x,y)| = |1 + y^2| < 1 + 25 = 26 = M.$$

We must now choose h, according to **A** of the theorem, so that the rectangle R of dimensions (here $x_0 = 0$ and $y_0 = 0$) $|x| \leqq h$, $|y| < 26h$ which has the point $(0,0)$ at its center lies in S. Remember S is now bounded by the lines $y = 5$ and $y = -5$. If therefore we choose $h =$

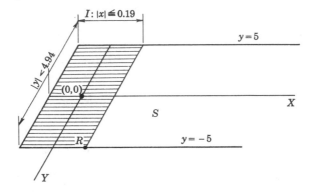

Figure 58.64

0.19, then the rectangle R of dimensions $|x| \leqq 0.19$, $|y| < 4.94$ which has $(0,0)$ at its center will lie in S. The interval I of (58.58) and the rectangle R are shown in Fig. 58.64. Hence the sequence of functions $y_0, y_1, \cdots, y_n$ will converge on the interval $-0.19 < x < 0.19$, to a function $y(x)$ which, on this interval, is the actual solution of (a) satisfying $y(0) = 0$.

Comment 58.65. The actual solution of (a) satisfying $y(0) = 0$ is

(d) $$y = \tan x.$$

Its Maclaurin series is

(e) $$\tan x = x + \frac{x^3}{3} + \frac{2x^5}{15} + \frac{17x^7}{315} + \frac{62x^9}{2835} + \cdots,$$

which converges for $|x| < \pi/2$. A comparison of (e) with $y_3(x)$ of (b) shows that the first three terms of each series are the same. If we had obtained additional Picard approximations, subsequent functions in the sequence would contain more and more terms in agreement with (e) and would approach the actual solution $\tan x$. Note, however, how much smaller the interval I, obtained from the theorem, is than the interval $(-\pi/2, \pi/2)$ for which the series (e) is valid. In the absence of a solution, we would know by the theorem only, that for $|x| < 0.19$, the sequence of approximating functions $y_0(x)$, $y_1(x)$, $\cdots$, $y_n(x)$ converges to a function $y(x)$ which is the particular solution of (a) satisfying $y(0) = 0$. We would thus know, for example, that $y_n(0.1)$ is an approximation to $\tan (0.1)$. However, we could not know from the theorem whether $y_0(0.5)$,

$y_1(0.5), \cdots, y_n(0.5)$ converges to the actual solution $\tan (0.5)$ unless we could find a larger h. *The theorem therefore does not always give us a maximum interval of convergence.*

EXERCISE 58

1. Show that the sequence of functions $x, x^2, x^3, \cdots, x^n$ converges on I: $0 \leq x < 1$, to the function $f(x) = 0$, but not uniformly. What is the limiting function of this sequence if I is the interval $0 \leq x \leq 1$? Is the limiting function continuous?

2. Show that the sequence of functions $\dfrac{x}{1 + x}, \dfrac{2x}{1 + 2x}, \cdots, \dfrac{nx}{1 + nx}$ converges on I: $0 < x < \infty$ to the function $f(x) = 1$, but not uniformly. What is the limiting function of this sequence if I is the interval $0 \leq x < \infty$? Is this limiting function continuous?

3. Show that the sequence of functions $1/x, 1/2x, 1/3x, \cdots, 1/nx$ converges uniformly on I: $1 \leq x < \infty$ to the continuous function $f(x) = 0$.

4. Show that the series

$$\frac{x^2}{1 + x^2} + \left(\frac{x^4}{1 + x^4} - \frac{x^2}{1 + x^2} \right) + \left(\frac{x^6}{1 + x^6} - \frac{x^4}{1 + x^4} \right) + \cdots,$$

whose sequence of partial sums is

$$\frac{x^2}{1 + x^2}, \frac{x^4}{1 + x^4}, \cdots, \frac{(x^2)^n}{1 + (x^2)^n},$$

converges to the function

$$f(x) = 0, \quad |x| < 1,$$
$$= \tfrac{1}{2}, \quad |x| = 1,$$
$$= 1, \quad |x| > 1.$$

5. In comment 58.59, we stated that the first Picard approximation need not be the function $y_0(x) = y_0$, where y_0 is the given initial value. In Example 58.63, our first approximation was $y_0(x) = 0$. Show that $y_0(x) = x$ would have been a better first approximation. Solve the problem starting with this new first approximation.

6. Solve the problem in Exercise 57,1, starting with the approximation $y_0(x) = 1 - x$. See 5 above. Is it a better or poorer approximation?

7. Find an interval of convergence for each sequence of Picard approximations obtained in Exercise 57,1, Exercise 57,2, Exercise 57,5.

ANSWERS 58

1. $f(x) = 0, 0 \leq x < 1; f(x) = 1, x = 1$. Limiting function is discontinuous.
2. $f(x) = 1, 0 < x < \infty; f(x) = 0, x = 0$. Limiting function is discontinuous.
6. Better approximation. Actual solution is $y = 2e^{-x} + x - 1 = 1 - x + x^2 - \dfrac{x^3}{3} + \dfrac{x^4}{12} - \dfrac{x^5}{60} \cdots.$

7. (1) $|f(x,y)| \leqq |x| + |y| < M$. Any values of x and y are permissible for region S. Choose h so that rectangle of dimensions $|x| \leqq h$, $|y - 1| < Mh$ which has $(0,1)$ at its center lies in S. The interval is $|x| < h$.
(2) $|f(x,y)| \leqq |x^2| + |y| < M$. Any values of x and y are permissible for region S. Choose h so that rectangle of dimensions $|x - 1| \leqq h$, $|y - 3| < Mh$ which has $(1,3)$ at its center lies in S. The interval is $|x - 1| < h$.

LESSON 59. The Ordinary and Singular Points of a First Order Differential Equation $y' = f(x,y)$.

In Lesson 5, we introduced the concept of an ordinary point and of a singular point of a first order equation

$$(59.1) \qquad y' = f(x,y).$$

For convenience we repeat these Definitions 5.4 and 5.41.

Definition 59.11. An ordinary point of a first order differential equation (59.1) is a point in the plane which lies on one and only one member of a 1-parameter family of solutions, i.e., it lies on one and only one integral curve.

Definition 59.12. A singular point of the first order differential equation (59.1) is a point in the plane which meets the following two requirements.

1. It lies on none or more than one integral curve of (59.1).
2. If a circle of arbitrarily small radius is drawn about this point, there is at least one ordinary point in its interior.

Comment 59.13. By Definition 59.11 and Comment 58.61, all points (x,y) of a region S, in which $f(x,y)$ satisfies (58.51) and (58.56), are ordinary points of (59.1). They lie on one and only one integral curve of its family of solutions. If the region S contains points which do not fulfill (58.51) or (58.56), then by Definition 59.12 and Comment 58.61, they *may* be singular. Hence in hunting for singular points, we need only examine those which fail to fulfill (58.51) or (58.56). There is certainly no need to look for them among those which do.

Example 59.2. Determine whether the differential equation

$$(a) \qquad y' = 2x$$

has singular points.

Solution. Here $f(x,y)$ of Theorem 58.5 is $2x$ and $\partial f(x,y)\partial y = 0$. Therefore in any bounded region S, $f(x,y)$ is continuous and satisfies (58.56). Its partial derivative with respect to y is also continuous and bounded in S. Therefore by Comment 58.24, $f(x,y)$ satisfies (58.51). Hence, by Comment 59.13, each point of S is an ordinary point. There are no singular points.

Remark. The solution of (a) is

(b) $$y = x^2 + c,$$

which is a family of parabolas with its vertices on the y axis. For each point (x_0, y_0) in the plane, (b) will give an unique particular solution of (a) satisfying the point (x_0, y_0). Hence, by Comment 58.62, we now also know that (b) is a general solution of (a).

Example 59.21. Determine whether the differential equation

(a) $$y' = -\sqrt{1 - y^2}, \quad -1 \leqq y \leqq 1,$$

has singular points.

Solution. Since $\sqrt{1 - y^2}$ is defined only for values of y for which $-1 \leqq y \leqq 1$, we confine our attention to a region which is contained between these lines. Here $f(x,y)$ of Theorem 58.5 is $-\sqrt{1 - y^2}$ and $\partial f(x,y)/\partial y = y/\sqrt{1 - y^2}$. Let us take for S the region defined by $-1 < y < 1$, $|x| < A$, where A is a positive number, Fig. 59.22. Let

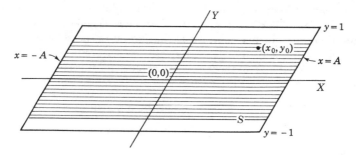

Figure 59.22

(x_0, y_0) be a point of S. Since it is an interior point of S, it can be surrounded by a rectangle contained in S. In this rectangle, $f(x,y)$ is continuous and satisfies (58.56). Its partial derivative with respect to y is also continuous and bounded. Therefore, by Comment 58.24, $f(x,y)$ satisfies (58.51). Hence each point in this rectangle, by Comment 59.13, is an ordinary point. Since each point of S can be similarly surrounded by a rectangle, all points of S are ordinary points.

If we enlarge S to include the lines $y = 1$ and $y = -1$, then for these values of y, (58.51) is not satisfied. [In (58.51), take $\bar{y} = 1$. Then as $y \to 1$, $|(-\sqrt{1 - y^2} + 0)/(y - 1)| \to \infty$.] Hence the points on these lines *may*, by Comment 59.13, be singular points of (a). That they are in fact singular points can be seen from the solutions of (a). These are

(b) $$y = \cos(x + c)$$

where c is the usual arbitrary constant, and the functions

(c) $y = \pm 1$.

Each point on the line $y = 1$, i.e., each point $(x_0, 1)$ lies on the integral curves $y = \cos(x - x_0)$ and $y = 1$; each point on the line $y = -1$, i.e., each point $(x_0, -1)$, lies on the curves $y = \cos(x + \pi - x_0)$, and $y = -1$. For example, the point $(0,1)$ lies on the integral curves $y = \cos x$ and $y = 1$, the point $(-2,1)$ lies on the integral curves $y = \cos(x + 2)$ and $y = 1$, the point $(3,-1)$ lies on the integral curves $y = \cos(x + \pi - 3)$ and $y = -1$. Hence, by Definition 59.12, the points on the lines $y = 1$ and $y = -1$ are singular points.

Comment 59.23. If the region S in the above example excludes the lines $y = \pm 1$, then each point of S, as remarked above is an ordinary point. Hence by Definition 59.11, there is one and only one particular solution through each point of S. Since for each point of S the family (b) yields a particular solution through it, we now know, by Comment 58.62, that the 1-parameter family (b) is, in S, a general solution of (a). If, however, S includes the lines $y = \pm 1$, then in the absence of further information, we cannot know whether (b) is a general solution of (a), i.e., whether it contains every particular solution. For since $(x, \pm 1)$ may be singular points, there may or may not, by Definition 59.12, be solutions through these points. Actually, as we saw above, $y = \pm 1$ are particular solutions of (a) not obtainable from the family (b). Hence for this enlarged region S, (b) is not the general solution of (a).

Example 59.24. Determine whether the differential equation

(a) $y' = (y + 1)^2/y, \quad y \neq 0$,

has singular points.

Solution. Here $f(x,y)$ of Theorem 58.5 is $(y + 1)^2/y$ and $\partial f(x,y)/\partial y = (y^2 - 1)/y^2$. Let us take for S the region defined by $-B \leqq y < 0$, $|x| < A$, where A, and $B > 1$, are positive numbers, Fig. 59.241. Let (x_0, y_0) be a point of S. Since it is an interior point of S, it can be surrounded by a rectangle contained in S. In this rectangle $f(x,y)$ is continuous and satisfies (58.56). Its partial derivative with respect to y is also continuous and bounded. Therefore, by Comment 58.24, $f(x,y)$ satisfies (58.51). Hence, each point in this rectangle, by Comment 59.13, is an ordinary point. Since each point of S can be similarly surrounded by a rectangle, all points of S are ordinary points. Therefore, by Definition 59.11, each point in S lies on one and only one integral curve of (a).

The n-parameter family of solutions of (a) is

(b) $\dfrac{1}{y + 1} + \log|y + 1| = x + c, \quad y \neq -1$.

We now note that (b) excludes particular solutions of (a) that lie on points whose y coordinate is -1. Hence we now also know that (b) cannot be the general solution of (a) and that there must be an additional solution (or solutions) of (a) that lie on the points $(x, -1)$. This particular solution

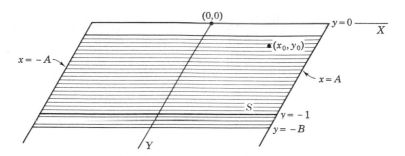

Figure 59.241

is, in fact, the line $y + 1 = 0$. This solution together with (b) contain every solution of (a). There are no others. Note the role played by the existence theorem in warning us that (b) cannot be the general solution of (a).

EXERCISE 59

Determine the singular points, if there are any, of each of the following equations.

1. $y' = x - y$.

2. $y' = \dfrac{2y - 1}{x}$, $x \neq 0$.

3. $yy' = \sqrt{1 - y^2}$, $|y| \leq 1$, $y \neq 0$.

4. $y' = \sqrt{y}$, $y \geq 0$.

5. $y' = \dfrac{1}{x - y}$, $x - y \neq 0$.

ANSWERS 59

1. None.   2. None in each half plane $x > 0$, $x < 0$.
3. Lines $y = \pm 1$.   4. Line $y = 0$.   5. None in each of the two regions divided by the line $x - y = 0$.

LESSON 60. Envelopes.

In this text, we have described various techniques for finding a 1-parameter family of solutions $f(x,y,c) = 0$, for special types of first order differential equations of the form $F\left(x, y, \dfrac{dy}{dx}\right) = 0$. If this 1-parameter family of solutions is not a general one, then the problem of finding particular solutions, if there are any, that are not obtainable from the family is usually a complex and difficult one. However, there is one case where a standard method exists for finding such particular solutions. The method we shall describe is closely associated with the notion of an "envelope" of a family of curves. Hence we shall first discuss the meaning of an envelope of a family of curves.

LESSON 60A. Envelopes of a Family of Curves. Consider the family of circles $(x - c)^2 + y^2 = 1$, with centers at $(c,0)$ and radius 1, Fig. 60.01. Each circle is tangent to the lines $y = \pm 1$. Since these two

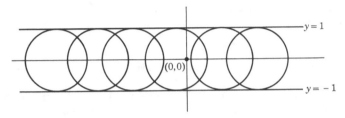

Figure 60.01

lines may be looked at as enclosing the family of circles, and since the word envelope means a wrapper for enclosing something, each has been called an envelope of the given family of circles. More generally, we define the envelope of a family of curves as follows.

Definition 60.1. Let $f(x,y,c) = 0$ be a given family of curves, Fig. 60.2. A curve C will be called an **envelope** of the family if the following two properties hold.

1. At each point of the envelope there is a unique member of the family tangent to it.

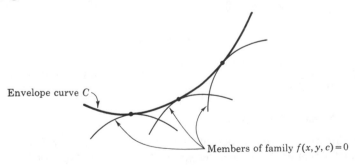

Figure 60.2

2. Every member of the family is tangent to the envelope at a distinct point of the envelope.

Definition 60.21. A function $f(x,y,c)$ is said to be **twice differentiable** if all its first and second partial derivatives exist.

We now state without proof* a theorem which gives a *sufficient* condition for the existence of an envelope of a family $f(x,y,c) = 0$.

Theorem 60.22. *If $f(x,y,c)$ is a twice differentiable function defined for a set of values of x,y,c, and if, for this set of values,*

$$(60.23) \qquad f(x,y,c) = 0, \qquad \frac{\partial f(x,y,c)}{\partial c} = 0,$$

and

$$(60.24) \qquad \begin{vmatrix} \dfrac{\partial f}{\partial x} & \dfrac{\partial f}{\partial y} \\[2ex] \dfrac{\partial^2 f}{\partial x \partial c} & \dfrac{\partial^2 f}{\partial y \partial c} \end{vmatrix} \neq 0, \qquad \frac{\partial^2 f}{\partial^2 c} \neq 0,$$

then the family of curves $f(x,y,c) = 0$ has an envelope whose parametric equations are given by (60.23).

Assume that $f(x,y,c)$ is a function which satisfies the hypothesis of Theorem 60.22 so that we can be sure the family of curves

$$(60.25) \qquad f(x,y,c) = 0$$

has an envelope. In place of the difficult proof of the above theorem which we have omitted, we shall now describe a formal method of finding this envelope. At the same time, we shall discover how the equations in (60.23) originated. No attempt will be made to justify each step rigorously. For each value of c, we obtain a member of the family (60.25). Hence if $c = c_0$ and $c = c_0 + \Delta c$, where $\Delta c \neq 0$ is small, then

$$(60.26) \qquad f(x,y,c_0) = 0 \quad \text{and} \quad f(x,y,c_0 + \Delta c) = 0$$

are two neighboring members of the family (60.25). By Definition 60.1, each of these two curves is tangent to the envelope of the family (60.25). Let us call P_1, P_2 their respective points of tangency, Fig. 60.29. The two curves themselves will usually intersect in a point, called P in the figure, which is not too far distant from P_1 and P_2. Since the coordinates of the point P satisfy each of the equations of (60.26), it will also satisfy the equation

$$(60.27) \qquad f(x,y,c_0 + \Delta c) - f(x,y,c_0) = 0.$$

---

*A proof can be found in William F. Osgood, *Advanced Calculus*.

(The point P lies in each curve; its coordinates will therefore make each term zero.) Since $\Delta c \neq 0$, we may divide (60.27) by it to obtain

(60.28)
$$\frac{f(x,y,c_0 + \Delta c) - f(x,y,c_0)}{\Delta c} = 0.$$

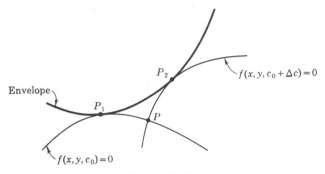

Figure 60.29

By the definition of a derivative, which we assumed exists,

(60.3)
$$\lim_{\Delta c \to 0} \frac{f(x,y,c_0 + \Delta c) - f(x,y,c_0)}{\Delta c} = \frac{\partial f(x,y,c)}{\partial c},$$

evaluated at $c = c_0$.

Therefore as $\Delta c \to 0$:

(a) The expression on the left of (60.28) approaches $\dfrac{\partial f(x,y,c)}{\partial c}$.

(b) The points P_2 and P approach P_1.

(c) The curve $f(x,y,c_0 + \Delta c) = 0$ approaches the curve $f(x,y,c_0) = 0$ so that both are tangent to the envelope at P_1.

Since c_0 is an arbitrary value of c, we have thus proved that if a family of curves has an envelope, then each point of the envelope, by (c) and (a) above and (60.28), must satisfy the two equations

(60.31)
$$f(x,y,c) = 0, \qquad \frac{\partial f(x,y,c)}{\partial c} = 0.$$

These two equations may be regarded as the parametric equations of an envelope of the family (60.25). Note that they are the same as (60.23).

Comment 60.32. It may not always be possible to eliminate the parameter c between the two equations in (60.31), but if it can be eliminated, then the resulting Cartesian equation $y = g(x)$ is called the **eliminant** of (60.31). An eliminant, however, may introduce additional loci which are not part of the envelope. For example,

(a) $$x = \sqrt{2 + c}, \qquad y = \sqrt{2 - c}$$

are the parametric equations of that part of a circle of radius two that lies in the first quadrant (since x and y are positive). Their eliminant is $x^2 + y^2 = 4$, which is the entire circle.

Comment 60.33. Not every family has an envelope. For example, the family of concentric circles $x^2 + y^2 = c^2$ has no envelope.

Comment 60.34. Theorem 60.22 gives only a sufficient condition for the existence of an envelope, not a necessary one. This means that if a family satisfies the hypotheses of Theorem 60.22, it must have an envelope; if it does not, it may or may not have an envelope.

Example 60.35. Find the envelopes, if there are any, of the family of curves

(a) $$y = \cos (x + c).$$

Solution. Here $f(x,y,c) = y - \cos (x + c)$ and $\dfrac{\partial f(x,y,c)}{\partial c} = \sin (x + c)$. The determinant (60.24) is therefore

(b) $$\begin{vmatrix} \sin (x + c) & 1 \\ \cos (x + c) & 0 \end{vmatrix} = -\cos (x + c).$$

Also $\partial^2 f/\partial c^2 = \cos (x + c)$. If $x + c \neq \pi/2$, then $\cos (x + c) \neq 0$. Hence, by Theorem 60.22, the sets of values, excluding $x + c = \pi/2$, that satisfy the parametric equations

(c) $$y - \cos (x + c) = 0,$$
$$\sin (x + c) = 0,$$

are envelopes of the family (a). Writing the first equation of (c) as $\cos (x + c) = y$, then squaring both equations of (c) and adding, we obtain the eliminant,

(d) $$y^2 = 1, \qquad y = \pm 1.$$

The two curves $y = 1$ and $y = -1$ are the envelopes of the family (a).

Comment 60.36. If we write (a) as

(e) $$\text{Arc} \cos y = x + c,$$

then $f(x,y,c) = \text{Arc} \cos y - x - c$ and $\partial f/\partial c = -1$. Equations (60.23) therefore become

(f) $$\text{Arc} \cos y - x - c = 0,$$
$$-1 = 0.$$

Since $-1 \neq 0$, the parametric equations are not satisfied. Yet we showed above that (a) has two envelopes, $y = \pm 1$. The trouble is that now $\partial^2 f / \partial c^2 = 0$ for all x,y, so that one of the hypotheses of Theorem 60.22 is not fulfilled. The fact that the family (e) has an envelope shows that the theorem gives only a sufficient condition, not a necessary one.

Example 60.37. Find the envelopes, if there are any, of the family of circles

(a) $$(x - c)^2 + y^2 = 1,$$

where c is a parameter.

Solution. Here $f(x,y,c) = (x - c)^2 + y^2 - 1$, $\partial f / \partial c = -2(x - c)$. The determinant (60.24) is therefore

(b) $$\begin{vmatrix} 2(x - c) & 2y \\ -2 & 0 \end{vmatrix} = 4y \neq 0 \quad \text{if } y \neq 0 \text{ and } \frac{\partial^2 f}{\partial c^2} = 2 \neq 0.$$

Hence by Theorem 60.22, the sets of values, excluding $y = 0$, which satisfy the parametric equations

(c) $$(x - c)^2 + y^2 - 1 = 0,$$
$$-2(x - c) = 0, \quad x - c = 0,$$

are envelopes of the family (a). Substituting the second equation of (c) in the first, we obtain the eliminant

(d) $$y^2 = 1, \quad y = \pm 1.$$

Therefore by Theorem 60.21, $y = 1$ and $y = -1$ are envelopes of (a). See Fig. 60.01.

Comment 60.38. A trouble similar to that mentioned in comment 60.36, occurs if we write (a) as

(e) $$c = x \pm \sqrt{1 - y^2}.$$

Equations (60.23) therefore become

(f) $$c - x \pm \sqrt{1 - y^2} = 0,$$
$$1 = 0.$$

Since $1 \neq 0$, (f) cannot be satisfied. But here $\partial^2 f / \partial c^2 = 0$ for all x,y,c, so that one of the hypotheses of Theorem 60.22 is not fulfilled. The fact that the family (e) has an envelope shows again that the theorem gives only a sufficient condition, not a necessary one.

Example 60.39. Find the envelopes if there are any of the family

(a) $$y^2 = 2cx - c^2.$$

Solution. You can verify that (60.24) is satisfied if $y \neq 0$. The parametric equations (60.23) are:

(b)
$$y^2 - 2cx + c^2 = 0$$
$$-2x + 2c = 0, \quad x = c.$$

Substituting the second equation of (b) in the first, we obtain the eliminants

(c) $y^2 - 2x^2 + x^2 = 0, \quad y^2 - x^2 = 0, \quad x - y = 0, \quad x + y = 0.$

Hence $x = y$ and $x = -y$, $y \neq 0$, are envelopes of (a). See Fig. 60.4.

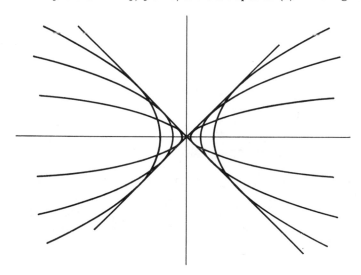

Figure 60.4

Example 60.41.   Find the envelopes, if there are any, of the family of curves

(a) $(x - c)^2 = 3y^2 - y^3.$

Solution.   Here $f(x,y,c) = (x - c)^2 - 3y^2 + y^3$, $\partial f/\partial c = -2(x - c)$. The determinant (60.24) is therefore

(b) $\begin{vmatrix} 2(x - c) & -6y + 3y^2 \\ 2 & 0 \end{vmatrix} \neq 0, \quad \text{if } y \neq 0 \text{ and } 2; \quad \dfrac{\partial^2 f}{\partial c^2} = 2 \neq 0.$

By Theorem 60.22, sets of values, excluding $y = 0$ and 2, that satisfy the parametric equations

(c) $(x - c)^2 - 3y^2 + y^3 = 0,$
$$2(x - c) = 0, \quad (x - c) = 0$$

are envelopes of (a). Substituting the second equation in the first, we obtain

(d) $$-3y^2 + y^3 = 0, \qquad y = 0, \qquad y = 3.$$

The eliminant $y = 0$ is excluded by (b). However, since Theorem 60.22 gives only a sufficient condition, we cannot assert without further study that $y = 0$ is not an envelope. (Actually it is not, see Example 60.53.) However, $y = 3$ is an envelope.

Example 60.42. Find the envelopes, if there are any, of the family of curves

(a) $$ce^y = x - y - 1.$$

Solution. Here $f(x,y,c) = ce^y - x + y + 1$, $\partial f/\partial c = e^y$. The determinant (60.24) is therefore

(b) $$\begin{vmatrix} -1 & ce^y + 1 \\ 0 & e^y \end{vmatrix} = -e^y \neq 0, \quad \text{but } \frac{\partial^2 f}{\partial c^2} = 0.$$

Hence one of the hypotheses of Theorem 60.22, namely $\partial^2 f/\partial c^2 \neq 0$, is not satisfied. The family (a), therefore, may or may not have envelopes. (It actually has none, see Example 60.54.)

LESSON 60B. Envelopes of a 1-Parameter Family of Solutions. Let $f(x,y,c)$ be a 1-parameter family of solutions of the differential equation $y' = f(x,y)$. The family may or may not have an envelope. If it has, then by Definition 60.1, at each point of the envelope, there is a member of the family of solutions that is tangent to it. Hence the envelope, at each of its points, has the same slope y' as an integral curve. It therefore follows that each envelope of the family of solutions, excluding those points of the envelope where the tangent is vertical, also satisfies the given differential equation $y' = f(x,y)$. *Hence an envelope of a family of solutions of $y' = f(x,y)$ is also a solution of this differential equation.*

Unless an envelope is coincident with an integral curve of a family, it is, by Definition 59.12, a locus of singular points. For each point of the envelope lies on two integral curves: the envelope itself and an integral curve of the family. Hence an envelope is a particular solution of a differential equation $y' = f(x,y)$ that cannot be obtained from a 1-parameter family of solutions by assigning a value to the arbitrary constant. The converse, however, need not be true. A particular solution not obtainable from a 1-parameter family of solutions is not necessarily an envelope. If, therefore, we can find an envelope of a 1-parameter family of solutions of a differential equation $y' = f(x,y)$, we shall at the same time have succeeded in finding a particular solution of the equation, not obtainable from the family.

Comment 60.43. By Theorem 60.22, we know that if $f(x,y,c) = 0$ is a twice differentiable function and satisfies (60.24) over a set of values of x,y,c, then the parametric equations of its envelope are given by

$$(60.44) \qquad f(x,y,c) = 0, \qquad \frac{\partial f(x,y,c)}{\partial c} = 0.$$

If the parameter c can be eliminated between these two equations, the resulting equation $y = g(x)$, which we called the eliminant, gives an envelope. However, as noted in Comment 60.32, the eliminant may introduce extraneous loci. It is essential, therefore, to verify by actual substitution in a given differential equation whether each factor of the eliminant satisfies the equation. If it does, it is a particular solution not obtainable from the family. See Example 60.53 below for an eliminant which is not a solution.

Example 60.5. Find, by means of envelopes, particular solutions, if there are any, of the differential equation

(a) $$y' = \sqrt{1 - y^2},$$

not obtainable from the family of solutions

(b) $$y = \cos{(x + c)}.$$

Solution. In Example 60.35, we found that $y = 1$ and $y = -1$ are envelopes of the family (b). You can verify by direction substitution that each is a solution of the given differential equation (a). Hence each envelope is a particular solution of (a) not obtainable from (b).

Example 60.51. Find by means of envelopes, particular solutions, if there are any, of the differential equation

(a) $$y' = -\frac{\sqrt{1 - y^2}}{y},$$

not obtainable from the family of solutions

(b) $$(x - c)^2 + y^2 = 1.$$

Solution. In Example 60.37, we found that $y = 1$ and $y = -1$ are envelopes of (b). Since each function satisfies (a), each is a particular solution of (a), not obtainable from the family (b).

Example 60.52. Find, by means of envelopes, particular solutions, if there are any, of the differential equation

(a) $$y' = \frac{x + \sqrt{x^2 - y^2}}{y},$$

not obtainable from the family of solutions

(b) $$y^2 = 2cx - c^2.$$

Solution. In Example 60.39, we found that $y = \pm x$ are envelopes of (b). Since these functions satisfy (a), they are particular solutions of (a), not obtainable from the family (b).

Example 60.53. Find, by means of envelopes, particular solutions, if there are any, of the differential equation

(a) $$9(y')^2(2 - y)^2 = 4(3 - y),$$

not obtainable from the family of solutions

(b) $$(x - c)^2 = 3y^2 - y^3.$$

Solution. In Example 60.41, we found two eliminants, namely $y = 3$ and $y = 0$. The first satisfies (a), and is therefore a particular solution of (a), not obtainable from (b). The second $y = 0$, however, does not satisfy (a). Hence, we now also know that $y = 0$ is not an envelope of (b). (In example 60.41, we remarked that in the absence of further information, we could not know whether or not $y = 0$ is an envelope.)

Example 60.54. Find, by means of envelopes, particular solutions, if there are any, of the differential equation

(a) $$y' = \frac{1}{x - y},$$

not obtainable from the family of solutions

(b) $$ce^y = x - y - 1.$$

Solution. In Example 60.42, we found that (60.24) of Theorem 60.22 was not satisfied. Hence we did not know whether or not the family (b) had an envelope. However, if we apply the existence Theorem 58.5 to the function $f(x,y) = 1/(x - y)$ of (a), we find that the only possible singular points lie on the line $x - y = 0$. Since the function $y = x$ does not satisfy (a), it cannot be an envelope of the family (b). In this example, therefore, a particular solution not obtainable from (b), if there is one, cannot be found by means of envelopes.

EXERCISE 60

Find the envelopes, if there are any, of each of the following family of curves.

1. $y = x + c.$

2. $y = x + \frac{1}{4}(x + c)^2.$

3. $y = (x - c)^3.$

4. $y^{1/3} = x - c.$

5. $y = cx + 3\sqrt{1 + c^2},\ y > 0$

6. $(y + x - c)^2 = 4xy.$

7. $y = \sin[(x - c)^2].$

Find, by means of envelopes, particular solutions, if there are any, of each of the following differential equations, not obtainable from the given family of solutions.

8. $y' = \sqrt{y - x} + 1$, $y = x + \frac{1}{4}(x + c)^2$, $x \leqq y \leqq x + 1$.

9. $y' = 3y^{2/3}$, $y = (x - c)^3$.

10. $y = xy' + 3\sqrt{1 + (y')^2}$, $y = cx + 3\sqrt{1 + c^2}$, $y > 0$, $x^2 < 9$.

11. $y' = 2\sqrt{(1 - y^2)} \sqrt{\text{Arc sin } y}$, $x = \sqrt{\text{Arc sin } y} + c$.

ANSWERS 60

1. None.
2. $y = x$.
3. $y = 0$.
4. $y = 0$. See problem 3 and read Comments 60.36, 60.38.
5. $y = \sqrt{9 - x^2}$.

6. $xy = 0$.
7. $y = \pm 1$.
8. $y = x$.
9. $y = 0$.
10. $y = \sqrt{9 - x^2}$.
11. $y = 0$, $y = \pm 1$.

LESSON 61. The Clairaut Equation.

The simplest type of family of curves is a family of straight lines. From analytic geometry we know that

$$(61.1) \qquad\qquad y = mx + b,$$

represents such a family. The slope of each line of the family is m, and its y intercept is b. The family (61.1) is thus a 2-parameter family. We can form a 1-parameter family from it by requiring that b be a function of the slope m, i.e., that $b = f(m)$. Hence (61.1) becomes

$$(61.12) \qquad\qquad y = mx + f(m).$$

To find a differential equation whose 1-parameter family of solutions is the family of lines (61.12), we proceed as we did in Lesson 4B. Differentiation of (61.12) gives $y' = m$. Substituting this value of m in (61.12), we obtain the differential equation

$$(61.13) \qquad\qquad y = y'x + f(y').$$

Equation (61.13) is called **Clairaut's equation.***

Comment 61.14. If we start with the differential equation (61.13), its solution (61.12) is easily found. Replace y' by the parameter m (or, if you prefer, by c). For example, the solution of the Clairaut equation $y = y'x + (y')^2$ is the family of nonvertical lines $y = mx + m^2$ or $y = cx + c^2$.

---

*Named after the French mathematician, Alex Claude Clairaut (1713–1765).

As we pointed out in Lesson 60B, if the family (61.12) has an envelope, then this envelope is a particular solution of (61.13) not obtainable from the family of solutions (61.12). By Theorem 60.22, a sufficient condition, among others, for the existence of an envelope of the family (61.12) is that points of the envelope satisfy the parametric equations

(61.15) $$y - cx - f(c) = 0,$$
$$x + f'(c) = 0.$$

The first equation in (61.15) has the same form as (61.12). It therefore satisfies Clairaut's equation (61.13). The eliminant of (61.15) also satisfies this first equation. (A solution of two simultaneous equations satisfies each equation.) Hence, if (61.15) yields an eliminant, then this eliminant by Comment 60.43, may be a particular solution of (61.13). It is if it satisfies (61.13).

Example 61.2.   Find a 1-parameter family of solutions of the Clairaut equation

(a) $$y = y'x + (y')^2.$$

Also investigate for envelopes of the family of solutions.

Solution.   First we note by (61.13), that (a) is a Clairaut equation with $f(y') = (y')^2$. Hence by Comment 61.14, its 1-parameter family of solutions is

(b) $$y = cx + c^2.$$

You can verify that (b) satisfies (60.24). Therefore, by Theorem 60.22, a set of values that satisfies the parametric equations (60.23),

(c) $$y - cx - c^2 = 0,$$
$$x + 2c = 0, \quad c = -x/2,$$

is an envelope of (b). Substituting the second equation of (c) in the first, we obtain the eliminant

(d) $$y = -\frac{x^2}{2} + \frac{x^4}{4} = -\frac{x^2}{4},$$

which is an envelope of the family of nonvertical lines (b). Since this function (d) satisfies (a), it is a particular solution of (a) not obtainable from the family (b).

Example 61.21.   Find a 1-parameter family of solutions of the Clairaut equation

(a) $$y = y'x + \log y'.$$

Also investigate for envelopes of the family of solutions.

Solution. First we note by (61.13) that (a) is a Clairaut equation with $f(y') = \log y'$. Hence by Comment 61.14,

(b) $$y = cx + \log c,$$

is a family of solutions of (a). You can verify that (b) satisfies (60.24). Therefore, by Theorem 60.22, the set of values that satisfies the parametric equations (60.23),

(c) $$y - cx - \log c = 0$$
$$x + \frac{1}{c} = 0, \quad x = -\frac{1}{c}$$

is an envelope of (b). Substituting the second equation of (c) in the first, we obtain

(d) $\quad y + 1 - \log(-x^{-1}) = 0, \qquad y + 1 + \log(-x) = 0, \quad x < 0,$

which is an envelope of the family of nonvertical lines (b). Since this function (d) satisfies (a), it is a particular solution of (a) not obtainable from the family (b).

Example 61.3. A source of light or sound which strikes a curve in the same plane with it, is reflected in a fixed direction. Find the equation of this curve.

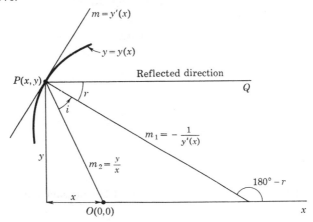

Figure 61.31

Solution. (See Fig. 61.31.) We place the origin at the source of the light or sound, and let the x axis be parallel to the fixed reflected direction. Let $y = y(x)$ be the equation of the curve we seek. We assume $y(x)$ is defined and differentiable on an interval I.

Call:

(a) i the angle the incident ray OP makes with the normal to the required curve,

   r the angle the reflected ray PQ makes with the normal to the required curve,

   y' the slope of the tangent to the required curve $y = y(x)$ at $P(x,y)$,

   m_1 the slope of the normal to the required curve at $P(x,y)$, $(= -1/y')$,

   m_2 the slope of OP $(=y/x)$.

By a law of physics

(b) $$\tan i = \tan r,$$

and by a formula of analytic geometry

(c) $$\tan i = \frac{m_1 - m_2}{1 + m_1 m_2} = \frac{-\dfrac{1}{y'} - \dfrac{y}{x}}{1 - \dfrac{1}{y'}\dfrac{y}{x}} = \frac{-x - yy'}{xy' - y}.$$

From Fig. 61.31, we see that $\tan (180° - r) = -1/y'$. Hence, $\tan r = 1/y'$. Therefore by (b), $\tan i = 1/y'$. Substituting this value of $\tan i$ in (c), we obtain

(d) $$\frac{1}{y'} = \frac{-x - yy'}{xy' - y}, \qquad y = 2xy' + y(y')^2.$$

The equation is almost like a Clairaut equation, but not quite. If, however, we multiply it by y, we obtain

(e) $$y^2 = 2xyy' + y^2(y')^2.$$

Now let

(f) $$u = y^2, \qquad u' = 2yy', \qquad (y')^2 = \frac{(u')^2}{4y^2}.$$

Substituting (f) in (e), we have

(g) $$u = u'x + \tfrac{1}{4}(u')^2,$$

which is a Clairaut equation in u. Hence its 1-parameter family of solutions is

(h) $$u = cx + \frac{c^2}{4}.$$

Substituting in (h) the value of u as given in (f), we obtain the family of parabolas

(i) $$y^2 = cx + \frac{c^2}{4}.$$

Geometric Problems Giving Rise to a Clairaut Equation. In Lessons 13 and 36D, we solved geometric problems which gave rise, respectively, to a first order equation and to a special type of second order equation. A curve whose tangents have properties which are independent of the point at which the tangent is drawn will lead to a Clairaut equation. Since the solution of a Clairaut equation is a family of straight lines, and since these straight lines will have the properties of the tangent lines, the envelope of the family will give the curve we seek.

Example 61.4. Every tangent to a curve has the property that the sum of its intercepts has a constant value k. Find the curve.

Solution. By Exercise 13,1(a) and 1(b), the x and y intercepts of a tangent line are respectively $x - y/y'$ and $y - xy'$. Hence the curve must satisfy the condition

(a) $\qquad x - \dfrac{y}{y'} + y - xy' = k, \qquad xy' - y + yy' - x(y')^2 = ky',$

$\qquad\qquad y(y' - 1) = xy'(y' - 1) + ky', \qquad y = xy' + \dfrac{ky'}{y' - 1}.$

The last equation in (a) is, by (61.13), a Clairaut equation. Hence its solution, by Comment 61.14, is

(b) $\qquad y = cx + \dfrac{kc}{c - 1}, \qquad xc^2 - (x + y - k)c + y = 0,$

which is a family of straight lines. You can verify that (b) satisfies (60.24). Therefore, by Theorem 60.22, the c-eliminant of

(c) $\qquad\qquad xc^2 - (x + y - k)c + y = 0$

$\qquad\qquad 2xc - (x + y - k) = 0, \qquad c = \dfrac{x + y - k}{2x},$

may be an envelope of (b). Substituting the second equation of (c) in the first, we obtain the eliminant

(d) $\qquad \dfrac{(x + y - k)^2}{4x} - \dfrac{(x + y - k)^2}{2x} + y = 0,$

$\qquad (x + y - k)^2 - 4xy = 0, \qquad x + y - k = \pm 2x^{1/2}y^{1/2},$

$\qquad x \pm 2x^{1/2}y^{1/2} + y = k, \qquad x^{1/2} \pm y^{1/2} = \pm k^{1/2}.$

Since it satisfies (a), it is the required solution.

EXERCISE 61

Find a 1-parameter family of solutions of each of the following Clairaut equations 1–6. Also investigate for envelopes.

1. $y = xy' - (y')^2$.

2. $y = xy' + [1 + (y')^2]$.

3. $y = xy' - (y')^{2/3}$.

4. $y = xy' - (y')^3$.

5. $y = xy' + \sqrt{1 + (y')^2}$.

6. $y = xy' - e^{y'}$.

7. Solve the equation $y = 3xy' + 6y^2(y')^2$. *Hint.* Multiply by y^2. Then make the substitution $u = y^3$, $du/dx = 3y^2y'$ to obtain the Clairaut equation

$$u = x\frac{du}{dx} + \frac{2}{3}\left(\frac{du}{dx}\right)^2.$$

In problems 8–13, every tangent to a curve has the property indicated. Find the curve.

8. The sum of its intercepts has a constant value 9.

9. The product of its intercepts has a constant value k.

10. The length of the segment intercepted by the coordinate axes has a constant value k^2. *Hint.* Set the square root of the sum of the squares of its intercepts equal to k^2.

11. Its distance from the origin has a constant value k. *Hint.* First show that the equation of a tangent line at a point (x,y) on the required curve is given by $y'X - Y + y - xy' = 0$, where (X,Y) is a point on the tangent line. Then use the fact that the distance of a point (A,B) to the line $aX + bY + c = 0$ is $\pm(aA + bB + c)/\sqrt{a^2 + b^2}$. Here $A = 0$, $B = 0$.

12. The sum of its distances from the points $(a,0)$ and $(-a,0)$ has a constant value k. In the distance formula from a point to a line, as given in 11, take the positive square root.

13. The segment intercepted by the coordinate axes forms with these axes a right triangle whose area has a constant value k^2.

In each of problems 14–16, find the Clairaut equation whose family of solutions of straight lines has the envelope indicated. *Hint.* First find x and y in terms of y' and then solve (61.13) for $f(y')$.

14. $y^2 = 2x$.   **15.** $2y = x^2$.   **16.** $xy = -k$.

ANSWERS 61

1. $y = cx - c^2$, $y = x^2/4$.

2. $y = cx + 1 + c^2$, $x^2 + 4y = 4$.

3. $y = cx - c^{2/3}$, $27x^2y + 4 = 0$.

4. $y = cx - c^3$, $27y^2 = 4x^3$.

5. $y = cx + \sqrt{1 + c^2}$, $y = \sqrt{1 - x^2}$, $y > 0$.

6. $y = cx - e^c$, $y = x\log x - x$.

7. $y^3 = cx + \frac{2}{3}c^2$.

8. $x^{1/2} \pm y^{1/2} = \pm 3$.

9. $4xy = k$.

10. $x^{2/3} + y^{2/3} = k^{4/3}$.

11. $x^2 + y^2 = k^2$.

12. $4(x^2 + y^2) = k^2$.

13. $2xy = \pm k^2$.

14. $y = y'x + \dfrac{1}{2y'}$.

15. $y = y'x - \dfrac{(y')^2}{2}$.

16. $y = y'x \pm 2\sqrt{ky'}$.

Chapter 12

Existence and Uniqueness Theorems for a System of First Order Differential Equations and for Linear and Nonlinear Differential Equations of Order Greater Than One. Wronskians.

LESSON 62. An Existence and Uniqueness Theorem for a System of n First Order Differential Equations and for a Nonlinear Differential Equation of Order Greater Than One.

LESSON 62A. The Existence and Uniqueness Theorem for a System of n First Order Differential Equations. In Theorem 39.12 we gave a sufficient condition for the existence of a *power series* solution of a system of n first order equations

$$(62.1) \qquad \frac{dy_1}{dt} = f_1(t, y_1, y_2, \cdots, y_n),$$

$$\frac{dy_2}{dt} = f_2(t, y_1, y_2, \cdots, y_n),$$

$$\cdots \cdots \cdots \cdots \cdots$$

$$\frac{dy_n}{dt} = f_n(t, y_1, y_2, \cdots, y_n),$$

satisfying the initial conditions,

$$(62.11) \qquad y_1(t_0) = a_1, \qquad y_2(t_0) = a_2, \cdots, y_n(t_0) = a_n.$$

Compare it with the theorem we now state, which gives a sufficient condition for the existence and uniqueness of a solution of (62.1) satisfying (62.11).

Theorem 62.12. *Let the functions* $f_1, f_2, \cdots, f_n$ *of the first order system* *(62.1) be continuous in a region S defined by*

$$(62.13) \qquad |t - t_0| \leqq k_0, \qquad |y_1 - a_1| \leqq k_1,$$
$$|y_2 - a_2| \leqq k_2, \cdots, |y_n - a_n| \leqq k_n.$$

In S, let each function satisfy a Lipschitz condition

$$(62.14) \quad |f_i(t,y_1,y_2, \cdots, y_n) - f_i(t,\overline{y}_1,\overline{y}_2, \cdots, \overline{y}_n)|$$
$$\leqq N(|y_1 - \overline{y}_1| + |y_2 - \overline{y}_2| + \cdots + |y_n - \overline{y}_n|), \quad i = 1, 2, \cdots n,$$

where $(t,y_1,y_2, \cdots, y_n)$ *and* $(t,\overline{y}_1,\overline{y}_2, \cdots, \overline{y}_n)$ *are any two points in S.*

Then an interval I_0: $|t - t_0| < h, h > 0$, *exists in which there is one and only one set of continuous functions* $y_1(t), y_2(t), \cdots, y_n(t)$ *with continuous derivatives in* I_0 *satisfying the given system (62.1) and the initial conditions (62.11).*

NOTE. The theorem does not always give a maximum interval of convergence; read Comment 58.65.

Remark. When $n = 1$, the system (62.1) reduces to a first order equation $y_1' = f_1(t,y_1)$. Theorem 58.5, which we proved previously for this first order equation, is thus a special case of the above more general theorem.

The proof of Theorem 62.12, which we have omitted, although more complicated than the proof of Theorem 58.5, follows its pattern exactly. We find the interval I_0 of Theorem 62.12, for example, in practically the same manner we found the interval I_0 of Theorem 58.5. Since the region S consists of closed intervals and the functions $f_1, f_2, \cdots, f_n$ are each continuous in S, each is bounded in S. Hence there is a positive number M such that

$$(62.15) \qquad |f_i(t,y_1,y_2, \cdots, y_n)| < M, \quad i = 1, 2, \cdots, n,$$

for every point $(t,y_1,y_2, \cdots, y_n)$ in S. By (62.13), $(t_0,a_1, \cdots, a_n)$ is an interior point of S. It is therefore possible to pick an h such that the $n + 1$ dimensional rectangle R,

$$(62.16) \quad |t - t_0| \leqq h \leqq k_0, \qquad |y_1 - a_1| < Mh \leqq k_1, \cdots,$$
$$|y_n - a_n| < Mh \leqq k_n,$$

which has this point $(t_0,a_1,a_2, \cdots, a_n)$ at its center, lies entirely in S. The M is given by (62.15); the k's are given in (62.13).

We then find sets of sequences of Picard approximations

$$y_{11}, y_{12}, \cdots, y_{1n},$$
$$y_{21}, y_{22}, \cdots, y_{2n},$$
$$\cdots \cdots \cdots \cdots$$
$$y_{n1}, y_{n2}, \cdots, y_{nn},$$

in a manner similar to that given in (57.33) for a system of two first order equations. The graph of each function in each approximating set, it is then proved, lies in the $n + 1$ dimensional rectangle R, and each set of sequences of approximating functions converges uniformly on I: $|t - t_0|$ $\leqq h$ to the respective set of functions $y_1(t)$, $y_2(t)$, $\cdots$, $y_n(t)$, which on I_0 is a solution of (62.1) satisfying (62.11). Finally we show that this set of functions y_1, y_2, $\cdots$, y_n is unique.

Example 62.17. Find an interval, for which the approximating functions obtained in Example 57.35 converge to the actual solution.

Solution. In this Example $f_1(t,x,y) = ty$, $f_2(t,x,y) = xy$. Comparing the initial conditions of this example with (62.11) we see that $t_0 = 0$, $a_1 = 1$, $a_2 = 1$. Since f_1 and f_2 are continuous for all t,x,y, we may take for S any bounded region, say one defined by $|t| \leqq 5 = k_0$, $|x - 1| \leqq 10 = k_1$, $|y - 1| \leqq 15 = k_2$. Hence $|x| \leqq 11$, $|y| \leqq 16$ and

(a) $\qquad\qquad |f_1| \leqq 80, \qquad |f_2| \leqq 176 = M.$

We must now choose h according to (62.16) so that

(b) $\quad |t| \leqq h \leqq 5, \qquad |x - 1| < 176h \leqq 10, \qquad |y - 1| < 176h \leqq 15.$

If $h = 0.056$, all three inequalities in (b) will be satisfied. The Picard approximations obtained in Example 57.35, will therefore converge to the actual solutions $x(t)$, $y(t)$ for at least $|t| < 0.056$.

LESSON 62B. Existence and Uniqueness Theorem for a Nonlinear Differential Equation of Order n. In Theorem 39.32, we gave a sufficient condition for the existence of a *power series* solution of the nth order nonlinear differential equation

(62.2) $\qquad\qquad \dfrac{d^n y}{dx^n} = f(x,y,y',y'', \cdots, y^{(n-1)})$

satisfying the initial conditions

(62.21) $\quad y(x_0) = a_0, \qquad y'(x_0) = a_1, \qquad y''(x_0) = a_2, \cdots,$
$$y^{(n-1)}(x_0) = a_{n-1}.$$

Compare it with the theorem we now state which gives a sufficient condition for the existence and uniqueness of a solution of (62.2) satisfying (62.21).

Theorem 62.22. *Let the function f in the nth order differential equation (62.2) be a continuous function of its arguments $x,y,y', \cdots, y^{(n-1)}$ in a region S defined by*

(62.23) $\quad |x - x_0| \leqq k_0, \qquad |y - a_0| \leqq k_1, \qquad |y' - a_1| \leqq k_2,$
$$|y'' - a_2| \leqq k_3, \cdots, |y^{(n-1)} - a_{n-1}| \leqq k_n,$$

and let f satisfy a Lipschitz condition,

(62.24) $\quad |f(x,y_1,y_1', \cdots, y_1^{(n-1)}) - f(x,y_2,y_2', \cdots, y_2^{(n-1)})|$
$$\leqq N(|y_1 - y_2| + |y_1' - y_2'| + |y_1'' - y_2''| + \cdots$$
$$+ |y_1^{(n-1)} - y_2^{(n-1)}|),$$

where $(x,y_1,y_1', \cdots, y_1^{(n-1)})$ and $(x,y_2,y_2', \cdots, y_2^{(n-1)})$ are any two points in S.

Then an interval I_0: $|x - x_0| < h$, $h > 0$, exists in which there is one and only one continuous function $y(x)$ with a continuous derivative of order n satisfying (62.2) and the initial conditions (62.21).

Proof. Assume that $y(x)$ is a solution of (62.2). We now define new functions $y_1(x)$, $y_2(x)$, $\cdots$, $y_n(x)$ by the following relations

(62.25) $\quad y(x) = y_1(x),$
$$y'(x) = y_1'(x) = y_2(x),$$
$$y''(x) = y_1''(x) = y_2'(x) = y_3(x),$$
$$\cdots \cdots \cdots \cdots \cdots \cdots \cdots \cdots \cdots \cdots \cdots \cdots \cdots \cdots \cdots$$
$$y^{(n-1)}(x) = y_1^{(n-1)}(x) = y_2^{(n-2)}(x) = \cdots = y_{n-1}'(x) = y_n(x).$$

Differentiating the function in the last line of (65.25), we obtain

(62.251) $\quad y^{(n)}(x) = y_1^{(n)}(x) = y_2^{(n-1)}(x) = \cdots = y_{n-1}''(x) = y_n'(x).$

By (65.251) and (62.25), we can therefore write (62.2) as

(62.26) $\qquad y_n'(x) = f(x,y_1,y_2,y_3, \cdots, y_n).$

Hence, because of (62.25) and (62.26), the set of functions $y_1,y_2, \cdots, y_n$ satisfies *the system of first order differential equations*

(62.27) $\qquad y_1'(x) = y_2(x),$
$$y_2'(x) = y_3(x),$$
$$y_3'(x) = y_4(x),$$
$$\cdots \cdots \cdots \cdots \cdots \cdots \cdots \cdots$$
$$y_{n-1}'(x) = y_n(x),$$
$$y_n'(x) = f(x,y_1,y_2,y_3, \cdots, y_n).$$

By (62.25), $y_1(x_0) = y(x_0)$, $y_2(x_0) = y'(x_0)$, $\cdots$, $y_n(x_0) = y^{(n-1)}(x_0)$. We can therefore replace the initial conditions (62.21) by the conditions

(62.28) $\quad y_1(x_0) = a_0, \qquad y_2(x_0) = a_1, \qquad y_3(x_0) = a_2, \cdots,$
$$y_n(x_0) = a_{n-1}.$$

Conversely, if we start with the system (62.27) and the initial conditions (62.28), and define

(62.281) $$y(x) = y_1(x),$$

then the relations in (62.25), (62.251), and (62.26) hold; the last equation of (62.27) becomes identical with (62.2), and the initial conditions (62.28) become (62.21). Hence if there exists a set of functions $y_1, y_2, \cdots, y_n$ which is a solution of the system (62.27) satisfying (62.28), then, by (62.281), the function $y_1(x)$ of this set is the solution $y(x)$ of (62.2) satisfying (62.21).

You can verify that the system (62.27) with initial conditions (62.28) satisfies the hypotheses of Theorem 62.12. Hence by the theorem, an interval I_0 about x_0 exists on which there is one and only one set of particular solutions $y_1, y_2, \cdots, y_n$ satisfying (62.27) and (62.28). Since $y_1(x)$ exists and is unique, it follows, by (62.281), that its equal $y(x)$ exists and is the unique solution of (62.2) satisfying (62.21).

Comment 62.282. We wish to emphasize that it is always possible, by means of (62.25) and (62.26), to replace a single differential equation of order $n > 1$ by an equivalent system of n first order equations.

Example 62.29. Set up a system of first order equations equivalent to the third order nonlinear equation

(a) $$y'''(x) = 3xy' - y^2 y''$$

for which

(b) $$y(0) = 1, \quad y'(0) = -1, \quad y''(0) = 2.$$

Solution. Let $y(x)$ be a solution of (a). As suggested in (62.25), we define

(c) $$y(x) = y_1(x),$$
$$y'(x) = y_1'(x) = y_2(x),$$
$$y''(x) = y_1''(x) = y_2'(x) = y_3(x).$$

Differentiating the last equation in (c), we obtain

(d) $$y'''(x) = y_1'''(x) = y_2''(x) = y_3'(x).$$

By (d) and (c), we can write (a) as

(e) $$y_3'(x) = 3xy_2 - y_1^2 y_3.$$

Hence, by (c) and (e), we can replace the single equation (a) by the system of first order equations

(f)
$$y_1'(x) = y_2(x),$$
$$y_2'(x) = y_3(x),$$
$$y_3'(x) = 3xy_2 - y_1{}^2 y_3,$$

and the initial conditions (b) by

(g) $y_1(0) = y(0) = 1,$ $y_2(0) = y'(0) = -1,$ $y_3(0) = y''(0) = 2.$

The function $y_1(x)$ of the solution of the system (f) satisfying (g) will be the solution $y(x)$ of (a) satisfying (b).

LESSON 62C. Existence and Uniqueness Theorem for a System of n Linear First Order Equations. In Theorem 62.12, no restrictions were placed on the degree of the dependent variables in the functions $f_1, f_2, \cdots, f_n$ of (62.1). In Theorem 62.3 below, each of the dependent variables $y_1, y_2, \cdots, y_n$ appears linearly, i.e., each of these variables has exponent one.

Theorem 62.3. *Given a system of n **linear** first order equations*

(62.31) $\dfrac{dy_1}{dt} = f_{11}(t)y_1 + f_{12}(t)y_2 + \cdots + f_{1n}(t)y_n + Q_1(t),$

$\dfrac{dy_2}{dt} = f_{21}(t)y_1 + f_{22}(t)y_2 + \cdots + f_{2n}(t)y_n + Q_2(t),$

$\cdot\ \cdot$

$\dfrac{dy_n}{dt} = f_{n1}(t)y_1 + f_{n2}(t)y_2 + \cdots + f_{nn}(t)y_n + Q_n(t),$

where all functions f_{ij} and Q_i, $i = 1, 2, \cdots, n$, $j = 1, 2, \cdots n$, are continuous on a common interval I. Then there is, on I_0, i.e., I without its end points, one and only one set of continuous functions $y_1(t)$, $y_2(t)$, $\cdots$, $y_n(t)$ with continuous derivatives, satisfying the system (62.31) and the initial conditions

(62.32) $y_1(t_0) = a_1, \cdots, y_n(t_0) = a_n,$

where t_0 is a point in I_0.

Remark. In Theorem 39.22, we gave a sufficient condition for the existence of a *power series* solution of (62.31) satisfying (62.32). Compare it with the above theorem which gives a sufficient condition for the existence and uniqueness of a solution of (62.31) satisfying (62.32).

Proof of Theorem. The proof of the theorem will consist in showing that the continuity requirement of the functions f_{ij} on I is equivalent to the requirement that each function satisfy a Lipschitz condition (62.14) in I, provided I is finite and closed. If I is not finite and closed, then it is sufficient, as we shall show, that each function satisfy this Lipschitz condition on a slightly smaller closed subinterval I' contained in I. And since all functions on the right of (62.31) are, by hypothesis, continuous on this common, closed interval I or I', the hypotheses of Theorem 62.12 will be satisfied. Hence its conclusion will follow. Let

(a) $\quad f_1(t,y_1,y_2, \cdots, y_n) = f_{11}(t)y_1 + f_{12}(t)y_2 + \cdots + f_{1n}(t)y_n + Q_1(t),$

$\qquad f_2(t,y_1,y_2, \cdots, y_n) = f_{21}(t)y_1 + f_{22}(t)y_2 + \cdots + f_{2n}(t)y_n + Q_2(t),$

$$\cdots \cdots \cdots \cdots \cdots \cdots \cdots \cdots \cdots \cdots \cdots \cdots \cdots \cdots$$

$\qquad f_n(t,y_1,y_2, \cdots, y_n) = f_{n1}(t)y_1 + f_{n2}(t)y_2 + \cdots + f_{nn}(t)y_n + Q_n(t).$

We assume the interval I of Theorem 62.3 is finite and includes its end points. If it is not, we take a finite closed subinterval I' of I. Hence in I', whether it is a subinterval of I or the whole interval I, each function $f_{ij}(t)$ and $Q_i(t)$ of (62.31) is, by hypothesis, continuous. All are therefore bounded in I'. This means that the absolute value of each function for all t in I' is less than some positive number. Let N be the largest of these numbers. Then for each function $f_{ij}(t)$ of (62.31) and for every t in I',

(b) $\qquad |f_{ij}(t)| \leqq N, \quad i = 1, 2, \cdots, n, \quad j = 1, 2, \cdots, n.$

Let $y_1, y_2, \ldots, y_n$ and $\bar{y}_1, \bar{y}_2, \cdots, \bar{y}_n$ be two sets of values of the dependent variables. Therefore by (a)

(c) $\quad f_i(t,y_1,y_2, \cdots, y_n) = f_{i1}(t)y_1 + f_{i2}(t)y_2 + \cdots + f_{in}(t)y_n + Q_i(t),$

$\qquad f_i(t,\bar{y}_1,\bar{y}_2, \cdots, \bar{y}_n) = f_{i1}(t)\bar{y}_1 + f_{i2}(t)\bar{y}_2 + \cdots + f_{in}(t)\bar{y}_n + Q_i(t),$

for each $i = 1, 2, \cdots, n$. Subtracting the second equation in (c) from the first, we obtain

(d) $\quad |f_i(t,y_1,y_2, \cdots, y_n) - f_i(t,\bar{y}_1,\bar{y}_2, \cdots, \bar{y}_n)|$

$$= |f_{i1}(t)(y_1 - \bar{y}_1) + f_{i2}(t)(y_2 - \bar{y}_2) + \cdots + f_{in}(t)(y_n - \bar{y}_n)|,$$

$$i = 1, 2, \cdots, n.$$

Since $|a - b| \leqq |a| + |b|$, and since, by (b), $|f_{ij}| \leqq N$, we obtain from (d)

(e) $\quad |f_i(t,y_1, \cdots, y_n) - f_i(t,\bar{y}_1, \cdots, \bar{y}_n)| \leqq |f_{i1}(t)| \, |y_1 - \bar{y}_1|$

$$+ |f_{i2}(t)| \, |y_2 - \bar{y}_2| + \cdots + |f_{in}(t)| \, |y_n - \bar{y}_n|$$

$$\leqq N\big(|y_1 - \bar{y}_1| + |y_2 - \bar{y}_2| + \cdots + |y_n - \bar{y}_n|\big).$$

If you will compare (e) with the Lipschitz condition (62.14), you will find that both are alike. Hence we have proved that for the special system of linear equations (62.31), the continuity requirement of the functions f_{ij} in I' is equivalent to the Lipschitz condition (62.14) of Theorem 62.12.

If the interval I of the theorem is finite and closed, then $I = I'$. If it is not, then since it is always possible to surround each t in I by a finite, closed subinterval I' contained in I, in which the theorem is valid, it follows that the theorem is valid on I. In this last case $I = I_0$.

EXERCISE 62

The following problems refer to the solutions of the systems in Exercise 57. For each find an interval for which the approximating solution converges.

1. Problem 6.   **2.** Problem 7.   **3.** Problem 8.   **4.** Problem 9.

Set up a system of first order equations with appropriate initial conditions equivalent to each of the following equations.

5. $y''(x) = 3x^2y' - y^2$, $y(0) = 0$, $y'(0) = 1$.
6. $y'''(x) = 2x(y')^2 - 3yy'' + xy$, $y(0) = 1$, $y'(0) = -1$, $y''(0) = 2$.

ANSWERS 62

1. In Theorem 62.12, $f_1 = t + y$, $f_2 = t - x^2$, $t_0 = 0$, $a_1 = 2$, $a_2 = 1$. S can be any bounded region: $|t| \leq k_0$, $|x - 2| \leq k_1$, $|y - 1| \leq k_2$, where k_0, k_1, k_2 are arbitrary. M is the larger of M_1 and M_2 in $|f_1| \leq |t| + |y| < M_1$, $|f_2| \leq |t| + |x^2| < M_2$. Choose h so that $|t| \leq h \leq k_0$, $|x - 2| < Mh \leq k_1$, $|y - 1| < Mh \leq k_2$.
2. See answer to problem 1 for method.
3. This system is a first order linear one. See Theorem 62.3: $f_{11}(t) = t$, $f_{21}(t) = 1$, $Q_2(t) = -e^t$. Each function is continuous for all t. Therefore the approximating functions converge for every t.
4. In Theorem 62.12, $f_1 = y^2$, $f_2 = x + z$, $f_3 = z - y$; $t_0 = 0$, $a_1 = 1$, $a_2 = 0$, $a_3 = 1$. S can be any bounded region: $|t| \leq k_0$, $|x - 1| \leq k_1$, $|y| \leq k_2$, $|z - 1| \leq k_3$, where k_0, k_1, k_3 are arbitrary. M is the larger of M_1, M_2, M_3 in $|f_1| < y^2 < M_1$, $|f_2| \leq |x| + |z| < M_2$, $|f_3| \leq |z| + |y| < M_3$. Choose h so that $|t| \leq h \leq k_0$, $|x - 1| < Mh \leq k_1$. $|y| < Mh \leq k_2$, $|z - 1| < Mh \leq k_3$.
5. $y_1'(x) = y_2$, $y_2'(x) = 3x^2y_2 - y_1^2$; $y_1(0) = 0$, $y_2(0) = 1$.
6. $y_1'(x) = y_2$, $y_2'(x) = y_3$, $y_3'(x) = 2xy_2^2 - 3y_1y_3 + xy_1$; $y_1(0) = 1$, $y_2(0) = -1$, $y_3(0) = 2$.

LESSON 63. Determinants. Wronskians.

LESSON 63A. A Brief Introduction to the Theory of Determinants. In this lesson, we shall outline only those essentials of the theory of determinants that we shall need for our purposes. We have already given the definition of a 2×2 and a 3×3 determinant in Comment 31.35 and (31.72). For convenience we recopy them here.

(63.1)
$$\begin{vmatrix} a_1 & b_1 \\ a_2 & b_2 \end{vmatrix} = a_1b_2 - a_2b_1,$$

(63.11)

$$\begin{vmatrix} a_1 & b_1 & c_1 \\ a_2 & b_2 & c_2 \\ a_3 & b_3 & c_3 \end{vmatrix} = a_1b_2c_3 + a_2b_3c_1 + a_3b_1c_2 - a_3b_2c_1 - a_2b_1c_3 - a_1b_3c_2.$$

We observe that each term in a determinant, as written on the right, contains one and only one element from each row and each column. Hence we conclude:

Theorem 63.12. *If each element in a row or in a column of a determinant is zero, then the determinant is zero.*

Consider a system of two equations in two unknowns.

(63.2) $$a_1x + b_1y = c_1,$$
$$a_2x + b_2y = c_2.$$

Multiplying the first equation by b_2, the second by $-b_1$, adding the two and then solving for x, we obtain

(63.21) $$x = \frac{c_1b_2 - c_2b_1}{a_1b_2 - a_2b_1}.$$

By (63.1), we can write (63.21) as

(63.22) $$x = \frac{\begin{vmatrix} c_1 & b_1 \\ c_2 & b_2 \end{vmatrix}}{\begin{vmatrix} a_1 & b_1 \\ a_2 & b_2 \end{vmatrix}}.$$

Note that the elements of the denominator determinant are the coefficients of x and y in (63.2); the elements of the numerator determinant are obtained by replacing the coefficients of x by the constants c_1 and c_2 and recopying the coefficients of y. Similarly you will find, if you solve (63.2) for y, that

(63.23) $$y = \frac{\begin{vmatrix} a_1 & c_1 \\ a_2 & c_2 \end{vmatrix}}{\begin{vmatrix} a_1 & b_1 \\ a_2 & b_2 \end{vmatrix}},$$

where the numerator determinant is obtained by replacing the coefficients of y by c_1 and c_2, and the denominator is the same as that in (63.22).

Comment 63.24.   From (63.22) and (63.23), we infer the following:

1. If $\begin{vmatrix} a_1 & b_1 \\ a_2 & b_2 \end{vmatrix} \neq 0$, the system of equations (63.2) has one and only one solution.

2. If the determinant in 1 equals zero and the numerators of (63.22) and (63.23) are $\neq 0$, the system has no solutions.

3. If the determinant in 1 equals zero and the numerators of (63.22) and (63.23) also equal zero, the system has an infinite number of solutions.

Each of these three possibilities is illustrated in the examples below.

Example 63.25.   Solve the system

(a)
$$2x + 3y = 1,$$
$$x - y = 4.$$

Solution.   By (63.22) and (63.23),

(b) $x = \dfrac{\begin{vmatrix} 1 & 3 \\ 4 & -1 \end{vmatrix}}{\begin{vmatrix} 2 & 3 \\ 1 & -1 \end{vmatrix}} = \dfrac{-1 - 12}{-2 - 3} = \dfrac{13}{5},\qquad y = \dfrac{\begin{vmatrix} 2 & 1 \\ 1 & 4 \end{vmatrix}}{\begin{vmatrix} 2 & 3 \\ 1 & -1 \end{vmatrix}} = \dfrac{7}{-5}.$

Hence (a) has an unique solution. Note that the denominator determinant $\neq 0$.

Example 63.26.   Solve the system

(a)
$$2x + 3y = 1,$$
$$4x + 6y = 3.$$

Solution.   By (63.22) and (63.23),

(b) $x = \dfrac{\begin{vmatrix} 1 & 3 \\ 3 & 6 \end{vmatrix}}{\begin{vmatrix} 2 & 3 \\ 4 & 6 \end{vmatrix}} = \dfrac{-3}{0},\qquad y = \dfrac{\begin{vmatrix} 2 & 1 \\ 4 & 3 \end{vmatrix}}{\begin{vmatrix} 2 & 3 \\ 4 & 6 \end{vmatrix}} = \dfrac{2}{0}.$

Hence there are no solutions for x and y. Note that the denominator determinant is zero but the numerator determinants are not. [The lines in (a) are actually parallel lines.]

Example 63.27.   Solve the system

(a)
$$2x + 3y = 1,$$
$$4x + 6y = 2.$$

Solution. By (63.22) and (63.23)

(b)
$$x = \frac{\begin{vmatrix} 1 & 3 \\ 2 & 6 \end{vmatrix}}{\begin{vmatrix} 2 & 3 \\ 4 & 6 \end{vmatrix}} = \frac{0}{0}, \qquad y = \frac{\begin{vmatrix} 2 & 1 \\ 4 & 2 \end{vmatrix}}{\begin{vmatrix} 2 & 3 \\ 4 & 6 \end{vmatrix}} = \frac{0}{0}.$$

Note that both numerator and denominator determinants are zero. As you can easily verify, the system (a) has an infinite number of solutions. Give x any value you wish in the first equation of (a). Solve it for y. You will find that these values of x and y will satisfy the second equation of (a). [The lines in (a) are in fact coincident.]

Comment 63.24 is also applicable to a system of n equations in n unknowns. For $n = 3$, the system becomes

(63.3)
$$a_1x + b_1y + c_1z = d_1,$$
$$a_2x + b_2y + c_2z = d_2,$$
$$a_3x + b_3y + c_3z = d_3.$$

Its solution is

(63.31)
$$x = \frac{\begin{vmatrix} d_1 & b_1 & c_1 \\ d_2 & b_2 & c_2 \\ d_3 & b_3 & c_3 \end{vmatrix}}{\begin{vmatrix} a_1 & b_1 & c_1 \\ a_2 & b_2 & c_2 \\ a_3 & b_3 & c_3 \end{vmatrix}}, \quad y = \frac{\begin{vmatrix} a_1 & d_1 & c_1 \\ a_2 & d_2 & c_2 \\ a_3 & d_3 & c_3 \end{vmatrix}}{\begin{vmatrix} a_1 & b_1 & c_1 \\ a_2 & b_2 & c_2 \\ a_3 & b_3 & c_3 \end{vmatrix}}, \quad z = \frac{\begin{vmatrix} a_1 & b_1 & d_1 \\ a_2 & b_2 & d_2 \\ a_3 & b_3 & d_3 \end{vmatrix}}{\begin{vmatrix} a_1 & b_1 & c_1 \\ a_2 & b_2 & c_2 \\ a_3 & b_3 & c_3 \end{vmatrix}}.$$

The system will have an unique solution for x, y, and z if the denominator $\neq 0$; no solution if the denominator equals zero and at least one numerator is not equal to zero; an infinite number of solutions if denominator and all numerators are equal to zero.

Our primary purpose in introducing determinants is to enable you to prove the following two important theorems. *Although for convenience, we have stated the theorems for three unknowns, they are applicable to a system of n equations in n unknowns.*

Theorem 63.4. *The system of equations*

(63.41)
$$a_1x + b_1y + c_1z = d_1,$$
$$a_2x + b_2y + c_2z = d_2,$$
$$a_3x + b_3y + c_3z = d_3,$$

has one and only one solution for x, y, z if the determinant formed by their coefficients is not equal to zero. It has none or an infinite number of solutions if this determinant equals zero.

Theorem 63.42.   *The system of equations*

(63.43)
$$a_1x + b_1y + c_1z = 0,$$
$$a_2x + b_2y + c_2z = 0,$$
$$a_3x + b_3y + c_3z = 0,$$

always has the trivial solution $x = 0$, $y = 0$, $z = 0$. It has an infinite number of nontrivial solutions, i.e., it has an infinite number of sets of values of x, y, and z satisfying (62.43), where at least one of x, y, z is not zero, if and only if the determinant formed by their coefficients is zero.

LESSON 63B.   Wronskians.   In Lesson 19A, when we discussed the linear dependence and independence of a set of functions, we stated that we would in time produce theorems by which their linear dependence or independence could be determined. Since these theorems are closely associated with the concept of a Wronskian, we shall first discuss this subject below.

Definition 63.5.   The **Wronskian*** of a set of functions $f_1(x)$, $f_2(x)$, $\cdots$, $f_n(x)$, each of which possesses derivatives of order $n - 1$, is defined to be the following determinant.

(63.51)
$$\begin{vmatrix} f_1(x) & f_2(x) & \cdots f_n(x) \\ f_1'(x) & f_2'(x) & \cdots f_n'(x) \\ f_1''(x) & f_2''(x) & \cdots f_n''(x) \\ \cdots \cdots \cdots \cdots \cdots \cdots \cdots \cdots \\ f_1^{(n-1)}(x)f_2^{(n-1)}(x) \cdots f_n^{(n-1)}(x) \end{vmatrix}$$

Note that the elements of the determinant are the given set of functions and their derivatives through order $n - 1$. We denote the Wronskian (63.51) by

(63.52)
$$W(f_1, f_2, \cdots f_n; x).$$

If $n = 2$, the Wronskian of f_1 and f_2 is, by (63.52) and (63.51),

(63.53)
$$W(f_1, f_2; x) = \begin{vmatrix} f_1 & f_2 \\ f_1' & f_2' \end{vmatrix} = f_1f_2' - f_2f_1'.$$

---

*Pronounced Vronskian and named for Hoëné Wronski, a Polish mathematician.

Definition 63.54. A function $f(x)$ is said to **vanish identically on an interval** I: $a \leqq x \leqq b$, or **to be identically zero on** I, if for *every* x in I, $f(x) = 0$. We indicate this fact by writing $f(x) \equiv 0$.

Theorem 63.55. *If a set of functions f_1, f_2, $\cdots$, f_n, each of which possesses a derivative of order $n - 1$, is linearly dependent on an interval I: $a \leqq x \leqq b$, then its Wronskian vanishes identically on I.*

Proof. Since the set of given functions f_1, f_2, $\cdots$, f_n is linearly dependent on I, there exist, by Definition 19.1, constants c_1, c_2, $\cdots$, c_n not all zero such that

(a) $$c_1 f_1 + c_2 f_2 + \cdots + c_n f_n = 0,$$

for every x in I. By differentiating (a) $(n - 1)$ times, we obtain

(b) $$f_1 c_1 + f_2 c_2 + \cdots + f_n c_n = 0,$$
$$f_1' c_1 + f_2' c_2 + \cdots + f_n' c_n = 0,$$
$$\cdots\cdots\cdots\cdots\cdots\cdots\cdots\cdots\cdots$$
$$f_1^{(n-1)} c_1 + f_2^{(n-1)} c_2 + \cdots + f_n^{(n-1)} c_n = 0.$$

The system (b) is a set of n equations, each equal to zero for every x in I. The c's in each equation are not *all* zero. There is therefore a set of c's not *all* zero that satisfies each equation in (b). Hence this set of c's is a solution of the system of equations (b). It therefore follows by Theorem 63.42, that the determinant

(c) $$\begin{vmatrix} f_1 & f_2 & \cdots & f_n \\ f_1' & f_2' & \cdots & f_n' \\ \cdots & \cdots & \cdots & \cdots \\ f_1^{(n-1)} & f_2^{(n-1)} & \cdots & f_n^{(n-1)} \end{vmatrix},$$

which is the Wronskian of the given set of functions, is zero.

Comment 63.56. Theorem 63.55 gives a necessary condition for the dependence of a set of functions. It says that if a set of functions is dependent on I, *then* its Wronskian is identically zero on I. Unfortunately it is not a sufficient condition. If the Wronskian of the set of functions is identically zero on I, one cannot conclude that the set is linearly dependent on I. It need not be. Here is an example. Let $f_1(x) = x^2$, $f_2(x) = x|x|$, Fig. 63.57. Let the interval I include $x = 0$. By (63.53), the Wronskian of the functions is

(a) $$\begin{vmatrix} x^2 & x|x| \\ 2x & x|x|' + |x| \end{vmatrix} = x^3 |x|' + x^2 |x| - 2x^2 |x|.$$

When $x > 0$, $|x| = x$. Therefore $|x|' = x' = 1$. The right side of (a) thus becomes $x^3 + x^3 - 2x^3 = 0$. When $x = 0$, $|x| = x = 0$, and again the right side of (a) is zero. When $x < 0$, $|x| = -x$ and therefore $|x|' = -x' = -1$. The right side of (a) now becomes $x^3(-1) - x^3 + 2x^3 = 0$. We have thus shown that the Wronskian of the two given func-

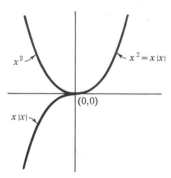

Figure 63.57

tions is identically zero on an interval I which includes $x = 0$. However, we shall now show that the two functions are linearly independent on I. Form with them the linear combination $c_1x^2 + c_2x|x| = 0$. If $x = 1$, then $c_1 + c_2 = 0$, and if $x = -1$, then $c_1 - c_2 = 0$. The only solution of this pair of equations is $c_1 = 0$, $c_2 = 0$. Hence by Definition 19.1, the functions x^2 and $x|x|$ are linearly independent.

On the other hand, if the Wronskian of a set of functions is not equal to zero for *only one* x in an interval I, then the set, by Theorem 63.55, must be linearly independent on I. In short, no conclusion can be drawn as regards the linear dependence or independence of a set of functions on I, from the fact that its Wronskian is identically zero on I. But if the Wronskian of the set is not equal to zero for only one value of x in I, then the set must be linearly independent on I.

In Exercise 63,5, you will find a necessary and *sufficient* condition for the linear independence of a set of functions on an interval.

EXERCISE 63

1. Determine, for each of the following systems, whether it has one, none, or infinitely many solutions.

(a) $x + 4y = 1$,
$2x - y = 3$.

(b) $x + y = 5$,
$2x + 2y = 10$.

(c) $2x + y = -2$,
$-2x - y = 3$.

(d) $x + 2y = 0$,
$2x + 4y = 0$.

(e) $-x + y = 0$,
$x - y = 0$.

(f) $5x + y = 0$,
$x - 3y = 0$.

2. Set up the Wronskian of each of the following set of functions.

(a) $1, e^x$. (b) $1, x$. (c) x, e^x. (d) x, xe^x. (e) $x, xe^x, 3x$.
(f) $\sin x, \sin 2x$. (g) $\sin x, \cos x$. (h) $ax, bx + c, c \neq 0$.
(i) $e^{ax}, e^{bx}, a \neq b$. (j) $x - \cos 2x, 2x - 2\cos 2x$. (k) $1, x, x^2$.
(l) $x, e^x, xe^x, 2e^x$. (m) $1, e^x, e^{2x}$. (n) x, e^x, xe^x. (o) $1, \sin^2 x$,
$\cos^2 x$, (p) $\sin^3 x, \sin x - \frac{1}{3}\sin 3x$.

3. Determine, by Comment 63.56, which of the sets in 2 are linearly independent over a suitable interval. For those sets whose Wronskian is zero, use Definition 19.1 to determine which are dependent and which are independent.

4. Prove that any two nonconstant functions that differ by a constant are linearly independent.

5. Theorem 63.55 gives only a necessary condition for the linear independence of a set of functions. The following theorem gives a *necessary and sufficient* condition for the linear independence of a set of functions. *A set of functions* $f_1(x), f_2(x), \cdots, f_n(x)$, *each continuous on an interval* $I: a \leq x \leq b$, *is linearly dependent on* I, *if and only if the determinant*

$$G = \begin{vmatrix} \int_a^b f_1^2\, dx & \int_a^b f_1 f_2\, dx & \cdots & \int_a^b f_1 f_n\, dx \\ \int_a^b f_2 f_1\, dx & \int_a^b f_2^2\, dx & \cdots & \int_a^b f_2 f_n\, dx \\ \cdots\cdots\cdots\cdots\cdots\cdots\cdots\cdots \\ \int_a^b f_n f_1\, dx & \int_a^b f_n f_2\, dx & \cdots & \int_a^b f_n^2\, dx \end{vmatrix} = 0$$

on I. The determinant G is called the **Grammian** of the set of functions. Use this theorem to prove that

(a) $x, 2x, 3x$ is a linearly dependent set on $I: -1 \leq x \leq 1$,
(b) e^x, e^{2x} is a linearly independent set on $I: -1 \leq x \leq 1$,
(c) $x, |x|$ is a linearly independent set on $I: -1 \leq x \leq 1$.

6. Prove Theorem 63.4.
7. Prove Theorem 63.42.

ANSWERS 63

1. (a) one. (b) Infinitely many. (c) None. (d) Infinitely many. (e) Infinitely many. (f) $x = 0, y = 0$.

2. (a) $\begin{vmatrix} 1 & e^x \\ 0 & e^x \end{vmatrix} = e^x$.

(d) $\begin{vmatrix} x & xe^x \\ 1 & xe^x + e^x \end{vmatrix} = x^2 e^x$.

(b) $\begin{vmatrix} 1 & x \\ 0 & 1 \end{vmatrix} = 1$.

(e) $\begin{vmatrix} x & xe^x & 3x \\ 1 & xe^x + e^x & 3 \\ 0 & xe^x + 2e^x & 0 \end{vmatrix} = 0$.

(c) $\begin{vmatrix} x & e^x \\ 1 & e^x \end{vmatrix} = (x - 1)e^x$.

(f) $\begin{vmatrix} \sin x & \sin 2x \\ \cos x & 2\cos 2x \end{vmatrix} = -2\sin^3 x$.

(g) $\begin{vmatrix} \sin x & \cos x \\ \cos x & -\sin x \end{vmatrix} = -1.$

(l) $W(x, e^x, xe^x, 2e^x; x) = 0.$

(h) $\begin{vmatrix} ax & bx + c \\ a & b \end{vmatrix} = -ac.$

(m) $2e^{3x}.$

(i) $\begin{vmatrix} e^{ax} & e^{bx} \\ ae^{ax} & be^{bx} \end{vmatrix} = (b - a)e^{(a+b)x}.$

(n) $e^{2x}(x - 2).$

(j) $\begin{vmatrix} x - \cos 2x & 2x - 2\cos 2x \\ 1 + 2\sin 2x & 2(1 + 2\sin 2x) \end{vmatrix} = 0.$

(o) $0.$

(k) $\begin{vmatrix} 1 & x & x^2 \\ 0 & 1 & 2x \\ 0 & 0 & 2 \end{vmatrix} = 2.$

(p) $0.$

3. Linearly independent sets are (a), (b), (c), (d), (f), (g), (h), (i), (k), (m), (n). Linearly dependent sets are (e), (j), (l), (o), (p)—see Example 21.32.

LESSON 64. Theorems About Wronskians and the Linear Independence of a Set of Solutions of a Homogeneous Linear Differential Equation.

In the previous lesson, we proved that if a set of functions is linearly dependent on I, then their Wronskian is identically zero on I. We also showed by an example, that the converse need not be true. If, however, the set of functions are n *solutions of a homogeneous linear differential equation of order* n, namely of

(64.1) $\quad f_n(x)y^{(n)} + f_{n-1}(x)y^{(n-1)} + \cdots + f_1(x)y' + f_0(x)y = 0,$

where $f_0(x), f_1(x), \cdots, f_n(x)$ are each continuous functions on an interval I, and $f_n(x) \neq 0$ when x is in I, then, as we shall show (see Comment 64.15 below) the vanishing of their Wronskian is a *necessary* and *sufficient* condition for the linear dependence of the set on I.

Theorem 64.11. *If each function* $y_1, y_2, \cdots, y_n$ *is a solution of (64.1) on an interval* $I: a \leq x \leq b$, *then the Wronskian of this set of functions is either identically zero on* I, *or it is not zero for any* x *in* I.

Proof. We shall prove the theorem only for $n = 2$. Call y_1, y_2 the two solutions of (64.1) when $n = 2$. Since each function satisfies a linear equation of the second order, their first and second derivatives exist. By (63.53),

(a) $$W(y_1, y_2; x) = \begin{vmatrix} y_1 & y_2 \\ y_1' & y_2' \end{vmatrix} = y_1 y_2' - y_2 y_1'.$$

Differentiating both sides of (a) with respect to x, we obtain

(b) $\quad W'(y_1, y_2; x) = y_1 y_2'' + y_1' y_2' - y_2' y_1' - y_2 y_1'' = y_1 y_2'' - y_2 y_1''.$

By hypothesis, each of the functions y_1 and y_2 is a solution of (64.1) with $n = 2$. Hence,

(c) $\qquad\qquad f_2(x) y_1'' + f_1(x) y_1' + f_0(x) y_1 = 0,$
$\qquad\qquad f_2(x) y_2'' + f_1(x) y_2' + f_0(x) y_2 = 0.$

Since we assumed $f_2 \neq 0$, we may divide each equation in (c) by it to obtain

(d) $\qquad y_1'' = -\dfrac{f_1 y_1' + f_0 y_1}{f_2}, \qquad y_2'' = -\dfrac{f_1 y_2' + f_0 y_2}{f_2}.$

Substituting these values in (b) gives, with the help of (63.53),

(e) $\quad W'(y_1, y_2; x) = -y_1 \left(\dfrac{f_1 y_2' + f_0 y_2}{f_2} \right) + y_2 \left(\dfrac{f_1 y_1' + f_0 y_1}{f_2} \right)$

$\qquad\qquad\qquad = -\dfrac{f_1}{f_2} (y_1 y_2' - y_2 y_1') = -\dfrac{f_1}{f_2} W(y_1 y_2; x).$

By (e), therefore

(f) $\qquad\qquad\qquad W' + \dfrac{f_1}{f_2} W = 0,$

an equation which is linear in W. Its solution is

(64.12) $\qquad\qquad W(y_1, y_2; x) = c e^{F(x)},$

where $F(x) = -\int (f_1 \, dx / f_2)$. Since, by (18.86), the exponential function $e^{F(x)}$ is never zero, we conclude from (64.12) that, if $c = 0$, then $W = 0$ for every x in I; if $c \neq 0$, then $W \neq 0$ for every x in I.

Remark. Formula (64.12) is known as **Abel's** formula.

Theorem 64.13. *If each function $y_1, y_2, \cdots, y_n$ is a solution of (64.1) in an interval $I: a \leqq x \leqq b$, and if*

(64.14) $\qquad\qquad W(y_1, y_2, \cdots, y_n; x) = 0,$

at a point $x = x_0$ in I, then the set of functions is linearly dependent on I.

Proof. We shall prove the theorem only for $n = 2$. By hypothesis, there is a point x_0 in I, for which

(a) $\qquad\qquad\qquad W(y_1, y_2; x_0) = 0,$

where y_1 and y_2 are two solutions of (64.1) with $n = 2$. Hence, by (63.53) and (a),

(b)
$$\begin{vmatrix} y_1(x_0) & y_2(x_0) \\ y_1'(x_0) & y_2'(x_0) \end{vmatrix} = 0.$$

Using the elements of the determinant (b) as coefficients, we form the two equations

(c)
$$y_1(x_0)c_1 + y_2(x_0)c_2 = 0,$$
$$y_1'(x_0)c_1 + y_2'(x_0)c_2 = 0.$$

Since the determinant whose elements are the coefficients of c_1 and c_2 of (c) is, by (b), zero, it follows, by Theorem 63.42, that there is an infinite number of pairs of nontrivial solutions of (c) for c_1 and c_2. With any one such pair, form the linear combination

(d)
$$y_3(x) = c_1y_1(x) + c_2y_2(x),$$

where y_1 and y_2 are the two given solutions. Therefore, by Theorem 19.3, $y_3(x)$ is also a solution of (64.1) with $n = 2$. By (d) and (c), $y_3(x_0) = c_1y_1(x_0) + c_2y_2(x_0) = 0$ and $y_3'(x_0) = c_1y_1'(x_0) + c_2y_2'(x_0) = 0$.

We have thus proved that $y_3(x)$ of (d) is a solution of (64.1), with $n = 2$, i.e., it is a solution of

(e)
$$f_2(x)y'' + f_1(x)y' + f_0(x)y = 0,$$

satisfying the conditions

(f)
$$y_3(x_0) = 0, \qquad y_3'(x_0) = 0.$$

But you can easily verify that $y(x) \equiv 0$ is also a solution of (e) and that this solution satisfies $y(x_0) = 0$, $y'(x_0) = 0$. By the uniqueness Theorem 19.2 (which we still have not proved but shall soon, see Theorem 65.2) the two solutions $y_3(x)$ of (d) and $y(x) = 0$ are identical. Hence by (d)

(g)
$$c_1y_1(x) + c_2y_2(x) = 0.$$

We remarked after (c), that c_1 and c_2 are nontrivial solutions of (c) and hence not both are zero. Therefore by Definition 19.1, the functions y_1 and y_2 are linearly dependent.

Comment 64.141. If the Wronskian of a set of solutions of (64.1) is zero at a point x_0 in I, then it is easy to show, by Theorem 64.13, that the Wronskian is identically zero on I. We prove the statement, which is equivalent to Theorem 64.11, for $n = 2$. Let y_1, y_2 be two solutions of (64.1). Since $W = 0$ at a point x_0 in I, the solutions, by Theorem 64.13

are linearly dependent. Therefore

(a) $c_1 y_1 + c_2 y_2 \equiv 0$

on I, where c_1 and c_2 are not both zero. Differentiation of (a) gives

(b) $c_1 y_1' + c_2 y_2' \equiv 0.$

Since c_1 and c_2 are not both zero, it follows by Theorem 63.42, that

(c)
$$\begin{vmatrix} y_1 & y_2 \\ y_1' & y_2' \end{vmatrix} \equiv 0.$$

But the determinant in (c) is, by (63.53), the Wronskian of y_1 and y_2.

Note. Although the above proof is much shorter than the one in Theorem 64.11, the latter proof was given in order to obtain (64.12), an equation which will be needed later.

Comment 64.15. Theorems 63.55 and 64.13 together give a necessary and sufficient condition for the linear dependence of a set of functions that are *solutions* of the linear differential equation (64.1). The first says that if the set is dependent on I, then its Wronskian is $\equiv 0$ on I. The second says that if its Wronskian is zero for only *one* x in I (and therefore by Theorem 64.11 it is $\equiv 0$ on I) then the set is linearly dependent on I. Hence you now have a test to determine whether a set of n *solutions* of a homogeneous linear differential equation (64.1) is linearly dependent or independent. If its Wronskian is zero for one x in I, the set is linearly dependent; if its Wronskian is not equal to zero for one x in I, the set is linearly independent.

Example 64.2. Show that if m_1, m_2, m_3 are distinct, then the set of functions

(a) $y_1 = e^{m_1 x}, \qquad y_2 = e^{m_2 x}, \qquad y_3 = e^{m_3 x},$

which are three solutions of a third order homogeneous linear equation, are linearly independent on $I\colon -\infty < x < \infty$.

Solution. By (63.51), the Wronskian of the given functions is

(b)
$$\begin{vmatrix} e^{m_1 x} & e^{m_2 x} & e^{m_3 x} \\ m_1 e^{m_1 x} & m_2 e^{m_2 x} & m_3 e^{m_3 x} \\ m_1^2 e^{m_1 x} & m_2^2 e^{m_2 x} & m_3^2 e^{m_3 x} \end{vmatrix}$$

By (63.11), (b) is equal to

(c) $e^{(m_1+m_2+m_3)x}[(m_2 m_3^2 -- m_3 m_2^2) - (m_1 m_3^2 - m_3 m_1^2)$
$$+ (m_1 m_2^2 - m_2 m_1^2)],$$

which can be written as

(d) $$e^{(m_1+m_2+m_3)x}(m_1 - m_2)(m_2 - m_3)(m_3 - m_1).$$

Since we have assumed m_1, m_2, m_3 are distinct and since, by (18.86), the exponential function $e^{(m_1+m_2+m_3)x}$ is never zero, it follows that (d) cannot be zero for any x. Hence the Wronskian (b) is not zero for any x. By Comment 64.15, the set of solutions in (a) is therefore linearly independent on $I: -\infty < x < \infty$.

Remark. The above proof can be extended to show that if m_1, $m_2, \cdots, m_n$ are distinct, then the set of n functions $y_1 = e^{m_1 x}$, $y_2 = e^{m_2 x}, \cdots, y_n = e^{m_n x}$, each of which is a solution of (64.1), is linearly independent on $I: -\infty < x < \infty$.

Example 64.21. Show that the set of functions

(a) $$y_1 = e^{ax}, \qquad y_2 = xe^{ax},$$

which are the solutions of $y'' - 2ay' + a^2y = 0$, are linearly independent on $I: -\infty < x < \infty$.

Solution. By (63.53), the Wronskian of the given functions is

(b) $$\begin{vmatrix} e^{ax} & xe^{ax} \\ ae^{ax} & axe^{ax} + e^{ax} \end{vmatrix} = axe^{2ax} + e^{2ax} - axe^{2ax} = e^{2ax}.$$

Since $e^{2ax} \neq 0$ for every x, the given functions are independent on $I: -\infty < x < \infty$.

Example 64.22. Show that the second solution $y_2(x)$ of (23.28) obtained by the reduction of order method described in Lesson 23B, is independent of the solution $y_1(x)$.

Solution. Let $F(x) = f_1(x)/f_2(x)$ and

(a) $$g(x) = \int \frac{e^{-\int F(x)dx}}{y_1^2} \, dx.$$

Then, by (23.28), $y_2 = y_1 g$ and, by (63.53), the Wronskian of the two solutions is

(b) $$W(y_1, y_2; x) = \begin{vmatrix} y_1 & y_1 g \\ y_1' & y_1 g' + y_1' g \end{vmatrix}$$

$$= y_1^2 g' + y_1 y_1' g - y_1' y_1 g$$

$$= y_1^2 g'.$$

By (a) and Theorem 58.3

(c) $$g'(x) = \frac{e^{-\int F(x)dx}}{y_1{}^2}.$$

Substituting (c) in the right of (b), we obtain

(d) $$W(y_1,y_2; x) = e^{-\int F(x)dx}.$$

Since the exponential function is never zero, it follows, by Comment 64.15, that the given functions are linearly independent solutions on any interval I over which the functions y_1 and y_2 are defined.

EXERCISE 64

1. Show that the set of functions 1, x, x^2 which are three solutions of $y''' = 0$ are linearly independent.
2. Show that each of the functions $y_1 = \sin x - \frac{1}{3}\sin 3x$ and $y_2 = \sin^3 x$, is a solution of $y'' + (\tan x - 2\cot x)y' = 0$, but that $y = c_1y_1 + c_2y_2$ is not the general solution of the differential equation. *Hint.* Show that the two functions are not linearly independent; see Exercise 63.3(p).
3. Prove that $y = c_1 \sin x + c_2 \cos x$ is a general solution of $y'' + y = 0$. *Hint.* Show that each function satisfies the equation and that the two functions are linearly independent.
4. Prove that $y = (c_1 + c_2x)e^x$ is a general solution of $y'' - 2y' + y = 0$.
5. Prove that $y = c_1e^x + c_2e^{2x} + c_3xe^{2x}$ is a general solution of $y''' - 5y'' + 8y' - 4y = 0$.

LESSON 65. Existence and Uniqueness Theorem for the Linear Differential Equation of Order n.

We are ready at last to prove the existence and uniqueness Theorem 19.2 for the linear differential equation of order n. Since on the interval I in which we shall be interested, $f_n(x) \neq 0$, we can divide (19.21) by $f_n(x)$ to obtain a linear differential equation

(65.1) $$y^{(n)}(x) + f_{n-1}(x)y^{(n-1)} + \cdots + f_1(x)y' + f_0(x)y = Q(x).$$

Theorem 65.2. *If the coefficients $f_0(x)$, $f_1(x)$, $\cdots$, $f_{n-1}(x)$ and $Q(x)$ in the linear differential equation (65.1) are each continuous functions of x on a common interval I, then for each point x_0 in I and for each set of constants $a_0, a_1, \cdots, a_{n-1}$, there is one and only one function $y(x)$ that satisfies (65.1) and the initial conditions*

(65.21) $$y(x_0) = a_0, \quad y'(x_0) = a_1, \cdots, y^{(n-1)}(x_0) = a_{n-1}.$$

Remark 65.22. In Theorem 37.51, we gave a sufficient condition for the existence of a *power series* solution of (65.1) satisfying (65.21). Compare it with the above theorem, which gives a sufficient condition for the existence and uniqueness of a solution of (65.1) satisfying (65.21).

Proof of the Theorem. Assume $y(x)$ is a solution of (65.1). We now define new functions $y_1(x)$, $y_2(x)$, $\cdots$, $y_n(x)$ by the following relations.

(65.23)
$$y(x) = y_1(x),$$
$$y'(x) = y_1'(x) = y_2(x),$$
$$y''(x) = y_1''(x) = y_2'(x) = y_3(x),$$
$$\cdots\cdots\cdots\cdots\cdots\cdots\cdots\cdots\cdots\cdots\cdots$$
$$y^{(n-1)}(x) = y_1^{(n-1)}(x) = y_2^{(n-2)}(x) = y_3^{(n-3)}(x)$$
$$= \cdots = y_{n-1}'(x) = y_n(x).$$

Differentiating the function in the last line of (65.23), we obtain

(65.231) $\quad y^{(n)}(x) = y_1^{(n)}(x) = y_2^{(n-1)}(x) = \cdots = y_{n-1}''(x) = y_n'(x).$

By (65.231) and (65.23), we can therefore write (65.1) as

(65.24) $\quad y_n'(x) + f_{n-1}(x)y_n(x) + \cdots + f_1(x)y_2(x) + f_0(x)y_1(x) = Q(x).$

Solving (65.24) for $y_n'(x)$, we obtain

(65.25) $\quad y_n'(x) = -f_{n-1}y_n - \cdots - f_1y_2 - f_0y_1 + Q.$

Hence because of (65.23) and (65.25), the set of functions y_1, y_2, $\cdots$, y_n, satisfies the system of linear first order equations

(65.26)
$$y_1'(x) = y_2(x),$$
$$y_2'(x) = y_3(x),$$
$$\cdots\cdots\cdots\cdots\cdots\cdots\cdots\cdots\cdots\cdots\cdots\cdots\cdots\cdots$$
$$y_{n-1}'(x) = y_n(x),$$
$$y_n'(x) = -f_{n-1}(x)y_n(x) - \cdots - f_1(x)y_2(x) - f_0(x)y_1(x) + Q(x).$$

By (65.23), $y_1(x_0) = y(x_0)$, $y_2(x_0) = y'(x_0)$, $\cdots$, $y_n(x_0) = y^{(n-1)}(x_0)$. We can therefore replace the initial conditions (65.21) by the conditions

(65.27) $\quad y_1(x_0) = a_0, \qquad y_2(x_0) = a_1, \cdots, y_n(x_0) = a_{n-1}.$

Conversely if we start with the system of equations (65.26) and the initial conditions (65.27), and define

(65.28) $$y(x) = y_1(x),$$

then the relations in (65.23), (65.231), and (65.24) hold; the last equation of (65.26) becomes identical with (65.1), and the initial conditions (65.27) become (65.21). Hence if there exists a set of functions $y_1(x)$, $y_2(x)$, $\cdots$, $y_n(x)$ that is a solution of the system (65.26) satisfying (65.27), then, by (65.28), the function $y_1(x)$ of this set is the solution $y(x)$ of (65.1) satisfying (65.21).

All the functions $f_0(x)$, $f_1(x)$, $\cdots$, $f_{n-1}(x)$, $Q(x)$ in the last equation of (65.26), are, by assumption, continuous on I. The functions y_2, y_3, $\cdots$, y_n are also continuous since their derivatives exist. Therefore the hypotheses of Theorem 62.3 are satisfied. Hence, by the theorem, there is, on I, or I_0 if I is closed, one and only one set of particular solutions $y_1(x)$, $y_2(x)$, $\cdots$, $y_n(x)$ that satisfies (65.26) and (65.27) where x_0 is a point in I_0. And since $y_1(x)$ exists and is unique, it follows that its equal $y(x)$ also exists and is the unique solution of (65.1) and (65.21).

Example 65.3. Set up a system of linear first order equations for the second order linear equation

(a) $$y'' - 3y' + 2y = x$$

and show that the solution $y_1(x)$ of the resulting system is the same as the solution $y(x)$ of (a).

Solution. Let $y(x)$ be a solution of (a). As suggested in (65.23), we define

(b) $$y(x) = y_1(x),$$
$$y'(x) = y_1'(x) = y_2(x).$$

Differentiating the last equation of (b), we obtain

(c) $$y''(x) = y_1''(x) = y_2'(x).$$

By (c) and (b), we can write (a) as

(d) $$y_2'(x) - 3y_2(x) + 2y_1(x) = x.$$

Hence, by (b) and (d), we can replace the single equation (a) by the system of linear first order equations

(e) $$y_1'(x) = y_2(x),$$
$$y_2'(x) = 3y_2 - 2y_1 + x.$$

In operator notation we can write (e) as

(f) $$Dy_1 - y_2 = 0,$$
$$2y_1 + (D - 3)y_2 = x.$$

Multiplying the first equation in (f) by $D - 3$ and adding it to the second, we obtain

(g) $$(D^2 - 3D + 2)y_1 = x, \qquad y_1'' - 3y_1' + 2y_1 = x.$$

Comparing the second equation in (g) with (a), we see that the solution $y_1(x)$ of the system (f) will be the same as the solution $y(x)$ of (a).

In Theorem 19.3, we stated that a homogeneous linear differential equation of order n has n linearly independent solutions, but proved this part of the theorem only for the case when the coefficients in the equation are constants. We now prove the statement for nonconstant coefficients.

Theorem 65.4. *If $f_0(x), f_1(x), \cdots, f_n(x)$ are each continuous functions of x on a common interval I, then the homogeneous linear differential equation*

$$(65.41) \qquad y^{(n)} + f_{n-1}(x)y^{(n-1)} + \cdots + f_1(x)y' + f_0(x)y = 0,$$

has n linearly independent solutions $y_1, y_2, \cdots, y_n$.

Proof. We consider first a *special set of solutions*, $g_1(x), g_2(x), \cdots, g_n(x)$ of (65.41), each satisfying respectively the initial conditions

(a) $\quad g_1(x_0) = 1, \quad g_1'(x_0) = 0, \quad g_1''(x_0) = 0, \cdots, g_1^{(n-1)}(x_0) = 0,$

$\qquad g_2(x_0) = 0, \quad g_2'(x_0) = 1, \quad g_2''(x_0) = 0, \cdots, g_2^{(n-1)}(x_0) = 0,$

$\qquad g_3(x_0) = 0, \quad g_3'(x_0) = 0, \quad g_3''(x_0) = 1, \cdots, g_3^{(n-1)}(x_0) = 0,$

$\qquad \cdots \cdots \cdots \cdots \cdots \cdots \cdots \cdots \cdots \cdots \cdots \cdots \cdots$

$\qquad g_n(x_0) = 0, \quad g_n'(x_0) = 0, \quad g_n''(x_0) = 0, \cdots, g_n^{(n-1)}(x_0) = 1,$

where x_0 is a point in I. By Theorem 65.2, each function $g_1, g_2, \cdots, g_n$ exists. We form a linear combination of this special set of solutions, set it equal to zero, and take its successive derivatives. We thus obtain

(b) $\qquad c_1 g_1(x) + c_2 g_2(x) + \cdots + c_n g_n(x) = 0,$

$\qquad c_1 g_1'(x) + c_2 g_2'(x) + \cdots + c_n g_n'(x) = 0,$

$\qquad c_1 g_1''(x) + c_2 g_2''(x) + \cdots + c_n g_n''(x) = 0,$

$\qquad \cdots \cdots \cdots \cdots \cdots \cdots \cdots \cdots \cdots \cdots \cdots \cdots \cdots$

$\qquad c_1 g_1^{(n-1)}(x) + c_2 g_2^{(n-1)}(x) + \cdots + c_n g_n^{(n-1)}(x) = 0.$

Let $x = x_0$. Then the first equation in (b), by the first column of values in (a), simplifies to $c_1 = 0$; the second equation in (b), by the second column of values in (a), simplifies to $c_2 = 0, \cdots$, the last equation in (b), by the last column of values in (a), simplifies to $c_n = 0$. We have thus shown that each constant $c_1, c_2, \cdots, c_n$ in (b) and in particular each such constant in the first equation of (b) is zero for x_0 in I. Therefore by Definition 19.1, the set of functions $g_1, g_2, \cdots, g_n$ is, on I, a linearly independent set. Since each function is, by assumption, a solution of (65.41), they form collectively a set of n linearly independent solutions of (65.41).

Let $y(x)$ be a solution of (65.41), and let x_0 be a point in I. By Theorem 19.3, the function

(c) $\qquad h(x) = y(x_0)g_1(x) + y'(x_0)g_2(x) + \cdots + y^{(n-1)}(x_0)g_n(x),$

is also a solution of (65.41) since it is a linear combination of n independent solutions. [Remember the coefficients in (c) are constants.] Taking successive derivatives of (c), we obtain

(d)
$$h'(x) = y(x_0)g_1'(x) + y'(x_0)g_2'(x) + \cdots + y^{(n-1)}(x_0)g_n'(x),$$
$$h''(x) = y(x_0)g_1''(x) + y'(x_0)g_2''(x) + \cdots + y^{(n-1)}(x_0)g_n''(x),$$
$$\cdots\cdots\cdots\cdots\cdots\cdots\cdots\cdots\cdots\cdots\cdots\cdots$$
$$h^{(n-1)}(x) = y(x_0)g_1^{(n-1)}(x) + y'(x_0)g_2^{(n-1)}(x) + \cdots + y^{(n-1)}(x_0)g_n^{(n-1)}(x).$$

Let $x = x_0$. Then (c), by the first column of values in (a), simplifies to $h(x_0) = y(x_0)$; the first equation in (d), by the second column of values in (a), simplifies to $h'(x_0) = y'(x_0)$, $\cdots$, the last equation in (d), by the last column of values in (a), simplifies to $h^{n-1}(x_0) = y^{(n-1)}(x_0)$. We have thus shown that the two solutions $h(x)$ and $y(x)$ of (65.41), and their first $n - 1$ derivatives are equal when $x = x_0$. Hence, by the uniqueness Theorem 65.2, the two solutions $h(x)$ and $y(x)$ are identical. We can therefore replace $h(x)$ by $y(x)$ in (c) to obtain

(e) $$y(x) = y(x_0)g_1(x) + y'(x_0)g_2(x) + \cdots + y^{(n-1)}(x_0)g_n(x),$$

where x_0 is a point in I. Since $y(x)$ is an arbitrary solution of (65.41), it follows that every solution of (65.41) can be expressed as a linear combination of the *special* set of n linearly independent solutions $g_1, g_2, \cdots, g_n$.

Let $y_1(x), y_2(x), \cdots, y_n(x)$ be n solutions of (65.41), and let x_0 be a point in I. Therefore by (e),

(f) $$y_1(x) = y_1(x_0)g_1(x) + y_1'(x_0)g_2(x) + \cdots + y_1^{(n-1)}(x_0)g_n(x),$$
$$y_2(x) = y_2(x_0)g_1(x) + y_2'(x_0)g_2(x) + \cdots + y_2^{(n-1)}(x_0)g_n(x),$$
$$\cdots\cdots\cdots\cdots\cdots\cdots\cdots\cdots\cdots\cdots\cdots\cdots$$
$$y_n(x) = y_n(x_0)g_1(x) + y_n'(x_0)g_2(x) + \cdots + y_n^{(n-1)}(x_0)g_n(x),$$

are each valid equations. The set of functions $g_1, g_2, \cdots, g_n$ are linearly independent. Hence the system (f) can be solved for $g_1, g_2, \cdots, g_n$ in terms of $y_1, y_2, \cdots, y_n$. This means, by Theorem 63.4, that the determinant formed by the coefficients of the g's is not zero. Therefore,

(g) $$\begin{vmatrix} y_1(x_0) & y_1'(x_0) & \cdots & y_1^{(n-1)}(x_0) \\ y_2(x_0) & y_2'(x_0) & \cdots & y_2^{(n-1)}(x_0) \\ \cdots & \cdots & \cdots & \cdots \\ y_n(x_0) & y_n'(x_0) & \cdots & y_n^{(n-1)}(x_0) \end{vmatrix} \neq 0.$$

A theorem of determinants now permits us to interchange rows and columns to obtain from (g) the equivalent determinant

(h)
$$\begin{vmatrix} y_1(x_0) & y_2(x_0) & \cdots & y_n(x_0) \\ y_1{}'(x_0) & y_2{}'(x_0) & \cdots & y_n{}'(x_0) \\ \cdots\cdots\cdots\cdots\cdots\cdots\cdots\cdots \\ y_1{}^{(n-1)}(x_0) & y_2{}^{(n-1)}(x_0) & \cdots & y_n{}^{(n-1)}(x_0) \end{vmatrix} \neq 0.$$

By Definition 63.5, the determinant (h) is the Wronskian of a set of solutions y_1, y_2, $\cdots$, y_n of (65.41) evaluated at $x = x_0$, where x_0 is a point in I. And since this Wronskian is not zero for one x in I, the set of solutions, by Comment 64.15, is linearly independent on I. We have thus proved the existence of a set of n linearly independent solutions of (65.41).

Theorem 65.5. *If y_1, y_2, $\cdots$, y_n are n linearly independent solutions of (65.41), then*

(65.51) $$y_c = c_1 y_1 + c_2 y_2 + \cdots + c_n y_n$$

is a general solution of (65.41), i.e., every solution of (65.41) can be obtained from (65.51) by a proper choice of the constants c_1, c_2, $\cdots$, c_n.

Proof. We prove the theorem for $n = 2$. Let y_1, y_2 be two linearly independent solutions of

(a) $$y'' + f_1(x)y' + f_0(x)y = 0.$$

Therefore, by Theorem 19.3,

(b) $$y_c = c_1 y_1 + c_2 y_2$$

is also a solution of (a). Assume y_3 is a solution of (a) not obtainable from (b). In the proof of Theorem 64.11, we showed that for each two solutions of (a), equation (64.12) is valid. Therefore, by (64.12),

(c) $$W(y_1, y_2; x) = c_{12} e^{F(x)},$$
$$W(y_1, y_3; x) = c_{13} e^{F(x)},$$
$$W(y_2, y_3; x) = c_{23} e^{F(x)},$$

where c_{12}, c_{13} and c_{23} are constants. Multiplying the first equation in (c) by y_3, the second by y_2, the third by y_1, and adding the resulting equations, we obtain

(d) $$y_3 W(y_1, y_2; x) + y_2 W(y_1, y_3; x) + y_1 W(y_2, y_3; x)$$
$$= (y_3 c_{12} + y_2 c_{13} + y_1 c_{23}) e^{F(x)}.$$

By (63.53), the left side of (d) is

(e) $$y_3(y_1 y_2{}' - y_2 y_1{}') + y_2(y_1 y_3{}' - y_3 y_1{}') + y_1(y_2 y_3{}' - y_3 y_2{}'),$$

which reduces to zero. Hence the right side of (d) must also be zero. And since, by (18.86), $e^{F(x)} \neq 0$ for every x, it follows that

(f) $$y_3 c_{12} + y_2 c_{13} + y_1 c_{23} = 0.$$

The solutions y_1, y_2 are, by assumption, linearly independent. Therefore, by Comment 64.15, the first Wronskian in (c) is not zero, i.e., $c_{12} \neq 0$. Hence dividing (f) by c_{12}, we obtain

(g) $$y_3 = -\frac{c_{13}}{c_{12}} y_2 - \frac{c_{23}}{c_{12}} y_1,$$

from which we see that y_3 is obtainable from (b) by a proper choice of the constants c_1 and c_2 in (b). Hence our assumption that y_3 is not obtainable from (b) is false.

Theorem 65.6. *Let y_1, y_2, $\cdots$, y_n be n linearly independent solutions of the homogeneous linear differential equation (65.41), and let*

(65.61) $$y_c = c_1 y_1 + c_2 y_2 + \cdots + c_n y_n$$

be its complementary function. Let y_p be a particular solution of the non-homogeneous linear differential equation (65.1).
Then

(65.62) $$y = y_c + y_p$$

is a general solution of (65.1), i.e., (65.62) includes every solution of (65.1).

Proof. By Theorem 65.2, at least one particular solution y_p of (65.1) exists. Assume g is a solution of (65.1) not obtainable from (65.62). Since y_p and g are each solutions of (65.1), we have

(a) $$y_p^{(n)} + f_{n-1}(x) y_p^{(n-1)} + \cdots + f_0(x) y_p = Q(x),$$
$$g^{(n)} + f_{n-1}(x) g^{(n-1)} + \cdots + f_0(x) g = Q(x).$$

Subtracting the first equation from the second, we obtain

(b) $$(g - y_p)^{(n)} + f_{n-1}(x)(g - y_p)^{(n-1)} + \cdots + f_0(x)(g - y_p) = 0.$$

Hence $y = g - y_p$ is a solution of the homogeneous linear equation (65.41). Therefore, by Theorem 65.5, $(g - y_p)$ can be obtained from (65.61) by a proper choice of the constants $c_1, c_2, \cdots, c_n$. Therefore,

(c) $$g - y_p = a_1 y_1 + a_2 y_2 + \cdots + a_n y_n,$$
$$g = y_p + (a_1 y_1 + a_2 y_2 + \cdots + a_n y_n).$$

We have thus shown that the solution g can be obtained from (65.62). Hence our assumption that g is not obtainable from (65.62) is false.

All statements made in Theorems 19.2, 19.3, and Comment 19.41, in connection with a linear differential equation of order n, have now been proved.

EXERCISE 65

1. Set up an equivalent system of linear first order equations for each of the following. Then show that solution $y_1(x)$ of the resulting system will be identical with the solution $y(x)$ of the given equation.

(a) $y'' - y' + 3y = x^2 + 1$. 　　(c) $y'' + 2y' + 3y = \tan x$.
(b) $y'' + 3y' - 2y = e^x + x$. 　　(d) $y''' - 5y'' + 8y' - 4y = 0$.

2. Set up an equivalent system of linear first order equations for each of the following.

(a) $y''' - 4y' + 2y = x + x^2$, $\quad y(0) = 1, y'(0) = -1$.
(b) $y''' + 3y'' - y' + y = e^x$, $\quad y(0) = 1, y'(0) = 0, y''(0) = 1$.
(c) $y^{(4)} - 3y''' + y'' + 5y' - 6y = 0$, $\quad y(0) = 0, y'(0) = 1, y''(0) = 2$,
$\quad y'''(0) = 3$.

ANSWERS 65

1. (a) $y_1'(x) = y_2$,
$\quad y_2'(x) = y_2 - 3y_1 + x^2 + 1$.
(b) $y_1'(x) = y_2$,
$\quad y_2'(x) = -3y_2 + 2y_1 + e^x + x$.

　　(c) $y_1'(x) = y_2$,
$\quad y_2'(x) = -2y_2 - 3y_1 + \tan x$.
(d) $y_1'(x) = y_2$,
$\quad y_2'(x) = y_3$,
$\quad y_3'(x) = 5y_3 - 8y_2 + 4y_1$.

2. (a) $y_1'(x) = y_2$,
$\quad y_2'(x) = 4y_2 - 2y_1 + x + x^2$, $\quad y_1(0) = 1, \quad y_2(0) = -1$.
(b) $y_1'(x) = y_2$,
$\quad y_2'(x) = y_3$,
$\quad y_3'(x) = -3y_3 + y_2 - y_1 + e^x$, $\quad y_1(0) = 1, \quad y_2(0) = 0, \quad y_3(0) = 1$.
(c) $y_1'(x) = y_2$, $\qquad y_3'(x) = y_4$,
$\quad y_2'(x) = y_3$, $\qquad y_4'(x) = 3y_4 - y_3 - 5y_2 + 6y_1$,
$\quad y_1(0) = 0, y_2(0) = 1, y_3(0) = 2, y_4(0) = 3$.

Bibliography

Intended for students who wish to advance beyond the material of this text.

Bessel Functions—G. N. Watson. *A Treatise on the Theory of Bessel Functions*. New York: Macmillan Co., 1944.

Celestial Mechanics—F. R. Moulton. *An Introduction to Celestial Mechanics*. New York: Macmillan Co., 1914. (Dover, 1970)

General Theory—E. L. Ince. *Ordinary Differential Equations*. London: Longmans, Green & Co., 1927. (Dover, 1956)

Laplace Transforms—R. V. Churchill. *Modern Operational Mathematics in Engineering*. New York: McGraw-Hill, 1944.

Legendre Functions—E. W. Hobson. *The Theory of Spherical and Ellipsoidal Harmonics*. Cambridge, England: Cambridge University Press, 1931.

Mechanics—W. F. Osgood. *Mechanics*. New York: Macmillan Co., 1946.

Numerical Methods—F. B. Hildebrand. *Introduction to Numerical Analysis*. New York: McGraw-Hill, 1956.

Perturbation Theory—F. R. Moulton. *An Introduction to Celestial Mechanics*. New York: Macmillan Co., 1914. (Dover, 1970)

INDEX

Index

Note: A boldfaced number refers to a problem in the exercises.

A CATALOG OF SELECTED
DOVER BOOKS
IN SCIENCE AND MATHEMATICS

A CATALOG OF SELECTED
DOVER BOOKS
IN SCIENCE AND MATHEMATICS

Astronomy

BURNHAM'S CELESTIAL HANDBOOK, Robert Burnham, Jr. Thorough guide to the stars beyond our solar system. Exhaustive treatment. Alphabetical by constellation: Andromeda to Cetus in Vol. 1; Chamaeleon to Orion in Vol. 2; and Pavo to Vulpecula in Vol. 3. Hundreds of illustrations. Index in Vol. 3. 2,000pp. 6⅛ x 9¼.
23567-X, 23568-8, 23673-0 Pa., Three-vol. set $46.85

THE EXTRATERRESTRIAL LIFE DEBATE, 1750–1900, Michael J. Crowe. First detailed, scholarly study in English of the many ideas that developed between 1750 and 1900 regarding the existence of intelligent extraterrestrial life. Examines ideas of Kant, Herschel, Voltaire, Percival Lowell, many other scientists and thinkers. 16 illustrations. 704pp. 5⅜ x 8½.
40675-X Pa. $19.95

A HISTORY OF ASTRONOMY, A. Pannekoek. Well-balanced, carefully reasoned study covers such topics as Ptolemaic theory, work of Copernicus, Kepler, Newton, Eddington's work on stars, much more. Illustrated. References. 521pp. 5⅜ x 8½.
65994-1 Pa. $15.95

AMATEUR ASTRONOMER'S HANDBOOK, J. B. Sidgwick. Timeless, comprehensive coverage of telescopes, mirrors, lenses, mountings, telescope drives, micrometers, spectroscopes, more. 189 illustrations. 576pp. 5⅜ x 8¼. (Available in U.S. only)
24034-7 Pa. $13.95

STARS AND RELATIVITY, Ya. B. Zel'dovich and I. D. Novikov. Vol. 1 of *Relativistic Astrophysics* by famed Russian scientists. General relativity, properties of matter under astrophysical conditions, stars and stellar systems. Deep physical insights, clear presentation. 1971 edition. References. 544pp. 5⅜ x 8½.
69424-0 Pa. $14.95

Chemistry

A SHORT HISTORY OF CHEMISTRY (3rd edition), J. R. Partington. Classic exposition explores origins of chemistry, alchemy, early medical chemistry, nature of atmosphere, theory of valency, laws and structure of atomic theory, much more. 428pp. 5⅜ x 8½. (Available in U.S. only)
65977-1 Pa. $12.95

CHEMICAL MAGIC, Leonard A. Ford. Second Edition, Revised by E. Winston Grundmeier. Over 100 unusual stunts demonstrating cold fire, dust explosions, much more. Text explains scientific principles and stresses safety precautions. 128pp. 5⅜ x 8½.
67628-5 Pa. $5.95

THE DEVELOPMENT OF MODERN CHEMISTRY, Aaron J. Ihde. Authoritative history of chemistry from ancient Greek theory to 20th-century innovation. Covers major chemists and their discoveries. 209 illustrations. 14 tables. Bibliographies. Indices. Appendices. 851pp. 5⅜ x 8½.
64235-6 Pa. $24.95

CATALYSIS IN CHEMISTRY AND ENZYMOLOGY, William P. Jencks. Exceptionally clear coverage of mechanisms for catalysis, forces in aqueous solution, carbonyl- and acyl-group reactions, practical kinetics, more. 864pp. 5⅜ x 8½.
65460-5 Pa. $19.95

THE HISTORICAL BACKGROUND OF CHEMISTRY, Henry M. Leicester. Evolution of ideas, not individual biography. Concentrates on formulation of a coherent set of chemical laws. 260pp. 5⅜ x 8½. 61053-5 Pa. $8.95

GENERAL CHEMISTRY, Linus Pauling. Revised 3rd edition of classic first-year text by Nobel laureate. Atomic and molecular structure, quantum mechanics, statistical mechanics, thermodynamics correlated with descriptive chemistry. Problems. 992pp. 5⅜ x 8½. 65622-5 Pa. $19.95

Engineering

DE RE METALLICA, Georgius Agricola. The famous Hoover translation of greatest treatise on technological chemistry, engineering, geology, mining of early modern times (1556). All 289 original woodcuts. 638pp. 6¾ x 11. 60006-8 Pa. $21.95

FUNDAMENTALS OF ASTRODYNAMICS, Roger Bate et al. Modern approach developed by U.S. Air Force Academy. Designed as a first course. Problems, exercises. Numerous illustrations. 455pp. 5⅜ x 8½. 60061-0 Pa. $12.95

DYNAMICS OF FLUIDS IN POROUS MEDIA, Jacob Bear. For advanced students of ground water hydrology, soil mechanics and physics, drainage and irrigation engineering and more. 335 illustrations. Exercises, with answers. 784pp. 6⅛ x 9¼.
65675-6 Pa. $19.95

ANALYTICAL MECHANICS OF GEARS, Earle Buckingham. Indispensable reference for modern gear manufacture covers conjugate gear-tooth action, gear-tooth profiles of various gears, many other topics. 263 figures. 102 tables. 546pp. 5⅜ x 8½.
65712-4 Pa. $16.95

ADVANCED STRENGTH OF MATERIALS, J. P. Den Hartog. Superbly written advanced text covers torsion, rotating disks, membrane stresses in shells, much more. Many problems and answers. 388pp. 5⅜ x 8½. 65407-9 Pa. $11.95

MECHANICS, J. P. Den Hartog. A classic introductory text or refresher. Hundreds of applications and design problems illuminate fundamentals of trusses, loaded beams and cables, etc. 334 answered problems. 462pp. 5⅜ x 8½. 60754-2 Pa. $12.95

MECHANICAL VIBRATIONS, J. P. Den Hartog. Classic textbook offers lucid explanations and illustrative models, applying theories of vibrations to a variety of practical industrial engineering problems. Numerous figures. 233 problems, solutions. Appendix. Index. Preface. 436pp. 5⅜ x 8½. 64785-4 Pa. $13.95

STRENGTH OF MATERIALS, J. P. Den Hartog. Full, clear treatment of basic material (tension, torsion, bending, etc.) plus advanced material on engineering methods, applications. 350 answered problems. 323pp. 5⅜ x 8½. 60755-0 Pa. $10.95

A HISTORY OF MECHANICS, René Dugas. Monumental study of mechanical principles from antiquity to quantum mechanics. Contributions of ancient Greeks, Galileo, Leonardo, Kepler, Lagrange, many others. 671pp. 5⅜ x 8½.
65632-2 Pa. $18.95

STATISTICAL MECHANICS: Principles and Applications, Terrell L. Hill. Standard text covers fundamentals of statistical mechanics, applications to fluctuation theory, imperfect gases, distribution functions, more. 448pp. 5⅜ x 8½.
65390-0 Pa. $14.95

THE VARIATIONAL PRINCIPLES OF MECHANICS, Cornelius Lanczos. Graduate level coverage of calculus of variations, equations of motion, relativistic mechanics, more. First inexpensive paperbound edition of classic treatise. Index. Bibliography. 418pp. 5⅜ x 8½.
65067-7 Pa. $14.95

THE VARIOUS AND INGENIOUS MACHINES OF AGOSTINO RAMELLI: A Classic Sixteenth-Century Illustrated Treatise on Technology, Agostino Ramelli. One of the most widely known and copied works on machinery in the 16th century. 194 detailed plates of water pumps, grain mills, cranes, more. 608pp. 9 x 12.
28180-9 Pa. $24.95

ORDINARY DIFFERENTIAL EQUATIONS AND STABILITY THEORY: An Introduction, David A. Sánchez. Brief, modern treatment. Linear equation, stability theory for autonomous and nonautonomous systems, etc. 164pp. 5⅜ x 8¼.
63828-6 Pa. $6.95

ROTARY-WING AERODYNAMICS, W. Z. Stepniewski. Clear, concise text covers aerodynamic phenomena of the rotor and offers guidelines for helicopter performance evaluation. Originally prepared for NASA. 537 figures. 640pp. 6⅛ x 9¼.
64647-5 Pa. $16.95

INTRODUCTION TO SPACE DYNAMICS, William Tyrrell Thomson. Comprehensive, classic introduction to space-flight engineering for advanced undergraduate and graduate students. Includes vector algebra, kinematics, transformation of coordinates. Bibliography. Index. 352pp. 5⅜ x 8½.
65113-4 Pa. $10.95

HISTORY OF STRENGTH OF MATERIALS, Stephen P. Timoshenko. Excellent historical survey of the strength of materials with many references to the theories of elasticity and structure. 245 figures. 452pp. 5⅜ x 8½.
61187-6 Pa. $14.95

CONSTRUCTIONS AND COMBINATORIAL PROBLEMS IN DESIGN OF EXPERIMENTS, Damaraju Raghavarao. In-depth reference work examines orthogonal Latin squares, incomplete block designs, tactical configuration, partial geometry, much more. Abundant explanations, examples. 416pp. 5⅜ x 8¼.
65685-3 Pa. $10.95

INCOMPRESSIBLE AERODYNAMICS, edited by Bryan Thwaites. Covers theoretical and experimental treatment of the uniform flow of air and viscous fluids past two-dimensional aerofoils and three-dimensional wings; many other topics. 654pp. 5⅜ x 8½.
65465-6 Pa. $16.95

Mathematics

HANDBOOK OF MATHEMATICAL FUNCTIONS WITH FORMULAS, GRAPHS, AND MATHEMATICAL TABLES, edited by Milton Abramowitz and Irene A. Stegun. Vast compendium: 29 sets of tables, some to as high as 20 places. 1,046pp. 8 x 10½. 61272-4 Pa. $29.95

CALCULUS REFRESHER FOR TECHNICAL PEOPLE, A. Albert Klaf. Covers important aspects of integral and differential calculus via 756 questions. 566 problems, most answered. 431pp. 5⅜ x 8½. 20370-0 Pa. $9.95

ASYMPTOTIC EXPANSIONS OF INTEGRALS, Norman Bleistein & Richard A. Handelsman. Best introduction to important field with applications in a variety of scientific disciplines. New preface. Problems. Diagrams. Tables. Bibliography. Index. 448pp. 5⅜ x 8½. 65082-0 Pa. $13.95

FAMOUS PROBLEMS OF GEOMETRY AND HOW TO SOLVE THEM, Benjamin Bold. Squaring the circle, trisecting the angle, duplicating the cube: learn their history, why they are impossible to solve, then solve them yourself. 128pp. 5⅜ x 8½. 24297-8 Pa. $5.95

VECTOR AND TENSOR ANALYSIS WITH APPLICATIONS, A. I. Borisenko and I. E. Tarapov. Concise introduction. Worked-out problems, solutions, exercises. 257pp. 5⅝ x 8¼. 63833-2 Pa. $9.95

THE ABSOLUTE DIFFERENTIAL CALCULUS (CALCULUS OF TENSORS), Tullio Levi-Civita. Great 20th-century mathematician's classic work on material necessary for mathematical grasp of theory of relativity. 452pp. 5⅜ x 8½. 63401-9 Pa. $11.95

AN INTRODUCTION TO ORDINARY DIFFERENTIAL EQUATIONS, Earl A. Coddington. A thorough and systematic first course in elementary differential equations for undergraduates in mathematics and science, with many exercises and problems (with answers). Index. 304pp. 5⅜ x 8½. 65942-9 Pa. $9.95

FOURIER SERIES AND ORTHOGONAL FUNCTIONS, Harry F. Davis. An incisive text combining theory and practical example to introduce Fourier series, orthogonal functions and applications of the Fourier method to boundary-value problems. 570 exercises. Answers and notes. 416pp. 5⅜ x 8½. 65973-9 Pa. $13.95

COMPUTABILITY AND UNSOLVABILITY, Martin Davis. Classic graduate-level introduction to theory of computability, usually referred to as theory of recurrent functions. New preface and appendix. 288pp. 5⅜ x 8½. 61471-9 Pa. $8.95

ASYMPTOTIC METHODS IN ANALYSIS, N. G. de Bruijn. An inexpensive, comprehensive guide to asymptotic methods–the pioneering work that teaches by explaining worked examples in detail. Index. 224pp. 5⅜ x 8½. 64221-6 Pa. $7.95

ESSAYS ON THE THEORY OF NUMBERS, Richard Dedekind. Two classic essays by great German mathematician: on the theory of irrational numbers; and on transfinite numbers and properties of natural numbers. 115pp. 5⅜ x 8½. 21010-3 Pa. $6.95

THE GEOMETRY OF RENÉ DESCARTES, René Descartes. The great work founded analytical geometry. Original French text, Descartes's own diagrams, together with definitive Smith-Latham translation. 244pp. 5⅜ x 8½. 60068-8 Pa. $9.95

APPLIED COMPLEX VARIABLES, John W. Dettman. Step-by-step coverage of fundamentals of analytic function theory–plus lucid exposition of five important applications: Potential Theory; Ordinary Differential Equations; Fourier Transforms; Laplace Transforms; Asymptotic Expansions. 66 figures. Exercises at chapter ends. 512pp. 5⅜ x 8½. 64670-X Pa. $14.95

INTRODUCTION TO LINEAR ALGEBRA AND DIFFERENTIAL EQUATIONS, John W. Dettman. Excellent text covers complex numbers, determinants, orthonormal bases, Laplace transforms, much more. Exercises with solutions. Undergraduate level. 416pp. 5⅜ x 8½. 65191-6 Pa. $11.95

MATHEMATICAL METHODS IN PHYSICS AND ENGINEERING, John W. Dettman. Algebraically based approach to vectors, mapping, diffraction, other topics in applied math. Also generalized functions, analytic function theory, more. Exercises. 448pp. 5⅜ x 8¼. 65649-7 Pa. $12.95

THE THIRTEEN BOOKS OF EUCLID'S ELEMENTS, translated with introduction and commentary by Sir Thomas L. Heath. Definitive edition. Textual and linguistic notes, mathematical analysis. 2,500 years of critical commentary. Unabridged. 1,414pp. 5⅜ x 8½. Three-vol. set. Vol. I: 60088-2 Pa. $10.95
Vol. II: 60089-0 Pa. $10.95
Vol. III: 60090-4 Pa. $12.95

CALCULUS OF VARIATIONS WITH APPLICATIONS, George M. Ewing. Applications-oriented introduction to variational theory develops insight and promotes understanding of specialized books, research papers. Suitable for advanced undergraduate/graduate students as primary, supplementary text. 352pp. 5⅜ x 8½. 64856-7 Pa. $9.95

COMPLEX VARIABLES, Francis J. Flanigan. Unusual approach, delaying complex algebra till harmonic functions have been analyzed from real variable viewpoint. Includes problems with answers. 364pp. 5⅜ x 8½. 61388-7 Pa. $10.95

AN INTRODUCTION TO THE CALCULUS OF VARIATIONS, Charles Fox. Graduate-level text covers variations of an integral, isoperimetrical problems, least action, special relativity, approximations, more. References. 279pp. 5⅜ x 8½. 65499-0 Pa. $8.95

CATASTROPHE THEORY FOR SCIENTISTS AND ENGINEERS, Robert Gilmore. Advanced-level treatment describes mathematics of theory grounded in the work of Poincaré, R. Thom, other mathematicians. Also important applications to problems in mathematics, physics, chemistry and engineering. 1981 edition. References. 28 tables. 397 black-and-white illustrations. xvii + 666pp. 6⅛ x 9¼. 67539-4 Pa. $17.95

INTRODUCTION TO DIFFERENCE EQUATIONS, Samuel Goldberg. Exceptionally clear exposition of important discipline with applications to sociology, psychology, economics. Many illustrative examples; over 250 problems. 260pp. 5⅜ x 8½. 65084-7 Pa. $10.95

UNBOUNDED LINEAR OPERATORS: Theory and Applications, Seymour Goldberg. Classic presents systematic treatment of the theory of unbounded linear operators in normed linear spaces with applications to differential equations. Bibliography. 199pp. 5⅜ x 8½. 64830-3 Pa. $7.95

DIFFERENTIAL GEOMETRY, Heinrich W. Guggenheimer. Local differential geometry as an application of advanced calculus and linear algebra. Curvature, transformation groups, surfaces, more. Exercises. 62 figures. 378pp. 5⅜ x 8½. 63433-7 Pa. $11.95

NUMERICAL METHODS FOR SCIENTISTS AND ENGINEERS, Richard Hamming. Classic text stresses frequency approach in coverage of algorithms, polynomial approximation, Fourier approximation, exponential approximation, other topics. Revised and enlarged 2nd edition. 721pp. 5⅜ x 8½. 65241-6 Pa. $16.95

POPULAR LECTURES ON MATHEMATICAL LOGIC, Hao Wang. Noted logician's lucid treatment of historical developments, set theory, model theory, recursion theory and constructivism, proof theory, more. 3 appendixes. Bibliography. 1981 edition. ix + 283pp. 5⅜ x 8½. 67632-3 Pa. $10.95

INTRODUCTION TO NUMERICAL ANALYSIS (2nd Edition), F. B. Hildebrand. Classic, fundamental treatment covers computation, approximation, interpolation, numerical differentiation and integration, other topics. 150 new problems. 669pp. 5⅜ x 8½. 65363-3 Pa. $16.95

THE FUNCTIONS OF MATHEMATICAL PHYSICS, Harry Hochstadt. Comprehensive treatment of orthogonal polynomials, hypergeometric functions, Hill's equation, much more. Bibliography. Index. 322pp. 5⅜ x 8½. 65214-9 Pa. $12.95

THREE PEARLS OF NUMBER THEORY, A. Y. Khinchin. Three compelling puzzles require proof of a basic law governing the world of numbers. Challenges concern van der Waerden's theorem, the Landau-Schnirelmann hypothesis and Mann's theorem, and a solution to Waring's problem. Solutions included. 64pp. 5⅜ x 8½. 40026-3 Pa. $4.95

THE PHILOSOPHY OF MATHEMATICS: An Introductory Essay, Stephan Körner. Surveys the views of Plato, Aristotle, Leibniz & Kant concerning propositions and theories of applied and pure mathematics. Introduction. Two appendices. Index. 198pp. 5⅜ x 8½. 25048-2 Pa. $8.95

INTRODUCTORY REAL ANALYSIS, A.N. Kolmogorov, S. V. Fomin. Translated by Richard A. Silverman. Self-contained, evenly paced introduction to real and functional analysis. Some 350 problems. 403pp. 5⅜ x 8½. 61226-0 Pa. $12.95

APPLIED ANALYSIS, Cornelius Lanczos. Classic work on analysis and design of finite processes for approximating solution of analytical problems. Algebraic equations, matrices, harmonic analysis, quadrature methods, much more. 559pp. 5⅜ x 8½. 65656-X Pa. $16.95

AN INTRODUCTION TO ALGEBRAIC STRUCTURES, Joseph Landin. Superb self-contained text covers "abstract algebra": sets and numbers, theory of groups, theory of rings, much more. Numerous well-chosen examples, exercises. 247pp. 5⅜ x 8½. 65940-2 Pa. $10.95

SPECIAL FUNCTIONS, N. N. Lebedev. Translated by Richard Silverman. Famous Russian work treating more important special functions, with applications to specific problems of physics and engineering. 38 figures. 308pp. 5⅜ x 8½. 60624-4 Pa. $9.95

QUALITATIVE THEORY OF DIFFERENTIAL EQUATIONS, V. V. Nemytskii and V.V. Stepanov. Classic graduate-level text by two prominent Soviet mathematicians covers classical differential equations as well as topological dynamics and ergodic theory. Bibliographies. 523pp. 5⅜ x 8½. 65954-2 Pa. $14.95

NUMBER THEORY AND ITS HISTORY, Oystein Ore. Unusually clear, accessible introduction covers counting, properties of numbers, prime numbers, much more. Bibliography. 380pp. 5⅜ x 8½. 65620-9 Pa. $10.95

THEORY OF MATRICES, Sam Perlis. Outstanding text covering rank, nonsingularity and inverses in connection with the development of canonical matrices under the relation of equivalence, and without the intervention of determinants. Includes exercises. 237pp. 5⅜ x 8½. 66810-X Pa. $8.95

OPTIMIZATION THEORY WITH APPLICATIONS, Donald A. Pierre. Broad spectrum approach to important topic. Classical theory of minima and maxima, calculus of variations, simplex technique and linear programming, more. Many problems, examples. 640pp. 5⅜ x 8½. 65205-X Pa. $17.95

INTRODUCTION TO ANALYSIS, Maxwell Rosenlicht. Unusually clear, accessible coverage of set theory, real number system, metric spaces, continuous functions, Riemann integration, multiple integrals, more. Wide range of problems. Undergraduate level. Bibliography. 254pp. 5⅜ x 8½. 65038-3 Pa. $9.95

MODERN NONLINEAR EQUATIONS, Thomas L. Saaty. Emphasizes practical solution of problems; covers seven types of equations. ". . . a welcome contribution to the existing literature...."–*Math Reviews*. 490pp. 5⅜ x 8½. 64232-1 Pa. $13.95

MATRICES AND LINEAR ALGEBRA, Hans Schneider and George Phillip Barker. Basic textbook covers theory of matrices and its applications to systems of linear equations and related topics such as determinants, eigenvalues and differential equations. Numerous exercises. 432pp. 5⅜ x 8½. 66014-1 Pa. $12.95

GEOMETRY OF COMPLEX NUMBERS, Hans Schwerdtfeger. Illuminating, widely praised book on analytic geometry of circles, the Moebius transformation, and two-dimensional non-Euclidean geometries. 200pp. 5⅝ x 8¼. 63830-8 Pa. $8.95

MATHEMATICS APPLIED TO CONTINUUM MECHANICS, Lee A. Segel. Analyzes models of fluid flow and solid deformation. For upper-level math, science and engineering students. 608pp. 5⅜ x 8½. 65369-2 Pa. $14.95

ELEMENTS OF REAL ANALYSIS, David A. Sprecher. Classic text covers fundamental concepts, real number system, point sets, functions of a real variable, Fourier series, much more. Over 500 exercises. 352pp. 5⅜ x 8½. 65385-4 Pa. $11.95

AN INTRODUCTION TO MATRICES, SETS AND GROUPS FOR SCIENCE STUDENTS, G. Stephenson. Concise, readable text introduces sets, groups, and most importantly, matrices to undergraduate students of physics, chemistry, and engineering. Problems. 164pp. 5⅜ x 8½. 65077-4 Pa. $7.95

SET THEORY AND LOGIC, Robert R. Stoll. Lucid introduction to unified theory of mathematical concepts. Set theory and logic seen as tools for conceptual understanding of real number system. 496pp. 5⅜ x 8¼. 63829-4 Pa. $14.95

LECTURES ON CLASSICAL DIFFERENTIAL GEOMETRY, Second Edition, Dirk J. Struik. Excellent brief introduction covers curves, theory of surfaces, fundamental equations, geometry on a surface, conformal mapping, other topics. Problems. 240pp. 5⅜ x 8½. 65609-8 Pa. $9.95

ORDINARY DIFFERENTIAL EQUATIONS, Morris Tenenbaum and Harry Pollard. Exhaustive survey of ordinary differential equations for undergraduates in mathematics, engineering, science. Thorough analysis of theorems. Diagrams. Bibliography. Index. 818pp. 5⅜ x 8½. 64940-7 Pa. $19.95

INTEGRAL EQUATIONS, F. G. Tricomi. Authoritative, well-written treatment of extremely useful mathematical tool with wide applications. Volterra Equations, Fredholm Equations, much more. Advanced undergraduate to graduate level. Exercises. Bibliography. 238pp. 5⅜ x 8½. 64828-1 Pa. $8.95

FOURIER SERIES, Georgi P. Tolstov. Translated by Richard A. Silverman. A valuable addition to the literature on the subject, moving clearly from subject to subject and theorem to theorem. 107 problems, answers. 336pp. 5⅜ x 8½. 63317-9 Pa. $11.95

DISTRIBUTION THEORY AND TRANSFORM ANALYSIS: An Introduction to Generalized Functions, with Applications, A. H. Zemanian. Provides basics of distribution theory, describes generalized Fourier and Laplace transformations. Numerous problems. 384pp. 5⅜ x 8½. 65479-6 Pa. $13.95

TENSOR CALCULUS, J.L. Synge and A. Schild. Widely used introductory text covers spaces and tensors, basic operations in Riemannian space, non-Riemannian spaces, etc. 324pp. 5⅜ x 8¼. 63612-7 Pa. $11.95

CALCULUS OF VARIATIONS, Robert Weinstock. Basic introduction covering isoperimetric problems, theory of elasticity, quantum mechanics, electrostatics, etc. Exercises throughout. 326pp. 5⅜ x 8½. 63069-2 Pa. $9.95

THE CONTINUUM: A Critical Examination of the Foundation of Analysis, Hermann Weyl. Classic of 20th-century foundational research deals with the conceptual problem posed by the continuum. 156pp. 5⅜ x 8½. 67982-9 Pa. $8.95

CHALLENGING MATHEMATICAL PROBLEMS WITH ELEMENTARY SOLUTIONS, A. M. Yaglom and I. M. Yaglom. Over 170 challenging problems on probability theory, combinatorial analysis, points and lines, topology, convex polygons, many other topics. Solutions. Total of 445pp. 5⅜ x 8½. Two-vol. set.
Vol. I: 65536-9 Pa. $8.95
Vol. II: 65537-7 Pa. $7.95

A SURVEY OF NUMERICAL MATHEMATICS, David M. Young and Robert Todd Gregory. Broad self-contained coverage of computer-oriented numerical algorithms for solving various types of mathematical problems in linear algebra, ordinary and partial, differential equations, much more. Exercises. Total of 1,248pp. 5⅜ x 8½. Two volumes.
Vol. I: 65691-8 Pa. $16.95
Vol. II: 65692-6 Pa. $16.95

INTRODUCTION TO PARTIAL DIFFERENTIAL EQUATIONS WITH APPLICATIONS, E. C. Zachmanoglou and Dale W. Thoe. Essentials of partial differential equations applied to common problems in engineering and the physical sciences. Problems and answers. 416pp. 5⅜ x 8½.   65251-3 Pa. $11.95

THE THEORY OF GROUPS, Hans J. Zassenhaus. Well-written graduate-level text acquaints reader with group-theoretic methods and demonstrates their usefulness in mathematics. Axioms, the calculus of complexes, homomorphic mapping, p-group theory, more. Many proofs shorter and more transparent than older ones. 276pp. 5⅜ x 8½.   40922-8 Pa. $12.95

GENERALIZED INTEGRAL TRANSFORMATIONS, A.H. Zemanian. Graduate-level study of recent generalizations of the Laplace, Mellin, Hankel, K. Weierstrass, convolution and other simple transformations. Bibliography. 320pp. 5⅜ x 8½.
65375-7 Pa. $8.95

Math–Decision Theory, Statistics, Probability

ELEMENTARY DECISION THEORY, Herman Chernoff and Lincoln E. Moses. Clear introduction to statistics and statistical theory covers data processing, probability and random variables, testing hypotheses, much more. Exercises. 364pp. 5⅜ x 8½.   65218-1 Pa. $12.95

STATISTICS MANUAL, Edwin L. Crow et al. Comprehensive, practical collection of classical and modern methods prepared by U.S. Naval Ordnance Test Station. Stress on use. Basics of statistics assumed. 288pp. 5⅜ x 8½.   60599-X Pa. $8.95

SOME THEORY OF SAMPLING, William Edwards Deming. Analysis of the problems, theory and design of sampling techniques for social scientists, industrial managers and others who find statistics increasingly important in their work. 61 tables. 90 figures. xvii + 602pp. 5⅜ x 8½.   64684-X Pa. $16.95

STATISTICAL ADJUSTMENT OF DATA, W. Edwards Deming. Introduction to basic concepts of statistics, curve fitting, least squares solution, conditions without parameter, conditions containing parameters. 26 exercises worked out. 271pp. 5⅜ x 8½.
64685-8 Pa. $9.95

LINEAR PROGRAMMING AND ECONOMIC ANALYSIS, Robert Dorfman, Paul A. Samuelson and Robert M. Solow. First comprehensive treatment of linear programming in standard economic analysis. Game theory, modern welfare economics, Leontief input-output, more. 525pp. 5⅜ x 8½.   65491-5 Pa. $17.95

DICTIONARY/OUTLINE OF BASIC STATISTICS, John E. Freund and Frank J. Williams. A clear concise dictionary of over 1,000 statistical terms and an outline of statistical formulas covering probability, nonparametric tests, much more. 208pp. 5⅜ x 8½.   66796-0 Pa.$8.95

PROBABILITY: An Introduction, Samuel Goldberg. Excellent basic text covers set theory, probability theory for finite sample spaces, binomial theorem, much more. 360 problems. Bibliographies. 322pp. 5⅜ x 8½.   65252-1 Pa. $10.95

GAMES AND DECISIONS: Introduction and Critical Survey, R. Duncan Luce and Howard Raiffa. Superb nontechnical introduction to game theory, primarily applied to social sciences. Utility theory, zero-sum games, n-person games, decision-making, much more. Bibliography. 509pp. 5⅜ x 8½. 65943-7 Pa. $14.95

FIFTY CHALLENGING PROBLEMS IN PROBABILITY WITH SOLUTIONS, Frederick Mosteller. Remarkable puzzlers, graded in difficulty, illustrate elementary and advanced aspects of probability. Detailed solutions. 88pp. 5⅜ x 8½.
65355-2 Pa. $4.95

PROBABILITY THEORY: A Concise Course, Y. A. Rozanov. Highly readable, self-contained introduction covers combination of events, dependent events, Bernoulli trials, etc. Translation by Richard Silverman. 148pp. 5⅜ x 8¼.
63544-9 Pa. $8.95

STATISTICAL METHOD FROM THE VIEWPOINT OF QUALITY CONTROL, Walter A. Shewhart. Important text explains regulation of variables, uses of statistical control to achieve quality control in industry, agriculture, other areas. 192pp. 5⅜ x 8½. 65232-7 Pa. $8.95

THE COMPLEAT STRATEGYST: Being a Primer on the Theory of Games of Strategy, J. D. Williams. Highly entertaining classic describes, with many illustrated examples, how to select best strategies in conflict situations. Prefaces. Appendices. 268pp. 5⅜ x 8½. 25101-2 Pa. $8.95

Math–History of

A SHORT ACCOUNT OF THE HISTORY OF MATHEMATICS, W. W. Rouse Ball. One of clearest, most authoritative surveys from the Egyptians and Phoenicians through 19th-century figures such as Grassman, Galois, Riemann. Fourth edition. 522pp. 5⅜ x 8½. 20630-0 Pa. $13.95

THE HISTORICAL ROOTS OF ELEMENTARY MATHEMATICS, Lucas N. H. Bunt, Phillip S. Jones, and Jack D. Bedient. Fundamental underpinnings of modern arithmetic, algebra, geometry and number systems derived from ancient civilizations. 320pp. 5⅜ x 8½. 25563-8 Pa. $9.95

GAMES, GODS & GAMBLING: A History of Probability and Statistical Ideas, F. N. David. Episodes from the lives of Galileo, Fermat, Pascal, and others illustrate this fascinating account of the roots of mathematics. Features thought-provoking references to classics, archaeology, biography, poetry. 1962 edition. 304pp. 5⅜ x 8½.
(USO) 40023-9 Pa. $9.95

HISTORY OF MATHEMATICS, David E. Smith. Nontechnical survey from ancient Greece and Orient to late 19th century; evolution of arithmetic, geometry, trigonometry, calculating devices, algebra, the calculus. 362 illustrations. 1,355pp. 5⅜ x 8½. Two-vol. set.
Vol. I: 20429-4 Pa. $13.95
Vol. II: 20430-8 Pa. $14.95

A CONCISE HISTORY OF MATHEMATICS, Dirk J. Struik. The best brief history of mathematics. Stresses origins and covers every major figure from ancient Near East to 19th century. 41 illustrations. 195pp. 5⅜ x 8½. 60255-9 Pa. $8.95

THE HISTORY OF THE CALCULUS AND ITS CONCEPTUAL DEVELOP-
MENT, Carl B. Boyer. Origins in antiquity, medieval contributions, work of Newton,
Leibniz, rigorous formulation. Treatment is verbal. 346pp. 5⅜ x 8½. 60509-4 Pa. $9.95

Math-Topology

ELEMENTARY CONCEPTS OF TOPOLOGY, Paul Alexandroff. Elegant, intu-
itive approach to topology from set-theoretic topology to Betti groups; how concepts
of topology are useful in math and physics. 25 figures. 57pp. 5⅜ x 8½.
60747-X Pa. $4.95

COMBINATORIAL TOPOLOGY, P. S. Alexandrov. Clearly written, well-orga-
nized, three-part text begins by dealing with certain classic problems without using
the formal techniques of homology theory and advances to the central concept, the
Betti groups. Numerous detailed examples. 654pp. 5⅜ x 8½. 40179-0 Pa. $18.95

EXPERIMENTS IN TOPOLOGY, Stephen Barr. Classic, lively explanation of one
of the byways of mathematics. Klein bottles, Moebius strips, projective planes, map
coloring, problem of the Koenigsberg bridges, much more, described with clarity
and wit. 43 figures. 210pp. 5⅜ x 8½. 25933-1 Pa. $8.95

CONFORMAL MAPPING ON RIEMANN SURFACES, Harvey Cohn. Lucid,
insightful book presents ideal coverage of subject. 334 exercises make book perfect
for self-study. 55 figures. 352pp. 5⅜ x 8¼. 64025-6 Pa. $11.95

CURVATURE AND HOMOLOGY: Enlarged Edition, Samuel I. Goldberg.
Revised edition examines topology of differentiable manifolds; curvature, homology
of Riemannian manifolds; compact Lie groups; complex manifolds; curvature,
homology of Kaehler manifolds. New Preface. Four new appendixes. 416pp. 5⅜ x 8½.
40207-X Pa. $14.95

TOPOLOGY, John G. Hocking and Gail S. Young. Superb one-year course in classical topology. Topological spaces and functions, point-set topology, much more.
Examples and problems. Bibliography. Index. 384pp. 5⅜ x 8¼. 65676-4 Pa. $11.95

THE FOUR-COLOR PROBLEM: Assaults and Conquest, Thomas L. Saaty and
Paul G. Kainen. Engrossing, comprehensive account of the century-old combinatorial topological problem, its history and solution. Bibliographies. Index. 110 figures.
228pp. 5⅜ x 8½. 65092-8 Pa. $7.95

Meteorology

PRINCIPLES OF METEOROLOGICAL ANALYSIS, Walter J. Saucier. Highly
respected, abundantly illustrated classic reviews atmospheric variables, hydrostatics,
static stability, various analyses (scalar, cross-section, isobaric, isentropic, more). For
intermediate meteorology students. 454pp. 6⅛ x 9¼. 65979-8 Pa. $14.95

LIGHTNING, Martin A. Uman. Revised, updated edition of classic work on the
physics of lightning. Phenomena, terminology, measurement, photography, spectroscopy, thunder, more. Reviews recent research. Bibliography. Indices. 320pp.
5⅜ x 8¼. 64575-4 Pa. $8.95

Physics

OPTICAL RESONANCE AND TWO-LEVEL ATOMS, L. Allen and J. H. Eberly. Clear, comprehensive introduction to basic principles behind all quantum optical resonance phenomena. 53 illustrations. Preface. Index. 256pp. 5⅜ x 8½.
65533-4 Pa. $10.95

ULTRASONIC ABSORPTION: An Introduction to the Theory of Sound Absorption and Dispersion in Gases, Liquids and Solids, A. B. Bhatia. Standard reference in the field provides a clear, systematically organized introductory review of fundamental concepts for advanced graduate students, research workers. Numerous diagrams. Bibliography. 440pp. 5⅜ x 8½.
64917-2 Pa. $11.95

QUANTUM THEORY, David Bohm. This advanced undergraduate-level text presents the quantum theory in terms of qualitative and imaginative concepts, followed by specific applications worked out in mathematical detail. Preface. Index. 655pp. 5⅜ x 8½.
65969-0 Pa. $15.95

ATOMIC PHYSICS (8th edition), Max Born. Nobel laureate's lucid treatment of kinetic theory of gases, elementary particles, nuclear atom, wave-corpuscles, atomic structure and spectral lines, much more. Over 40 appendices, bibliography. 495pp. 5⅜ x 8½.
65984-4 Pa. $13.95

AN INTRODUCTION TO HAMILTONIAN OPTICS, H. A. Buchdahl. Detailed account of the Hamiltonian treatment of aberration theory in geometrical optics. Many classes of optical systems defined in terms of the symmetries they possess. Problems with detailed solutions. 1970 edition. xv + 360pp. 5⅜ x 8½.
67597-1 Pa. $10.95

HYDRODYNAMIC AND HYDROMAGNETIC STABILITY, S. Chandrasekhar. Lucid examination of the Rayleigh-Benard problem; clear coverage of the theory of instabilities causing convection. 704pp. 5⅜ x 8¼.
64071-X Pa. $17.95

INVESTIGATIONS ON THE THEORY OF THE BROWNIAN MOVEMENT, Albert Einstein. Five papers (1905-8) investigating dynamics of Brownian motion and evolving elementary theory. Notes by R. Fürth. 122pp. 5⅜ x 8½.
60304-0 Pa. $5.95

THE PHYSICS OF WAVES, William C. Elmore and Mark A. Heald. Unique overview of classical wave theory. Acoustics, optics, electromagnetic radiation, more. Ideal as classroom text or for self-study. Problems. 477pp. 5⅜ x 8½.
64926-1 Pa. $14.95

THIRTY YEARS THAT SHOOK PHYSICS: The Story of Quantum Theory, George Gamow. Lucid, accessible introduction to influential theory of energy and matter. Careful explanations of Dirac's anti-particles, Bohr's model of the atom, much more. 12 plates. Numerous drawings. 240pp. 5⅜ x 8½.
24895-X Pa. $7.95

ELECTRONIC STRUCTURE AND THE PROPERTIES OF SOLIDS: The Physics of the Chemical Bond, Walter A. Harrison. Innovative text offers basic understanding of the electronic structure of covalent and ionic solids, simple metals, transition metals and their compounds. Problems. 1980 edition. 582pp. 6⅛ x 9¼.
66021-4 Pa. $19.95

PHYSICAL PRINCIPLES OF THE QUANTUM THEORY, Werner Heisenberg. Nobel Laureate discusses quantum theory, uncertainty, wave mechanics, work of Dirac, Schroedinger, Compton, Wilson, Einstein, etc. 184pp. 5⅜ x 8½.
60113-7 Pa. $8.95

ATOMIC SPECTRA AND ATOMIC STRUCTURE, Gerhard Herzberg. One of best introductions; especially for specialist in other fields. Treatment is physical rather than mathematical. 80 illustrations. 257pp. 5⅜ x 8½.   60115-3 Pa. $7.95

AN INTRODUCTION TO STATISTICAL THERMODYNAMICS, Terrell L. Hill. Excellent basic text offers wide-ranging coverage of quantum statistical mechanics, systems of interacting molecules, quantum statistics, more. 523pp. 5⅜ x 8½.
65242-4 Pa. $13.95

THEORETICAL PHYSICS, Georg Joos, with Ira M. Freeman. Classic overview covers essential math, mechanics, electromagnetic theory, thermodynamics, quantum mechanics, nuclear physics, other topics. First paperback edition. xxiii + 885pp. 5⅜ x 8½.   65227-0 Pa. $21.95

BOUNDARY VALUE PROBLEMS OF HEAT CONDUCTION, M. Necati Özisik. Systematic, comprehensive treatment of modern mathematical methods of solving problems in heat conduction and diffusion. Numerous examples and problems. Selected references. Appendices. 505pp. 5⅜ x 8½.   65990-9 Pa. $12.95

PROBLEMS AND SOLUTIONS IN QUANTUM CHEMISTRY AND PHYSICS, Charles S. Johnson, Jr. and Lee G. Pedersen. Unusually varied problems, detailed solutions in coverage of quantum mechanics, wave mechanics, angular momentum, molecular spectroscopy, scattering theory, more. 280 problems plus 139 supplementary exercises. 430pp. 6½ x 9¼.   65236-X Pa. $14.95

THEORETICAL SOLID STATE PHYSICS, Vol. 1: Perfect Lattices in Equilibrium; Vol. II: Non-Equilibrium and Disorder, William Jones and Norman H. March. Monumental reference work covers fundamental theory of equilibrium properties of perfect crystalline solids, non-equilibrium properties, defects and disordered systems. Appendices. Problems. Preface. Diagrams. Index. Bibliography. Total of 1,301pp. 5⅜ x 8½. Two volumes.   Vol. I: 65015-4 Pa. $16.95
Vol. II: 65016-2 Pa. $16.95

A TREATISE ON ELECTRICITY AND MAGNETISM, James Clerk Maxwell. Important foundation work of modern physics. Brings to final form Maxwell's theory of electromagnetism and rigorously derives his general equations of field theory. 1,084pp. 5⅜ x 8½. Two-vol. set.   Vol. I: 60636-8 Pa. $14.95
Vol. II: 60637-6 Pa. $12.95

OPTICKS, Sir Isaac Newton. Newton's own experiments with spectroscopy, colors, lenses, reflection, refraction, etc., in language the layman can follow. Foreword by Albert Einstein. 532pp. 5⅜ x 8½.   60205-2 Pa. $13.95

THEORY OF ELECTROMAGNETIC WAVE PROPAGATION, Charles Herach Papas. Graduate-level study discusses the Maxwell field equations, radiation from wire antennas, the Doppler effect and more. xiii + 244pp. 5⅜ x 8½. 65678-0 Pa. $9.95

INTRODUCTION TO QUANTUM MECHANICS With Applications to Chemistry, Linus Pauling & E. Bright Wilson, Jr. Classic undergraduate text by Nobel Prize winner applies quantum mechanics to chemical and physical problems. Numerous tables and figures enhance the text. Chapter bibliographies. Appendices. Index. 468pp. 5⅜ x 8½. 64871-0 Pa. $12.95

METHODS OF THERMODYNAMICS, Howard Reiss. Outstanding text focuses on physical technique of thermodynamics, typical problem areas of understanding, and significance and use of thermodynamic potential. 1965 edition. 238pp. 5⅜ x 8½. 69445-3 Pa. $8.95

TENSOR ANALYSIS FOR PHYSICISTS, J. A. Schouten. Concise exposition of the mathematical basis of tensor analysis, integrated with well-chosen physical examples of the theory. Exercises. Index. Bibliography. 289pp. 5⅜ x 8½. 65582-2 Pa. $10.95

RELATIVITY IN ILLUSTRATIONS, Jacob T. Schwartz. Clear nontechnical treatment makes relativity more accessible than ever before. Over 60 drawings illustrate concepts more clearly than text alone. Only high school geometry needed. Bibliography. 128pp. 6⅛ x 9¼. 25965-X Pa. $7.95

THE ELECTROMAGNETIC FIELD, Albert Shadowitz. Comprehensive undergraduate text covers basics of electric and magnetic fields, builds up to electromagnetic theory. Also related topics, including relativity. Over 900 problems. 768pp. 5⅜ x 8¼. 65660-8 Pa. $19.95

GREAT EXPERIMENTS IN PHYSICS: Firsthand Accounts from Galileo to Einstein, edited by Morris H. Shamos. 25 crucial discoveries: Newton's laws of motion, Chadwick's study of the neutron, Hertz on electromagnetic waves, more. Original accounts clearly annotated. 370pp. 5⅜ x 8½. 25346-5 Pa. $11.95

RELATIVITY, THERMODYNAMICS AND COSMOLOGY, Richard C. Tolman. Landmark study extends thermodynamics to special, general relativity; also applications of relativistic mechanics, thermodynamics to cosmological models. 501pp. 5⅜ x 8½. 65383-8 Pa. $15.95

LIGHT SCATTERING BY SMALL PARTICLES, H. C. van de Hulst. Comprehensive treatment including full range of useful approximation methods for researchers in chemistry, meteorology and astronomy. 44 illustrations. 470pp. 5⅜ x 8½. 64228-3 Pa. $12.95

STATISTICAL PHYSICS, Gregory H. Wannier. Classic text combines thermodynamics, statistical mechanics and kinetic theory in one unified presentation of thermal physics. Problems with solutions. Bibliography. 532pp. 5⅜ x 8½. 65401-X Pa. $14.95

Prices subject to change without notice.

Available at your book dealer or write for free Dover Mathematics and Science Catalog (59065-8) to Dept. Gl, Dover Publications, Inc., 31 East 2nd St., Mineola, N.Y. 11501. Dover publishes more than 250 books each year on science, elementary and advanced mathematics, biology, music, art, literature, history, social sciences, and other subjects.